Methods in Enzymology

Volume XXIII

PHOTOSYNTHESIS

Part A

METHODS IN ENZYMOLOGY

EDITORS-IN-CHIEF

Sidney P. Colowick   Nathan O. Kaplan

Methods in Enzymology

Volume XXIII

Photosynthesis

Part A

EDITED BY

Anthony San Pietro
DEPARTMENT OF BOTANY
INDIANA UNIVERSITY
BLOOMINGTON, INDIANA

1971

ACADEMIC PRESS New York and London

ACADEMIC PRESS, INC.
111 Fifth Avenue, New York, New York 10003

United Kingdom Edition published by
ACADEMIC PRESS, INC. (LONDON) LTD.
Berkeley Square House, London W1X 6BA

LIBRARY OF CONGRESS CATALOG CARD NUMBER: 54-9110

PRINTED IN THE UNITED STATES OF AMERICA

Table of Contents

Section I. Isolation and Culture Techniques

Section II. Preparation and Properties of Mutants

Section III. Cellular and Subcellular Preparations

Section IV. Components

Contributors to Volume XXIII

Article numbers are in parentheses following the names of contributors.
Affiliations listed are current.

C. Freeman Allen (48), *Department of Chemistry, Pomona College, Claremont, California*

Mary Belle Allen (14), *Institute of Marine Science, University of Alaska, College, Alaska*

Daniel I. Arnon (39), *Department of Cell Physiology, University of California, Berkeley, California*

Assunta Baccarini-Melandri (50), *Istituto Botanico, Universita degli Studi-Bologna, Bologna, Italy*

Rita Barr (37), *Department of Biological Sciences, Purdue University, Lafayette, Indiana*

Robert G. Bartsch (34, 62), *Chemistry Department, University of California at San Diego, La Jolla, California*

D. S. Bendall (33), *Department of Biochemistry, Cambridge University, Cambridge, England*

John Biggins (17), *Division of Biological and Medical Sciences, Brown University, Providence, Rhode Island*

Norman I. Bishop (3, 11), *Department of Botany, Oregon State University, Corvallis, Oregon*

Clanton C. Black, Jr. (24, 55), *Graduate Studies Research Center, Department of Biochemistry, University of Georgia, Athens, Georgia*

N. K. Boardman (26), *Division of Plant Industry, C.S.I.R.O., Canberra, Australia*

Peter Böger (21), *Ruhr-Universität, Institut für Biochemie der Pflanzen, Bochum, Germany*

J. S. Brown (43), *Department of Plant Biology, Carnegie Institution of Washington, Stanford, California*

Bob B. Buchanan (39), *Department of Cell Physiology, University of California, Berkeley, California*

Roderick K. Clayton (69), *Division of Biological Sciences, Cornell University, Ithaca, New York*

J. R. Cook (5), *Department of Zoology, University of Maine, Oregon, Maine*

Benjamin T. Cope (42), *Chemistry Division, Argonne National Laboratory, Argonne, Illinois*

F. L. Crane (37), *Department of Biological Sciences, Purdue University, Lafayette, Indiana*

Michael A. Cusanovich (32), *Department of Chemistry, University of Arizona, Tucson, Arizona*

W. M. Darley (7), *Department of Botany, The University of Georgia, Athens, Georgia*

H. E. Davenport (33), *A.R.C. Food Research Institute, Norwich, England*

Barrie Entsch (46), *Plant Physiology Unit, C.S.I.R.O. Division of Food Research and School of Biological Sciences, Macquarie University, North Ryde, Sydney, Australia*

William R. Evans (52), *Charles F. Kettering Research Laboratory, Yellow Springs, Ohio*

Giorgio Forti (41), *Istituto Botanico, Universita de Napoli, Napoli, Italy*

Albert W. Frenkel (25), *Department of Botany, University of Minnesota, Minneapolis, Minnesota*

Yoshihiko Fujita (58), *Ocean Research Institute, University of Tokyo, Nakano, Tokyo, Japan*

Augusto Garcia (31), *Faculty of Exact and Natural Sciences, University of Buenos Aires, Buenos Aires, Argentina*

Pearl Good (48), *Department of Chemistry, Pomona College, Claremont, California*

D. Graham (20), *Plant Physiology Unit,*

C.S.I.R.O. Division of Food Research and School of Biological Sciences, Macquarie University, North Ryde, Sydney, Australia

S. GRANICK (13), *The Rockefeller University, New York, New York*

ELAINE GREENBERG (68), *Department of Biochemistry and Biophysics, University of California, Davis, California*

R. R. L. GUILLARD (9), *Woods Hole Oceanographic Institute, Woods Hole, Massachusetts*

R. H. HAGEMAN (45), *Department of Agronomy, University of Illinois, Urbana, Illinois*

EIJI HASE (6), *Institute of Applied Microbiology, University of Tokyo, Bunkyo-ku, Tokyo, Japan*

ROBERT HILL (33), *Department of Biochemistry, Cambridge University, Cambridge, England*

T. HORIO (63, 64), *Division of Enzymology, Institute for Protein Research, Osaka University, Osaka, Japan*

Y. HORIUTI (63, 64) *Research Laboratories, Toyo Jozo Company, Ltd., Shizuoka, Japan*

D. P. HUCKLESBY (45), *Long Ashton Research Station, University of Bristol, Bristol, England*

G. JACOBI (28), *Lehrstuhl für Biochemie der Pflanze der Universität Göttingen, Untere Karspüle, Germany*

ARNE JENSEN (56), *Norwegian Institute of Seaweed Research, Trondheim-NTH, Norway*

JOSEPH S. KAHN (51), *Department of Biochemistry, North Carolina State University, Raleigh, North Carolina*

SAKAE KATOH (38), *Department of Biochemistry and Biophysics, Faculty of Science, University of Tokyo, Hongo, Tokyo, Japan*

BACON KE (60), *Charles F. Kettering Research Laboratory, Yellow Springs, Ohio*

B. KOK (47), *Research Institute for Advanced Studies, Baltimore, Maryland*

W. M. LAETSCH (8), *Department of Botany, University of California, Berkeley, California*

M. DANIEL LANE (53), *Department of Physiological Chemistry, Johns Hopkins University School of Medicine, Baltimore, Maryland*

R. P. LEVINE (10), *The Biological Laboratories, Harvard University, Cambridge, Massachusetts*

SYNNØVE LIAAEN-JENSEN (56), *Organic Chemistry Laboratories, Norwegian Institute of Technology, University of Trondheim, Trondheim, Norway*

STEPHEN LIEN (49), *Department of Microbiology, Indiana University, Bloomington, Indiana*

MANUEL LOSADA (44), *Instituto de Biologia Celular, C.S.I.C., Facultad de Ciencias, Universidad de Sevilla, Sevilla, Spain*

HARVARD LYMAN (12), *Department of Biological Sciences, State University of New York at Stony Brook, Stony Brook, New York*

RICHARD E. MCCARTY (23, 30) *Section of Biochemistry and Molecular, Biology, Cornell University, Ithaca, New York*

T. V. MARSHO (47), *Department of Biology, University of Maryland, Baltimore, Maryland*

BRUNO A. MELANDRI (50), *Istituto Botanico, Universita degli Studi-Bologna, Bologna, Italy*

AKIRA MITSUI (36), *Department of Botany, Indiana University, Bloomington, Indiana*

YUJI MORIMURA (6), *Institute of Applied Microbiology, University of Tokyo, Bunkyo-ku, Tokyo, Japan*

JACK MYERS (58), *Department of Zoology, University of Texas, Austin, Texas*

RODGER A. NELSON (25), *Department of Botany, University of Minnesota, Minneapolis, Minnesota*

K. NISHIKAWA (63, 64), *Division of Enzymology, Institute for Protein*

Research, Osaka University, Osaka, Japan

JOHN M. OLSON (61), *Biology Department, Brookhaven National Laboratory, Upton, New York*

ANTONIO PANEQUE (44), *Instituto de Biologia Celular, C.S.I.C., Facultad de Ciencias, Universidad de Sevilla, Sevilla, Spain*

R. B. PARK (22), *Department of Botany, University of California, Berkeley, California*

JACK PREISS (59, 68), *Department of Biochemistry and Biophysics, University of California, Davis, California*

EFRAIM RACKER (49), *Section of Biochemistry and Molecular Biology, Cornell University, Ithaca, New York*

GILLES RIBÉREAU-GAYON (59), *Department of Molecular Biology, Albert Einstein College, Bronx, New York*

ALBERT W. RUESINK (16), *Department of Botany, Indiana University, Bloomington, Indiana*

GEORGE K. RUSSELL (12), *Department of Biology, Adelphi University, Garden City, New York*

JEROME A. SCHIFF (12), *Department of Biology, Brandeis University, Waltham, Massachusetts*

GEORG H. SCHMID (15), *Max-Planck-Institut für Züchtungsforschung, Köln-Vogelsang, West Germany*

P. SCHOPFER (54), *Biological Institute II, University of Freiburg, Freiburg, Germany*

HORST SENGER (3), *Botanical Institute, University of Marburg, Marburg/Lahn, Germany*

ELWOOD R. SHAW (27), *Charles F. Kettering Research Laboratory, Yellow Springs, Ohio*

KAZUO SHIBATA (29), *Laboratory of Plant Physiology Rikagaku Kenkyusho, The Institute of Physical and Chemical Research, Wakohshi, Saitama Prefecture, Japan*

MASATERU SHIN (40), *Kobe Yamate Women's College, Kobe, Japan*

H. W. SIEGELMAN (9, 54), *Biology Department, Brookhaven National Laboratory, Upton, New York*

ROBERT M. SMILLIE (20, 46), *Plant Physiology Unit, C.S.I.R.O. Division of Food Research and School of Biological Sciences, Macquarie University, North Ryde, Sydney, Australia*

RICHARD C. STARR (2), *Department of Botany, Indiana University, Bloomington, Indiana*

C. R. STOCKING (19), *Department of Botany, University of California, Davis, California*

HAROLD H. STRAIN (42), *Chemistry Division, Argonne National Laboratory, Argonne, Illinois*

STEFAN SURZYCKI (4), *Botany Department, University of Iowa, Iowa City, Iowa*

WALTER A. SVEC (42), *Chemistry Division, Argonne National Laboratory, Argonne, Illinois*

ATUSI TAKAMIYA (57), *Department of Biology, Faculty of Science, Toho University, Narashino City, Chiba, Japan*

J. PHILIP THORNBER (31, 66, 67), *Department of Botanical Sciences, University of California, Los Angeles, California*

N. E. TOLBERT (65), *Department of Biochemistry, Michigan State University, East Lansing, Michigan*

C. B. VAN NIEL (1), *Hopkins Marine Station, Stanford University, Pacific Grove, California*

LEO P. VERNON (27, 31), *Department of Chemistry, Brigham Young University, Provo, Utah*

B. E. VOLCANI (7), *Scripps Institution of Oceanography, University of California at San Diego, La Jolla, California*

D. A. WALKER (18), *Department of Botany, The University, Sheffield S10 2TN, Yorkshire, England*

RICHARD T. WANG (69), *Department of Zoology, University of Texas, Austin, Texas*

MARCIA WISHNICK (53), *Department of*

Biochemistry, Public Health Research Institute of the City of New York, New York, New York

EIJIRO YAKUSHIJI (35), *Department of Biology, Faculty of Science, Toho University, Izumicho, Narashino, Chiba-Ken, Japan*

N. YAMAMOTO (64), *Division of Enzymology, Institute for Protein Research, Osaka University, Osaka, Japan*

Preface

The discovery of the chloroplast reaction by Hill in 1937, commonly known as the Hill reaction, heralded the concerted enzymological attack on the mechanism of photosynthesis. In the ensuing three and one-half decades, a tremendous but diverse wealth of knowledge has accumulated from the multiplicity of experimental approaches used to study this problem. The aim of this (and the succeeding) volume is to provide as comprehensive as possible a coverage of these methodological approaches, namely, biochemical, biophysical, genetic, and physiological.

The presentations in this volume consider isolation and culture techniques of algae, bacteria, and diatoms; plant tissue culture; the preparation and properties of mutants; cellular and subcellular preparations from algae, bacteria, and plants; and the purification and properties of components of the photosynthetic systems. Volume XXIV will cover methodology both chemical and physical, buffers and inhibitors, synthesizing capabilities of the photochemical systems, and a short section devoted to the analytical techniques and isolation of components of nitrogen fixation. Where appropriate, reference is provided to articles published previously in this series.

I wish to express my deepest gratitude to all the authors for their fine contributions, advice, and suggestions. The continued patience and valuable cooperation of the staff of Academic Press are acknowledged with gratitude. Excellent secretarial assistance was provided by Mrs. Virginia Flack and Mrs. Cheryl Fisher.

ANTHONY SAN PIETRO

METHODS IN ENZYMOLOGY

EDITED BY

Sidney P. Colowick and Nathan O. Kaplan

VANDERBILT UNIVERSITY
SCHOOL OF MEDICINE
NASHVILLE, TENNESSEE

DEPARTMENT OF CHEMISTRY
UNIVERSITY OF CALIFORNIA
AT SAN DIEGO
LA JOLLA, CALIFORNIA

METHODS IN ENZYMOLOGY

EDITORS-IN-CHIEF

Sidney P. Colowick   Nathan O. Kaplan

VOLUME XX. Nucleic Acids and Protein Synthesis (Part C)
Edited by KIVIE MOLDAVE AND LAWRENCE GROSSMAN

VOLUME XXI. Nucleic Acids (Part D)
Edited by LAWRENCE GROSSMAN AND KIVIE MOLDAVE

VOLUME XXII. Enzyme Purification and Related Techniques
Edited by WILLIAM B. JAKOBY

VOLUME XXIII. Photosynthesis (Part A)
Edited by ANTHONY SAN PIETRO

Methods in Enzymology

Volume XXIII

PHOTOSYNTHESIS

Part A

Section I

Isolation and Culture Techniques

A. ENRICHMENT TECHNIQUES

Articles 1 and 2

B. SYNCHRONOUS CULTURES

Articles 3 through 7

C. TISSUE CULTURE

Article 8

D. LARGE-SCALE GROWTH OF ALGAE

Article 9

[1] Techniques for the Enrichment, Isolation, and Maintenance of the Photosynthetic Bacteria

By C. B. van Niel

The photosynthetic bacteria are characterized by their ability to grow in the absence of air when exposed to light. Unlike the algae and higher plants, they do not produce oxygen, and they require an external supply of an oxidizable substrate whose nature varies with the species.

One large group, the photosynthetic *sulfur* bacteria, can use one or more reduced inorganic sulfur compounds as oxidizable substrate; it comprises the green and brown Chlorobacteriaceae[1] and the brown, red, and purple Thiorhodaceae. All of them can grow when supplied with H_2S, and some representatives can also use elementary sulfur, sulfite, or thiosulfate. The numerous types are found widely distributed in aqueous environments where H_2S is present, either through volcanic emanations (sulfur springs) or as a result of microbial activities, mainly sulfate reduction. They often occur as mass developments ("blooms") easily visible to the naked eye; such blooms may extend over large areas. Photosynthetic sulfur bacteria have been encountered in bodies of water whose salt content ranges from near zero to saturation (salt ponds), and at temperatures from below zero to about 80° (hot springs).

The second group, collectively known as the Athiorhodaceae, is composed of the brown and red bacteria that grow preferentially at the expense of a variety of simple organic compounds. Even though they, too, are common inhabitants of bodies of water, they have not to my knowledge been found as mass developments in nature, nor have they thus far been reported growing in strongly saline environments or at high temperatures.

[1] H. G. Trüper and N. Pfenning (personal communication, 1969) have recently submitted a proposal to the Judicial Commission of the International Commission on Bacterial Nomenclature of the International Association of Microbiological Societies, advocating a change of the name of this taxon to Chlorobiaceae. The arguments advanced in favor of the proposal seem compelling, and it is therefore probable that the proposal will be accepted.

Meanwhile, the Judicial Commission, which met during the Tenth International Congress for Microbiology in Mexico City, August, 1970, has declined a proposal for the conservation of the family names Thiorhodaceae and Athiorhodaceae [N. Pfennig and H. G. Trüper, *Int. J. Syst. Bacteriol.* **20,** 31 (1970)], and Pfennig and Trüper (personal communication, 1970) have since submitted a proposal to substitute the names Chromatiaceae and Rhodospirillaceae, respectively, for these taxa. In the present chapter, written before September, 1969, the older and more familiar terms have been used.

Early Enrichment Culture Methods

Winogradsky Columns

Engelmann[2] was the first to recognize the importance of light for the development of the photosynthetic bacteria, but it was Winogradsky[3] who stressed the role of H_2S in the metabolism of the Thiorhodaceae and devised an effective method for establishing cultures of these organisms in the laboratory. Its principle consists in creating an environment that ensures a continuous supply of H_2S over a long period of time. This is accomplished by putting a mixture of mud, $CaSO_4$, and roots of water plants, such as *Butomus*, in the bottom of a tall glass cylinder, filling the cylinder with water, and exposing it to light. The organic matter in the mud layer undergoes a slow decomposition under the influence of various kinds of anaerobic bacteria, resulting in the production of CO_2, H_2, and simple organic substances, such as alcohols, fatty and hydroxy acids. These, in turn, are there oxidized by the sulfate-reducing bacteria, which use sulfate as the final electron acceptor and reduce it to H_2S. Its concentration being greatest near the place of its formation, i.e., in the mud, the first development of the photosynthetic sulfur bacteria becomes visible here, usually in the form of green or reddish spots on the side of the cylinder exposed to the light. As the activities of the sulfate-reducing bacteria increase, the H_2S concentration in the bottom layer of the cylinder may rise to a level too high for the growth of Thiorhodaceae. At that stage the green sulfur bacteria tend to become predominant in that region, while various kinds of Thiorhodaceae appear in the water column, either as swarms at different distances from the mud–water interfase, or as pink, red, or purplish patches that adhere to the glass.

Instead of plant roots, other kinds of insoluble and slowly decomposable organic matter, such as cellulose, may be used. In that case it is advisable to supplement the contents of the cylinder with a source of nitrogen in order to enhance the development of the microbial population; the addition of a small amount of NH_4MgPO_4 to the mud–cellulose mixture is particularly suitable for the purpose.

Sometimes the development of photosynthetic sulfur bacteria in a Winogradsky column may be delayed owing to a slow initiation of the decomposition processes in the mud. This may result in the early growth of algae which, by virtue of their oxygen production, seriously interfere with the growth of photosynthetic bacteria, and allow various aerobic,

[2] T. W. Engelmann, *Bot. Ztg.* **46,** 661 (1888).

[3] S. Winogradsky, "Beiträge zur Morphologie und Physiologie der Bacterien," Helft I: "Zur Morphologie und Physiologie der Schwefelbacterien." Arthur Felix, Leipzig, 1888.

nonphotosynthetic microorganisms to become predominant. In order to avoid such undesirable results, it is advisable to use a rich mud containing an ample supply of anaerobic bacteria, including sulfate reducers. Algal growth can also be retarded or prevented by keeping the cylinder in the dark for the first few days and exposing it to light only after the decomposition is well under way. As a source of radiant energy incandescent bulbs are preferable to daylight, because their emission favors the growth of photosynthetic bacteria over that of algae. Even better results are obtained, as Gaffron[4] showed long ago, by exposing the cultures to radiation in the near infrared region (720–1000 nm). This entirely prevents algal growth, while the photosynthetic bacteria, whose pigment systems absorb maximally in the near infrared, can develop rapidly. Exposure of the cylinders to light from "daylight" (fluorescent) tubes, which emit little infrared radiation, has the opposite effect, and should be avoided.

For the enrichment of green and red sulfur bacteria from brackish, marine or strongly saline environments, both the mud and the water used to fill the cylinders should, of course, contain salt in corresponding concentrations; thermophilic representatives can be obtained by incubating the columns at appropriately high temperatures.

A Winogradsky column may remain functional over a long period of time and provide a succession of different representatives of green and red sulfur bacteria long after the initially introduced organic matter has been decomposed. If oxygen is rigorously excluded from such cultures, they actually represent a closed system in which a more primitive cycle of matter occurs than that which now perpetuates life on earth through the influx of solar radiation. The contrast between the two kinds of cycles is clear from the following diagrams:

Primitive cycle of matter

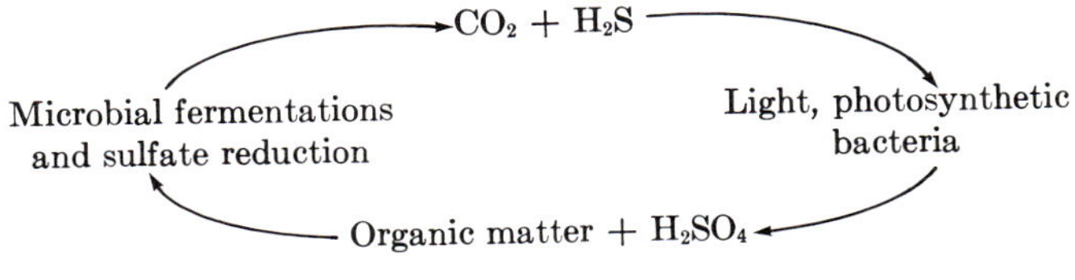

Terrestrial cycle of matter

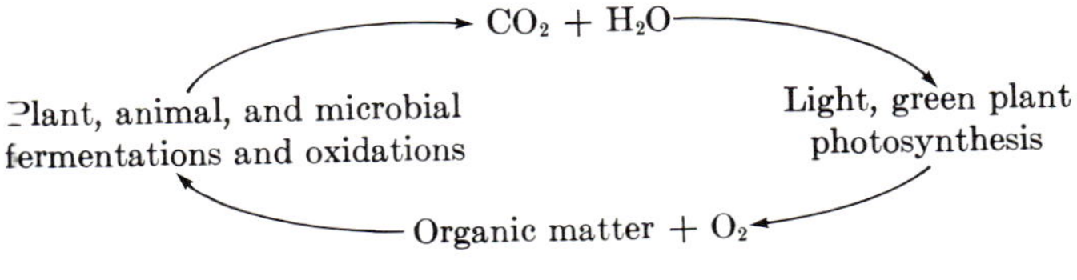

[4] H. Gaffron, *Biochem. Z.* **275,** 301 (1935).

Molisch Columns

A dead animal, or a piece of animal tissue, placed with some mud at the bottom of a cylinder filled with water and exposed to light generally undergoes a much more rapid anaerobic decomposition than does a husky plant root or a mass of cellulose. Although in both cases the decomposition products may be quite similar in kind, their rate of formation during protein decomposition is apt to be far greater than that at which they can be subsequently oxidized by the sulfate reducing bacteria. Consequently, these decomposition products will accumulate in the medium; and this gives rise to a rapid development of Athiorhodaceae—photosynthetic bacteria for which simple organic substances rather than H_2S are the oxidizable substrate *par excellence*.

This was first observed by Molisch,[5] who thus procured the material for his studies on this group of organisms. The method is still useful for the preparation of mass cultures in the laboratory. However, the putrefaction of proteinaceous substances yields, in addition to a variety of simple organic compounds, also H_2S, from sulfhydryl groups rather than from sulfate reduction. In the long run such cultures, even if prepared without the addition of sulfate, are therefore likely to become populated by a mixture of Thio- and Athiorhodaceae, particularly because some Thiorhodaceae can use organic substrates as well as H_2S.

Improved Enrichment Culture

General Remarks

The outcome of a Winogradsky or Molisch column depends on the gradual modification of the primary environment through the activities of numerous kinds of microorganisms, and they are largely uncontrollable. Consequently this will apply also to the kinds of photosynthetic bacteria that eventually make their appearance. A better understanding of the metabolic properties of the latter[6] made it possible to improve upon these methods and to design more specific procedures for the selective enrichment of particular representatives by the use of appropriate synthetic media and conditions of incubation. Nevertheless, the success of these methods is contingent upon the availability of inocula that contain the desired organisms in sufficient numbers, so that a small inoculum suffices to induce their growth; if a large amount of inoculum were re-

[5] H. Molisch, "Die Purpurbakterien nach neuen Untersuchungen." G. Fischer, Jena, 1907.

[6] C. B. van Niel, *Arch. Mikrobiol.* **3**, 1 (1931); **7**, 323 (1936); *Bacteriol. Rev.* **8**, 1 (1944).

quired, it would also introduce enough foreign matter to modify the composition of the medium and thereby render it less specific.

Usually a few milliliters of mud or stagnant water in which organic matter undergoes decomposition suffice to yield cultures of most if not all of the hitherto known species of Athiorhodaceae, and of the commonest among the green and red sulfur bacteria. For many representatives of the latter groups one may, however, have to depend on naturally occurring mass developments in which their presence can be ascertained by microscopic examination; alternatively, one can resort to the use of a Winogradsky column to provide a satisfactory primary inoculum, as Larsen[7] found when trying to isolate green sulfur bacteria.

Photosynthetic Sulfur Bacteria

As mentioned earlier, all members of this group can use H_2S as oxidizable substrate; therefore they can be selectively enriched in sulfide-containing media, in the absence of air, when exposed to light. Without known exceptions, they are obligatory anaerobes and phototrophs; attempts to cultivate them in darkness have so far failed.

A strictly inorganic medium suffices for the cultivation of some of the green sulfur bacteria (*Chlorobium* species) and of small Thiorhodaceae (*Chromatium, Thiocystis,* and *Thiococcus* species). But such a medium is inadequate for growing the large and more conspicuous Thiorhodaceae as well as many strains belonging to the Chlorobacteriaceae, because they all require a supply of cyanocobalamine (vitamin B_{12}).[8,9]

A medium that permits growth of nearly all photosynthetic sulfur bacteria thus far isolated was developed by Pfennig[10]; it has subsequently been slightly modified by Pfennig and Lippert.[9] Because its preparation is somewhat complicated, it is convenient to make up a fairly large batch at one time and to store it in completely filled and tightly closed screw-cap bottles in the dark, where it keeps for many months.

The directions for the preparation of enough medium to fill about 30 bottles of 100-ml capacity call for the use of a number of solutions that can be separately sterilized.

Solution 1: 0.83 g of $CaCl_2 \cdot 2\ H_2O$ in 2.5 liters of H_2O; for organisms living in salt water environments, NaCl should be added in a concentration 1.5 times as high as the one desired in the final medium.

[7] H. Larsen, *J. Bacteriol.* **64,** 187 (1952); *Kgl. Nor. Vidensk. Selsk. Skr.* 1953 Nr. 1.

[8] H. G. Schlegel and N. Pfennig, *Arch. Mikrobiol.* **38,** 1 (1961).

[9] N. Pfennig and K. D. Lippert, *Arch. Mikrobiol.* **55,** 245 (1966).

[10] N. Pfennig, *Zentralbl. Bakteriol. Parasitenk. Infectionskr. Hyg., Abt. 1, Supplementh.* **1,** 179 (1965).

Solution 2: H_2O, 67 ml; KH_2PO_4, 1 g; NH_4Cl, 1 g; $MgCl_2 \cdot 2\ H_2O$, 1 g; KCl, 1 g; 3 ml of a vitamin B_{12} solution, containing 2 mg of cyanocobalamine per 100 ml of H_2O[11]; and 30 ml of a heavy metal solution which contains, per liter H_2O, EDTA (disodium salt), 500 mg; $FeSO_4 \cdot 7\ H_2O$, 200 mg; $ZnSO_4 \cdot 7\ H_2O$, 10 mg; $MnCl_2 \cdot 4\ H_2O$, 3 mg; H_3BO_3, 30 mg; $CoCl_2 \cdot 6\ H_2O$, 20 mg; $CuCl_2 \cdot 2\ H_2O$, 1 mg; $NiCl_2 \cdot 6\ H_2O$, 2 mg; $Na_2MoO_4 \cdot 2\ H_2O$, 3 mg; the ingredients are severally dissolved in H_2O before adding them to the EDTA solution; the final pH of the heavy metal solution is adjusted to about 3.

Solution 3: 3 g of Na_2CO_3 in 900 ml of H_2O; this solution is autoclaved in a cotton-plugged container supplied with a cotton-plugged inlet tube for subsequent gassing of the solution with CO_2.

Solution 4: 3 g of $Na_2S \cdot 9\ H_2O$ in 200 ml of H_2O. It is sterilized in a flask with a Teflon-covered magnetized stirring rod.

The bulk of solution 1 (2000 ml) is dispensed in 67-ml aliquots over 30 bottles of 100-ml capacity, and these are autoclaved with the screw caps only loosely screwed on. The remainder of this solution, which is eventually used to fill the bottles to capacity, is separately sterilized, as are solutions 2, 3, and 4.

Following removal from the autoclave, the partially filled bottles and the other solutions are rapidly cooled to room temperature by placing them in a bath with running cold water; this prevents prolonged exposure to air, which would have the undesirable effect of causing more than a minimum amount of oxygen to dissolve in the liquids.

When cool, solution 3 is gassed with CO_2 to saturation; its pH should then be about 6.2. To it is added the cooled solution 2, and 33-ml aliquots of the combined solutions are dispensed aseptically into the bottles containing solution 1.

The flask with solution 4 is placed on a magnetic stirrer, and its contents are partially neutralized by the dropwise addition of 1.5 ml of sterile, 2 *M* H_2SO_4 while being agitated. It is added to the bottles in 5-ml portions each, though in some cases this amount must be reduced (see below). Finally, the bottles are filled completely with the surplus of solution 1 and tightly closed; a small gas bubble, about the size of a pea, should remain in order to accommodate small changes in volume and pressure during storage and incubation.

The pH of the final medium should be between 6.7 and 7.2; the con-

[11] Stock solutions must be stored in the dark to prevent photodecomposition of the vitamin.

tents of one bottle is used to check this and, if necessary, to determine the amount of acid or alkali needed for proper pH adjustment.

The freshly prepared medium is not entirely free of oxygen and, because many of the photosynthetic sulfur bacteria are strict anaerobes, it is advisable to store the bottles in the dark for at least 24 hours before inoculating them. Should the medium be needed immediately, a small amount of sterile ascorbic acid (final concentration not more than 0.05%) may be added to lower the redox potential sufficiently.

It should be mentioned that the sulfide concentration of the complete medium (0.075% $Na_2S \cdot 9\ H_2O$) is too high to allow some species of the photosynthetic sulfur bacteria to grow; these must be cultivated in a medium with only one-half the amount of sulfide ordinarily used (2.5 instead of 5 ml of the neutralized solution 4 per bottle).

Because the total crop of bacteria at full growth is small, even at the higher sulfide concentrations, it is often desirable or even necessary periodically to "feed" them. This should be done at the time when the medium no longer contains sulfide and the bacteria have used up most of the sulfur stored as droplets in the cells, which can easily be determined by microscopic examination, or after some experience has been gained, by a simple visual inspection of the cultures. The difference in appearance of a culture in which the organisms are stuffed with, or free of, sulfur droplets is quite striking; in the former case the cultures look chalky; in the latter they are clear. Feeding is done by withdrawing 2.5 or 5 ml from the bottles and refilling them with an equal amount of the neutralized Na_2S solution. After feeding, it is usually advisable to place the cultures in darkness for a few hours, after which they are returned to their appropriate position in the light.

As long as the aim of an enrichment culture is restricted to the procurement of a culture in which the desired organism is growing, it is obviously superfluous to use sterile media and aseptic techniques in the preparation of the primary cultures, especially because the inoculum for such cultures generally contains a mixture of many kinds of microbes. Sterile media are required only if one wishes to establish beyond any doubt that the organisms developing in these cultures cannot have been introduced initially other than with the inoculum. This applies whenever studies on the distribution of the organisms in question are undertaken. During subsequent purification procedures, the use of sterile media and aseptic precautions is of course, imperative.

Prior to the introduction of Pfennig's standard medium, it had been possible to obtain fairly specific enrichment cultures of a few green and red sulfur bacteria with the aid of simpler media, merely by varying the sulfide and hydrogen ion concentrations.[6] Owing to the extensive studies of Pfennig and co-workers, it has become possible to define the conditions under which a far greater number of species can be grown selectively; they mainly involve variations in the wavelength and intensity of the radiant energy source to which the cultures are exposed, and in the temperature of incubation.

The photosynthetic activities of the organisms under discussion are mediated by pigment systems which comprise specific chlorophylls and carotenoids. Thus far, four different bacteriochlorophylls have been isolated from various members of the group; they can be readily distinguished by their absorption characteristics. Following the proposal of Jensen *et al.*,[12] they will here be designated as bacteriochlorophylls *a*, *b*, *c*, and *d*.

Bacteriochlorophyll *a* is the principal pigment of most Thiorhodaceae, as well as of most Athiorhodaceae, and a minor component of the pigment system of all Chlorobacteriaceae (Chlorobiaceae). Its *in vivo* absorption maxima in the infrared region are at about 800–810 and 880–890 nm. The relative heights of the 800 and 880 peaks differ among species.

Bacteriochlorophyll *b*, with a minor *in vivo* absorption maximum at about 820–830 nm and a major one at 1020–1040 nm, has thus far been encountered in only one species of Thiorhodaceae, a *Thiococcus* sp. discovered by Eimhjellen *et al.*[13]; it also occurs in some green *Rhodopseudomonas* strains, members of the Athiorhodaceae (see below).

On the other hand, bacteriochlorophylls *c* and *d* have been found exclusively in the Chlorobacteriaceae (Chlorobiaceae). These pigments were first isolated from strains of *Chlorobium* species and designated as chlorobium chlorophylls 660 and 650, respectively; the numbers refer to the position of the longwave absorption maxima of the extracted pigments in diethyl ether solution. The *in vivo* absorption maxima of bacteriochlorophylls *c* and *d* are situated at approximately 750 and 730 nm, respectively. It has now been established that different strains of a particular species of green sulfur bacteria, indistinguishable on the basis of several other criteria, may contain either bacteriochlorophyll *c* or *d*, although this does not necessarily apply to all species. Further information on this point will be found in the next section.

The growth rate of a photosynthetic bacterium is a function of its

[12] A. Jensen, O. Aasmundrud, and K. E. Eimhjellen, *Biochim. Biophys. Acta* **88**, 466 (1964).

[13] K. E. Eimhjellen, H. Steensland, and J. Traetteberg, *Arch. Mikrobiol.* **59**, 82 (1967). This organism has recently been renamed *Thiocapsa pfennigii* [K. E. Eimhjellen, *Arch. Mikrobiol.* **73**, 193 (1970)].

photosynthetic activity, and the latter depends—within limits—upon the amount of radiant energy absorbed by its chlorophyll. Therefore, exposure to radiation at a wavelength corresponding to the major absorption maximum of a particular chlorophyll can act as a powerful selective agent and lead to the eventual predominance of such organisms in cultures initiated with a mixed inoculum. This, along with differences in substrate specificity and H_2S tolerance, permits a convenient selective cultivation of the Chlorobacteriaceae (Chlorobiaceae) and Thiorhodaceae, respectively.

Chlorobacteriaceae (Chlorobiaceae). The currently known representatives of this group are severally allocated to a number of different genera that are readily distinguishable on the basis of morphological features.

The organisms belonging to the genus *Chlorobium* occur as small, nonmotile, ovoid to short rod-shaped or curved cells, sometimes growing in chains resembling streptococci; the curved members have been isolated from brackish water and, in media prepared with sea salt instead of pure NaCl, may grow in the form of tightly wound coils. The presently recognized species are the green-colored *Chlorobium limicola, C. thiosulfatophilum,* and *C. chlorochromatii* and the brown-colored *C. phaeobacteroides* and *C. phaeovibrioides.* All of them grow well in Pfennig's standard medium at 25°–30°, even if the sulfide concentration is increased to 0.1–0.2% $Na_2S \cdot 9\ H_2O$ and the pH lowered to 6.6, and at high light intensity (700–2000 lux, corresponding to exposure to the radiation emitted by a 60-W bulb at 35–15 cm distance). Under these conditions they can successfully compete with the Thiorhodaceae that may be present in the original inoculum, so that cultures essentially free from the latter organisms can easily be obtained by repeatedly transferring cultures at an early stage of development to fresh media.

With the exception of *C. chlorochromatii,* for which only bacteriochlorophyll *c* has been reported, all species are represented by strains that contain either bacteriochlorophyll *c* or *d*. Consequently their growth can also be selectively favored by exposing the enrichment cultures to radiation in the 720–760 nm wavelength range. It is even reasonable to expect that a specific selection of strains with either of the two chlorophylls can be achieved by exposing cultures to radiation in the 740–760 nm and 720–735 nm wavelength regions, respectively. As far as I am aware, pertinent experiments of this kind have not yet been reported.

Chlorobium limicola cannot use thiosulfate as electron donor; as the name implies, *C. thiosulfatophilum* can do so. Enrichment cultures in media containing both H_2S and thiosulfate, the latter in a concentration of 0.1–0.2%, will therefore permit *C. thiosulfatophilum* to multiply after the sulfide supply has been exhausted, and in transfers to media with both substrates it will thus eventually become predominant.

The two brown-colored *Chlorobium* species, *C. phaeobacteroides* and *C. phaeovibrioides,* with rod-shaped and slightly curved cell shapes, respectively, have been isolated by Pfennig[14] and Trüper and Genovese[15] from samples of fresh, brackish, and salt lake and pond water, where they have been found to occur in large numbers at depths of 7–13 m, that is, well below those harboring the green species. According to Trüper and Genovese, this can be attributed to the fact that little or no radiation at wavelengths greater than 700 nm penetrates to such depths, whereas the chocolate-brown *Chlorobium* species contain an abundance of carotenoids which maximally absorb in the 450–470 nm range—also the wavelength penetrating most deeply—and permit the organisms to photosynthesize there. If this explanation is correct, it should be possible to enrich specifically these species in Pfennig's standard medium illuminated with light in this wavelength range.

In contrast to *Chlorobium, Chloropseudomonas* comprises the small green sulfur bacteria that are motile; they are polarly flagellated. Only one species, *Chloropseudomonas ethylicum,* has thus far been described; in addition to H_2S it can also use some simple organic compounds, particularly ethanol, as oxidizable substrate. It can therefore be enriched selectively by using the standard medium supplemented with 0.1–0.2% ethanol and exposing the cultures to light of high intensity. Bacteriochlorophyll *d* has not been found as a component of the pigment system, but enrichment cultures exposed to light at a wavelength of 725–735 nm may well turn up strains with this bacteriochlorophyll.

From crude cultures containing both *Chlorobium* and *Chloropseudomonas,* the latter can also be specifically enriched by taking advantage of its motility, which allows the cells to migrate in a light gradient. A simple way to achieve a separation is to use a glass tube, nearly completely filled with the standard medium plus ethanol and rendered semisolid by the addition of 0.2% agar, which prevents disturbances due to convection currents. A few drops of the mixed culture are pipetted on top of the column, and the open end is closed off so as to exclude the entrance of air; whereupon the tube is exposed to a beam of light directed at the other end of the tube. The nonmotile chlorobia can continue to grow only in the liquid on top of the agar column, while the

[14] N. Pfennig, *Arch. Mikrobiol.* **63,** 224 (1968).

[15] H. G. Trüper and S. Genovese, *Limnol. Oceanogr.* **13,** 225 (1968).

motile chloropseudomonads migrate through it toward the light and develop throughout the tube.

Yet another member of the group of green sulfur bacteria is *Pelodictyon clathratiforme*. Its cells are rod-shaped, larger that those of *Chlorobium* and *Chloropseudomonas*, and bifurcated at one end. They grow as three-dimentional networks whose meshes are composed of closed chains of cells.

In enrichment cultures prepared as described for other members of the Chlorobiaceae, this unusual bacterium cannot compete with *Chlorobium* species, which will soon outgrow it. But a medium with a low sulfide concentration (0.03% $Na_2S \cdot 9\ H_2O$), exposure to light of low intensity, and incubation at a temperature below 20° provide conditions that severely limit the growth of chlorobia while permitting the continued development of *Pelodictyon*. Pfennig,[10,16] the first to grow *Pelodictyon* in the laboratory, specifies the conditions for its cultivation as follows: initial sulfide concentration not over 0.07%, later kept at or below 0.03% $Na_2S \cdot 9\ H_2O$; light intensity 200–500 lux (30–20 cm from a 25-W bulb), with diurnal alterations of light and darkness; incubation temperatures between 10° and 20°. In such cultures, *Pelodictyon* grows at the bottom of the culture vessel in the form of small, discrete flocks which are used for the inoculation of subcultures. If cultures are incubated for some weeks at a temperature between 4° and 8°, the cells of *Pelodictyon clathratiforme* produce gas vacuoles; the flocks then rise to the surface of the medium, where they accumulate immediately below the stopper of the culture bottle. This behavior provides another means for obtaining an incoulum rich in *Pelodictyon* and relatively free from other organisms that may be present in the culture.

Further purification can be effected by using the washing procedure that Pringsheim[17] has introduced and perfected for the isolation of algae. Some of the bottom deposit or surface scum containing flocks of *Pelodictyon* is dispersed in a few milliliters of sterile medium in a watch crystal, supported on a wire triangle in a petri dish. Individual flocks (colonies) are withdrawn with as little of the liquid as possible and transferred to another watch crystal with a few milliliters of sterile medium; the procedure is repeated several times. The operations are conveniently performed while being viewed through a good dissecting microscope. The consecutive washings eliminate many of the contaminating microorganisms; the progressive purification can be determined by examining the contents of the liquid left behind in the watch crystals under high

[16] N. Pfennig and G. Cohen-Bazire, *Arch. Mikrobiol.* **59,** 226 (1967).

[17] E. G. Pringsheim, "Pure Culture of Algae—Their Preparation and Maintenance." Cambridge Univ. Press, London and New York, 1946.

magnification. Finally the washed flocks are transferred to small culture vessels, e.g., screw-capped tubes with a capacity of 10 ml and completely filled with sterile medium; if inoculated into a large amount of liquid, growth may be considerably delayed.

It must be realized that the medium is unstable and apt to undergo marked changes in composition while being exposed to air in the watch crystals. Escape of CO_2 and H_2S into the atmosphere will cause an increase in pH, and the entrance of O_2 will create more or less aerobic conditions. These undesirable effects can be minimized by using a medium supplemented with ascorbate; by not dispensing it into the watch crystals until immediately prior to use; and by performing the operations as speedily as possible.

At the time of writing, pure cultures of *Pelodictyon clathratiforme* had not yet been obtained; the Pfennig isolates were still contaminated with sulfate reducing bacteria when last reported on. Thus it is not yet possible to determine whether increased knowledge of its physiological properties might suggest improvements in culture conditions that would render enrichment cultures still more specific.

Chlorobiaceae also occur in nature growing in symbiotic association with nonphotosynthetic bacteria. These "consortia," known as *Chlorochromatium aggregatum* (*Chloronium mirabile*) and *Pelochromatium roseum*, are composed of a large, rod-shaped, colorless, motile "inner bacterium"—which has not yet been isolated in pure culture—enveloped in a coating of green or brown *Chlorobium* species, respectively. The consortia multiply by the synchronous division of the components.

Both *Chlorochromatium* and *Pelochromatium* have been grown by Pfennig[10,18] in the standard medium under the same conditions described above for the cultivation of *Pelodictyon*. The "inner bacterium" confers motility on the complex, which responds phototactically to a light gradient, so that the organisms accumulate at the side of the culture vessel turned toward the light. From here they can be selectively removed by careful pipetting and transferred to fresh media for subculturing.

Thiorhodaceae. In Pfennig's standard medium exposed to daylight or incandescent bulbs, *Chlorobium* and *Chloropseudomonas* species grow faster than most, if not all, representatives of the red and purple sulfur bacteria, and thus tend to overgrow the latter in cultures started with a mixed inoculum. In order to establish enrichment cultures of members of the Thiorhodaceae it is therefore necessary either to use an inoculum in which these organisms are far more abundant than the green sulfur bacteria, or else to inhibit or prevent the growth of Chlorobiaceae by

[18] N. Pfennig, *Annu. Rev. Microbiol.* **21**, 285 (1967).

using light passed through a filter that transmits only radiation at wavelengths beyond 780 nm as a source of radiant energy. Additional refinements, which permit the specific enrichment of species of Thiorhodaceae that contain either bacteriochlorophyll *a* or *b*, involve exposure of the cultures to radiation in the 800–900 nm or in the 1000–1050 nm range, respectively. This was the approach used by Eimhjellen *et al.*[13] for isolation of *Thiocapsa pfennigii*, the first red sulfur bacterium known to produce bacteriochlorophyll *b*.

As in the case of the Chlorobiaceae, it has become possible to establish environmental conditions favoring the growth of particular species of Thiorhodaceae by modifying the sulfide concentration of the medium, the intensity of the incident light, and the temperature of incubation. Most of the work so far accomplished has served to define the optimum conditions for the cultivation of various types that occur in fresh, brackish, or seawater; it is to be hoped that this will be extended to the Thiorhodaceae commonly found in strongly saline environments. Despite the fact that here, too, many forms are encountered, including a variety of typical *Thiospirillum* species, only a single species has been isolated from this habitat until now (see below).

In the early studies with pure cultures of Thiorhodaceae,[6] it was observed that the organisms displayed considerable morphological variability. Recent studies have made it clear that this variability must be attributed to the use of media deficient in vitamin B_{12}[8,19,20]; in cultures of marine species, media with a low NaCl concentration may also induce abnormal cell shapes.[20] In media of adequate composition the morphology of all the species so far studied is remarkably constant. Specific directions for their cultivation, based largely on the work of Pfennig,[10,18] are found in the following paragraphs.

As mentioned above, *Thiocapsa pfennigii*, isolated by Eimhjellen *et al.*[13] and characterized by its production of bacteriochlorophyll *b*, was enriched in bottle cultures exposed to radiation in the 1000–1050 nm range. Light from a 100-W tungsten filament lamp was passed first through a water filter and next through an infrared filter opaque to radiation of wavelengths below 900 nm. The medium was Pfennig's standard medium with 0.075% $Na_2S \cdot 9\ H_2O$ at pH 6.6; the inoculum was river or brackish water mud. Since this organism was found not to require any vitamins for growth, the medium supplied more than the minimum requirements; a B_{12}-free solution should have been equally satisfactory. It is possible that a concentrated search for other bacterio-

[19] E. A. Petrova, *Mikrobiologiya* **28**, 814 (1959).

[20] H. G. Trüper and H. W. Jannasch, *Arch. Mikrobiol.* **61**, 363 (1968).

chlorophyll *b*-containing Thiorhodaceae will turn up strains that do need B_{12}-supplementation; in that event the use of media with and without B_{12} should permit specifically to enrich the different organisms. In view of the selective effects of light intensity and temperature of incubation on various Thiorhodaceae that produce bacteriochlorophyll *a*, these factors may also serve to discover a larger variety of bacteriochlorophyll *b* containing Thiorhodaceae.

All other species of this group that have been examined to date produce bacteriochlorophyll *a*. They can conveniently be grown in Pfennig's medium in bottle cultures illuminated by incandescent bulbs; their emission in the 800–900 nm wavelength range is fully adequate. In enrichment cultures exposed to continuous illumination at high intensity (700–2000 lux, equivalent to exposure to a 60-W bulb at a distance of 35–15 cm) and incubated at 25°–30°, small *Chromatium, Thiocapsa, Thiocystis, Rhodothece,* and *Amoebobacter* are predominant species. At a lower light intensity (300–700 lux; 25–15 cm distant from a 25-W bulb) and an incubation temperature of 20°–25°, *Chromatium warmingii* flourishes; while intermittent illumination (e.g., 16 hours light, 8 hours dark) at still lower light intensity (100–300 lux; 45–25 cm distant from a 25-W bulb), and an incubation temperature of 15°–20° favors the growth of the large *Chromatium okenii, C. weissei,* and the marine *C. buderi.* The last-mentioned species, discovered by Trüper and Jannasch,[20] requires a medium with about 3% NaCl for normal growth.

Essentially the same conditions (intermittent light of low intensity and low incubation temperature) are used for the cultivation of *Thiodictyon* and the large *Thiospirillum jenense;* but for these organisms the sulfide concentration must be reduced to 0.04% $Na_2S \cdot 9\ H_2O$, and the pH adjusted to 6.5–6.8.

From salt lakes and salt flats which contain H_2S, small vibrio- or spirillum-shaped Thiorhodaceae can readily be isolated with the aid of enrichment cultures in Pfennig's medium with a corresponding salt content[6,21–23]; they grow well under continuous illumination and at 30°; the sulfide concentration can be raised to 0.1–0.2% $Na_2S \cdot 9\ H_2O$, and the pH to 7.6–9.0. These organisms, for which Trüper[22] has proposed the generic name *Ectothiorhodospira,* do not store sulfur droplets inside their cells. As long as the medium still contains sulfide, the sulfur formed

[21] J. C. Raymond and W. R. Sistrom, *Arch. Mikrobiol.* **59**, 255 (1967).

[22] H. G. Trüper, *J. Bacteriol.* **95**, 1910 (1968).

[23] C. C. Remsen, S. W. Watson, J. B. Waterbury, and H. G. Trüper, *J. Bacteriol.* **95**, 2374 (1968).

as the primary oxidation product remains in solution as polysulfide, which imparts a distinctly yellow color to the medium. When all the sulfide has been oxidized, the cultures suddenly change appearance; they become turbid and milky white, owing to the presence of small sulfur droplets in the liquid. Some of the strains that have been studied in pure culture are obligate halophiles, requiring media with 14–25% NaCl for optimal growth and failing to develop in media with less than 5% NaCl. Others have lower salt requirements and cannot grow at high salt concentrations.

As is to be expected, the first enrichment cultures for purple sulfur bacteria usually contain mixtures of different representatives. A preliminary separation of particular types can often be achieved by taking advantage of the fact that the motile species respond chemo- and phototactically, and that some of the nonmotile ones produce gas vacuoles under appropriate conditions.

For an easy separation of motile from nonmotile cells, one uses a tube, bent upward at one end, filled with Pfennig's standard medium, and placed in a horizontal position. A sample of a thriving mixed culture is then introduced at the open end. Owing to their strongly negative aerotaxis, the motile cells rapidly migrate toward the closed end of the tube, whence they can be collected free from the nonmotile forms which perforce remain in the inoculated area. If, after inoculation, the tube is placed in a dark room with the closed end exposed to a not too bright light source (25-W bulb at a distance of about 10 cm), so that a light gradient is established over the length of the tube, the motile forms also respond phototactically and move toward the region of optimum light intensity. Since the rate of locomotion differs for various species, a further separation takes place inside the tube in the course of time. A flattened capillary tube with a broader open end to facilitate inoculation often permits one to follow the migration and distribution of the organisms by direct microscopic examination. Regions of the capillary in which the cells of a particular kind are found at a given time are then cut out and used as inoculum for new cultures, preferably on a small scale (screw-capped tubes of 10-ml capacity).

Cells of nonmotile species forming gas vacuoles accumulate at the surface of liquid cultures and thus can be readily obtained as a concentrated suspension for subculturing.

Photosynthetic Nonsulfur Bacteria

In contrast to the photosynthetic sulfur bacteria, the members of this group cannot use reduced inorganic sulfur compounds as oxidizable substrate.[24] Instead, their photosynthetic metabolism depends on the presence of organic substrates or, in some cases, of molecular hydrogen, as oxidizable substrates. Some species, such as *Rhodospirillum fulvum*, *R. molischianum*, and *R. photometricum*, and *Rhodopseudomonas viridis*, are obligate anaerobes and obligate phototrophs; others can be grown in darkness in the presence of air. However, under the latter conditions they cannot successfully compete with various nonphotosynthetic bacteria, so that satisfactory enrichment cultures can be obtained only with media incubated anaerobically and exposed to a source of radiant energy.

Bottles completely filled with solutions of complex organic materials, such as peptone or yeast extract, inoculated with some mud or stagnant water, incubated at 25°–30°, and exposed to continuous illumination from an incandescent light bulb, generally yield cultures of Athiorhodaceae in 4–7 days. But the outcome is quite unpredictable because nonphotosynthetic bacteria of various kinds are usually the first to develop, and they cause changes in the medium that are impossible to control. The use of sugar-containing media should be entirely avoided; these soon become infested with bacteria that ferment the sugar with the production of acid, which renders the media unfit for the growth of Athiorhodaceae.

Much better results are obtained with media that do not permit the growth of nonphotosynthetic organisms, while satisfying the minimum requirements of the photosynthetic ones. Since the latter can use a variety of oxidizable substrates that are not subject to fermentative decomposition (primary and secondary alcohols, fatty and aromatic acids, succinic acid), these are the substrates of choice for enrichment cultures of the nonsulfur purple bacteria. It is true that these substrates can also be oxidized under anaerobic conditions by nitrate-, sulfate-, and carbonate-reducing (methane producing) bacteria. But the development of the first two groups of organisms can be eliminated by using media without nitrate and sulfate, while the methane producers are not likely to cause difficulties on account of their limited distribution and low growth rate.

For many purposes a satisfactory base is a solution composed of H_2O; $NaHCO_3$, 0.2%; NH_4Cl, 0.1%; KH_2PO_4, 0.05%; $MgCl_2$, 0.05%; trace metals (see page 8); and NaCl in appropriate concentrations for organisms living in brackish or marine environments. It provides the necessary inorganic ingredients except sulfur. The addition of sulfate

[24] An exception is *Rhodopseudomonas palustris*, which can grow at the expense of thiosulfate. Besides, the author has obtained growth of *Rhodomicrobium vannielii* in inorganic media with sulfide as the oxidizable substrate, but this still requires confirmation.

should be avoided because it allows sulfate-reducing bacteria to grow and produce H_2S, which is toxic to Athiorhodaceae in very low concentrations while it encourages the development of photosynthetic sulfur bacteria. A small amount of $Na_2S \cdot 9\ H_2O$ (0.01–0.02%), supplied at the time of inoculation, is recommended for primary enrichment cultures; for subsequent transfers, when Arthiorhodaceae have already established themselves, $MgSO_4$ may be substituted for $MgCl_2$, and Na_2S may be omitted.

Most Athiorhodaceae require one or more B vitamins for growth, though none have yet been found to need B_{12}. To meet this requirement, it has been customary to supplement the mineral medium with 0.01–0.02% yeast extract. Higher concentrations are not recommended since they tend to promote the development of various nonphotosynthetic bacteria. Ethanol and acetate appear to be used by all species as oxidizable substrate, and if added in a concentration of 0.2%, either one yields cultures of considerable density. The pH of the complete medium should be adjusted to 6.8–7.0.

Though fully adequate for the cultivation of all the presently known representatives of the Athiorhodaceae, this medium is not selective for any of its members. But the important advances in our knowledge of the physiological properties of the different species have made it possible to specify more rigorous conditions that enable the specific enrichment of particular types. They are based upon the use of media in which one or more B vitamins are substituted for yeast extract; different organic compounds for ethanol or acetate; and exposure of the cultures to radiation in a limited wavelength region.

Vitamins. The studies of Hutner[25] have revealed that the various species of Athiorhodaceae have markedly different vitamin requirements, as follows: *Rhodopseudomonas palustris*, *p*-aminobenzoic acid; *Rhodopseudomonas capsulata*,[26] thiamine; *Rhodopseudomonas gelatinosa*, thiamine and biotin; *Rhodopseudomonas spheroides*, thiamine, biotin, and nicotinic acid; *Rhodospirillum rubrum*, biotin; *Rhodospirillum fulvum*, *p*-aminobenzoic acid; *Rhodospirillum molischianum*, unknown[27]; *Rhodomicrobium vannielii*, none.

Thus, in an ethanol or acetate medium supplemented with biotin as the sole B vitamin, only *Rhodospirillum rubum* can grow abundantly;

[25] S. H. Hutner, *J. Bacteriol.* **52**, 213 (1946); *J. Gen. Microbiol.* **4**, 286 (1950).

[26] The specific epithet *capsulatus*, which had earlier been used, is in conflict with the internationally accepted rules of nomenclature, and has accordingly been changed to *capsulata*.

[27] N. Pfennig, K. E. Eimhjellen, and S. Liaaen-Jensen, *Arch. Mikrobiol.* **51**, 258 (1965).

similarly, a thiamine-containing medium selects for *Rhodopseudomonas capsulata;* and one with *p*-aminobenzoic acid for *Rhodospirillum fulvum* and *Rhodopseudomonas palustris.* An ethanol medium with 0.03% $Na_2S \cdot 9\ H_2O$, but devoid of vitamins is useful for the enrichment of *Rhodomicrobium;* however, it also supports growth of some of the red sulfur bacteria.

Substrates. The range of organic substances that can serve as oxidizable substrates also varies with the species, so that the mineral medium supplemented with yeast extract and an organic compound used by only one or a few kinds of Athiorhodaceae can also serve for the specific enrichment of those organisms.

Glutarate in a concentration of 0.2%, and benzoate, *p*-hydroxybenzoate, or cinnamate (0.05%) frequently produce cultures in which *Rhodopseudomonas palustris* in the predominant species, but Pfennig[27] has occasionally found *Rhodospirillum fulvum* in benzoate enrichment cultures. If a primary enrichment culture containing both species is subcultured in a medium with 0.2% thiosulfate as the substrate, only *Rhodopseudomonas palustris* continues to grow, while *Rhodospirillum fulvum* comes to the fore in subcultures in a medium with 0.02–0.04% pelargonate.

Secondary alcohols, particularly *i*-propanol in a concentration of 0.2%, often yield cultures rich in *Rhodopseudomonas gelatinosa*[28]; propiqnate in the same concentration is a preferred substrate for *Rhodopseudomonas capsulata.* This species can also be enriched in a vitamin-supplemented medium incubated in the light in an atmosphere of 80% H_2 and 20% CO_2.[29] *Rhodopseudomonas spheroides,* which is nearly always present in primary enrichment cultures with ethanol or acetate, can be further enriched by transferring such cultures to a medium with 0.3% tartrate[30]; and *Rhodospirillum rubrum* to one with 0.3% alanine.

Caprylate (maximum concentration 0.05%) and pelargonate (0.02–0.04%) serve for the enrichment of *Rhodospirillum molischianum;* according to Drews[31] this species in inhibited by yeast extract in concentrations above 0.02% but can tolerate sulfide in a concentration of 0.05% $Na_2S \cdot 9\ H_2O$. Pelargonate has also been used for the enrichment of *Rhodospirillum fulvum.*

[28] J. W. Foster, *J. Gen. Physiol.* **24,** 123 (1940); *J. Bacteriol.* **47,** 355 (1944).

[29] J. Klemme, Wasserstoff-Stoffwechsel und Hydrogenase-Reaktionen bei Athiorhodaceae, Ph.D Thesis, Göttingen Univ., 1968.

[30] Tartrate is not a satisfactory substrate for primary enrichment cultures of this species because it usually undergoes a rapid decomposition under the influence of representatives of the Enterobacteriaceae.

[31] G. Drews, *Zentralbl. Bakteriol. Parasitenk. Infectionskr. Hyg., Abt. 1. Supplementh.* **1,** 170 (1965).

The most effective enrichment cultures for particular species are, of course, obtained by using media containing a combination of type-specific vitamins and substrates.

Pfennig[32,33] has recently described two new species of Athiorhodaceae, neither of which depends on an external supply of B vitamins for growth. *Rhodopseudomonas acidophila*[32] can be effectively enriched in a vitamin-free mineral medium with 0.15–0.2% succinate at an initial pH of 5.1. *Rhodospirillum tenue*[33] develops preferentially in a neutral medium containing 0.02–0.04% pelargonate where, in the absence of B vitamins, it can successfully compete with *Rhodospirillum fulvum*.

Radiant Energy Source. The green *Rhodopseudomonas viridis*[34,35] is the only currently known member of the Athiorhodaceae that produces bacteriochlorophyll *b*. It can therefore be specifically enriched by exposing cultures in an ethanol or acetate medium to radiant energy in the 1000–1050 nm range, as described above for the isolation of *Thiocapsa pfennigii*. Different strains have been found to differ in their vitamin requirements; Pfennig[18] mentions that one of his strains can grow in the absence of vitamins, while another requires both biotin and *p*-aminobenzoic acid.

Isolation of Pure Cultures

As described in the previous section, the improved enrichment culture methods for photosynthetic bacteria, based upon the use of chemically defined media, make it possible to obtain cultures in which one particular species predominates. Its isolation in pure culture can be accomplished in various ways.

Dilution Cultures

In some cases the most convenient method is the preparation of subcultures in liquid media inoculated with progressive dilutions of the mother culture; it obviates the possibility of damaging the organisms though manipulations that involve a temporary exposure to air or to high temperatures. The media used should be properly "aged" beforehand; this is particularly important for those species that are obligate anaerobes (green sulfur bacteria, many of the large Thiorhodaceae, *Rhodospirillum fulvum*, *Rhodospirillum molischianum*, *Rhodospirillum photometricum*, *Rhodopseudomonas viridis*). It is advisable to use screw-capped tubes

[32] N. Pfennig, *J. Bacteriol.* **99**, 597 (1969).

[33] N. Pfennig, *J. Bacteriol.* **99**, 619 (1969).

[34] K. E. Eimhjellen, O. Aasmundrud, and A. Jensen, *Biochem. Biophys. Res. Commun.* **10**, 232 (1963).

[35] G. Drews and P. Giesbrecht, *Arch. Mikrobiol.* **53**, 255 (1966).

of about 10-ml capacity, filled completely with the appropriate medium, so that the growth of a small number of cells in the inoculum is not unduly retarded.

In principle, it is possible to obtain a pure culture by determining the cell density of a mother culture, preparing a suspension diluted to the point where a conveniently measurable fraction contains a single cell, and using this as the inoculum for a subculture. In practice this is, however, complicated by the fact that the population density cannot be estimated with sufficient accuracy. Besides, only viable cells will grow in subcultures, and it is therefore necessary that the relative numbers of living and dead cells be known.

If microscopic examination of the enrichment culture shows the vast majority of organisms to be motile, it may, of course, be safely concluded that viability is close to 100%. In such cases one prepares a suspension in sterile growth medium containing about 100 cells/ml and inoculates a series of, say, 10 tubes with 1 ml each. One of these, after careful mixing, is used to inoculate a second series with 1-ml aliquots each; a tube of this series is similarly used to inoculate a third series; and the process is repeated once more. Because the tubes of the third series presumably were inoculated with one cell each, most of these, and only a single tube of the final series, are apt to show growth.

If the viability of the population of the mother culture cannot be judged by simple microscopic examination, a larger primary inoculum is advisable; it is then also necessary to prepare a larger series of cultures at progressive, 10-fold dilutions.

In either case, the final series should comprise a number of tubes in which growth does not occur; the positive cultures in that set may then be assumed to have received a single viable cell, and thus to represent pure cultures.

A fairly reliable estimate of the number of cells in the mother culture can usually be made by direct microscopy. A volume of 0.01 ml of the culture is placed on a slide, spread over an area of 1 cm^2, and viewed under oil immersion. A culture containing 10^6 cells/ml then shows approximately one cell per field. More accurate determinations can be made by using a standard counting chamber.

Shake Cultures

The use of solid media in which separate and spatially fixed cells can grow out into colonies is often preferable for pure culture isolation and, where applicable, should be the method of choice. For the isolation of many kinds of photosynthetic bacteria, so-called "shake cultures" offer advantages over plate cultures; they are prepared as follows.

A series of sterile tubes, held in a water bath at 45°, is filled about two-thirds with a liquefied agar medium of the desired composition at the same temperature. The first tube is inoculated with a few drops of the mother culture, and the contents thoroughly mixed by gently rolling the tube and at the same time tilting it up and down.[36] About one-tenth to one-twentieth is quickly poured into the second tube; the first tube is placed in a vertical position in cold water so as rapidly to cause the agar to cool and solidify; the second tube, after mixing its contents, is similarly used to inoculate the third tube; and the procedure is repeated until 8–10 tubes, representing 10- to 20-fold consecutive dilutions, have been prepared. After the medium has solidified, the agar is covered with an approximately 2-cm deep layer of a melted, sterile 1:1 mixture of paraffin and paraffin oil; this greatly reduces the rate of diffusion of air into, and the escape of gaseous components from, the agar medium. During its solidification the paraffin plug contracts, and this is frequently accompanied by its partial retraction from the glass wall, leaving air channels that would defeat its purpose. It is therefore necessary to inspect the tubes after some time and to perfect the seal by mild, local heating and tapping to drive out the air bubbles; this may have to be repeated a number of times. The tubes are incubated with exposure to a light source such as a 25-W incandescent bulb at a distance of about 25 cm.

Only a small number of colonies should develop in the tubes representing the higher dilutions. These colonies are then carefully examined for uniformity of shape and contents, and the process is repeated by using a suspension of a single, apparently homogeneous colony as the inoculum for a second series of shake cultures. All the colonies in the tubes of this series must be identical before it can be concluded that a pure culture has been obtained.

Removal of the agar column from the tubes is best accomplished by first melting and discarding the paraffin seal, and then slowly pushing a sterile capillary pipette, provided with a piece of rubber tubing which permits the maintenance of a slight air pressure, between the glass wall and the agar down to the bottom of the tube. Plugging of the capillary can be prevented by frequent up-and-down movements of the pipette. When the tip has reached the bottom, further application of pressure causes the agar column to be pushed out of the tube; it is col-

[36] The term "shake culture," although in common use, is misleading: the mixing should definitely not be done by shaking the contents of the inoculated agar. This would introduce air bubbles which remain in the solidified agar and may prevent the growth of colonies.

lected in a sterile dish, where it can be dissected so as to permit ready access to individual colonies.

Attention must here be drawn to a phenomenon that can often be observed in cultures of this kind. A colony developing in the agar in close proximity to the glass wall may, during its further expansion, abut on the film of liquid between the agar column and the tube wall, where it is formed as the result of syneresis associated with the solidification of the agar. If this happens, the growth is apt to spread out in the liquid film and produce a continuous sheet of cells on the outside of the agar column over a considerable area. Inevitably, these cells are distributed still more widely when the column is eventually extruded and dissected. Consequently, any discrete colony inside the agar that has been made accessible to further manipulations will always be "contaminated." Fortunately, the extent of this contamination is usually too small to eliminate the colony from being used as a promising inoculum for a new series of shake culures.

The agar medium used for the isolation of the photosynthetic sulfur bacteria should contain the same ingredients, and in the same proportions, as the liquid medium described earlier (see pages 7–9). It is best prepared in small batches by mixing 100 ml of solution 1, now supplemented with 2.5–3% agar, 50 ml of solutions 2 plus 3, and 5–7 ml of solution 4. The final amount suffices for filling a set of 8–10 tubes At the time of mixing, all solutions should be at a temperature of 45°; at lower temperatures the medium prematurely solidifies, while at higher temperatures a considerable amount of CO_2 and H_2S are lost to the atmosphere. Because it is not feasible to let the medium age before it is inoculated, the addition of ascorbic acid to a final concentration of about 0.05% is recommended; this ensures anaerobic conditions from the start. In order to minimize the escape of CO_2 and H_2S, all operations, including the inoculation of the consecutive tubes, should be performed as quickly as possible.

The isolation of Athiorhodaceae can be similarly accomplished with agar media of a composition corresponding to those used for their enrichment. But in many cases it is more convenient to use a single complex medium in which most if not all members of the group can grow. A 1% yeast extract or peptone, 0.2% sodium malate, 1.5% agar medium has been found eminently satisfactory for this purpose. It has the advantage that the addition of CO_2 is unnecessary, so that the medium can be prepared and sterilized in bulk, stored in the gel state, and melted when needed without undergoing changes in composition. Addition of ascorbate prior to inoculation is necessary for the cultivation of the strictly anaerobic species.

Unfortunately, agar shake cultures cannot be used for the isolation of *Pelodictyon* and most of the larger Thiorhodaceae. Thus far it has not been possible to make these organisms grow in a solid medium. Their purification and isolation must therefore be accomplished with the aid of the liquid dilution technique.

Plate Cultures

Instead of leaving the agar media, inoculated with progressive dilutions of an enrichment culture, to solidify in the tubes in which they have been prepared, they can also be poured into sterile culture dishes. Here the developing colonies are more or less immediately accessible to subsequent operations, and examination of the colonies in a plate by direct microscopy under low to intermediate power is greatly facilitated. Besides, a plate with as many as a few hundred evenly spaced colonies is still quite satisfactory for pure culture isolation, which is not so in the case of an agar column where they would be more crowded together and more difficult to reach individually. In consequence, one can use a greater dilution rate (100-fold) for consecutive inocula, and thus save on materials.

For anaerobic incubation the plates are stacked right side up in a container which is repeatedly evacuated and filled with an O_2-free atmosphere. During the evacuation the air that has been trapped between the agar and the bottom of the culture dish tends to escape, causing the agar to be broken into bits; this can be avoided by making a few slits through the agar, down to the glass.

The evacuation also removes dissolved gases, such as CO_2 and H_2S, which may be necessary constituents of the medium, thereby radically changing its composition. For this reason the method is not suitable for the isolation of photosynthetic sulfur bacteria. It is, however, satisfactory for Athiorhodaceae because they can be grown in the abovementioned complex medium which is not subject to such alterations.

Those members of the latter group that are not strict anaerobes can develop colonies in pour plates exposed to air, or even on the surface of agar plates, which may therefore be inoculated by streaking. This eliminates the need to prepare consecutive dilutions and simplifies the examination and removal of colonies even more. These organisms are also facultative phototrophs, and can be grown in darkness on the same media that permit anaerobic growth in the light. It must, however, be mentioned that one and the same species is likely to produce colonies of rather dissimilar appearance when grown in darkness or in light. Under the former conditions, the production of photosynthetically active pigments is greatly reduced, so that the colonies appear much paler. Some species, such as *Rhodopseudomonas capsulata* and *Rhodopseudomonas spheroides,* when grown in the light, develop as brown-colored or deep red colonies under anaerobic and aerobic conditions, respectively; when incubated in darkness, the colonies are nearly colorless with reddish centers.

Criteria for Determining the Purity of Cultures

Even the most careful microscopic examination of a culture is insufficient to establish its purity beyond all doubt. Not only is it impossible thus to detect the presence, among thousands or millions of cells, of a single, microscopically distinct organism; the problem is further aggravated by the fact that cells of quite different bacterial species may so resemble each other as to render them indistinguishable under the microscope. Yet, for many kinds of experiments the use of pure cultures is imperative because an occasional contaminant may become predominant in cultures exposed to different environmental conditions. A simple example may illustrate the point.

Suppose that a flourishing culture of a *Chlorobium* species, growing anaerobically in a sulfide medium exposed to radiation in the 730–760 nm range, contains a few cells of a *Chromatium*, a *Rhodopseudomonas,* and a nonphotosynthetic *Pseudomonas* species, which escape detection by microscopic examination. If this culture were transferred to a sterile sulfide medium of the same composition as the original one, and the new culture exposed to radiation in the 800–900 nm region, the *Chromatium* would rapidly outgrow the *Chlorobium.* In transfers to a sulfide-free medium supplied with an organic substrate, either the *Pseudomonas* or the *Rhodopseudomonas* would soon become predominant, depending upon the conditions of incubation—the former in dark-aerobic, the latter in light-anaerobic cultures. This may suffice to show that far more rigorous criteria are needed to determine whether a culture is pure.

Those organisms that can be grown in or on agar media are easiest to deal with; here a sole requirement suffices. It is that a suspension, prepared from a single colony growing on or in an agar medium, when subcultured in or on a medium of the same composition and incubated under fully comparable conditions, must yield only one kind of colony, identical with the original one. If so, any colony from these subcultures may be confidently considered pure; if more than one type of colony appears, repetition of the procedure is necessary.

The examination of the subcultures must, however, be quite thorough, for colonies of different bacteria may appear indistinguishable to the naked eye. Nevertheless, they show differences in shape, color, structure, and texture, and even the slightest of these can be observed by careful inspection of plates or tubes with the aid of a microscope.

If the organism in question is unable to grow in or on solid media, this criterion obviously cannot be applied. Then it is necessary to resort to subcultures in liquid media. These should be of widely different composition, simple as well as complex, to provide conditions under which various kinds of contaminants would be enriched; and for the same reason the cultures should be incubated both aerobically and anaerobically, in light and in darkness, and at different temperatures. Only if all such subcultures fail to reveal growth of other organisms may the initial culture be presumed to be pure.

As mentioned earlier, the photosynthetic sulfur bacteria are usually closely associated in their natural habitat with sulfate reducing bacteria, which can persist in cultures of the former. Consequently, cultures of the sulfur bacteria should be tested particularly for the presence of sulfate reducers. This can best be done by inoculating a bottle, completely filled with an aqueous solution containing 0.1% peptone, 0.1% $(NH_4)_2SO_4$, 0.1% K_2HPO_4, 0.05% $MgSO_4$, 0.05% sodium ascorbate, 0.5% sodium lactate, a small crystal of $FeSO_4$, NaCl as needed, and adjusted to a final pH of 6.8–7.2, with the culture to be tested, and incubating the bottle at 25°–30° in darkness.

Maintenance of Pure Cultures

All the photosynthetic sulfur bacteria are best maintained in liquid cultures; for some of them it is the only method so far available. Cultures in 100-ml bottles, filled with Pfennig's standard sulfide medium, incubated at a temperature of 10°–20°, and alternately exposed to radiation from a 25-W incandescent bulb and to darkness, require relatively little attention because growth and sulfide consumption are slow. At intervals, replenishment of the sulfide supply is necessary (see page 9); but this does not become an excessive burden. More than four consecutive "feedings," at about 1-week intervals, is not advisable because of the progressive increase in density of the cultures. After that, the cultures should be transferred to fresh media. This kind of maintenance ensures the instant availability of cultures growing under conditions closely approximating those in which the organisms live in nature.

The Athiorhodaceae can also be maintained in liquid cultures; but here a wider choice of media is possible. The various chemically defined media in which they had been initially enriched may be replaced either by a single synthetic medium which contains an organic substrate, such as acetate or malate in a concentration of 0.2%, that can be used by all of the various species, supplemented with a mixture of B vitamins required by some of them; or by a complex medium, such as a 1% yeast extract or peptone solution, in which all the representatives can grow.

The cell density reached in any of these media is such that subsequent feeding is unnecessary, and transfers need be made no more often than at monthly intervals. In fact, cultures of *Rhodospirillum rubrum* in bottles with a yeast extract medium have been found to contain viable cells even after a year when the cultures had been stored on a shelf exposed to daylight from a north window at room temperature (15°–22°). Nevertheless, more frequent transfers are recommended because growth of the organisms, especially in media with salts of organic acids, is accompanied by an appreciable increase in alkalinity.

A disadvantage of cultivation in a complex medium is that spontaneously appearing mutants may there find better opportunities for development, so that in the course of time a pure culture tends to contain a mixture of organisms with different properties. This is less likely to happen if a minimal medium is used.

Instead of perpetuating cultures of Athiorhodaceae in liquid media, they can also be maintained as stab cultures. These are prepared by inoculating tubes, filled for about two-thirds of their length with an appropriate agar medium solidified in a vertical position, with a straight needle which carries the inoculum at its tip and is stabbed into the agar column over the greatest part of its length. For cultures of strictly anaerobic species the space above the agar is then filled with sterile paraffin oil, and the tube is tightly closed with a screw cap or sterile rubber stopper. This is superfluous for organisms that are not inhibited by exposure to air. After a few days' incubation with continuous illumination, when good growth along the stabs has become visible, they can be stored in dim daylight. Monthly transfers are recommended.

Many of the smaller sulfur and nonsulfur photosynthetic bacteria have been successfully preserved in a lyophilized state; and some of the most sensitive types, which have not remained viable during lyophilization, have responded well to storage in liquid nitrogen.

The most extensive collection of photosynthetic sulfur bacteria is being maintained by Professor N. Pfennig, Institute for Microbiology of the University, 34 Göttingen, Grisebachstrasse 8 (Germany). The American Type Culture Collection carries a large number of strains of Athiorhodaceae.

[2] Algal Cultures—Sources and Methods of Cultivation

By RICHARD C. STARR

Natural populations of algae are rarely species pure (contain a single species), clonal (composed of descendants from a single cell through asexual reproduction), or axenic (devoid of other microorganisms such as bacteria), yet for many types of investigations it is absolutely necessary that the population used possess all three of these attributes. Methods for establishing such species pure, clonal, axenic cultures of algae have been developed by many workers in the 25 years since the publication of the review "Cultivation of the Algae" by Bold[1] and the small book "Pure Cultures of Algae" by Pringsheim.[2] No longer are investigators limited to such laboratory weeds as *Chlorella*. The investigations of Provasoli, Pintner, Droop, Hutner, Pringsheim, and many others have defined the sophisticated requirements of the media, such as trace metals, vitamins, and pH, and many of the more exotic forms such as *Ochromonas, Euglena,* and *Phaeodactylum* can now be grown successfully. Thus, algae exhibiting unique pigment systems, varied heterotropic nutrition, sexual systems, etc., are now available for investigation.

Algae may be isolated from freshly collected natural populations or from enrichment cultures designed to favor the growth and development of certain physiological types if they are present. Populations of algae occur in a great variety of habitats—hot and cold; wet and dry; high and low osmotic pressures, light intensities, or pH; dilute and enriched watery environments. The growth of a natural population under any one set of environmental conditions does not, however, necessarily indicate that the alga grows best under those conditions; rather, the successful growth of the alga may be due to its tolerance rather than preference in combination with a lack of competition among the microflora. On the other hand, an investigator interested in an alga with a special set of physiological attributes may profit often by searching for it in habitats with the requirements under which it must, or might, grow. One of the most detailed and useful accounts of methods used in the isolation and purification of algal cultures has been given by Lewin.[3] It should be remembered, however, that one may successfully free an alga of bacteria

[1] H. C. Bold, *Bot. Rev.* **8,** 69 (1942).

[2] E. G. Pringsheim, "Pure Cultures of Algae," 119 pp. Cambridge Univ. Press, Cambridge, England, 1946.

[3] R. A. Lewin, *Rev. Algol.* **4,** 181 (1959).

and, as a result, find that some vital organic constituent normally provided by bacterial action must now be supplied in the medium. It is this stumbling block that continues to prevent the successful establishment of axenic cultures of many algae.

Culture Collections of Algae

A wide variety of algal strains is now available from established culture collections of algae. Many are small private collections where an investigator working with a particular alga or on a particular process has accumulated stocks as a working collection. Such collections tend to be temporary in nature and limited in scope. On the other hand, there are now five major collections of algae which are organized to serve the general scientific community and which supply cultures at nominal cost. With no restrictions on the importation of algal cultures into most of the countries of the world, these collections are a source for a wide variety of algal strains, many of which have served in the past in classic research.

The five major collections are: Culture Centre of Algae and Protozoa, 36 Storey's Way, Cambridge CB3 ODT, England; Sammlung von Algenkulturen, Pflanzenphysiologischen Institut der Universität Göttingen, Nikolausbergerweg 18, Göttingen, Germany; Sbirka kultur autotrofnich organismu CSAV, Vinicna, 5, Praha 2, Czechoslovakia; Culture Collection of Algae, Institute of Applied Microbiology, University of Tokyo, Tokyo, Japan; Culture Collection of Algae, Department of Botany, Indiana University, Bloomington, Indiana 47401.

Lists of the cultures maintained in each Collection are available upon request. Some lists have appeared in scientific publications.[4-6]

With few exceptions, speciation in the algae is based on morphological criteria with little concern or knowledge of the physiological characteristics. It is therefore most important that strains of algae used in physiological and biochemical research be identified more explicitly than by the species name. Investigators often consider the name of a laboratory or other investigator as proper identification of a strain, but this is no solution; witness the frequent reference in the literature to the "Emerson" strain[7] of *Chlorella* and the many strains attributed to him

[4] W. Koch, *Arch. Mikrobiol.* **47**, 402 (1964).

[5] M. Baslerova and J. Dvorakova, "Algarum, Hepaticarum, Muscorumque in culturis collectio," 59 pp. Ceskoslovenke akademie ved. Praha., 1962.

[6] R. C. Starr, *Amer. J. Bot.* **51**, 1013 (1964).

[7] "The Emerson strain of *Chlorella* referred to in the paper in *Nature* **187**, 613–614, 1959, was obtained from Jack Myers, Austin, Texas. I asked Jack about its origin myself. He obtained it directly from Emerson, probably before the Indiana Collection was established, and he does not know if it can be identified with any number of Indiana's or any other collection. I am not able to tell you if it is

in the various culture collections. It is most important that strains be identified by numbers, preferably that of an established culture collection.

In the following section certain genera of algae are listed, and the directions for their growth are given. The directions should be considered as a point of departure, for in many instances the particular requirements of the experimental procedure may require some alteration of medium or physical conditions.

It is recommended that glass-distilled water be used in compounding all media. Deionized water may be suitable in some instances, but experience has shown that this is not always the case.

Growth Conditions

Anabaena: The toxic strains of *Anabaena flos-aquae,* a planktonic filamentous blue green alga, have been grown by Gorham *et al.*[8] in the following medium known as ASM-1:

	(μmole/liter)
$NaNO_3$	2000
$MgSO_4$	200
$MgCl_2$	200
$CaCl_2$	200
K_2HPO_4	100
Na_2HPO_4	100
$FeCl_3$	4
H_3BO_3	40
$MnCl_2$	7
$ZnCl_2$	3.2
$CoCl_2$	0.08
$CuCl_2$	0.0008
Na_2EDTA	20

The alga was grown at 22 ± 1° in 125-ml flasks, continuously shaken with illumination at approximately 1700 lux.

Anacystis: The high-temperature strain of blue green algae known as *Anacystis nidulans* (Bloomington 625; Cambridge 1405/1; Göttingen

identical with *Chlorella vulgaris* as used in the paper in *Plant Physiol.* **33,** 109–113 (1958) since, as I understand, Myers, from whom I obtained the Emerson strain, does not know himself." (Letter from C. Sorokin to R. C. Starr, July 28, 1970.)

[8] P. R. Gorham, J. McLachlan, U. T. Hammer, and W. K. Kim, *Verh. Int. Verein. Limnol.* **15,** 796 (1964).

1402-1) has been used extensively in research. It is generally agreed that this strain is not unicellular, but forms short chains of cells, and thus the identification is not correct. The strain continues to be used under this name inasmuch as phycologists have not yet agreed as to its proper designation. Pringsheim[9] has discussed the nomenclature and proposed *Lauterbornia nidulans* as a new genus and species. The formula for medium C of Kratz and Myers[10] is as follows:

	(g/liter)
$MgSO_4 \cdot 7H_2O$	0.25
K_2HPO_4	1.00
$Ca(NO_3)_2 \cdot 4H_2O$	0.025
KNO_3	1.0
Na_3 citrate $\cdot$ $2H_2O$	0.165
$Fe_2(SO_4)_3 \cdot 6H_2O$	0.004
A_5 microelements	1.0 ml
Microelements stock solution	
H_3BO_3	2.86
$MnCl_2 \cdot 4H_2O$	1.81
$ZnSO_4 \cdot 7H_2O$	0.222
MoO_3 (85%)	0.0177
$CuSO_4 \cdot 5H_2O$	0.079

The optimum temperature for growth is 39°.

Van Baalen[11] eliminated some problems of precipitation after autoclaving by modifying the above medium in the following manner:

Van Baalen's C_g-10 medium

Kratz and Myers medium C with the following changes:

1. K_2HPO_4 lowered to 0.050 g/l;
2. Glycylglycine added at 1.0 g/l;
3. Na_2EDTA (0.010 g/l) in place of sodium citrate.

Van Baalen pointed out that this modified medium gave good results only when the alga was grown at 39°.

Allen[12] was able to grow such blue green algae as *Anacystis nidulans* and *Gloeocapsa alpicola* on the surface of agar plates using a modified

[9] E. G. Pringsheim, *Arch. Mikrobiol.* **63,** 1 (1968).

[10] W. A. Kratz and J. Myers, *Amer. J. Bot.* **42,** 282 (1955).

[11] C. Van Baalen, *J. Phycol.* **3,** 154 (1967).

[12] M. M. Allen, *J. Phycol.* **4,** 1 (1968).

medium of Hughes *et al.*[13] solidified with 1.5% agar which had been sterilized separately from the mineral components of the medium.

Modified medium[12] of Hughes *et al.*[13]

	(g/liter)
$NaNO_3$	1.5
K_2HPO_4	0.039
$MgSO_4 \cdot 7H_2O$	0.075
Na_2CO_3	0.02
$CaCl_2$	0.027
$Na_2SiO_3 \cdot 9H_2O$	0.058
EDTA	0.001
Citric acid	0.006
Fe citrate	0.006
Microelements	1 ml

Microelements solution

H_3BO_4	2.86
$MnCl_2 \cdot 4H_2O$	1.81
$ZnSO_4 \cdot 7H_2O$	0.222
$Na_2MoO_4 \cdot 2H_2O$	0.391
$CuSO_4 \cdot 5H_2O$	0.079
$Co(NO_3)_2 \cdot 6H_2O$	0.0494

The pH of the medium was 7.8.

To prepare the solid medium, equal volumes of double-strength mineral base and double-strength Difco Bacto-agar were separately sterilized and combined after cooling at 48°.

When the algae were grown in liquid culture, the medium was continuously bubbled with a 0.5% CO_2–air mixture.

Ankistrodesmus: As in *Chlorella*, different strains of *Ankistrodesmus* with the same morphology may exhibit different physiological traits. The biochemical properties of 18 strains of *Ankistrodesmus* as reported by Kessler and Czygan[14] and Kessler[15] are shown in Table I. The comparable numbers of the strains in the four major Collections are given.

Kessler and Czygan used the following medium, as devised earlier by Kessler *et al.*[16]:

[13] E. O. Hughes, P. R. Gorham, and A. Zehnder, *Can. J. Microbiol.* **4**, 225 (1958).

[14] E. Kessler and F. Czygan, *Arch. Mikrobiol.* **55**, 320 (1967).

[15] E. Kessler, *Arch. Mikrobiol.* **58**, 270 (1967).

[16] E. Kessler, W. Langner, I. Ludewig, and H. Wiechmann, *Stud. Microalgae Photosyn. Bact.* pp. 7–20 (1963).

TABLE I
BIOCHEMICAL PROPERTIES OF 18 STRAINS OF *Ankistrodesmus*

Species names	Nos. of algal collections				Color of N deficiency	Secondary carotenoid[a]	Hydro-genase activity	Gelatine digestion	pH limits
	Göttingen	Bloomington	Cambridge	Tokyo					
A. ammalloides Chodat et Oettli	202-1	190	202/1		orange	Ast., (Can.)	+	+	6.0
A. braunii (Naegeli) Collins	202-7a	244	202/7a		orange	Ast., (Can.)	++	++	3.5
A. braunii (Naegeli) Collins	202-7b	245	202/7b	C96	orange	Ast., (Can.)	++	++	4.0
A. braunii (Naegeli) Collins	202-7c				orange	Ast., (Can.)	++	++	4.0
A. braunii (Naegeli) Collins	202-7d				orange	Ast., (Can.)	++	++	3.5
A. braunii (Naegeli) Collins	202-7e				orange	Ast., (Can.)	++	+	6.0
A. braunii (Naegeli) Collins		750			orange	Ast., (Can.)	++	++	3.5
A. densus Korschikoff	202-12				orange	Ast., (Can.)	+	+	5.5
A. falcatus (Corda) Ralfs		749			orange	Ast., (Can.)	+	+	5.0

A. falcatus var. *acicularis* (A. Br.) G. S. West		101		C88	orange	Ast., (Can.)	++	+	5.0
A. falcatus var. *acicularis* (A. Br.) G. S. West		748			orange	Ast., (Can.)	++	+	5.0
A. falcatus var. *duplex* (Kützing) G. S. West	202-9	187	202/9a	C90	orange	Ast., (Can.)	++	+	5.5
A. falcatus var. *spirilliformis* G. S. West	202-2[b]	189	202/2	C91	orange	Ast., (Can.)	+	+	5.5
A. falcatus var. *spirilliformis* G. S. West	202-3	188	202/3		orange	Ast., (Can.)	+	+	5.5
A. falcatus var. *spirilliformis* G. S. West	202-4[c]	241	202/4		orange	Ast., (Can.)	+	+	4.5
A. falcatus var. *stipitalus* (Chodat) Lemm.	202-5	242	202/5		orange	Ast., (Can.)	+	+	5.0
A. nannoselene Skuja	202-6	243	202/6		orange	Ast., (Can.)	++	+	4.5
A. stipitatus	202-11[d]				orange	Ast., (Can.)	—	—	3.5

[a] Ast. = Astaxanthin-Ester; Can. = Canthaxanthin.
[b] Also known as *A. angustus*.[6]
[c] Also known as *A. falcatus* var. *mirabilis*.[6]
[d] Also known as *Keratococcus bicaudatus* (A. Braun) Boye-Peteroen, [B. Fott and E. Truncová, Acta Univ. Carol., Biol. 1964, p.97].

	(g/liter)
KNO_3	0.81
NaCl	0.47
$NaH_2PO_4 \cdot 2H_2O$	0.47
$Na_2HPO_4 \cdot 12H_2O$	0.36
$MgSO_4 \cdot 7H_2O$	0.25
$CaCl_2 \cdot 6H_2O$	0.02
$FeSO_4 \cdot 7H_2O$	0.010
$MnCl_2 \cdot 4H_2O$	0.0005
H_3BO_3	0.0005
$ZnSO_4 \cdot 7H_2O$	0.0002
$(NH_4)_6Mo_7O_{24} \cdot 4H_2O$	0.00002

The final pH was adjusted to 6.3.

The cultures were illuminated with light of an intensity of 4000 lux and maintained at a temperature of 23°. Sterile air containing 2% CO_2 was bubbled through the cultures.

Chlamydomonas: Sueoka[17] and Levine's group at Harvard, who use *Chlamydomonas reinhardi,* employ the following medium or a modification of it as indicated:

	Minimal medium	High salt minimal medium
NH_4Cl (g)	0.05	0.50
$MgSO_4 \cdot 7H_2O$ (g)	0.02	0.02
$CaCl_2 \cdot 2H_2O$ (g)	0.01	0.01
K_2HPO_4 (g)	0.72	1.44
KH_2PO_4 (g)	0.36	0.72
Hutner's trace elements (ml)	1	1
Distilled water (liter)	1	1

Hutner's trace elements solution[18]

	(g)
EDTA	50.0
$ZnSO_4 \cdot 7H_2O$	22.0
H_3BO_3	11.4
$MnCl_2 \cdot 4H_2O$	5.1
$FeSO_4 \cdot 7H_2O$	5.0
$CoCl_2 \cdot 6H_2O$	1.6
$CuSO_4 \cdot 5H_2O$	1.6
$(NH_4)_6Mo_7O_{24} \cdot 4H_2O$	1.1

Boil the medium in 750 ml of distilled water, cool slightly, and bring to pH 6.5–6.8 with KOH (do not use NaOH). The clear solution is diluted to 1000 ml with distilled water and should have a green color which changes to purple on standing. It is stable for at least one year.

Surzycki[19] recommends two modifications of Sueoka's media: (1) Tris–minimal-phosphate medium and (2) minimal high-magnesium medium.

For heterotrophic acetate mutants the media may be supplemented with sodium acetate at a concentration in the medium of 0.20%.

Chlorella: Chlorella was one of the first algae isolated into pure culture by Beijerinck in the 1890's and since that time many other investigators have isolated new strains.

Chlorella has few distinctive morphological characteristics which can be used in the classical taxonomic fashion; studies of the strains in culture have resulted in monographs of the genus based largely on morphological characteristics,[20] on morphological and physiological characteristics,[21] and on physiological criteria.[22-24]

There is obvious lack of agreement between the several investigators as to the proper designation of species. In view of the difficulty of positive identification, it is strongly urged that any investigator using *Chlorella* should identify his strain not so much by the species name, which may be highly suspect in many instances, but more importantly by a strain number, preferably that of an established Culture Collection. Many strains occur in more than a single Collection; therefore, to clarify partially the problem of cross identification, Table II includes not only morphological and physiological data but also cross listings of the many strains from the Cambridge, Göttingen, Bloomington, and Tokyo Collections, using species names according to Kessler.[24]

A high-temperature strain of *Chlorella* (Tx 7-11-05) was isolated by Sorokin and Myers[25] and has been used extensively by Sorokin and

[17] N. Sueoka, *Proc. Nat. Acad. Sci.* (*Wash.*) **46,** 83 (1960).

[18] S. K. Bose, *in* "Bacterial Photosynthesis" (H. Gest, A. San Pietro, and L. P. Vernon, eds.), p. 501. Antioch Press, 1963.

[19] S. Surzycki, this volume, [4].

[20] B. Fott and M. Novakova, *in* "Studies in Phycology" (B. Fott, ed.). Prague, 1969.

[21] C. J. Soeder, *in* "Studies on Microalgae and Photosynthetic Bacteria," pp. 21–34. Japanese Society of Plant Physiologists 1963.

[22] I. Shihira and R. W. Krauss, "*Chlorella,* Physiology and Taxonomy of Forty-one Isolates." University of Maryland, College Park, Maryland, 1965.

[23] E. Kessler, *Arch. Mikrobiol.* **52,** 291 (1965).

[24] E. Kessler, *Arch. Mikrobiol.* **55,** 346 (1967).

[25] C. Sorokin and J. Myers, *Science* **117,** 330 (1953).

TABLE II

Chlorella Strains in Four Major Culture Collections of Algae (Speciation According to Kessler[24])

Species	Numbers in culture collection				Species characteristics				
	Gött-ingen	Bloom-ington	Cam-bridge	Tokyo	Hetero-trophic	Hydro-genase	Second-ary carote-noids	pH limit	Cell form
C. luteoviridis Chodat	211-2a	21	211/2a	C97	–	–	–	3	Large, round
	211–2b	248	211/2b						
	211-3	22	211/3						
	211-4	23	211/4						
	211-5a	24	211/5a						
	211-5b	490	211/5b						
C. saccharophila	211-1a	20	211/1a	C87	–	–	–	2–3	Ellipsoidal
	211-1b	246	211/1b	C102					
	211-1c	247	211/1c						
	211-1d								
			211-1f						
	211-9a	27	211-9a						
	211-9b								
C. paramecii Loefer	211-6	130	211/6		+				
C. protothecoides Kruger	211-7a	25	211/7a	C99	+				
	211-7b	249	211/7b						
	211-7c	250	211/7c						
	211-7d	411	211/7d						
C. variegata Beijerinck	211-10a	28	211/10a						
	211-10b	255	211/10b						
	211-10c	256	211/10c						
	211-10d	257	211/10d						
	211-10e	258	211/10e						

C. xanthella Beijerinck	211-13	31	211/13		+				
C. vulgaris Beijerinck	211-1e				–	–	–	4–4.5	Small, round
	211-8l								
	211-8m								
	211-11b	259	211/11b						
	211-11c	260	211/11c						
	211-11f								
	211-11j	265	211/11j						
	211-11l								
	211-11p								
	211-11q								
	211-11r								
	211-11s								
	211-11t								
	211-12	30							
	211-19								
			211/21						
	211-30								
		395							
		396							
C. zopfingiensis Donz	211-14a	32	211/14	C111	–	–	++	4.5–5.5	Large, round
	211-14b								
	211-14c								
Chlorella I[a]	211-8a	26	211/8a		–	++	+	3.5–4	Large, slightly ellipsoidal
(= "pyrenoidosa")	211-8b	251	211/8b	C104					
	211-8c	252	211/8c	C105					
	211-8e								
			211/8g						
			211/8h						
	211-11m								
	211-11n								
	211-15								
		343		C101					

(*Continued*)

TABLE II (*Continued*)

Species	Numbers in culture collection				Species characteristics				
	Gött-ingen	Bloom-ington	Cam-bridge	Tokyo	Hetero-trophic	Hydro-genase	Second-ary carote-noids	pH limit	Cell form
Chlorella II	211-11g	262	211/11g		–	+	–	3	Large, round
	211-11h	263	211/11h						
		397							
		398							
Chlorella III	211-8k	1230[b]	211/8k		–	+	–	4–4.5	Small, round
		261	211/11d						
	211-11k								
	211-31								
	211-32[c]								
	211-33[c]								
	211-34[c]								
	211-40a								
	211-40c								

[a] The description of *Chlorella pyrenoidosa* Chick does not differ from that of *C. vulgaris* by Beijerinck, and therefore, Soeder[21] prefers to use Roman numbers as a temporary measure.
[b] Sorokin and Myers Tx7-11-05; thermophilic.
[c] Thermophilic.

others. Comparison of the characteristics of low- and high-temperature strains of *Chlorella* by Sorokin[26] is given in Table III.

Ellsworth and Aronoff[27,28] in their investigations on the biogenesis of chlorophyll *a* have isolated mutants of the high temperature strain of *Chlorella* (Tx 7-11-05). These have been deposited in the Collection at Bloomington under the following numbers:

IU 1663	Wild type (Tx 7-11-05) received by Aronoff from Sorokin
IU 1664	AJ mutant; protoporphine mutant of wild type
IU 1665	BE mutant; vinyl pheoporphine mutant of wild type
IU 1666	CA mutant; pheophorbide mutant of wild type
IU 1667	CA-AC-AZ mutant; protoporphine submutant of CA
IU 1668	CA-CA mutant; subisolate of CA mutant accumulating increased amounts of CA pigments
IU 1669	Ca-B_2C-B_2F mutant; divinyl pheoporphine submutant of CA
IU 1670	B_1E-B_2P mutant; divinyl pheoporphine submutant of BE
IU 1671	B_1E-B_1D-B_1N mutant; subisolate of BE with increased production of divinyl pheoporphine

Sorokin and Krauss[29] used the following medium in their work on the effects of light intensity on the growth rates of green algae:

	(g/liter)
KNO_3	1.25
KH_2PO_4	1.25
$MgSO_4 \cdot 7H_2O$	1.00
$CaCl_2$	0.0835
H_3BO_3	0.1142
$FeSO_4 \cdot 7H_2O$	0.0498
$ZnSO_4 \cdot 7H_2O$	0.0882
$MnCl_2 \cdot 4H_2O$	0.0144
MoO_3	0.0071
$CuSO_4 \cdot 5H_2O$	0.0157
$Co(NO_3)_2 \cdot 6H_2O$	0.0049
EDTA	0.5

The pH of the medium was 6.8.

For their work on the effects of temperature and illuminance on

[26] C. Sorokin, *Nature* **184,** 613 (1959).

[27] R. K. Ellsworth and S. Aronoff, *Arch. Biochem. Biophys.* **125,** 35 (1968).

[28] R. K. Ellsworth and S. Aronoff, *Arch. Biochem. Biophys.* **125,** 269 (1968).

[29] C. Sorokin and R. W. Krauss, *Plant Physiol.* **33,** 109 (1958).

TABLE III
TEMPERATURE CHARACTERISTICS OF TWO STRAINS OF *Chlorella pyrenoidosa*

Characteristic	Unit of measurement	Strain of Emerson[7]	*Chlorella* 7-11-05[a]
(1)	(2)	(3)	(4)
Temperature optimum for:			
growth	°C	25–26	38–39
photosynthesis	°C	32–35	40–42
endogenous respiration	°C	30	40–42
glucose respiration	°C	30	40–42
Upper temperature limit for:			
growth	°C	29	42
photosynthesis	°C	39	45
endogenous respiration	°C	Above 45	Above 50
glucose respiration	°C	Above 45	Above 50
Growth rate at light saturation:			
at 25°C	number of doublings per day	3.1	3.0
at 39°C	number of doublings per day	—	9.2
Growth rate at half saturating light intensity:			
at 25°C	number of doublings per day	2.4	2.3
at 39°C	number of doublings per day	—	7.0
Rate of apparent photosynthesis at light saturation:			
at 25°C	mm^3 O_2/mm^3 cells/hr	43	47
at 39°C	mm^3 O_2/mm^3 cells/hr	Rapidly declining	170
Rate of apparent photosynthesis at half saturating light intensity:			
at 25°C	mm^3 O_2/mm^3 cells/hr	32	30
at 39°C	mm^3 O_2/mm^3 cells/hr	—	115
Rate of endogenous respiration:			
at 25°C	mm^3 O_2/mm^3 cells/hr	1.8	2.0
at 39°C	mm^3 O_2/mm^3 cells/hr	1.4	5.5
Rate of glucose respiration:			
at 25°C	mm^3 O_2/mm^3 cells/hr	4.5	8
at 39°C	mm^3 O_2/mm^3 cells/hr	1.6	18
Saturating light intensity for growth:			
at 25°C	fc	500	500
at 39°C	fc	—	1400
Saturating light intensity for photosynthesis:			
at 25°C	fc	400	500
at 39°C	fc	—	1600

[a] Göttingen 211-8k; Bloomington 1230; Cambridge 211/8k.

Chlorella, Sorokin and Krauss[30] used a medium with the following composition (in grams per liter): KH_2PO_4, 1.31; $MgSO_4 \cdot 7H_2O$, 0.5; and urea, 0.44. The concentration of minor elements (in parts per million) was Ca, 5; Fe, 2; Mn, Zn, and B, 0.5; Cu, 0.04; Mo, 0.02; Co and V, 0.01 The Fe, Mn, Zn, Cu, and Co were used as compounds chelated by ethylenediaminetetraacetic acid (EDTA). The pH of the medium was 6.0.

Euglena gracilis: All the strains below were used by Pringsheim and Pringsheim[31] in experimental elimination of chloroplasts and eyespots using heat. Further data on their nutrition is given by Hutner *et al.*[32]:

Göttingen	Bloomington	Cambridge	Pringsheim (1952)	Remarks
1224-5/11	1716	1224/5k	No. 11	Grows at 34°; does not bleach
1224-5/16	1717	1224/5q	No. 16	Grows at 34°; does not bleach
1224-5/19	1718	1224/5i	No. 19	Grows at 34°; does not bleach
1224-5/20	1719	1224/5v	No. 20	Grows at 34°; does not bleach
1224-5/15	160	1224/7	No. 15	Grows and bleaches at 34°
1224-5/1	367	1224/5a	No. 1	No growth above 28°
1224-5/2	368	1224/5b	No. 2	No growth above 28°
1224-5/9	752	1224/5t	No. 9	Grows and bleaches at 34°; uses glucose

For growth and maintenance of stock cultures of *Euglena gracilis,* the following medium is used routinely at the Culture Collection of Algae at Cambridge and at Bloomington[33]:

To 1000 ml of Pyrex-distilled water add:

Sodium acetate	1.0 g
Beef extract	1.0 g
Tryptone	2.0 g
Yeast extract	2.0 g
Calcium chloride	0.01 g

If desired, the above medium may be solidified by adding 15 g of agar.

Böger and San Pietro[34] in studying ferredoxin and cytochrome *f* used the following medium for autotropic growth of *E. gracilis:*

[30] C. Sorokin and R. W. Krauss, *Plant Physiol.* **37,** 37 (1962).
[31] E. G. Pringsheim and O. Pringsheim, *New Phytol.* **51,** 65 (1952).
[32] S. H. Hutner, M. K. Bach, and G. I. M. Ross, *J. Protozool.* **3,** 101 (1956).
[33] R. C. Starr, *Amer. J. Bot.* **51,** 1013 (1964).
[34] P. Böger and A. San Pietro, *Z. Pflanzenphysiol.* **58,** 70 (1967).

	(mg/liter)	Stock solution (mg/100 cc)	(cc/liter)
$(NH_4)_2SO_4$	0.135×10^3	1.35×10^3	10
NH_4Cl	0.90×10^3	9.0×10^3	10
KH_2PO_4	1.0×10^3	10×10^3	10
Na citrate · $2H_2O$	0.51×10^3	5.0×10^3	10
$MgCl_2 \cdot 6H_2O$	0.35×10^3	3.5×10^3	10
$FeCl_2 \cdot 4H_2O$	4.72		1
NH_4 molybdate	0.20	20	1
$NaVO_4 \cdot 16H_2O$	0.10	10	1
H_3BO_3	0.50	50	1
$ZnCO_3$ ($ZnSO_4 \cdot 6H_2O$)	0.18 (0.38)	18 mg (38 mg)	1
$CaCl_2$	0.20	20	1
$MnCl_2 \cdot 4H_2O$	1.8	180	1
$Co(NO_3)_2 \cdot 6H_2O$ ($CoCl_2$)	1.3 (1.35)	130 mg (135 mg)	1
$CuSO_4 \cdot 5H_2O$	0.02	2	1
Thiamine	1	1 (mg/ml)	1
Vitamin B_{12}	0.005	50 (mg/liter) (dil. 1:10)	1

The pH is 4.4.

Hutner *et al.*[35] recommend the following heterotrophic medium for the vitamin B_{12} assay using the "Z" strain of *Euglena gracilis*. The "Z" strain is found in the various collections as Bloomington No. 753; Cambridge 1224/5z; Göttingen 1224-5/25.

"Dry mix" basal medium for B_{12} assay

	Concentration (mg %) in final medium	Wt. (g) for 10 liters of final medium
KH_2PO_4	30	3.0
$MgSO_4 \cdot 7H_2O$	40	4.0
L-Glutamic acid	300	30.0
$CaCO_3$	8	0.8
Sucrose	1500	150.0
DL-Aspartic acid	200	20.0
DL-Malic acid	100	10.0
Glycine	250	25.0
Ammonium succinate*	60	6.0
Thiamine HCl†	0.06	0.6
"Metals 45"	2.2	0.22

* This appears to be no longer commercially available in the U.S.A.; it may be replaced by an equivalent amount of succinic acid + NH_4HCO_3.

† Triturate (0.6 g) containing 1 g of thiamine HCl + 99 g of sucrose ("1:100" triturate).

The basal medium is made up double strength by dissolving the ingredients in distilled water by steaming for 15 minutes. Cool; check pH—it should be pH 3.6; if adjustment is necessary, use NaOH or H_2SO_4. Filter. Bottle. Sterilize by autoclaving at 10 pounds for 15 minutes.

For assay of B_{12} in sera of patients treated with sulfonamides, add *p*-aminobenzoic acid (PABA) 0.125 mg % to the final medium. PABA does not alter the growth curve in the absence of sulfonamides.

"Metals 45" dry mix

Concentration (mg %) in final medium	Compound used	Gravimetric factor	Wt. for 1000 liters of final medium (g)
Fe 0.2	$FeSO_4 \cdot (NH_4)_2SO_4 \cdot 6H_2O$	7.02	14.0
Zn 0.1	$ZnSO_4 \cdot 7H_2O$	4.4	4.4
Mn 0.05	$MnSO_4 \cdot H_2O$	3.1	1.55
Cu 0.008	$CuSO_4 \cdot 5H_2O$	3.9	0.31
Co 0.01	$CoSO_4 \cdot 7H_2O$	4.8	0.48
B 0.01	H_3BO_3	5.6	0.57
Mo* 0.035	$(NH_4)_6Mo_7O_{24} \cdot 4H_2O$	1.8	0.64
V 0.001	$Na_3VO_4 \cdot 16H_2O$	9.3	0.093
		Total	22.043

* The value for Mo is almost certainly needlessly high; in practice no inhibition was observed even at much higher levels.

Notes on the metal mix: The ingredients are ground together and stored in a dry place. No direct evidence is at hand that *Euglena* requires B, Mo, V, or Co beyond that in B_{12}; they are added for the sake of completeness in the light of data for other algae.

Ochromonas: Ochromonas is a unicellular flagellate belonging to the Chrysophyceae. In addition to being photosynthetic, it is able to ingest particulate matter or to absorb dissolved materials as a source of carbon. It grows at acidic, basic, and neutral pH's. Its photosynthetic apparatus exhibits a reversible light–dark, appearance–disappearance, a feature permitting useful experimentation as to compounds associated with the photosynthetic apparatus.[36]

[35] S. H. Hutner, M. K. Bach, and G. I. M. Ross, *J. Photozool.* **3,** 101 (1956).
[36] S. Miyachi, S. Miyachi, and A. A. Benson, *J. Photozool.* **13,** 76 (1966).

Ochromonas malhamensis (Bloomington 1297; Cambridge 933/1a; Göttingen 933-1a) and *O. danica* (Bloomington 1298; Cambridge 933/2; Göttingen 933-7) have been studied in greatest detail. Hutner *et al.*[37] described in detail the nutrition of *O. malhamensis*, especially the changes in requirements when the alga is grown at temperatures above 35°. *Ochromonas danica* later proved to be a better experimental material for studies involving cytotoxic compounds; Hutner *et al.*[38] used the following medium for its growth:

	(g/liter)
DL-Malic acid	0.4
KH_2PO_4	0.3
$MgCO_3$	0.4
$MgSO_4 \cdot 3H_2O$	1.0
L-Glutamic acid	3.0
Na_2 succinate · $6H_2O$	0.1
NH_4Cl	0.5
Glucose	10.0
$CaCO_3$	0.05
Minor elements	(mg/liter)
Fe	1.0
Mn	3.0
Zn	3.0
Mo	0.37
Cu	0.18
B	0.037
V	0.037
Vitamins	(mg/liter)
Thiamine HCl	10.0
Biotin	0.01

The pH of the medium is 3.6.

Phaeodactylum: Phaeodactylum tricornutum Bohlin has been used

[37] S. H. Hutner, H. Baker, S. Aaronson, H. A. Nathan, E. Rodriguez, S. Lockwood, M. Sanders, and R. A. Petersen, *J. Protozool.* **4**, 259 (1957).

[38] S. H. Hutner, A. C. Zahalsky, S. Aaronson, and R. M. Smillie, *in* "Biochemistry of Chloroplasts" (T. W. Goodwin, ed.), Vol. **2**, pp. 703–720. Academic Press, New York, 1967.

widely in the last 50 years for investigations dealing with the physiology and biochemistry of the diatoms, but the strain has been known as *Nitzschia closterium* f. *minutissima*. The Plymouth strain (Bloomington 642; Cambridge 1052/1a; Göttingen 1090-1a) and others are discussed by Lewin,[39] and their taxonomic relationship to *P. tricornutum* are described in detail.

Mann and Myers[40] used the Droop strain (Bloomington 646; Cambridge 1052/6) for a study of the pigments, growth, and photosynthesis of *P. tricornutum*. The medium which they used was modified from the ASP-2 medium of Provasoli *et al.*[41]:

	(g/liter)
NaCl	5.0
$MgSO_4 \cdot 7H_2O$	1.2
$NaNO_3$	1.0
KCL	0.60
$CaCl_2$	0.30
K_2HPO_4	0.10
Tris	1.0
Micronutrient stock (added at 10 ml/liter)	
Na_2EDTA	3.0
H_3BO_3	0.60
$FeSO_4 \cdot 7H_2O$	0.20
$MnCl_2$	0.14
$ZnSO_4 \cdot 7H_2O$	0.033
$Co(NO_3)_2 \cdot 6H_2O$	0.0007
$CuSO_4 \cdot 5H_2O$	0.0002

Optimum growth was obtained at 18°, aeration with 1–2% CO_2, and a moderate light intensity (one 20-W fluorescent lamp on each side, 6 inches from the test tube cultures).

Porphyridium: The unicellular red alga *Porphyridium cruentum* is maintained in three Collections (Bloomington 161; Cambridge 1380/1a;

[39] J. C. Lewin, *J. Gen. Microbiol.* **18,** 427 (1958).
[40] J. E. Mann and J. Myers, *J. Phycol.* **4,** 349 (1968).
[41] L. Provasoli, J. J. A. McLaughlin, and M. R. Droop, *Arch. Mikrobiol.* **25,** 392 (1957).

Göttingen 1380-1a). Brody and Emerson[42] used the following medium:

Major components	(mole/liter)
KCL	0.215
NaCl	0.214
KNO_3	0.0123
$MgSO_4 \cdot 7H_2O$	0.0101
K_2HPO_4	0.00287
$Ca(NO_3)_2 \cdot 4H_2O$	0.00106
KI	0.000301
KBr	0.000420
$FeSO_4$	0.00002*
Micronutrients	**(μmole/liter)**
A_5: H_3BO_3	50.0
$MnSO_4$	13.0
$ZnSO_4$	1.0
$(NH_4)_6Mo_7O_{24} \cdot 4H_2O$	0.5
$CuSO_4$	0.5
B_9: $Al_2(SO_4)_3K_2SO_4 \cdot 2H_2O$	1.0
KBr	1.0
KI	0.5
$Cd(NO_3)_2 \cdot 4H_2O$	0.5
$CO(NO_3)_2 \cdot 6H_2O$	0.5
$NiCl_2 \cdot 6H_2O$	0.5
$Cr(NO_3)_3 \cdot 7H_2O$	0.1
$V_2O_4(SO_3)_3 \cdot 16H_2O$	0.1
$Na_2WO_4 \cdot 2H_2O$	0.1

* Use 1 ml of stock per liter of final medium. $FeSO_4$ stock 0.02 *M* in 0.01 *M* HNO_3.

The precipitate which forms after sterilization and cooling is dissolved by bubbling commercial carbon dioxide through the medium for a few minutes.

Pyrobotrys (*Chlamydobotrys*): The heterotrophic nutrition of this green colonial flagellate, incapable of autotrophic growth, was studied by Pringsheim and Wiessner[43,44] using the following defined medium:

[42] M. Brody and R. Emerson, *Amer. J. Bot.* **46,** 433 (1959).
[43] E. G. Pringsheim and W. Wiessner, *Arch. Mikrobiol.* **40,** 231 (1961).
[44] W. Wiessner, *Arch. Mikrobiol.* **43,** 402 (1962).

	(mole/liter)
NH_4Cl	1.5×10^{-3}
K_2HPO_4	3.0×10^{-4}
$MgSO_4 \cdot 7H_2O$	4.0×10^{-5}
$CaSO_4$	7.0×10^{-5}
Asparagine ($H_2OC—CH(NH_2)—CH_2—CO—NH_2 \cdot H_2O$)	1.0×10^{-2}
Na acetate ($CH_3—COONa \cdot 3H_2O$)	4.0×10^{-3}

Minor elements

$FeSO_4 \cdot 7H_2O$	2.0×10^{-4}
$ZnSO_4 \cdot 7H_2O$	3.5×10^{-5}
$MnSO_4 \cdot 4H_2O$	1.0×10^{-5}
H_3BO_3	2.0×10^{-4}
$Co(NO_3)_2 \cdot 6H_2O$	3.5×10^{-6}
$CuSO_4 \cdot 5H_2O$	4.0×10^{-9}
$Na_2MoO_4 \cdot 2H_2O$	4.0×10^{-6}

Vitamins:

B_{12} 0.01% stock; use 0.2 ml/liter
B_1 0.2% stock; use 0.1 ml/liter

Na-EDTA 200 mg/liter

The solution of vitamin B_1 should be sterilized by Seitz or Millipore filtration and added separately after the medium has been autoclaved and cooled. The pH is adjusted to 7.0 with NaOH or HCl.

Stock cultures were maintained by Pringsheim and Wiessner in:

	(g/liter)
Na acetate	5
Beef extract	5
Tryptone	10
Yeast extract	10

Minor elements and salts as in previous defined medium

Scenedesmus: Kessler and Czygan[14] reported on certain biochemical properties of *Scenedesmus* strains and showed that, as in *Ankistrodesmus* and *Chlorella,* strains with the same morphology may differ physiologically. The strains and their cross listing among the several major culture

TABLE IV
BIOCHEMICAL PROPERTIES OF 26 SCENEDESMUS STRAINS[14,15]

Species names	Nos. of algal collections				Color of N deficiency	Secondary carotenoids[a]	Hydrogenase activity	Gelatin-digestion	pH limit
	Göttingen	Bloomington	Cambridge	Tokyo					
S. acuminatus (Lagerh.) Chodat	276-12	415	276/12		orange	Ast., Can.	++	+	4.0
S. acutiformis Schröder	276-11	416	276/11		orange	Ast., (Can.)	++	+	4.0
S. basiliensis Vischer	276-1	83	276/1a		yellow	Ast., Can., Ech.	++	+	4.0
S. bijugatus var. *seriatus* Chodat		413	276/14		orange	(Ast.), (Can.)	+	++	4.0
S. dimorphus Kützing	276-10	417	276/10		orange	Ast., Can., (Ech.)	++	+	4.5
S. dimorphus Kützing		746			yellow	Ast., Can., Ech.	++	+	4.0
S. dimorphus Kützing		1237			yellow	Ast., Can., (Ech.)	+	+	4.0
S. dispar Bréb.	276-13	414	276/13		orange	Ast., Can., (Ech.)	++	++	5.5
S. longus Meyen		1236			orange	Ast., Can., Ech.	+	++	4.0
S. naegelii Chodat	276-2	74	276/2		yellow	Ast., Can., Ech.	++	+	4.0

S. obliquus (Turp.) Krüger	276-3a	72	276/3a		yellow	Ast., Can.	++	+	4.0
S. obliquus (Turp.) Krüger	276-3b	78	276/36		yellow	Ast., Can., (Ech.)	++	+	4.0
S. obliquus (Turp.) Krüger	276-3c				yellow	Ast., Can., (Ech.)	++	+	3.5
S. obliquus (Turp.) Krüger	276-3d				yellow	Ast., Can.	++	+	4.0
S. obliquus (Turp.) Krüger	276-3e				yellow	Ast., Can., (Ech.)	++	+	4.0
S. quadricauda Bréb.	276-4a	77	276/4a		orange	Ast., Can.	++	++	5.5
S. quadricauda Bréb.	276-4b	76	276/4b		orange	Ast., Can., (Ech.)	+	++	5.0
S. quadricauda Bréb.	276-4c				orange	Ast., Can., (Ech.)	+	++	5.0
S. quadricauda Bréb.	276-4d				orange	Ast., Can., (Ech.)	++	++	5.0
S. quadricauda Bréb.	276-4e				orange	Ast., Can., Ech.	+	++	5.0
S. spec.	276-5				yellow	Ast., Can.	++	+	3.5
S. spec.	276-6[b]	393		C72	yellow	Ast., Can.	++	+	4.0
S. spec.	276-7				yellow	Ast., Can.	++	+	4.0
S. spec.	276-8				orange	Ast., Can.	++	+	4.5
S. spec.	276-9				orange	Ast., Can., Ech.	+	+	4.5
S. spec.	276-20				orange	Ast., Can., Ech.	++	++	4.0

[a] Ast. = Astaxanthin; Can. = Canthaxanthin; Ech. = Echinenon.
[b] *Gaffron D* 3; also known as *S. obliquus*.

collections are given in Table IV. The medium used by Kessler and Czygan has been given earlier (see *Ankistrodesmus*).

Volvulina: Volvulina is a green colonial flagellate. The more common species, *V. steinii,* requires an exogenous carbon source, while *V. pringsheimii* is autotrophic. Carefoot[45] studied the nutrition of these two species, utilizing the following media:

	Stock solution (g/400 ml)	(ml/liter)
$NaNO_3$	10	10
$CaCl_2$	1	10
$MgSO_4 \cdot 7H_2O$	3	10
K_2HPO_4	3	10
KH_2PO_4	7	10
NaCl	1	10
PIV metal mix (See Volvox medium)		5

This medium would support growth of *V. pringsheimii,* but for *V. steinii* it was necessary to add an exogenous carbon source. Carefoot showed that *V. steinii* would respond to a wide variety of carbon sources, but sodium acetate at a concentration in the medium of 2.5×10^{-3} *M* was routinely used.

Volvox: In 1959 Provasoli and Pintner devised a medium for growing *Volvox globator*. More recently in the writer's laboratory this medium has proved to be excellent for growing many species of *Volvox,* other colonial green algae, and various other algae. It is absolutely necessary that cultures be axenic when using this medium. The medium is dilute, and the glycylglycine does not buffer the medium sufficiently to permit aeration with air containing added CO_2. Aeration with sterile air is permissible. The medium is prepared by Starr[46] in the following manner:

For each 1000 ml of medium required, stock solutions in the amounts indicated are added to 981 ml of glass-distilled water:

(ml)	Stock solution	(g/100 ml)
1	$Ca(NO_3)_2 \cdot 4H_2O$	11.8
1	$MgSO_4 \cdot 7H_2O$	4.0
1	Na_2 glycerophosphate	5.0
1	KCl	5.0
10	Glycylglycine	5.0
1	Biotin	25.0×10^{-6}
1	B_{12}	15.0×10^{-6}
3	PIV metal solution	(as given below)

[45] J. R. Carefoot, *J. Protozool.* **14,** 15 (1967).
[46] R. C. Starr, *Arch. Protistenk.* **111,** 204 (1969).

PIV METAL SOLUTION. To 500 ml of glass-distilled water, add 0.750 g of Na_2-EDTA. After this chelating agent is dissolved, add the following salts in the amounts indicated:

	(g)
$FeCl_3 \cdot 6H_2O$	0.097
$MnCl_2 \cdot 4H_2O$	0.041
$ZnCl_2$	0.005
$CoCl_2 \cdot 6H_2O$	0.002
Na_2MoO_4	0.004

Adjust final *Volvox* solution to pH 7.0 using 1 *N* NaOH. For some strains of *Volvox* and other algae, thiamine is added to an amount equal to 1 mg/liter.

[3] Preparation and Photosynthetic Properties of Synchronous Cultures of *Scenedesmus*

By NORMAN I. BISHOP and HORST SENGER

In 1919 Warburg[1] introduced the unicellular alga *Chlorella* into photosynthesis research. Since then, mass cultures of green algae have been increasingly used for plant physiology studies, but little or no attention has been paid to the influence of different developmental stages on their photosynthetic behavior. There was no way to approach this problem until the techniques of producing synchronous cell cultures became available and the synchronous cell development was understood. Since a synchronous culture represents a parameter for the development of a single cell, it also provides the tool to study any inherent photosynthetic changes of a single cell.

When Tamiya and co-workers[2] first obtained synchronized *Chlorella* cultures, they recognized that a change in the photosynthetic capacity occurred during the life cycle. Subsequently many synchronous cultures have been examined for their changes in photosynthetic behavior, but the initial results appeared confusing. Most investigators found an increase of photosynthetic capacity during the first part of the life cycle, followed by a decrease to a minimal level just prior to cell division and subsequently an additional small increase. Others noted either a steady increase or a continuous decline of photosynthetic capacity during the

[1] O. Warburg, *Biochem. Z.* **100,** 230 (1919).

[2] H. Tamiya, T. Iwamura, K. Shibata, E. Hase, and T. Nihei, *Biochim. Biophys. Acta* **12,** 23 (1953).

life cycle. Most of the disagreements in these results can be traced back to different and inadequate synchronizing methods.[3] Therefore, the most desirable conditions for synchrony of a culture should be obtained prior to assessment of the possible effects of developmental stages on photosynthetic activity.

Comparison of Random and Synchronized Cultures

Every green microalga in a mass culture follows its own life cycle, each of which probably starts at a different time. In an ideal *random* culture, the initiation of the cell's cycle depends only upon statistical distribution (Fig. 1). Distribution curves of cell size should remain the same at any time during the culture period.

A random culture can be obtained only when all growth factors are kept constant. This implies that the self-shading of the cells and the selective uptake of individual compounds from the nutrient medium must be compensated for by dilution with fresh nutrient medium. This can be accomplished by using a photoelectrically controlled dilution device.[4-6] Ideally, such a culture is in a steady state and continuously in a logarithmic growth phase.

To obtain a *synchronized* culture requires only that all life cycles be initiated simultaneously (Fig. 1). This requires that at any time during the life cycle of the cells, the distribution curves of cell size will be different. Distribution curves for both culture types are given in Fig. 1. The development of a synchronized culture requires at least one discontinuous growth factor which causes the synchronization. In the case of green algae, the most useful parameter is light. The dilution with fresh nutrient medium is mostly done once during a life cycle, but might be performed, for certain experimental reasons, continuously during the light period.[7,8] A synchronized culture is never in a steady state and only temporarily in a logarithmic growth phase.

A third type of culture is the so-called *batch* culture. Its cell population is the most difficult to define, but also is the most commonly used one for photosynthetic studies. The external growth conditions are the same as for the random culture, but self-shading and selective uptake from the nutrient medium are not compensated for by dilution. Thus the culture has only a short period of logarithmic growth during which the population may have any stage between random and synchrony, depend-

[3] H. Senger, *Planta* **90,** 243 (1970).
[4] J. N. Philips and J. Myers, *Plant Physiol.* **29,** 148 (1954).
[5] J. A. Bassham and M. Kirk, *Biochim. Biophys. Acta* **43,** 464 (1960).
[6] H. Senger and H. J. Wolf, *Arch. Mikrobiol.* **48,** 81 (1964).
[7] C. Sorokin, *Physiol. Plant.* **10,** 659 (1957).
[8] C. Sorokin, *Physiol. Plant.* **13,** 20 (1960).

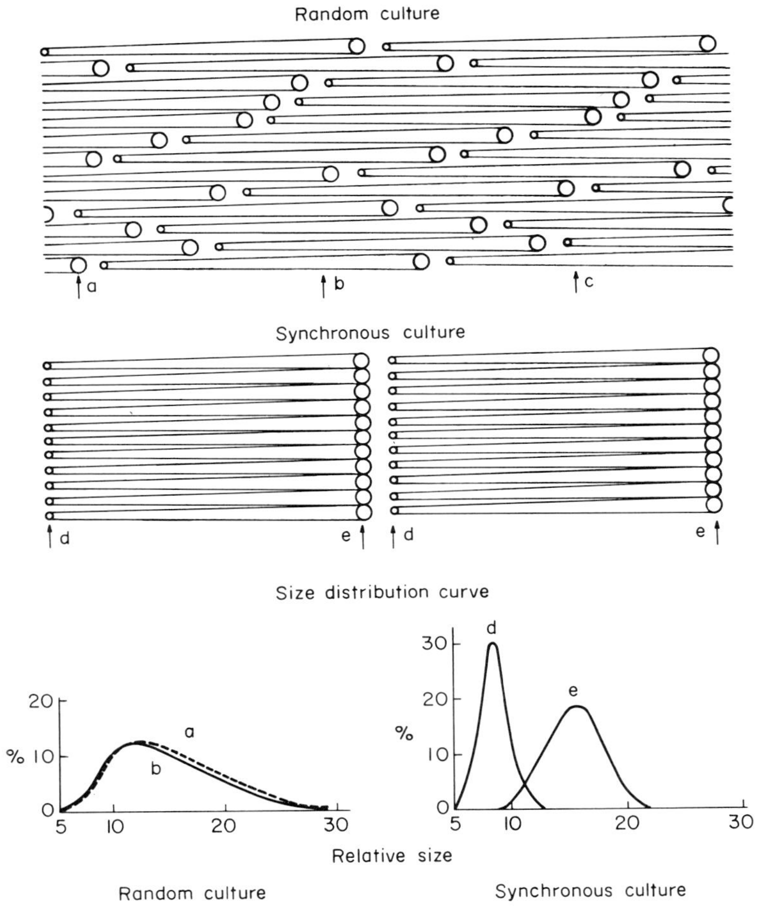

FIG. 1. Schematic representation of the distribution of the life cycles (o———O) and the cell size distribution curves for random and synchronous cultures of unicellular algae.

ing upon the synchrony status of the inoculum. After the logarithmic growth phase, this type of culture will become more or less stagnant; such a culture never reaches a steady state.

The Synchronizing Procedure

Among the different procedures for synchronizing green algae, the light-dark regime combined with a dilution to a constant cell number after each life cycle is the most appropriate.[9-11]

[9] H. Tamiya, Y. Morimura, Y. Yokota, and R. Kunieda, *Plant Cell Physiol.* **2,** 383 (1961).

Before starting the synchronization of a culture, the optimal growth conditions should be established. The following factors should be observed: composition of the nutrient medium, temperature, intensity and spectral composition of the light, and CO_2 content as well as flow rate of the aerating gas.

Nutrient Media. Several nutrient media for *Scenedesmus* have been published.[12-14] We have employed a modification of Kessler's medium in our studies; its composition (*M*) is as follows:

KNO_3	8.0×10^{-3}
NaCl	8.0×10^{-3}
$Na_2HPO_4 \cdot 2H_2O$	1.0×10^{-3}
$NaH_2PO_4 \cdot 2H_2O$	3.0×10^{-3}
$CaCl_2 \cdot 2H_2O$	1.0×10^{-4}
$MgSO_4 \cdot 7H_2O$	1.0×10^{-3}
H_3BO_3	4.5×10^{-5}
$MnCl_2 \cdot 4H_2O$	8.0×10^{-6}
$ZnSO_4 \cdot 7H_2O$	7.0×10^{-7}
MoO_3	1.0×10^{-7}
$CuSO_4 \cdot 5H_2O$	3.0×10^{-7}
$Fe_2(SO_4)_3 \cdot 6H_2O$	7.5×10^{-6}
Sodium citrate$\cdot 2H_2O$	5.5×10^{-4}

In addition, 1 ml of A_5 microelement solution[15] is added per liter of growth medium.

The algae are grown in this medium in glass culture tubes (length, 42 cm; diameter, 3.7 cm) under sterile conditions. The conical bottom of the tube ends in a capillary glass tube through which the algal suspension is aerated, and the algae are thereby kept in suspension.

Temperature Control. Constant temperature can be maintained by placing the culture tubes in a water bath with an automatically controlled heating and cooling system. The water bath must allow irradiation from one or two sides. Commercially available "light thermostats" have been described.[16] Culture vessels can also be kept in a growth chamber, but since cultures absorb radiated heat from the light sources, sufficient air circulation or a cooling system for the light sources should be provided in order to maintain a uniform temperature.

[10] H. L. Tamiya, *Annu. Rev. Plant Physiol.* **17**, 1 (1966).
[11] A. Pirson and H. Lorenzen, *Annu. Rev. Plant Physiol.* **17**, 439 (1966).
[12] O. Kandler, *Z. Naturforsch. B* **5**, 423 (1950).
[13] E. Kessler, W. Arthur, and J. E. Brugger, *Arch. Biochem. Biophys.* **71**, 326 (1957).
[14] H. M. Müller, *Planta* **56**, 555 (1961).
[15] W. A. Kratz and J. Myers, *Amer. J. Bot.* **42**, 282 (1955).
[16] A. Kuhl and H. Lorenzen, *in* "Methods in Cell Physiology" (D. M. Prescott, ed.), p. 159. Academic Press, New York, 1964.

Light Source. Sufficient light intensity for the growth of the algae can be obtained with a battery of 3 warm-white (KEN-RAD) and 3 grow-Lux (Sylvania) fluorescent lamps mounted in front of the culture tubes. This combination of lamps provides a light field of uniform spectral distribution in the visible region of the spectrum. The intensity can be adjusted by varying the distance between the culture tubes and the fluorescence lamps. The maximum intensity at the surface of the culture tubes will approach 10,000 lux.

Aeration. Aeration with air does not provide sufficient carbon dioxide for the rapidly growing culture. Therefore, the air must be enriched with additional CO_2. A constant partial pressure of CO_2 can be achieved by mixing air with CO_2 of constant pressure via two flowmeters. A gas mixture of 2–3% CO_2 in air seems to be optimal when the flow rate is in the range of 20 liters/hour.

Light-Dark Regime. After optimal growth conditions have been established, the synchronizing light-dark regime most suitable for the particular algal strain must be established. This regime should be set such that the dark period begins shortly before the separation of the mother cells into daughter cells and should last until all division is completed. The completion of this latter step is essential since the beginning of the light period is the timer for the next cell division.[17]

The synchronizing light-dark regime can be established for a new algal strain in the following way: The algae are grown as a random culture to a moderate density and then the light is turned off for 12 hours. During his dark period, all cells ready for cell division will divide and remain in the stage of young daughter cells. At the end of this dark period, the culture is diluted so that excess shading of cells is not a problem. After the light has been turned on, the cell number of the culture is followed under the microscope. After a certain time, a burst of division occurs in those cells which had divided in the previous dark period. Knowing the approximate length of the period from the onset of illumination until cell division, the appropriate dark period can be selected as mentioned above. Little correction of the light-dark period is usually required later. If the light-dark regime is chosen correctly, a random culture should be fully synchronized after 3 light-dark regimes.[18]

Dilution. Growing cultures of unicellular algae decrease the average incident light intensity per cell by mutual shading.[6] But since each life cycle of the synchronous culture should start with the same growth conditions, the culture has to be diluted to a constant density or cell num-

[17] H. Senger and N. I. Bishop, *in* "The Cell Cycle" (G. M. Padilla, G. L. Whitson, and I. L. Cameron, eds.), p. 179. Academic Press, New York, 1969.

[18] H. Senger, *Arch. Mikrobiol.* **40,** 47 (1961).

ber at the end of each light-dark regime.[19-21] In addition, the dilution provides the culture with fresh nutrient medium needed for the rapid growth of the new cells. Normally, the rate of dilution is calculated by counting the cells in a hemacytometer under the microscope. Recently the automatic dilution device for a homocontinuous culture[6] has been modified to dilute the synchronous culture at the beginning of each light period to a constant density.[22] This device allows the maintenance of a synchronous culture over many life cycles. The number of cells normally used for starting a synchronous culture of unicellular green algae runs around 10^6 cells/ml.

Considering the general growth conditions and synchronization methods mentioned above, *Scenedesmus* can be synchronized under the following conditions: light-dark regime of 14 hours light (10,000 lux) and 10 hours dark with dilution to a cell number of 1.56×10^6 cells/ml at the beginning of each light period; The temperature of the growth chamber at 25° and the cultures aerated with 3% CO_2 in air at a rate of 20 liters/hour. Under these conditions, the cells follow a simple developmental cycle: the young "daughter cells" grow larger and become denser; nuclear division takes place and transforms them into "mother cells." Sixteen hours after the beginning of the light period, the mother cells divide into a coenobium of 8 cells in the form of a zipper. The turbulence in the culture, caused by the aeration, soon breaks down the coenobium into single daughter cells (see Fig. 2, upper part).

It should be mentioned that *Scenedesmus* can also be synchronized as a mixotrophic culture. The addition of 0.5% glucose (and beef and yeast extract) to the inorganic medium causes an increase in the growth rate and a doubling of the number of daughter cells per mother cell, as observed earlier for synchronous cultures of *Chlorella*.[17,23]

Requirements for "Ideal" Synchronization

If synchronous cultures are the basis for studying developmental changes—like photosynthetic behavior—of microalgae, highest demands are required for the quality of synchrony. Recently we have compiled the following demands for an "ideal" synchronous culture.[17]

1. Complete synchronization
2. Homogeneity

[19] H. Lorenzen, *Flora (Jena)* **144,** 473 (1957).

[20] H. Lorenzen, *in* "Synchrony in Cell Division and Growth" (E. Zeuthen, ed.) p. 571. Wiley, New York, 1964.

[21] H. Metzner and H. Lorenzen, *Ber. Deut. Bot. Ges.* **73,** 410 (1960).

[22] H. Senger and J. Pfau, *Arch. Mikrobiol.* in press.

[23] H. Senger, *Vortr. Bot., Deut. Bot. Ges.* [N.F.] **1,** 205 (1962).

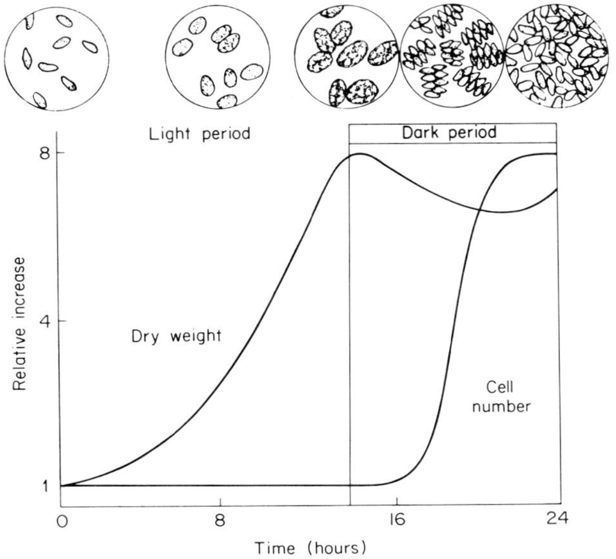

FIG. 2. Characterization of a synchronous culture of *Scenedesmus* by its curves of cell number and dry weight. The microscopic pictures of typical developmental stages of the cells are represented according to the time scale.

3. Exponential growth and shortest life cycle
4. Nonsusceptibility of the life cycle to the synchronizing procedure

That cultures of *Scenedesmus obliquus* synchronized in the above-mentioned way fulfill or well approximate the requirements for an "ideal" synchronous culture has been amply demonstrated.[3]

Growth Parameters of the Synchronous Cultures

The most significant parameter for the synchrony of a culture is the time course of the *cell number* (Fig. 2). A period of constant cell number is followed by the rapid burst of cell division. The *dry weight* of a synchronous culture increases exponentially during most of the light period. Due to respiration, it decreases during the dark period and increases slightly after separation of daughter cells (Fig. 2). *Packed cell volume* and *pigment concentration* demonstrate an almost parallel increase to the dry weight during the light period,[3,24] but in contrast to the dry weight, these two factors stay nearly constant during the dark period.

For computation of the rate of photosynthesis, one of these growth

[24] H. Senger and N. I. Bishop, *in* "Progress in Photosynthesis Research" (H. Metzner, ed.), Vol. I, p. 425. Tübingen, 1969.

parameters has to be chosen. Since dry weight, packed cell volume, and the pigment concentration demonstrate a similar time course during the synchronous growth of *Scenedesmus*, and since packed cell volume is the easiest to measure, it has been chosen as a parameter for calculation of the rates of different photosynthetic measurements. That values expressed on the basis of equal packed cell volume could as well be considered on the basis of equal dry weight, chlorophyll, and absorption is again clearly demonstrated from the data of Table I. In addition, it should be mentioned that the ratio *a/b* changes only within the limits of the experimental error.

TABLE I
COMPARISON OF THE RATIOS OF DRY WEIGHT, TOTAL CHLOROPHYLL AND ABSORPTION TO THE PACKED CELL VOLUME (PCV) OF SYNCHRONIZED *Scenedesmus* CELLS OF DIFFERENT DEVELOPMENTAL STAGES

Hour of the life cycle	Dry weight (mg/l PCV)	Chlorophyll (μl/l PCV)	Absorption at $\lambda = 680$ nm (%)
0	21 ± 3	6.2 ± 0.4	80.0
6	23 ± 5	6.3 ± 0.7	80.9
8	25 ± 2	6.8 ± 0.8	78.8
12	27 ± 1.5	6.9 ± 0.6	79.6
16	24 ± 3	6.5 ± 0.8	78.0
20	21 ± 2	6.6 ± 0.8	80.5
24	24 ± 3	6.2 ± 0.4	80.5

Photosynthetic Reactions during the Life Cycle of *Scenedesmus*

Total Photosynthesis. Photosynthetic capacity measured in saturating white light (either manometrically or polarographically as O_2 evolution during the life cycle of *Scenedesmus*) increases from the onset of the light period for about 8 hours, then decreases until the 16th hour, i.e., just before release of the daughter cells.[3,24] As the number of young daughter cells increase during the dark period, the photosynthetic capacity increases again slightly. The same pattern of photosynthetic behavior has been reported for *Scenedesmus* sp.[25] and for different strains of *Chlorella*.[7,19,21,26,27] The relative difference in rate of O_2 evolution between the 8th and the 16th hour of the life cycle is the same in the light-limiting region as under light-saturated conditions[3] (Fig. 3A). A

[25] L. H. J. Bongers, *Landbouwhogeschool Wageningen* **58,** 1 (1958).
[26] C. Sorokin, "Photosynthetic Mechanisms in Green Plants." *Nat. Acad. Sci., Publ.* 1145, p. 742.
[27] B. Gerhardt, *Planta* **61,** **101** (1964).

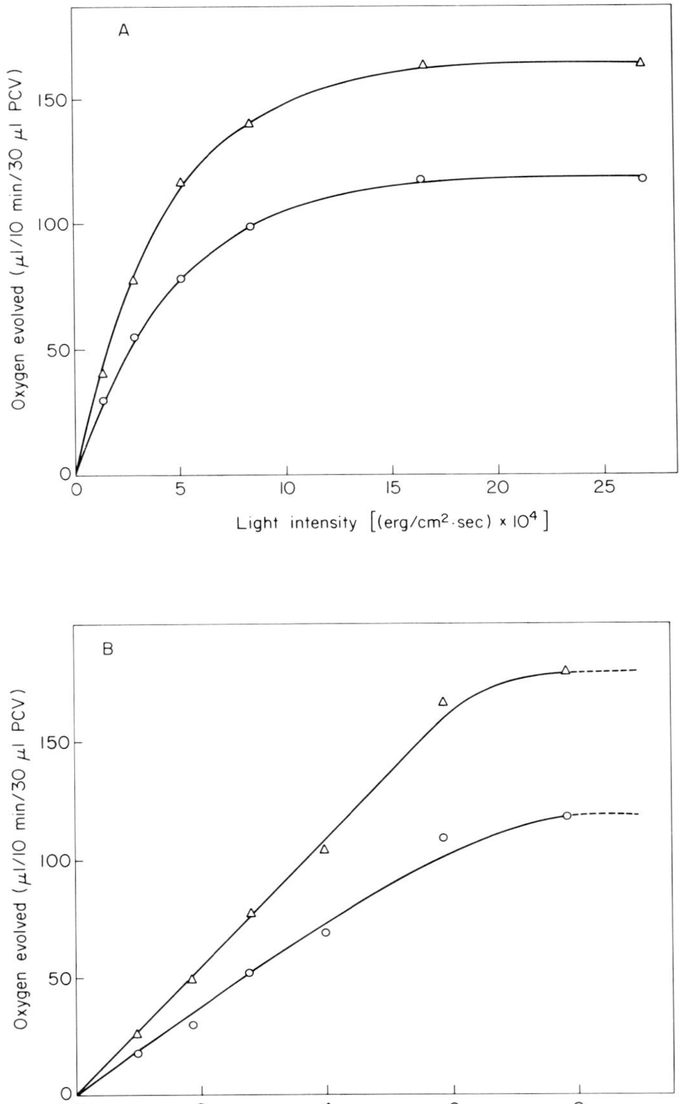

FIG. 3. Light intensity curves of photosynthesis (A), photoreduction (B) and Hill-reaction (C, see p. 62) for *Scenedesmus* cells of the 8th (○) and 16th (△) hour of the synchronous life cycle [in part from H. Senger and N. I. Bishop, *Nature* **214**, 140 (1967)].

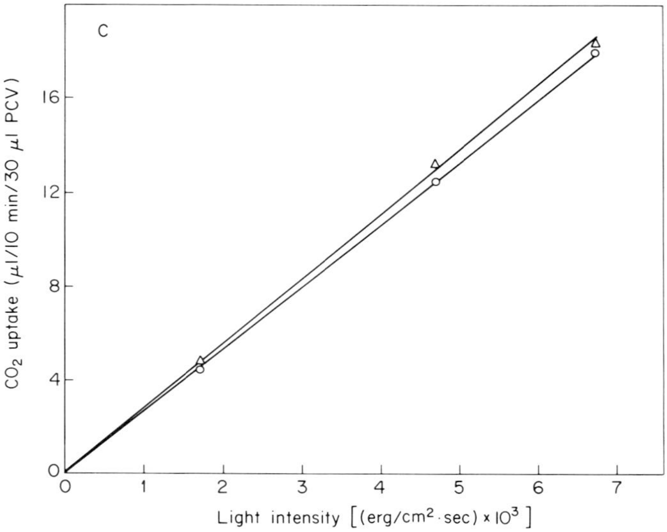

FIG. 3 (*continued*).

comparable observation was made for synchronized *Chlorella* cultures.[28] The observed changes in photosynthetic activity are not the result of the light-dark regime itself since a similar response occurs if the cells are placed in continuous light instead of a subsequent dark period (see Senger[3] for further discussion of this point).

The photosynthetic quotient as measured with Warburg's "two vessel method"[29] remains around 1 throughout the life cycle of *Scenedesmus* in both the light-limiting and the light-saturating region.[3]

The quantum yield of O_2 evolution, measured in monochromatic light of $\lambda = 640$ nm, follows the same pattern as photosynthetic capacity (Fig. 4). Methodological details are reported elsewhere.[30] Similar results were obtained for the relative quantum yield of synchronous *Chlorella pyrenoidosa*.[28]

To extend our observations on the quantum yield, the fluorescence yield of a suspension of cells was measured during the life cycle of *Scenedesmus*. Using monochromatic light ($\lambda = 436$ nm) at different intensities (maximal 200 erg/cm^2 sec), the relative yield of fluorescence at $\lambda = 686$ nm was determined (Fig. 4). As anticipated, the relative yield of fluorescence shows just the inverse pattern of the quantum yield of

[28] C. Sorokin and R. W. Krauss, *Biochim. Biophys. Acta* **48,** 314 (1961).

[29] O. Warburg, *Biochem. Z.* **152,** 51 (1924).

[30] H. Senger and N. I. Bishop, *Nature (London)* **214,** 140 (1967).

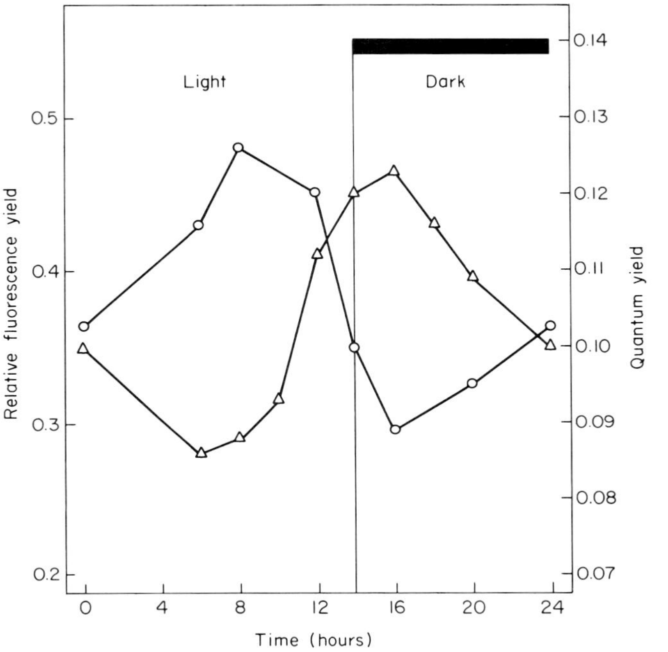

FIG. 4. Relative fluorescence yield (△—△) and quantum yield (○—○) of photosynthesis during the synchronous life cycle of *Scenedesmus*. For experimental details see text.

photosynthesis. Dohler reported a similar pattern of fluorescence in synchronous cultures of *Chlorella pyrenoidosa*.[31]

Separate Reactions of Photosystems I and II. Hydrogen-adapted cells of *Scenedesmus obliquus* have the ability to assimilate CO_2 through only a photosystem I reaction, called photoreduction.[32-34] Photoreduction was measured manometrically as CO_2 uptake in monochromatic light (λ = 691 nm) in 8- and 16-hour-old cells of a synchronous culture of *Scenedesmus* (Fig. 3C). (For further details of the method, see Senger and Bishop[30] and Senger.[35]) As demonstrated in Fig. 3, the rate of photoreduction is the same in the two extreme stages of photosynthetic capacity. Accordingly, the quantum yield of photoreducion remains constant for the cells of the 8th and 16th hours.[30]

[31] G. Dohler, *Beitr. Biol. Pflanz.* **40,** 1 (1964).
[32] H. Gaffron, *Biol. Zentralbl.* **59,** 302 (1939).
[33] N. I. Bishop and H. Gaffron, *Biochem. Biophys. Res. Commun.* **8,** 471 (1962).
[34] N. I. Bishop, *Photochem. Photobiol.* **6,** 621 (1967).
[35] H. Senger, *Planta* **92,** 327 (1970).

TABLE II

RELATIONSHIP BETWEEN DIFFERENT PHOTOSYNTHETIC REACTIONS AT THE TWO EXTREME STAGES OF PHOTOSYNTHETIC ACTIVITY DURING THE SYNCHRONOUS LIFE CYCLE OF *Scenedesmus*

Reaction measured	Comparison of the rate between the 8th and 16th hour[a]	
Photosynthesis (white light)	ll	Increased
Photosynthesis (white light)	ls	Increased
Photosynthesis (640, 680 nm)	ll	Increased
Quantum yield (640, 680 nm)	ll	Increased
Fluorescence (686 nm)	ll	Decreased
Photoreduction (680 nm)	ll	Same
Quantum yield of PR (680 nm)	ll	Same
Action spectrum of photosystem I	ll	Same
Hill reaction (above 620 nm)	ll	Increased
Hill reaction (above 620 nm)	ls	Increased
Action spectrum of photosystem II	ll	Increased
Emerson enhancement effect	ll	Increased
Cyclic photophosphorylation	ls	Decreased

[a] ls, light-saturated region; ll, light-limiting region.

The Hill reaction, measured with *p*-benzoquinone as the oxidant, is believed to be entirely a photosystem II reaction. It was measured polarographically as O_2 evolution in a Gilson-Oxygraph with a micro-Clark electrode over the life cycle of *Scenedesmus*. To prevent photooxidation of the quinone, light above $\lambda = 620$ nm (Plexi-glass 2423) was used for irradiation. The oxygen content of the samples was always brought to a low and predetermined level at the beginning of each measurement by blowing nitrogen through the cell suspension. The curve obtained in this way[35] is similar to the one for photosynthetic capacity. The light-intensity curves for cells of the 8th and 16th hours (Fig. 3B) demonstrate as well in the light-limiting as in the light-saturated region a difference similar to that of the photosynthetic capacity.

Action spectra of O_2 evolution for cells of the 8th and 16th hour are also different.[24,35] Curve analysis of these action spectra demonstrated that the peak at $\lambda = 691$ nm (photosystem I) has the same heights at the different times, whereas the peaks around 676 and 657 nm (photosystem II) decrease considerably from the 8th to the 16th hour.

Emerson-Enhancement Effect and Cyclic Photophosphorylation. According to the different activities of photosystems I and II, the Emerson-enhancement effect changes considerably during the life cycle of *Scenedesmus*.[36] As actinic light for photosystem I 704 nm and for photosystem

[36] H. Senger and N. I. Bishop, *Nature (London)* **221**, 975 (1969).

II 650 nm were chosen. Oxygen evolution was measured manometrically in an all-glass differential manometer in a Gilson photorespirometer. The curve for the Emerson-enhancement in synchronous cultures of *Scenedesmus* follows closely that for the photosynthetic capacity. It reaches about 50% at the 8th hour and drops below zero at the 16th hour.[36]

Anaerobic photoassimilation of glucose is a representative measurement of the cell's capacity for cyclic photophosphorylation.[37] In addition, cyclic photophosphorylation is believed to be driven only through a photosystem I reaction.[38-40] Anaerobic photoassimilation of glucose was estimated in cells of the 8th and 16th hour of the synchronous life cycle of *Scenedesmus* by measuring the uptake of labeled glucose in saturated light under argon atmosphere in the presence of DCMU.[35] The amount of glucose uptake in the 8th hour was about 30% *lower* than in the 16th hour.

Conclusion. In synchronized cultures of *Scenedesmus obliquus,* the photosynthetic fluctuates between the maximum around the 8th and a minimum around the 16th hour. At the 8th hour, both photosystems appear to be most efficiently coupled. Quantum yield and Emerson-enhancement effect are accordingly high. At the 16th hour, photosystem II drops in activity and the capacity of photosystem I thereby becoming free is most probably used for additional cyclic photophosphorylation.

Comparison of the Photosynthetic Capacity of *Scenedesmus* from Different Culture Types

It has been demonstrated that synchronous cultures of green algae provide an excellent tool for studying the photosynthetic behavior in the different developmental stages of these organisms. And it was also shown that the developmental stage of the cells can cause quite different results in photosynthetic experiments. To stress the importance of the culture conditions for photosynthetic research even more, we carried out the following experiments. Over a period of 5 days, the three different culture types discussed above were tested for their photosynthetic capacity. From the synchronous culture the samples were always taken at the 8th hour. The random culture was maintained at the density of the synchronous culture (8th hour) by using a turbidostat for automatic dilution.[6] Samples from this culture were harvested each day at the same time. A batch culture was obtained by dividing a random culture equally

[37] O. Kandler and W. Tanner, *Ber. Deut. Bot. Ges.* **79,** Suppl., 48 (1966).
[38] W. Tanner, V. Zinecker, and O. Kandler, *Z. Naturforsch. B* **22,** 358 (1967).
[39] W. Tanner, E. Loos, W. Kolb, and O. Kandler, *Z. Pflanzenphysiol.* **59,** 301 (1968).
[40] W. Wiessner, *Nature (London)* **212,** 403 (1966).

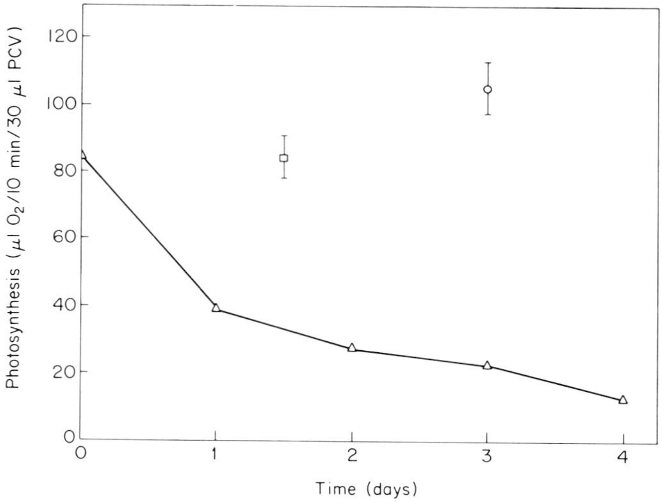

FIG. 5. Decay of photosynthetic capacity in a batch culture after turning off the continuous dilution device. For comparison the mean values and standard deviations of a random (□) and a synchronous (○) culture (each day measured at the eighth hour) for the 5 days are given. Photosynthetic capacity was measured in red light (above $\lambda = 620$ nm) with an intensity of 1×10^5 erg/cm^2 sec.

and leaving one portion in continuous light without dilution. Samples for photosynthesis analysis from this culture were taken at the same time as from the random culture. In the samples from the three cultures the photosynthetic capacity of cells from the three different culture types was measured manometrically in saturating white light. The results are shown in Fig. 5. The photosynthetic capacity of the synchronous and the random culture vary only within the experimental error. However, the synchronous culture at the eighth hour always showed higher maximal capacity than the random culture. In contrast to these two culture types, the batch culture shows a striking and continuous decrease in its photosynthetic capacity. Parallel measurements of the fluorescence yield showed it to be constant in the samples from the synchronous and random culture, but to increase (reciprocally to the decrease in photosynthetic capacity) in the samples from the batch culture.

Acknowledgments

The preparation of this manuscript and the support of research described was sponsored in part by the United States Public Health Service (GM-11745) and the United States Atomic Energy Commission AT(45-1)-1783.

[4] Synchronously Grown Cultures of *Chlamydomonas reinhardi*

By STEFAN SURZYCKI

Information concerning the regulation of the synthesis of cell components has been derived largely from studies with exponentially growing cultures of unicellular organisms. These cultures contain a mixture of cells at various stages of their life cycle, and the analysis of such cells yields information representing the average values of the individual members of the population. In this kind of study, however, little information can be obtained regarding the way in which processes of cellular biosynthesis and regulation are integrated into cell division and growth. In synchronously dividing cultures, on the other hand, the events occurring in single cells can be amplified and the investigator is provided with a large amount of cellular material which exists in more or less the same morphological and physiological condition.

Mass cultures of green algae can be easily synchronized by means of a light-dark cycle. The following method of synchronization of division of the unicellular green alga *Chlamydomonas reinhardi* is a modification of a method developed by Bernstein[1] and by Kates and Jones.[2] It involves a 12 hour light-12 hour dark cycle with the temperature maintained at 21°.

Method

Stock Solutions and Reagents. The salt solution contains in 1 liter of distilled water: NH_4Cl, 8 g; $MgSO_4 \cdot 7H_2O$, 2 g; $CaCl_2 \cdot H_2O$, 1 g. The first two salts are dissolved in 700 ml of distilled water. $CaCl_2$ is dissolved separately in 300 ml of distilled water and then added to the first solution. If the three salts are dissolved together, a precipitate forms and the solution cannot be used.

Phosphate buffer I contains per liter of distilled water: K_2HPO_4, 14.34 g; KH_2PO_4, 7.26 g.

Phosphate buffer II contains per liter of distilled water: K_2HPO_4, 93.5 g; KH_2PO_4, 63.0 g.

The trace elements[3] solution contains: EDTA disodium salt, 50 g; H_3BO_3, 11.4 g; $ZnSO_4 \cdot 7H_2O$, 22 g; $MnCl_2 \cdot 4H_2O$, 5.06 g; $FeSO_4 \cdot 7H_2O$,

[1] E. Bernstein, *Science* **131,** 1528 (1960).

[2] J. R. Kates and R. F. Jones, *J. Cell. Comp. Physiol.* **63,** 157 (1964).

[3] S. H. Hutner, L. Provosoli, A. Schatz, and C. P. Haskins, *Proc. Amer. Phil. Soc.* **94,** 152 (1950).

4.99 g; $CoCl_2 \cdot 6H_2O$, 1.61 g; $CuSO_4 \cdot 5H_2O$, 1.57 g; $(NH_4)_6Mo_7O_{24} \cdot 4H_2O$, 1.1 g.

All the above salts except EDTA are dissolved in 550 ml of distilled water, one at a time, in the order indicated above. EDTA is dissolved in 250 ml of distilled water and heated until it is dissolved. The first solution is heated to 100°, and the 250 ml solution of EDTA is added to it. The entire solution is again brought to 100° and then allowed to cool slightly (80°–90°) before the pH is adjusted to 6.5–6.8 with 20% KOH. Slightly less than 100 ml of KOH is required. The temperature setting on the pH meter should be adjusted with hot standard buffer (75°). The temperature of the trace elements solution should not be allowed to drop below 70° during this process. The solution is adjusted to a final volume of 1 liter and aged for 2 weeks at room temperature in a 2-liter Erlenmeyer flask which is stoppered with a loosely fitting cotton plug. During this time the solution should change color from green to purple. The resulting rust-colored precipitate is removed by several filtrations under suction with a Büchner funnel using three layers of Whatman No. 1 filter paper. The clear purple solution can be stored at —20° for more than a year.

Reagents

$MgSO_4$, 1 *M*
Trisma base (reagent grade), Sigma Chemical Company
Concentrated hydrochloric acid

Preparation of Media. Two kinds of minimal media are used: (1) Tris·minimal phosphate (TMP) and (2) minimal-high magnesium (MM).[4]

TMP medium contains: 2.42 g of Trisma base, 50 ml of salt solution, 1 ml of phosphate buffer II, 1 ml of trace elements, 1.5 ml of concentrated hydrochloric acid, and 1 ml of 1 *M* $MgSO_4$ added to 950 ml of distilled water. The final pH of the medium before autoclaving is 7.0.

The minimal-high magnesium medium (MM) contains: 50 ml of salt solution, 50 ml of phosphate buffer I, 1 ml of trace elements, and 1 ml of 1 *M* $MgSO_4$ added to 900 ml of distilled water. The pH of this medium before autoclaving is between 6.7 and 6.8. After autoclaving, the medium will have a white precipitate which should almost completely dissolve upon the slow cooling of the medium. The precipitate which remains does not affect the synchronous growth of the cells.

TMP medium is used when exact control of the pH is required of the medium. The cells growing on this medium will also reach a higher cell

[4] N. Sueoka, *Proc. Nat. Acad. Sci. U.S.* **46,** 83 (1960).

concentration at the last synchronous division (5 to 6 $\times$ 10^6 cells/ml). Otherwise the properties of the synchronous cultures (synthesis of macromolecules, mitotic division, cell division, etc.) are identical when grown on either medium.

Strain of C. reinhardi Used for Synchronous Growth. Synchronously dividing cultures can be obtained using the 137c wild type strain of either mating type (+ or —). Chiang[5] and Kates and Jones[2] obtained synchronous growth with the 89 and 90 strains using a method similar to that described here. Any mutant strain which requires acetate for its growth cannot be synchronized by this method, but phototrophic strains requiring amino acids or vitamins have been synchronized. The use of strains that are selected for a high degree of synchronization of cell division is not recommended, since this selection usually gives strains that are incapable of synthesizing chlorophyll in the dark (yellow mutants). The properties of synchronous cultures of yellow mutants differ significantly from those of wild type in the time of synthesis of macromolecules (chlorophyll, RNA, etc.). It is desirable, on the other hand, to select a strain in which the daughter cells are liberated from the mother cell wall after cytokinesis during the early dark period. Such a strain, when grown on either TMP or MM media, will, achieve a higher degree of synchronous cell division after fewer light-dark cycles (2 cycles) than wild type cells that are not selected in this way.

Selection for a strain suitable for synchronization is carried out in the following way. Wild type cells are plated onto 1.5% agar plates (about 100 colonies per plate) containing minimal medium (MM) supplemented with 0.4% vitamin-free casamino acids and allowed to grow (25°, 4000 lux) for 6 days. The largest green colonies are selected and at least ten of these are mixed together to form a new strain with a genetic background as close as possible to the original wild type. The strain selected in this way usually liberates its daughter cells almost immediately after cytokinesis and does not require light for this process. The strains are maintained in stock tubes or on plates as described by Levine (this volume [10]).

Inoculation of Cultures for Synchronous Growth. Three-liter volumes of medium (TMP or MM) contained in 4-liter Erlenmeyer flasks equipped with bubbler tubes are used. A 300-ml "seed" culture of cells is first grown to stationary phase in continuous light (3 days, 25°). Cell number is determined, and a volume containing 1.5 $\times$ 10^8 cells is then inoculated into the 3 liters of medium. An amount of medium

[5] K. S. Chiang, Doctoral Thesis, Department of Biology, Princeton University, Princeton, New Jersey, 1965.

equivalent to that to be inoculated is removed from the 3 liters of medium so as to maintain exactly a 3-liter volume after inoculation. The final cell concentration should be 5×10^4 cells/ml. Cultures should be inoculated 6–7 hours before the dark part of the cycle is initiated. The cultures are bubbled with 5% CO_2 in air in order to agitate the cells and to supply carbon dioxide for growth.

Illumination. Cells are synchronized using a 24-hour light-dark cycle (12 hours light-12 hours dark). The light part of the cycle is carried out from 0 to 12 hours; the dark, from 12 to 24 hours. The light intensity is 6000 lux at the level of the cultures. Illumination is from the bottom as well as from two sides of the cultures. Satisfactory results have been obtained only when white and "daylight" fluorescent lamps are used in a ratio of three white lamps to one "daylight" lamp for illumination from the sides. Five "daylight" lamps are used to illuminate from the bottom.

Temperature. The temperature is maintained at 21° as measured inside a culture. An illuminated glass water bath or a Belco water-jacketed culture bottle can be used for this purpose. Variation in the temperature of more than 1° results in poor synchrony.

Cell Division. Cell division in synchronous cultures occurs during the dark period (12–24 hours) between 16 and 19 hours. Cells are inoculated at 4–5 hours (during the light). No cell division occurs during the dark period that follows inoculation. The following light period is referred to as the first light period of synchrony, and the ensuing dark period, as the first dark period. The first cell division occurs during this dark period. At this time every cell divides to four daughter cells; the cell concentration increases from 5×10^4 to 2×10^5 cells/ml. During the second dark period, the cell number also increases four times; every cell divides to four daughter cells, from 2×10^5 to 8×10^5 cells/ml. During the third dark period, the total increase in cell number is only three times; from 8×10^5 to 2.4×10^6 cells/ml. This is the result of half of the cell population dividing to two daughter cells and half, to four. During the fourth dark period, the cell number increases only two times since every cell divides to two daughter cells; from 2.4×10^6 to 4.8×10^6 cells/ml. Synchronous cell division will continue indefinitely as long as the cells are diluted with fresh medium to a concentration which after division will not exceed 5×10^6 cells/ml.

The extent of synchronous cell division in the third and fourth cycles depends upon light intensity. As mentioned above, the light intensity which is usually used results in a 3-fold increase in cell number during the third cycle, and 2-fold during the fourth. However, when the light intensity is increased, each cell will divide to four daughter cells; and lowering the light intensity results in the division of each cell to only two daughter cells.

The extent of synchronous cell division also depends on the length of the light period. No division occurs if the light period is less than 8 hours. The extent of increase in cell number increases proportionally with the length of the light period from 8 to 12 hours.

The first and second cell divisions are not fully synchronized since some of the cells (less than 10%) are dividing during the light period of the cycle. The first fully synchronized division occurs during the third dark period of synchronous growth. The cells used to study properties of synchronous cultures (photosynthetic capacity, respiration, synthesis of macromolecules, etc.) are taken from the third or fourth light period of synchronous cultures.

Photosynthesis and Respiration. Photosynthetic capacity, as measured on a Clark-type oxygen electrode, increases in a stepwise[6,7] fashion from 5 to 11 hours after the beginning of the light period. The final extent of cell division that occurs during a given dark period is proportional to the increase of cell components during the previous light period. Cell number will increase 3-fold when the photosynthetic capacity of the cells triples. Respiration increases (stepwise)[6,7] from 11 to 12 hours, just before the dark part of the cycle.

Synthesis of Macromolecules

Chlorophyll Synthesis. Chlorophyll synthesis, as measured by a modification[8] of the method of Mackinney,[9] increases in parallel with the increase of photosynthetic capacity (from 5 to 11 hours). It also increases in a stepwise fashion and is a good indication as to how well the cells are synchronized. Any gradual increase in chlorophyll during the first 5 hours of the light period may mean that some of the cells are escaping synchronization.

Synthesis of Nucleic Acids. RNA synthesis during the light part of the cycle proceeds in an exponential fashion. Chloroplast ribosomal RNA is synthesized only during the light period.[10] Cytoplasmic ribosomal RNA synthesis also occurs during the light period but continues slowly in the

[6] S. J. Surzycki, U. W. Goodenough, R. P. Levine, and J. J. Armstrong, *Symp. Soc. Exp. Biol.* **24,** 13 (1970).

[7] R. P Levine, J. J. Armstrong, and S. J. Surzycki, *Tenth Annu. Meet. Amer. Soc. for Cell Biol.,* San Diego, p. 120a, 1970.

[8] D. I. Arnon, *Plant Physiol.* **24,** 1 (1949).

[9] G. Mackinney, *J. Biol. Chem.* **140,** 315 (1941).

[10] P. J. Hastings and S. J. Surzycki (unpublished results).

dark for about 9 hours and then stops.[10] At the beginning of the light period, the synthesis of chloroplast ribosomal RNA precedes the synthesis of cytoplasmic ribosomal RNA by 30 to 60 minutes. There is some evidence that the chloroplast ribosomal RNA synthesis is under stringent control, whereas, the cytoplasmic ribosomal RNA synthesis appears to be under a relaxed type of control.[10,11]

DNA has two distinctive periods of synthesis. Chloroplast DNA (β-band, $\rho = 1.694$ g/cm^3) is synthesized in a stepwise fashion between 6 and 8 hours (in the light).[12] Nuclear DNA (α-band, $\rho = 1.724$ g/cm^3[3,12,13]) is synthesized during the dark period from 14 to 16 hours when the cells are preparing to divide to two daughter cells, and from 11 to 12 and 14 to 16 hours when the ensuing cell division is to four daughter cells. It was shown[13] that when cells are preparing to, divide to four daughter cells, both the chloroplast and the nuclear DNA go through two consecutive rounds of replication. When the cells are preparing to divide to only two daughter cells (the fourth day of synchrony), there is only one round of DNA replication (chloroplast and nuclear). Under the described culture conditions, the increase in DNA is always proportional to the following increase in cell number. This is not the case with synchronous cultures of the 89 and 90 strains of *C. reinhardi,* in which the DNA per cell shows a strong dependence on the light intensity.[13]

Kates *et al.*[13] have shown that when vegetative cells divide to four daughter cells, a round of DNA replication precedes each of the two consecutive nuclear divisions. An excellent study with the electron microscope of nuclear division and the following cytokinesis with the 137c strain of *C. reinhardi* was done by Johnson and Porter.[14] They showed that this organism possesses a "closed" mitotic apparatus; the nuclear envelope does not break down during mitotic division. However, the spindle microtubules appear to extend into the cytoplasm through a large dilation of the nuclear envelope. After the first mitotic division, the two daughter nuclei do not usually completely reconstitute their interphase appearance. The spindle microtubules disappear from the nucleoplasm, and septa are seen across the nuclear pores, but most nuclei contain condensed chromatin and nucleoli are not visible. The claim of Kates *et al.*[13] that after the first mitotic division there is reconstitution of the nucleus to its interphaselike appearance does not seem to be justified in view of this study, at least for the 137c strain of *C. reinhardi.*

In conclusion, the synchronized vegetative cultures of *C. reinhardi*

[11] S. J. Surzycki and P. J. Hastings, *Nature* (*London*) **220,** 786 (1969).
[12] K. S. Chiang and N. Sueoka, *Proc. Nat. Acad. Sci. U.S.* **57,** 1506 (1967).
[13] J. R. Kates, K. S. Chiang, and R. F. Jones, *Exp. Cell Res.* **49,** 121 (1968).
[14] U. G. Johnson and K. R. Porter, *J. Cell Biol.* **38,** 403 (1968).

can be considered to have a G1 phase with a duration of 14 to 15 hours; the S period lasts about 4 hours, and partially overlaps the division (D period) which also lasts about 4 hours. The synchronized cells do not appear to have a true G2 period. However, Kates *et al.*[13] found with the 89 and 90 strains that only under very high light intensities (15,000 lux), was the amount of DNA synthesized during a synchronous cycle proportional to the number of daughter cells produced in the ensuing division. Under low light intensity (5000 lux), the cells contained twice the normal G1 level of DNA and underwent their next division without an intervening S period. Thus, under these conditions, an artificial prolongation of the G2 period has been achieved. The same phenomenon was not observed with the 137c strain of *C. reinhardi* in which the increase in the amount of DNA under low light intensity (6000 lux) occurs during every synchronous cycle and is proportional to the ensuing cell division to two daughter cells. The increase in the amount of DNA under high light intensity also occurs during every synchronous cycle and is proportional to the ensuing cell division to four daughter cells. The only difference in the effect of high and low light intensity is that, with high light the time of the first round of nuclear DNA replication occurs between 11 and 12 hours; and of the second, between 14 and 16 hours. The amount of DNA per cell under both conditions of illumination is the same ($1.38\ \mu g \times 10^{-7}$/cell) for interphase cells.

Protein and Enzyme Synthesis. The net increase in protein per cell occurs only during the light part of the cycle and increases linearly from 1 to 12 hours. This increase in the amount of protein is proportional to the following increase in cell number. The synthesis of several enzymes during the light period has been studied. It was shown that the following enzymes exhibit a "stepwise" pattern of synthesis: aspartate carbamoyltransferase,[15] ornithine carbamoyltransferase,[15] phosphoenolpyruvate carboxylase,[15] alanine dehydrogenase,[15] glutamate dehydrogenase,[15] ribulose-1,5-diphosphate carboxylase,[6,7] Fd-NADP reductase,[7] ribulose-5′-phosphate kinase,[16] NADP-dependent glyceraldehyde-3-phosphate dehydrogenase,[16] NAD-dependent glyceraldehyde-3-phosphate dehydrogenase,[16] and EDTA-resistant fructose-1,6-diphosphate aldolase.[16]

[15] J. R. Kates and R. F. Jones, *Biochim. Biophys. Acta* **145,** 153 (1967).

[16] B. Moll, Doctoral Thesis, Harvard Biological Laboratories, Harvard University.

[5] Synchronous Cultures: *Euglena*[1]

By J. R. Cook

Populations of the flagellate protozoan *Euglena gracilis* display periodicity of cell division when grown on a light-dark cycle.[2,3] When the light-dark periods are appropriately chosen with respect to incubation temperature, divisions are synchronized with approximately 100% of the cells dividing in each "burst."[4] Phototrophic nutrition is essential; population expansion becomes asynchronous if a utilizable carbon source is present in the medium.[5]

For reasons of convenience the only studies with synchronized *Euglena* have used environmental conditions leading to a 24-hour generation time. Since one cell divides by longitudinal fission to yield two filial cells (without sex), one expects a doubling of population number at the division burst. This occurs during each dark period, the length of which differs depending on the strain of *Euglena* used; in an obligate phototrophic strain, 8 hours of dark in each 24 hours yielded the best synchrony,[4] whereas the more popular *E. gracilis* Klebs strain Z (Pringsheim) requires 10 hours of dark and 14 of light.[6,7] A method of estimating the proper ratio of light to dark for 24 hour doubling times has been described.[4]

Synchronizing Procedure

Except where noted, the following account is confined to work with *E. gracilis* Z. This strain multiplies most rapidly at 29°–30°, but for synchronization a temperature of 21.5° ± 0.5° is required.[8]

The medium used, described by Cramer and Myers,[9] is given in the table. Citrate, not used as a carbon source by *Euglena,* is added as a chelating agent. The medium is phosphate-buffered at pH 6.8, which is near the upper limit for growth of *Euglena:* it may be adjusted to as low

[1] Supported by N.I.H. Grant GM-12179. The author is a Research Career Development Awardee of the Public Health Service (1-K4-GM-9395).

[2] G. F. Leedale, *Biol. Bull.* **116,** 162 (1959).

[3] R. T. Huling, *Trans. Amer. Microsc. Soc.* **79,** 384 (1960).

[4] J. R. Cook and T. W. James, *Exp. Cell Res.* **21,** 583 (1960).

[5] L. N. Edmunds, Jr., *J. Cell. Comp. Physiol.* **66,** 147 (1965).

[6] J. R. Cook and M. Hess, *Biochim. Biophys. Acta* **80,** 148 (1964).

[7] J. R. Cook, *Plant Physiol.* **41,** 821 (1966).

[8] J. R. Cook, *Biol. Bull.* **131,** 83 (1966).

[9] M. Cramer and J. Myers, *Arch. Mikrobiol.* **17,** 384 (1952).

CRAMER-MYERS[9] MEDIUM FOR PHOTOTROPHIC GROWTH OF *Euglena*

Component	(mg/liter)
$(NH_4)_2HPO_4$	1000
KH_2PO_4	1000
$MgSO_4 \cdot 7H_2O$	200
$CaCl_2$	20
$Fe_2(SO_4)_3 \cdot nH_2O$	3
$MnCl_2 \cdot 4H_2O$	1.8
$CoSO_4 \cdot 7H_2O$	1.5
$ZnSO_4 \cdot 7H_2O$	0.4
$Na_2MoO_4 \cdot 2H_2O$	0.2
$CuSO_4 \cdot 5H_2O$	0.02
Citrate (as $Na_3C_6H_5O_7 \cdot 2H_2O$)	800
Vitamin B_1	0.1
Vitamin B_{12}	0.0005

as pH 3.0 with H_2SO_4 prior to autoclaving. $(NH_4)_2HPO_4$ is the nitrogen source; the vitamins B_1 and B_{12} are required.

The culture vessel found most convenient is a 9×50 cm Pyrex cylinder, (about 3.5-liter capacity) fitted with a Pyrex water jacket. The vessel is plugged with cotton, through which pass the necessary tubes. The vessel is autoclaved with the reservoir of medium. Inoculation is made with a syringe through a port sealed with a syringe stopper. Air enriched to 5% CO_2 is bubbled from the bottom of the vessel (a tube packed with cotton sterilizes the air). A Teflon-covered magnetic stirring bar further serves to keep the culture well mixed. Cell suspension is removed by a simple siphon; the tip may be kept in a tube of dilute acid to prevent contamination. If a large reservoir is autoclaved with the vessel, several runs may be made with the same setup by appropriate dilutions. Carboys fitted with tubulature at the bottom are most convenient for holding the fresh media.

Daylight type fluorescent tubes or incandescent flood lamps may be used; more even lighting is possible with the former. Light intensities should be between 300 and 1000 fc. Lower intensities do not support optimum growth, and higher intensities are inhibitory.[10] Any 24-hour clock timer can be used to turn the lights on or off. At 21.5°, the program calls for 14 hours of light followed by 10 hours of dark, offered repetitively. The program may be set at any time with respect to the solar day; synchrony is established almost at once, and is maintained until the population approaches the stationary phase.

[10] J. R. Cook, *J. Protozool.* **10,** 436 (1963).

Biochemical Properties

In continuous light biochemical properties of *Euglena* depend heavily on the light intensity. At 1000 fc, *Euglena* contains large amounts of paramylum, the polysaccharide reserve; at 300 fc, near the lower limit for multiplication, little or no paramylum is accumulated, and the cells have somewhat smaller amounts of protein than at 1000 fc.[10] These patterns are also seen in synchronized populations. When 1000 fc is used, paramylum is synthesized continuously in the light period, and is virtually all consumed in the following dark period.[7] Presumably this paramylum is used as the carbon source for protein and RNA synthesis, which continues throughout most of the dark period as well as the light period. At 300 fc, on the other hand, there is no paramylum synthesis, and the synthesis of protein and RNA is confined to the light period.[11,12] At either light intensity, synthesis of the photosynthetic pigments is

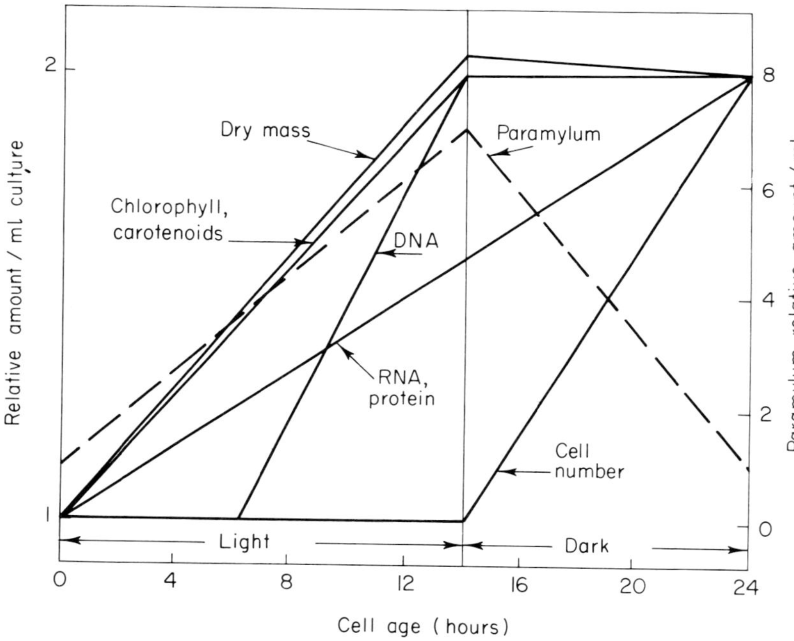

FIG. 1. Patterns of biosynthesis in synchronized *Euglena* (21.5°, 1000 fc). Paramylum (right ordinate) increases about 7-fold in the light and is utilized in the dark. Other constituents (left ordinate) double in characteristic patterns.

[11] J. R. Cook, *Biol. Bull.* **121, 277** (1961).

[12] L. N. Edmunds, Jr., *J. Cell. Comp. Physiol.* **66,** 159 (1965).

found only in the light period. The light intensity is also without effect on the timing of total cellular DNA replication, which occurs in the latter half of the light period.[6,13] The replication of cytoplasmic DNA, not detectable by ordinary chemical analysis, is also discontinuous, with the most vigorous synthetic activity occurring very early and very late in the light period.[14] All other synthetic activity is linear with respect to time. The biochemical growth patterns of *E. gracilis* Z synchronized with 1000 fc light are summarized in Fig. 1.

Gas Exchange

Oxygen consumption of synchronized *Euglena* (1000 fc), measured in the dark immediately after removal from the culture vessel, increases fairly rapidly early in the light period, leveling off (after a doubling in rate) by the end of the light period. The activities of at least two of the respiratory enzymes show no clear relationship to the overall pattern of oxygen consumption. Referred to protein, the specific activity of isocitrate dehydrogenase (NADP, EC 1.1.1.42) increases steadily during the

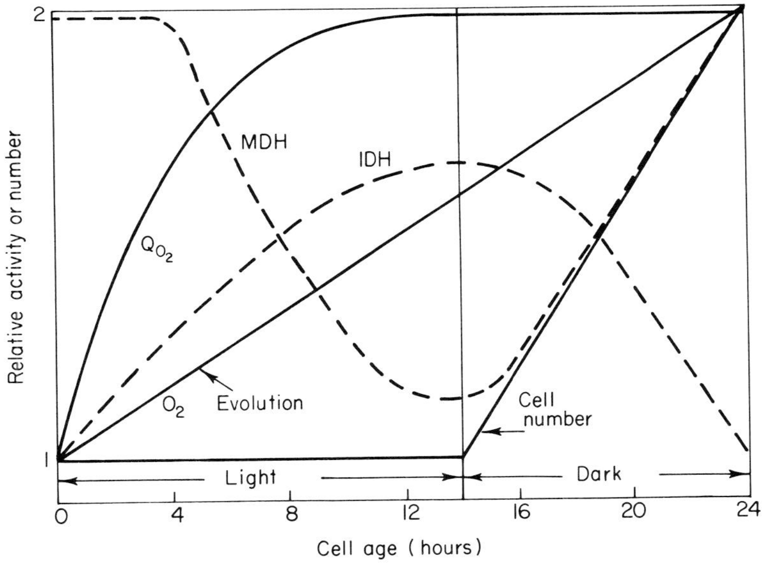

FIG. 2. Relative rates of oxygen consumption and production referred to unit volume of culture (and to individual cells during the light period). At age 0, oxygen consumption is 13 μl O_2/hour/10^6 cells, and oxygen production is 80 μl O_2/hour/10^6 cells. Relative activity of the two enzymes is referred to protein.

[13] L. N. Edmunds, Jr., *Science* **145,** 266 (1964).
[14] J. R. Cook, *J. Cell Biol.* **29,** 369 (1966).

light period, then declines to its initial value during the dark period. Malate dehydrogenase (EC 1.1.1.37) decreases steadily during the light period, then rises during the dark period to its former value. Since total protein doubles during this time, the total measured activity of the two enzymes also doubles, as expected.

Oxygen production (corrected for consumption) is nearly as great at 1000 fc as in saturating light (2000 or 3000 fc), indicating that the photosynthetic apparatus is operating near maximum capacity at all ages. The capacity for oxygen production increases steadily throughout the 24-hour light-dark cycle, proportional to protein but not to chlorophyll content.[7] The gas exchange data for synchronized *Euglena* are summarized in Fig. 2.

[6] Synchronous Culture of *Chlorella*

By EIJI HASE and YUJI MORIMURA

There are two methods of synchronous culture which have been most widely in use for *Chlorella*. One is the method developed by Tamiya and his collaborators[1-3] in their first study in 1953 with *Chlorella ellipsoidea*.[4] The cultures are started from a homogeneous population of small young cells—called Ds cells[3]—which are separated from less homogeneous populations by differential centrifugation. When the cells reach a mature stage—called L_3 cells[1]—after which they do not require light for further development to complete cellular division, the light is turned off. When all the cells have completed division, the light is turned on to start the second cell cycle. In the other method,[5] which may be called the method of programmed light-dark regimes, a random culture is subjected, without being fractionated beforehand into a homogeneous population, to a programmed light and dark alternation, in which the lengths of the light and dark periods are adequately chosen according to the algal species and culture conditions. The synchrony of growth and cell division is attained

[1] H. Tamiya, *in* "Synchrony in Cell Division and Growth" (E. Zeuthen, ed.), p. 247. Wiley (Interscience), New York, 1964.

[2] Y. Morimura, *Plant Cell Physiol.* **1,** 49 (1959).

[3] H. Tamiya, Y. Morimura, M. Yokota, and R. Kurieda, *Plant Cell Physiol.* **2,** 383 (1961).

[4] H. Tamiya, T. Iwamura, K. Shibata, E. Hase, and T. Nihei, *Biochim. Biophys. Acta* **12,** 23 (1953).

[5] H. Lorenzen, *in* "Synchrony in Cell Division and Growth" (E. Zeuthen, ed.), p. 571. Wiley (Interscience), New York, 1964.

on repeating the light and dark regimes several times. The advantages and disadvantages of the two methods have been discussed by several authors.[1,3,5-7] Herein the technical procedures and some precautions to be taken in running the synchronous cultures by the former method are described. It would be easy for the investigator to apply the latter method if he refers to a part of the procedures described below since it is in many aspects similar to the first method.

Culture Apparatus

According to the scale of the experiment, oblong flat flasks (glass) of relatively small capacity or large chambers (acrylate resin) are used as containers. The typical shapes of the containers are shown in Fig. 1 and their features in the table. The flask has an opening with a cotton plug carrying a glass tube, the lower end of which almost reaches the bottom of the flask. The upper end of this tube is widened and filled with cotton to serve as an air filter. In the upper opening of the tube is fitted a rubber stopper carrying a glass tube that serves to conduct CO_2-

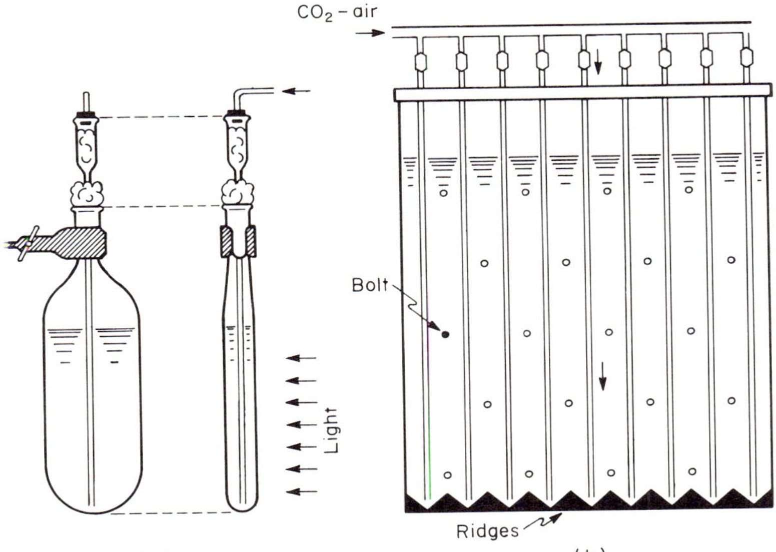

FIG. 1. An example of oblong flat flask (a) and a culture chamber made of clear acrylate resin (b).

[6] H. Tamiya, *Annu. Rev. Plant Physiol.* **17**, 1 (1966).
[7] R. R. Schmidt, *in* "Cell Synchrony—Studies in Biosynthetic Regulation" (I. L. Cameron and G. M. Padilla, eds.), p. 189. Academic Press, New York, 1966.

FEATURES OF CULTURE CONTAINERS

A. Oblong flat flasks

Flask	Size of the main body of the flask (cm)			Volume of culture solution (ml)
	Length	Width	Inner thickness	
Small	15	7	1.5	50
Medium	37	12	2.8	500
Large	52	17	3.5	1000

B. Flat culture chamber made of clear acrylate resin[a]

Size[b] (cm)			Volume of culture solution (ml)
Height	Width	Inner thickness	
120	100	2	20,000

[a] Y. Morimura, S. Yanagi, and H. Tamiya, *Plant Cell Physiol.* **5,** 281 (1964).

[b] The plates of acrylate resin are 0.5 cm in thickness. By changing the height and width, one can construct chambers of different capacities. The inner thickness in these cases is kept at about 2 cm to maintain the uniformity of light intensity available for individual cells in the culture. This, of course, depends upon the cell concentration of the culture.

enriched air into the culture medium. The flasks filled with culture solutions are autoclaved, inoculated, and placed in a thermostat with a glass wall on one side. They are illuminated from outside, in a direction perpendicular to the flat walls of the culture containers, either by a bank of 20–40 W daylight fluorescent lamps or of 150–250 W reflector food lamps. The chamber is a rectangular vessel with plane-parallel walls made of clear acrylate resin.[8] At the bottom there are triangular ridges to prevent sedimentation of the algal cells. To prevent deformation of the vessel caused by water pressure, the plane-parallel walls are fixed with bolts. These bolts, as well as the ridges, are also made of acrylate resin. In a large-scale experiment, two chambers of this type are used, one placed on either side of a light source chamber, which is also made of acrylate resin, with separate compartments for parallel fluorescent lamps. The culture and light source chambers are set in a large water thermostat. The cultures for these chambers are prepared without special precautions to keep them under aseptic conditions. However, no serious contamination by other organisms is usually encountered during rela-

[8] Y. Morimura, S. Yanagi, and H. Tamiya, *Plant Cell Physiol.* **5,** 281 (1964).

tively short periods when the culture media contain only inorganic nutrients or in the experiments of repetitive cell cycle in which the culture is diluted with fresh medium at the end of each cycle, as described later.

Homogeneous Cell Population

To obtain starting cells that are as uniform in size and properties as possible, their preculture has to be effected under constant conditions with inocula having as constant a prehistory as possible. The culture is, therefore, performed in three steps: (1) culturing of the inoculum for the preculture, (2) preculturing of the starting cells, and (3) the synchronous culture of the main experiment. The procedures described below are those employed for *Chlorella ellipsoidea,* but they would be also applicable, with some modifications, to other strains of *Chlorella*. The culture medium used in the three steps is of the following composition per liter: 5.0 g of KNO_3; 2.5 g of $MgSO_4 \cdot 7H_2O$; 1.25 g of KH_2PO_4; 2.8 mg of $Fe_2SO_4 \cdot H_2O$; and 1 ml of Arnon's A5-solution, pH 5.6. Arnon's A5-solution contains per 1 ml: 2.85 mg of H_3BO_3; 1.81 mg of $MnCl_2 \cdot 4H_2O$; 0.22 mg of $ZnSO_4 \cdot 7H_2O$; 0.018 mg of $(NH_4)_6Mo_7O_{24} \cdot 4H_2O$; and 0.078 mg of $CuSO_4 \cdot 5H_2O$. Aeration of the culture is with air containing 2–3% CO_2 at a rate of about 50 ml/100 ml of culture solution per minute.[9]

Preparation of Inoculum for Preculture. To prepare the inoculum for preculture, the culture solution is dispensed in 50-ml portions into the small flasks (Fig. 1,a; see part A of the table); after being sterilized by autoclaving, it is inoculated with a pure algal culture taken from a 2- to 4-week-old slant culture which has been preserved at 20°–25° under illumination with daylight fluorescent lamps, about 5 kilolux in intensity. The cultures in these flasks are kept at 25° under illumination with daylight fluorescent lamps of 10 kilolux in intensity and with constant aeration of CO_2-enriched air.

Preculture and Separation of the Starting Cells for Synchronous Culture. After 7–8 days, 25-ml portions of inoculum culture are taken out with sterilized pipettes and poured into medium- or large-sized flasks, each containing 500 or 1000 ml, respectively, of the sterilized culture solution. The cultures are kept at 25° under illumination with incandescent lamps, the light intensity being kept at 10 kilolux for the first 4–5 days, after which it is reduced to 0.8 kilolux. Culturing under the lower light intensity lasts for 3–4 days; by this time the algal population

[9] Before it enters the culture solution, the CO_2-enriched air is brought to the same water-vapor pressure as the culture medium by being scrubbed through gas wash bottles containing a solution having the same composition and temperature as the culture medium.

density attains the level of about 10 ml of packed cell volume per liter, and more than 90% of the total population is in the form of small cells having diameters ranging from 2.5 to 3.5 μ. The culture is diluted 10- to 20-fold with fresh culture solution, and then centrifuged at 1200 g for 5–6 minutes to discard the larger, heavier cells. The supernatant cell suspension is further centrifuged, at 1200 g for 12–15 minutes; this time, to remove the cells in the supernatant which are comparatively smaller and lighter. The cells thus separated are 2.7–3.3 μ, in diameter, the maximum frequency (more than 80%) being 3.0 μ, and used as the starting material (Ds cells) for synchronous culture. The yield of the Ds cells is 10–20% of the total population (existing before centrifugation) in terms of packed cell volume.

The preculture can also be made in the following way.[8] After the inoculum culture is transferred to the medium- or large-sized flasks, the cultures are kept at 20° under illumination with incandescent lamps, the light intensity being maintained at 10 kilolux. The illumination is applied continuously during the initial 2 days, after which the cultures are subjected to two or three changes of light and dark periods lasting, respectively, 20 and 10 hours. Harvest of cells is made at the end of the last dark period, by which time the population of the cultures is about 5–6 ml packed cell volume per liter in density and is composed mostly of small cells. The subsequent procedures of separating the starting Ds cells are the same as those mentioned above.

To separate the starting cells in sufficient quantity for synchronous mass cultures, use is made of a refrigerated centrifuge of De Laval type.[8] The first centrifugation is made at 5000 g, the rate of flow being 700 ml/minute. The precipitate is discarded, and the supernatant solution is again centrifuged at 7700 g with a flow rate of 500 ml/minute. The precipitate thus obtained is a mass of the small cells which can be used as the starting material for synchronous culture. The yield is about 10–20% of the total population in terms of packed cell volume.

Synchronous Culture. The Ds cells obtained as above are immediately—or after storage in a refrigerator for not more than 24 hours—used as the starting material for synchronous culture. According to the purpose of the experiment, various types of flat flasks or acrylate resin chambers are used as shown in Fig. 1 and in the table. In most of the physiological studies of algal cell cycle, the medium- or large-sized flat flasks are convenient, but for biochemical studies, large chambers made of acrylate resin are preferable. In all cases, the initial inoculation of Ds cells in the culture medium is made at a concentration of about 7×10^9 cells or 0.15 ml of packed cell volume per liter. Maintenance of low cell concentrations during the culture is important for maintaining uniformity of the light intensity available for individual cells. As a criterion of the synchrony of the culture, the statistical distribution of cell size is measured by the optical method using an ocular micrometer or by using a Coulter counter.[10] Following the changes in size and appearance of cells occurring during the progress of cell development, the L_3 cells which do not require light for the completion of cellular division are recognized,[11] and the light is turned off. Since the number of cells increases 4-fold after the completion of cellular division of *Chlorella ellipsoidea* under the standard conditions (light intensity: 10 kilolux),[1]

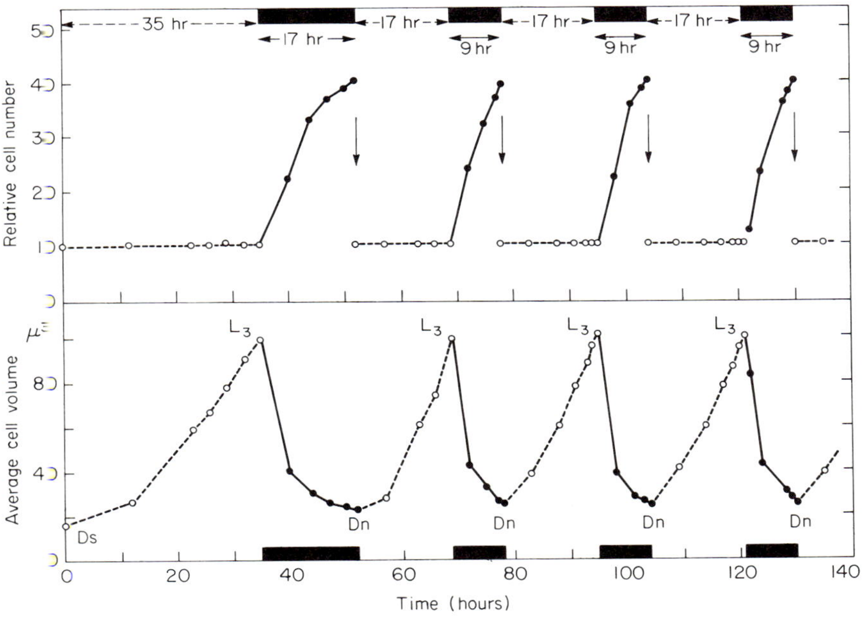

FIG. 2. Synchronous culture of *Chlorella ellipsoidea* by the method starting from a homogeneous cell population (Ds cells). The upper figure shows changes of relative cell number and the lower those of average cell volume (calculated from the statistical distribution of cell size) occurring in the DsLD cycle followed by repeated DLD cycles (see text). Temperature: 21°, light intensity: 10 kilolux. Black bars along the abscissa indicate the dark periods. [From H. Tamiya, Y. Morimura, M. Yokota, and R. Kunieda, *Plant Cell Physiol.* **2**, 383 (1961).]

[10] K. Shibata, Y. Morimura, and H. Tamiya, *Plant Cell Physiol.* **5**, 315 (1964).

[11] Initially this mature stage of cell development is recognized by transferring portions of the algal cells from light to darkness at different times during the culture and by examining whether or not they can completely divide in the dark incubation.[3] After several preliminary experiments, it would be easy to recognize this stage by the size and appearance of cells under the light microscope.

three quarters of the culture is taken out and the rest is diluted 4-fold to start the following cycle.

Various types of synchronous cultures started from Ds cells have been studied, and their advantages and disadvantages are discussed by Tamiya *et al.*[3] An example of synchronous cultures using the so-called DsLD and DLD methods giving the most homogeneous synchrony is shown in Fig. 2. The daughter cells produced from the L_3 cells in the dark are represented by Dn. The cycle starting from Ds and ending, via L_3, with Dn is called the DsLD cycle. When the Dn cells produced in the DsLD cycle are incubated in the light, they grow and, after attaining the L_3 cell stage, divide into Dn cells in the dark. This cycle beginning and ending with Dn is called the DLD cycle, which is beautifully repetitive, as seen in Fig. 2. In the repetitive DLD cycle, one can harvest portions of the culture at different times during one cycle. When large amounts of algal material at different stages are needed, the cells at respective stages are harvested in successive cell cycles. As demonstrated by Hare and Schmidt,[12] one can apply the continuous-dilution (and harvest) method, in which the culture is continuously harvested and diluted with fresh culture medium. In the DLD cycle shown in Fig. 2, the light period is 17 hours and the dark period 9 hours at a temperature of 21°. Recently Kanazawa and Kanazawa[13] showed that the DLD cycle can be equally successfully carried on under the conditions of 17 hours light (10 kilolux) and 7 hours dark at 23°

Programmed Light-Dark Regimes

The technical procedures in this method are practically the same as those described above, except that an adequate program of light-dark alternation is previously chosen according to the algal species and culture conditions, especially temperature and light intensity. For details of this method reference is made to an article by Lorenzen,[5] and also to the articles in this volume dealing with the synchronous cultures of other algae using this method.

Acknowledgment

The authors are grateful to Professor H. Tamiya for his valuable advice in the preparation of this manuscript.

[12] T. A. Hare and R. R. Schmidt, *Appl. Microbiol.* **16**, 496 (1968).
[13] T. Kanazawa and K. Kanazawa, *Plant Cell Physiol.* **10**, 495 (1969).

[7] Synchronized Cultures: Diatoms

By W. M. Darley and B. E. Volcani

In order to carry out an extensive investigation on the role of silicon in metabolism and shell formation in diatoms, it was necessary to devise methods for obtaining synchronized cultures that could be used for a wide variety of biochemical studies, and that would provide large amounts of cell material, at a high degree of synchrony. Two systems for inducing synchrony were developed: (a) silicon-starvation, which results in synchronized formation of the wall, and (b) a light-dark regime which produces a complete cycle of synchronized growth and division. These synchronies have been used for studies on changes in DNA, RNA, protein, carbohydrate, lipid, and pigment during the life cycle[1] and wall formation[2,3] as well as for studies on changes of nucleoside triphosphates,[4,5] photosynthesis, and respiration[6] during wall formation only.

Navicula pelliculosa

Culture and Media. The strain of *N. pelliculosa* used in both synchronies is listed as No. 668 in the Culture Collection of Algae at the Indiana University, Bloomington, Indiana. Recently cloned cultures, necessary for best results, are obtained by plating on a freshwater Tryptone medium (FWT) solidified with 1.5% agar composed of $Ca(NO_3)_2 \cdot 4H_2O$, 0.1 g; $K_2HPO_4 \cdot 3H_2O$, 0.0135 g; $MgSO_4 \cdot 7H_2O$, 0.025 g; $Na_2SiO_3 \cdot 9H_2O$, 0.1 g; Na_2CO_3, 0.02 g; trace elements, 1 ml; Bacto-Tryptone peptone (Difco) 1 g, in 1 liter of distilled water; pH (not adjusted) 8.3. The mixture of the trace elements[7] consists of H_3BO_3, 0.568 g; $ZnCl_2$, 0.624 g; $CuCl_2 \cdot H_2O$, 0.268 g; $Na_2MoO_4 \cdot 2H_2O$, 0.252 g; $CoCl_2 \cdot 6H_2O$, 0.42 g; $MnCl_2 \cdot 4H_2O$, 0.36 g; $FeSO_4 \cdot 7H_2O$, 2.50 g, sodium tartrate·2 H_2O, 1.76 g in 1 liter of glass-distilled water.

After 7–10 days growth at 20°, and exposure to a fluorescent light intensity of 5000 lux, colonies varying in size from small, compact and

[1] W. M. Darley, C. W. Sullivan, and B. E. Volcani, in preparation.

[2] J. Coombs, W. M. Darley, O. Holm-Hansen, and B. E. Volcani, *Plant Physiol.* **42,** 1601 (1967).

[3] F. P. Healey, J. Coombs, and B. E. Volcani, *Arch. Mikrobiol.* **59,** 131 (1967).

[4] J. Coombs, P. J. Halicki, O. Holm-Hansen, and B. E. Volcani, *Exp. Cell Res.* **47,** 302 (1967).

[5] J. Coombs, P. J. Halicki, O. Holm-Hansen, and B. E. Volcani, *Exp. Cell Res.* **47,** 315 (1967).

[6] J. Coombs, C. Spanis, and B. E. Volcani, *Plant Physiol.* **42,** 1607 (1967).

[7] P. R. Burkholder and L. G. Nickell, *Bot. Gaz.* **110,** 426 (1949).

smooth margined, to large and diffuse margined, can be distinguished. A colony of intermediate size is isolated into 8 ml of FWT liquid medium in 12.5 × 2 cm screw-cap tubes. After growth for about 2 weeks under the above conditions, the cultures are placed in dim light; they are transferred once every 2 months. It may be necessary to reclone the cultures periodically to improve the synchrony.

Cell Number. The cell samples (about 5 ml) are preserved with a drop of Lugol's solution (6 g of KI, 4 g of I_2 in 100 ml of H_2O). To separate clumped cells effectively without causing significant cell breakage, samples are hand-homogenized by plunging at least 20 times with a Teflon pestle in a Potter-Elvehjem tissue grinder. In the light-dark (L-D) synchrony, the homogenized suspension is subjected to mild sonication for 15 minutes in an ultrasonic cleaner (model System Forty, Ultrasonic Industries) and is again hand-homogenized. Then 0.5 ml is pipetted into 49.5 ml of Millipore-filtered 0.85% NaCl solution; the cell number is determined with a Coulter Model B electronic counter equipped with a 100 μ pore aperture tube and set at amplification 1/4, aperture current 1/4, upper threshold 100, and lower threshold 10. The settings should be established for each instrument.

Silicon-Starvation Synchrony

Principle. An exponential culture is grown in the light in a medium containing a limited amount of silicon, but an excess of other essential nutrients. At the period of silicon starvation, increase in cell number stops and initiation of cell wall occurs in most of the cells. Upon the addition of silicon, wall formation is completed with a resulting synchronous increase in cell number.[5,8,9]

Inoculum. Cultures of 50 ml of FWT medium in 125-ml Erlenmeyer flasks inoculated with 5 ml are grown on a reciprocal shaker (120 strokes/minute) for 3 days at 18°–20° under continuous illumination at 5000 lux with "cool white" and "warm white" fluorescent lamps. After two further subculturings under the same conditions, the inoculum consists of cells at a stationary stage of growth (about 6×10^6 cells/ml). Approximately 400 ml of culture are required for inoculation.

Culture Vessel. A 4-liter polycarbonate bottle (specially manufactured by Nalge Co. Inc., Rochester, New York) is adapted with three openings for a polypropylene aeration tube, a sampling tube, and the

[8] J. C. Lewin, B. E. Reimann, W. F. Busby, and B. E. Volcani, *in* "Cell Synchrony—Studies in Biosynthetic Regulation" (I. L. Cameron and G. M. Padilla, eds.), p. 169. Academic Press, New York, 1966.

[9] W. F. Busby and J. Lewin, *J. Phycol.* 3, 127 (1967).

FIG. 1. Vessels for silicon-starvation (left) and light-dark (right) synchronization cultures. 1, Aeration tube (5 mm, i.d.); 2, sampling tube (5 mm, i.d.); 3, inoculation port; 4, glass tubing plugged with cotton for suction of sample; 5, replaceable glass cap for aseptically removing sample; 6, magnetic bar, Teflon coated, octagonal surfaces, 5 cm long; 7, magnetic stirrer, solid state; 8, inlet for sterile dilution medium; 9, resin flask cover; 10, Teflon washer; 11, bank of fluorescent lamps.

introduction of inoculum (Fig. 1). The bottle is placed between two banks of 40-W fluorescent lamps, each bank containing two "cool white" and two "warm white" lamps, equally distributed; this provides a light intensity of 17,000 lux midway between the two banks. The temperature is maintained at 21° ± 1°; the culture is aerated at the rate of 10 liters/minute and stirred with a magnetic stirrer (a solid state speed control stirrer is used since it does not generate heat).

Synchronization Procedure. Three liters of FWT medium to which is added 9 ml of a sterile antifoam agent [polypropylene glycol (P-2000): water 2:1000 by volume], and 5 ml of a 5% filter-sterilized solution of thiosulfate are inoculated to an initial cell density of about 6.5×10^5 cells/ml. After 36 hours of growth, by which time silicon is depleted from the medium and cell division has ceased, the culture is starved for 14 hours. The synchrony is initiated by adding a solution of neutralized

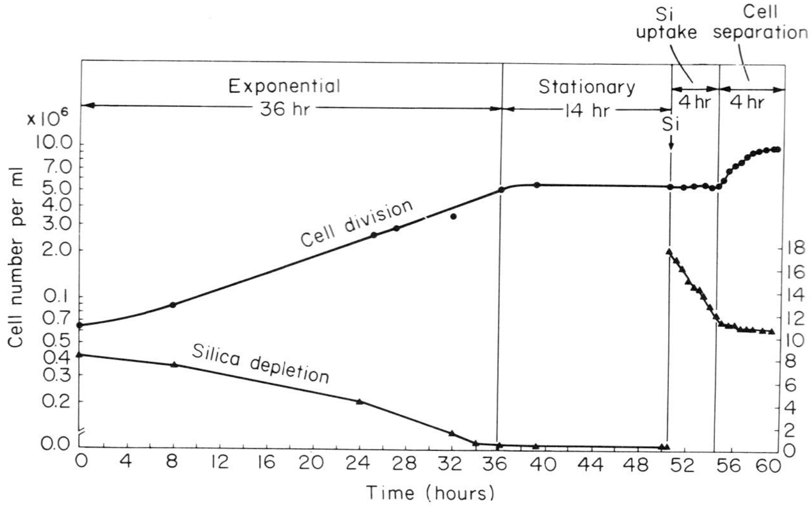

FIG. 2. Course of silicon-starvation synchrony of *Navicula pelliculosa,* and depletion of silicon from the medium [J. Coombs, P. J. Halicki, O. Holm-Hansen, and B. E. Volcani, *Exp. Cell Res.* **47,** 315 (1967)].

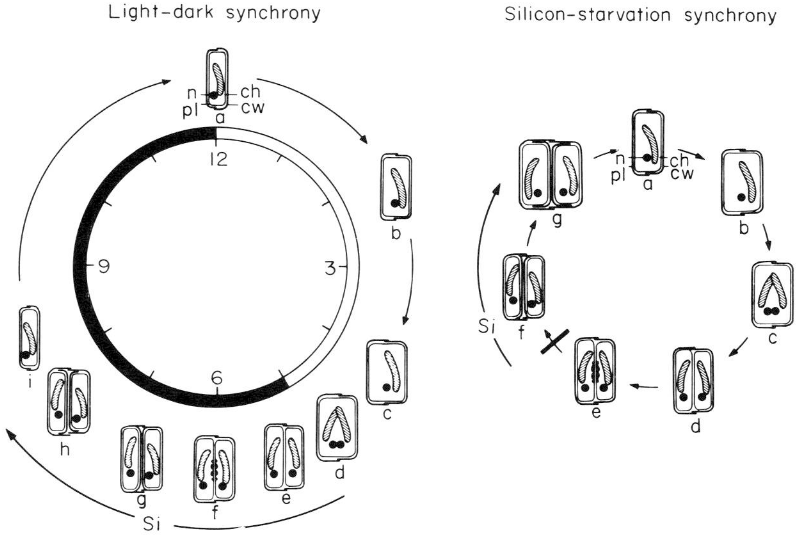

FIG. 3. Schematic representation of division cycle of *Navicula pelliculosa.* Left: Light-dark synchrony (W. M. Darley, C. W. Sullivan, and B. E. Volcani, in preparation). Right: Silicon-starvation synchrony [J. Coombs, P. J. Halicki, O. Holm-Hansen, and B. E. Volcani, *Exp. Cell Res.* **47,** 315 (1967)]. n, nucleus; ch, chloroplast; pl, plasmalemma; cw, cell wall.

sodium silicate to a concentration of 17 μg/ml silicon, together with L-methionine and L-cysteine, to concentrations of 10^{-5} and 10^{-4} *M*, respectively. If the pH has risen above 8 during the starvation period, it is adjusted to pH 7.5 by the addition of sterile HCl.

Properties of the Synchrony. Cell separation ceases once the medium is depleted of silicon (36 hours). However, cellular development continues until cytokinesis has taken place and the deposition of new walls has begun. Thus, after 14 hours the cell population consists of biprotoplastic cells, i.e., two daughter protoplasts, each surrounded by a new plasmalemma and separated by intercellular space, contained within the parent frustule (Fig. 3). When silicon is resupplied, a 3–4 hour period of rapid silicon uptake, during which new walls are completed, is followed by a 3–4 hour period of cell separation, at which time 80–95% of the cells divide (Figs. 2 and 3).

Light-Dark Synchrony

Principle. An exponential culture is placed in the dark for 24 hours. At the end of this period, cell division has ceased and the culture consists for the most part of small, young cells. The culture is then exposed to a repetitive light-dark cycle such that the light period provides enough energy for one division cycle and the dark period allows completion of the division cycle.

Inoculum. Prior to inoculation of the synchrony vessel, the FWT culture is subcultured at least twice in a defined freshwater glycylglycine medium (FWG); this consists of the medium described above in which Tryptone is replaced by glycylglycine 0.66 g/liter, and which is supplemented with vitamin B_{12}, 1μg; biotin, 2 μg; and thiamine·HCl, 0.5 mg/liter. The pH is adjusted to 8.2 with NaOH. Fifty-milliliter cultures in 125-ml Erlenmeyer flasks are grown at 20°, and illuminated with fluorescent lamps (5000 lux), on a reciprocal shaker (120 strokes/minute) for 24 hours to a cell count of 4 to 5 $\times$ 10^6 cells/ml.

Culture Vessel and Chamber. The culture vessel consists of a 4-liter Pyrex reaction kettle (14 cm in diameter and 39 cm in height). The glass cover, resting on a Teflon gasket, has four openings through which an aeration tube, a sampling tube, and an inlet for fresh medium enter the vessel through cotton plugs; the fourth opening is used for introduction of the inoculum (Fig. 1).

The synchrony is carried out in the glycylglycine medium to which is added sodium lactate (FWGL), 2 g/liter (6.6 ml of a 30% sodium lactate sterilized separately and added at inoculation). All media are autoclaved at least 24 hours prior to use. Medium to be used for dilution of the synchrony is aerated for 12 hours prior to use at the temperature of the synchrony, and transferred to the synchrony vessel through silicone rubber tubing.

The sychronization is maintained in a lightproof chamber (50 cm wide × 100 cm long × 75 cm high) provided with fans for ventilation. Illumination of the culture vessel is the same as that for the silicon-starvation synchrony, and the culture is aerated at the rate of 10 liters/minute. The light regime is controlled by a time-switch electric clock.

Synchronization Procedure. To initiate the synchrony, 3 liters of FWGL medium in the synchrony vessel are inoculated to a final concentration of 1.7 to 2.0×10^4 cells/ml. A 24-hour culture in FWG medium in the late exponential phase of growth (4 to 5×10^6 cells/ml) is used for inoculum. The synchrony vessel is placed in the synchrony chamber in the light for 32–36 hours; then the lights are turned off. After 24 hours of starvation in the dark, cell number is 0.5 to 0.75×10^6 cells/ml. Cells are then placed in a L-D regime of 5 hours light and 7 hours dark (one cycle) for as many cycles as desired. The cell number is approximately doubled during each of the first 3 L-D cycles, and reaches 4 to 6×10^6 cells/ml at the end of the third cycle (Fig. 4).

The culture is diluted by half with fresh medium at the beginning of

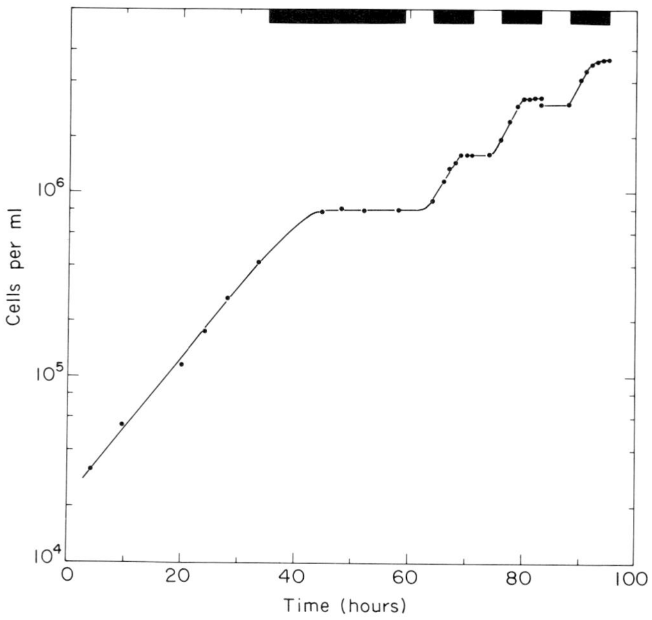

FIG. 4. Change in cell number of *Navicula pelliculosa* during exponential growth period, initial period of dark starvation, and first light-dark cycles. Solid bars at the top denote dark periods (W. M. Darley, C. W. Sullivan, and B. E. Volcani, in preparation).

the fourth cycle and every cycle thereafter, resulting in 2 to 3×10^6 cells/ml at the beginning of each cycle. Occasionally the culture is diluted slightly at the beginning of the third cycle if the cell number exceeds 3×10^4 cells/ml. To dilute the culture, 1.5 liters of culture is removed by suction through the sample tube into a large bottle, and 1.5 liters of medium is aseptically added from the reservoir of fresh medium.

Continuous-Light Synchrony. When studying light-dark synchronized cells, it is possible to avoid abnormal shifts in metabolism resulting from the light-dark changes, simply by leaving the already L-D synchronized culture in continuous light until cell division is completed. Dilution is carried out in the usual manner after the period of cell division. The light-dark synchronized culture retains a reasonable degree of synchrony for at least two successive cycles in continuous light.

Properties of the Synchrony. Cell separation occurs only during the dark period. During the first 3–4 cycles, however, the length of the separation burst steadily decreases from 6–7 hours during the first cycle to 4–5 hours during cycle 3. By cycle 5, however, the culture has reached a consistent degree of synchrony. On the average, the separation burst lasts for 3 hours (range 2.0–4.0), with the midpoint occurring between hour 7 and 10 of the L-D cycle; 75–100% of the cells divide at each burst, for an average of 88%. Experiments are therefore confined to the fifth and successive cycles. The culture can be maintained for as many as 19 cycles without loss of synchronization.

The sequence of events occurring both in the light-dark cycle and in the second cycle in continuous light of a synchronized culture consists of: cell growth culminating in mitosis and cytokinesis, followed by silicic acid uptake, wall formation, and separation of complete daughter cells. The division cycle is shown in Fig. 3.

Cylindrotheca fusiformis

Light-Dark Synchrony

Principle. The synchronizing principle for *C. fusiformis* is the same as that for the L-D synchrony of *N. pelliculosa*. However, in contrast to *N. pelliculosa* which requires a repetitive L-D cycle, *C. fusiformis* requires only a single period of dark starvation followed by illumination at high intensity to synchronize cell separation.[4,8]

Culture and Media. Axenic cultures of *Cylindrotheca fusiformis* (S. Watson's strain 13) are maintained and cloned on an enriched seawater medium (ESW) composed of $NaNO_3$, 0.25 g; $K_2HPO_4 \cdot 3H_2O$, 0.027 g; $Na_2SiO_3 \cdot 9H_2O$, 0.1 g; trace elements, 1 ml; vitamin B_{12}, 1 μg; thiamine·HCl, 0.5 mg; Tryptone, 1 g, in 1 liter filtered seawater.

In this synchrony it is essential to use a recently isolated clone for inoculum. The clone is obtained by plating on ESW medium solidified with 1.5% agar. Single colonies are transferred to test tubes containing 5 ml of ESW.

The synchrony is carried out in a medium prepared from commercial synthetic seawater (Utility Chemical Co., Paterson, New Jersey) of the following composition: NaCl, 27.5 g; $MgCl_2 \cdot 6H_2O$, 5.38 g; $MgSO_4 \cdot 7H_2O$, 6.77 g; KCl, 0.722 g; $NaHCO_3$, 0.20 g; $SrCl_2 \cdot 6H_2O$, 19.7 mg; $MnSO_4 \cdot H_2O$, 3.95 mg; $Na_2MoO_4 \cdot 2H_2O$, 0.987 mg; $Na_2HPO_4 \cdot 7H_2O$, 3.29 mg; LiCl, 0.987 mg; $CaCl_2$, 1.375 g; KI, 0.095 mg; KBr, 0.0285 mg; $Al_2(SO_4)_3$, 0.475 mg; $CoSO_4$, 0.0526 mg; RbCl, 0.157 mg; $CuSO_4 \cdot 5H_2O$, 0.488 mg; $ZnSO_4 \cdot 7H_2O$, 0.101 mg; and calcium gluconate, 0.659 mg.

To 40 g of the above is added: $NaNO_3$, 0.16 g; $Na_2SiO_3 \cdot 9H_2O$, 0.1 g; trace elements, 1 ml; 2 Na-EDTA (ethylenediamine tetraacetic acid, disodium), 0.012 g; thiamine hydrochloride, 0.5 mg; Tryptone, 1 g in 1 liter of glass distilled water. Before autoclaving, the pH is adjusted with NaOH to 8.0.

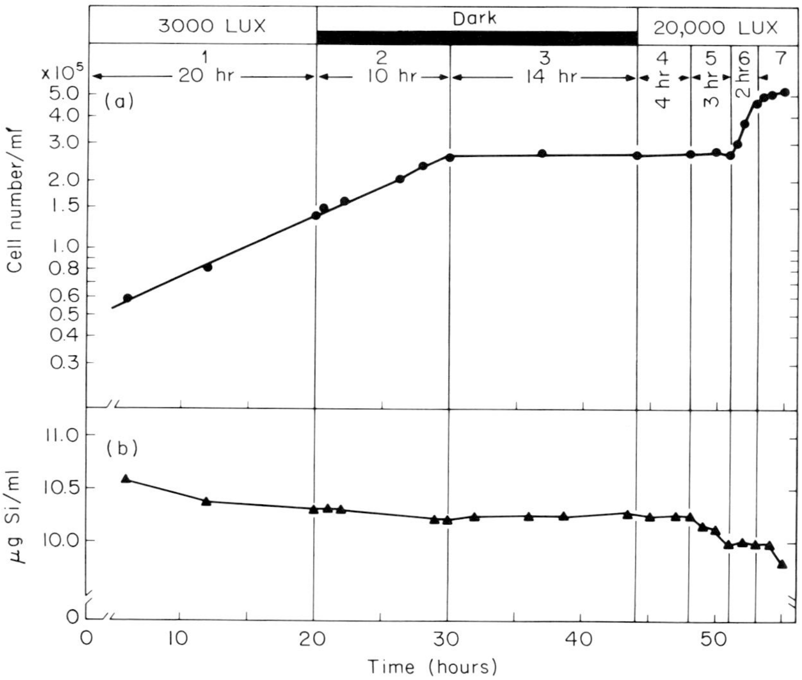

FIG. 5. Course of a synchronized culture of *Cylindrotheca fusiformis* (a), and depletion of silicon from the medium (b). [J. Coombs, P. J. Halicki, O. Holm-Hansen, and B. E. Volcani, *Exp. Cell Res.* **47**, 302 (1967)].

Cell Number. Samples are preserved and handled as in *N. pelliculosa.* Before counting, one drop of 5 *N* HCl is added to the suspension to dissolve any precipitated salts. Cells are counted in the light microscope with a Levy counting chamber, or in the Coulter counter set at amplification 1/4, aperture current 0.354, upper threshold 100, and lower threshold 10.

Culture Vessel. The culture vessel consists of a 2-liter Erlenmeyer flask adapted with an aeration and sampling assembly (as for *N. pelliculosa*) which enters through a cotton plug at the top. The flask is kept in a thermostatically controlled water bath at 25°. Illumination is provided either by two "warm-white" and two "cool-white" fluorescent lamps, providing about 3000 lux at the center of the flask, or by an iodine lamp ("Quartzline," General Electric Co.) which provides 20,000 lux. Air is bubbled through the culture at 6 liters/minute.

Inoculum. A 7–10-day-old clone is inoculated (2 ml) into 50 ml of ESW medium in a 125-ml Erlenmeyer flask and grown for 3 days on an illuminated reciprocal shaker as for *N. pelliculosa.* This culture is inoculated (5 ml) into identical medium; when it reaches 9 to 10 $\times$ 10^5 cells/ml (approximately 24 hours) it is used for inoculum; 40–50 ml of culture is required for inoculation.

Synchronization Procedure. Artificial seawater medium, 1.5 liters, in the synchrony vessel is inoculated to a final cell density of 4 $\times$ 10^4 cells/ml. After incubation at 3000 lux for about 20 hours, the culture

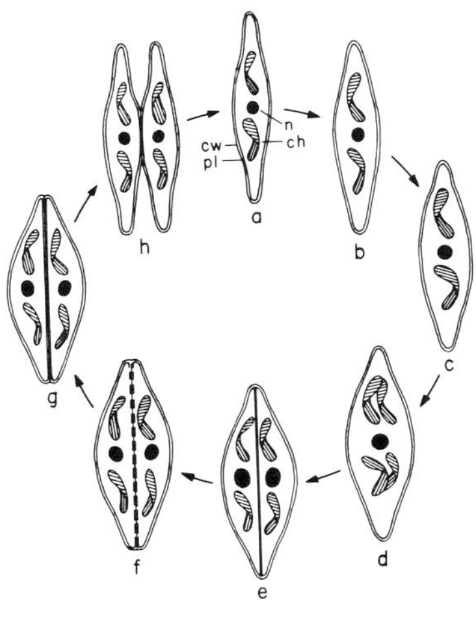

FIG. 6. Schematic representation of division cycle of *Cylindrotheca fusiformis.* n, nucleus; ch, chloroplast; pl, plasmalemma, cw, cell wall [J. Coombs, P. J. Halicki, O. Holm-Hansen, and B. E. Volcani, *Exp. Cell Res.* **47**, 302 (1967)].

reaches a cell number of 1.2×10^5 cells/ml; the flask is then placed in the dark by wrapping the synchrony vessel with aluminum foil. After 24 hours of darkness, the culture is exposed to the high intensity iodine lamp and at the same time filter-sterilized solutions of L-cysteine, L-methionine, and sodium lactate are added to final concentrations of 10^{-4}, 10^{-5}, and 2×10^{-2} *M*, respectively. The culture is left in the light for 10–12 hours for the completion of the synchronized division.

Properties of the Synchrony. The time course of increase in cell number during the entire synchrony procedure is shown in Fig. 5a. When the exponential culture is placed in the dark, cell division continues for about 10 hours, resulting in an 80% increase in cell number. When the culture is reexposed to light (20,000 lux) there is a 7-hour lag during which the cells increase in size and carbon content and take up silicic acid (Fig. 5b); cell separation then takes place during a 2–2.5 hour period. Cell number increases by 92% on the average. A schematic representation of the division cycle is shown in Fig. 6. As with *Navicula pelliculosa,* a sequence of events that may be grouped into 3 phases can be distinguished: (1) increase in carbon mass; (2) nuclear division, cytokinesis, silicon uptake, and cell wall formation; (3) cell separation (see phases 4, 5, and 6 respectively, in Fig. 5).

Light-Dark Synchrony of Other Diatoms

Light-dark regimes have been used to obtain synchronized cell division in four other species. Comparative data on seven L-D synchronies are presented in the table.

The varying degrees of synchronization and the applicability to biochemical research vary with the organism and reflect, to some extent, the purposes for which the synchronies were devised. The comparatively low cell concentrations in the last four synchronies are due to the fact that laboratory conditions were designed to simulate those of the natural environment.

Two different methods have been used to synchronize *Ditylum brightwellii.* In the first,[10] an 8:16 hour L-D regime produced a comparatively high degree of synchronization; this synchrony was used for morphological and metabolic studies. In another set of experiments[11] with *D. brightwellii* and *Nitzschia turgidula* using a variety of cycles, growth rate was measured as a function of light intensity. By interpolating from these growth curves, combinations of photoperiod and light intensity were chosen that gave a growth rate of one division per day. It was

[10] R. W. Eppley, R. W. Holmes, and E. Paasche, *Arch. Mikrobiol.* **56,** 305 (1967).
[11] E. Paasche, *Physiol. Plant.* **21,** 66 (1968).

Data on Light-Dark Synchronized Cultures of Diatoms

Organism	Habitat[a]	Type[b]	Carbon (μg/ml)[c]	Total cycle time (hr)	Division (%)	Burst time (hr)	Generation time of exponential culture[d] (hr)	Synchronization index[e]	Reference
Navicula pelliculosa	FW	P	12	12	88	3	8	0.55	*f*
Cylindrotheca fusiformis	M	P	16	1 step	92	2	13	0.78	*g,h*
Cyclotella cryptica	M	C	40	8	100	4	5	0.20	*i*
Ditylum brightwellii	M	C	0.5	24	95	4	30	0.82	*j*
Ditylum brightwellii	M	C	2.0	24	100	8	24	0.66	*k*
Nitzschia turgidula	M	P	1.0	24	90	6	10	0.36	*j*
Skeletonema costatum	M	C	0.8	24	100	9	10	0.10	*l*

[a] FW, fresh-water; M, marine.

[b] P, pennate; C, centric.

[c] A representative value obtained from the original paper (*C. fusiformis*[g] and *D. brightwellii*[k]) or estimated from data obtained from the literature.

[d] Doubling time in continuous light under the conditions of the synchrony.

[e] Calculated from the data in the 3 columns to its immediate left, according to the formula of O. H. Scherbaum [*J. Protozool.* **9,** 61 (1962)].

[f] W. M. Darley, C. W. Sullivan, and B. E. Volcani, in preparation.

[g] J. Coombs, P. J. Halicki, O. Holm-Hansen, and B. E. Volcani, *Exp. Cell Res.* **47,** 302 (1967).

[h] J. C. Lewin, B. E. Reimann, W. F. Busby, and B. E. Volcani, *in* "Cell Synchrony—Studies in Biosynthetic Regulation" (I. L. Cameron and G. M. Padilla, eds.), p. 169. Academic Press, New York, 1966.

[i] D. Werner, *Arch. Mikrobiol.* **55,** 278 (1966).

[j] E. Paasche, *Physiol. Plant.* **21,** 66 (1968).

[k] R. W. Eppley, R. W. Holmes, and E. Paasche, *Arch. Mikrobiol.* **56,** 305 (1967).

[l] E. G. Jørgensen, *Physiol. Plant.* **19,** 789 (1966).

found that short photoperiods of bright light yielded the highest degree of synchrony.

The synchrony developed for *Skeletonema costatum,* a familiar experimental species, was only slightly better than exponential growth.

[8] Tissue Culture: Plant

By W. M. Laetsch

The first true tissue cultures were apparently green,[1] but the use of such plant material for investigating photosynthesis and/or chloroplast development was essentially ignored until recent years.[2,3] This is unfortunate, because tissue cultures offer the prospect of providing the desirable experimental features of algal cultures while possessing the unique developmental patterns of higher plants.

The delineation of some of the advantages of this system will perhaps indicate some of the ways in which the following methods can be used. Techniques for handling cultured tissues are very similar to those for microorganisms, and the physical environment can be controlled in a common manner. The long-term control of temperature and illumination is, therefore, much more precise than is presently possible for either seedlings or mature plants. It is notoriously difficult, for example, to expose intact plants or detached organs to very high light intensities for an extended time period. This problem is minimized with cultured cells or tissues. The same is true for quantitative work involving light quality. Since most work on the development of higher plant chloroplasts centers on the light-induced etioplast to chloroplast transformation, the value of a well-defined illumination system cannot be underestimated. The control of the chemical environment is also susceptible to far greater precision than is possible with either whole plants or plant parts. Sterile conditions open the way to a variety of experiments, and the absence of a cuticle in cultured tissues lessens the problems of absorption of exogenous chemicals which so often plague those working with higher plant tissue. A defined substrate for cultured tissues offers opportunities which too often have been ignored. This permits the isolation of events in chloroplast development from general cell responses such as replication and growth. It is next to impossible to regulate the chemical imputs of

[1] R. J. Gautheret, *C. R. Acad. Sci.* **198,** 2195 (1934).
[2] W. M. Laetsch and D. A. Stetler, *Amer. J. Bot.* **52,** 798 (1965).
[3] L. Bergman and C. Berger, *Planta* **69,** 58 (1966).

normal leaf tissue, and a result is that "triggers" for chloroplast development usually "trigger" other processes as well. The difficulty in determining reciprocal influences usually results in most of them being ignored. In addition to a greater amount of physiological homogeneity, tissue cultures can possess greater anatomical homogeneity than do organs such as leaves and cotyledons. As a result, a host of variables encountered in "normal" tissues is minimized. A defined substrate also permits the regulation of tissue and organ differentiation. The ability to induce bud formation, for example, allows the examination of chloroplast development in "normal" tissue while still maintaining the advantages of sterility and precise control over the physical environment.

Angiosperms, unlike most of the algae, do not synthesize chlorophyll in the dark, but this phenomenon has not been fully exploited experimentally because of the inability to maintain plants or organs in the dark for extended periods. As a result, experiments have been confined to seedlings with large seeds. Tissue cultures can be maintained indefinitely in the dark, and they are the only system for effectively investigating chloroplast continuity in higher plants.

The electron microscope is a basic tool for studying chloroplast development, and cultured tissues have certain advantages for electron microscopy. The tissue generally responds well to chemical fixatives and is easy to section. The tissues of many leaves respond in just the opposite fashion.

It has frequently been stated that many problems in the development of higher plant chloroplasts will not be solved until chloroplasts can be cultured independently of the cell. While this breakthrough is awaited, cultured tissues offer the possibility of working with simpler, better defined systems than is available with attached or detached plant organs.

Summary of Investigations on Chloroplast Development in Cultured Tissues

This brief review will provide some examples of how tissue cultures have been used in studying plastid development, but it does not pretend to cover the literature. A number of papers have described aspects of fine structure of developing plastids,[2-5] while others have been more concerned with the effect of growth regulators, nutrients, and light quality on synthesis and/or chloroplast development.[6-17] Still others have investigated enzyme patterns during chloroplast development.[18,19]

[4] H. W. Israel and F. C. Steward, *Ann. Bot.* **31**, 1 (1967).

[5] S. J. Blackwell, W. M. Laetsch, and B. B. Hyde, *Amer. J. Bot.* **56**, 457 (1969).

[6] D. A. Stetler and W. M. Laetsch, *Science* **149**, 1387 (1965).

[7] E. M. J. Jaspars, *Physiol. Plant.* **18**, 933 (1965).

Comparisons of photosynthetic structure and function of developing chloroplasts in cultured versus intact tissues have been made to my knowledge only in tobacco.[2,20] There are differences in the two types of tissue, but these are mostly quantitative in nature. These various studies have established some of the basic parameters influencing chloroplast development in cultured tissue and have demonstrated that the system is amenable to experimental manipulation. Most of these authors have employed callus tissue grown on solid media, but Bergman has utilized liquid suspension cultures in his attempts to induce autotrophic growth of tobacco cells.[21] Most of the previous studies have been on problems similar to those which have been or could be conducted on leaf tissue, but some recent work has been concerned with problems that can be investigated only by means of tissue culture techniques. One such problem concerns the ability of mature chloroplasts in mature leaf tissue to dedifferentiate, divide, and develop again into mature chloroplasts.[22] The leaf is a determinate organ and studies on chloroplast development usually end with the death of the leaf, but the chloroplast life cycle can be continuously generated in cultured tissue. Another recent project concerns the relationship of tissue growth to organelle differentiation in dark-grown cultures. The etioplast characteristic of dark-grown leaves of seedlings has not been found in dark-grown callus. It has been determined in tobacco that the appearance of mature etioplasts can be controlled by regulating the growth rate of the callus.[23] The prolamellar body, for example, is induced by inhibiting tissue growth and is eliminated by stimulating tissue growth. It has previously been thought that the prolamellar body was intimately involved with the normal etioplast to choroplast transformation, but evidence from tissue cultures suggest

---

[8] S. Venketeswaran, *Physiol. Plant* **18,** 776 (1965).
[9] I. K. Vasil and A. C. Hildebrandt, *Planta* **68,** 69 (1966).
[10] G. Beauchesne and M. C. Poulain, *Photochem. Photobiol.* **5,** 157 (1966).
[11] L. Bergman and A. Bälz, *Planta* **70,** 285 (1966).
[12] R. Schantz, H. Duranton, and M. Peyrière, *C. R. Acad. Sci. Ser. D* **265,** 205 (1967).
[13] N. Sunderland, *Ann. Bot.* **30,** 253 (1966).
[14] N. Sunderland, *Ann. Bot.* **31,** 573 (1967).
[15] R. Boasson and W. M. Laetsch, *Experientia* **23,** 967 (1967).
[16] A. K. Stobart, I. McLaren, and D. R. Thomas, *Phytochemistry* **6,** 1467 (1967).
[17] P. K. Chen and S. Venketeswaran, *Physiol. Plant.* **21,** 262 (1968).
[18] A. K. Stobart and D. R. Thomas, *Phytochemistry* **7,** 1313 (1968).
[19] A. K. Stobart and D. R. Thomas, *Phytochemistry* **7,** 1963 (1968).
[20] D. A. Stetler, Ph.D. Thesis Univ. of California, Berkeley, 1967.
[21] L. Bergman, *Planta* **74,** 243 (1967).
[22] W. M. Laetsch, *Amer. J. Bot.* **54,** 639 (1967).
[23] D. A. Stetler and W. M. Laetsch, *Amer. J. Bot.* **55,** 709 (1968).

that its appearance is a function of the rate of cell division and that it does not play a causal role in chloroplast development.

Problems Encountered with Cultured Tissue

It is legitimate to ask why cultured tissues have not been more widely used for studies in chloroplast development and function. Aside from the fact that most people think it is easier to grow bean seedlings than to master tissue culture techniques, there are features of tissue cultures that make them unsuitable for certain studies. A greater appreciation for the potential of cultured tissues in studies on chloroplast development will be obtained if the difficulties are discussed in some detail.

Anatomical heterogeneity is generally less of a problem than in leaves, but genetical heterogeneity is a greater problem. Not only are the cells in a particular callus likely to have a variety of genotypes, but the longer the tissue is kept in culture the greater is the chance for increases in ploidy levels and chromosomal aberrations.[24,25] This presents problems in duplicating experiments, because the tissue used one month is often genetically very different tissue when used several months later. This genetic variability is at least one cause of the considerable variability in physiological response which many workers have found. In spite of what has been said above about the anatomy of callus, it is a mistake to think that all tissues are completely homogeneous in terms of cell type and cell size. Meristematic rates within callus vary and there are often considerable differences in cell size.

Differentiation in the form of xylemlike and phloemlike elements is frequent, and these elements are often localized in complex nodules. An important result of these various types of heterogeneity is the difference in developmental rates within the total plastid population. The chloroplasts in the cells near the surface of the callus often differentiate faster than those in the cells of the interior. The islands of meristematic cells commonly found in callus mean that division of proplastids is taking place in some cells while well developed chloroplasts are present in others. In other words, synchronous development of the chloroplast population is very rare in cultured tissues.[26] It might be thought that liquid suspension cultures would be the answer to some of these problems, but such cultures are generally very heterogeneous themselves since they consist of single cells and cell clumps of various sizes. Many of the cell clumps are actually smaller than many of the single cells. There are some

[24] J. G. Torrey, *Physiol. Plant.* **20,** 265 (1967).
[25] D. J Heinz, G. W. P. Mee, and L. G. Nickell, *Amer. J. Bot.* **56,** 450 (1969).
[26] M. M. Yeoman and P. K. Evans, *Ann. Bot.* **31,** 323 (1967).

cultures of tobacco which have a high proportion of free cells, but the cells will not synthesize chlorophyll.

Dark-grown bean or barley leaves turn green within a matter of hours upon exposure to light. Dark-grown cultured tissue turns green very slowly and may take as long as a week for the development of mature chloroplasts. While the concentration of chlorophyll per dry weight of green tissue varies between cultures, it is usually much less than that of the leaf tissue of the corresponding tissue. In tobacco, for example, the mature leaf can have twenty times more chlorophyll on a dry weight basis than callus.[2] This is mainly a result of fewer chloroplasts per cell. The low final concentration of chlorophyll means that small differences during chloroplast development are difficult to quantify. While the leisurely synthesis of chlorophyll in cultured tissue is a handicap for those in a hurry, it can be used to advantage to study the intermediate steps in development which sometimes happen so rapidly in leaf tissue they are difficult to detect.

The number of chloroplasts per unit volume of tissue is rather small, and rich yields of isolated chloroplasts are difficult to obtain. The chloroplasts are also full of starch, and, since the medium has sugar, the starch cannot be removed by placing the tissue in the dark. A solution to this problem might be effected by a continuation of the preliminary work on growing tissues on sugar-free medium.[21,27]

Initiation of Cultures

General methods for establishing and maintaining tissue cultures have been provided in detail by several authors.[28-30] A common impression is that tissue culture is a complex and difficult technique, but it is actually very simple. The basic rules of sterile technique must be followed, but other than that there is not a single "best way" to grow cultured tissue. The main problem is to place uncontaminated but live tissue on a medium that will induce cell division. Sterile tissue is obtained either by stripping away external tissue from an organ in order to excise internal uncontaminated tissue, or the external surface is sterilized and the whole organ is cut into explants. The latter method is used in the leaf sterilization method described below. Sterile transfer rooms and expensive inoculation hoods are nice to have, but are quite unnecessary for such work. The main problem in these operations is to eliminate air

[27] T. Fukami and A. C. Hildebrandt, *Bot. Mag. Tokyo* **80**, 199 (1967).

[28] R. J. Gautheret, "La Culture des Tissues Végétaux." Masson, Paris, 1959.

[29] P. R. White, "The Cultivation of Animal and Plant Cells." Ronald Press, New York, 1963.

[30] H. E. Street, *in* "Techniques for the Study of Development" (F. H. Wilt and N. K. Wessells, eds.), p. 425. Crowell, New York, 1967.

movements, and a few well arranged pieces of cardboard on a laboratory bench are as satisfactory as the glossiest transfer rooms.

Stem explants containing cambial tissue are used most commonly as starting material for cultures. Leaf tissue is often more difficult to culture, although in some species such as tobacco, callus is easy to obtain from leaves. Callus is readily obtained from the roots of many species, but such cultures are probably less likely to be green than are those from stems and leaves. Herbaceous dicotyledons have been used most extensively for tissue cultures, because they have proved relatively easy to grow. Monocotyledonous tissues, and especially grasses, have traditionally been difficult to grow in culture, but recent work suggests that many of these difficulties show signs of being solved.[31] It remains to be seen whether chlorophyll synthesis in these monocot tissues will be sufficient to use them for studies in chloroplast development. Certain gymnosperms have been cultured with considerable success, but pteridophytes have not been promising material.[28] With the exception of tissue cultures of *Ginko,* there has only been casual mention of the ability of cultured tissues of gymnosperms to form chlorophyll.[32]

Many types of culture vessels are available, and the most suitable type is more a function of the arrangement of the available light system than any other factor. Culture tubes and flasks are more suitable for vertical lamps, and petri dishes are best for overhead lamps. Their optical properties make plastic petri dishes preferable to glass. A fair amount of nonsense has been written about the necessity for special types of specialized (and expensive) culture vessels, but there is little evidence that such exotic ware produces any better results than those obtained with standard flasks and culture tubes.

Cultured tissues are generally defined as clumps of randomly proliferating cells, but there is no reason the term cannot apply to tissues of organs which are cultured in their "normal" state of differentiation under sterile conditions. This type of cultured tissue eliminates many of the disadvantages associated with callus tissue, and permits the exploitation of "normal" tissue in a sterile environment with a defined source of nutrition.[33] Small pieces of leaf tissue are examples of such cultures. The following method has been primarily used with tobacco tissue, but the same procedures work with bean leaves, and other species would undoubtedly provide suitable material. The same basic method can be used to surface sterilize stems, seed, cotyledons, or flowers, and the tissue can

[31] Y. Yamada, K. Tanaka, and E. Takahashi, *Proc. Jap. Acad.* **43,** 156 (1967).
[32] W. Tulecke, *Amer. J. Bot.* **54,** 797 (1967).
[33] R. Boasson and W. M. Laetsch, *Science* **166,** 749 (1969).

then be maintained on a minimal medium or callus can be induced with a complete medium.

Cut strips approximately 1 inch wide from the center portion of the leaf avoiding the main vein, the tip, and the base of the leaf. Etiolated leaves are treated intact. Soak the strips for 2–4 minutes in a beaker of water with a small amount of laboratory detergent. The subsequent steps should be conducted under a hood. Transfer the strips to a 4.0% hypochlorite solution (dilute commercial bleach 1:5 with water) with a small amount of laboratory detergent as wetting agent. The detergent is more effective than Tween. Soak the leaf strips for 5 minutes; stirring occasionally to remove air bubbles adhering to the leaf surface. Use a sterile forceps to transfer the strips to a beaker or flask containing sterile water and rinse by swirling the beaker. Transfer the leaf strips to petri dishes containing two pieces of sterile filter paper. Two strips are a convenient number for each dish. Use a sterile cork borer (preferably stainless steel) to cut disks from the strips. Large veins should be avoided. The disks will collect inside the cork borer and should be pushed out with a sterile glass rod. Never sterilize the disks after they have been cut from the strips, because the large cut surface will permit serious damage by the hypochlorite. When the disks are placed on medium in petri dishes, they should be randomized; that is, the disks from one leaf should not all be placed in the same culture dish. Disks 2 mm in diameter are a convenient size, and 10 disks will fit in a 5-cm petri dish.

Small disks are recommended because the cells rapidly lose stored nutrients and become dependent upon those in the medium. Large disks invariably mean the inclusion of major veins, and this means greater anatomical and physiological heterogeneity in the tissue. A minimal medium composed only of the inorganic constituents of the medium described below plus sucrose will maintain leaf tissue in the dark for as long as 3 months. It is also sufficient for normal rates of chlorophyll synthesis in disks of etiolated leaves. Cell proliferation resulting in callus formation can be induced in either etiolated or mature leaf tissue cultured on the minimal medium by transferring the tissue to a complete medium composed of the inorganic and organic components plus the growth regulators described below. Cell division will occur throughout the leaf disk, but it will be most active at the periphery. Enough callus will be formed in the latter region within 1–2 weeks to permit subculturing.

There are many types of culture media, and the tissues of each species and perhaps each organ will have their special requirements. A reasonable approach is to start with a medium on which a large number of different tissues will grow reasonably well and then gradually define the requirements for specific tissues. The following medium is basically that of Murashige and Skoog,[34] and it has proved satisfactory for a wide variety of dicotyledonous tissues.

The use of undefined or complex ingredients such as coconut water, casein hydrolyzate, or yeast extract is to be avoided if at all possible. If used for the reasons described below, they should be added to the other inorganic and organic constituents before addition of the agar. Coconut water is generally used at a concentration of 10–20% (v/v). If 100 ml of coconut water is used, then the amount of water added to the organic and inorganic constituents is correspondingly reduced. The fresh coconut water should be autoclaved for a few minutes, and the precipitated proteins filtered before use in the culture medium. It is convenient to freeze unused coconut water in 50- or 100-ml portions for future use.

The medium is dispensed into tubes or flasks and autoclaved at 15 lb for 15 minutes. Certain vitamins, antibiotics, etc., are filter-sterilized (the sterile plastic units with a membrane filter are most convenient) and added to the sterile medium.

Experiments with tissue cultures are usually measured in weeks rather than days and desiccation of the medium is frequently a problem. This is aggravated in the light because of the greenhouse effect. Petri dishes sealed around the bottom with masking tape will not dry out for many weeks and gas exchange does not seem to be seriously affected. Flasks and tubes with metal or plastic closures can be treated in a similar fashion. Cotton plugs are the least satisfactory type of closure. Sterile sheets of polyethylene are frequently used to cover culture vessels, but their effectiveness in inhibiting desiccation and permitting the transmittance of light is counterbalanced by their difficult manipulation during tissue transfer.

Control of Chloroplast Development

Chloroplast development is strongly influenced by the growth and physiological state of the host cells, so the ability to control the development of cells assumes great importance. Nutrients and growth regulators are very effective control agents.

Chlorophyll synthesis in light-grown tissue cultures can be markedly influenced by sugar concentration.[7,12] High concentrations inhibit synthesis and stimulate tissue growth. Sucrose also inhibits chlorophyll synthesis in cultured tobacco leaf tissue when it is present at concentrations of 2.0% and higher.[35] The standard sucrose concentration for tissue cultures

[34] T. Murashige and F. Skoog, *Physiol. Plant.* **15,** 473 (1962).

[35] W. M. Laetsch, *Amer. J. Bot.* **53,** 613 (1966).

Stock Solution 1

Dissolve the following one at a time in 5 liters of glass-distilled water:

NH_4NO_3	16.5 g
KNO_3	19.0 g
$CaCl_2$ 2 H_2O	2.2 g
$MgSO_4 \cdot 7\ H_2O$	1.85 g
KH_2PO_4	1.7 g
H_3BO_3	0.062 g
$MnSO_4 \cdot H_2O$	0.169 g
$ZnSO_4 \cdot 7\ H_2O$	0.106 g
KI	0.0083 g

Stock Solution 2

Dissolve in 100 ml of glass-distilled water:

$FeSO_4 \cdot 7\ H_2O$	5.57 g
Na_2EDTA	7.45 g

Stock Solution 3

Dissolve in 100 ml of glass-distilled water:

$CuSO_4 \cdot 5\ H_2O$	0.025 g
$CoCl_2 \cdot 6\ H_2O$	0.025 g

Stock Solution 4

Dissolve in 100 ml of glass-distilled water:

Na_2MoO_4	0.025 g

Stock Solution 5

Dissolve in 100 ml of glass-distilled water:

Glycine	0.2 g
Nicotinic acid	0.05 g
Pyridoxine·HCl	0.05 g
Thiamine·HCl	0.01 g

Medium

Combine the following to make 1 liter of medium:

Stock Sol. 1	500 ml
Stock Sol. 2	0.5 ml
Stock Sol. 3	0.1 ml
Stock Sol. 4	1.0 ml
Stock Sol. 5	1.0 ml

Add the following to the above:

Myoinositol	0.1 g
Sucrose	20 g

Growth regulators are added to the above (see discussion below). Adjust the pH of the solution to 5.7–6.0 with 0.1 *N* NaOH. A liter of solid medium is made by melting 8 g of agar in 497 ml of glass-distilled water and adding to the above solution. A liter of liquid medium is made by adding 497 ml of glass-distilled water to the above solution.

is 2.0–3.0%, so it is not surprising that chlorophyll synthesis in many cultures has not been very noticeable.

It was early noted that chloroplast development is inversely correlated with tissue growth rate,[2] and other studies have substantiated this observation. From the inception of tissue culture studies, major emphasis has been on optimum conditions for growth, and this emphasis has led to an unconscious selection against conditions promoting chlorophyll synthesis. The use of sugar concentrations optimum for growth is a case in point. The auxin, 2,4-dichlorophenoxyacetic acid (2,4-D), is an extremely effective promoter of cell division in cultured tissues and cultures grown on medium containing this auxin usually have little, if any, chlorophyll. A specific affect of this compound on chlorophyll synthesis has been suggested,[21] but it is more likely that this is the result of a primary effect on cell proliferation. Gibberellic acid also promotes growth in many cultured tissues, and chlorophyll synthesis in such cultures is usually inhibited. It is best, therefore, to use a medium that will not promote growth at the expense of chlorophyll synthesis. We have found α-naphthyleneacetic acid (NAA) to be the most effective auxin for such a medium, and a concentration of 0.5 mg of NAA/liter provides satisfactory results.

Cytokinins are important ingredients of culture media, since they are generally required for cell division and are often regulators of organ induction.[36] The most generally used cytokinins are kinetin and benezyladenine. The addition of either of these compounds to the above medium at 0.5 mg/liter completes the basic medium. Exogenously applied cytokinin has been found necessary for chlorophyll synthesis in light-grown tissue of at least one strain of tobacco callus.[6] This same strain will grow slowly without exogenously applied cytokinin, thus implying endogenous synthesis. If this strain is placed in the light on a medium without auxin or cytokinin it will not grow, but it will form chlorophyll.[15] It is felt that if growth is not limited by other factors, the endogenous cytokinin will be limiting for chloroplast development, but if growth is limited by other factors, then the available cytokinin is available to support chloroplast development. Again, the relationship between chloroplast development and cell replication and growth is extremely close. These events can be influenced by traces of cytokinin remaining after the tissue is transferred to cytokinin-free medium, so particular attention must be paid to ensuring the absence of residual growth regulator in such experiments. Kinetin can induce chloroplast replication without inducing cell replica-

[36] F. Skoog and C. O. Miller, *Symp. Soc. Exp. Biol.* **11**, 118 (1957).

tion,[33] so the involvement of cytokinins in chloroplast development in cultured tissue has a promising future.

Cytokinin and auxin interactions are also important because of their effect on bud and root induction in certain tissue cultures, and the ability to regulate tissue differentiation has obvious implications for investigations of chloroplast development. Buds can be induced in tobacco tissue by increasing the cytokinin concentration and lowering the auxin concentration. The inverse relationship results in roots induction. It is best to use indoleacetic acid (IAA) for such experiments. An IAA concentration of 0.1 mg/liter and a kinetin concentration of 2.0 mg/liter will promote extensive bud induction both in our callus cultures and in leaf disk cultures. This bud induction will proceed in either light or dark. Cultures isolated from different species or varieties will have their own requirements, and it is not suggested that the stated auxin–cytokinin ratio will provide positive results in all cases. Embryoids which develop into normal plants can be induced from undifferentiated carrot tissue cultures by eliminating the 2,4-D from the medium.[37] The effect of 2,4-D on bud induction and chloroplast development is seen in sugar cane cultures.[38] The undifferentiated tissues grown on high concentrations of 2,4-D do not form chlorophyll, but when the 2,4-D concentration is lowered, organized leaf primorida appear and chloroplast development commences. This relationship between tissue differentiation and the ability of proplastids to develop into chloroplasts is especially intriguing because of the opportunities it offers to control the development of dimorphic chloroplasts in sugar cane.[39]

The kinds and ratios of growth regulators will have to be adapted to the respective requirements of tissues when they are grown on either solid or liquid media. The aim of cell liquid suspension cultures is to have cells which do not stick together, but conditions promoting friability may not promote chloroplast development. In many cultures, for example, low cytokinin concentrations, and the use of 2,4-D and gibberellic acid promote tissue friability but inhibit chloroplast development. High cytokinin concentrations promote chloroplast development and tissue compactness.

The use of undefined media components, such as coconut water,[40] was described earlier, because a major purpose of tissue culture techniques is to minimize variables. On the other hand, coconut water and casein

[37] W. Halperin and D. F. Wetherell, *Science* **147,** 756 (1965).

[38] J. P. Nitsch.

[39] W. M. Laetsch and I. Price, *Amer. J. Bot.* **56,** 77 (1969).

[40] W. Tulecke, L. H. Weinstein, A. Rutner, and H. S. Laurencot, *Contrib. Boyce Thompson Inst.* **21,** 115 (1961).

hydrolyzate frequently stimulate chloroplast development, so their use is sometimes justified if chlorophyll synthesis cannot be induced by other means. These additives will often make it possible to culture tissues which only grow slowly, if at all, on defined media, but a tissue which requires such additives for either growth or chlorophyll synthesis must possess some unusual features of chloroplast development to offset the disadvantages of using a chemically undefined medium.

Fluorescent lamps are satisfactory light sources for inducing chloroplast development in cultured tissues. The advantage of incandescent lamps in combination with fluorescent lamps has not been established. Many cultured tissues prosper in continuous light, but the optimum photoperiod should be determined for specific tissues. A light intensity of 500–1000 fc is generally the optimum. Cultures tend to become bleached at higher intensities, but this might be as much the result of high temperature within the culture tubes induced by the high light intensity as of any direct effect of the illumination. Cultures are frequently maintained at 100–200 fc, and this intensity is too low for optimum chloroplast development in many tissues. It is often difficult to arrange suitable facilities for maintaining large numbers of cultures under banks of high intensity fluorescent lamps, because the compressor capacity of most temperature-controlled rooms will not handle the high heat output of the lamp ballasts. Mounting the ballasts outside the rooms will usually solve the problem. Most temperature-controlled rooms are fairly narrow, so vertical lamps behind shelves represent the most economical use of space for exposing cultures. The use of commercial growth chambers for growing tissue cultures in the light is very uneconomical in terms of both space and money. Light intensity is controlled either by wrapping the culture vessels in cheesecloth or paper tissues, or by inserting metal screens of various mesh size in front of the lamps. Cultures are commonly grown at 25°–27°, but some tissues might have unusual temperature requirements.

Chloroplast development in cultured tissues can be controlled by influencing the growth rate of the tissue, by manipulating various components of the medium, and by regulating light and temperature. Certain tissues will still not synthesize chlorophyll in spite of such maneuvers. One approach to overcoming this problem is to supply intermediates in prophyrin biosynthesis.[32] Cultured tissues are frequently impoverished, with respect to their biosynthetic potential, in comparison with differentiated tissues, and it is probably infrequent that chlorophyll synthesis is blocked at only one point. Even a refractory tissue, such as sugar cane callus, will "turn on" its chloroplast development machinery as soon as an organized meristem appears. It is not necessary, therefore, to discard a tissue which will not produce chloroplasts, because it might well yield interesting problems because of that fact.

Cytological Techniques

A number of standard techniques will be presented, because they are particularly useful for chloroplast studies with cultured tissue. The general anatomical features of cultured tissues are sometimes ignored, and this is unfortunate because an understanding of chloroplast development can be obtained only if there is adequate knowledge of cells and tissues. Histological preparations for light microscopy employ standard techniques.[41] The same can be said for electron microscopy, but since it is necessary to use the electron microscope to really see chloroplasts, the following schedule is given because it has yielded excellent preparations for a variety of cultured tissues. As with all such schedules, it is possible to vary many of the steps without courting disaster.

Small pieces of tissue (1 mm or less across the longest axis) are fixed in 2.0% glutaraldehyde and 0.05 *M* cacodylate buffer pH 6.8–7.0 for 3–10 hours. (Prepare 4.0% glutaraldehyde and add 1:1 to 0.1 *M* cacodylate buffer. Prepare the buffer with 21.4 g/liter of sodium cacodylate. Adjust the pH with HCl.) The fixation should be done at 4°. Wash the tissue in cold buffer three times with a 20-minute interval between each wash. Fix the tissue overnight at 4° in 2.0% OsO_4 in water. Dehydrate in the following acetone series: 30%, 20 minutes, 4°; 50%, 20 minutes, 4°; 70% (1.0% $UrNO_3$) overnight at 4°; 90%, 1 minute, 4°; 95%, 15 minutes, 4°; 100%, 20 minutes, room temperature, repeat 3 times. Embed tissue in Epon[42] in the following manner: 50% Epon: 50% acetone overnight (lid on vials); replace 50% Epon in vials with 100% Epon and let stand for 4–6 hours. Fill embedding capsules with Epon and transfer tissue. Keep at room temperature for 24 hours, and at 45° and 60° each for 24 hours, respectively.

The relationship between chloroplast development and the replication and growth of cells has been emphasized, and important aspects of this relationship can be obtained only by counting and measuring both chloroplasts and cells. The best method of counting chloroplasts is by making thick sections (about 40 μ) of paraffin-embedded material, so that at least one cell layer is included in each section. Chloroplasts can be counted by focusing through cells. Some cells in cultured tissues are very large and the cells must be separated to count the plastids. This can be done with pectinase.[43] The incubation period should be extended to

[41] W. A. Jensen, "Botanical Histochemistry." W. H. Freeman, San Francisco, 1962.
[42] J. H. Luft, *J. Biophys. Biochem. Cytol.* **9,** 409 (1961).
[43] E. C. Humphries and P. W. Wheeler, *J. Exp. Bot.* **11,** 81 (1960).

3 days, and the suspension should be forced through a fine pipette several times at the end of this period. Better cell separation is obtained by the chromic acid method[44] which is recommended for making cell counts. The chloroplasts cannot be counted in cells separated with chromic acid, because they are destroyed.

Chlorophyll Determinations

The method of Arnon[45] or of Winterman and DeMots[46] employing acetone and ethanol, respectively, as solvents are generally used. Callus tissues usually contain more water than leaf tissues, and this introduces the possibility of error when comparing chlorophyll in the two tissues. Some callus has so much water that the use of 80% acetone will result in too dilute a solution. Another problem is that callus tissue often contains so much starch that much longer centrifugation periods are required to obtain a clear supernatant solution than is customary with leaf tissue.

Photosynthetic Measurements

Standard methods are easily adapted to tissue cultures. Studies on the incorporation of $^{14}CO_2$ in liquid cultures can be conducted in the same manner as for unicellular algae. The same methods used for leaf tissue can be applied to callus. Very thin slices or small pieces of callus should be used since the very compact tissue is resistant to CO_2 diffusion relative to leaf tissue. This is one factor that probably limits completely autotrophic growth of tissue cultures. Actually there has been little quantitative work on photosynthesis in cultured tissue, and the instrumentation currently available could be used to good effect.

Conclusions

Many of the advantages and difficulties in using cultured tissue for investigating chloroplast development have been considered. The critical point is one that is not always realized by present or potential investigators in this field. It is that tissue cultures should be used only when they can provide a unique tool for investigating a basic problem in plastid development. They should not be used when "normal" tissues will do just as well. If cultured tissues are regarded as a potentially powerful technique for studying certain problems associated with the photosynthetic apparatus, rather than as a way of life, there is reasonable chance of the development of higher plant chloroplasts being explained in more satisfactory terms than at present.

[44] R. Brown and P. Rickless, *Proc. Roy. Soc. Ser. B.* **136,** 110 (1949).
[45] D. I. Arnon, *Plant Physiol.* **24,** 1 (1949).
[46] J. F. G. M. Wintermans and A. DeMots, *Biochim. Biophys. Acta* **109,** 448 (1965).

[9] Large-Scale Culture of Algae

By H. W. Siegelman[1] and R. R. L. Guillard[2]

Large quantities, greater than 100 g, of photoautotrophically grown algae, may be required for detailed studies of algal constituents and organelles. Methods are described here for growing such large quantities of algae using simple and inexpensive equipment. A description of the procedure has been given for the culture of *Euglena gracilis* var. *bacillaris,* strain Z.[3]

The method, with slight variations, has been employed for growing many kinds of marine and freshwater algae, both colonial and single celled. Included are many blue-green and green algae; the single-celled red alga *Porphyridium;* xanthophyte algae; cryptomonad flagellates; prasinophyte flagellates; haptophyte (chrysophyte) flagellates including coccolithophorids; many diatoms; and finally, certain dinoflagellates, both naked and armored. Various nutrient media have been used in the same basic apparatus.

Inoculum

Cultures are maintained in unshaken 500-ml Erlenmeyer flasks containing 200 ml of medium. (Certain species are sensitive to phenolic plastics used for screw caps, to certain plastic plugs or stoppers, and even to materials found in cotton or cheesecloth. Sensitivity can be detected experimentally; inverted beakers, glass wool plugs, washed cotton plugs, or stainless steel closures are generally satisfactory.)

Illumination of 100–500 fc is provided by banks of fluorescent "cool-white" lamps. In general, lower intensities are suited to blue-green and green algae; moderate intensities are better for diatoms, and dinoflagellates require highest intensities. Illumination can be continuous unless synchronous growth is required (a special problem) or unless the algae cannot tolerate continuous light (rare).

Most commonly used algae will tolerate temperatures of 18°–25°. Certain tropical species absolutely require temperatures in this range or even higher; 32° may be taken as a safe maximum for such forms. However, arctic or cold-temperature species often will not tolerate tempera-

[1] Research carried out at Brookhaven National Laboratory under the auspices of the U.S. Atomic Energy Commission.

[2] Contribution No. 2415 from the Woods Hole Oceanographic Institution. This work was supported by National Science Foundation Grant 7682.

[3] H. Lyman and H. W. Siegelman, *J. Protozool.* **14,** 297 (1967).

tures as high as 15°. For such species, special provisions for suitable low temperatures (aim for 5°) must be provided. Lighted and refrigerated water baths are generally most successful.

When the 500-ml culture flasks attain a dense algal growth (1–4 weeks), but before the cultures cease to grow rapidly, 2–4 flasks are used as inoculum for an 8-liter serum bottle containing 6 liters of medium. (New 200-ml cultures should be started at this time.) The 8-liter cultures are aerated through a single 6-mm glass aerator tube (see large-scale culture for other details). Aeration should be slow at first then increased to about 1 liter/minute as growth starts. Illumination is as described above, the growth should be dense by 4–14 days. As before, inoculum should be used just before the end of the period of rapid growth. Up to this point, sterile techniques are employed; see precautions associated with autoclaving seawater media.

Large-Scale Culture

The 8-liter bottle inoculum is transferred to a 50-gal polyethylene drum containing 160 liters of the nonsterile nutrient medium. The polyethylene drum is covered with a lid which is perforated to allow the insertion of fluorescent lamps and vinyl tubing for aeration, cooling, and CO_2 (Fig. 1). Aeration and stirring are accomplished by passing air at 10 liters/minute scrubbed by a "Koby Junior" disposable air purifier (Koby Corporation, Melrose, Massachusetts) in vinyl tubing (5 mm i.d. × 7 mm o.d.) through two cylindrical gas diffusion stones at the bottom of the drum. If required for freshwater media, CO_2 from a cylinder of 99.5% CO_2 is introduced through a separate vinyl tube at 100 ml/minute through a single gas diffusion stone at the bottom of the drum. Air and CO_2 are metered with needle valves, and flow rates are monitored with rotameters. CO_2 need not generally be added for marine media. If it is used, its flow rate should be slow enough to prevent the pH from dropping too low (7.2 should be a safe lower limit).

Illumination. Fluorescent lamps are positioned both external and internal to the culture vessel. A bank of 8-F96T12 "cool-white" fluorescent lamps attached to the wall provides the external source (Fig. 1). One to eight (as required) F48T12 "cool-white" lamps, contained in 47-mm glass tubes closed at one end, are inserted into the culture through the perforations in the cover of the drum.

Temperature Control. Primary temperature control is achieved by the ambient temperature of the culture room. A large loop (about 3 m) of vinyl tubing (11 mm i.d. × 17 mm o.d.) is coiled within the drum for circulation of cold or hot tap water as indicated by the temperature requirement of the algae being grown. Cooling will generally be required

ALGAL GROWTH MEDIA[a]

Compound	Medium				
	Kratz and Myers D[b] (mg/liter)	Ukai *et al.*[c] (mg/liter)	Modified from Chu No. 10[d] (mg/liter)	Modified from ASPM of McLachlan[e] (mg/liter)	"f/2' seawater[f] (mg/liter)
$NaNO_3$	1000	—	85.01	85	75
$Ca(NO_3)_2 \cdot 4\ H_2O$	10	—	—	—	—
$NaH_2PO_4 \cdot H_2O$	—	—	—	13.8	5
$Na_2HPO_4 \cdot 12\ H_2O$	—	500	—	—	—
K_2HPO_4	1000	—	8.71	—	—
KNO_3	—	3000	—	—	—
$Na_2SiO_3 \cdot 9\ H_2O$	—	—	28.4	56.8	15–30
$FeCl_3 \cdot 6\ H_2O$	—	—	3.15	0.54	3.15
$Fe_2(SO_4)_3 \cdot 6\ H_2O$	4	—	—	—	—
$FeSO_4 \cdot 7\ H_2O$	—	20	—	—	—
Na_2EDTA	50	—	4.36	11.2	4.36
KCl	—	—	—	746	—
$CaCl_2 \cdot 2\ H_2O$	—	20	36.8	1470	—
$MgCl_2 \cdot 6\ H_2O$	—	—	—	4067	—
$MgSO_4 \cdot 7\ H_2O$	150	500	37	4930	—
NaCl	—	—	—	23,400	—
$NaHCO_3$	—	—	12.6	168	—
H_3BO_3	—	2.86	4.6	12.3	—
$MnCl_2 \cdot 4\ H_2O$	1.44	1	—	—	—
$MnCl_3 \cdot 4\ H_2O$	1.44	1.81	0.18	1.39	0.18
$CuSO_4 \cdot 5\ H_2O$	1.57	0.079	0.01	0.00005	0.01
$ZnSO_4 \cdot 7\ H_2O$	8.82	0.222	0.022	0.23	0.022
$CoCl_2 \cdot 6\ H_2O$	0.40	—	0.01	0.005	0.01
$NaMoO_4 \cdot 2\ H_2O$	1.19	0.006	0.006	—	0.006
Thiamine·HCl	—	—	0.1	0.1	0.1
Biotin	—	—	0.005	0.005	0.005
Vitamin B_{12}	—	—	0.005	0.005	0.005
Water	Distilled	Distilled	Distilled	Distilled	Sea

[a] The Kratz and Myers D medium[b] is suitable for many freshwater blue-green and green algae. Vitamins are not called for but may be required by some species for which this medium is otherwise suitable.

The medium of Ukai *et al.*[c] was especially designed for *Tolypothrix tenuis*.

Chu's medium[d] No. 10 and numerous modifications are satisfactory for many freshwater algae, including species that will not tolerate concentrated media used for *Chlorella* and other hardy algae. The modification given here employs the Fe EDTA trace metal solution of the enriched seawater "f/2" plus added boron, and has vitamins added. A more usual trace metal solution for Chu No. 10 would have iron citrate plus citric acid, ≈ 3 mg per liter each, with the other elements about as given.

The artificial seawater ASPM[e] is suitable for many marine or brackish water algae. The vitamin solution employed here is that of "f/2" rather than the more complex if more than two lamps are in the drum. An external electric strip heater can be used as an alternate to hot circulating water.

Medium. Concentrated solutions of the ingredients are made and added to demineralized water or distilled water in the drum to give the desired final concentrations. Salts used in large amounts, such as NaCl

---

one of McLachlan. The very low copper level in the trace metal solution may reflect the presence of copper in the major salts. However, many algae will grow if the trace metal solution for "f/2" is substituted. The buffer glycylglycine has been omitted. The algae may grow well with no buffer, or Tris (0.5–1 g/liter) can be used if the algae will tolerate it.

The enriched seawater[f] "f/2" will be useful if seawater is easily available. Contamination by the native phytoplankton or zooplankton may be serious. If filtration (see Comments) is impractical, the water can be sterilized by heat or at least pasteurized in carboy lots. The nutrients should be added after the water has cooled.

In all these media the silicon need be added only for diatoms, as far as is known at present. Much silica dissolves from glass vessels, so its use is not critical even for diatoms if they are cultured in moderate density in glass, but in the plastic containers silicon addition is mandatory.

Some algae cannot use nitrate, in which case NH_4Cl should be used. Ammonia is toxic to many species at relatively low levels, so 0.1 m*M* is suggested as a starting level (5.3 mg/liter of NH_4Cl). The concentration may be raised to 1 m*M* for some species, which can favor dense growth.

There are a number of commercial formulations of artificial seawaters that are satisfactory for many marine algae. "Rila Marine Mix" (Rila Products, Teaneck, New Jersey) can be used 2–3% (w/v)[g] and treated essentially like natural seawater of equivalent salinity, with the "f/2" enrichment. "Instant Ocean" (Aquarium Systems, Inc., 1450 East 289 St., Wickliffe, Ohio) is another formulation. Note that some of the artificial seawaters contain EDTA or Tris buffer, which may have to be taken into account.

Detailed treatment of freshwater algal media,[h] and considerations applicable to marine media[e,i] are described.

The pH of seawater media will normally start at 7.2 to 7.8 and become higher as growth progresses. Aeration aids in keeping the pH low. For some species it may be necessary to adjust the initial pH of the freshwater media. If a buffer is desired, Tris should be tried because the more favorable glycylglycine is rapidly metabolized by bacteria.

[b] W. A. Kratz and J. Myers, *Amer. J. Bot.* **42,** 282 (1955).
[c] Y. Ukai, Y. Fujita, Y. Morimura, and A. Watanabe, *J. Gen. Appl. Microbiol.* **4,** 163 (1958).
[d] S. P. Chu, *J. Ecol.* **30,** 284 (1942).
[e] J. McLachlan, *Can. J. Microbiol.* **10,** 769 (1964).
[f] R. R. L. Guillard and J. H. Ryther, *Can. J. Microbiol.* **8,** 229 (1962).
[g] R. Ukeles, *Limnol. Oceanogr.* **10,** 492 (1965).
[h] L. Provasoli and I. J. Pintner, *in* "The Ecology of Algae" (C. A. Tryon, Jr. and R. T. Hartman, eds.), p. 84. Pymatuning Laboratory of Field Biology, Univ. of Pittsburgh, 1960.
[i] L. Provasoli, J. J. A. McLaughlin, and M. R. Droop, *Arch. Mikrobiol.* **25,** 392 (1957).

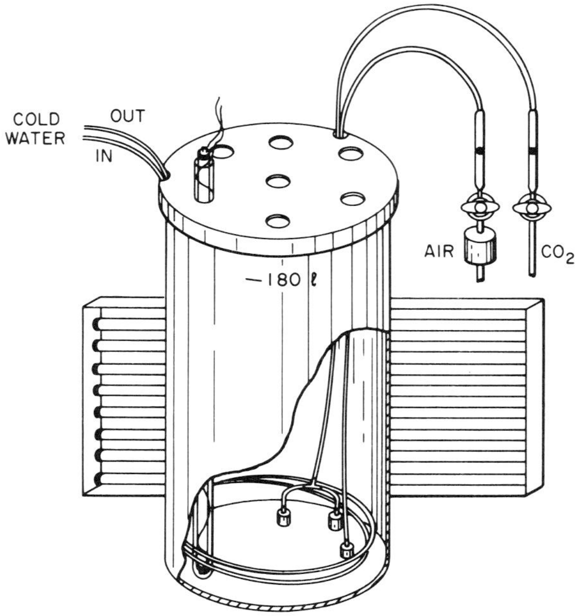

FIG. 1. Apparatus for large-scale culture of algae. The bank of external lamps is shown behind the polyethylene drum. One internal lamp is shown in place. The gas-diffuser stones from the air and CO_2 lines are shown at the bottom of the drum.

or $MgSO_4$, in artificial marine media are added directly. No attempt is made to sterilize the drum, medium, or water. However, the drums should be cleaned between batches with extremely hot water and allowed to dry. Troublesome contaminants that may have grown on the drum walls or bottom can be controlled by washing with ethanol or concentrated HCl; the latter must be removed completely with sterile water or a sterile buffer rinse.

Several media are given in the table, together with notes and references.

Harvesting of Cultures. Large cultures are grown for 6–14 days after inoculation. Cultures are best harvested by continuous centrifugation, e.g., in a Sorvall RC-2B refrigerated centrifuge having the Sorvall continuous-flow apparatus with an 8-channel distributor operated at 15,000 rpm with a flow rate of 1700 ml/min. A more simple method of harvest for most freshwater algae consists of adding a solution of 4 M $CaCl_2$ to the culture vessel to a final concentration of 5–10 mM. The actual concentration required is determined by adding 4 M $CaCl_2$ to a 100-ml graduated cylinder containing the algae until the algae are observed to aggregate. The algae aggregate and settle to the bottom of the culture vessel. A large pump, siphon, or aspirator is used to remove the bulk of the clear supernatant solution. The thick slurry of aggregated algae is collected by centrifugation in a large basket centrifuge lined with hardened filter paper, or in large-size centrifuge bottles. Marine algae may not aggregate with $CaCl_2$ but most will aggregate on addition of $KAl(SO_4)_2$ to a concentration of about 5 m*M*. All centrifugal operations can be simplified by appropriate use of a pump, electrical float valve, and reservoir to control flow rates to the centrifuge.

Yields. Yields will vary depending on the several factors regulating growth and upon the growth characteristics of any particular algae. In general, yields of 100–600 g of fresh weight per large culture vessel are achieved easily.

Comments

The large-scale culture procedure has been used to provide 40-kg lots (fresh weight) of *Phormidium luridum* var. *olivaceae* in 6 weeks using 10–180 liter culture vessels. Contamination of the cultures has not proved to be a serious problem although the medium and the polyethylene drum were not sterilized. The culture apparatus may be sterilized by placing germicidal lamps in the container and dispensing the culture medium through a large sterile Millipore R apparatus or a ceramic filter, such as the Selas No. FP-128-03, 0.6 μ, 20 psi filtration pressure.[4]

The practical problem of growing large amounts of algae is best handled by discontinuous culture. The continuous culture method is an elegant procedure, but the many problems inherent in algal culturing preclude its use for large yields.

Semicontinuous culture, in which most of the culture is removed periodically and replaced with fresh medium, has limited value because contamination becomes significant. If this method is employed, then harvest enough culture to keep growth *rate* maximal, though this will mean a yield less than maximal.[5,6]

[4] H. C. Davis and R. Ukeles, *Science* **134,** 562 (1961).

[5] B. H. Ketchum and A. C. Redfield, *Biol. Bull.* **75,** 165 (1935).

[6] B. H. Ketchum, L. Lillick, and A. C. Redfield, *J. Cell. Comp. Physiol.* **33,** 267 (1949).

Section II

Preparation and Properties of Mutants

[10] Preparation and Properties of Mutant Strains of *Chlamydomonas reinhardi*

By R. P. LEVINE

Methods are available for the induction, screening, and maintenance of mutant strains of the unicellular green alga, *Chlamydomonas reinhardi*, that are unable to carry out normal photosynthesis. These mutant strains can be used to study the nature and function of components of the photosynthetic apparatus in ways that are somewhat similar to those that have been used with auxotrophic mutant strains of microorganisms, for each of the single gene mutations has caused the alga to lose the ability to synthesize, at least in active form, a component of the photosynthetic apparatus. It is the purpose of this article to describe some of the properties of the mutant strains of *C. reinhardi* and methods for obtaining and culturing them.

Kinds of Mutations to be Expected

Mutations affecting chlorophyll and carotenoid biosynthesis are known in *C. reinhardi*.[1-5] Screening techniques also make it possible to select for mutations of genes that affect photosynthetic electron transport (and often, simultaneously, photosynthetic phosphorylation) as well as of those that affect enzymes of the photosynthetic carbon reduction cycle.[6,7] Mutations of genes that play a role in the architecture of the photosynthetic apparatus have also been identified.[5,8]

The Induction of Mutations

Common mutagenic agents are X rays, ultraviolet light, and a host of chemical mutagens; the choice depends to some extent on which is the most conveniently used in the laboratory. Microorganisms require relatively high doses of irradiation in order to mutate, and this in turn

[1] R. Sager and G. E. Palade, *J. Biophys. Biochem. Cytol.* **3**, 463 (1957).
[2] R. Sager, *Brookhaven Symp. Biol.* **11**, 101 (1958).
[3] G. A. Hudock and R. P. Levine, *Plant Physiol.* **39**, 889 (1964).
[4] G. A. Hudock, G. C. McLeod, J. Moravkova-Kiely, and R. P. Levine, *Plant Physiol.* **39**, 898 (1964).
[5] U. W. Goodenough, J. J. Armstrong, and R. P. Levine, *Plant Physiol.* **44**, 1001 (1969).
[6] R. P. Levine, *Science* **162**, 768 (1968).
[7] R. P. Levine, *Annu. Rev. Plant Physiol.* **20**, 523 (1969).
[8] U. W. Goodenough and R. P. Levine, *Plant Physiol.* **44**, 990 (1969).

requires the availability of a high-intensity source of X rays. Sources of ultraviolet light are more readily available in the form of germicidal lamps, and they are highly effective in inducing mutations in microorganisms. Among the chemical mutagens, the alkylating agents are known to be effective in *C. reinhardi*,[9] and they do not pose some of the serious health hazards that are encountered with potent carcinogens such as nitrosoguanidine.

Mutagenic agents can induce both "point" mutations and mutations that affect either chromosome structure or number. A mutation affecting a large portion of a chromosome may lead to the loss of more than one photosynthetic function whereas a point mutation is most likely to affect a single cistron and thus a single function. Of course, it must always be kept in mind that the loss of a single component can have a pleiotropic effect, meaning that it can cause an array of alterations in the phenotype. The problem in such cases is to determine which is the primary effect of the mutation. Genetic and cytological tests are necessary in order to distinguish a point mutation from a chromosomal mutation. Genetic analysis is possible with *Chlamydomonas*, whereas it is not yet possible with such algae as *Chlorella* or *Scenedesmus*. The genetics of *C. reinhardi* are particularly well known.

Methyl Methane Sulfonate

Methyl methane sulfonate has been used successfully to induce mutations in *C. reinhardi* in which the capacity to form different components of the photosynthetic apparatus has been lost,[10] and genetic analyses to date indicate that the mutations are point mutations. Because of the ease with which this mutagen can be used, a description of the method is given here.

The experiment is carried out under sterile conditions, as are all procedures with *C. reinhardi*. Cells in the logarithmic phase of growth are harvested and washed in 68 m*M* potassium phosphate buffer, pH 7.0. After washing they are resuspended in 10 ml of this buffer contained in a test tube, and the cell concentration is adjusted to about 1×10^7 cells/ml. The test tube is then placed in a water bath at 22°, and the cell suspension is agitated continually. After equilibration to the temperature of the water bath, methyl methane sulfonate is added to give a final concentration of 0.14 *M*. The cells are exposed to the mutagen for 3.5 hours, after which the action of the mutagen is stopped by diluting the

[9] R. Loppes, *Z. Verebungslehre* **98,** 143 (1966).
[10] P. Bennoun and R. P. Levine, *Plant Physiol.* **42,** 1284 (1967).

cell suspension 20-fold with the phosphate buffer. The diluted suspension is then washed 2 times with buffer and is adjusted to have approximately 1×10^4 cells/ml. Aliquots of 0.2 ml are then plated to a high salt minimal medium[11] supplemented with 0.2% sodium acetate to support heterotrophic growth, and colonies are allowed to form for subsequent screening (see below). Under the conditions described above, a typical experiment gives approximately 15% survivors. A survival rate commensurate with a reasonable yield of mutations is desired; this can be estimated by constructing a killing curve, which plots survivors vs. dose for a given mutagen and set of experimental conditions.

Ultraviolet Light

Another effective agent for inducing photosynthetic mutations in *C. reinhardi* is ultraviolet light.[12] A dose of ultraviolet light from two 15-W General Electric Germicidal lamps (equivalent to 97 ergs/cm² sec) is given to cells in the logarithmic phase of growth. This permits a survival of about 50% when given in a 4 ml suspension of cells (about 5×10^6 cells/ml) contained in a petri dish 60 mm in diameter. During the irradiation the cells are agitated with the aid of a magnetic stirrer. At the end of the treatment the cell suspension is diluted so that 0.2-ml aliquots contain about 100 surviving cells, and the aliquots are plated to minimal medium supplemented with sodium acetate. To prevent photoreactivation, all postirradiation procedures are carried out under the light from a General Electric "Bugaway" yellow lamp. The cells are incubated in the dark for 12 hours after plating, after which time they are placed in white light for subsequent growth.

The yield of acetate-dependent mutations of *C. reinhardi* from an experiment similar to the one just described was found to be 0.1%, and of these about 10% were found to be unable to carry out normal photosynthesis.

Screening for Mutations

Two methods have been devised to screen for or to select for mutations in *C. reinhardi* in which photosynthesis is affected. The first takes advantage of the fact that the mutations result in the inability of cells to fix carbon dioxide by photosynthesis,[13] and the second bases its selectivity on the fact that a block in the photosynthetic electron trans-

[11] N. Sueoka, *Proc. Nat. Acad. Sci. U.S.* **46,** 83 (1960).
[12] N. W. Gillham and R. P. Levine, *Nature* (*London*) **194,** 1165 (1962).
[13] R. P. Levine, *Nature* (*London*) **188,** 339 (1960).

port chain can lead to a fluorescence yield that is significantly higher than the normal yield.[10] The two methods can lead to the selection of different classes of mutant phenotypes.

Screening by ^{14}C Decay

Selecting for the inability to fix carbon dioxide by photosynthesis should theoretically lead to the widest spectrum of mutant phenotypes. This spectrum should include mutations that affect photosynthetic electron transport, photosynthetic phosphorylation, and the synthesis of enzymes of the photosynthetic carbon reduction cycle. So far, however, this technique as applied to *C. reinhardi* has yielded only mutant strains in which either electron transport or the carbon cycle have been affected. No mutations have yet been found that result in the loss of photosynthetic phosphorylation without affecting electron transport at the same time.

In screening for mutations by the carbon dioxide fixation method, cells that have been treated with a mutagen are plated to acetate-supplemented minimal medium. After colonies have formed, they are replica-plated to several plates of the same medium. When colonies have formed on the replica plates, one plate is selected for the screening test and the others are put aside in dim light to await the results of the screening. The plate selected for the screening test is inverted over a small watch glass containing a small volume (usually less than 1.0 ml) of ^{14}C-labeled sodium bicarbonate at a concentration that contains of the order of 1 μCi. Each replica plate to be screened is illuminated from above with light from reflector floodlights (150 W), and about 0.5 ml of a 1 *N* solution of acetic acid is introduced. All operations are carried out in a hood.

The lights are allowed to remain on for 5 minutes; during this time, a comparable series of plates can be exposed to $^{14}CO_2$ in the dark as controls. At the end of the exposure period, the colonies are replicated to filter paper and the paper is exposed to the fumes of concentrated HCl for 3 minutes in order to drive off any unfixed carbon dioxide. Radioautographs are then made of the filter paper. The film can usually be developed after a week's exposure to the colonies. The ability of the cells of a colony to fix carbon dioxide is seen by the darkening of the emulsion of the film at a position opposite the colony on the filter paper. Cells in colonies that cannot fix carbon dioxide do not affect the film.

It is now necessary to return to an untreated replica plate or, if colonies have re-formed upon it, to the master plate; those colonies that did not cause the film to be exposed are picked off and subcultured.

Cells are then grown in sufficient number to allow measurements of carbon dioxide fixation by more orthodox means (see section below dealing with culturing mutant strains).

Screening by Fluorescence

The fluorescence of colonies of wild-type cells of *C. reinhardi* can be detected by eye, and indeed the eye can detect an increase in the level of fluorescence when these colonies have been exposed to DCMU, an inhibitor of photosynthetic electron transport. Cells that are blocked in photosynthetic electron transport as a consequence of mutation will exhibit a similarly high level of fluorescence, and this fact permits a sensitive and rapid criterion for their detection.

The fluorescence level of 200–300 colonies on a petri dish is easily monitored in the following way. The fluorescence of the colonies is induced by illumination from a high-pressure mercury arc lamp (an Osram HBO 200, for example). The light from the lamp is passed through a water filter to remove ultraviolet and infrared light and then through a glass filter (Corning 4305) that cuts off all wavelengths above 640 nm. An intensity of some 4000 lux at the agar surface in the petri dish is sufficient to produce fluorescence that is detectable by eye. The fluorescence of the colonies may also be photographed using type 413 Polaroid film in a camera fitted with a glass filter (Corning 2030) that cuts off all wavelengths of light below 640 nm, with exposure times adjusted to give an optimum image of the fluorescent colonies. Putative mutants appear as colonies that are brighter than the surrounding wild-type colonies.

The major drawback with this technique is the need for an optimum amount of illumination and, more importantly, an evenness of illumination. Uniform colony size is also desirable since the level of fluorescence will be a function, in part, of the number of cells that fluoresce.

Testing the Mutants

After the cells in a colony selected by either of the two techniques described above have been subcultured, their photosynthetic capacity is "retested" to ascertain that they have undergone a heritable change. The retesting is most easily carried out by measuring either light-dependent carbon dioxide fixation or oxygen evolution of whole cells. If the measurements give values that are at least 5- to 10-fold less than those found for the wild-type strain, it can be assumed, at least for *C. reinhardi*, that a mutation has occurred which has had a significant effect on some aspect of photosynthesis.

If desired, the genetics of the mutant strains can be analyzed.[14–16] These analyses will reveal whether the mutation is inherited in the classical Mendelian manner of single nuclear gene mutations or whether it is inherited in an extranuclear fashion or in a manner suggesting that a nonchromosomal mutation has occurred. It is also possible to map the mutant genes to their sites in the genome and thus to determine whether two (or more) mutant strains having identical properties have arisen as the consequence of mutations at the same or at different loci.

Maintaining Mutant Strains in Culture

Stock cultures of mutant strains of *C. reinhardi* are maintained in dim light (~1000 lux) from daylight fluorescent lamps on agar medium contained in screw-cap culture tubes. Two types of media are used: either minimal medium[11] supplemented with sodium acetate or this medium with the additional supplement of 0.4% yeast extract. Carbon sources other than acetate have been found to be ineffective. Several stock cultures of different ages are maintained for each mutant strain. Certain strains can remain alive in the stock tubes for many months whereas others must be transferred at frequent intervals. There is at least one mutant strain of *C. reinhardi* (*ac-206*) that dies in stock tubes, and it must be cultured on media contained in petri dishes.

When mutant strains are to be used with any frequency, *working cultures* are maintained at low light intensity (~2000 lux) on agar medium in petri dishes. These cultures are transferred to fresh medium about every 3 days.

The culture of mutant strains for experiments is carried out in liquid medium. These *experimental cultures* are normally grown in 300 ml of Tris-acetate medium[17] in 500-ml Erlenmeyer flasks. The cultures are agitated by placing the flasks on rotary shakers. Illumination is from above. At the surface of the culture it is maintained at 4000 lux for wild type and for *ac-115*, *ac-141*, and F-34, and at 2000 lux for the remaining strains. Cells of the last group of strains tend to clump together at the higher light intensity. Logarithmic-phase cultures of wild type are obtained within 36–48 hours after inoculation whereas the mutant strains will be found in logarithmic phase between 72 and 84 hours.

Two to four such cultures usually provide ample material for the

[14] W. T. Ebersold and R. P. Levine, *Z. Verebungslehre* **90,** 74 (1959).

[15] W. T. Ebersold, R. P. Levine, E. E. Levine, and M. A. Olmsted, *Genetics* **47,** 531 (1962).

[16] R. P. Levine and W. T. Ebersold, *Annu. Rev. Microbiol.* **14,** 197 (1960).

[17] D. S. Gorman and R. P. Levine, *Proc. Nat. Acad. Sci. U.S.* **54,** 1665 (1965).

preparation of chloroplast fragments by the sonic disruption of cells.[18] If more cells are needed, cultures can be grown in large Erlenmeyer flasks or in carboys. The cells must be agitated continuously; vigorous aeration is a satisfactory method for the large cultures.

Stock, working, and experimental cultures should be maintained at 20°–27°.

Reversions and the Occurrence of Suppressor Mutations

Mutant strains in culture in the laboratory occasionally develop a phenotype that approaches or equals that of the wild-type strain. This can occur in one of two ways—reversion and suppression. Reversion, the reverse mutation of the gene in the direction of wild type, occurs rarely (at a frequency of less than 1 in 10^7 to 10^8). Suppression arises through the occurrence of a mutational event at a gene locus that is different from that of the original mutation and is a more common event in *C. reinhardi*. Several strains carrying suppressor mutations have been isolated from stocks of the mutant strains,[19] and genetic analyses have shown that the mutations occur randomly throughout the alga's genome. The suppressor mutations are of interest in themselves, but for the study of the photosynthetic properties of the original mutant strains they are a nuisance. Efforts have been made in this laboratory to determine whether there are conditions of culture that favor the accumulation of these mutations and whether there are conditions that would tend to select against them. So far, the only means that has been found to be at all effective against them is the frequent transfer of the cells in the working cultures. Even this procedure is only partially effective, at least for certain mutant strains.

There are two means of separating suppressed from unsuppressed mutant cells in *C. reinhardi:* by cloning and by crossing to the wild-type strain. The cloning procedure is the simplest and most rapid, and it is usually effective if the culture has not become heavily overrun with suppressor mutations. Cells are plated to minimal medium supplemented with sodium acetate and placed in the light. After colonies have formed, they are replica plated to minimal medium and to minimal medium supplemented with sodium acetate. When colonies have formed on the replica plates, the plates are examined and colonies are selected that have not grown on the minimal medium but have grown on the acetate-supplemented medium. These isolates or clones are subcultured and then tested to determine whether they have the properties of the parent mutant culture from which they were derived.

[18] R. P. Levine and D. S. Gorman, *Plant Physiol.* **41**, 1293 (1966).

[19] E. Cosbey, P. J. Hastings, and R. P. Levine, *Microbiol. Gen. Bull.* **23**, 20 (1965).

Occasionally the cloning procedure is unsatisfactory, and it then becomes necessary to cross cells[14-16] of the suppressed mutant strain to wild-type cells of opposite mating type. The meiotic products are separated on minimal medium supplemented with acetate. After the products have grown to form colonies, these are replica-plated to minimal medium and acetate-supplemented medium. Cells of the unsuppressed mutant will not grow on the minimal medium, and they can be isolated from the acetate-containing medium and tested to see whether their photosynthetic properties are those of the original mutant strain. Normally this process is resorted to only when essentially the entire culture has been taken over by the suppressor mutations so that the cloning procedure offers little probability of finding an isolate that is unsuppressed.

Properties of the Mutant Strains

The properties of the mutant strains of *C. reinhardi* that have been studied to date are listed below. Pertinent published references will be found next to each mutant symbol. When an unpublished observation is given, it too will be noted. Interpretations of the photosynthetic properties are found in the references that are cited.

ac-80a[8,20-22]: linkage group II; F-1: linkage group undetermined. Component affected: P700. Neither a light-induced nor a chemically induced absorbance change at 700 nm is detected in chloroplast fragments of the two mutant strains. The *ac-80a* strain is leaky, meaning that its genetic lesion is not entirely effective in eliminating photosynthetic activity. The rate of photosynthetic carbon dioxide fixation by whole cells can be as high as 25% of the wild-type rate. In chloroplast fragments, the rates of NADP photoreduction with either water or the DPIP-ascorbate couple are significantly less than wild-type rates. However, there is significant Hill activity with either DPIP or ferricyanide as the oxidants. Cytochromes 559 and 553 can be reduced by light but not oxidized. No light-induced change in cytochrome 564 has been detected although the cytochrome is present.[23] Neither cyclic photosynthetic phosphorylation with PMS as the electron carrier nor the fixation of carbon dioxide by photoreduction under hydrogen are detected in the mutant strain. However, noncyclic photosynthetic phosphorylation coupled to the photoreduction of ferricyanide occurs. Of the two light-

[20] A. L. Givan and R. P. Levine, *Plant Physiol.* **42,** 1264 (1967).

[21] N. H. Chua and R. P. Levine, *Plant Physiol.* **44,** 1 (1969).

[22] P. J. Hastings, E. E. Levine, E. Cosbey, M. O. Hudock, N. W. Gillham, S. J. Surzycki, R. Loppes, and R. P. Levine, *Microbiol. Gen. Bull.* **23,** 17 (1965).

[23] R. P. Levine, *Proc. Int. Congr. Photosyn. Res.,* Freudenstadt, Germany (1968).

induced pH changes found in chloroplast fragments of *C. reinhardi,* only the one associated with noncyclic electron flow is detected in F-1.[24] The 520-nm light-induced absorbance change associated with light absorbed by photochemical system II (PS II) is present in cells of F-1, but the change associated with photochemical system I (PS I) is missing.[21,25] At least four other independently occurring mutations have been found that have phenotypes similar to *ac-80a*. None as yet have been mapped to their respective sites in the genome. However, it is known that F-1 is not linked to the *ac-80* locus. F-1 has all the properties of *ac-80a* except it is not leaky. F-1 is also relatively easy to maintain free of suppressor mutations. Its growth rate is particularly slow.

ac-208[8,17,22,26]: linkage group III. Component affected: plastocyanin. No plastocyanin can be detected in extracts of acetone powders of cells that have been purified by chromatography on DEAE followed by chromatography on Sephadex. This mutant strain is presently the most difficult to maintain free of suppressors. Whole cells of the unsuppressed strain have a negligible rate of carbon dioxide fixation by photosynthesis and by photoreduction under hydrogen. Chloroplast fragments exhibit insignificant rates of NADP photoreduction with either water or the DPIP–ascorbate couple as electron donor. However, there are significant Hill reaction rates with ferricyanide or DPIP as the oxidants. P700 is present. Cytochromes 553 and 559 can be reduced by light but not oxidized. Their oxidation and the photoreduction of NADP can be restored if plastocyanin, purified from the wild-type strain, is added to the reaction mixtures. The light-induced absorbance change attributed to cytochrome 564 is not detected, although the cytochrome is present.[27] Neither cyclic nor noncyclic photosynthetic phosphorylation have been detected. Whole cells of the mutant strain exhibit the 520 nm absorbance change associated with PS II but not the change associated with PS I.[25] Only the light-induced pH change associated with noncyclic electron flow is detected in this mutant strain.[24]

ac-206[8,17,18,22,23,26]: linkage group XIV. Component affected: cytochrome 553. No cytochrome 553 can be detected in acetone powders of cells or in extracts of these powders that have been purified by chromatography on DEAE and Sephadex. This mutant strain is difficult to maintain in stock tubes, and is cultured on medium in petri dishes with frequent transfers to fresh medium. Cells in liquid cultures die relatively rapidly at the onset of the stationary phase, and they are usually

[24] P. Sokolove and R. P. Levine, unpublished data.

[25] N. H. Chua and R. P. Levine, unpublished data.

[26] D. S. Gorman and R. P. Levine, *Plant Physiol.* **41**, 1648 (1966).

[27] R. P. Levine, unpublished data.

harvested after 48 hours of growth. Whole cells have insignificant rates of carbon dioxide fixation by photosynthesis and photoreduction, and chloroplast fragments exhibit only a trace of the Hill reaction with NADP as the electron acceptor. However, there are significant rates of NADP reduction with the DPIP–ascorbate couple as the electron donor as well as significant rates of the Hill reaction with DPIP or ferricyanide as the electron acceptors. The light-induced oxidation of P700 is detected in *ac-206,* but of course no light or chemically induced absorbance changes are detected that can be attributed to cytochrome 553. The light-induced reduction of cytochrome 559 is seen, however. Although cytochrome 564 is present, its light-induced absorbance change is not detected. Cyclic photosynthetic phosphorylation occurs in *ac-206* but noncyclic photosynthetic phosphorylation coupled to ferricyanide reduction has not been detected. The 520-nm light-induced absorbance change that is associated with PS I is seen in whole cells of the mutant strain.[25] The light-induced pH changes associated with cyclic and noncyclic electron flow have been seen in *ac-620.*[24]

ac-21[8,18,22,23,28,29]: linkage group XI. Component affected: unknown. Cells of this strain have very low carbon dioxide fixation by photosynthesis and by photoreduction. Chloroplast fragments exhibit no significant NADP photoreduction with water as the electron donor but a relatively high rate of NADP photoreduction with the DPIP–ascorbate couple. There is a Hill reaction with DPIP or ferricyanide as electron acceptors. P700 is present as seen by its light-induced oxidation. Cytochrome 553 can be oxidized by PS I but it is not reduced by PS II. The reverse is seen for cytochrome 559; it is reduced by PS II, but it is not oxidized by PS I. This result has led to the conclusion that a component affected by mutation at the *ac-21* locus lies between the two cytochromes in the photosynthetic electron transport chain. The light-induced absorbance change of cytochrome 564 is not observed even though the cytochrome is present. Noncyclic photosynthetic phosphorylation coupled to ferricyanide reduction has not been seen in *ac-21,* but cyclic photosynthetic phosphorylation does occur at a significant rate[30] in properly prepared chloroplast fragments.[17] The 520-nm light-induced absorbance change associated with PS I is detected in whole cells of *ac-21.*[25] The light-induced pH change associated with cyclic electron flow is observed, but the change associated with noncyclic electron flow is absent.[24]

ac-115: linkage group I; *ac-141:* linkage group III; and F-34: link-

[28] R. P. Levine and R. M. Smillie, *J. Biol. Chem.* **238,** 4052 (1963).
[29] J. Lavorel and R. P. Levine, *Plant Physiol.* **43,** 1049 (1968).
[30] D. S. Gorman and R. P. Levine, unpublished data.

age group undetermined but not an allele of either *ac-115* or *ac-141*.[8,18,21–23,28,29] Components affected: cytochrome 559 and Q. Whole cells fix negligible amounts of carbon dioxide by photosynthesis, but they have a normal rate of carbon dioxide fixation by photoreduction. There is no Hill reaction with NADP, DPIP, or ferricyanide. The rate of NADP photoreduction with the DPIP–ascorbate couple is high, however. There is not a light- or chemically induced absorbance change attributable to cytochrome 559. The level of PS II fluorescence is high in weak measuring light, and it cannot be increased by actinic light. Cytochrome 553 and cytochrome 564 can be oxidized by PS I as can P700. The mutant strains exhibit cyclic but not noncyclic photosynthetic phosphorylation. The 520-nm absorbance change associated with PS II is missing, but the change associated with PS I is present.[21,25] Of the two light-induced pH changes, only the change associated with cyclic electron flow is seen.[24]

F-60: linkage group undetermined. Component affected: phosphoribulokinase. Cells of this strain do not fix carbon dioxide by photosynthesis or photoreduction. Reactions of the electron transport chain and photosynthetic phosphorylation are normal. Of the enzymes of the photosynthetic carbon reduction cycle only phosphoribulokinase appears to be affected.[31]

ac-20[22,32,35]: linkage group XIII. Component affected: chloroplast ribosomes. Cells grown on acetate-supplemented medium contain 5–10% of the chloroplast ribosomes found in wild-type chloroplasts. This is a conditional mutant strain, for it will also grow at a slow rate on minimal medium, and under this growth condition the level of chloroplast ribosomes is 25% of the wild-type level. When grown on an acetate-supplemented medium, rates of photosynthetic reactions are some 10- to 30-fold less than those of the wild-type, due to the absence of normal levels of ribulose-1,5-diphosphate carboxylase and cytochrome 559. When grown on a minimal medium, photosynthetic reactions occur at faster rates, and the aforementioned photosynthetic components are present at more normal levels.

[31] B. Moll, *Plant Physiol.*, in press.
[32] R. P. Levine and R. K. Togasaki, *Proc. Nat. Acad. Sci. U.S.* **53,** 987 (1965).
[33] R. K. Togasaki and R. P. Levine, *J. Cell Biol.* **44,** 531 (1970).
[34] R. P. Levine and A. Paszewski, *J. Cell Biol.* **44,** 540 (1970).
[35] U. W. Goodenough and R. P. Levine, *J. Cell Biol.* **44,** 547 (1970).

[11] Preparation and Properties of Mutants: *Scenedesmus*

By Norman I. Bishop

The induction and isolation of specific mutants as a tool for study of the mechanism of photosynthesis represents an extremely diverse methodology. It is apparent from the onset of such a program that a wide variety of mutations will be attained ranging from those strains with altered biosynthetic patterns for pigment synthesis to those blocked in a single essential step of the photosynthetic electron transport system. The obtainment of a desired mutant strain will be limited by the physiological restraints imposed by the organism selected for study, by the techniques available for the detection of the desired characteristic(s) of that strain, and by the imagination of the investigator.

General Considerations

The initial step in the preparation of a mutant strain must be the obtainment of a culture of *Scenedesmus obliquus* which is free from all other microorganisms. Since the continuation of a photosynthetic mutant will depend upon the ability of that strain to survive under heterotrophic growth conditions, all possible contaminants must be removed. For this purpose, aliquots of an algal suspension are spread uniformly on petri dishes containing a basal medium composed of a nitrate medium[1] plus 0.5% glucose, 0.25% yeast extract (Difco), and 2% agar (NGY medium). The inoculated plates are placed in an incubator (30°) and observed daily for a period of 10–12 days. Single clones are selected from plates that are free of bacterial and fungal contamination. This clone is subcultured onto the above medium in culture tubes and retained for later treatments for the induction of mutations.

The wild-type strain of *Scenedesmus* grows satisfactorily under autotrophic, mixotrophic, or heterotrophic conditions, producing normally pigmented cells that perform comparable rates of photosynthesis when selected at the appropriate time of the growth cycle.

Mutant Induction

X ray. Cultures of the purified *Scenedesmus* are transferred aseptically into sterile medium, agitated thoroughly, and a sample counted with a hemacytometer to determine the cell number prior to irradiation studies. An initial cell number of 6×10^6 cells/ml is chosen in order to

[1] E. Kessler, W. Arthur, and J. E. Brugger, *Arch. Biochem. Biophys.* **71,** **3263** (1957).

have a workable number of cells per culture plate at the end of the induction procedure. Suspensions of cells are placed in 2-ml volumetric flasks for subsequent irradiation with either a Philip 25-750 or a General Electric 300-kV X-ray machine; exposure times are adjusted such that 60 krad are delivered to the samples. This treatment, with *Scenedesmus,* produces about 95% kill, the general mutation frequency in the surviving cells being about 2%. After irradiation the algal suspension is diluted and an inoculum sufficient to give 25–30 viable cells per plate is spread onto petri dishes as described above.

Ultraviolet Irradiation. The use of ultraviolet irradiation as a mutagenic agent for *Scenedesmus* is limited because of the extreme killing effect produced and the low mutation rate among surviving cells. The few mutations obtained by ultraviolet irradiation have been less stable than those produced by other methods.

Chemical. An alternate and equally satisfactory technique for the induction of mutations in *Scenedesmus* and other related algae is through the use of a number of chemical mutagenic agents. Two such chemicals, ethyl methane sulfonate (EMS) and *N*-Methyl-*N′*-nitro-*N*-nitrosoguanidine have been successfully employed for mutant induction with *Scenedesmus* and *Chlorella.*

The techniques employed are similar to those described for mutant induction with X ray. The mutagenic and killing action of EMS is dependent upon the physiological status of the algal culture used; young actively growing cells are much more susceptible to EMS than are older cells. For this reason a certain degree of variability is encountered in the number of cells that survive treatment. From a sterile slant of *Scenedesmus* a loopful of cells is transferred to 10 ml of sterile H_2O and the cell number determined with a hemacytometer. An initial cell number of approximately 6.5×10^4 cells/ml is desirable. In order to have a countable number of cells on control tests the algal culture is serially diluted such that approximately 100 cells are placed on each petri dish.

Chemical treatment of the cells is performed at a cell concentration of 6.5×10^4 cells/ml. A sufficient amount of EMS is added to give a final concentration of 0.14 M and the cells are allowed to incubate for 1 hour at room temperature. The action of EMS is terminated by either centrifuging out the algal cells, followed by successive washings with sterile growth medium, or by the addition of sodium thiosulfate. The pelleted cells from the centrifugation step are resuspended in 200 ml of sterile water and 0.1-ml aliquots spread onto petri dishes. As mentioned above the killing effect of EMS is variable, but in general about 95% kill occurs.

If the action of EMS is terminated by the addition of sodium thiosulfate, the following procedure is used: 5 ml of treated cell suspension is pipetted into 5 ml of 0.4 *M* sodium thiosulfate and allowed to stand for 10 minutes at room temperature. The sample is then diluted 10-fold with water and 0.1-ml aliquots plated out and incubated as indicated above.

The techniques employed with *N*-Methyl-*N'*-nitro-*N*-nitrosoguanidine are essentially identical to those for EMS treatment except for the concentration of the mutagen employed. Successful mutant induction is obtained by incubating algal cells for 1 hour in a 20 m*M* solution. The action of this mutagen is terminated by repeated washings of the cells with sterile distilled water.

Mutant Isolation

Visual Examination for Pigment Mutants. Close examination of the petri plates when algal clones have reached a diameter of approximately 2 mm will allow for the selection of a broad category of pigment mutants. At this time the mutant clones are removed with a platinum loop and individual culture tubes containing the NGY medium inoculated. Such cultures when maintained at 23° remain healthy for a 3–4 month period; periodic subculturing is required in order to retain the mutant strain.

It is also possible at this point in the procedure to recognize colonies of cells which are limited in their growth rate and appear either as minute colonies or have a granulated, rather than the usual shiny, appearance. No detailed study has been made by the author on this type of mutation in *Scenedesmus*.

Further characterization of a pigment mutant is performed by comparative studies on the ability of the mutant to grow heterotrophically and mixotrophically, by examination of the absorption spectrum of a methanol extract of mutant cells, and by thin-layer chromatographic analysis of this extract. In general, those mutants which grow only heterotrophically usually do not synthesize a normal complement of carotenoids and are therefore light sensitive. Those strains which show identical growth and pigment development under the two growth conditions form normal amounts of the carotenoids but generally do not synthesize chlorophyll. A number of intermediate mutant types will also be obtained which fail to synthesize chlorophyll in the dark but undergo greening during mixotrophic growth. This last mutant form also grows autotrophically provided a supply of energy is available for the initial stages of chloroplast-membrane development.

The chloroplast pigments of *Scenedesmus* are not removed by the conventional extraction with acetone so commonly employed for chlorophyll estimation in isolated chloroplasts and other algae. It is necessary to extract *Scenedesmus* with hot methanol; a complete removal of pigments requires three extractions and sometimes more depending upon the age of the algal culture. Chlorophyll concentration determinations can be made directly on the methanol by employing the extinction coefficients of chlorophylls *a* and *b* in methanol.[2] Analysis of the absorption spectrum of the methanol extract, or even of the whole cells, generally reveals the basic nature of the pigment mutation. The accumulation of C_{40} polyenes, among them ζ-carotene, proneurosporene, prolycopene, and β-zeacarotene, is evident by the increased absorbancy that occurs at 378, 400, and 425 nm. The general isolation and characterization of similar mutants of *Chlorella vulgaris*, and their properties, have been previously described by Claes.[3,4]

More detailed analyses on the specific nature of pigment synthesis can be obtained by chromatography of the methanol extract. For this purpose the methanol extract is brought to 5% KOH by the addition of KOH pellets, heated to the boiling point, and cooled rapidly in an ice bath. The methanol concentration is brought to 80% by the addition of 5% NaCl, and the nonsaponifiable pigments are partitioned into petroleum ether. The partitioning is repeated three times, the petroleum ether fractions are combined and then evaporated to dryness under vacuum. For thin-layer chromatography (TLC) 20×20 cm silica gel plates are activated for 1 hour at 105°, streaked with the sample (dissolved in chloroform), preincubated for 15 minutes with a solvent composed of petroleum ether, isopropanol, and water (100:10:0.25), and developed. This system resolves all the carotenes and carotenoids of the normal strain of heterotrophically grown *Scenedesmus*. These include, in band order down the TLC plate, α- and β-carotene, α-cryptoxanthin, α-carotene epoxide, HO-echininone, lutein and zeaxanthin, violaxanthin, and neoxanthin. In mutants which form the acyclic C_{40} polyenes, compounds which are in general extremely mobile in the petroleum ether–isopropanol solvent, a better separation is obtained by employing a less polar solvent, i.e., 5–10% benzene in petroleum ether. For quantization of α- and β-carotene, and of lutein and zeaxanthin, it is necessary to rechromatograph these bands on TLC plates with an absorbent composed of $CaCO_3$, $Ca(OH)_2$, and MgO (6:1:1.2) and with a solvent consisting of either 1% or 10% isopropanol in heptane.

Screening for Photosynthetic Mutants: Uptake of $^{14}CO_2$. Following the retrieval of the desired pigment mutants the surviving colonies are

[2] M. Holden, *in* "Chemistry and Biochemistry of Plant Pigments" (T. W. Goodwin, ed.), p. 466. Academic Press, New York, 1965.

[3] H. Claes, *Z. Naturforsch. B* **12**, 401 (1957).

[4] H. Claes, *Z. Naturforsch. B* **14**, 4 (1959).

screened for mutants which do not perform photosynthesis. The technique employed involves exposure of the clones to light and carbon dioxide labeled with carbon-14 by a method previously described.[5] Following exposure to light and $^{14}CO_2$, a replicate plate is made for each petri dish with a sterilized velvetin pad. After thorough drying of the cloth pad, which must retain imprints of the colonies on each plate, it is stapled to Kodak No-Screen X-ray film. After exposure for 7–10 days, the film is developed and the ability of each clone to incorporate carbon dioxide is evaluated. Colonies that fail to incorporate $^{14}CO_2$ are located either on the original or on the replicate plate and transferred into individual culture tubes. After sufficient growth occurs larger scale culture flasks, containing either basal nitrate medium for autotrophic growth or NGY medium for heterotrophic growth, are inoculated. Only those strains which show little or no growth autotrophically are retained for further investigation. It is possible to obtain mutations which have a decreased rate of photosynthesis (as evidenced by a slower rate autotrophically but not heterotrophically), but no retrieval of such mutants have been attempted in the author's laboratory.

Screening for Photosynthetic Mutants: Fluorescence Analysis. An additional useful method of screening for photosynthetic mutants involves the detection of mutant clones which have excessively high fluorescence. This technique has been recently applied by Levine for isolation of mutants of *Chlamydomonas*[6] and is discussed elsewhere in this volume. It is well known that inhibition of photosynthesis at various stages ranging from early steps in photosystem II to some of the initial stages of photosystem I results in increased fluorescence yield. From mutant studies it is known that loss of P-700, of cytochrome *f*, and of lowered plastoquinone levels also produce such an effect.[7,8] Thus this technique is not only rapid but also specific for the types of mutants detected. Its major drawback is that it does not detect mutants whose photosynthesis is blocked in later steps in photosynthesis, nor will it detect certain types of mutants that are genetically blocked on the oxidizing side of photosystem II which would be unable, hypothetically, to show a variable yield fluorescence. A combination of the two techniques in searches for photosynthetic mutants is recommended.

Characterization of Photosynthetic Mutants

General Nature of Mutants. Further assessment of the nature of the block to photosynthesis in mutants of *Scenedesmus* is possible after it

[5] R. P. Levine, *Nature (London)* **188**, 339 (1960).
[6] P. Bennoun and R. P. Levine, *Plant Physiol.* **42**, 1264 (1967).
[7] R. Powls, J. Wong, and N. I. Bishop, *Biochim. Biophys. Acta* **180**, 490 (1969).
[8] N. I. Bishop and J. Wong, *Biochem. Biophys. Acta* (in press).

has been determined that heterotrophically cultured cells are comparable to the wild type in their growth pattern and pigmentation. Such cells can then be analyzed for their ability to perform a number of partial reactions of photosynthesis. The initial assumptions made are that the mutant may be blocked in photosystem II, in photosystem I, or in the electron transport system connecting the two photosystems. This rudimentary characterization can be implemented by *in vivo* studies on whole cells of the mutants and by various observations on reactions of isolated chloroplasts.

Generally, the photosynthetic capacity of heterotrophically grown cells is measured polarographically since it is possible to determine quantitatively the level of both photosynthesis and respiration and also the immediate effect of light upon apparent respiration. For an example, the photosynthetic capacity of wild type and three separate mutant types are summarized in Fig. 1. The strains are mutant 8, mutant 11, and mutant 50. These strains have been shown to be a system I mutant (void of P-700), a system II mutant (deficient in plastoquinone A), and an

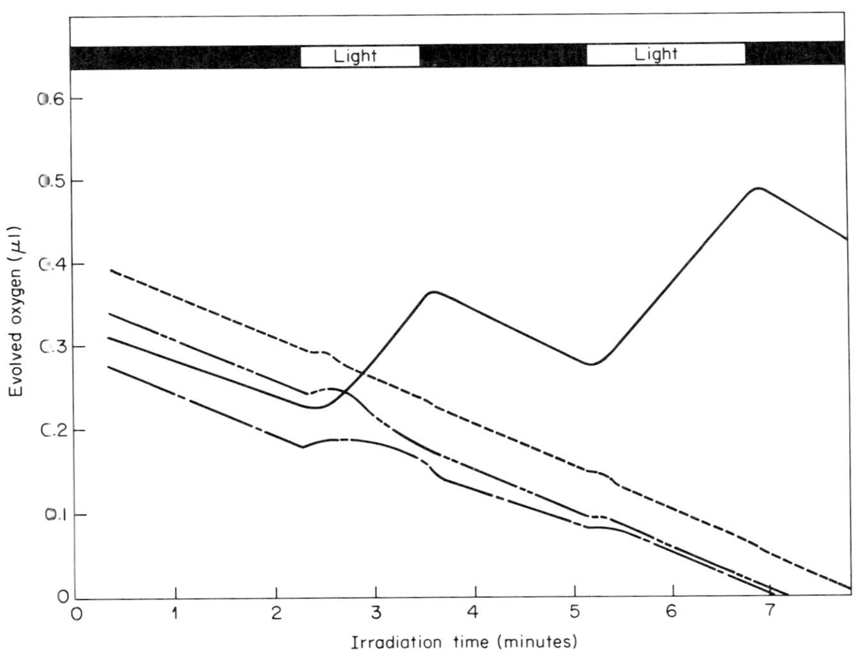

FIG. 1. Polarographic determination of oxygen consumption and production by heterotrophically grown cells of wild type (———), mutant 11 (- - -), mutant 8 (— - - —), and mutant 50 (— - —). Cells, 10 μl, were suspended in 2 ml of 50 m*M* phosphate buffer, pH 6.5; temperature, 25°. Red light intensity = 1×10^5 ergs/sec/cm^2.

electron transport mutant (lacking both bound and soluble cytochrome 552), respectively.[7] The legends of each of the following figures will provide the conditions under which the assays were performed. A suspension of normal cells shows normal respiration and an almost instantaneous production of oxygen upon illumination. None of the mutant strains show any marked response to light; often a slight amount of photosynthesis is detected during the first minute of illumination, but a subsequent exposure to light usually produces no response. In some mutant strains, usually pigment mutants, a brief exposure to light will cause a marked stimulation of the subsequent dark respiration. This has not been detected in photosynthetic mutants.

To assay for photosystem II activity the capacity of the mutant strains to perform the quinone–Hill reaction can be determined. The typical response obtained with the variety of mutant strains mentioned above is summarized in Fig. 2. Mutants which have a portion of photosystem I deleted or are mutated in the electron transport system show a decreased rate of the Hill reaction. Thus mutants 8, 26, and 50 have less

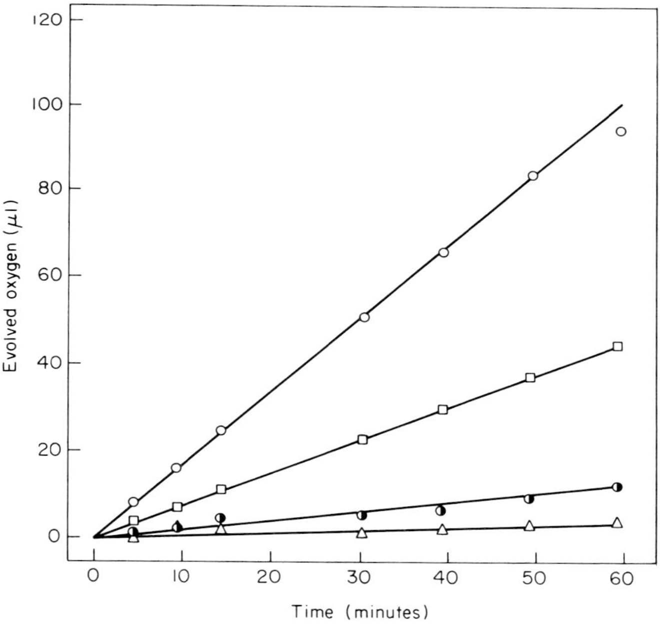

FIG. 2. Manometric determination of the Hill reaction capacity of wild type (○——○), mutant 8 (□——□), mutant 11 (△——△), and mutant 50 (◑——◑) of *Scenedesmus*. Cells, 50 μl, were suspended in 3.0 ml of 50 m*M* phosphate buffer, pH 6.5, in a gas phase of 96% N_2–4% CO_2; temperature, 25°. Manometric reaction vessels each contain 1 mg of *p*-benzoquinone. Light intensity $= 5 \times 10^5$ ergs/sec/cm^2.

than 50% of the activity of the wild type. A true photosystem II mutant such as mutant 11, should be unable to evolve oxygen regardless of the type of Hill oxidant employed.

The assay procedures discussed so far do not allow for a distinction between electron transport and system I mutants. *Scenedesmus* is one of the few green algae which has the capacity to form an hydrogenase whose action can be coupled to that of photosystem I such that a normal fixation of carbon dioxide will occur without the concomitant production of oxygen. This activity, photoreduction, is a strict photosystem I reaction by all criteria.[9] Mutants which are blocked only in photosystem II should perform a normal photoreduction but those altered in either photosystem I, or certain stages of the electron transport system, or photophosphorylation will have such activity inhibited. Some of the characteristic responses of photoreduction obtained with the various mutant types are depicted in Fig. 3.

A mutant with impaired photosynthesis and photoreduction, but good

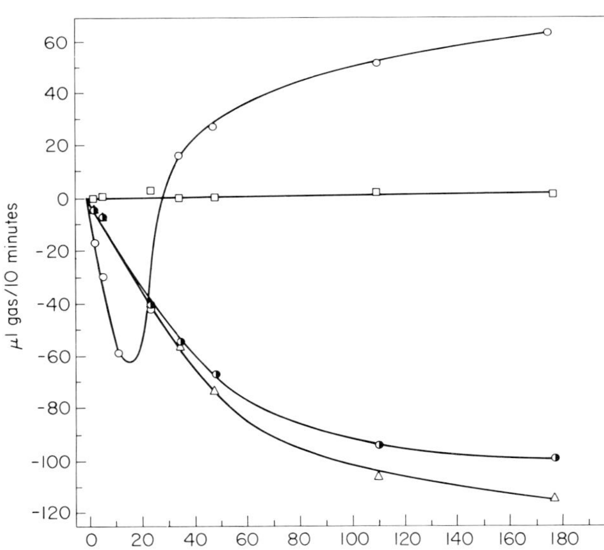

FIG. 3. Manometric determination of the photoreductive capacity of wild-type *Scenedesmus* and mutant strains Nos. 8, 11, and 50 at various white light intensities. Cells, 100 μl, were suspended in 3 ml of 50 m*M* KH_2–K_2HPO_4 buffer, pH 6.5, and incubated for 8 hours prior to illumination in an atmosphere of 96% H_2–4% CO_2; temperature, 25°. Wild type (○———○), mutants 8 and 50 (□———□), wild type + DCMU (2 μ*M*), (◑———◑), and mutant 11 with and without DCMU (△———△).

[9] N. I. Bishop, *Photochem. Photobiol.* **6**, 621 (1967).

Hill reaction activity, can be classified then as being deficient in either photosystem I activity or in some portion of the mechanism of carbon dioxide fixation. Mutants with normal photoreduction but inhibited Hill reaction and photosynthetic activities are clearly photosystem II deficient mutants. Mutants which are inhibited both in photoreduction and Hill reaction activity are most likely electron transport or photophorylation-type mutants. To determine the precise type of mutation in those strains which are not completely differentiated by the *in vivo* assays, requires the isolation of chloroplasts from individual mutants and subsequent assays of their capacity to utilize hydrogen donors, other than water, for the reduction of NADP and to perform photophosphorylation.

Isolation of Algal Chloroplasts

Previous attempts to isolate chloroplasts from *Scenedesmus* resulted in particles with low and irreproducible photochemical activity.[10] More thorough studies on the conditions necessary for heterotrophic growth and on the composition of chloroplast isolation medium have allowed for the isolation of algal chloroplasts fragments with activity comparable to that of isolated spinach chloroplasts.

Culture Conditions. When the medium described previously (NGY medium) is employed for heterotrophic growth of *Scenedesmus* (at 29°–30°) increase in cell number ceases after 4–5 days. At that time all cells have assumed the same stage of division, i.e., they are synchronized because of nutritional limitations. Consequently, cultures are utilized after 2–3 days of growth (i.e., during the logarithmic phase) after sufficient subculturing (usually 3 or 4) has provided a random culture of cells. For this purpose subculturing is carried out by inoculating 250 ml of the NGY medium with 50 μl of cells from a 3–4 day-old culture, allowing 60 hours of growth and then repeating the subculturing procedure several times. Chloroplasts isolated from such cultures show the highest photochemical activity.

Chloroplast Particle Preparation. Dark-grown cells are harvested in the centrifuge, washed once in a medium consisting of 20 m*M* tricine, pH 7.5, 10 m*M* KCl; 50 m*M* EDTA; and 0.5 m*M* DTT (dithiothreitol) and recentrifuged. This solution should be made fresh daily. The pellet is resuspended in approximately 5 ml of breaking solution, transferred to a 50-ml stainless-steel cup by the addition of 20 ml of breaking solution and the cup filled to within 0.5 cm of the top with 0.35-mm-diameter glass beads. The cells and glass beads are vibrated at full power on a Vibrogen-Zellmuhle for 5 minutes. The cup and contents are cooled prior

[10] L. H. Pratt and N. I. Bishop, *Biochim. Biophys. Acta* **153,** 664 (1968).

to and during the actual breaking process with ice water. The contents of the cup are then filtered through a coarse, fritted-glass filter into a suction flask to remove the glass beads which are subsequently washed three times with a solution of the following composition: 20 mM tricine, pH 7.5; 0.6 M sucrose; 10 mM KCl; and 25 μM DTT. This procedure removes most of the algal material from the glass beads. The filtrate is centrifuged at 500 g for 15 minutes, the supernatant is carefully poured into clean centrifuge tubes, and centrifuged for 15 minutes at 27,000 g. The supernatant is discarded and the top light-green bands of the pellet resuspended in a solution consisting of 0.4 M sucrose, 30 mM KCl, and bovine serum albumin (10 mg/ml). A glass stirring rod capped with cotton is used to remove only the top layer of the pellet. It is important throughout this procedure to maintain the samples cold.

The use of algal chloroplast fragments in subsequent tests follows the same procedure routinely employed in general Hill reaction analysis. For the determination of NADP-reducing capacity there is an absolute requirement for ferredoxin, ferredoxin-NADP reductase, and chloride ion. Particles having an activity of approximately 300 μmoles NADP reduced per hour per milligram of chlorophyll can be routinely obtained; these particles have shown no requirement for either plastocyanin or cytochrome *f*. Chloroplast isolated from the various mutant types can then be evaluated for their capacity to perform the numerous partial reactions known to be catalyzed by isolated spinach chloroplasts.

Analysis of Components of the Electron Transport System of *Scenedesmus*

Water-Soluble Cofactors. In mutants which are suspected of being electron transport type mutants several techniques are available for determining whether or not a given cofactor is present. These range from light-dark difference spectra analysis of whole cells to actual extraction and purification of individual components of the electron transport system. The latter technique has been employed in the author's laboratory in studies on *Scenedesmus, Chlorella,* and *Cyanidium*. The procedure requires an appreciable large amount of cell material (120 g fresh weight) which is thoroughly freed from any possible source of contamination. A thick suspension of cells in 0.2 M Tris·HCl, pH 7.5, to which has been added a small amount of deoxyribonuclease to degrade the DNA released upon cell rupture, is passed twice through a French press at 24,000 psi.[7] The broken cell suspension is made to 2% with respect to Triton X-100 and incubated overnight at 2°. The cell debris is removed by centrifugation (25,000 g for 15 minutes) and the dark-green supernatant brought to 35% saturation with solid $(NH_4)_2SO_4$. After standing for 10 minutes the suspension is centrifuged at 25,000 *g* for 30 minutes. The yellow-brown solution is then separated from a floating green lipoprotein matter by decantation and by filtration through glass wool. The supernatant is brought to 90% saturation with solid $(NH_4)_2SO_4$, allowed to stand for 1 hour at 2° and then centrifuged. The reddish brown pellet obtained is suspended in distilled water and dialyzed overnight to remove $(NH_4)_2SO_4$.

Further purification of the 90% ammonium sulfate pellet is accomplished by adsorption of the dialyzed extract onto DEAE-cellulose (2 × 15 cm, Whatman DE-32, microgranular) and the adsorbed protein is eluted with an increasing gradient of Tris·HCl (pH 7.5). For this purpose a mixing chamber containing 500 ml of distilled water is connected to a reservoir containing 1 liter of 1.0 *M* Tris·HCl (pH 7.5), and the outlet of the mixing chamber is connected to the column. Fractions of 5-ml volume are collected beginning with the application of the extract to the column. The following cytochromes are obtained by this procedure from wild-type strain of *Scenedesmus:* cytochrome *c* 549 (tubes 1–5), cytochrome *b* 562 (tubes 10–14), cytochrome *c* 551 (tubes 13–16), cytochrome *b* 558 (tubes 22–25), and cytochrome *c* 552 (*f*) (tubes 28–31). Ferredoxin-NADP reductase is collected in tube 13–17, plastocyanin in tubes 26–28, and ferredoxin in tubes 44–48. Many of these compounds are obviously not of chloroplast origin.

Electron transport components which remain bound to the insoluble lamellar protein fraction can be determined after exhaustive extraction of the green lamellar lipoprotein float (obtained in the 35% ammonium sulfate precipitation step) with acetone (—15°) until free of chlorophyll. The gray powder obtained is dissolved in 3% sodium dodecyl sulfate in 0.2 *M* Tris·HCl (pH 7.5) and the ferricyanide and dithionite difference spectra analyzed. This procedure reveals the presence of cytochrome *c* 552 and of cytochromes *b*. The *b*-type cytochromes are not too clearly distinguished, but the absorption spectra suggest the presence of cytochrome *b* 562 and cytochrome *b* 559.

A more thorough discussion of the purification of these various cofactors and some of their physical characteristics has appeared elsewhere.[8]

Lipid Soluble Cofactors. To determine the water-insoluble transport components of normal and mutant strains of *Scenedesmus* approximately 10–15 g wet weight of washed cells are extracted four times with 30-ml portions of hot methanol. The extracts are combined, evaporated to dryness, and washed onto a 2.5 × 15 cm column of 100-mesh silicic acid with two 50-ml washes of isooctane (2,2,5-trimethylpentane). Upon adsorption of the extract, the β-carotene is eluted from the column with 125 ml of 15% $CHCl_3$–85% isooctane. Subsequent elution with 150 ml of 75%

$CHCl_3$–25% isooctane removes most of the quinone-type molecules. This procedure is expedited by using a water-aspirator vacuum and collecting the fractions in 250-ml suction flasks. The quinone-containing fraction is evaporated to dryness for purification by TLC.

Separation and purification of the various chloroplastic quinones is achieved by a three-step thin-layer chromatographic procedure: (1) The extract is dissolved in chloroform, streaked onto a 20 × 20 cm silica gel plate and developed in a 60:40 benzene–heptane solvent. Vitamin E, vitamin K, and plastoquinone A are located on the plate as purple bands when the plate is sprayed with a 0.1% rhodamine solution (in absolute ethanol). (2) Further purification of plastoquinone and vitamin K is obtained by scraping off the respective bands, eluting the compound, streaking a paraffin-treated silica gel plate [immersed previously in a 5% solution of liquid paraffin in ligroine (b.r. 66°–75°)], and developing in 3% H_2O–acetone solvent. The various quinone bands are localized as indicated above and scraped from the TLC plate. (3) The quinones are eluted free of the dye from the silica gel–paraffin plate with redistilled $CHCl_3$, restreaked on 4 × 20 cm silica gel plates and developed in petroleum ether (b.r. 26°–43°) until all the paraffin is removed from the origin. The quinones, which remain at or near the origin, are removed from the silica gel with chloroform and then evaporated to dryness. The purified sample is taken up in buffered ethanol (10 m*M* ammonium acetate, pH 5.0, in absolute ethanol) for spectrophotometric analysis. This technique affords approximately 75% recovery of the original amount of plastoquinone A.

Additional Techniques for Determination of Properties of Mutants

Through the application of various photosynthetic mutants of *Scenedesmus* and *Chlamydomonas* to studies on the mechanism of photosynthesis, it has become apparent that a number of more sophisticated techniques can be applied to actual detailing of the nature of the mutant under study. Space does not permit extensive elaboration on the various techniques that are available; they will be mentioned only briefly.

Electron Paramagnetic Resonance (*EPR*) *Analysis.* Early attempts to detect unpaired electrons in photosynthetic tissue demonstrated two distinct light-dependent signals which have been characterized by differences in their kinetics of appearance and decay, hyperfine structure, and "*g* values." A number of studies have shown that the EPR signal with fast rise and decay kinetics (signal I or R signal) is associated with fast photochemical events while the one with slow rise and decay kinetics (signal II or S signal) is associated with reactions of photosystem II. Mutant studies have shown that the S signal is absent in mutants lacking photosystem II activity and the R signal is absent in a mutant strain deficient in P-700.[11] Although not commonly employed as a diagnostic tool, access to an EPR spectrometer would facilitate identification of mutant types.

Delayed Light Emission Studies. It is well known that in the process of energy conversion and storage by the photosynthetic apparatus that a small amount of the stored energy can be reemitted as a delayed fluorescence (delayed light emission) which is due to an inefficient reexcitation of chlorophyll by the stored energy. Bertsch and his colleagues have determined through mutant studies that the principle source of this energy originates from photosystem II. Electron transport mutants or photosystem I mutants, on the other hand, behave either normally or produce more delayed light emission than the wild-type strain of *Scenedesmus*.[12] Facilities to measure this phenomenon are not generally available, but, nevertheless, it is representative of a sophisticated technique for possible mutant study.

Fluorescence Yield Analysis. The fluorescence of green plants is believed to consist of two components; the first is a fluorescence of constant quantum yield, probably emanating from the bulk absorbing chlorophyll whose yield is not affected by changes in the efficiency of the photochemical reactions and the second is a fluorescence of variable quantum yield which is directly related to photosynthetic capacity. The latter form is presumed to emanate from the trapping centers of photosystem II. A mechanism has been proposed for the variable-yield fluorescence based upon the oxidation and reduction of an electron transport system between the two photosystems wherein the fluorescence of chlorophyll of the trapping center is quenched by the oxidized state of the hypothetical substance, compound Q, but not by its reduced form.[13] A number of other hypothetical intermediates have also been proposed as being involved in this reversible fluorescence-quenching phenomenon. Application of this technique to studies on mutant characteristics have demonstrated that mutants deficient in photosystem II activity possess an extremely high fluorescence without any variable-yield component; also the onset kinetics of fluorescence is altered. The fluorescence of photosystem I or electron transport mutants is typified by having an initial fluorescence yield greater than that of wild-type cells and which is strongly augmented by the addition of photosystem II light. This latter observation can be utilized as a good test for the presence of photosystem II

[11] E. C. Weaver and N. I. Bishop, *Science* **140,** 1095 (1963).
[12] W. Bertsch, J. R. Azzi, and J. B. Davidson, *Biochim. Biophys. Acta* **143,** 129 (1967).
[13] W. L. Butler, *Curr. Top. Bioenerg.* **1,** 49 (1966).

activity in a mutant strain.[7] Photosynthetic mutants which are blocked in later steps of photosynthesis, such as in the reductive pentose cycle, show fluorescence patterns basically indistinguishable from wild-type *Scenedesmus*.

A combination of these specialized techniques and the general biochemical ones mentioned earlier will allow for a rather specific delineation of the nature of the mutation. Additional techniques are being sought for the identification of more select genetic blocks within photosystem II.

[12] Isolation of Mutants from *Euglena gracilis*

By JEROME A. SCHIFF, HARVARD LYMAN, and GEORGE K. RUSSELL

Euglena gracilis has proven to be a very useful organism for studies of photosynthesis,[1-5] chloroplast development,[6,7] and other metabolic processes[8] because chloroplast development and replication are subject to environmental control by the investigator. Dark grown cells of *Euglena* contain proplastids which can be induced with light to form chloroplasts, and because of this, *Euglena* has been used extensively for studies on the regulatory mechanisms controlling chloroplast differentiation.[6] Proplastids and chloroplasts (and the DNA which they contain) can be selectively eliminated from the cells, yielding strains which provide useful comparisons for chloroplast-localized activities.[6,7,9] On suitable growth media the chloroplast is a gratuitous organelle since the organism grows equally well as an organotroph. The ease with which chloroplast-forming ability is lost from the cell is doubtlessly connected with this fact.

[1] J. M. Olson and R. M. Smillie, *in* "Photosynthetic Mechanisms of Green Plants" (B. Kok and A. Jagendorf, eds.), pp. 56–65. National Academy of Sciences, Washington, D.C., 1963.
[2] G. K. Russell and H. Lyman, *Plant Physiol.* **43,** 1284 (1968).
[3] G. K. Russell, H. Lyman, and R. Heath, *Plant Physiol.* **44,** 929 (1969).
[4] S. Katoh and A. San Pietro, *Arch. Biochem. Biophys.* **118,** 488 (1967).
[5] W. R. Evans, *in* "The Biology of Euglena" (D. Buetow, ed.), Vol. II, p. 73. Academic Press, New York, 1968.
[6] J. A. Schiff and H. T. Epstein, *in* "The Biology of Euglena" (D. Buetow, ed.), Vol. II, p. 285. Academic Press, New York, 1968.
[7] J. A. Schiff and M. Zeldin, *J. Cell Physiol.* **72,** Suppl. 1, 103 (1968).
[8] "The Biology of Euglena" (D. Buetow, ed.), Vols. I and II. Academic Press, New York, 1968.
[9] M. Edelman, J. A. Schiff, and H. T. Epstein, *J. Mol. Biol.* **2,** 769 (1965).

Genetic recombination[10] has not been observed in this organism, although various investigators have searched for meiosis,[11] sexual recombination,[12] transformation, or infective agents which might be used for genetic transduction. Part of the difficulty may lie in the apparent polyploidy of the two strains of *Euglena* commonly studied; target analysis of X-ray and uv killing curves of *Euglena gracilis* var. *bacillaris* suggest the presence of eight targets,[13] a result consistent with octaploidy. This also renders phenotypic expression of recessive chromosomal mutations unlikely since up to eight identical loci on separate chromosomes would have to be affected simultaneously in the same cell (unless some chromosomal mutations had already accumulated in those loci). In the absence of meiosis, mitosis ensures that the daughter cells each have a genetic complement identical to that of the parent, rendering it unlikely that chromosomal mutants will segregate during cell division. If some means were found to render these cells haploid or to interfere with the regularity of mitosis, recovery of chromosomal mutations might be possible. A recent report on the isolation of auxotrophic mutants in *Euglena*[14] suggests that it may be possible to isolate nuclear mutants, but further work will be needed to validate these findings.

In the absence of further information, it is reasonable to assume that most mutations observed in *Euglena* are nonchromosomal, leading to a consideration of the mitochondrial[15] and plastid[9,16–20] genomes. *Euglena* appears to be an obligate aerobe and apparently cannot grow by fermentative processes[21–23]; therefore, all mitochondrial mutations which affect function should be lethal, at least in the dark. The possibility that phosphorylation and photoreduction occurring in the chloroplasts might

[10] J. R. Preer, *in* "Research in Protozoology," Vol. 3, p. 133. Pergamon, New York, 1968.

[11] G. F. Leedale, *Arch. Mikrobiol.* **42,** 237 (1962).

[12] B. Biechler, *Co. R. Soc. Biol.* **124,** 1264 (1937).

[13] H. Z. Hill, J. A. Schiff, and H. T. Epstein, *Biophys. J.* **6,** 125 (1966).

[14] T. V. Come and D. M. Travis, *J. Heredity* **60**(1), 39 (1969).

[15] M. Edelman, H. T. Epstein, and J. A. Schiff, *J. Mol. Biol.* **17,** 463 (1966).

[16] J. Leff, M. Mandel, H. T. Epstein, and J. A. Schiff, *Biochem. Biophys. Res. Commun.* **13,** 126 (1963).

[17] M. Edelman, C. A. Cowan, H. T. Epstein, and J. A. Schiff, *Proc. Natl. Acad. Sci.* **52,** 1214 (1964).

[18] D. S. Ray and P. C. Hannawalt, *J. Mol. Biol.* **9,** 812 (1964).

[19] D. S. Ray and P. C. Hannawalt, *J. Mol. Biol.* **11,** 760 (1965).

[20] G. Brawerman and J. M. Eisenstadt, *Biochim. Biophys. Acta* **91,** 477 (1964).

[21] J. A. Gross and T. L. Jahn, *J. Protozool.* **5,** 126 (1958).

[22] J. R. Cook, *in* "The Biology of Euglena" (D. Buetow, ed.), Vol. I, p. 44. Academic Press, New York, 1968.

[23] G. Alexander, *Biol. Bull.* **61,** 165 (1931).

support the growth of a mutant strain possessing defective mitochondria cannot be excluded, but such mutants have not yet been reported. In addition, there are more than 500 mitochondria per cell, and the probability of segregation of a mutation present in any one mitochondrion is very small. Although mutation of the mitochondrial DNA undoubtedly occurs (aberrant mitochondria are occasionally seen in electron micrographs of mutagenized cells), it is unlikely that any of these could be recovered as mutant cell lines. These arguments also apply to any mutations in the chromosomal or mitochondrial genomes which produce gene products affecting the chloroplast.

Experimental evidence suggests that mutations of the chloroplast genome can be detected. Ultraviolet target analyses[13,24] and microbeam studies[25] have indicated that there are approximately 30 cytoplasmic sites affecting chloroplast replication. It is generally assumed that these are identical with the plastid DNA. Target analysis assumes that any one of the 30 entities is sufficient to render the cell competent to yield progeny containing chloroplasts. The very high sensitivity of chloroplast replication to various physical and chemical agents suggests the following hypothesis to account for the production of plastid mutants. Mutant strains might represent clones derived from cells in which all but one of the plastid genomes were prevented from replicating by the mutagenic treatment, and a mutation was induced in the one surviving genome. This mutated genome could then have replicated several times to produce a cell in which all of the chloroplast genomes carried the mutation.

From the foregoing discussion, it seems likely that the mutants expressed as plastid phenotypes in *Euglena* are mutations or deletions of the plastid DNA. This is consistent with the observation that the plastid of *Euglena* is more sensitive than the viability of the cell to alteration by most environmental agents (including many of the mutagens to be discussed below).

Types of Mutants

Most workers have restricted the naming of mutants to easily recognized phenotypic characteristics, regardless of the mechanisms which may later be found to determine them. This is a useful procedure, and we recommend its retention in naming *Euglena* mutants. The types of mutants which have been recovered, or which are potentially obtainable, include the following:

1. *Mutations Affecting Colony Color.* Any colony exhibiting a color

[24] E. Lyman, H. T. Epstein, and J. A. Schiff, *Biochim. Biophys. Acta* **50**, 301 (1961).

[25] A. Gibor and S. Granick, *J. Cell Biol.* **15**, 590 (1962).

other than the normal green of wild type is included here. These may be found to involve mutations affecting plastid pigment formation, plastid structure, plastid inheritance or deletion, plastid function, or other processes when characterized in more detail.

The following types all yield normal green colonies (like wild type):

2. *Mutations Affecting Photosynthetic Function.* These include all mutants having impaired photosynthetic capacity not already included in 1 above. It may be inferred that these mutants are probably blocked elsewhere than in pigment formation and are not photosynthetically deficient because of gross pigment limitations.[2,26]

3. *Drug-Resistant Mutants.* These mutants fail to exhibit sensitivity to drugs which affect the wild-type cell.

4. *Motility Mutants.* These are unable to swim and may be selected and tested by their loss of phototaxis.[27]

5. *Heat-Resistant Mutants.* These retain plastids when grown at temperatures above 32° under conditions where the wild-type cell rapidly loses plastids in an irreversible manner.[28]

6. *Auxotrophic Mutants.* These require one or more exogenous factors for growth in the usual media.[14]

7. *Mutants of Cell Morphology.* Mutants of this class show easily recognized changes in gross cell morphology at the light microscope level, e.g., loss of eyespot.

8. *Mutants in Colony Morphology.* Colonies appear to be different than those from wild type. (One must take into account the great variability of wild-type colony morphology, especially under different plating conditions.)

Definition of the Term "Mutant": Any strain possessing a heritable alteration of phenotype is defined as a mutant. In this context, a mutagen is any agent causing a heritable alteration in phenotype. In naming the mutant strains no assertion is made concerning the underlying genetic change, i.e., point mutation, deletion, etc.

Naming of Mutants

A method already employed in the published literature provides a convenient means of naming mutants and inferring their derivation at sight. The method consists of a four letter code, the letters of which describe, in sequence, the phenotype as isolated, the parent species ("B" for *bacillaris* and "Z" for the Z strain), the mutagen used, and whether

[26] R. P. Levine, *Nature* **188,** 339 (1960). Also this volume, article (10).
[27] R. A. Lewin, *Can. J. Microbiol.* **6,** 21 (1960).
[28] A. Uzzo and H. Lyman, *Biochim. Biophys. Acta* **180,** 573 (1969).

the organism treated with mutagen was light grown ("L") or dark grown ("D").

Phenotype as Isolated: (First symbol of code)

1. *Color:* Y = yellow; W = white; P = pale green; O = olive-green; H = halo or sectored-colony former; G = golden; O_r = orange.
2. *Photosynthetic Competence:*
   Ph^- = Will not grow in light on media lacking carbon compounds other than carbon dioxide, thiamine, and B_{12}.[28a] (Other phenotypes may be isolated, e.g., inability to fix CO_2; enhanced fluorescence.)[26]
3. *Drug-Resistant Mutations:*
   Dap^r = diaminopurine resistant
   Am^r = actinomycin resistant
   Sm^r = streptomycin resistant, i.e., does not lose plastids on Sm
   Nal^r = nalidixic acid resistant, i.e., does not lose plastids on Nal
4. *Motility Mutants:*
   M^- = lack of motility
   and further symbols to be invented as mutants become available.

All mutants are numbered within the type in order of isolation (or, in the future, in order of appearance in the published literature) by a numerical subscript on the first symbol (e.g., P_1, W_8, etc.); each new mutant receives a new number, so that no mutants are confused with those which might have been lost in the past. Differences between laboratories might be resolved by priority of publication. See Table I for a list of mutants in recent publications.

Strain (second symbol of code):

B = *Euglena gracilis* Klebs var. *bacillaris*, Pringsheim
Z = *Euglena gracilis* Klebs, Z strain, Pringsheim

Mutagen (third symbol of code):

S = Spontaneous; X = X ray; U = ultraviolet; H = heat; Sm = streptomycin; Ng = nitrosoguanidine; Nal = nalidixic acid.

Treated Cells (fourth symbol of code):

L = light-grown cells; D = dark-grown cells.

Proper designation would be as follows: Y_1BXL: yellow mutant number one isolated from *bacillaris* after X-ray treatment of light-grown cells. While this is the formal designation, the mutants become colloquially

[28a] H. Lyman and H. W. Siegelman, *J. Protozool.* **14,** 297 (1967).

TABLE I

MUTANTS OF *Euglena gracilis* DESCRIBED IN SOME RECENT PUBLICATIONS[a]

Mutant	Footnote	Reference
O_1BS	*b*	6, 9
O_2BX	*b*	6, 9
P_1BXL		6, 9, 13, *g–i*
P_4ZUL	*c*	2, 3
P_7ZNgL	*c*	2
P_8ZNgl	*c*	2
P_9ZNgL		*j*
$P_{10}ZNalL$	*c*	2
W_1BVL	*d*	50
W_1ZXL		*i*
W_2ZUL		*i*
W_3BUL		6, 7, 9, 16, 17, *g*, *h*, *k*
"W_3ZSL"	*e*	*i*
W_4BUL		*i*
W_8BHL		6, 9, *g*, *k*
$W_{10}BSmL$		6, 9, 15, *k*
$W_{30}BS$	*b*	9
$W_{33}ZUL$	*f*	*l*
Y_1BXD		6, 9, *h*, *k*, *m*
Y_2BUL		9, *h*, *i*
Y_3BUD		6, *g*, *h*, *k*
Y_6ZNalL		*n*

[a] A series of mutants was described by Gross and Jahn[18] before the present terminology was suggested. These mutants are listed in the Indiana Algae Culture Collection [R. C. Starr, *Amer. J. Bot.* **51,** 1013 (1964)]. Other mutant strains have been described by L. G. Moriber, B. Hershenov, S. Aaronson, and B. Bensky [*J. Protozool.* **10,** 80 (1963)], G. Brawerman, C. A. Rebman, and E. Chargaff [*Nature* **187,** 1037 (1960)], R. H. Neff [*J. Protozool.* **7,** 69 (1960)], A. Gibor and S. Granick [*J. Protozool.* **9,** 327 (1962)], and A. Gibor and Helen Herron *in* "the Biology of Euglena" [(D. Buetow, ed.), Vol. II, p. 335. Academic Press, New York, 1968].

[b] It is not known whether the source of these was light—or dark—grown cells.

[c] These four strains were originally called P_4, P_7, P_8, and P_9, respectively, by Russell and Lyman,[n] but the present designation is in agreement with the proposed nomenclature.

[d] This strain should probably be properly named W_1BVL, L indicating that it was isolated in the light. This conflicts with a strain called W_1BZX at an earlier date by Stern.[i]

[e] Since this conflicts with W_3BUL and since W_3ZSL has been lost, we recommend dropping W_3ZSL entirely.

[f] This strain was originally called ZUV-1 by Scott and Smillie.[m]

[g] S. C. Lewis, J. A. Schiff, and H. T. Epstein, *J. Protozool.* **12,** 281 (1965).

[h] A. I. Stern, J. A. Schiff, and H. T. Epstein, *Plant Physiol.* **39,** 220 (1964).

[i] A. I. Stern, J. A. Schiff, and H. P. Klein, *J. Protozool.* **7,** 52 (1960).

[j] H. Lyman and A. S. Jupp, *J. Cell Biol.* **43,** 83a (1969).

[k] M. Zeldin and J. A. Schiff, *Planta* (Berlin) **81,** 1 (1968).

[l] N. S. Scott and R. M. Smillie, *Biochem. Biophys. Res. Commun.* **28,** 598 (1967).

[m] Inadvertently cited as "Y_1BXL" in this reference.

[n] G. K. Russell and H. Lyman, *J. Protozool.* **14** (Suppl.), 13 (1967).

known as Y_1, W_8, etc. since the combination of first letter and number are unique.

Secondary Derivation of Mutants

If a second mutant is derived from the first either spontaneously or by treatment, the designation is as follows:

Reversion to wild-type phenotype: "R" with numerical subscript is added to the mutant designation, e.g., Y_2BULR_1, first wild-type revertant isolated from the yellow mutant Y_2BUL. All other secondary derivations receive a designation appropriate to the new mutant followed by the old mutant name in parentheses. In this context the parentheses mean "derived from," e.g., $W_{16}BUD(P_1BXL)$ is a white mutant derived from P_1BXL by treatment of dark-grown cells with uv. This ensures that the mutant lineages are not lost and that the derivation is shown for all strains.

Methodology

In the following discussion distilled water and reagent grade chemicals are used throughout (except as indicated). Sterilization by autoclaving is performed for 15 minutes at 15 lbs/inch.[2] For autoclaving very large volumes the time should be increased accordingly.

Growth Media: Euglena grows optimally from pH 3 to pH 8 on a variety of substrates.[29-31] Low pH media are popular since bacterial growth is unlikely, and the only contaminants are yeasts and other fungi. However, several inhibitors (e.g., streptomycin) do not penetrate the cell at low pH, and it is frequently useful to have media at neutral or slightly basic pH. Phototropic media (media from which utilizable carbon sources other than carbon dioxide are excluded) have been devised which yield good growth.[31] High pH media favor carbon dioxide solubility through the formation of HCO_3^- and CO_3^{2-}, but the low pH phototrophic medium is convenient as a counterpart to the low pH organotrophic medium. Growth in the low pH phototrophic medium can be enhanced by incubating the plates in 5% carbon dioxide in air or by bubbling a 5% CO_2–air mixture through liquid medium during growth. The composition of five media which have proven very useful in our hands are shown in Table II.

All liquid media may be converted to solid by the addition of agar.

[29] S. H. Hutner, A. C. Zahalsky, S. Aaronson, H. Baker, and O. Frank, *in* "Methods in Cell Physiology" (D. M. Prescott, ed.), Vol. II, p. 217. Academic Press, New York, 1966.

[30] J. J. Wolken, "*Euglena,* an experimental organism for biochemical and biophysical studies." Rutgers University Press, New Brunswick, New Jersey, 1961.

[31] M. Cramer and J. Myers, *Arch. Mikrobiol.* **17,** 384 (1952).

TABLE II
CULTURE MEDIA FOR *Euglena gracilis*[a]

Compound	(g/liter)
Hutner's acidic organotrophic[b] (pH ≈ 3.5[c,d])	
KH_2PO_4	0.4
$MgSO_4 \cdot 7H_2O$	0.5
$CaCO_3$	0.2
L-Glutamic acid	5.0
d-Malic acid	2.0
Trace metals mix	0.05
$FeCl_3$	0.005
Thiamine · HCl (vitamin B_1)	0.001
Cyanocobalamin (vitamin B_{12})	2×10^{-7}
$(NH_4)_2HPO_4$	0.2
Hutner's neutral organotrophic[29] (pH = 6.8)	
Potassium citrate · H_2O	1.0
Sodium acetate · $3H_2O$	1.0
Glycine ethyl ester · HCl	1.0
L-Glutamic acid (gamma ethyl ester · HCl)	1.0
L-Asparagine · H_2O	1.5
α-Glycerophosphoric acid (disodium salt)	0.5
$MgSO_4 \cdot 7H_2O$	0.5
$(NH_4)_2SO_4$	0.02
Thiamine · HCl (vitamine B_1)	0.01
Cyanocobalamin (vitamin B_{12})	4×10^{-6}
Trace metals mix	0.04
Adjust to pH 6.8 with Tris	
Hutner's alkaline organotrophic[b] (pH = 8.0)[d]	
K_2HPO_4	0.10
$CaCl_2 \cdot 2H_2O$	0.05
NH_4Cl	0.40
Sodium acetate · $3H_2O$	1.00
n-Butanol	3.00
EDTA (disodium salt)	0.50
$MgSO_4 \cdot 7H_2O$	0.50
$FeCl_3 \cdot 6H_2O$	0.002
Thiamine · HCl (vitamin B_1)	0.001
Cyanocobalamin (vitamin B_{12})	4×10^{-7}
Trace metals mix	0.05
Adjust to pH 8.0 with KOH	
Acidic phototrophic[d,28a] (pH 3.5)	
EDTA (disodium salt)	0.5
KH_2PO_4	0.3

Table II (*Continued*)

Compound	(g/liter)
$MgSO_4 \cdot 7H_2O$	0.5
$CaCO_3$	0.06
$(NH_4)_2SO_4$	1.0
Trace metals mix	0.065
Thiamine · HCl (vitamin B_1)	0.001
Cyanocobalamin (vitamin B_{12})	2×10^{-8}
Adjust to pH 3.5 with H_3PO_4 or H_2SO_4	0.02
Neutral phototrophic (pH = 6.8) Cramer and Myers[31] modified by Edmunds[e]	
$(NH_4)_2HPO_4$	1.0
KH_2PO_4	1.0
$MgSO_4 \cdot 7H_2O$	0.409
$CaCl_2 \cdot 2H_2O$	0.026
Sodium citrate · $2H_2O$	0.516
$FeSO_4 \cdot 7H_2O$[f]	3.283×10^{-3}
$MnCl_2 \cdot 4H_2O$[f]	1.800×10^{-3}
$CoCl_2 \cdot 6H_2O$[f]	1.063×10^{-3}
$ZnSO_4 \cdot 7H_2O$[f]	4.0×10^{-4}
$Na_2MoO_4 \cdot 2H_2O$[f]	2.99×10^{-4}
$CuSO_4 \cdot 5H_2O$[f]	2.0×10^{-5}
Thiamine · HCl (vitamin B_1)	1.0×10^{-4}
Cyanocobalamin (vitamin B_{12})	0.5×10^{-7}
Trace metals mix[g]	
$Fe(NH_4)_2(SO_4)_2 \cdot 6H_2O$	28.0
$MnSO_4 \cdot H_2O$	24.8
$ZnSO_4 \cdot 7H_2O$	52.8
$CuSO_4 \cdot 5H_2O$	0.8
$(NH_4)_6Mo_7O_{24} \cdot 4H_2O$	0.36
$CoSO_4 \cdot 7H_2O$	4.8
$Na_3VO_4 \cdot 16H_2O$	0.37
H_3BO_3	1.14

[a] Descriptions of other media and a discussion of their merits may be found in Hutner *et al.*,[29] Wolken,[30] and Cook.[22]

[b] C. L. Greenblatt and J. A. Schiff, *J. Protozool.* **6,** 23 (1959).

[c] *Euglena* media similar to the acidic organotrophic medium above (±agar) is available from the Difco Co.

[d] For *double-strength* medium (i.e., for mixing with agar) multiply these amounts by two. (Only strictly necessary for acidic media.)

[e] L. N. Edmunds, Jr., *J. Cellular Comparative Physiol.* **66,** 147 (1965).

[f] A proportional amount of the trace metal mix may be substituted for these salts.

[g] It is convenient to make up a highly concentrated stock solution of trace metals. If difficulty is encountered in dissolving high concentrations of these salts, addition of HCl will usually facilitate solution. Alternatively, the salts can be ground together and weighed as "trace metals."

For agar slants, 2–4% agar is convenient; for agar plates 2% is best. With low pH media, however, autoclaving hydrolyzes the agar. It is therefore necessary to prepare double-strength medium and a double-strength agar solution, autoclave them separately, and mix them after cooling somewhat but before the agar solidifies. For example, an 8-liter batch of medium should cool 30–60 minutes before mixing with a simultaneously autoclaved 8-liter batch of double-strength agar. This admonition applies to any agar media prepared from low pH media.

Materials for Plating: Euglena grows well on agar plates with plating efficiencies approaching 100% when the overlay technique is employed (See below).

Plates: Petri dishes (100 × 25 mm) filled to a depth of about 8 mm are employed. The deep dishes are necessary since the incubation times are a week or more and thin layers of media tend to desiccate. Either plastic or glass dishes may be employed. The plates are filled with 2% agar medium. When using low pH media they are prepared by sterilizing equal amounts of double-strength medium and 4% agar separately and mixing the two solutions under sterile conditions after cooling. One liter of 2% agar medium fills about 30 plates. After solidification the plates are inverted to avoid the dripping of condensate onto the agar surface and are stored in a dust-free environment either at room temperature (lab drawers are adequate) or at 4°.

Overlay Agar: Method I. To make low pH 1.2% agar medium for overlay, autoclave separately equal amounts of 2.4% agar and double strength medium. Mix under sterile conditions and pipette 3.5–4 ml into capped Wasserman tubes (12 × 100 mm) (loose-fitting plastic or aluminum caps are best). One tube is required for each plate previously sterilized by autoclaving. The tubes are stored at 4° until used.

Method II. An alternate method requires a somewhat different preparation of overlay. Equal volumes (80 ml) of double-strength low pH medium and of 2.4% agar are prepared in separate Erlenmeyer flasks and are autoclaved. These media may then be stored at 4°. Prior to use, the flask of agar is heated in a boiling water bath to melt it completely, and the double strength medium, previously warmed to 45°, is added. The mixture is placed immediately into a 45° water bath. This provides enough overlay for 30 plates. This agar is dispensed using wide-bore 10-ml pipettes.

Dilution Tubes: Loosely capped Wasserman tubes (12 × 100 mm) are sterilized by autoclaving. They are then filled with a convenient volume of sterile single strength liquid medium to serve as a diluent. Volumes which have been found convenient for serial dilution are: 0.9 ml (0.1 ml of cell suspension is added to make 1.0 ml, a 10-fold dilution); 4.0 ml and other sizes as needed.

Discard Bins: It is convenient to have an empty agar can on hand to receive the tops of Wasserman tubes after use, as well as a plastic basin or wire basket to receive the discarded tubes themselves.

Pipettes: Cotton-plugged pipettes are used except for 0.1– and 0.2-ml sizes where this is unnecessary. For overlay method II above, the wide-bore 10-ml pipettes used may be prepared by cutting off the ends of conventional 10-ml pipettes. Wide bore pipettes are also available commercially. Pipettes are either sterilized in conventional pipette canisters or are packed in two's in glassine paper envelopes. The pipette canisters are autoclaved and dried in an oven somewhat above 100°. It should be noted that since many microorganisms grow much faster than *Euglena,* conventional bacteriological techniques used with *Escherichia coli,* for example, are often inadequate. If several small experiments are to be done, it is safer to use glassine envelopes with a few pipettes in each rather than to continually open and close a canister of pipettes.

Growth of Cells in Liquid Culture: For culturing cells in liquid, 500-ml portions of culture medium are dispensed into 1-liter Erlenmeyer flasks or 1000 ml into 2-liter flasks, and the flasks are plugged with gauze-covered cotton, foam plastic, or inverted plastic beakers held in place with masking tape applied after autoclaving. For smaller cultures the volumes used are 50–75 ml of medium in a 125-ml Erlenmeyer flask, or other convenient volumes. The volume should be small compared to the flask volume to avoid wetting the plugs when the flasks are shaken. After autoclaving and cooling, the flasks are innoculated using conventional aseptic techniques. They are then incubated on rotary shakers under 150–200 foot-candles of light provided by cool white and red fluorescent tubes, when light-grown cells are desired, or in darkness when dark-grown cells are used. For dark growth either a dark room is employed or flasks are taped with several layers of black plastic electrician's tape (Scotch brand) and are carefully checked for light leaks. The plugs are covered with aluminum foil, and all manipulations are performed under a dim green safelight.[32,33] Growth takes place at 26°, both in liquid and on solid media.

Mutagenesis

The mutagens commonly employed and the techniques for their use are as follows:

[32] R. B. Withrow and L. Price, *Plant Physiol.* **32**(3), 244 (1957).

[33] See J. A. Schiff, this series, Volume XXIV, for a cheap and convenient safelight (1969).

1. X Rays

Polyethylene caps or glass dishes make convenient irradiation dishes when covered with aluminum foil. These dishes can be sterilized by exposing the open dish and aluminum foil to prolonged uv, by prolonged X-ray treatment, or by autoclaving. Cell suspensions ($1–2 \times 10^6$ cells/ml) in growth medium are placed in the dishes and are exposed to 30–50,000 R which provides about 99% killing.[34] The killing curve should be determined for each experimental situation. A Phillips X-ray machine at 100 kV and 6.7 mA filtered by aluminum, delivering 6000 R/min, has been used. After treatment, the cells are diluted appropriately and plated.[34]

2. Ultraviolet Irradiation

Ultraviolet light eliminates the ability of cells to transmit plastids to their progeny and produces strains in which chloroplast DNA is undetectable.[9,24] These white colonies can be isolated as uv bleached mutants. A dose of 100 erg/mm^2 from a low-pressure mercury arc (15-W GE germicidal lamp, major output at 2537 Å) is generally sufficient to yield greater than 99% loss of green-colony forming ability without impairing cell viability. At this dose, mutants can also be found among the few remaining pigmented colonies, if sufficient numbers of cells are plated. Care must be taken to avoid reversal of the uv effect by longer wavelength light[35-37] (PR = photoreactivation). This is easily accomplished by performing all experiments under a dim green safelight[32,33] or under red or yellow incandescent safelights, as the peak for PR effectiveness extends from the near uv through the blue region of the spectrum.[36] The same precaution must be taken in growing colonies on plates. Cultures may be incubated in the dark for five to seven days and then placed in the light (PR decays with cell division), or the plates may be incubated in light from red fluorescent tubes (with ends taped) which allows chlorophyll and chloroplast formation but does not bring about photoreactivation.[38]

For uv irradiation, a 6.0-ml aliquot of cell suspension (2×10^6 cells/ml) is placed in a sterile 100×25 mm petri dish, the cover is removed, and the suspension is agitated during exposure to uv. The cells are then suitably diluted and plated under nonphotoreactivating conditions.

[34] J. Miller, J. A. Schiff, and H. T. Epstein, unpublished work, 1962.

[35] P. C. Hannawalt, *in* "Photophysiology" (A. C. Giese, ed.), Vol. IV, p. 203. Academic Press, New York, 1968.

[36] J. A. Schiff, H. Lyman, and H. T. Epstein, *Biochim. Biophys. Acta* **50,** 310 (1961).

[37] A. Kelner, *J. Gen. Physiol.* **34,** 835 (1951).

[38] J. A. Schiff, H. Lyman, and H. T. Epstein, *Biochim. Biophys. Acta* **51,** 340 (1961).

3. Nalidixic Acid (Nal)

A stock solution is prepared by dissolving 5 mg of nalidixic acid in 1.0 ml of 1.0 N NaOH and bringing the volume to 25 ml with distilled water. The solution is sterilized by millipore (0.45 μ, pore size) filtration. The solution must be stored at $-15°$.

Nal is used at a concentration of 50 μg/ml in low pH organotrophic medium. This medium is inoculated with log phase cells to a final concentration of 1–2 $\times$ 10^4 cells/ml and is then incubated in the light for 24 hours. Visible light is absolutely necessary for the bleaching action of Nal.[39] Upon plating, these cells give rise to greater than 99% white colonies with no loss of cell viability. Mutants can be found at reasonably high frequencies among the 1% nonwhite colonies. It is not certain whether these mutants are induced by Nal or are simply resistant to the bleaching action of Nal. Several recently isolated photosynthetically deficient strains of *Euglena* have been shown[40] to be Nal^r; treatment with Nal may therefore represent a means of isolating photosynthetic mutants of *Euglena*.

4. Nitrosoguanidine (Ng)[41,42]

N-methyl-N'-nitro-N-nitrosoguanidine is a powerful carcinogen and should be handled with caution. Solutions of Ng are not very stable, particularly at neutrality, and should be freshly prepared. The powder should be kept desiccated in the dark at low temperature. Three different procedures have been employed for Ng mutagenesis in *Euglena*. In all cases, Ng solutions are sterilized shortly before use by millipore filtration.

Method I. Washed *Euglena* cells are suspended in 50 mM citrate buffer (pH 5.0) containing Ng (50 μg/ml) and a final cell concentration of 10^5/ml. After 20 minutes the cells are diluted 1000-fold and plated by the overlay technique. This procedure results in very little loss of cell viability, but about 60% of the resulting colonies are permanently bleached. The remaining 40% nonwhite colonies contain many mutant strains. This technique has been used for the isolation of several photosynthetically deficient mutants of *Euglena*.[2]

Method II. According to the method of McCalla,[43] Ng is dissolved in sterile low pH organotrophic medium at concentrations of 2–8 μg/ml. Washed cells of *Euglena* are inoculated into the medium to a final con-

[39] H. Lyman, *J. Cell Biol.* **35**, 726 (1967).
[40] H. Lyman, *J. Cell Biol.* **39**, 83 (1968).
[41] E. A. Adelberg, M. Mandel, and C. C. C. Chen, *Biochem. Biophys. Res. Commun.* **18**, 788 (1965).
[42] N. W. Gillham, *Genetics* **52**, 529 (1965).
[43] D. R. McCalla, *J. Protozool.* **13**(3), 472 (1966).

centration of 10^5 cells/ml and incubated for periods of 8–24 hours. At the end of this time the cells are washed, diluted, and plated. This procedure has been used to study the bleaching effects of Ng.[43]

Method III.[44] To an aliquot of cells in resting medium (mannitol, 1%, K_2HPO_4, 0.01 *M*; KH_2PO_4, 0.01 *M*; $MgCl_2$, 0.01 *M*, added after separate autoclaving) an equal volume of Ng solution in resting medium (50 μg/ml) is added to yield a final concentration of 25 μg/ml. After 60 minutes the cells are washed, resuspended in fresh sterile resting medium, diluted, and plated.

5. *Streptomycin (Sm)*

Cells growing in medium containing Sm at pH 7 or higher rapidly lose the ability to form green colonies when plated.[45,46] Some of the white colonies yield strains in which chloroplast DNA is undetectable.[9] Other mutants may be found among the nonwhite colonies. Treatment with Sm is carried out in pH 8 medium (Table II) under growing conditions in the light. A concentrated streptomycin sulfate solution is freshly prepared and filtered in a sterile millipore filter apparatus. This solution is dispensed into flasks of sterile pH 8 medium to yield a final concentration of 0.05–0.1% (w/v). After inoculation, the flasks are incubated under ordinary growth conditions in the light, and after several generations white colonies can be recovered by plating. It is advantageous to plate on low pH medium since the further entry of Sm is retarded. At these concentrations of Sm the pH of the medium should not change appreciably. If higher concentrations are employed, the Sm solution should be neutralized before sterile filtration since the effective concentration varies considerably with the pH. Dihydrostreptomycin is much less effective in *Euglena*, and higher concentrations must be used.[44,47]

6. *Heat (i.e., temperatures above 32°)*

Temperatures below 32° have little effect on green colony formation in *Euglena*. Above 32° there is rapid loss of green-colony forming ability.[28,48] Temperatures approaching 37° are inhibitory or lethal to the cells, however, and a temperature of 34° is usually selected as a good compromise where little cell death occurs. Under these conditions, grow-

[44] J. Diamond and J. A. Schiff, unpublished, 1969.
[45] S. W. Chang and J. A. Schiff, in preparation.
[46] L. Provasoli, S. H. Hutner, and A. Schatz, *Proc. Soc. Exptl. Biol. Med.* **69**, 279 (1948).
[47] J. Mego and D. Buetow, cited by J. Mego, *in* "The Biology of Euglena" (D. Buetow, ed.), Vol. II, p. 363. Academic Press, New York, 1968.
[48] E. G. Pringsheim and O. Pringsheim, *New Phytol.* **51**, 65 (1952).

ing cells rapidly lose the ability to form green colonies. From the white colonies, strains can be cloned in which chloroplast DNA is undetectable.[9]

Cells are grown in the light at 34° under the usual nutritional and illumination conditions. After about six generations more than 90% of the cells give rise to white colonies on plating. The effect of elevated temperatures is much less in the dark.[28]

7. Other Mutagenic Treatments

These include theophylline and theobromine,[14] 5-Bromouracil,[49] high intensity visible light,[50] high pressure,[51] and pyribenzamine,[21] nitrofurans,[52] and macrolide antibiotics.[53] Since these agents were not specifically employed for the production of mutants or the techniques and mutants have not been described completely, experimental details will be omitted here, and the reader is referred to the original publications.

Isolation of Mutants by Plating: After treatment with a mutagen, the cells can be plated to produce colonies. Assuming all materials to be prepared in advance, as described above, the following are convenient and rapid plating methods:

The aliquot of diluted cells to be plated should contain 100–300 cells. Numbers higher than this are difficult to count on plates. Convenient numbers can be achieved through serial dilution using the dilution tubes described above. The initial cell concentrations to be diluted can be determined either by counting an aliquot in a precalibrated Coulter Electronic Cell Counter with 100 μ aperture, or in a hemocytometer chamber under a conventional light microscope.[22] In the latter procedure it is convenient to add a minute drop of saturated mercuric chloride to immobilize the cells. To detect rare color mutants against a background of white colonies, a lawn may be plated consisting of 5×10^5 to 1×10^6 cells. Similarly, where selective media are used (e.g., phototrophic medium which selects against nonphotosynthetic cells), large numbers may be plated since only those capable of growth on that medium will form colonies.

The overlay agar must be prepared before plating by heating the tubes in boiling water or steam for the minimum time needed to produce complete melting (usually 1–2 minutes) (Pour out one tube to be sure).

[49] S. Scher and J. C. Collinge, *Nature* **205,** 828 (1965).
[50] J. Leff and N. I. Krinsky, *Science* **158,** 1332 (1967).
[51] J. A. Gross, *Science* **147,** 741 (1965).
[52] L. Ebringer, R. Krkoska, M. Macor, A. Jurasek, and R. Kada, *Arch. Mikrobiol.* **57,** 61 (1967).
[53] L. Ebringer, *Naturwissenschaften* **24,** 1 (1965).

Prolonged heating must be avoided, especially with low pH medium, since the overlay will fail to solidify. Keep the overlay agar molten until needed by placing the tubes in a 45° water bath. They may be kept at 45° for up to six hours.

An aliquot of cells (0.1–0.5 ml) is pipetted directly upon the agar suface of a plate. After flaming the lip, pour the contents of one tube of overlay agar directly upon the drop of cells. Grip the covered plate at one edge and shake carefully using both reciprocal and rotational motion. Quickly tilt the plate back and forth, gripping it at various quadrants. This serves to spread the overlay agar and cells evenly before the overlay hardens. The plate is then allowed to stand for about 30 minutes and is inverted for incubation.

A useful modification of the overlay technique is employed when many plates are to be handled in sequence. For the necessary overlay, a flask of 2.4% agar (see above) is heated in a boiling water bath until melted, and about 4 ml of double-strength medium prewarmed to 45° is then added. The mixture is placed in a 45° water bath. The molten agar is dispensed from a flask of overlay agar onto the plates (which have previously received an aliquot of cell suspension) with a sterile 10-ml pipette from which the tip has been removed to provide a wide bore. The pipette is controlled with a propipette bulb. With this device it is possible to dispense overlay agar onto plates in quick succession, but each plate should be agitated as above before proceeding to the next.

Plates are conveniently incubated in the light at 26° on glass or Plexiglass shelves over alternating red and white fluorescent tubes (400–600 foot-candles). (If a constant temperature room is used, the ballasts for the fluorescent tubes should be located outside of the room since most of the heat dissipation occurs from the ballasts.) Colonies can be detected (particularly with a binocular microscope) after about five days but are not scoreable before seven to nine days. It should be noted that mutant colonies often grow more slowly and may change their appearance with time. Dark incubation can take place on covered shelves at 26°. Colonies appear almost the same time as with light grown cells. If pigment production is to be scored, the plates must be incubated in the light for at least two days following the dark incubation.

Replica Plating: For routine work, the overlay method is best since the cells are immobilized and a distinct colony is obtained for each cell plated. Due to cell motility, conventional spreading techniques are not routinely used. In the case of replica plating, however, the overlay interferes, and the spreading technique must be used. The master plate is prepared from a conventional dilution where about 0.1 ml contains 100 cells or less. The 0.1-ml aliquot is pipetted onto the agar surface and is spread with a bent glass rod (sterilized by dipping in alcohol and flaming). Fifteen minutes is allowed to elapse before inverting the plates for incubation. Best results are obtained when plates containing little or no excess moisture are employed. Replica plating is particularly useful for recovering mutants which are potentially lethal unless supplied with a nutrient or particular set of conditions (e.g., for the recovery of photosynthetic mutants which would be lost on phototrophic medium).

A cylindrical wooden block a bit smaller than the inner diameter of the petri dish is necessary. The block is washed with methanol before use. Velveteen pads (or filter paper circles) larger than the block are sterilized by autoclaving in a suitable container. Rubber bands are also sterilized at the same time.

In a reasonably clean area, free from drafts, the sterile velveteen or filter paper is stretched over the block and held fast with a sterile rubber band (or a methanol-washed brass ring). The plate to be replicated is marked at the edge with a reference point whose relation to the block is noted. The plate is then inverted and pressed firmly and evenly against the pad. After removing the master plate, fresh plates (a maximum of 4–5) lacking or containing the selective agent and marked with a reference point which is aligned with the reference point of the block derived from the master plate are pressed against the pad in succession. After covering and inverting the plates, the replicas are incubated as usual, while the master plate is kept at 4° for later comparison.

Patch Plating: A convenient method of checking mutants and comparing them with the wild-type is to patch plate them upon selective media. A grid somewhat smaller than a petri dish is prepared using graph paper. The interstices of the grid lines should be far enough apart to prevent colonies from overlapping when fully grown and the interstices should be numbered in sequence. A petri dish containing medium is placed over the grid, and an index mark is made to register it with the numbered grid. Cells to be tested are either applied as a small drop of suspension, or can be transferred with the aid of sterile toothpicks to the numbered interstices. In this way 12–16 mutants can be compared with the wild type on several different media or under several different sets of conditions.

Euglena will form colonies on filter pads (e.g., Millipore filters) placed on the agar surface of plates.[22] This technique is potentially useful in the isolation of mutants, but our experience with it is too limited for inclusion here.

Stability of Mutants: When a mutant colony is obtained from primary plating, it should be rigorously checked for stability before it is named and placed in the collection. Since most of these mutants will be chloroplastic, the regularities of mitosis are not available to ensure equal distribution of genomes to progeny during cell division. Thus mutant colonies frequently yield many different genetic types including reversions to wild type. For this reason, mutants, when first obtained, should be grown extensively on liquid medium and replated several times to obtain stable substrains, if possible. After being carried on slants, they should be checked from time to time to be sure that their characteristics have not been altered. Reversion can be minimized by maintaining the mutants on selective media or under selective conditions.

Pretreatment and Selection for Photosynthetic Mutants: A. Shneyour and M. Avron of the Weizmann Institute have kindly provided us with their novel unpublished method for obtaining mutants of *Euglena* which are incapable of evolving oxygen photosynthetically. Using a previous observation[54] that phototrophic growth on 3-(3,4-dichlorophenyl)-1,1-dimethyl urea (DCMU), a powerful and selective inhibitor of photosynthetic oxygen evolution, would bring about a decrease in the number of chloroplasts per cell, they utilized this pretreatment to speed up the selection process for mutants. Their method is as follows.

Phototrophically grown cells were transferred to phototrophic medium (one liter containing 5×10^4 cells/ml) containing 20 μM DCMU added in a minimum volume of methanol. The culture was grown for seven days without gas bubbling through. This treatment reduces the average number of chloroplasts per cell by about two-thirds and a significant proportion of the cells contain only one chloroplast per cell.[54] The cells were collected by centrifugation (1000 *g* for 10 minutes), washed once with sterile 50 m*M* sodium citrate buffer, pH 5, and were then suspended in the same buffer to a final concentration of 10^6 cells/ml. The cells were then subjected to nitrosoguanidine mutagenesis, essentially by method I (see above). After washing with fresh phototrophic medium the cells are resuspended in 9 ml of this medium in 25 ml flasks and are grown phototrophically with gas bubbling for five days. This step seems to be necessary to allow sufficient time for phenotypic expression before selection.

To provide conditions for the selection of photosynthetic mutants, 1 ml of 250 m*M* sodium arsenate was added to each of these cultures, and the cells were grown phototrophically for two more days under the same conditions. They were then washed with phototrophic medium, diluted in pH 3.5 organotrophic medium and plated on organotrophic medium. The green and pale-green colonies which formed after 10 days of growth were transferred to liquid organotrophic medium. Each liquid culture was

[54] A. Schneyour, Y. Ben-Shaul, and M. Avron, *Exper. Cell Res.* **58,** 1 (1969).

checked polarographically for oxygen evolution when the cells were in the early log phase (about 10 μg chlorophyll/ml). Some mutants were leaky and regained the ability to evolve oxygen when measured in the late log phase or later. Between 5 and 15% of the green cultures were photosynthetic mutants. They either did not evolve O_2 in the light or possessed a much reduced capacity for oxygen evolution. They are preserved on slants of organotrophic medium and are occasionally recloned.

Appendix: Commercial Sources for Materials Used in the Culture and Mutagenesis of *Euglena**

Agar
- Difco Laboratories, 920 Henry Street, Detroit, Michigan 48201
- Meer Corporation, 318 West 46th Street, New York, New York 10036

Foam plugs (for flasks, tubes, etc.)
- Gaymar Industries, 701 Seneca Street, Buffalo, New York

Plexiglass (This material is made by Rohm and Haas, Philadelphia, Pennsylvania 19105. It is distributed by franchised dealers such as the firm listed here.)
- Commercial Plastics, 353 McGrath Highway, Somerville, Massachusetts

Glassine envelopes for sterilization [No. P 5330 "dispowrap (pipette)"]
- Scientific Products, Evanston, Illinois

Plastic petri dishes
- Aloe Scientific, 1831 Olive Street, St. Louis, Missouri 63103
- Fisher Scientific, 633 Greenwich Street, New York, New York 10014 (Fisher maintains other offices in Atlanta, Boston, Chicago, Houston, St. Louis, Washington, Pittsburgh, and other cities.)
- Falcon Plastics, 550 West 83rd Street, Los Angeles, California 90045

"Wasserman" tubes and tube closures
- Bellco Glass Inc., P.O. Box B, Vineland, New Jersey 08360
- Kimble Laboratory Glassware, Owens-Illinois, Toledo, Ohio (Kimble sells its glassware through local franchised dealers)

Streptomycin
- Nutritional Biochemicals Corp., 26201 Miles, Cleveland, Ohio 44128
- Calbiochem, P.O. Box 54282, Los Angeles, California 90054
- Mann Research Laboratories, 136 Liberty Street, New York, New York 10006
- Sigma Chemical Company, 3500 DeKalb Street, St. Louis, Missouri 63118

* The firms listed here are by no means the only suppliers of materials mentioned in this article. This is only a representative listing made from references at hand and implies no endorsement of the listed companies.

Nitrosoguanidine

Aldrich Chemical Co., 2371 North 30th Street, Milwaukee, Wisconsin 53210

Nalidixic Acid

Mann Research Laboratories, 136 Liberty Street, New York, New York 10006

Calbiochem, P.O. Box 54282, Los Angeles, California 90054

Ultraviolet (uv) lamp

"Germicidal" low-pressure mercury lamp, 15 W, General Electric Co., 1 River Road, Schenectady, New York 12305

Euglena medium

Difco Laboratories, 920 Henry Street, Detroit, Michigan 48201

[13] Preparation and Properties of *Chlorella* Mutants in Chlorophyll Biosynthesis

By S. Granick

Chlorella vulgaris is a unicellular green alga, 3–10 μ in diameter, and possesses a single cup-shaped chloroplast containing the pigments—chlorophylls and carotenoids. Division of the cell into 8–16 "autospores" occurs within the original celluose wall; the cell wall then breaks, releasing the autospores, which enlarge and begin the cycle again.[1] Unlike higher plants, *Chlorella* can multiply and synthesize chlorophyll in the dark when grown on a glucose plus inorganic salt medium (see the table). The normal clone we used was obtained in 1946 from Professor S. Trelease of Columbia University.

General Procedure for Obtaining Mutants of Chlorella. Irradiate a suspension of the cells in distilled water with X rays, plate out the cells on a solid agar medium thinly enough so that the individual surviving cells will multiply and each will form a colony, and select those colonies which are paler green, yellow, white, or brown.[2] Such mutant colonies are further purified by making a thin suspension of a colony, plating it out an agar, and selecting among the resulting new colonies. This step may have to be repeated several times to obtain "stable" clones. Finally, cells of the stable clone are preserved as stocks by transferring them to several test tubes, growing them slowly at 11° in the dark on agar–glucose–inorganic salt medium and retransferring once a year. A stable

[1] H. Tamiya, *Annu. Rev. Plant Physiol.* **17**, 1 (1966).

[2] S. Granick, *J. Biol. Chem.* **172**, 717 (1948).

CULTURE MEDIA

1. *Inorganic salts medium*	
KNO_3	1.5 g/liter
Na_2HPO_4	1.0
$NaH_2PO_4 \cdot H_2O$	3.1
$MgSO_4 \cdot 7H_2O$	0.24
EDTA disodium salt	0.012
$FeSO_4 \cdot 7H_2O$	0.007[a]
$CaCl_2$ anhydrous	0.080[a]
Trace metals stock solution	20 ml
Trace metals stock solution (g/liter)	
$MnSO_4 \cdot 4H_2O$	0.050
H_3BO_3	0.050
$CuSO_4 \cdot 5H_2O$	0.005
$ZnSO_4 \cdot 7H_2O$	0.005
2. *Inorganic salts–glucose medium (g/liter)*	
Glucose[b] added to medium 1	7.5
Inorganic salts–glucose–agar medium[c] *(g/liter)*	
Difco Bacto-Agar (1.5%) added to medium 2	15.0

[a] Dilute the salts to 800 ml; add the $FeSO_4$ in 20 ml of water; add the $CaCl_2$ in 100 ml of water, with stirring. EDTA was not used originally. It helps to keep heavy metals in solution.

[b] Preferably as sterile filtered 50% glucose solution added to sterile inorganic salt medium.

[c] The agar is added to the inorganic medium before sterilizing. The medium is heated on a boiling water bath 15–20 minutes until the agar dissolves. It is then distributed into test tubes or flasks, plugged with cotton, and autoclaved for 17 minutes at 15 psi at 121°.

clone may be further mutated by ultraviolet irradiation if desired. The reason for first irradiating with X-rays is to cause irreversible mutations by breaking chromatin strands.

Most of the *Chlorella* mutants obtained may be classified as "leaky" mutants, that is, mutants incompletely blocked at a particular step. In such a mutant the activity of some enzyme is greatly diminished leading not only to an accumulation of one or more precursor substrates, but also to the formation of small amounts of substrates beyond the mutated step.

X Irradiation. The wild-type cells are first grown in an inorganic medium in the light, concentrated by centrifugation, washed, and diluted with H_2O to 5×10^8 cells/ml and placed in 5-cm-wide thin-walled petri dishes. Two X-ray tubes placed immediately below and above the covered dishes provide an intense radiation. The radiation is delivered by 180-kV tungsten-target tubes at a rate of 2000 roentgens per minute with a characteristic half-value layer of 0.19 mm of Cu for 10–20 minutes.

Considering that about 1 in 10,000 cells has survived, dilutions are made over a 1000-fold range in steps of 1:10, 1:100, and 1:1000. In this way a suitable dilution of cells will be obtained in one of the dilution steps. Of each dilution 0.5 ml are plated out onto 10-cm-diameter petri dishes, each containing 20 ml of inorganic salts–glucose–agar medium. A glass spreader and a turntable are used to spread the cells evenly. Each plate is wrapped separately in Saran Wrap to prevent contamination and grown at room temperature in the dark for about 2 weeks.

Ultraviolet Irradiation. Cells in a concentration of 10^5/ml are suspended in distilled water in a dish to form a layer 1 cm high. The dish is covered with a quartz plate and irradiated for 6 minutes with light from a 15-W General Electric germicidal lamp at a distance of 30 cm above the dish. In this way W_5B was converted to W_5B-17, a mutant which leaked out soluble hematoporphyrin into the medium.[3]

Preservation of Clones. The following precautions have been found necessary to prevent fungal contamination of stock cultures which are transferred yearly. Selected clones are preserved on inorganic salts–glucose–agar medium in 20-ml screw-cap culture tubes. Heated medium, 7 ml, is pipetted into the tubes using an automatic pipettor. The tubes are plugged with cotton and autoclaved. They are allowed to cool in slanting racks. Transfers are made from the clones to the slants using sterile wire loops. The new slants are kept in the dark for 2–3 weeks until the cultures are partially grown. Then the cotton plug is cut to the level of the glass tube, and the tube is flamed at the top and capped with a sterile bakelite cap. A 2-inch square of aluminum foil is wrapped tightly around each cap and then flamed. The tubes are finally placed in a refrigerator at 11° in the dark. The refrigerator is precleaned by wiping the walls and wire support screens with a 1% solution of cetyl trimethyl ammonium bromide. When dry, the inside is wiped with a solution of 1 g of cetyl trimethyl ammonium bromide, 5×10^5 units of mycostatin, and 10 ml of glycerin in 100 ml of water. This leaves a thin film on all surfaces. A sticky glass floss air filter strip is attached to the underside of one of the lower shelves of the refrigerator to trap dust and spores.

Isolation of a Protoporphyrin Mutant as Specific Example.[2] After X irradiation, several colorless colonies were found on the agar. One of these, W_5, was unstable and gave rise to colorless and pale-green colonies, especially when grown in liquid media. However on solid media, both the pale-green and colorless cells often gave rise to brown colonies within

[3] L. Bogorad and S. Granick, *J. Biol. Chem.* **202,** 793 (1953).

one to several weeks. From one of these brown colonies, W_5B clones were derived; these became brown when grown on an agar medium. W_5B, when grown in a liquid medium even with continual aeration, appeared only faintly brown when centrifuged into a thick pellet. For large numbers of cells and rapid development of the brown color, the following procedure was used. An inoculum, taken from brown cells on a solid medium or from almost colorless cells grown on a liquid medium, was sown on the surface of solid medium in Blake bottles each containing 1 liter of solidified agar medium. At 25°, growth was maximal on the fourth day, but the cells were colorless. At this time the bottles were tilted slightly to drain away excess surface liquid, and within the next 2 or 3 days the cells became brownish. If the liquid was not drained away some 4–6 days after inoculation, only a slight browning resulted within the next few days. By water addition to the agar surface, the cells could be shaken off and the cell suspension decanted. About 7–10 ml of packed cells were obtained per bottle. It became apparent that the brown pigment developed most rapidly in the resting stage at the end of the logarithmic period of growth. Under these growth conditions the cells were often enlarged to 2 to 3 times their normal diameter and contained 5–20 large starch grains which appeared to be localized in a vesicle making up most of the cell volume; a thin crescent-shaped area of cytoplasm containing brown granular material lay on one side of the cell. The brown material fluoresced intensely red in a fluorescence microscope, indicating the presence of porphyrins. These cells were easily broken by merely shaking them in distilled water. Small cells were seldom observed to contain this brown material. From large batches of cells, in addition to large amounts of protoporphyrin, traces of chlorophyll could be detected.

By ultraviolet irradiation of W_5B, mutant W_5B-17 was obtained. This mutant could be grown in a liquid medium in 20-liter bottles using vigorous aeration, an inorganic salts–glucose medium, and the addition of 5 ml of sterile Mazola oil as a defoaming agent. At 29° production was maximal in 9 days and the contents of the bottles appeared chocolate brown. From 15 liters of liquid, 50 g dry weight of cells were obtained and about 1 g of porphyrins.[3]

Chlorophyll Biosynthetic Chain. A number of the *Chlorella* mutants were isolated, and their pigments were identified as porphyrins. The structures of these porphyrins suggested that they were intermediates in chlorophyll biosynthesis. These porphyrins could be arranged into a biosynthetic chain sequence that was reasonable both organochemically and biochemically (Fig. 1).

Comments on the Biosynthetic Chain as Revealed by Chlorella

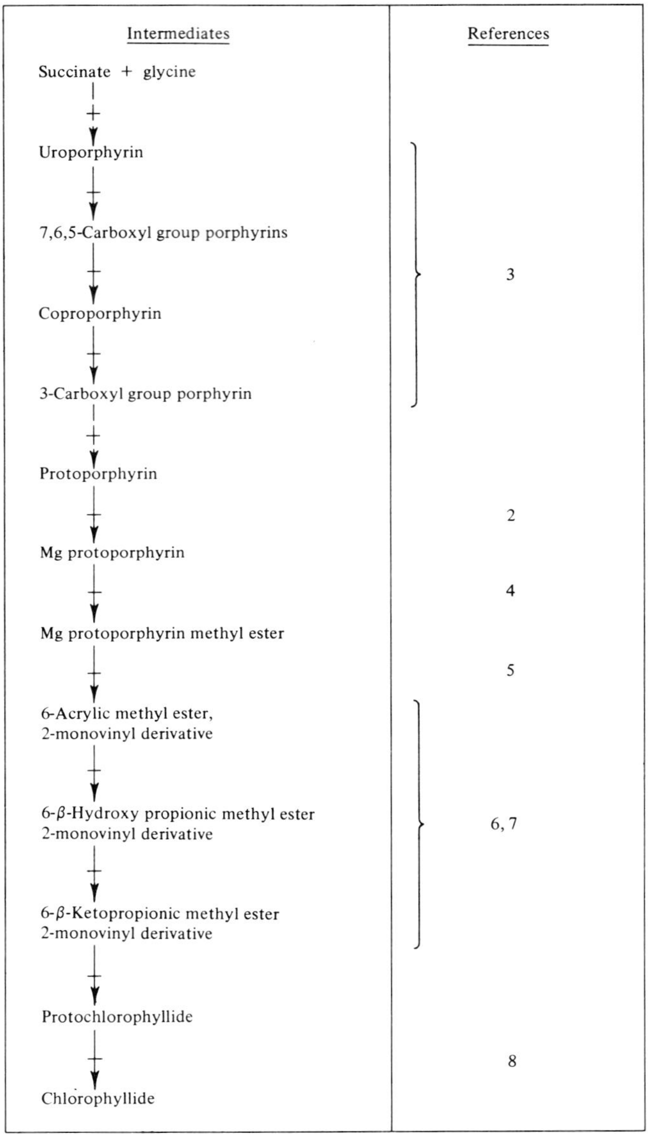

FIG. 1. Porphyrins isolated from mutant strains of *Chlorella vulgaris* and arranged in a biosynthetic sequence.

Mutants. The enzyme synthesizing δ-aminolevulinic acid from succinyl-CoA and glycine has been detected as yet only in *Rhodopseudomonas spheroides,* but not in higher plants. The porphyrins, uroporphyrin and its decarboxylated derivatives[3] are the oxidized forms of colorless porphyrinogens, which are the true intermediates. From protoporphyrin onward, the pigments are all in their oxidized colored states. The finding of Mg protoporphyrin[4] was the first direct clue that heme and chlorophyll arose from the same biosynthetic chain.

From mutant 60A, the monomethyl ester of Magnesium protoporphyrin was isolated.[5] The esterification is on the propionic acid group in side chain position 6 around the ring. The oxidation of this propionic acid methyl ester side chain appears to take place by a β-oxidation. This idea is supported by the recent studies of Ellsworth and Aronoff,[6,7] who prepared *Chlorella* mutants by ultraviolet irradiation and used mass spectrometry to identify the isolated pigments. From a red-orange mutant, A.J., they obtained evidence for the presence of the acrylic ester derivative. From a green-yellow mutant, B.E., they obtained evidence for the presence of the β-hydroxypropionic ester derivative and the β-ketopropionic ester derivative. During the oxidation of the propionic ester side chain, there also occurs a reduction of the vinyl side chain at position 4 (but not at position 2) to an ethyl group. It is suggested by these authors that there is a parallel pathway in which both mono- and divinyl analogs undergo β-oxidation of the propionic acid side chain; however, the monovinyl derivatives appear to be the predominant ones.

Protochlorophyllide can be converted by normal *Chlorella* in the dark to chlorophyllide by an enzymatic reaction; in this reaction 2 H atoms are added to pyrrole ring IV. In mutant 31,[8] which appears yellow when dark grown, protochlorophyllide can be converted to chlorophyllide only in the light, as is the case in higher plants. Presumably wild-type *Chlorella* has both an enzyme that does not require light and also an enzyme that requires light, either one of which can form the chlorophyllide. In mutant 31, the gene governing the formation of the dark enzyme is thought to have been mutated and inactivated.

Two pigments have been isolated from W_5B-17,[9] which do not fit into the sequence. These are hematoporphyrin and the corresponding monovinyl, monohydroxyethyl porphyrin. The lack of optical activity in these

[4] S. Granick, *J. Biol. Chem.* **175,** 333 (1948).
[5] S. Granick, *J. Biol. Chem.* **236,** 1173 (1961).
[6] R. K. Ellsworth and S. Aronoff, *Arch. Biochem. Biophys.* **125,** 269 (1968).
[7] R. K. Ellsworth and S. Aronoff, *Arch. Biochem. Biophys.* **130,** 374 (1969).
[8] S. Granick, *J. Biol. Chem.* **183,** 713 (1950).
[9] S. Granick, L. Bogorad, and H. Jaffe, *J. Biol. Chem.* **202,** 801 (1953).

pigments suggests that they are artifacts, perhaps derived from some unstable intermediate between coproporphyrinogen and protoporphyrinogen.

Mutations and Genes. It is not known which genes of the chlorophyll biosynthetic chain reside in the nucleus, the plastids, or the mitochondria.[10] The use of new antibiotics specific for one or other of these organelles may soon make such distinctions possible. At present, the best guess is that these genes are localized in the nucleus. Studies on the enzyme δ-aminolevulinic acid synthetase of mitochondria suggests that this enzyme is coded by a nuclear gene. In the human Mendelian dominant disease, acute porphyria, it is inferred that the synthetase of δ-aminolevulinic acid is coded for by a nuclear gene which forms a mRNA, the mRNA provides the information to synthesize the enzyme in the cytoplasm, and the newly formed enzyme migrates into and lodges on the inner mitochondrial cristae membrane, where it functions to combine succinyl-CoA, produced by the citric acid cycle, with glycine to form δ-aminolevulinic acid.[11] However, ultraviolet damage to the plastid DNA of *Euglena* can result in plastids devoid of all pigments, even though the nucleus had been shielded from irradiation.[12] Such an experiment indicates that the expression of nuclear genes in forming the pigments of the plastid requires the proper functioning of plastid genes.

[10] A. Gibor and S. Granick, *J. Cell Biol.* **15**, 599 (1962).
[11] S. Granick and A. Kappas, *Proc. Nat. Acad. Sci. U.S.* **57**, 1463 (1967).
[12] S. Granick and A. Gibor, *Progr. Nucl. Acid Res. Mol. Biol.* **6**, 143 (1967).

[14] *Chlorella* Mutants

By MARY BELLE ALLEN

Mutant strains of various species of *Chlorella* have been used in studies of a number of topics related to photosynthesis, including the synthesis of chlorophylls[1-5] and of plastids,[6] the degradation of chlorophyll,[7] the absorption spectra of different chlorophyll complexes in the living cell,[8-10] the role of carotenoids[11,12] and of chlorophyll *b*[13] in photo-

[1] S. Granick, *J. Biol. Chem.* **172,** 717; **175,** 333 (1948).
[2] S. Granick, *J. Biol. Chem.* **183,** 715 (1950).
[3] S. Granick, *Annu. Rev. Plant Physiol.* **2,** 115 (1951).
[4] L. Bogorad and S. Granick, *J. Biol. Chem.* **202,** 793 (1953).
[5] R. K. Ellesworth and S. Aronoff, *Arch. Biochem. Biophys.* **125,** 35, 269–277 (1968).
[6] G. W. Bryan, A. H. Zadylak, and C. F. Ehrel, *J. Cell Sci.* **2,** 513 (1967).
[7] R. Ziegler and S. H. Schanderl, *Photosynthetica* **3,** 45 (1968).
[8] C. S. French, *Brookhaven Symp. Biol. No. 11,* p. 65. Upton, New York, 1958.

synthesis, the size of the photosynthetic unit,[14,15] and the possible role of photosynthetic pigments in synchronization of growth.[16] The methods for production and isolation of the mutants are similar to the well known ones for handling mutants of other microorganisms, so that only a brief treatment seems necessary.[17]

Production and Isolation of Mutants. X rays and ultraviolet light are the mutagens which have most commonly been used. A mutagenic effect of streptomycin has been reported.[18] Other chemical mutagens would no doubt be effective.

Since a large number of mutations in *Chlorella* involve the pigment system and since many of these render the organisms nonphotosynthetic, it is desirable to use a medium containing glucose for the initial culture of the mutants. These should be incubated in the dark, at least for the first few hours, to avoid photoreactivation effects. Dark incubation is also desirable because many pigment mutants are photosensitive.[11,19] Since auxotrophic mutants may be formed, inclusion of a growth factor supplement (yeast extract is convenient) is desirable. Any of a number of mineral media may serve as the basal medium; *Chlorella* is very tolerant in this respect. Two sample experimental protocols are briefly presented.

1. *Chlorella vulgaris* was grown on a mineral medium, concentrated by centrifugation, suspended in a small drop of the same medium in a thin-walled Carrel flask, and exposed to X rays from a 180-kV tungsten target tube at a rate of 2050 roentgens per minute for 10–20 minutes.

The irradiated cells were diluted and grown in the dark on a solid medium consisting of 3% agar, 0.7% glucose, and inorganic salts.[1]

2. *Chlorella pyrenoidosa* was grown on a maintenance medium of mineral salts supplemented with beef extract and glucose (BAD, see table). A suspension of the cells in water was placed in an open petri dish and stirred while irradiating with a Hanovia ultraviolet lamp. The time of

---

[9] M. B. Allen, C. S. French, and J. S. Brown, *in* "Comparative Biochemistry of Photoreactive Systems" (M. B. Allen, ed.), p. 33. Academic Press, New York, 1960

[10] C. S. French, *Carnegie Inst. Wash. Yearbook* **66,** **177** (1966/67).

[11] H. Claes, *Z. Naturforsch.* **96,** 461 (1954); **11b,** 260 (1956); **12b,** 401 (1957); **13b,** 222 (1958); **14b,** 4 (1959).

[12] H. Claes and T. O. M. Nakayama, *Z. Naturforsch. B* **14,** 746 (1960).

[13] M. B. Allen, *Brookhaven Symp. Biol. No. 11,* p. 337. Upton, New York, 1958.

[14] A. Wild and K. Egle, *Planta* **42,** 73 (1968).

[15] A. Wild and K. Egle, *Photosynthetica* **2,** 253 (1968).

[16] H. Metzner, H. Rau, and H. Sangen, *Planta* **65,** 186 (1965).

[17] See also this volume [13].

[18] V. V. Tugarinov and V. V. Kuznetsov, *Dokl. Akad. Nauk SSSR* **166,** **722** (1966).

[19] S. Bendix and M. B. Allen, *Arch. Mikrobiol.* **41,** 115–141 (1962).

COMPOSITION OF BAD MEDIUM

Macronutrients		Micronutrients (mg/liter)	
KNO_3	4 m*M*	Fe (as NaFeEDTA or NaFe-diethylene–triamine–pentaacetate)	4
K_2HPO_4	1 m*M*	B (as H_3BO_3)	0.5
$MgSO_4$	1 m*M*	Mn (as $MnSO_4 \cdot H_2O$)	0.5
Beef extract	0.1%	Zn (as $ZnSO_4 \cdot 7H_2O$)	0.05
Glucose	0.2%	Cu (as $CuSO_4 \cdot 5\ H_2O$)	0.02
		Mo (as MoO_3 85%)	0.01
		V (as NH_4VO_3)	0.01

Composition of Screening Media

	BGM[a]	
KNO_3	20 m*M*	BGM + YE = BGM + 0.05% yeast extract
$MgSO_4$	1 m*M*	M = BGM + 1% glucose
NaCl	4 m*M*	C = BGM + 1% glucose + 0.05% yeast extract
$CaCl_2$	0.5 m*M*	
K_2HPO_4	2 m*M*[b]	
Micronutrients as in BAD		

[a] M. B. Allen and D. I. Arnon, *Plant Physiol.* **30,** 372 (1953).
[b] Added aseptically after autoclaving.

irradiation was determined from previous measurements of the time required to kill more than 99% of the population (since different strains vary in sensitivity to radiation, this must be determined for each culture).

Portions of the suspension were then plated on M or C (see table) and incubated in the dark. Spreading a drop of the cell suspension on the surface of an agar plate (2% agar was used to solidify the media of the table) was found to be the most effective method for obtaining colonies for examination.[19]

Screening of Mutants. Pigment mutants, the most common type, are detected by inspection. An illuminated viewing box is a helpful aid to this. Many pigment mutants show a high degree of phenotypic variability, which may persist through many transfers. Sectored colonies are common, and even after repeated isolations of one color type, a whole spectrum of colors may be obtained on subsequent transfer. Fortunately for biochemical experiments, not all pigment mutants are this variable. As mentioned, many pigment mutants are photosensitive and must be cultured in the dark. Many others show differences in pigmentation depending on whether they are grown in the light or the dark.

Other types of mutants that have been reported include auxotrophs, antibiotic producers,[19] streptomycin-resistant strains,[18] cells with different temperature relations from wild type, and strains that differ from the parent only in growth rate.[19]

The details of the screening procedure to be employed will depend on the particular type of mutant to be isolated. A general method which proved useful in the author's laboratory consists of transferring cells from individual colonies from the initial plating to a C (see table) agar plate placed over a 44-square grid drawn on the frosted glass top of a viewing box. One colony is transferred to each square. When colonies have grown up, cells can be quickly transferred in an identical array to a series of plates of other media or to the same medium for incubation under different growth conditions.[19]

The probability of back mutation to wild type appears to be rather high with *Chlorella* mutants, so that cultures which have been maintained for some time should be checked to see that they retain their original properties. It may be necessary to reisolate the original form. An interesting, if puzzling case of chemically directed back mutation has been reported by Czygan,[20] who found that 50 m*M* chloramphenicol caused several *C. pyrenoidosa* mutants to revert to wild type.

[20] F. C. Czygan, *Nature* **212**, 960 (1966).

[15] Origin and Properties of Mutant Plants: Yellow Tobacco

By GEORG H. SCHMID

Color variation in green tobacco leaves caused by chlorophyll deficiency may be the result of changes in environment, e.g., temperature, chemicals, such as numerous commercial herbicides and insecticides, or virus and fungus infections. Chlorophyll deficiency in tobacco may be induced by mutation. These mutations either occur spontaneously or are induced by chemicals or irradiation. On the other hand, chlorophyll-rich mutants can also be induced artificially: Colchicine treatment was successfully used in obtaining polyploid forms, mostly by species crossings. Seeds or shoots of *Nicotiana tabacum* are treated with 0.8% colchicine. This treatment leads to genome doubling and to an increase of the chloroplast number per cell, which in turn may cause a tendency toward especially chlorophyll-rich plants. Considerable work has been done on the induction of mutations by irradiation (X rays, radioactive substances). The latter method causes chromosome aberration and chromosome deletion. The relative ease of detection of chlorophyll mutants in tobacco is a function of a discernible color difference and the large population of small seedlings which characterize experimental sowings. One plant of *N. tabacum* produces 200,000–400,000 seeds. Simple aging of the seed lot increases the probability of finding mutants.

Description of Different Aurea Phenotypes of Tobacco

A number of chlorophyll mutants of tobacco with aurea or yellow appearance have been used in photosynthesis research.

Recessive Chlorophyll Mutants. Recessive chlorophyll mutants have been frequently observed in tobacco. It is probable that the degree of recessive chlorophyll deficiency is influenced by environmental conditions during seed development. The best known mutants are *Nicotiana tabacum* derivatives with yellow-green (*yg*) character from No. 63, Dixie Shade, Florida 15 and Rg. The so-called glaucous or yellow character in *N. tabacum,* first reported by Nolla[1] in 1934, is controlled by a single recessive gene pair (*yg/yg*). The mutant phenotype is commonly termed "consolation." These plants are chlorophyll deficient in the seedling stage and may regain almost normal pigmentation at later stages of development; the plants grow vigorously.

Nicotiana tabacum L. Tobaccos with "White Burley" Character (Burley 21, Ky 16, Ky 151, etc.). White Burley tobacco originated in 1864 or 1865, presumably as a mutation from the green variety "Little Burley,"[2] in Brown County, Ohio. White Burley seedlings are only somewhat lighter green than seedlings of green varieties. A few weeks after transplantation of the seedlings the color difference with green varieties almost disappears. With the approach to maturity, however, White Burley tobaccos lose most of their pigmentation, particularly in the lower leaves and the stem. If the usual commercial practice is followed and the plants are "topped" below the seed head, the loss of chlorophyll is increased and the plants become light yellow. Except for a slight mottling of the leaves as they ripen, green varieties retain their color under this treatment.

Dominant Chlorophyll Mutants. Nicotiana tabacum L. Aurea mutant Su/su. Burk and Menser found in 1964 an unusual aurea mutation of tobacco in a single, initially variegated plant in a population derived from an 18-year-old seed lot of the Connecticut cigar variety John Williams Broadleaf.[3] They called the aurea factor "*Su*." When present in a heterozygous condition (*Su/su*) this factor resulted in plants with the yellow green or aurea appearance. The plant was the first recorded dominant

[1] J. A. B. Nolla, *J. Agr. Univ. Puerto Rico* **18,** 443 (1934).
[2] E. H. Mathewson, *U.S. Dep. Agr. Bw. Plant. Ind. Bull.* **244** (1912).
[3] L. Burk and H. A. Menser, *Tobacco Sci.* **8,** 101 (1964).

aurea mutation in tobacco. The name *Su* was chosen by these authors because of a similarity to sulfur deficiency. Under special environmental conditions, such as high light intensity, long-day conditions, warm temperatures, and high humidity, the plants are very chlorophyll deficient, especially during the first weeks after germination. At later stages of development the plants are distinctly greener, though still very chlorophyll deficient. Grown at low light intensity, or short-day conditions, growth is extremely poor. The plants are distinctly greener from the beginning. Grown under either condition, the plants have sometimes the ability of becoming spontaneously greener. This process is apparently induced by various environmental conditions and cannot be controlled yet.

Cytoplasmic Inheritance in N. tabacum L. Variegated Plants from NC 95. Variegated NC 95 contains a mixture of normal and chlorophyll-deficient cells distributed between adjacent histogens. The mode of inheritance is maternal.

All tobacco mutants described above grow autotrophically within reasonable time to maturity just as normal green tobacco.

X-Ray Mutants in Nicotiana tabacum L. Virescent 402. There are several lines of virescent tobacco. Virescent 402 has been induced by X irradiation; the mode of inheritance is not known. These plants grow poorly and are injured by high light intensities.

Chemical Properties

Chlorophyll Content per Leaf Area. Pigments are extracted from the leaves with 100% methanol at 60° for 30 minutes. The mixture is centrifuged at 20,000 *g* for 10 minutes. The supernatant solution is adjusted to 90% methanol and 10% water and examined with a Zeiss spectrophotometer. The amount of chlorophyll is determined from the extinction values at 663 (Ex 663) and 645 (Ex 645) nm. For a 1-cm light path, chlorophyll *a* and chlorophyll *b* are calculated from the formula:

$$\frac{1.16 \times (\text{Ex } 663 \times 45.6 - \text{Ex } 645 \times 9.27) \times \text{volume (ml)}}{3585.75} = \text{mg chlorophyll } a \quad (1)$$

$$\frac{1.07 \times (\text{Ex } 645 \times 82.04 - \text{Ex } 663 \times 16.75) \times \text{volume (ml)}}{3585.75} = \text{mg chlorophyll } b \quad (2)$$

Without the factors 1.16 [Eq. (1)] and 1.07 [Eq. (2)] the formula can be used for chlorophyll solutions in 80% acetone–20% water. Leaves from yellow tobacco mutants contain much less chlorophyll than the green siblings. The carotenoid content is reduced too in these yellow

TABLE I
AVERAGE PIGMENT CONTENT PER LEAF AREA OF DIFFERENT TOBACCO VARIETIES[a]

Nicotiana tabacum	Chlorophyll ($\mu g/cm^2$)	Chl *a*:Chl *b*	Total Chl:total carotenoid
John Williams Broadleaf	27 ± 12	2.9 ± 0.4	3.8 ± 0.5
Mutant *Su/su*	7.6 ± 5.2	3.5 ± 0.6[b]	2.7 ± 0.6
Rg green	25 ± 5	2.5 ± 0.4	4.0 ± 0.5
Rg derivative with yellow-green character	9 ± 4	3.0 ± 0.2	3.0 ± 0.7
NC 95 green leaf patch	32 ± 6	2.5 ± 0.6	3.7 ± 0.8
NC 95 yellow-green leaf patch	4.2 ± 1	3.4 ± 0.6	3.1 ± 0.1

[a] At 6–12 weeks after germination.
[b] Under conditions of spontaneous greening the average chlorophyll content reaches 9 ± 2 μg Chl/cm^2 with a Chl *a*:Chl *b* ratio of 5.1 ± 1.5.

leaves, but considerably less in proportion. Thus the total chlorophyll: total carotenoids ratio is smaller in the aurea mutant than in the green control (Table I). Chlorophyll-deficient tobacco mutants are yellow because this ratio, carotenoids:chlorophyll, is greater than in green plants. The chlorophyll content per leaf area varies in the aurea mutants as a result of age and growing conditions.

Chlorophyll Content per Chloroplast. Determination of the chlorophyll content per chloroplast is achieved by using the Coulter technique. A suitable Coulter counter is the Model F with a 70-μ capillar from Coulter Electronics Inc., Hialeah, Florida, fitted with a mean cell volume computer. Good chloroplast preparations give a counting plateau between the attenuation settings 1 and 0.5 at threshold 10, and aperture 8 or 16.

A countable chloroplast suspension is prepared by the following procedure: 500 g of leaves are disrupted in a roller mill at 5° with 1.5 liters of buffer containing 50 m*M* Tris·HCl, pH 7.8; 0.25 *M* sucrose; and 100 mg/liter of bovine serum albumin (Buffer I). The brei is filtered through a folded filter (Schleicher and Schüll), and the filtrate is centrifuged for 20 minutes at 600 *g*. The supernatant solution is discarded. The sediment is suspended in 100 ml of buffer I and centrifuged at 100 *g* for 2 minutes. The sediment is discarded and the supernatant solution is diluted with 100 ml of buffer I and recentrifuged at 100 *g* for 2 minutes. This procedure is repeated once more, and the sediment is discarded. Thereafter the supernatant solution is centrifuged at 600 *g* for 20 minutes. The sediment is placed on a sucrose gradient containing layers of 40%, 45%, 50%, and 55% sucrose from top to bottom. The gradient is centrifuged for 60 minutes at 2200 *g*. In general, the chloroplasts are found in the

TABLE II
CHLOROPHYLL CONTENT OF DIFFERENT MUTANT CHLOROPLASTS OF *Nicotiana tabacum*[a]

N. tabacum	Total chlorophyll per chloroplast (10^{-12} g)	Chloroplast volume (μ^3)
John Williams Broadleaf	1 ± 0.16	75 ± 28
Mutant *Su/su*	0.5 ± 0.28	54 ± 13
NC 95 yellow-green leaf patch	0.05 ± 0.02	52 ± 10

[a] The absolute variations are induced by different average chloroplast sizes of the individual preparations. All plants grown under low light intensity.

45% sucrose layer. The layer is separated and diluted with 100 ml of buffer I and centrifuged again at 600 *g* for 20 minutes. The sediment is washed with buffer I and recentrifuged at 600 *g* for 20 minutes. The sediment is suspended finally in a 10-ml-volume flask in a buffer containing 0.1 *M* K_2HPO_4, 0.25 *M* sucrose, and 0.1 g/liter bovine serum albumin (Buffer II). For the counting procedure, 0.1 ml of this suspension is usually diluted to 100 ml with buffer II.

The Coulter technique offers the possibility of calculating pigment content and other chemical properties per chloroplast (Table II). It is clearly seen (Table II) that chlorophyll deficiency in the mutant plants is caused by chlorophyll deficiency of the chloroplast.

Biological Properties

Chloroplast Structure. The decrease of chlorophyll content of the mutant chloroplasts is accompanied by a change of the chloroplast structure.[4,5] The well-known typical grana structure where thylakoids are highly stacked is found only in the green sibling. The dominant tobacco mutant *Su/su* contains chloroplasts which have low stacked grana.[4] Depending above all on the light conditions, the plants may contain only chloroplasts with grana of two to three stacked thylakoids. Grown at low light intensity, or under other unfavorable conditions, the chloroplast structure changes in favor of more highly stacked grana. Correlated with this structural change is an increase of the chlorophyll content. At a later stage of development the mutant chloroplast structure approaches that of the normal green chloroplast.

Chloroplasts from yellow-green patches of the variegated *N. tabacum* L. NC 95 contain chloroplasts with widely separated thylakoids, which

[4] G. H. Schmid, J. M. Price, and H. Gaffron, *J. Microsc.* (*Paris*) **5**, 205 (1966).
[5] G. H. Schmid, *J. Microsc.* (*Paris*) **6**, 485 (1967).

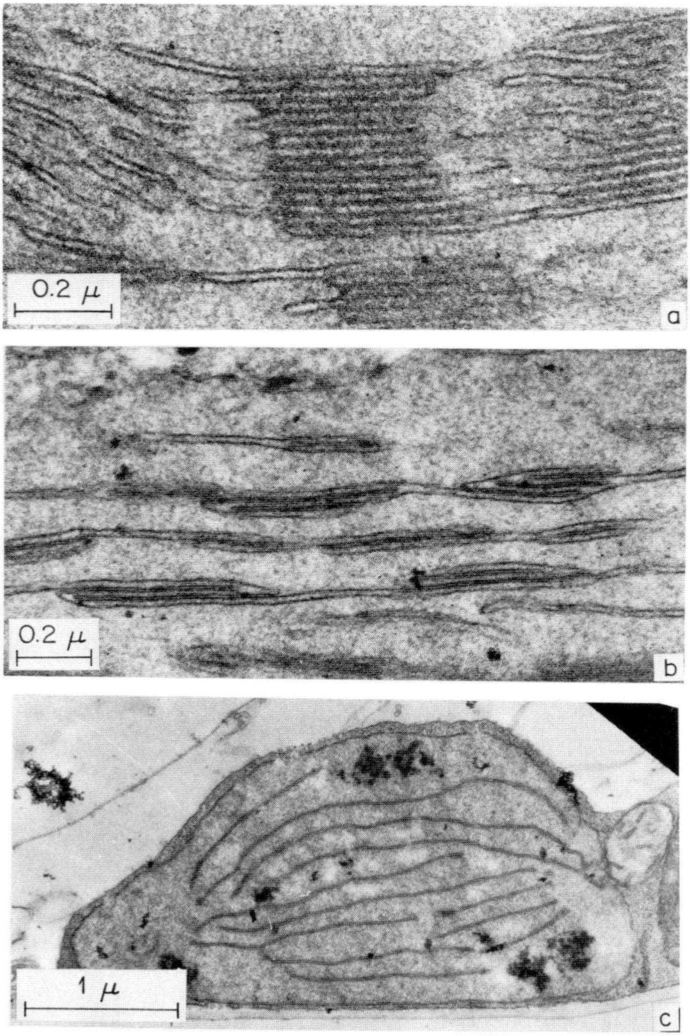

FIG. 1. Chloroplasts from *Nicotiana tabacum* L. (a) John Williams Broadleaf (normal green tobacco). (b) Aurea mutant *Su/su*. (c) NC 95 yellow-green leaf patch of a variegated plant.

do not form stacks or doublings.[5] It is seen (Table II) that NC95 chloroplasts are very chlorophyll deficient. This observation together with the structural appearance (Fig. 1) demonstrates the well-known fact that grana formation parallels chlorophyll synthesis, and vice versa.[6] In nor-

[6] D. von Wettstein, *Ber. Deut. Bot. Ges.* **74**, 221 (1961).

mal green plants, grana are apparently a device for packing chlorophyll. But chlorophyll is also present in isolated thylakoids, which is in contrast to the opinion that chlorophyll is only located in the grana. There is no precise description in the literature as to which kind of thylakoids and chloroplast structure is really indispensable for photosynthesis. There is unanimity only in the fact that there have to be thylakoids and that the active lamellar system must contain chlorophyll.

Physical Properties and Light Metabolism

Photosynthesis Measured Manometrically as Oxygen Evolution. The leaf sections are floated on a carbonate–bicarbonate buffer in a Warburg vessel, or supported over the buffer on a wire spiral. The latter arrangement prevents submersion and is mandatory with leaves that are easily wetted. Depending on the ratio of the mixture carbonate–bicarbonate, CO_2 concentrations in the gas phase of 0.45–3.5% can be obtained. The area of the leaf section has to be adjusted so that no more than 20 μl of O_2 are evolved in 10 minutes. Depending on the light intensity, this area varies for the yellow tobacco mutants and the green control between 2 and 8 cm.[2] Whenever the rate of O_2 evolution is faster than 20 μl/10 minutes, CO_2 becomes limiting due to problems of gas diffusion.[7] In such an event the values of the saturation rate of photosynthesis obtained by this method are lower than those obtained by measuring $^{14}CO_2$ fixation in a gas chamber. The light-dependent rate stays the same, however, and is not affected by too low CO_2 concentrations.

Photosynthesis Measured as $^{14}CO_2$ Fixation in the Gas Phase. When photosynthesis is measured as $^{14}CO_2$ incorporation (0.45–5% CO_2 in air), the temperature of the illuminated leaf area may rise above that of the thermostated gas phase, despite agitation of the latter. This causes a tendency toward higher values of photosynthesis. The temperature effect on the saturation rate can be considerable at high CO_2 concentrations.[7] With leaves floated on buffer, however, this is a minor problem. In addition the stomata probably respond differently to the conditions in either method, and this may also affect the saturation level. In the extreme case, a low saturation value obtained by manometry and a high value obtained by $^{14}CO_2$ fixation, both in the gas phase with 5% CO_2, may differ by a factor of 2. The correct values will probably lie between these extremes. Either method, manometry or $^{14}CO_2$ fixation, gives the same relative results, however, for the mutants and the green control. The $^{14}CO_2$-fixation method generally gives higher values, but manometry is sufficient for routine experiments. Because of a lower temperature sensi-

[7] P. Chartier, *Ann. Physiol. Veg.* **8**, 167 (1966).

tivity,[8] the rate of photosynthesis of the *Su/su* mutant is not affected as much by local heating as is the rate in the control John Williams Broadleaf. That is, these differences of the method used preponderantly affect the green control. The data of Table III and Fig. 2 contain maximal rates of photosynthesis with all variations of both methods just described. Despite the known transfer of energy from accessory chloroplast pigments, particularly carotenes to chlorophyll, light absorbed by such pigments is not responsible for the high rates of photosynthesis in *Su/su* and consolation, because the rates presented in Table III and Fig. 2 are observed in red light.

Maximal Rates of Photosynthesis on the Basis of Leaf Area and Chlorophyll Content in Nicotiana tabacum var. John Williams Broadleaf and the Aurea Mutant Su/su. Owing to age or growing conditions, the chlorophyll content per leaf area may vary because leaves are of different thicknesses. The active chlorophyll which happens to be present

TABLE III
SATURATION RATES FOR DIFFERENT PLANTS IN RED LIGHT AT 25°[a]

Plant	Chlorophyll content per leaf area ($\mu g/cm^2$)	Rate of photosynthesis: O_2 evolution or CO_2 fixation (μl/mg chlorophyll/min)
Medicago polymorpha L.	63 ± 14	38 ± 12
Cassia obtusifolia L.	45 ± 12	40 ± 13
Nicotiana tabacum L. var. John W. Broadleaf	27 ± 12	45 ± 17
N. tabacum L. var. Rg green	25 ± 5	53 ± 12
N. tabacum L. var. Kentucky 16	20 ± 6	41 ± 12
N. tabacum L. var. Dixie Shade green	20 ± 5	34 ± 12
N. tabacum L. var. Virescent 402	12 ± 6	55 ± 15
N. tabacum L. Rg derivative with yellow-green character 526-1-2y	9 ± 4	140 ± 40
N. tabacum L. yellow leaf patch of variegated NC 95	4.2 ± 1.0	8.5 ± 7

[a] Averages of at least 50 independent determinations for each species. Red filter 575 nm < λ < 750 nm; CO_2 concentration was varied between 0.45% and 5% in air.

[8] G. H. Schmid and H. Gaffron, *J. Gen. Physiol.* **50**, 563 (1967).

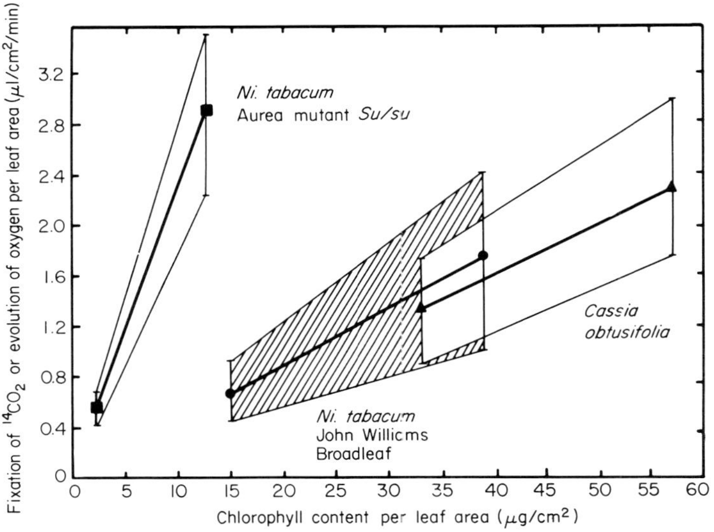

FIG. 2. Maximal rate of photosynthesis measured as $^{14}CO_2$ fixation or as O_2 evolution for *Nicotiana tabacum Su/su,* John Williams Broadleaf and *Cassia obtusifolia.*

in a certain area will determine the photosynthetic rate per leaf area. For instance, leaves of both the mutant and the green control are often thicker when grown in a liquid medium[9] than in soil. The mutant plant, when soil grown, often produces comparatively thicker, smaller leaves than the green control (dry weight of Table IV). Thus, the concept of chlorophyll deficiency on the basis of leaf area is rather relative (Table IV). The mutant responds to environmental changes in a way which differs from the responses of the green control. The maximal rate of photosynthesis per leaf area of the mutant may be either greater or lower or the same as in the green control, depending on growth conditions and chlorophyll content. As has been pointed out in the paragraph on chloroplast structure, in the seedling stage the *Su/su* mutant contains chloroplasts which are very chlorophyll deficient and which have very poorly stacked grana. With this structure and pigment content, the rate of photosynthesis based on chlorophyll is approximately five times higher than that of the green control (Tables III and IV; Fig. 2). At later stages of development, or if the plants have a higher chlorophyll content due to unfavorable growing conditions, or if the effect of spontaneous greening has taken place, the chloroplast structure approaches that of the green control. With this more normalized chloroplast structure and

[9] G. H. Schmid, *Planta* **77,** **77** (1967).

TABLE IV
METABOLIC TRENDS IN THE *Su/su* MUTANT AND JOHN WILLIAMS BROADLEAF TOBACCO[a]

Growth conditions	Plant	Chlorophyll per leaf area (μg/cm²)	Dry weight per leaf area (mg/cm²)	Saturation rate of photosynthesis: O_2 evolved/hr/10 cm² (μl)	Saturation rate of photosynthesis: O_2/mg chlorophyll/min (μl)
Growth chamber plants, 40 days after sowing	*Su/su*	4	1.3	411	170
	JWB	18	1.3	415	39
Soil-grown seedlings, weeks after germination	*Su/su*	5.3	1.2	555	175
	JWB	29	2.0	835	48
Soil-grown mature plants, 4 months after germination, shortly before blooming	*Su/su*	18	2.3	1080	100
	JWB	44	1.4	750	29
Water-culture plants 4 months after germination, shortly before blooming	*Su/su*	25	3.1	1040	70
	JWB	54	4.1	860	27

[a] Rate of photosynthesis measured by manometry at 25° in red light 575 nm $< \lambda < 700$ nm, energy 36,000 erg/cm²/sec¹. Each value is the average of at least 20 determinations.

the higher pigment content, the rate of photosynthesis, on the basis of chlorophyll, approaches more and more the maximal values observed in green plants (Table IV).

In contrast to the saturation rates of photosynthesis, where the yellow tobacco mutant *Su/su* exhibits on the basis of leaf area the same or sometimes even higher rates of photosynthesis (Fig. 2) as the green sibling, the rate is much below par at low light intensities.

In blue light, 380 nm $< \lambda <$ 575 nm, mutants described in this article derive no special advantage of the increased carotenoid:chlorophyll ratio.[8]

Maximal Rates of Photosynthesis on the Basis of Leaf Area and Chlorophyll Content in Nicotiana tabacum var. Kentucky 16 (Tobacco with "White Burley Character"). Plants in narrow stands of "White Burley" tobaccos may show, at the stage when the plant has produced only six–eight leaves, a distinct color difference from the bottom to the top. The bottom leaf is distinctly lighter in color, sometimes it looks

TABLE V
METABOLIC TRENDS IN ASCENDING LEAVES OF *Nicotiana tabacum* Ky 16 AT 25°[a,b]

Leaf number and color	Total leaf area (cm^2)	Dry weight leaf (mg)	Total chlorophyll (μg/leaf)	Apparent rate of photosynthesis as oxygen evolved: per leaf (μl/hr)	Apparent rate of photosynthesis as oxygen evolved: per 10 cm^2 (μl/hr)	Rate of dark respiration as oxygen uptake: per leaf (μl/hr)	Rate of dark respiration as oxygen uptake: per 10 cm^2 (μl/hr)
1. White	45.5	79	73	690	151[c]	56	13.4
2. Yellow green	29	74	250	540	187	80.5	28
3. Green	28	81	384	372	133	129	46
4. Green	26	65	630	440	170	190	73
5. Green	21	66	420	296	141	140	62

[a] A tobacco with "White Burley" character.

[b] The rate of photosynthesis was measured in red light 575 nm < λ < 700 nm of 20,000 ergs/sec[1]/cm[2].

[c] Except for the white leaf, photosynthetic rates were at saturation level at this intensity.

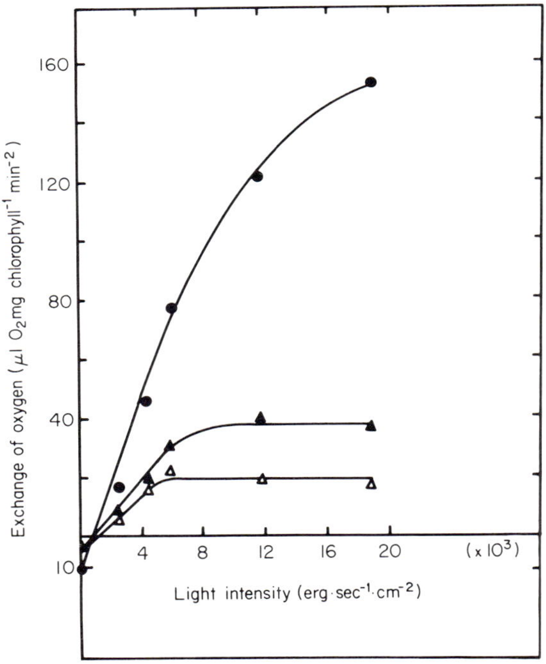

FIG. 3. *Nicotiana tabacum* L. Ky 16. Light intensity curve for oxygen gas exchange. Red light. Rates on the basis of chlorophyll: ●, bottom leaf; ▲, middle leaf; △, upper leaf.

almost white, whereas the top leaves are dark green. Between these extremes there is a continuity of transient colored leaves. The metabolic trends found in these leaves are given in Table V. It shows an increase of chlorophyll content per unit of leaf area from the bottom to the top leaves. Because of different leaf sizes and different chlorophyll content, the maximal rate of photosynthesis per leaf decreases from bottom to the top. The saturation rates of photosynthesis calculated on the basis of chlorophyll decrease by roughly one order of magnitude whereas the saturation rate per unit leaf area remains fairly constant. The light intensity curve (Fig. 3) for red light shows that the least pigmented bottom leaf required the highest intensity for saturation of photosynthesis. At low light intensity the chlorophyll-deficient leaf performed very poor photosynthesis. As discussed in the last paragraph of this chapter, in certain stages of development the light-harvesting chlorophyll is reduced in these leaves, thus proportionally increasing the amount of reaction center chlorophyll.

Respiration

Dark Respiration. The O_2 consumption and CO_2 evolution in dark respiration is the reversal of photosynthesis. Therefore, all measurements of apparent rates of photosynthesis must be corrected for respiration in order to obtain the net rates of photosynthesis. Respiration in the tobacco mutants is generally very high but at the same time extremely variable. Despite standardized growing conditions the rate of dark respiration varies from experiment to experiment. Leaves from Ky 16 (Table V) are an example of this unsatisfactory experimental condition.

The Effect of Light on Respiration. Recently the glycolate metabolism in green plants has been examined by several investigators.[10,11] In green plants glycolic acid, an early product of photosynthesis induces the formation of the enzyme glycolate oxidase. Plants with a high content of this enzyme show a high photorespiration (enhanced O_2 consumption and CO_2 evolution in the light. This so-called photorespiration is apparently one of the factors, as Zelitch pointed out, which limits net photosynthesis. It has been shown that the *Su/su* mutant has a higher rate of photorespiration than the green sibling. This result explains why the *Su/su* mutant grown in air (0.03% CO_2) at high light intensities cannot fully compete in growth vigor with the green sibling, the latter having a 40% greater leaf mass.[9] At low light intensities, the *Su/su* mutant was not viable. Indeed, the *Su/su* mutant contains the

[10] I. Zelitch and P. R. Day, *Plant Physiol.* **43,** 1883 (1968).

[11] N. E. Tolbert and M. S. Cohan, *J. Biol. Chem.* **204,** 639 (1953).

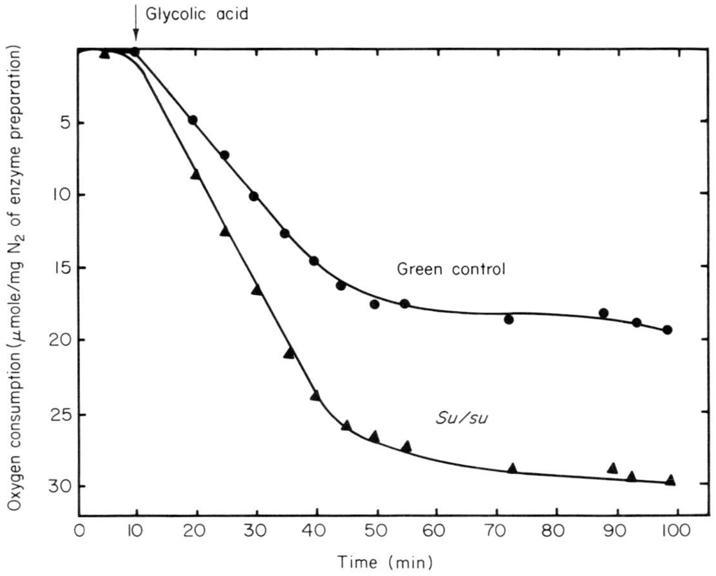

FIG. 4. Glycolate oxidase activity (μmoles oxygen uptake per milligram of N) in *Nicotiana tabacum* L. at 30°. Green control: John Williams Broadleaf; *Su/su:* aurea mutant.

enzyme glycolate oxidase in higher concentrations than the green sibling,[12] which confirms the Zelitch experiments (Fig. 4).

Quantum Requirements for Photosynthesis and Emerson Enhancement

The quantum requirements are quite independent of any conspicuous difference between dark-green and yellow plants. The optimal quantum requirement for the evolution of one O_2 (or fixation of one $^{14}CO_2$) remains the same as in all reliable measurements reported in the literature for all kinds of plants. As seen in Table VI, all plants tested required 9–12 quanta of red light for the evolution of one O_2. The fact that the tobacco mutant *Su/su* has a quantum requirement of 13 is easily explained (see below).

In blue light the quantum requirement for the evolution of one O_2 is between 18 and 20. This is due to the loss of light energy absorbed by inactive pigments (Table I).

Principle. The quantum requirement is determined by comparing the absorbed light quanta within the linear range of the light intensity curve

[12] G. H. Schmid, *Hoppe Seyler's Z. Physiol. Chem.* **350**, 1035 (1969).

TABLE VI
CHLOROPHYLL CONTENT AND QUANTUM NUMBER FOR OXYGEN EVOLUTION IN RED AND FAR-RED LIGHT AT 25°

Plant species	Number of experiments	Chlorophyll averages ($\mu g/cm^2$)	Quanta/O_2 evolved
Cassia obtusifolia	2	57	11
Lespedeza sp.	7	45	8
Nicotiana tabacum var. Burley	5	35	10
N. tabacum JWB green	3	20	9
N. tabacum X-ray 402	2	18.6	10
N. tabacum Rg.Y.Gr.	9	8	12
N. tabacum *Su/su* Y.Gr.	16	7.6	13
N. tabacum Burley 21	10	4.6	10
N. tabacum Jap. br. yellow	3	2.1	11
Scenedesmus D_3[a]	16	—	11

[a] For *Scenedesmus* the light absorption by 15, 30, and 45 μl of cells was measured.

of photosynthesis, with the amount of oxygen evolved (or CO_2 fixed). The amount of oxygen evolved (apparent rate) has to be corrected for respiration. This is done by adding the mean rate of respiration observed in the dark before and after illumination to the measured apparent rate of photosynthesis. What is obtained is "net" photosynthesis. Since there is no reasonable way of eliminating the source of error induced by the so-called photorespiration, all values for quantum requirement reported in the literature are not entirely correct, provided the investigated plants really exhibit photorespiration under the respective experimental conditions. In the tobacco mutant *Su/su*, a plant with very high photorespiration,[10] it is therefore not surprising that the quantum requirement for oxygen evolution is found to be 13. The question whether 8 or 12 quanta are really required is under these circumstances irrelevant.

Light Absorption Measurements. A large surface bolometer (H. Röhrig, Berlin, Germany) connected with a null detector, in a set up of the kind shown in Fig. 5 is calibrated with the use of a standard carbon filament lamp (E 6410, The Eppley Laboratory Inc., Newport, Rhode Island). Calibration is done at least at three different light intensities or socket voltages of the lamp. The light absorption of a leaf section is measured under the exact experimental light conditions of the measurement of O_2 evolution. The light absorption itself is measured in an Ulbricht sphere which contains in parallel-connected silicon photocells. The photocurrent is measured with the use of a sensitive ammeter. These ammeter readings are calibrated against the bolometer readings (R_2 and R_3 in Fig. 5). The difference of the ammeter reading without leaf

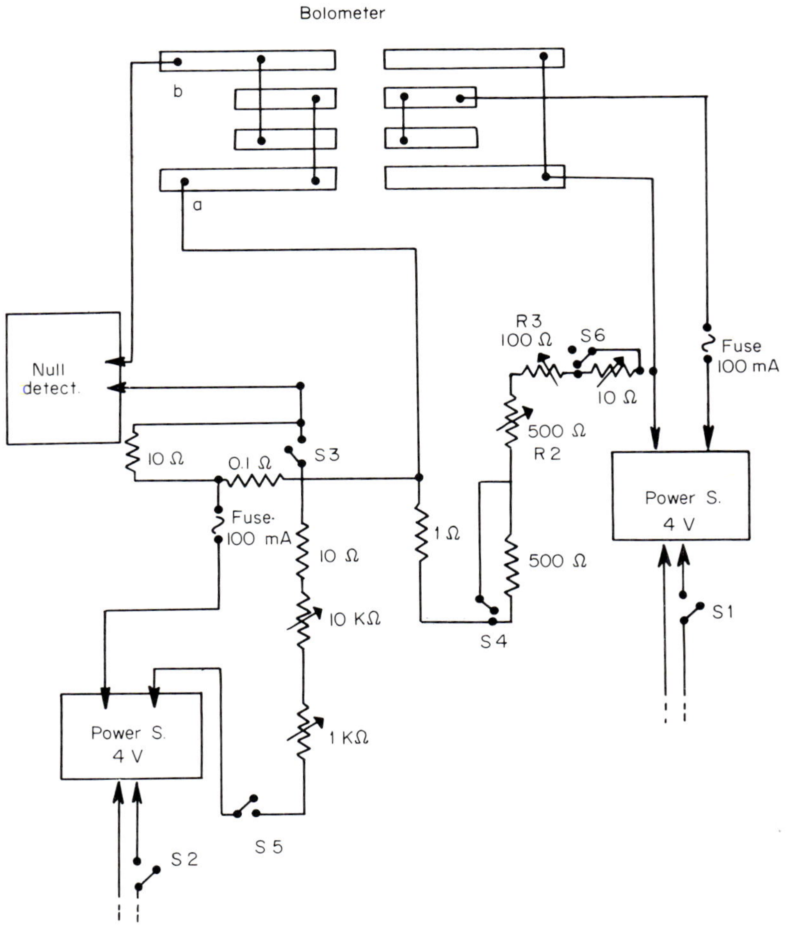

FIG. 5. Diagram of a bolometer set up. The upper part of the diagram shows the connection of the bolometer legs.

section and with leaf section is converted to the value of the absorbed light energy. Errors are induced by light scattering and measurements in the nonlinear light intensity range of the bolometer characteristic. Measurements at too low or too high light intensities both lead to incorrect quantum requirements. There is no correlation between chlorophyll content and relative light absorption in leaves of yellow tobacco mutants.[13] Light will be reflected more often in a pale leaf than in a strongly absorbing dark-green one. The relative light absorption in a chlorophyll-deficient

[13] G. H. Schmid and H. Gaffron, *J. Gen. Physiol.* **50,** 9, 2131 (1967).

leaf is much higher than is expected from the chlorophyll content. Although the principle of measuring quantum requirements for photosynthesis is rather simple, good measurements depend on a great number of factors. This is especially valid for leaf sections of yellow tobacco mutants. For the success of the method, the following points must be observed:

1. The bolometer or the thermopile must be accurately calibrated and sufficiently sensitive. Here errors may arise from uncertainties in the standard.

2. The light should be monochromatic, or if a wide band filter or a Plexiglass filter is used, the main wavelength for which the energy content per light quantum is calculated, must be carefully determined.[13]

3. The incident light must be parallel.

4. In weakly absorbing leaf samples, light scattering even when an Ulbricht sphere is used may lead to faulty measurements.

5. Light absorption for far-red is so low in the leaf sections that measurements are often in the region of the instrumental limit and depend heavily on points 1 and 4.

6. The light measurements must be done above the compensation point of the photosynthesis–light-intensity curve.

In the paragraph on quantum requirement for photosynthesis in yellow tobacco mutants, it is reassuring to see that as long as all components of the complete system are activated, the thermodynamics of the overall process, not the average concentration of pigments and enzymes, determine the quantum yield.

Emerson Effect. The Emerson effect is caused by the peculiar power of far-red light to enhance rates of photosynthesis in other parts of the spectrum. The effect may mean that the plant world has adapted itself to white light, not to colored light, and makes use of radiation of the active spectrum for more than one purpose. Besides the assimilation of carbon there are other light-induced reactions such as photoperiodism which depend on the red end and phototropism which depends on the blue end of the spectrum. These two may exert an indirect control on the state of the lamellar system. But quite directly the two spectrally not identical chlorophyll systems and their accessory pigments have to share the incoming radiation in a balanced way for the best efficiency. From this point of view, monochromatic light taken from any place in the spectrum can hardly be expected to produce optimal rates of photosynthesis. The Emerson effect means that the most serious unbalanced absorption in monochromatic or colored light can be cured by far-red light, which might be absorbed by a chlorophyll system whose task it is to produce ATP.[13] The Emerson enhancement has been discussed in the literature

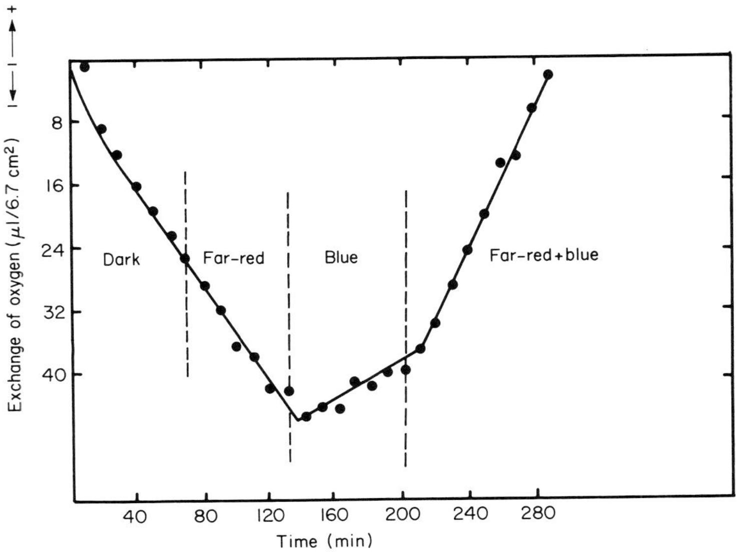

FIG. 6. Time course of Emerson enhancement in *Nicotiana tabacum* Virescent 402. Far-red, 25,300 erg/sec/cm^2; blue, 3080 erg/sec/cm^2. Letaf area 6.7 cm^2; 7.2 μg total chlorophyll/cm^2. Age of tobacco: 4 months after germination.

from the point of view that the low efficiency of photosynthesis in the far-red, the so-called red drop, is avoided or diminished if light of shorter wavelength (e.g., blue light) is added. It appears that this idea is trivial, because far-red light of the suitable wavelength does not promote photosynthesis in the X-ray mutant of tobacco (Fig. 6). But the same far-red light enhances photosynthesis in the blue half of the spectrum. This confirms what has been suggested above.

Correlation of Structure and Function

If the photosynthetic reactions of the mutant chloroplasts are correlated with the lamellar structure of the chloroplasts as seen in the electron micrographs (Fig. 1), it can be established that a fully active photosystem II apparently requires the close packing of at least two thylakoids.[14] A fully active photosystem I, however, is already associated with single unfolded thylakoids. Chloroplasts of types *a* and *b* evolve oxygen and catalyze the ferricyanide Hill reaction. Chloroplasts of type *c* from *N. tabacum* NC95, however, show no photoreactions involving system II but have a perfectly active photosystem I. The rates for

[14] P. H. Homann and G. H. Schmid, *Plant Physiol.* **42,** 1619 (1967).

$NADP^+$ photoreduction with ascorbate–DCPIP and for PMS-mediated photophosphorylation are, on the basis of chlorophyll, as good as those of type *a*. Moreover, chloroplasts of type *c* from yellow patches of variegated NC95 contain chlorophyll that is not associated with system I. This is shown by comparing the quantum yield for ATP formation in photophosphorylation in the far-red with the quantum yield in red light. In chloroplasts from yellow patches of variegated *N. tabacum* NC95, this quantum efficiency in the far-red is four times better than in red light. This structure and function relationship was extended by Levine[15] and confirmed by Wiessner[16] with algae.

Photosynthetic Units in Yellow Tobacco Mutants

The properties of the yellow tobacco mutants just described show that in relation to their maximal photosynthetic capacity most green plants contain a great excess of chlorophyll. At high light intensities most of the light energy absorbed by the chloroplast pigment in a green plant is wasted. The accepted explanation is that there are not enough "enzymes" to make use of the flux of primary photochemical products. This excess of light-capturing capacity over light-utilizing capacity in normal green plants does not mean, however, that only part of the chlorophyll has access to the enzyme system. As Emerson and Arnold pointed out, at low light intensity light quanta captured anywhere inside the chlorophyll aggregates are fully effective.[17] This paradoxical property of the chlorophyll system has led to the theoretical concept of the "photosynthetic unit," in which several hundred chlorophyll molecules can equally well feed the energy of captured quanta into an active site or reduction center.[18] Because all green plants tested have similar maximal rates of photosynthesis on the basis of chlorophyll (Table III), they have approximately the same ratio of effective pigments to "enzyme." The *Su/su* mutant and the consolation mutants (like RG yg 526-1-2y, Dixie Shade 525-10-9y, Florida 15, 528-7-4y, No. 63, 527-3-2y) reach maximal rates of photosynthesis on a chlorophyll basis which are higher than in normal green plants, and thus they permit a better photochemical utilization of high light intensities. At the time of Willstätter and Stoll it was obvious why a higher light intensity was needed to saturate a pigment-rich rather than a pigment-poor photochemical mechanism. From this point of view, the fact that a pigment-poor mechanism requires a five-times-higher light intensity for saturation would be not understandable. Today we

[15] R. P. Levine, *Science* **162,** 768 (1968).
[16] W. Wiessner and F. Amelunxen, *Arch. Mikrobiol.* **66,** 14 (1969).
[17] R. Emerson and W. Arnold, *J. Gen. Physiol.* **16,** 191 (1932).
[18] H. Gaffron and K. Wohl, *Naturwissenschaften* **24,** 81 (1936).

distinguish between the light-harvesting and light-converting pigments, e.g., the photosynthetic unit and its reaction center. Plenty of light-harvesting and little reaction center chlorophyll has been the accepted explanation for low saturation levels in photosynthesis.[18] The high efficiency of the aurea mutants is the first hint that in the yellow mutant chloroplasts there were more reaction centers that have to be kept light saturated than in a green plant. The concept of the photosynthetic unit arose from the classical flashing light experiments of Emerson and Arnold.[17]

Definition of the Unit. The metabolic photosynthetic unit is defined by the number of chlorophyll molecules involved in the fixation of one molecule of carbon dioxide or the evolution of one molecule of oxygen as the consequence of a very strong light flash, which activates all available reaction centers at once. The flash has to be so short as to make a repetition of the ensuing normal photosynthetic sequence of reactions during its lifetime unlikely.

Method for the Determination of the Photosynthetic Unit in Leaf Sections

The principle of these measurements consists of the following steps:

1. The measurement of $^{14}CO_2$ incorporation in a leaf section in dim light and in dim light plus a selected number of saturating flashes.
2. The determination of the chlorophyll content in the same leaf section.
3. Check for flash saturation of $^{14}CO_2$ fixation by cutting down the flash intensity (for instance with a wire screen) and repeating steps 1 and 2.
4. The exact measurement of the radioactive $^{14}CO_2$ and the total CO_2 concentration of the gas phase in which the $^{14}CO_2$ incorporation is carried out.

The detailed method is described elsewhere.[19]

The average photosynthetic unit in the yellow tobacco mutant *Su/su* or in aurea from Japanese Bright Yellow tends to come out smaller than for the green plant (Fig. 7). This means that in these mutants the amount of light-harvesting chlorophyll is reduced. Moreover, the unit size varies in all plants tested between 300 and 5000 (Figs. 7 and 8). All unit sizes which are found in the yellow plant are also present in the green controls. Only the distribution of values is different in green and yellow plants. Hence a small unit is not a characteristic feature of a

[19] G. H. Schmid and H. Gaffron, *J. Gen. Physiol.* **52,** 212 (1968).

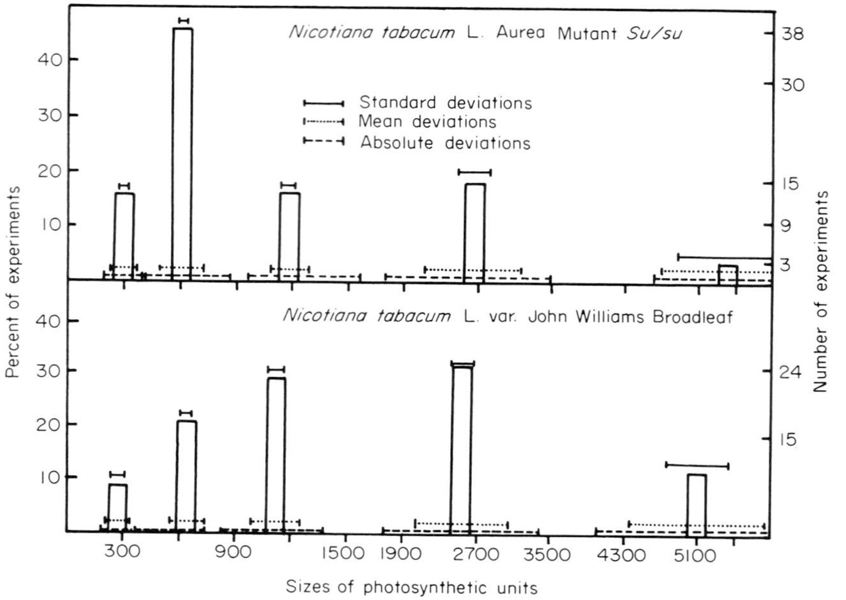

FIG. 7. *Top: Nicotiana tabacum* L. aurea mutant *Su/su*, young plants 6–8 weeks old; 82 flashed samples, 147 controls. Leaf area for the assay 3 cm²; 229 determinations of chlorophyll in 3 cm². Highest total chlorophyll content 13 μg/cm²; average 5.9 μg/cm²; standard deviation $\pm$0.14 μg/cm². *Bottom: N. tabacum* L. var. John Williams Broadleaf, young plants, 6–8 weeks old; 77 flashed samples, 143 controls. Leaf area for the assay, 2.25 cm²; 220 determinations of chlorophyll in 1.5 cm². Highest total chlorophyll content 76.3 μg/cm²; lowest total chlorophyll content 8.6 μg/cm²; average 22.1 μg/cm²; standard deviation $\pm$0.563 μg/cm².

yellow tobacco mutant. In the distribution pattern of unit sizes in young *Su/su* leaves (Fig. 7) the peak is asymmetrically located at 600 chlorophylls/CO_2 fixed/flash whereas the green control gives the Emerson and Arnold value of 2500 in a symmetrical distribution pattern. In Fig. 8 it is seen that the distribution pattern for *N. tabacum* aurea from Japanese Bright Yellow yields a unit peak at 300, *Cassia obtusifolia* at 2400, and Swiss chard at 600. The distribution of unit sizes and the location of the peaks suggest that there are unit groups which can be arranged in the following series of values for n Chl/CO_2: $\approx$300, $\approx$600, $\approx$1200, $\approx$2400 and $\approx$5000. Most interestingly, unit sizes vary much less in algae. The Emerson and Arnold value of 2400 Chl/CO_2 fixed/flash is the value found with the highest probability under the experimental conditions where unit sizes vary between 300 and 5000 in leaves. The variations do not extend significantly farther than 1200 Chl/CO_2 fixed from this peak.

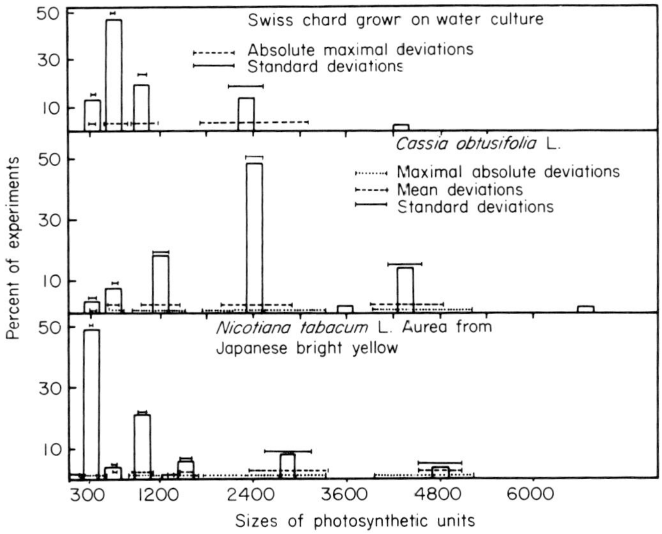

FIG. 8. *Top:* Swiss chard. Growing leaves of water culture plants; 35 flashed samples, 31 controls, leaf area for the assay 2.25 cm^2; 66 determinations of chlorophyll in 1.5 cm^2. Highest value 71.6 $\mu g/cm^2$; lowest value 27.1 $\mu g/cm^2$; average value 43.8 $\mu g/cm^2$; standard deviation ± 1.245 $\mu g/cm^2$. *Middle: Cassia obtusifolia* L., 47 flashed samples, 52 controls, leaf area for the assay 2.25 cm^2; 98 determinations of chlorophyll in 1.5 cm^2. Highest total chlorophyll content 78.4 $\mu g/cm^2$; lowest total chlorophyll content 37.5 $\mu g/cm^2$; average 58 $\mu g/cm^2$; standard deviation ± 1 $\mu g/cm^2$. *Bottom: Nicotiana tabacum* L. aurea from Japanese Bright Yellow; 46 flashed samples, 55 controls, leaf area for the assay 3 cm^2; 101 determinations of chlorophyll in 3 cm^2. Highest total chlorophyll content 8.5 $\mu g/cm^2$; lowest total chlorophyll content 0.7 $\mu g/cm^2$; average value 3.6 $\mu g/cm^2$; standard deviation ± 0.19 $\mu g/cm^2$.

Synchronously grown algae are no different. This means that cells in well-selected tobacco leaves are in a much better synchronous condition than any algal culture. If the term of forbidden unit sizes is applied, forbidden numbers occur with higher probability in algae. Twenty-five individual unit sizes determined for *Anacystis nidulans* are shown in Table VII. Statistical treatment yields the same peaks as in tobacco plants.

The results of Menke,[20] Kreutz,[21] and Hosemann[22] offer a possibility of explaining why units vary in the way shown in Figs. 7 and 8. The main result of this research is that a group of oriented pigments exists besides

[20] W. Menke, *Kolloid Z.* **85**, 256 (1938).
[21] W. Kreutz, Habilitationsschift, Berlin, 1968.
[22] R. Hoseman and W. Kreutz, *Naturwissenschaften* **53**, 298 (1966).

TABLE VII
VARIABILITY OF PHOTOSYNTHETIC UNIT SIZES IN *Anacystis nidulans*[a]

Range of unit sizes	Individual unit sizes	Average unit size	Standard deviation
760–1300	1230, 855, 1211, 1043, 760, 975, 1300, 1300, 1021, 1203	1090	±60
1536–3067	2404, 2604, 2028, 1592, 2500, 2427, 1536, 1550, 3000, 1825, 3011, 3067, 1555, 2667, 2132	2250	±145

[a] Chlorophylls per CO_2 fixed per flash. Note that the mean deviation within one group is symmetrical.

a group of unoriented chlorophyll. From measurements of polarized fluorescence of oriented chloroplasts after irradiation with unpolarized light and measurements of the dichroism it was postulated that chlorophyll must exist in two different physical states.[23] It is concluded that

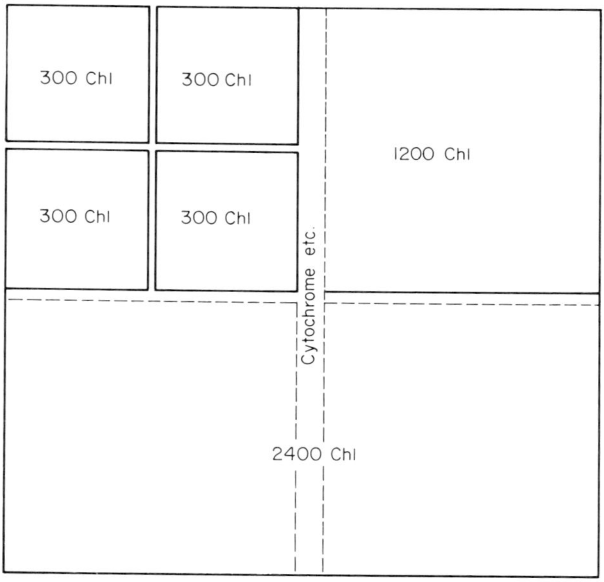

FIG. 9. Model of photosynthetic units according to Kreutz (Habilitationsschrift, Berlin, 1968). This is applied to a smallest unit size of 300 chlorophylls.

[23] R. A. Olson, W. H. Jennings, and W. L. Butler, *Biochim. Biophys. Acta* **54,** 615 (1961); **58,** 144 (1962); **88,** 331 (1964).

FIG. 10. Model of photosynthetic units. Each cube represents 300 chlorophylls with the white dot as reaction center of photosystem II. Expansion of the lattice cell distance between the 300 chlorophyll units makes new centers.

the absorbed light is transferred from the unoriented (*c* 673, *c* 683, *c* 705) to the oriented chlorophyll (*c* 695).

In the very elaborate Kreutz model a crystalline quadratic lattice consists of 16 mass centers, the dimensions of the crystallite being 165 × 165 × 38 Å. These crystallites represent the dominant part of the protein component of the formerly so-called quantosomes. To these crystallites, a pigment and a lipid layer is attached. Each crystallite consists of four subparticles (42 Å × 42 Å × 38 Å), each containing 16 chlorophyll dimers, that is, 32 chlorophylls. The whole crystallite contains 128 chlorophylls. Each subparticle of the structural protein contains a hydrophobic hole, the shape and dimension of which fit the space requirement of porphyrin rings. Another 32 chlorophylls can be packed into these holes per crystallite according to Kreutz. The crystallite thus contains maximally 160 chlorophylls if all the holes are occupied, a case which seems to be realized in *Euglena*. The number of 160 chlorophylls is very close to one-half of the smallest unit found in tobacco. In general, however, the crystallite contains only 139 chlorophylls because not all "holes" are occupied. In his model the *c* 695, that is, the dichroic chlorophyll, is located within the protein subparticle. In the Kreutz model, the chlorophylls, whatever the number assumed to be the photosynthetic unit, are organized in an overall quadratic or plain lattice, which, since the lattice energy must be at a minimum and the 4-fold symmetry requirement must be respected, yields multiple units which differ by the factors of 2, 4, 8, 16, etc. (Fig. 9). The functioning of such a unit would be limited by this overall structure (Fig. 10). Any change—for example, the expansion of the overall crystal lattice dimensions by far-red—will affect the functioning of the unit.

Section III

Cellular and Subcellular Preparations

A. CELLULAR

Articles 16 and 17

B. SUBCELLULAR

1. Chloroplasts and grana

   Articles 18 through 24

2. Chromatophores

   Article 25

3. Subchloroplast fragments

   Articles 26 through 30

4. Subchromatophore fragments

   Articles 31 and 32

[16] Protoplasts of Plant Cells

By ALBERT W. RUESINK

Protoplasts of plant cells consist of ordinary plant cells devoid of their largely polysaccharide cell wall. Although protoplasts may arise naturally in some cases (certain egg cells, floating cells in some fruit juices), those of most importance to the physiologist or biochemist are produced artificially by removal of the cell wall. Several reviews on microbial protoplasts have appeared,[1-3] including one dealing specifically with techniques.[4] There exist two major reviews of the plant protoplast literature,[5,6] but neither emphasizes techniques, the subject of this review. Plant protoplasts have been used to study membrane permeability,[7,8] hormone action,[9-11] protein uptake,[12,13] and the nature of the plasma membrane surface.[14] Further, they were used to show that the site of action of the toxin of *Helminthosporium victoriae* is the plasma membrane of susceptible host cells.[15] Research areas for which the use of protoplasts shows great promise include cell wall synthesis, somatic hybridization, and organelle isolation.

As shown in Fig. 1, protoplasts assume a spherical shape upon isolation from a restraining cell wall into an appropriate osmoticum. The membrane surface is very soft and may be distorted markedly by currents within the solution or by micromanipulation.[16] The types of protoplasts that can be seen in the phase contrast microscope, representing

[1] C. Weibull, *Ann. Rev. Microbiol.* **12,** 1 (1958).
[2] K. McQuillen, *in* "The Bacteria" (I. C. Gunsalus and R. V. Stanier, eds.), Vol. I, p. 249 Academic Press, New York, 1960.
[3] "Microbial Protoplasts, Spheroplasts, and L-Forms" (L. B. Guze, ed.). Williams and Wilkins Co., Baltimore, Maryland, 1968.
[4] J. Spizizen, this series, Vol. V, p. 122.
[5] E. C. Cocking, *in* "Viewpoints in Biology" (J. D. Carthy and C. L. Duddington, eds.), Vol. 4, p. 170. Butterworth, London, 1965.
[6] E. C. Cocking, *Int. Rev. Cytol.* **28,** 89 (1970).
[7] J. Levitt, G. W. Scarth, and R. D. Gibbs, *Protoplasma* **26,** 237 (1936).
[8] D. Vreugdenhil, *Acta Bot. Neer.* **6,** 472 (1957).
[9] E. C. Cocking, *Nature* **193,** 998 (1962).
[10] D. W. Gregory and E. C. Cocking, *J. Exptl. Bot.* **17,** 68 (1966).
[11] A. W. Ruesink and K. V. Thimann, *Science* **154,** 280 (1966).
[12] E. C. Cocking, *Biochem. J.* **95,** 28P (1965).
[13] E. C. Cocking, *Planta* **68,** 206 (1966).
[14] A. W. Ruesink and K. V. Thimann, *Proc. Nat. Acad. Sci.* **54,** 56 (1965).
[15] K. R. Samaddar and R. P. Scheffer, *Plant Physiol.* **43,** 21 (1968).
[16] J. Q. Howe, *Protoplasma* **12,** 196, 221 (1931).

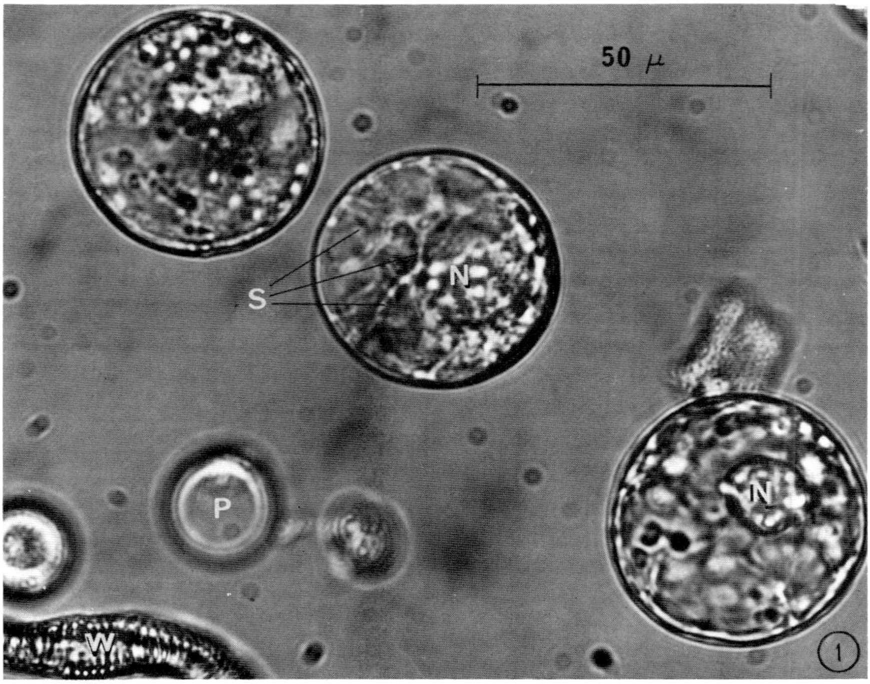

FIG. 1. Photoplasts released from Avena coleoptile cells. N, nucleus; S, transvacuolar strands; P, partial or subprotoplast; W, remaining secondary wall of a xylem cell.

various degrees of viability, are illustrated in Fig. 2. Arrows designate typical directions of cyclosis. Type B is often seen immediately after pipetting the protoplasts; most of the cytoplasm is at one side of the vacuole, and waves of cytoplasmic streaming move around the vacuole. In a few minutes, type B will usually convert to type A with strands across the vacuole. Cells from some tissues will eventually convert into type G, apparently the most viable of all, with the nucleus and plastids suspended by transvacuolar strands in a small amount of cytoplasm in the center of the vacuole. A striking amount of cyclosis occurs in the strands. Type E represents an occasionally seen protoplast in which there are no transvacuolar strands but in which cyclosis is very vigorous, sometimes occurring as concentric rings of movement beneath the vacuole as depicted here.

The other types represent distinctly unhealthy conditions—danger signals for the investigator working with protoplasts. Type C occurs on occasions when the protoplasts are shrunken by a high external osmotic

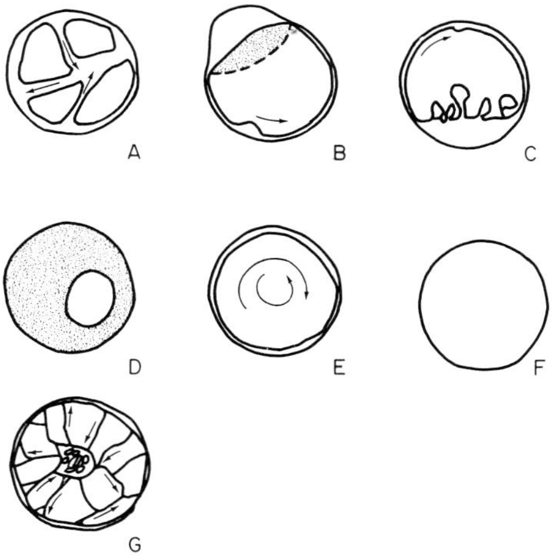

FIG. 2. Types of vacuolated photoplasts. Arrows indicate directions of cytoplasmic streaming. Complete description in text.

concentration. Vesicles which appear to be empty bleb off from the cytoplasm and wave around in the vacuole. Cyclosis is often maintained. Bolhar-Nordenkampf has reported similar structures in intact onion cells exposed to water stress.[17] Type D illustrates the protoplast equivalent of cap or tonoplast plasmolysis, in which the vacuole is shrunken and the cytoplasm is swollen. The only cytoplasmic motion is Brownian motion, and recovery of a viable condition has never been observed. Type F represents the naked vacuole that is left when the plasma membrane bursts and the cytoplasm is lost. Such structures are easily recognized by their lack of phase contrast and by the occasional remnants of cytoplasm stuck on their exterior surface. They never are stable for more than a few hours.

Although several investigators through the last half of the 19th century saw a few higher plant protoplasts removed from their wall, the first record of numerous protoplasts comes from Klercker.[18] He tore up plasmolyzed leaf tissue of *Stratiotes aloides* under a microscope, extruding protoplasts. The method was improved by subsequent workers,[7,19,20] and has been used to prepare protoplasts from onion, apparently the most suitable tissue, as recently as 1957.[8]

[17] H. Bolhar-Nordenkampf, *Protoplasma* **61,** 85 (1966).
[18] J. von Klercker, *Öfvers. Kungl. Vetensk. Förh, Stockh.* **49,** 463 (1892).
[19] E. Küster, *Wilhelm Roux Arch. Entwicklungsmech. Organismen* **30,** 351 (1910).
[20] W. Seifriz, *Protoplasma* **3,** 191 (1927).

In recent years, enzymes have been used to remove the wall; the first success was reported by Cocking using an enzyme mixture containing predominantly cellulase.[21] Later he was the first to demonstrate that for some tissues, enzyme mixtures containing predominantly pectinase work best.[22] The three major advantages of an enzymatic over a cutting method are: (1) Larger numbers of protoplasts may be obtained; (2) less osmotic shrinking of the cytoplasm is required; and (3) no cells are broken by the cutting tool. On the other hand, not all tissues release protoplasts under the enzyme treatments thus far tried,[11] and many of the enzyme solutions derived mainly from fungal sources have contained substances deleterious to protoplast viability.[11,23]

Cell Wall Removal Techniques

Cutting the Wall of Plasmolyzed Cells: This is the simplest method of preparing protoplasts and will work with any tissue whose cells can be plasmolyzed and whose walls can be cut cleanly. As with any protoplasts, it is necessary to have a nonmetabolizable, nonpenetrable osmoticum present. Here the initial concentration must be high enough to cause considerable plasmolysis, usually about 1 *M*. Walls are cut with an instrument causing low shear; a honed razor blade is a good choice.[20] Sometimes the protoplasts are extruded relatively spontaneously when the cells are cut; in contrast, in other cases it is necessary to tease them out by reducing the osmotic concentration of the surrounding medium to about 0.5 *M* and/or by pulling on the protoplasts with a glass needle to which they stick. A similar technique, but with somewhat more chopping, has been used to isolate small vacuoles from root meristems.[24]

Enzymatic Digestion: Two distinct types of enzyme preparations have been used to prepare protoplasts—cellulases and pectinases. It has never been shown explicitly that only one enzyme is present in the preparation, and the possibility that more than one enzyme is essential to wall removal must certainly remain open in every case. Many polysaccharidase preparations are contaminated with "hemicellulase" activity, and its contribution to the wall-removing ability is completely unknown. Usually, it appears that the type of enzyme mentioned in any given case is primarily responsible for the effect observed.

Cellulase was the first enzyme to be used to remove cell walls,[21] and it has continued to be used for a number of tissues (e.g., see Ruesink and Thimann[11]). The most successful source of the enzyme has been the

[21] E. C. Cocking, *Nature* **187**, 962 (1960).
[22] D. W. Gregory and E. C. Cocking, *Biochem. J.* **88**, 40P (1963).
[23] R. U. Schenk and A. C. Hildebrandt, *Crop. Sci.* **9**, 629 (1969).
[24] P. Matile, *Z. Naturforsch. B* **21**, 871 (1966).

fungus *Myrothecium verrucaria*,[11,14] which is not used by any firm producing commercially available cellulase. However, treating the commercial enzymes, Takamine 20,000 (Miles Laboratories), Cellulase 36 Concentrate (Rohm and Haas), and Cellulase Onozuka 1500 (All-Japan Biochemical), with acid to remove presumptive peroxidases yielded enzyme mixtures capable of releasing viable protoplasts from a number of different tissue cultures.[23] On a dry weight basis, the *Myrothecium* enzyme preparation contains from 3–20 times the cellulase activity of any of the other three enzyme preparations just mentioned. The difference in activity is greater with powdered cellulose as the substrate than with the artificial substrate carboxymethyl cellulose.

Myrothecium and other microbes excrete cellulase into the culture medium as an adaptive enzyme with cellulose as the best carbon source.[25] The growth medium for *Myrothecium* is given in Table I. Inoculation is accomplished by growing a 7-day culture of *Myrothecium verrucaria* QM 460 (Quartermaster Laboratories, Natick, Mass.) on a slant of potato-dextrose agar.[26] A suspension of spores and pieces of mycelium is formed by adding sterile distilled water and stirring with a glass rod; the suspension is poured into the liter of medium in a 4-liter Fernbach culture flask. The culture is grown for 14 days at 22° with continuous agitation. The solid material is removed by filtering through glass fiber paper (Whatman GF/A). Since the contents of the filtrate that are essential for removing walls are heat stable, the filtrate can be reduced 10-fold in volume by flash evaporation at 40°. The fraction precipitating between 30 to 70% saturation with $(NH_4)_2SO_4$ is redissolved in a very small volume of water and desalted on Biogel P-4 or Sephadex G-25. The active

TABLE I
Myrothecium verrucaria GROWTH MEDIUM[a]

$NaNO_3$	3.8 g	H_3BO_3	0.06 mg
NH_4NO_3	0.6 g	$CuCl_2$	0.11 mg
$MgSO_4 \cdot 7H_2O$	0.3 g	$CaCl_2 \cdot 4H_2O$	0.75 mg
KH_2PO_4	0.4 g	$FeSO_4 \cdot 7H_2O$	0.05 mg
Glucose	1.0 g	$MnCl_2 \cdot 4H_2O$	0.08 mg
Cellulose powder[b]	10.0 g	$ZnSO_4 \cdot 7H_2O$	2.00 mg
		$Co(NO_3)_2 \cdot 6H_2O$	0.49 mg
		$(NH_4)_6Mo_7O_{24} \cdot 4H_2O$	0.04 mg

[a] The solution is adjusted to pH 6.3 with NaOH and made up to 1 liter.
[b] Whatman Cellulose Powder, standard grade, is satisfactory, but Munktell's Cellulose Powder No. 410 (A. H. Thomas) produces a better cellulase preparation.

[25] M. Mandels and E. T. Reese, *J. Bacteriol.* **79,** 819 (1960).
[26] H. E. Gruen, *Plant Physiol.* **38,** 652 (1963).

enzymes are eluted with brown pigments ahead of the salt, the color serving as a convenient marker for the location of the enzyme.[11] Such an enzyme solution is ready for use, or it may be kept frozen for several months. One can lyophilize the material with little loss of activity. To release protoplasts, the plant tissue is exposed to a 3–8% (weight/volume) solution of the cellulase preparation for 1 to 2 hours. For detailed information on cellulase see Gascoigne and Gascoigne,[27] Norkrans,[28] or Reese.[29]

Cellulase releases protoplasts from *Avena* coleoptiles and primary leaves, some root tissues, and many but not all tissue cultures.[11,23,30] Also suitable are tomato cotyledons[31] and tomato fruits.[32] Mature green tissues are often not susceptible to cellulase digestion,[11] although walls of isolated tobacco leaf mesophyll cells can be removed by cellulase[33] and a snail gut preparation has released protoplasts from several kinds of leaves.[34] Although cellulose is a major component of cell walls, the other components may be untouched by the cellulase preparation and will preserve wall integrity in some cases. Using a purified cellulase, a correlation has been noted between the ease of protoplast release and the amount of glucose present in the wall.[30] Two other factors that may limit the removal of cell walls by cellulase, neither of which has been studied in relation to protoplast formation, are (1) natural inhibitors of cellulase that are present in some plant tissues,[35] and (2) end products of the cellulase reaction, especially cellobiose, that inhibit cellulase activity.[36]

Pectinase enzyme preparations are suitable for releasing protoplasts from placental tissues of solanaceous fruits.[22,37-39] Pectinol R-10 (Rohm and Haas) is the preparation usually used, often at concentrations up to 20% without further purification. Little information is available on the

[27] J. A. Gascoigne and M. M. Gascoigne, "Biological Degradation of Cellulose." Butterworth, London, 1960.

[28] B. Norkrans, *Ann. Rev. Phytopathol.* **1**, 325 (1963).

[29] "Advances in Enzymic Hydrolysis of Cellulose and Related Materials" (E. T. Reese, ed.). Pergamon Press, New York, 1963.

[30] W. A. Keller, B. Harvey, O. L. Gamborg, R. A. Miller, and D. E. Eveleigh, *Nature* **226**, 280 (1970).

[31] E. C. Cocking, *Biochem. J.* **82**, 12P (1961).

[32] E. C. Cocking and D. W. Gregory, *J. Exptl. Bot.* **14**, 504 (1963).

[33] I. Takebe, Y. Otsuki, and A. Aoki, *Plant Cell Physiol.* **9**, 115 (1968).

[34] P. G. Pinto da Silva, *Naturwissenschaften* **56**, 41 (1969).

[35] M. Mandels and E. T. Reese, *Ann. Rev. Phytopathol.* **3**, 85 (1965).

[36] E. T. Reese, *Appl. Microbiol.* **4**, 39 (1956).

[37] D. W. Gregory and E. C. Cocking, *J. Cell Biol.* **24**, 143 (1965).

[38] E. Pojnar, J. H. M. Willison, and E. C. Cocking, *Protoplasma* **64**, 460 (1967).

[39] B. Raj and J. M. Herr, *Protoplasma* **69**, 291 (1970).

ability of this preparation to release protoplasts from other tissues. When treated with pectinase, some tissues release single cells with the walls relatively intact.[33,40–42]

The method yielding the largest number of protoplasts was developed by Gregory and Cocking.[37] With pectinase, the walls of fruit placental tissue were weakened, but not completely dissolved. A device much like a wool carder teased out protoplasts from the larger, thinner-walled cells. To separate tissue fragments from single protoplasts, a stainless steel sieve was used. Protoplasts were floated free from the debris on a sucrose gradient.

Mixtures of enzyme preparations seem to be useful in some situations. In one case,[33] pectinase was used to isolate single cells, and subsequently cellulase would remove their walls and yield protoplasts. This technique has been improved, by using a mixture of a *Trichoderma* cellulase and a pectinase to procure protoplasts from leaf cells.[43] Fully expanded tobacco leaves were stripped of their lower epidermis and 2 g of pieces of the leaf were floated on 20 ml of a solution of 0.4% Macerozyme pectinase and 4% Onozuka cellulase (both from All-Japan Biochemical) with 20% (w/v) sucrose at pH 5.5. Following an initial vacuum filtration of 30 minutes, the enzyme solution was replaced by an identical 20 ml of enzyme mixture. Incubation lasted 4 hours at room temperature without shaking. Digestion was terminated by pouring the suspension through coarse wire gauze to remove remaining chunks of tissue. Subsequent flotation of the protoplasts through fresh 20% sucrose freed them of the enzymes. Mixtures of enzyme preparations are not, however, effective in all situations; pectinases inhibit the release of protoplasts from *Avena* coleoptiles by cellulase.[11]

Although snail gut juice from Industrie Biologique Francaise has been found to be quite toxic to higher plant cells and slow in digesting their walls, one report indicates that the snail gut preparation Glusulase (Endo Laboratories, Garden City, N. J.) will release protoplasts from a number of leaf and root tissues.[34]

Intact, isolated vacuoles have been released from tomato fruit protoplasts by reducing the sucrose concentration of the medium to 10%. Such vacuoles were considerably less stable with time than were the protoplasts.[10]

[40] J. Chayen, *Nature* **170,** 1070 (1952).
[41] M. Zaitlin and D. Coltrin, *Plant Physiol.* **39,** 91 (1964).
[42] W. H. Jyung, S. H. Wittwer, and M. J. Bukovac, *Plant Physiol.* **40,** 410 (1965); *Nature* **205,** 921 (1965).
[43] J. B. Power and E. C. Cocking, *J. Exptl. Bot.* **21,** 64 (1970).

Osmotic Stabilization of Protoplasts

Protoplasts are exceedingly sensitive to their osmotic environment. Ideally, the osmoticum maintaining the water balance of the protoplasts should be inert, nontoxic, nonviscous, and nonpermeable, with a density causing protoplasts either to sink or to float as desired by the investigator. The osmoticum must not interfere with assays using protoplasts or protoplast derivatives. Three types of osmotica are commonly used: (1) *mannitol*—not metabolized, protoplasts sink in it[11,14]; (2) *sucrose*—metabolized in some cases, protoplasts float in it, may inhibit cellulase activity[21,37]; and (3) *mixtures of salts*, usually $CaCl_2$ and either KCl or NaCl—neither taken up nor metabolized, protoplasts sink in it, produces a highly charged external environment.[7,8,11] Mannitol or sucrose with some ionic addition has been used occasionally.[10,11] A divalent cation improves considerably the stability of protoplasts. That the osmoticum does affect the plasma membrane is evidenced by the different residue left when a protoplast bursts in various osmotica. In mannitol, the membrane and cytoplasm disperse into tiny fragments barely visible in the light microscope. In sucrose, stringy aggregates of membrane and cytoplasm remain. In an ionic osmoticum, the plasma membrane remains coherent, yielding a ghost with much adherent cytoplasm.

In one case, sorbitol with a small amount of sucrose provided osmotic stabilization.[44] This provided the only report to date of nuclear division occurring in protoplasts. Nuclear division was described only on the basis of light microscopy of living protoplasts, and a more rigorous proof of division is needed.

Since naked protoplasts function somewhat as miniature osmometers,[7,8] the osmotic concentration of the surrounding medium is obviously very important. Concentrations varying from 0.3 to 5.0 osmolal have been used, with values of 0.5 to 0.6 used most routinely.[5,11,16,45] Little information is available on the effect of osmotic strength of the surrounding medium on the metabolic competence of the protoplasts.

When an object of any shape is converted into a sphere without a volume change, the surface area is decreased. Therefore it should be possible to convert any cell into a protoplast without changing the vacuolar or cytoplasmic volume and without rupturing the plasma membrane. What happens to the excess membrane is unknown, although observations of numerous small vesicles within the cytoplasm (Fig. 3) suggest that membrane components from the surface are found within the cell. Whether these are formed by pinocytosis or a reaggregation of

[44] T. Eriksson and K. Jonasson, *Planta* **89,** 85 (1969).

[45] R. U. Schenk and A. C. Hildebrandt, *Phyton* **26,** 155 (1970).

membrane components that have left the protoplast surface as individual molecules is unknown, although the former is suggested by the observation that tobacco mosaic virus is indeed taken up by pinocytosis of the protoplast surface.[13]

pH Effects on Protoplasts

The most extensive work on pH effects showed that initial protoplast release from tissue cultures with cellulase was optimum at pH 5.0 or below, while subsequent survival was best at pH 5.5–6.0.[45] The survival optimum was corroborated by other investigators, who found that release was enhanced at pH's below 5.8.[44] Some workers have achieved excellent release with cellulase using no buffer, so long as the pH of the incubation solutions was about 5.5.[11,43]

Techniques for Handling Protoplasts

The handling of protoplasts for research purposes involves special problems. Protoplasts are fragile; even such slight agitation as a few quick shakes of a tube containing protoplasts with large vacuoles is sufficient to rupture most of them. These protoplasts are manyfold larger than the relatively stable protoplasts of bacteria and yeasts, and the plasma membrane will not withstand physical shock. Divalent cations such as magnesium[10] and calcium[11] improve membrane stability to some extent, but biochemical studies are still difficult to perform because of the loss of protoplasts.

During the preparation of protoplasts, very gentle motion at 15-minute intervals will often hasten the release of protoplasts, apparently by enhancing the penetration of the enzyme and/or popping protoplasts free from weakened walls. Removing the enzyme preparation from the protoplasts can be accomplished by washing gently several times in the osmoticum, either by centrifugation (see below) or by allowing the protoplasts to settle to the bottom of a conical centrifuge tube between washes. A comparison of the number of protoplasts left in a washed preparation after each of several washes compared to the number left in an unwashed preparation is given in Fig. 4. Although about 90% of the enzyme is removed at each wash, almost 50% of the protoplasts are lost. Even this relatively gentle operation lyses many protoplasts. An alternative method[37] involves preparing the protoplasts in 20% sucrose and then overlayering 10% sucrose, through which the protoplasts float to the surface. Such a flotation method of removing enzyme will work only with sucrose or an equally dense osmoticum.

Pouring a solution containing protoplasts must be done carefully and with a short falling distance. Gentle pipetting can usually be executed

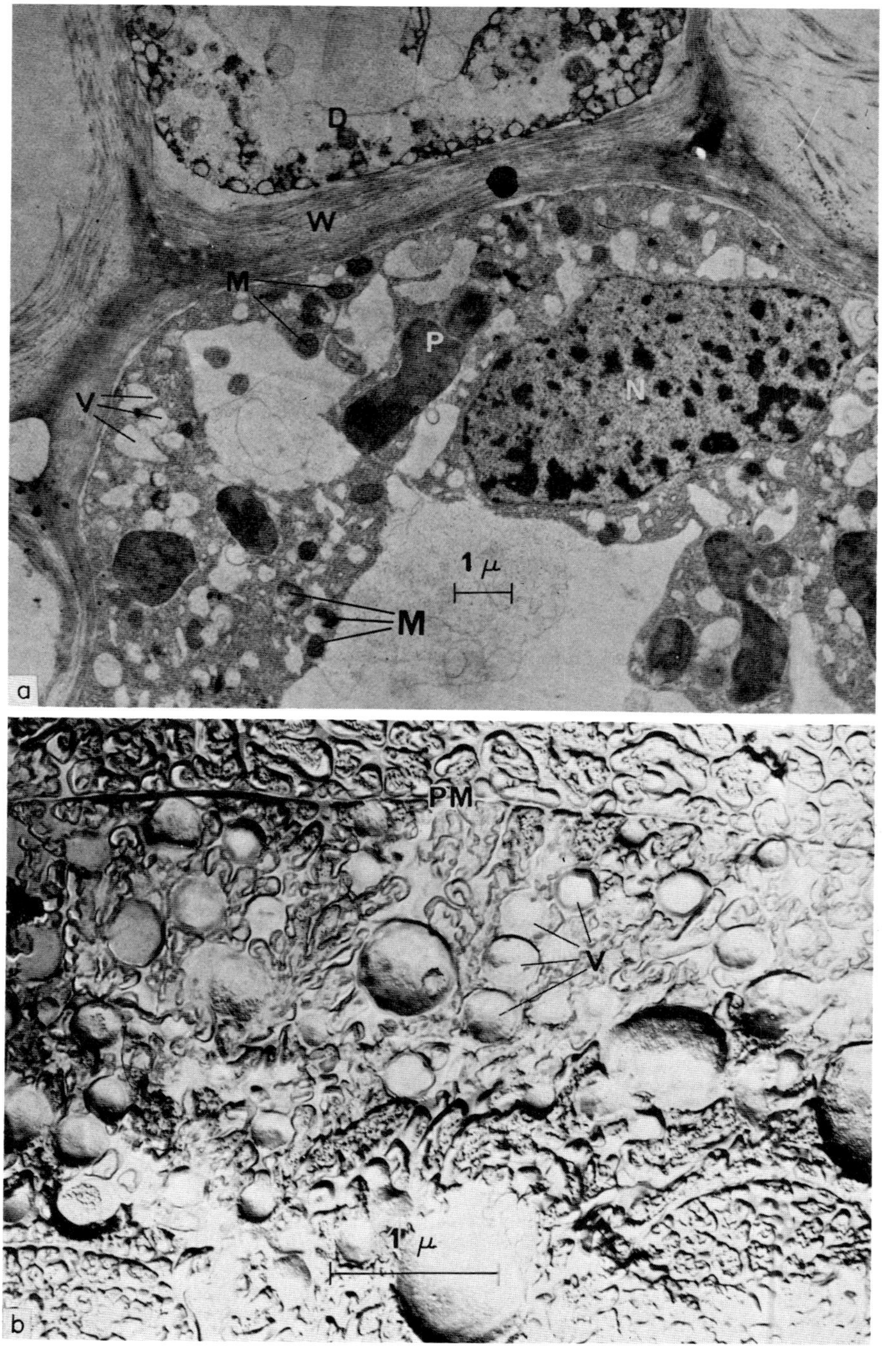
D
W
M
P
N
V
1 μ
M
a
PM
V
1 μ
b

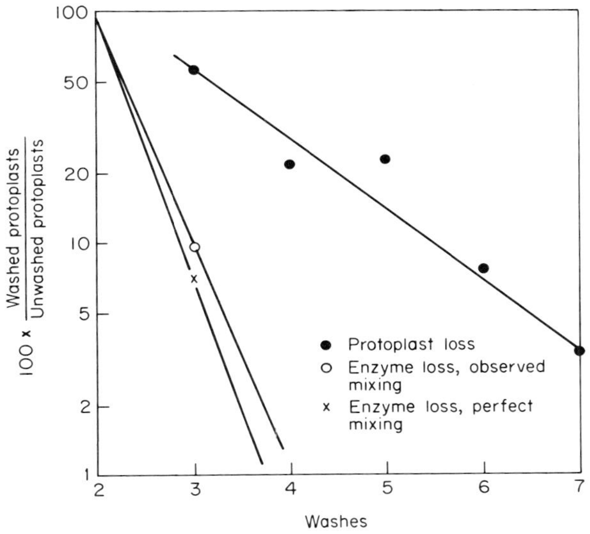

FIG. 4. Loss of protoplasts during washing. Protoplasts were prepared in 12-ml, conical centrifuge tubes containing 100 μl of 5% cellulase and 0.5 *M* mannitol. Washes consisted of adding 1.3 ml of 0.5 *M* mannitol, stirring gently, and letting the protoplasts settle before pipetting off the supernatant solution to 100 μl. The point "enzyme loss, observed mixing" was determined by treating a tube of dye in exactly the same manner and measuring the decrease in absorbance. Note that survival of the protoplasts would have been higher had a divalent cation been present during this experiment.

with a minimum of breakage. For small numbers of protoplasts, Carlsberg or Lang–Levy micropipets are useful.

The centrifugation of higher plant protoplasts to concentrate them can be done only moderately successfully. Forces large enough to form a genuine pellet succeed only in collapsing all protoplasts into an homogenate. In an ionic osmoticum, it is possible, however, to spin down protoplasts to the bottom of a tube at forces up to 500 *g* without unusually high breakage. Higher forces may be used in density gradients where the protoplasts are not forced against a hard glass surface, with less danger of physical lysis.

No simple method exists for determining the number of protoplasts in a given solution. Because protoplasts are fragile and settle rapidly

---

FIG. 3. Numerous vesicles within the cytoplasm of *Avena coleoptile* protoplasts. (a) Electron micrograph of protoplasts still within remaining cell wall: V, vesicles in cytoplasm; W, cell wall loosened and swollen by the cellulase preparation; D, dead cytoplasm of a cell disorganized by the enzyme treatment; P, plastid; N, nucleus; M, mitochondrion. (b) Freeze-etch electron micrograph of a portion of an isolated protoplast: V, vesicles in cytoplasm; PM, plasma membrane.

out of suspension, it is difficult even to get a representative sample into a pipet. Haemocytometers are not suitable instruments for counting numbers of protoplasts because they settle quickly, preventing an accurate distribution of protoplasts on the grid when a coverslip is applied. Two methods have been used to count protoplasts. Firstly, protoplasts on a microscope slide can be photographed, and a count can be made of the photograph at leisure.[9] Unfortunately, small currents often move some of the protoplasts around on the slide, and it is difficult to be sure that one is observing exactly the same field of protoplasts when doing time-course experiments. Secondly, they can be placed on a depression slide and the whole depression scanned systematically for a count.[9,14] Since commercial depression slides have slightly rounded bottoms, even when called flat, more convenient slides can be made by placing rings of parafilm four layers thick on a warmed slide, to which they will stick. The inside of the chamber can be ringed with Vaseline to prevent water being drawn under the edge. After 10–15 μl of solution is placed on the slide with a micropipet, the chamber is covered with an ordinary thin coverslip. Such chambers are particularly well suited for phase contrast observations. It should be noted that some kinds of protoplasts may split into subprotoplasts, causing inaccuracies in such counts.[32] This method does, however, allow a determination of the fraction of the protoplasts showing vigorous cyclosis, i.e., a measurement of viability.

Metabolic Competence

Several lines of evidence suggest reasonably active metabolism within protoplasts, although definitive experiments have not been done. Cyclosis is readily apparent in protoplasts, and it has been observed on several occasions to continue for weeks.

In many cases, wall regeneration has been the best evidence of metabolic competence. Binding noted that moss protoplasts would regenerate a wall and even resume polar growth in many cases.[46] Yoshida showed that *Elodea* cells which were strongly plasmolyzed, but not actually isolated from their wall, would regenerate a wall around pieces of cytoplasm containing a nucleus, but not around pieces not containing a nucleus.[47] Differences in the chloroplasts were also noted, with those of nucleated portions of cytoplasm losing starch and necrosing faster than those of enucleate portions.[48]

Wall regeneration has also been reported around protoplasts prepared

[46] H. Binding, *Z. Naturforsch. B* **19,** 775 (1964).

[47] Y. Yoshida, *Plant Cell Physiol.* **2,** 139 (1961).

[48] Y. Yoshida, *Protoplasma* **54,** 476 (1962).

enzymatically with either pectinase[6,38,49] or cellulase.[50] Obviously, a protoplast able to resynthesize a wall must have many of its vital functions unimpaired.

General Perspectives

Ideally, enzymatic preparations of protoplasts should provide large quantities of unbroken protoplasts for many physiological studies. Two major factors render this difficult: (1) the relatively large amount of cellulase required to remove walls from cells with a high cellulose content and the limited number of tissues releasing protoplasts after pectinase treatment; and (2) the extreme physical fragility of protoplasts of highly vacuolated, mature higher plant cells. The latter point retards biochemical work since the remains of broken protoplasts interfere significantly with analyses of protoplast metabolism. Additions to the osmotic medium may in the future be found to strengthen membranes; however, one would then have an altered membrane, and many of the uses for which protoplasts are suitable (e.g., studies on the nature and permeability of the plasma membrane) would be lost. Further work will undoubtedly see significant contributions from the use of protoplasts for organelle isolation, somatic hybridization, and wall synthesis, the areas to which the use of protoplasts is now most vigorously being applied.

[49] K. Mishra and J. R. Colvin, *Protoplasma* **67,** 295 (1969).
[50] R. K. Horine and A. W. Ruesink, *Plant Physiol.* **46,** Suppl., Abstr. no. 70 (1970).

[17] Protoplasts of Algal Cells

By John Biggins

Naked cells prepared by the enzymatic removal of their cell walls offer unique opportunities for the study of ion and water movement, gross conformational changes, and the creation of membrane preparations by relatively mild physical procedures, such as osmotic lysis. Certain blue-green algae are sensitive to muramidase (EC 3.2.1.17), and, following the suggestion of Crespi, Mandeville, and Katz,[1] we have successfully prepared metabolically active protoplasts from *Phormidium luridum*.

[1] H. L. Crespi, S. E. Mandeville, and J. H. Katz, *Biochem. Biophys. Res. Commun.* **9,** 569 (1962).

Preparation

Photoautotrophically grown cells are harvested by centrifugation in the mid or late logarithmic phase of growth and washed in the buffered stabilizing osmoticum: 0.5 *M* mannitol, 30 m*M* potassium phosphate, pH 6.8. The washed cells are suspended in fresh mannitol–phosphate at a cell concentration of approximately 5% (w/v) and solid egg-white muramidase (Worthington Biochemical Corporation, Freehold, New Jersey) is added to give a final enzyme concentration of 0.5 g/liter. The cells are incubated in a sealed flask at 35° for 2 hours with occasional swirling. Toward the end of the incubation period it is advisable to check the progress of the reaction by microscopic examination of the filaments. Progress is indicated by (1) the change in shape of cells within the filament from rectangular to spherical and (2) the disintegration of the filament into units of smaller length.

The suspension is then cooled in an ice bath, and the protoplasts are separated from the untreated and clumped filaments by filtering through glass wool. The filtrate is centrifuged at a low speed (500 *g* for 5 minutes) to sediment the protoplasts.

The protoplasts are gently resuspended in fresh mannitol–phosphate or alternately osmoticum of similar tonicity. Resuspension is most conveniently accomplished by slow Vortex speeds or careful use of a homogenizer with a Teflon pestle. These techniques usually result in a protoplast preparation consisting of many single cells and very short filaments. The protoplasts can then be recentrifuged to remove residual muramidase.

Properties and Stability

If kept ice-cold, the protoplasts are stable for several hours. Cell breakage is reliably indicated by the release of the soluble accessory pigment, phycocyanin, which appears in the supernatant after a short test centrifugation at low speed. We have not found the progress of the enzyme digestion or subsequent stability of the preparation to be improved by addition of EDTA or Mg^{2+} as is the case with certain bacteria.[2]

The protoplasts are capable of endogenous respiration and photoassimilation of carbon dioxide at rates that are very similar to those of the untreated cells.[3] Membrane preparations derived by osmotic lysis of the protoplasts are very active in cyclic and noncyclic photophosphorylation, the Hill reaction, NADP photoreduction,[4] and the oxidation of certain respiratory substrates.[5]

[2] J. Spizizen, this series, Volume V [11].

[3] J. Biggins, *Plant Physiol.* **42,** 1442 (1967).

[4] J. Biggins, *Plant Physiol.* **42,** 1447 (1967).

[5] J. Biggins, *J. Bacteriol.* **99,** 570 (1969).

Comments

Our investigations have been restricted to *P. luridum,* which is readily available as a bacteria-free culture from the Culture Collection of Algae, Indiana University, Bloomington. Crespi *et al.,*[1] however, report that the blue-green algae *Plectonema calothricoides, Oscillatoria tenuis,* and *Fremyella diplosiphon* are similarly affected by muramidase, and therefore, these algae should also serve as suitable starting material for similar protoplast preparations. Much longer incubation periods with muramidase are necessary for other species of blue-green algae, such as *O. formosa, O. amoena,*[1] and *Anacystis nidulans.*[6]

Cyanocyta korschikoffiana is a pigmented alga (a so-called cyanelle) existing in a symbiotic relationship within a cryptomonad, *Cyanophora paradoxa.* The cyanelles are structurally similar to blue-green algae but lack the cell wall and sheath, and they have been isolated by differential centrifugation following breakage of the host protozoan.[7] Such isolated cyanelles are expected to have properties similar to protoplasts and, therefore, to be of considerable utility in studies of photosynthesis.

These are no reports concerning the preparation of protoplasts from the other algal divisions. The cell walls of the higher algae are markedly different in composition from the Cyanophyta, and they require cellulases for degradation. However, since investigators have successfully prepared protoplasts from yeasts[8] and higher plants[9] using snail gut and fungal enzymes, it should be possible to prepare protoplasts from the higher algae.

[6] C. J. Ludlow and R. B. Park, *Plant Physiol.* **44,** 540 (1969).
[7] M. Edelman, D. Swinton, J. A. Schiff, H. T. Epstein, and B. Zeldin, *Bacteriol. Rev.* **31,** 315 (1967).
[8] A. A. Eddy and D. H. Williamson, *Nature (London)* **179,** 1252 (1958).
[9] A. W. Ruesink and K. V. Thimann, *Proc. Nat. Acad. Sci. U.S.* **54,** 56 (1965).

[18] Chloroplasts (and Grana): Aqueous (Including High Carbon Fixation Ability)

By D. A. Walker

In the past, chloroplasts have been isolated (see Gorham[1]) for three principal purposes: (a) studies on electron transport, photophosphorylation and oxygen evolution in the presence of oxidants other than CO_2; (b) carbon assimilation; and (c) examination of enzymic and structural

[1] P. R. Gorham, this series, Vol. 1, p. 22.

components. The purpose for which the chloroplasts are to be used governs the material, the extraction, and the assay.

Preparation

Material: Spinach (*Spinacea oleracea*) is favored because in many latitudes it is freely available for most of the year, it is readily macerated, its vacuoles are largely free of endogenous inhibitors, and it often contains only small quantities of starch and calcium oxalate. For these reasons it normally yields chloroplasts of exceptionally high activity which behave in a predictable and relatively reproducible fashion. Its widespread use allows ready comparison of results with those of most other workers. It should not be confused with various cultivars of beet (e.g., perpetual spinach, Swiss chard, spinach beet, i.e., varieties of *Beta vulgaris*) which are sometimes grown or sold as spinach, but which are as unsatisfactory from a biochemical as from a culinary standpoint. The most useful related species include *Chenopodium album* and various species of *Atriplex*. Of these only *Chenopodium album* gives high yields of intact chloroplasts using current techniques. The principal disadvantages of spinach are that in northern latitudes it is unobtainable for many months of the winter *and* summer, and it is only grown really successfully under glass if it is possible to combine high light intensity with short days, moderate temperatures, and high humidity.

The most useful plant which can be grown easily in the laboratory is the garden pea (*Pisium sativum*). A dwarf variety (such as Feltham First, Laxton Superb, or Progress No. 9) with uniform germination is to be preferred. The principle advantage of the pea is that in appropriate conditions it can be grown for 2–3 weeks in an inert medium such as vermiculite (heat-expanded mica) without showing signs of mineral deficiency so that if light, temperature, and humidity are kept constant, then more or less invariable plant material is obtained. Starch production can be avoided by keeping the light intensity low (1500–2000 lux, e.g., five 40 W fluorescent tubes at 2–3 feet) and provided that low light is combined with relatively low temperature (15–20°) then healthy plants are readily obtained. If 150 gm of seed are soaked in water overnight and then sown at a depth of 1–2 cm in moist vermiculite of area 750 cm^2 and depth 10 cm, it will yield 100 gm or more of leaf and shoot after 12–14 days (11 hours light) at 18° and 1500 lux. Fifty grams of leaf and shoot yields a chloroplast suspension containing about 2 mg of chlorophyll. It is not necessary to separate the leaves from the shoots prior to maceration. Intact chloroplasts from peas will assimilate CO_2 at about 60–70% of the rate of comparable preparations from spinach, whereas for the majority of species from which CO_2-fixing chloroplasts have been separated the rates are 10%, or less, of those achieved by spinach chloroplasts.

Other species which have been successfully used for work on electron transport and photophosphorylation (but not yet for CO_2 assimilation) include lettuce, flax, barley, maize, poke weed (*Phytolacca americana*), New Zealand spinach (*Tetragonia expansa*), *Claytonia perfoliata*, *Lamium album*, *Stellaria media*, *Impatiens biflora*, and Good King Henry (*Chenopodium bonus henricus*).

Standard techniques applied unmodified to the vast majority of species yield chloroplasts which retain poor, uncoupled, electron, transport only. Some species, e.g., Dogs Mercury (*Mercurialis perennis*), are of interest because their chloroplasts are almost entirely inactive due to the presence of a potent endogenous inhibitor.[2]

Principles: Leaves from many species quickly pulverized in a mortar with a little salt or sugar solution will yield a green juice which, after filtration through moist cotton wool, may be seen to contain large numbers of discrete green particles capable of supporting the Hill reaction (cf. Gorham[1]). In essentials the methods first employed by Hill[3] in the 1930's have not really been improved on (cf. Walker[4] and Kalberer *et al.*[5]). However electron transport often persists when all else is lost. For this reason it may seem desirable to prepare the best chloroplasts possible, even if they are to be subsequently degraded, rather than degrade them carelessly during extraction simply because they may be wanted for nothing more than electron transport. The problem then resolves itself into three parts: (1) how to open the cells without damaging the chloroplasts; (2) how to provide, in a simple mixture, conditions in which the chloroplasts can survive extraction, and (3) how to separate the chloroplasts from other cellulose components. Other things being equal, the more simple the procedure the better (cf. Cockburn *et al.*[6]).

Maceration of the Tissue: Relatively undamaged chloroplasts have been obtained by cutting open individual cells of *Acetabularia mediterranea* with scissors.[7] For some purposes these may be extremely useful, but the procedure is often laborious, the plants may be somewhat difficult to grow, and the photosynthetic activity of the parent tissue may

[2] R. Hill and D. A. Walker, unpublished.
[3] R. Hill, *Essays Biochem.* **1,** 121 (1965).
[4] D. A. Walker, *Biochem. J.* **92,** 22 (1964).
[5] P. P. Kalberer, B. B. Buchanan, and D. I. Arnon, *Proc. Nat. Acad. Sci. U.S.* **57,** 1542 (1967).
[6] W. Cockburn, D. A. Walker, and C. W. Baldry, *Plant Physiol.* **43,** 1415 (1968).
[7] D. C. Shephard, W. B. Levin, and R. G. S. Bidwell, *Biochem. Biophys. Res. Commun.* **32,** 413 (1968).

be low. For more usual species such as spinach, leaves may be pounded (pestle and mortar, without sand). A Waring blendor or domestic homogenizer provides, however, a more efficacious and reproducible technique, and there is no indication that preparations obtained in this way are of poorer quality. For intact chloroplasts, it is necessary to restrict the grinding procedure to a minimum (3–5 seconds blending). Longer blending, improves yields of recovered chlorophyll but increases the proportion of fragmented chloroplasts. Most blendors work by cavitation as much as by actual cutting so that excessive frothing, an indication of denatured protein, is to be avoided. It is necessary to avoid freezing, but to keep as close to the freezing point as possible. To this end it is desirable to use the grinding medium as a semifrozen slush (like melting snow). The latent heat of melting then prevents local increases in temperature related to the heat generated by the mechanical action of the blendor. The initial viscosity of the semifrozen medium also facilitates the cutting action of the blades by providing a degree of resistance to their tendency to act as paddles rather than knives.

Higher yields of intact chloroplasts may be obtained by using less conventional blendors such as the Polytron.

Separation: Many of the earlier techniques incorporated a preliminary low-speed centrifugation to remove whole cells and cell debris followed by a second spin at low speed to recover the chloroplasts. This is undesirable for three reasons: (1) It lengthens the time during which the chloroplasts are exposed to various substrates inhibitors, etc., from which they are normally sequestered in the cell; (2) it lengthens exposure to the grinding medium which may contain protective but otherwise undesirable additives; and (3) it lowers the yield of intact chloroplasts, many of which will come down with the cell debris. These drawbacks may be avoided by substituting an initial rapid filtration followed by a quick centrifugation. One method which has been used is to squeeze the juice from the pulp manually through two layers of muslin and then to filter through eight layers. If this is done, it is important to select a hairy cloth containing cotton or wool, rather than nylon, which is less retentive. Alternatively, or additionally, the juice may be drained through a pad of cotton wool previously moistened with grinding medium and sandwiched between fine nylon mesh to retain the cotton wool fibres. At this stage the juice is free of whole cells and debris so that the chloroplasts may be recovered by a quick centrifugation. It is desirable to use a centrifuge with rapid acceleration and braking such that a 50- to 100-ml tube may be taken to 6000–7000 rpm and back to rest within 60–90 seconds. Longer spins may be necessary with some centrifuges, and various combinations of times and speeds have been successfully used, but in general it is desirable to aim for the shortest centrifugation possible. In some circumstances it is possible to recover the chloroplats within 2–3 minutes of starting maceration of the leaf.

A swing-out centrifuge gives a better pellet and may prevent damage which could occur in an angle-head as chloroplasts are dragged, under gravity, along the upper surface of the centrifuge tubes. Detergent and scouring agents should be avoided in cleaning centrifuge tubes.

Blending Volume: If appreciable contamination by cytoplasmic or vacuolar constituents is to be avoided, the volume of the initial grinding medium should be kept relatively high (by weight, about 4–6 times that of the tissue).

Resuspension: Once removed, chloroplasts are best suspended initially in small volume, using a glass rod and cotton wool soaked in chilled medium. For washing or breaking (see below), the volume is then increased and the chloroplasts again recovered by any appropriate centrifugation. Storage is also best in small volume at 0°.

Grinding and Resuspending Media

The grinding media which have been used are almost as legion as the people who have employed them. For particular purposes it is obviously desirable to use precisely the same procedure as that employed by some other worker. For this reason I shall not attempt to provide a catalogue of media but simply to give one or two examples together with some evaluation of the rationale behind them.

Osmoticum: In his original experiments Hill[3] used a sugar (often 12% sucrose or 6% glucose) to keep the osmotic pressure of the medium approximately similar to that of the cytoplasm of the parent tissue. Anxious to avoid a potential metabolite Arnon and co-workers[5,8] used 0.35 *M* NaCl in their equally classic experiments on photophosphorylation. Possibly because it penetrates the outer envelope more readily then sugars, salt is less satisfactory for obtaining intact chloroplasts. For this reason Walker[4] returned to the original practice of Hill. There is some evidence that glucose, fructose, and sucrose (which are otherwise entirely satisfactory) may act as metabolites. Accordingly 0.33 *M* sorbitol was introduced[4,9] and was subsequently adopted by Jensen and Bassham.[10] It should be noted that sorbitol partially inhibits cyclic photophosphorylation by envelope-free chloroplasts and also interferes with estimation of inorganic phosphate by some methods. Sugar polymers of various sorts

[8] D. I. Arnon, *Physiol. Rev.* **47,** 317 (1967).

[9] D. A. Walker, *Biochem. Chloroplasts Proc.* **2,** 53 (1967).

[10] R. G. Jensen and J. A. Bassham, *Proc. Nat. Acad. Sci. U.S.* **56,** 1095 (1966).

(e.g., dextrans, Ficoll) have also been used[11,12] in the hope of simulating cytoplasmic vicosity. Choline chloride has been used when it has seemed necessary to avoid both sugars and cations such as Na^+, K^+, and Mg^{2+}.

Buffer: Even when the pH of the unbuffered homogenate is close to neutrality, it is desirable to work at a constant pH. Hill used phosphate or pyrophosphate as a buffer. For work on carbon assimilation, orthophosphate may lead to complications (see below). Arnon and Whatley (cf. Arnon[8]) concerned with photophosphorylation used tris-(2-amino-2-hydroxymethylpropane-1:3-diol), although this has subsequently been found to act as a mild uncoupling agent. Good *et al.*[13] have developed an extensive range of new buffers [such as MES, 2-(*N*-morpholino)-ethane sulfonic acid] which are excellent for chloroplasts separation and assay and may be selected according to their various characteristics. For many purposes inorganic pyrophosphate (cf. Cockburn *et al.*[6] and Moore *et al.*[14]) is useful because it is cheap, does not enter the chloroplast freely, and may provide a measure of protection during extraction by acting as a chelating agent. Buffers are usually used in the range 0.1–0.005 *M* according to the nature of the leaf material. Some workers[4,9,10] find that it helps to extract at a slightly acid pH (6.5–6.8), but the advantages to be derived are probably marginal (cf. Kalberer *et al.*[5]) and any pH between about 6.3 and 8.5 may be safely selected for specific needs. Good's caution[13] about adjusting the pH of the buffer at the temperature at which it is to be used should be borne in mind. It should be noted that many of Good's buffers and compounds such as EDTA will support a rapid nonbiological oxygen uptake if illuminated by white light in the presence of flavin nucleotides and their degradation products.

Ascorbate: In the summer many chloroplasts (from field-grown plants) contain as much ascorbate (by weight) as chlorophyll, and there is then no need to add it to the grinding medium. In other circumstances the endogenous ascorbate concentration may be low, and chloroplast performance may be strikingly improved by its addition. The exact role of ascorbate is uncertain, although it seems clear that in the absence of a suitable coupling agent and DCMU it does not readily contribute to electron transport. It may act as a poising agent (cf. Arnon[8] and Brant and Whatley[15]) and almost inevitably as an antioxidant and hence

[11] R. M. Leech, *Biochem. Chloroplasts Proc.* **1**, 65 (1967).
[12] S. I. Honda, T. Hongladarom, and G. G. Laties, *J. Exp. Bot.* **17**, 460 (1966).
[13] N. E. Good, G. D. Winget, W. Winter, T. N. Conolly, S. Izawa, and R. M. M. Singh, *Biochemistry* **5**, 467 (1966).
[14] R. E. Moore, H. Springer-Lederer, H. C. J. Ottenheym, and J. A. Bassham, *Biochim. Biophys. Acta* **180**, 368 (1969).
[15] B. R. Brant and F. R. Whatley, *Biochem. Chloroplasts Proc.* **2**, 505 (1967).

preservative. Because of its instability, ascorbate should be added to solutions immediately prior to use. The use of sodium *iso*-ascorbate was introduced[4] because the commercial product is more stable and more easily available in the United Kingdom than its epimer. Even when ascorbate will subsequently interfere in an assay (e.g., with 2,6-dichlorophenolindophenol or ferricyanide as an oxidant), it may be useful to have it present during grinding and subsequently remove it by washing.

$MgCl_2$: Electron transport may become uncoupled from photophosphorylation in the absence of $MgCl_2$. It is unlikely that the magnesium concentration will fall below the critical value during grinding, but it is usually added as a safeguard. Sodium or potassium chloride may also be added as a stabilizing agent, but again the concentration of chloride and monovalent cations is unlikely to fall below the critical value during grinding, especially if they are present as a component of some other additive such as the buffer.

Other Additives: A variety of additional additives have been used, e.g.:

1. Cysteine or reduced glutathione (10^{-3}–10^{-4} M) to protect –SH groups
2. Carbowax (polyethylene glycol), e.g., carbowax 4000 at 0.6%,[16] principally as a protection against tannins
3. EDTA, e.g., at about 10^{-3} M, as a chelating agent
4. BSA (Bovine serum albumin) at about 0.1% is thought to act as a stabilizing agent for a variety of reasons.[11,12]
5. PVP (polyvinyl pyrollidone) as an absorbant for tannins and sodium metabisulphite at about 2×10^{-3} M as an inhibitor of polyphenoloxidases (which promote tannin formation) have been used in preparing mitochondria[17,18] and may therefore find an application in chloroplast work.

Where appropriate, these compounds may also be incorporated in resuspending media.

Other Resuspending Media: Occasionally it may be thought necessary to resuspend chloroplasts in some medium other than grinding medium or the assay medium (cf. Walker[4]) possibly as an aid to storage. However unless the chloroplasts are to be washed or processed further they may be conveniently resuspended directly in the assay medium, or one similar to it.

[16] J. Miflin and R. H. Hageman, *Plant Physiol.* **38,** 66 (1963).
[17] J. D. Jones, A. C. Hulme, and L. S. C. Wooltorton, *Phytochem.* **4,** 659 (1965).
[18] D. M Stokes, J. W. Anderson, and K. S. Rowan, *Phytochem.* **7,** 1509 (1968).

Carbon Assimilation and Associated Oxygen Evolution

The most useful medium so far devised for this assay[4,6,9,10,19] contains the following reagents, together with 10 mM $NaHCO_3$ and 100 μg/ml of chlorophyll:

0.33 M sorbitol
2 mM EDTA
1 mM $MgCl_2$
2 mM Na isoascorbate
0.5 mM K_2HPO_4
5.0 mM $Na_4P_2O_7$
50 mM HEPES (N-2-hydroxyethylpiperazine-N'-2-ethanesulfonic acid), all adjusted to pH 7.6 with NaOH

This medium is similar to that used by Jensen and Bassham,[10] and it differs in only one important respect from essentially identical mixtures used previously (e.g., by Walker[4,9]). This is that it contains inorganic pyrophosphate in addition to orthophosphate. The facts relating to orthophosphate and pyrophosphate are still far from resolved but the following points are pertinent.

(a) Orthophosphate is necessary for carbon assimilation, but as the concentration is increased beyond 10^{-5} M, the initial induction period is extended and the maximum rate depressed until at about 10^{-2} M (with chloroplasts prepared from dark stored tissues) carbon assimilation is almost completely suppressed.[20] This inhibition may be reversed by the addition of various cycle intermediates such as glyceraldehyde-3-phosphate.[21]

(b) Pyrophosphate does not appear to penetrate the chloroplast envelope, but it is externally hydrolyzed by a pyrophosphatase released from damaged chloroplasts. At high pyrophosphate concentrations this hydrolysis proceeds slowly, presumably because of inhibition by excess substrate.

(c) In the presence of pyrophosphate, orthophosphate is less inhibitory.

The total effect of these three factors is that inorganic pyrophosphate is able to provide the orthophosphate required for photosynthesis in a noninhibitory form.[5] For this reason, chloroplasts assayed in this mix-

[19] D. A. Walker, C. W. Baldry, and W. Cockburn, *Plant Physiol.* **43,** 1419 (1968).
[20] W. Cockburn, C. W. Baldry, and D. A. Walker, *Biochim. Biophys. Acta* **143,** 614 (1967).
[21] W. Cockburn, D. A. Walker, and C. W. Baldry, *Biochem. J.* **107,** 89 (1968).

ture show no requirement for added cycle intermediates in order to achieve their maximal rates.

Carbon Assimilation: If in addition to the reagents listed above, the mixtures contain bicarbonate as $NaH^{14}CO_3$ and if samples are removed during illumination and acidified with HCl to allow subsequent determination of CO_2 fixed, distribution of counts fixed may be determined after chromatography. Phosphate incorporation may be followed with ^{32}P.

Associated Oxygen Evolution: At the present time measurements of the oxygen evolution associated with carbon assimilation under aerobic conditions have been made entirely with the oxygen electrode. Assimilation and evolution have been followed simultaneously in the same mixture.[19]

Rates: Although rates of carbon assimilation in excess of 200 μmoles/mg of chlorophyll/hour have been reported, the value for a typical preparation is probably nearer to 60. An average figure for an intact plant is about 100.

Chlorophyll: In many types of apparatus a chlorophyll concentration of 100 μg/ml is optimal; more concentrated suspensions are only adequately illuminated in thin films. The chlorophyll concentration may be determined by suspending 50 μl of chloroplast suspension in 20 ml of 80% acetone and filtering (in the dark). If the optical density at 652 nm (1-cm light path) is then multiplied by a factor of 100/9 (i.e., 11.11 recurring), the value so obtained approximates closely to the chlorophyll content of the suspension in micrograms per microliter.[22,23] Alternatively, $(1/\text{O.D.} \times 9 = \mu l)$ of suspension required to give 100 μg of chlorophyll.

Envelope-Free Chloroplasts

Using the methods described above preparations may be obtained from spinach or peas in which 50–95% of the chloroplasts will have intact envelopes and therefore retain the ability to assimilate CO_2. Since the intact envelope presents a permeability barrier to a variety of compounds (e.g., NADP and ribulose-1,5-diphosphate), it may be necessary to remove the envelope. This is readily accomplished by osmotic shock (a 10-fold dilution of grinding or resuspending medium produces virtually 100% envelope-free chloroplasts in less than 1 minute). For purposes of comparison this may be done within the reaction mixture as follows:

1. "Intact" chloroplasts + water + double strength assay medium → envelope free
2. Double strength assay medium + water + "intact" chloroplasts → "intact"

[22] D. I. Arnon, *Plant Physiol.* **24**, 1 (1949).
[23] J. Bruinsma, *Biochim. Biophys. Acta* **52**, 576 (1961).

These two reaction mixtures now contain exactly the same additives at identical concentrations, but by simply varying the order in which the additions were made the chloroplasts are converted from 50–95% intact to 100% envelope free.

Free-Lamellar Bodies: When "intact" chloroplasts are suspended in a large volume of 1/10 dilution grinding medium (or grinding medium in which the concentration of osmoticum is correspondingly reduced), envelope loss is followed by appreciable loss of soluble components including the enzymes, coenzymes, and substrates concerned in carbon traffic. If the chloroplasts are exposed to dilute medium for only 5 minutes (or less) before they are returned to full strength medium, there is little apparent expansion of the thylakoids, and the preparation comprises, in essentials, chloroplasts without stroma or envelopes (i.e., free lamellae).

Expanded Lamellae: If free lamellae (above) are left in dilute media, endosmosis of water causes the thylakoids to separate, producing expanded vacuoles bounded by thylakoid membranes (expanded lamellae).

Chloroplast Particles: Thylakoid fragments (sometimes called "grana" but not necessarily composed of the stacked thylakoids which the microscopist refers to as "grana") are present as a contaminant in all of the above preparations. They may be prepared from intact chloroplasts or lamellar bodies by ultrasonic disintegration. Further fractionation may also be obtained by treatment with digitonin or detergents such as Triton X-100, followed by differential or density gradient centrifugation (see Boardman[24]).

Purified Chloroplast Fractions: Both intact chloroplasts and free lamellar bodies (envelope-free chloroplasts minus stroma) have been purified[25] by sucrose density gradient centrifugation (see also Price and Hirvonen[26]) to a point at which they are essentially free of contamination by other cellular components, such as mitochondria and nuclei.

Other Assays: Only intact chloroplasts will support carbon assimilation and the reduction of 3-phosphoglycerate at appreciable rates. It should be noted, however, that to obtain maximal rates of electron transport, cyclic, and noncyclic photophosphorylation and to obtain the largest light-reduced pH shifts and some other phenomena, it is necessary to work with envelope-free chloroplasts or lamellar bodies. The assay mixtures and methods of assay for these purposes are, of course, far too numerous to list here, but they include the use of radioactive tracers, manometry, spectrophotometry, mass spectrometry, electron spin resonance, chemical analysis, and a variety of electron measurements.

[24] N. K. Boardman, *Advan. Enzymol.* **30,** 1 (1968).

[25] R. M. Leech, *Biochem. Chloroplasts Proc.* **1,** 65 (1967).

[26] C. A. Price and A. P. Hirvonen, *Plant Physiol. Suppl.* **9,** 41 (1966).

[19] Chloroplasts: Nonaqueous

By C. R. STOCKING

Principles

Since many of the enzymes, cofactors, activators, substrates, and products of metabolism are water soluble, a determination of their intracellular location by means of aqueous methods of isolating cell organelles often is extremely difficult. Two major sources of error are adsorption of water-soluble materials onto the organelles and leaching of the materials from the organelles during isolation. The Behrens[1] nonaqueous isolation procedure and various modifications of it[2-4] avoid these difficulties. Since the tissue, which is rapidly frozen and dried without thawing, is ground in nonpolar solvents and the density gradient separation of the organelles takes place in the absence of water, the movement of water-soluble substances during the isolation is prevented.

Freeze Drying Leaves

In order to deplete the starch in the chloroplasts, young vigorously growing plants are kept in the dark for 24–48 hours immediately before they are used. Starch grains increase the density of chloroplasts and make difficult their separation from other cellular components by density gradient techniques. On the day of the freeze-drying, the plants are placed in the light for 1–1.5 hours before the leaves are harvested. The leaves are cut into strips (about 1 cm wide), and the major veins are removed. The strips are frozen rapidly by placing them in a cheesecloth bag and submerging it in liquid nitrogen or in isopentane at liquid-nitrogen temperature. An alternate method[2] is to submerge the bag containing the strips in a mixture of petroleum ether and dry ice.

The frozen leaves are transferred, without thawing, to a lyophilization flask kept at about —78° by immersing its base in a dry-ice–acetone mixture. The frozen leaf strips may be broken with a glass rod and are dried under vacuum at —20° to —25°. This temperature can be maintained by placing the lyophilization apparatus in a deep freezer. Depending upon the amount of leaves dried, about 2 or 3 days are required. The freeze-dried material may be stored under vacuum over P_2O_5 at —25° for several weeks.

[1] M. Behrens, *Z. Physiol. Chem.* **209,** 59 (1932).
[2] R. Thalacker and M. Behrens, *Z. Naturforsch. B* **14,** 443 (1959).
[3] U. Heber, *Ber. Deut. Bot. Ges.* **70,** 371 (1957).
[4] C. R. Stocking, *Plant Physiol.* **34,** 56 (1959).

Grinding

Since the purity of the chloroplast preparation is dependent on a separation of the dried cytoplasm from the chloroplasts,[5] the method of grinding is critical. Grinding must be done in nonaqueous solvents precooled to about —10°.

Isolation of Chloroplasts

Procedure A

Direct Density Gradient. Freeze-dried leaves, 0.4 g, are ground at about —10° in 30 ml of hexane–carbon tetrachloride with a density of 1.26 measured at 20°. The exact density used may have to be chosen by trial since the density of chloroplasts varies depending on species and growth conditions. The grinding is done in a Virtis homogenizer at full speed for 30 seconds, with 8 g of acid-washed glass homogenizing beads (approximately 0.2 mm diameter) in the grinding vessel. An alternate method is to grind 0.1 g samples in a Ten Broeck homogenizer. The ground material is filtered through Nylon cloth, 270 mesh (53-μ opening). Since the grinding medium has been in contact with the entire sample, in studies of the lipid composition of the cell,[6] it is necessary to have a complete grind. In this case, the residue that remains on the nylon cloth is washed with a small volume of grinding medium, into a graduated cylinder. After the glass beads have settled, the supernatant is transferred to a Ten Broeck hand homogenizer and ground and refiltered through the nylon cloth. This operation is repeated until the complete sample passes through the cloth. The filtrates are combined.

Five-milliliter aliquots of the ground sample are layered on top of 22-ml linear density gradients in Spinco SW 25.1 swinging-bucket centrifuge tubes. The gradients are prepared with carbon tetrachloride and hexane mixtures so that the density at the top of the tube is 1.27 and at the bottom of the tube is 1.40. The preparation is centrifuged for 2.5 hours at 20,000 rpm (35,000 *g*) in a swinging-bucket head (Spinco SW 25.1).

The light chloroplasts collect in a band near the top of the tube and may be separated from the rest of the material by any of several gradient sampling devices.[7]

The samples are diluted with an equal volume of cold hexane to lower

[5] C. R. Stocking, L. K. Shumway, T. E. Weier, and D. Greenwood, *J. Cell Biol.* **36**, 270 (1968).
[6] A. Ongun, W. W. Thomson, and J. B. Mudd, *J. Lipid Res.* **9**, 409 (1968).
[7] V. Allfrey, *in* "The Cell" (J. Brachet and A. E. Mirsky, eds.), Vol. 1, p. 193. Academic Press, New York, 1959.

the density and then centrifuged for 10 minutes at 3000 *g*. The supernatant is decanted from the pellets which are then dried under vacuum and stored under vacuum over P_2O_5 at —20°.

This procedure yields three fractions: (1) nonaqueously isolated chloroplasts (which may be further purified as described below); (2) a chloroplast-depleted sample which still contains chloroplasts as well as cell walls, cytoplasm, vacuolar material, and nuclei; and (3) supernatant liquid with some dissolved lipids.

Reverse Density Gradient. In some instances, a purer preparation of chloroplasts is obtained by grinding the freeze-dried leaves in a hexane–carbon tetrachloride mixture of a density equal to the heaviest density to be used in the gradient. In this case, the ground and filtered sample is placed at the bottom of the centrifuge tube, and the density gradient is layered on top of it. The chloroplasts move up through the density gradient during the centrifugation and consequently are washed continuously during this process.

Further Purification of Chloroplasts. To obtain the purest chloroplast preparation, the chloroplasts isolated by the density gradient centrifugation are resuspended in a light hexane–carbon tetrachloride mixture (60:40 v/v) and separated from small particles by repeated (8–10 times) short-term centrifugations (70 seconds at 3500 rpm) in this medium.[8,9]

Procedure B[2]

Two to three grams of freeze-dried leaves are ground in 250 ml of precooled petroleum ether (b.p. 30°–60°) + carbon tetrachloride (80:20 v/v) for five minutes (in 1-minute bursts) in a cooled Waring Blendor. The brei is filtered through nylon cloth and centrifuged for 10 minutes at 1400 *g*. The supernatant contains small particles, and these can be recovered by diluting the supernatant with an equal volume of petroleum ether and centrifuging it for 10 minutes at 3000 *g*.

The precipitate from the original centrifugation (containing chloroplasts, cell wall fragments, cytoplasm, cell membranes, vacuolar materials, and nuclei) is resuspended in 50 ml of petroleum ether, transferred to a 50-ml centrifuge tube, and centrifuged at 800 *g* for 2.5 minutes. In this process the chloroplasts and all the larger heavy particles go to the bottom while some particles smaller than chloroplasts remain in the supernatant. This process is repeated (four or five times) until the supernatant is only weakly turbid.

The chloroplasts, larger particles, and heavier cell material are sus-

[8] U. Heber and E. Tyszkiewicz, *J. Exp. Bot.* **38,** 185 (1962).
[9] U. Heber and J. Willenbrink, *Biochim. Biophys. Acta* **82,** 313 (1964).

pended in 50 ml of petroleum ether, and 5-ml aliquots are layered on discontinuous density gradients in 50-ml centrifuge tubes and centrifuged for 15 minutes at 2000 *g*. Care must be used not to overload the gradient with plant material. The gradients are prepared by layering into 50-ml centrifuge tubes four 10-ml layers of carbon tetrachloride–petroleum ether from bottom to top (densities 1.40, 1.35, 1.30, 1.25).

The purest chloroplasts are found in a dark-green band with a density of less than 1.30 below the upper green petroleum ether layer. The chloroplast fraction is removed and diluted with an equal volume of petroleum ether, mixed, and centrifuged at 3000 *g* for 10 minutes. The precipitated plastids may be further purified by repeated short-time centrifugations in petroleum ether–carbon tetrachloride (60:40 v/v) as previously described. After purification, they are dried under vacuum over P_2O_5 and stored in vacuum at −25°.

The rest of the density gradient sample is mixed with an equal volume of petroleum ether and centrifuged; the resulting plastid-depleted cellular material is dried under vacuum and stored.

Precautions

1. All operations in which the organic solvents are in contact with plant material should be carried out at −5° to −10° with prechilled solvents and equipment. This reduces the extraction of chlorophyll and other lipids, reduces the inactivation of enzymes, and reduces the danger of handling toxic and flammable solvents.

2. Once the plant material is frozen, it must be kept at a low temperature, preferably below −20°, during the drying step. A slight thawing before or during drying will negate the results of the isolation.

3. Once the material is freeze-dried, all traces of moisture must be kept from it. Care must be exercised to prevent the condensation of small amounts of moisture on the glassware, centrifuge tubes, and equipment that have been precooled and that will come intact with the sample. The solvents should be dried by adding anhydrous sodium sulfate or other suitable desiccant to the reagent bottles.

4. Since the removal of cytoplasm that surrounds the chloroplasts in the freeze-dried tissue is dependent on a separation during the grinding step, the grinding is especially critical and must be done in the solvents. This apparently weakens the chloroplast envelope and allows a cleavage to occur between the chloroplasts and cytoplasm. The grinding must be sufficiently complete to accomplish this cleavage and a subsequent separation.

5. Since the density of chloroplasts varies within one species under different growth conditions and from species to species, a preliminary

trial may be necessary to determine the correct densities to use in the isolation procedure. Chloroplasts containing starch have a density near that of other cellular material and are not easily separated from it.

6. Accurate determination of densities of the isolation media is essential. The usual procedure is to use a set of hydrometers that read to the third decimal place. The range 1.000–1.620 in steps of 0.001 unit is satisfactory.

Calculations

Procedure A. A simplified method of calculating the fraction (y) of a particular cellular compound that is present in chloroplasts is based on the quantitative recovery of both the purified chloroplasts and the chloroplast-depleted fraction from the same sample of leaf homogenate. Let U = the total chlorophyll in the isolated plastid fraction; V = the total chlorophyll in the isolated plastids and in the plastid depleted fractions; A = the total amount of the particular cellular compound in the isolated plastid fraction; C = the total amount of particular cellular compound in the isolated plastid and the plastid depleted fractions.

Then

$$y = \frac{A \times V \times 100}{U \times C} = \text{amount (\%) of compound present in chloroplasts.}$$

This calculation is based on the assumptions that the chloroplast fraction is pure and that the extraction of chlorophyll by the solvent is the same for plastids in all fractions.

Procedure B.[9] The ratio (z) of the chloroplast weight to cell weight of dried leaf cells may be calculated from the chlorophyll content per unit weight of the purified chloroplasts (y) and the chlorophyll content per unit weight of dried leaf material (x).

$$z = x/y \tag{1}$$

The percent yield of isolated chloroplasts (w) may be calculated from the total chlorophyll in the isolated plastids (u) and the total chlorophyll in the original tissue (v).

$$w = 100 \times u/v \tag{2}$$

The fraction (y) of a particular cellular compound that is present in the chloroplasts is given by the equation

$$y = z \times a/c \tag{3}$$

where a and c are the amounts on a unit weight basis, of this compound in chloroplasts and the total cell, respectively.

The amount (c) of the substance on a unit weight basis in the total cell may be determined directly or may be calculated from the amount (d) of the substance in the chloroplast-depleted cell residue (measured on a unit weight basis)

$$c = d + (a - d)(z \times w/100) \tag{4}$$

By substituting Eq. (4) into Eq. (3), the fraction of the compound in the cell may be calculated.

$$y = \frac{z \times a}{d + (a - d)(z \times w/100)}$$

The fraction of the compound present in the nonchloroplast part of the cell $= 1 - y$.

Estimation of Purity. The cytoplasmic contamination of the chloroplast preparation can be determined by measuring the distribution, in the various fractions, of an enzyme known to be located only in the cytoplasm *in vivo*. Pyruvate kinase is a suitable enzyme under most conditions for this purpose.[10,11] However, in some cases pyruvate kinase is inhibited in the total leaf extracts, and another cytoplasmic enzyme must be used for the purity test. Well prepared chloroplast preparations should contain less than 8% of the total leaf pyruvate kinase.

In those instances where anthocyanin is contained in leaf cell vacuoles, the distribution of this pigment may be used to estimate vacuolar contamination of nonaqueously isolated chloroplasts. Chloroplasts free of at least 92% of the vacuolar anthocyanin have been prepared from *Irisene* leaves.[4]

A qualitative evaluation of the purity of the chloroplast preparation may be made by observation of the samples with the electron microscope. Since nonaqueously prepared plant material swells rapidly in water, it is necessary in preparing samples for the electron microscope to fix the samples in a buffered glutaraldehyde–acetone mixture (3% glutaraldehyde:acetone, 50:50 v/v) and not allow the plastids to come in contact with too polar a solution.[5]

Limitations

Lipid Extraction. While the nonaqueous method prevents the leaching of water-soluble materials from chloroplasts during chloroplast isolation, it results in a loss of some lipid components to the supernatant solution. If low temperatures (—10° to —20°) are maintained during the

[10] U. Heber, *Z. Naturforsch. B* **15,** 95 (1960).
[11] R. M. Smillie, *Can. J. Bot.* **41,** 123 (1963).

isolation, the lipid loss is minimized. Nevertheless, as much as 4% of the total chlorophyll and significant amounts of other lipids,[12] especially monogalactosyl diglyceride, phosphatidylchlorine,[6] and plastoquinone, are lost to the supernatant. Nonaqueously isolated chloroplasts when placed in buffer swell and lose their water-soluble constituents.[5] They are unable to carry out the Hill reaction unless supplemented with plastoquinone.[13]

Purity of Fractions. The usefulness of the method is limited by the fact that the chloroplast fraction is the only subcellular fraction that has been isolated nonaqueously in relatively pure form from leaves. The fate of microbodies, mitochondria, and nuclei (in leaf samples) have not been adequately studied in nonaqueously fractionated leaf material. Consequently only the distribution or movement of a compound between the chloroplast and the rest of the cell can be obtained.

Since the chloroplast envelope is destroyed during the isolation, the intracellular location of any enzyme specifically associated with this membrane cannot readily be determined by this procedure.

Enzyme Inactivation. Only some of the enzymes of the cell can resist freeze-drying and exposure to organic solvents. The validity of results of experiments involving a particular enzyme rests on the demonstration that the enzyme is not inactivated during the nonaqueous isolation. This may be tested by comparing, on a unit dry weight basis, the enzyme activity in fresh tissue and in freeze-dried tissue subjected to all of the steps in the nonaqueous isolation. Many of the enzymes do retain their activity when subjected to freeze-drying and exposure to hexane and carbon tetrachloride.[4,10,11]

The localization of activators and inhibitors within specific parts of the cell may present difficulties. For example, the specific occurrence of a cytoplasmic located aldolase inhibition has been reported.[4]

Usefulness

Most of the problems of adsorption and transfer of polar components, of pH changes, osmotic balance, and autolysis that often are associated with aqueous methods of isolating chloroplasts are not problems in the nonaqueous method. Consequently this method has been very useful in studying the intracellular distribution of metabolic compounds. In particular the method is useful in studying the intracellular distribution of

[12] R. Thalacker and M. Behrens, *Separation Experientia* **16**, 165 (1960).

[13] W. L. Ogren and D. W. Krogmann, *in* "Photosynthetic Mechanism of Green Plants" (B. Kok and A. T. Jagendorf, eds.), *Nat. Acad. Sci. Publ.* **1145**, 684 (1963).

enzymes,[4,10,11,14] the intracellular distribution of ions,[15,16] and nucleotides,[17] and the movement of the early products of photosynthesis between plastids and cytoplasm.[18]

[14] I. F. Bird, H. K. Porter, and C. R. Stocking, *Biochim. Biophys. Acta* **100,** 366 (1965).
[15] C. R. Stocking and A. Ongun, *Amer. J. Bot.* **49,** 284 (1962).
[16] A. W. D. Larkum, *Nature (London)* **218,** 447 (1968).
[17] U. Heber and K. A. Santarius, *Biochim. Biophys. Acta* **109,** 390 (1965).
[18] C. R. Stocking, G. R. Williams, and A. Ongun, *Biochem. Biophys. Res. Commun.* **10,** 416 (1963).

[20] Chloroplasts (and Lamellae): Algal Preparations

By D. GRAHAM and ROBERT M. SMILLIE

In the preparation of chloroplasts and chloroplast fragments (lamellae), much less work has been done with algae than with higher plants. One reason for this is the difficulty often encountered in obtaining active preparations from algae. This can frequently be attributed to the drastic treatment required to break the cell walls of many unicellular algae. A feature of some of the most popular of the algal species used in the laboratory such as *Chlorella, Scenedesmus, Chlamydomonas,* and *Ochromonas* is that each cell contains a single, relatively large chloroplast which is difficult to expel intact from the enclosing cell wall. The algal chloroplast appears to be at least as fragile as its higher plant counterpart, so that the methods used for breaking the cell wall can result in considerable damage to the plastid. A reasonable degree of success has, however, been achieved with *Euglena* and *Chlamydomonas,* particularly where the preparation of chloroplasts has been required for characterization of macromolecular components such as DNA and RNA and proteins of the electron transport chain.

No one technique for chloroplast isolation has been shown to be suitable for a range of algal species. The techniques employed are mostly the result of trial and error and vary with the particular attribute of the chloroplast that is to be monitored.

General Considerations. The preparation and assay of chloroplasts usually consists of three phases: (1) disruption of the cells and production of a homogenate with as little damage to the chloroplast as possible (unless chloroplast fragments are especially required); (2) fractionation of the homogenate to concentrate the chloroplasts (or their fragments) and to remove nonchloroplast components; and (3) assay of the properties of the isolated chloroplasts.

The particular properties which are to be examined will determine the methods that are permissible under (1) and (2). Where freedom from contamination by components of cytoplasmic origin is necessary, this is invariably achieved at the expense of yield. Conversely, where quantitative analytical considerations are paramount, high yields of chloroplasts must be achieved at the expense of purity. It is therefore necessary to be quite clear which of these alternatives is required and design the methods accordingly. The purity of the isolated chloroplasts can be determined by morphological examination with the light and electron microscopes and by biochemical analysis.

All preparative operations described below should be carried out at about 2° to minimize enzymatic degradation of the cell materials.

Methods for Disruption of Algae

A review by Hughes and Cunningham[1] describes the methods generally employed for cellular disruption. The methods used for algae are more closely allied to those used for bacteria than to those used for higher plant or animal tissues. Algal cell walls are usually tough and elastic, often requiring considerable forces for their disruption.

No systematic survey of methods of disruption of algae appear to have been made, and most research groups seem to have developed an empirical method to suit their own particular requirements. Thus a number of variations of similar methods have been described for the disruption and fractionation of algae.

Solid Shear Methods

These usually involve grinding by hand using a pestle and mortar and a thick paste of about 1 g wet weight of cells with 12 g of acid-washed sand. Glass beads (ballitini) can be substituted for sand in which case it is generally desirable to choose beads about the same size as the organisms. The period of grinding depends on the endurance of the operator and the toughness of the cells. Four minutes of "vigorous" grinding is usually adequate for *Chlamydomonas,* whereas up to 30 minutes may be required for *Chlorella.* Hand-grinding procedures are difficult to make reproducible, and the use of a mechanically operated pestle and mortar improves reproducibility.

[1] D. E. Hughes and V. R. Cunningham, *Biochem. Soc. Symp.* **23,** 8 (1963).

Liquid Shear Methods

A variety of liquid shear methods are available. These are used on dense suspensions of algae, usually in the proportion of 1-ml packed cell volume to 2–4 ml of suspending medium.

French Pressure Cell.[2] This apparatus employs a plunger driven by an hydraulic press at pressures up to 20×10^3 psi (1.41×10^3 kg/cm^2) to force the suspension through a small orifice. The rate of flow, controlled by a needle valve, is often difficult to regulate, and the reduction gear system introduced by Wimpenny (see reference 1) offers better control. If the hydraulic press is manually operated, it is difficult to maintain a given pressure and uniform rate of flow. This difficulty is largely overcome by using a mechanically operated pump. The pressures employed vary with the organism, the growth conditions and age of the culture, and the composition of the suspending medium used. (Hydraulic presses commonly used in the laboratory usually have a 2-inch diameter piston, whereas the French pressure cell has a 1-inch diameter piston. In this case the actual pressure applied to the cell suspension is four times that shown on the pressure gauge of the hydraulic press. It is generally not made clear in published reports which of these two values is quoted. It is suggested that the actual pressure in the cell suspension be given.) The method has been employed for breaking mainly *Euglena* and *Chlamydomonas*, but it is applicable to a variety of unicellular organisms.

Ultrasonic Oscillator. This method employs ultrasonic generation of gaseous cavitation as the principal disruptive force. Hughes and Cunningham[1] discuss the principles involved. The commercial instruments provide a variety of titanium probe sizes of which 0.5-inch diameter is generally useful. In practice, the probe is adjusted to give a narrow clearance above the lower end of the vessel containing the algal suspension. Since considerable heat is generated during operation, the suspension is chilled in ice water to maintain a low temperature. It is necessary to operate the ultrasonic oscillator in short periods of, say, 30–60 seconds and to raise the probe out of the suspension for a few minutes so that adequate cooling of the suspension can be maintained. This method is particularly suitable for producing chloroplast fragments, but it is generally too vigorous for preparing whole chloroplasts.

Waring Blendor. This apparatus may be used to disrupt algae when used in conjunction with glass beads about the size of the organism. Sand or polyethylene pellets may also be effective. Special seals are

[2] H. W. Milner, N. S. Lawrence, and C. S. French, *Science* **111,** 633 (1950).

necessary to prevent the sand or glass from entering the bearing sleeves of the drive shaft, and a stainless-steel vessel with heavily convoluted walls is required to ensure rapid breakage. Operating times vary with the organism but can range up to several minutes. Frothing is a serious problem with this technique and may result in denaturation of proteins.

Oscillatory Shakers. A number of commercial devices have been designed to give very rapid shaking of a cell suspension containing glass beads (ballitini) which are about the size of the organism. Cooling can be achieved either with an external ice bath or with a direct stream of liquid CO_2 into the chamber. The method is destructive, but it is useful for preparing fragments of organelles, for example particles containing photosystem I or photosystem II.[3]

Other Methods. Several other treatments have been used to weaken algal cell walls and membranes prior to breaking the cells by one of the methods mentioned above. These include freezing and thawing of cells and incubation with proteolytic enzymes. Examples using these treatments are given below.

Medium and Fractionation

The choice of suspending medium used and the fractionation procedure employed in the isolation of algal chloroplasts depends upon the purpose for which the chloroplast or chloroplast fragments are required. The preparation of chloroplasts with good biosynthetic capabilities requires a relatively gentle method for breaking the cell and use of a buffered isotonic or hypertonic medium containing high molecular weight carbohydrates or bovine serum albumin. Such preparations do not necessarily exhibit high photochemical activities; these, however, can often be obtained by using fragmented chloroplasts.

Differential centrifugation and density gradient centrifugation using either discontinuous or continuous gradients may be used to fractionate algal homogenates. Zonal rotors employing continuous density gradients make possible the large-scale preparation of chloroplasts and chloroplast particles. This technique, which should have a wide application in the fractionation of organelles, has been used to isolate *Euglena* chloroplasts.[4]

The isolation of pure spinach chloroplasts by free-flow electrophoresis has been reported,[5] and this method should also be applicable to the isolation of algal chloroplasts.

[3] J. S. Brown, *Carnegie Inst. Washington Yearb.* **68** (1970).
[4] G. J. Knight and C. A. Price, *Biochim. Biophys. Acta* **158,** 283 (1968).
[5] W. Klofat and K. Hannig, *Hoppe-Seyler's Z. Physiol. Chem.* **348,** 739 (1967).

Specific Methods for Isolation of Chloroplasts from Different Organisms

Euglena gracilis

The method described by Eisenstadt and Brawerman[6] is the one most commonly used to prepare chloroplasts from *E. gracilis*. The method is applicable to both *E. gracilis* var. *bacillaris* and the Z strain, the two strains most frequently employed in biochemical studies.

Isolation of Chloroplasts from E. gracilis. Procedure of Eisenstadt and Brawerman[6] (Slightly Modified)

Medium

Sucrose, 10% (w/v)
Tris·HCl buffer, 10 m*M*, pH 7.6
$MgCl_2$, 4 m*M*
2-Mercaptoethanol, 1 m*M*

Procedure. Cells grown photoautotrophically are harvested and washed free of growth medium by centrifugation at 2500 *g* for 2 minutes. They are suspended [1 vol packed cells in 20–40 vol of 10 m*M* Tris·HCl buffer (pH 7.6)] and recentrifuged. The cells are given a final wash, this time in the medium shown above and suspended in the same medium (1 vol packed cells to 3–4 vol medium at 0°–2°). The cell suspension is then passed through an ice-chilled French pressure cell. A range of values for the correct pressure to be used has been given in the literature. The optimum pressure for breakage is influenced by a number of factors including culturing conditions and the composition of the suspending medium. Hence it is recommended that small portions of the suspension of cells be passed individually through the cell, a range of pressures being used, and then examined for cell breakage in the light microscope. The lowest pressure giving over 90% breakage of the cells should be used for the bulk of the suspension. After passage through the pressure cell, the suspension is immediately diluted with an equal volume of medium, stirred, and centrifuged at 500 g for 10 minutes. The pellet is retained and is suspended in medium (approximately 100 ml per 10 ml of the original packed cell volume), allowed to stand 10 minutes, and then filtered through four layers of gauze. The filtered suspension is centrifuged at 500 *g* for 10 minutes, and the resulting pellet is again suspended in medium (20–30 ml per 10 ml of the original packed cell volume). The suspension is then thoroughly mixed with 2 vol of 75% (w/v) sucrose

[6] J. M. Eisenstadt and G. Brawerman, *J. Mol. Biol.* **10**, 392 (1964).

and centrifuged at 23,000 *g* for 10 minutes, preferably in a swinging-bucket rotor. The surface layer of chloroplasts is removed, diluted with medium, and centrifuged at 3000 *g* for 5 minutes to recover the chloroplasts. The flotation procedure may be repeated if highly purified chloroplasts are required. The yield of chloroplasts is about 50%.

Comments. The method can also be used to isolate chloroplasts from cells grown heterotrophically in the light. It is not recommended for isolating proplastids from dark-grown cells, or small developing chloroplasts from greening cells during the first 12 hours of illumination.

Properties. Under light or phase contrast microscopy the preparation of chloroplasts appears to be free of other particles. The chloroplasts look intact and are active photochemically, but they have lost most of their *c*-type cytochrome-552 and ferredoxin (unpublished experiments). Presumably, many other soluble proteins are also lost. The chloroplasts retain DNA,[7,8] DNA polymerase,[8] and ribosomes.[6] Ribosomal RNA extracted from these chloroplasts is relatively free of contamination by cytoplasmic ribosomal RNA.[9] The isolated chloroplasts can incorporate amino acids into peptides.[6]

Chloroplasts for Photochemical Measurements

In experiments involving only photochemical measurements, it is usually sufficient to break the cells in a suitable medium and isolate the chloroplasts by differential centrifugation. Katoh and San Pietro[10] have described a method which is similar to that of Brawerman and Eisenstadt[6] described above, except that the flotation step is omitted.

Medium[10]

K phosphate buffer, 50 m*M*, pH 7.0
Sucrose, 0.4 *M*
NaCl, 10 m*M*

Procedure.[10] Autotrophic cells are washed by centrifugation in 50 m*M* phosphate buffer (pH 7.0) and suspended in the above medium (1 vol packed cells to 2–3 vol of medium). The cells are disrupted by passage through a French pressure cell (see above). The resulting suspension is centrifuged at 100 *g* for 5 minutes and then at 270 *g* for 5 minutes to remove whole cells and the larger cell fragments. The chloroplasts are then collected by centrifugation at 1000 *g* for 10 minutes. The chloro-

[7] M. Edelman, J. A. Schiff, and H. T. Epstein, *J. Mol. Biol.* **11,** 769 (1965).
[8] N. S. Scott and R. M. Smillie, *J. Cell Biol.* **38,** 151 (1968).
[9] N. S. Scott and R. M. Smillie, *Curr. Mod. Biol.* **2,** 339 (1969).
[10] S. Katoh and A. San Pietro, *Arch. Biochem. Biophys.* **118,** 488 (1967).

plasts are washed once in medium and finally suspended in a few milliliters of medium.

Properties. The preparation consists mainly of chloroplasts and some fragments, but is free of whole cells and larger cell fragments. The photochemical activities of the preparation are summarized in the final section.

Methods Employing Freezing and Thawing of Cells

Freezing and thawing apparently weakens the pellicle of *E. gracilis* so that cells are more easily broken by mechanical means. Kahn[11] has made use of this observation in isolating chloroplasts showing photophosphorylating activity. In a second method described below, a combination of freezing and thawing and partial digestion with proteolytic enzymes is used to weaken the pellicle.

In the procedure devised by Kahn,[11,12] the harvested cells are washed once with 1 mM NaCl, once with water, and then suspended in 4–6 times their wet weight of a medium consisting of 400 mM mannitol, 25 mM Tricine·NaOH buffer (pH 8.0), and 10 mM NaCl. The cell suspension is rapidly frozen in a low temperature bath (—12°), partially thawed and then homogenized with one-eighth volume of 3–5-mm polyethylene pellets (Marlex high density polyethylene resin, the Phillips Petroleum Company, U.S.A.) in a chilled Waring Blendor for 1.0–1.5 minutes. The suspension is filtered through eight layers of gauze, and centrifuged at 1000 g for 45 seconds. The chloroplasts in the supernatant fluid are collected by centrifugation at 1000 g for 10 minutes and washed once with medium.

When suspended in medium to which bovine serum albumin (10 mg/ml) has been added, the chloroplasts show a photophosphorylating activity (with pyocyanine) of 80–100 micromoles ATP formed per milligram of chlorophyll per hour.[12] This preparation retains activity for several hours at 2°–4°.

The chloroplasts can be further purified by centrifugation in a gradient of sucrose,[13] but photophosphorylating activity is reduced.

Rawson and Stutz[14] have modified a procedure[15] employing proteolytic enzymes to prepare chloroplasts from *E. gracilis*. Cells are harvested, washed twice in 10 mM Tris·HCl (pH 7.5) and frozen to —60°. They are thawed slowly and incubated at 4° for 12 hours in a medium consisting of 0.34 M sucrose, 10 mM Tris·HCl (pH 7.5), 4 mM $MgCl_2$,

[11] J. S. Kahn, *Biochem. Biophys. Res. Commun.* **24,** 329 (1966).
[12] I. C. Chang and J. S. Kahn, personal communication (1970).
[13] E. F. Carrel and J. S. Kahn, *Arch. Biochem. Biophys.* **108,** 1 (1964).
[14] J. R. Rawson and E. Stutz, *Biochim. Biophys. Acta* **190,** 368 (1969).
[15] C. A. Price and M. F. Bourke, *J. Protozool.* **13,** 474 (1966).

40 mM KCl, 5 mM 2-mercaptoethanol, and trypsin (1.0 mg per 1 g of packed cells). The cells are diluted with 7.5 vol of medium, except that the trypsin is omitted, and centrifuged at about 1000 g to sediment the cells, which are washed once more with the same medium. The cells are disrupted by suspending in a medium consisting of 0.7 M sucrose, 10 mM Tris·HCl (pH 7.5), 4 mM $MgCl_2$, 40 mM KCl, and 5 mM 2-mercaptoethanol (7 ml per 1 g of packed cells) and homogenizing in a Waring Blendor for 50 seconds. The homogenate is diluted with 1 vol of the same medium, filtered through 10 layers of gauze and centrifuged at 400 g for 2 minutes. The upper three-quarters of the supernatant fluid is withdrawn and centrifuged at 700 g for 12 minutes and the resulting pellet of chloroplasts is washed once with medium. These chloroplasts retain their ribosomes. Analysis of chloroplast and cytoplasmic ribosomal RNA indicated that the chloroplasts are relatively free of cytoplasmic RNA contamination.

Comment. Procedures employing freezing and thawing may be useful in specific instances, for example, in isolating chloroplasts showing high rates of photophosphorylation, but they do not appear to offer advantages over the procedure of Eisenstadt and Brawerman[6] as a general method for isolating chloroplasts from *E. gracilis.* Freezing and thawing results in leakage of soluble proteins from the cells, including some chloroplast proteins, e.g., cytochrome 552, and it obviously causes damage to the outer membrane of the chloroplast as well as the outer membrane of the cell.

Chlamydomonas reinhardi

Chloroplast Fragments

Several methods have been used to disrupt the cells of *C. reinhardi* to obtain chloroplast fragments suitable for photochemical studies. These include sonication,[16] grinding with sand,[17] passage through the French pressure cell,[18,19] freezing and thawing,[20] and disintegration in a Nossal shaker,[20] Eppenbach colloid mill,[20] or Braun mechanical cell homogenizer[3] using a mixture of algal paste and glass beads. Chloroplast fragments in the homogenate can be separated by differential centrifugation. They can be further fractionated into particles containing different ratios

[16] R. P. Levine and D. S. Gorman, *Plant Physiol.* **41,** 1293 (1966).

[17] D. S. Gorman and R. P. Levine, *Proc. Nat. Acad. Sci. U.S.* **54,** 1665 (1965).

[18] R. P. Levine and D. Volkmann, *Biochem. Biophys. Res. Commun.* **6,** 264 (1961).

[19] J.-M. Michel and M.-R. Michel-Wolwertz, *Carnegie Inst. Washington Yearb.* **67,** 508 (1969).

[20] T. Hiyama, M. Nishimura, and B. Chance, *Plant Physiol.* **44,** 527 (1969).

of chlorophyll *a* to chlorophyll *b* by centrifugation in a sucrose gradient.[3,21]

Two methods for preparing chloroplast fragments from *C. reinhardi* are described below. The first method employs sonication to break the cells and yields particles showing a good rate of photoreduction of NADP and other acceptors but poor photophosphorylation activity. Particles showing a reasonable rate of photophosphorylation can be obtained by the second method.

Procedure I.[16] Cells are harvested, washed once in 10 m*M* potassium phosphate buffer (pH 7.0) containing 20 m*M* KCl and 2.5 m*M* $MgCl_2$, and suspended in 5 ml of the same solution. The cells are broken by sonic disintegration for 30 seconds at 0°–2° using a Mullard ultrasonic disintegrator. The sonicate is centrifuged at 480 *g* for 6 minutes to sediment unbroken cells. Chloroplast fragments are sedimented from the supernatant fluid by centrifugation at 20,000 *g* for 20 minutes. The fragments are resuspended in 1–2 ml of buffer, leaving behind the starch, and transferred to a Ten Broeck hand homogenizer for complete dispersal of the fragments.

Procedure II.[17] The cells are broken in a medium consisting of 10 m*M* potassium phosphate buffer (pH 7.5), 20 m*M* KCl, 2.5 m*M* $MgCl_2$, 1 m*M* $MgNa_2EDTA$, and 1 m*M* reduced glutathione. Disruption of the cells is accomplished by grinding a paste of cells and clean sand (0.7 g wet weight of packed cells and 8.5 g of dry sand) using a chilled pestle and mortar. The paste is diluted with medium, the suspension containing broken cells is decanted from most of the sand and then centrifuged to remove whole cells and residual sand. The supernatant fluid containing chloroplast fragments is centrifuged at 20,000 *g* for 15 minutes to collect the fragments which are resuspended in 1–2 ml of medium.

Chloroplasts

Sucrose gradient centrifugation has been used to isolate chloroplasts from a homogenate of *C. reinhardi* cells. The method outlined below[22] is similar to an earlier procedure described by Sager and Ishida.[23]

Buffers

Buffer I: Sucrose, 0.25 *M*
Tris·HCl, 50 m*M*, pH 8.0

[21] J. S. Brown, this volume [43].

[22] S. J. Surzycki, *Proc. Nat. Acad. Sci. U.S.* **63,** 1327 (1969); and personal communication.

[23] R. Sager and M. R. Ishida, *Proc. Nat. Acad. Sci. U.S.* **50,** 725 (1963).

MgNa$_2$EDTA, 1 mM
MgSO$_4$, 20 mM

Buffer II: Sucrose, 0.5 M
Tris·HCl, 50 mM, pH 8.0
MgSO$_4$, 20 mM
Bovine serum albumin (fraction V, Bacteriological, from Nutritional Biochemicals Corp.), 0.25% (w/v)
2-Mercaptoethanol, 2 mM

Procedure. Cells are grown photoautotrophically to the stationary phase of growth. The cells from 6 liters of culture are harvested and washed three times with cold buffer I, and two times with cold buffer II. The cells are suspended in buffer II at a ratio of 1 vol of buffer to 2 vol of packed cells and broken in a French pressure cell. The breaking pressure should be adjusted so as to give mostly protoplasts and chloroplasts, (ratio of about 4:1) and as few fragments as possible. The suspension is immediately diluted with 3 vol of cold buffer II, and centrifuged in 30-ml glass tubes (acid-washed) in 20-ml volumes for 1.5 minute at 500 g. The supernatant fluid containing chloroplasts and protoplasts is centrifuged for 5 minutes 1900 g. By means of an aspirator, the supernatant fluid is discarded together with any membrane fragments sticking to the walls of glass tubes. This centrifugation is repeated three times, each time discarding the supernatant fluid and membrane fragments adhering to the walls of the glass tubes and resuspending the pellet in 20 ml of buffer II. The membrane fragments tend to stick to the glass tubes provided the tubes have been acid washed and clean tubes are used for each centrifugation. As a result of these centrifugations the pellet consists mostly of a mixture of chloroplasts and protoplasts. These are separated on a discontinuous sucrose gradient prepared in buffer II. The gradients contain four 5-ml layers: 1.5, 1.75, 2, and 2.5 M sucrose. Five milliliters of the chloroplast–protoplast mixture is layered on top of the gradient which is centrifuged at 8300 g using a swinging-bucket rotor for 60 minutes. Three distinct green layers are formed. The first, located at the border of the 1.5 and 1.75 M sucrose layers, contains chloroplasts with damaged outer membranes. The second band is located between the 1.75 and 2.0 M sucrose layers and contains intact chloroplasts. The third band is formed on the top of the 2.5 M sucrose layer and contains protoplasts.

Chloroplasts collected from the second band incorporate nucleotides into RNA and amino acids into peptides. They show low Hill reaction activity compared with chloroplast fragments prepared by Procedure I. The degree of retention of soluble chloroplast proteins by chloroplasts prepared by this method has not been determined. Chloroplasts isolated from *C. reinhardi* cells by using gradient centrifugation contain chloroplast DNA[23] and 69 S chloroplast ribosomes.[24]

Scenedesmus and Chlorella

Procedures for isolating chloroplast fragments from *Scenedesmus* have been published by Pratt and Bishop[25] and by Kok and Datko.[26] The cells are broken by shaking with glass beads and the chloroplast fragments are isolated by differential centrifugation. These fragments are active in photoreduction and cyclic photophosphorylation.

A procedure for isolating chloroplasts from *Chlorella protothecoides* based on the method developed by Sager and Ishida[23] for isolation of chloroplasts from *Chlamydomonas* cells has been described by Mihara and Hase.[27] The isolated chloroplasts contain chloroplast ribosomal 30 S and 50 S subunits. Photochemical and other activities have not been investigated.

Acetabularia mediterrania

The procedure of Bidwell, Levin, and Shephard[28] (given below) yields chloroplasts from *A. mediterrania* that fix CO_2 photosynthetically. Gibor[29] has also described a method for isolating DNA-containing chloroplasts from this marine alga.

Media

Medium I: Mannitol, 0.6 *M*
Na$_2$EDTA, 1.0 m*M*
Bovine serum albumin, 0.1% (w/v)
n-Tris(hydroxymethyl)methyl 2-aminoethane sulfonic acid (TES)–KOH buffer, 0.1 *M*, pH 7.8
Dithiothreitol, 1 m*M*

Medium II: Mannitol, 0.6 *M*
Na$_2$EDTA, 1.0 m*M*
Bovine serum albumin, 0.1% (w/v)
TES–KOH buffer, 5 m*M*, pH 7.2
Dithiothreitol, 0.1 m*M*

[24] R. Sager and M. G. Hamilton, *Science* **157,** 709 (1967).
[25] L. H. Pratt and N. I. Bishop, *Biochim. Biophys. Acta* **153,** 664 (1968).
[26] B. Kok and E. A. Datko, *Plant Physiol.* **40,** 1171 (1965).
[27] S. Mihara and E. Hase, *Plant Cell Physiol.* **10,** 465 (1969).
[28] R. G. S. Bidwell, W. B. Levin, and D. C. Shephard, *Plant Physiol.* **44,** 946 (1969).
[29] A. Gibor, *in* "Biochemistry of Chloroplasts" (T. W. Goodwin, ed.), Vol. 2, p. 321. Academic Press, New York, 1967.

Procedure.[28] Cells in the elongating phase of growth, 15–25 mm in length, are collected (total of 1–5 g fresh weight) and minced to small pieces with scissors into medium I. The homogenate is strained through 173-mesh bolting silk. The debris is rinsed by mixing with medium II and straining. The strained suspensions are combined and centrifuged at 500 *g* for 5 minutes. The pellet containing the chloroplasts is suspended gently in 1 ml of medium II and layered on top of a bilayer consisting of 3 ml of medium II containing 2.5% (w/v) of dextran on 2 ml of medium II containing 5% (w/v) of dextran. After centrifugation at 50 *g* for 5 minutes, the upper two layers are removed and diluted with 6 ml of medium II.

The chloroplasts are concentrated by centrifugation at 500 *g* for 5 minutes and can be suspended in medium II or directly in an assay medium.

Properties. Examination by electron microscopy indicates that most of the isolated chloroplasts retain their outer membrane and are intact. The chloroplasts fix CO_2 into photosynthetic products. The rate of fixation is stated to be comparable to rates of photosynthesis shown by the whole cells.[30]

Fractionation of Chloroplasts in Nonaqueous Medium

Smillie[31] has described a method for isolating chloroplasts from *E. gracilis* using a nonaqueous medium consisting of *n*-hexane and carbon tetrachloride. The method is useful for preparing chloroplasts containing water-soluble chloroplast proteins which are lost using isolation procedures involving aqueous media. Since chloroplast lipids are partially extracted by the nonaqueous medium, the photochemical properties of the chloroplasts are modified.

Photochemical Activities of Algal Chloroplast Preparations

Hill Reaction and Photophosphorylation

The photochemical activities shown by various algal preparations are listed in Tables I and II. Some of these rates approach those obtained with higher plant chloroplasts, but in other cases the rates reported so far are very low.

CO_2 Fixation by Algal Chloroplast Preparations

Few results have been published on CO_2 fixation by isolated algal chloroplasts. No significant incorporation of $^{14}CO_2$ could be obtained

[30] D. C. Shephard, W. B. Levin, and R. G. S. Bidwell, *Biochem. Biophys. Res. Commun.* **32,** 413 (1968).

[31] R. M. Smillie, *Can. J. Bot.* **41,** 123 (1963).

TABLE I

HILL REACTIONS BY ALGAL CHLOROPLAST PREPARATIONS

Organism	Method of preparation of chloroplasts	Electron acceptor[a]							Reference
		Dichlorophenol indophenol	Ferricyanide	NADP[b]	*p*-Benzoquinone	Cytochrome *c*[b]	Methyl viologen	NADP[b,c]	
Chlamydomonas reinhardi	Sonication and differential centrifugation	269	—	169	—	—	—	81	*e*
		40–125	556	60–110	63	—	—	30–60	*f*
Scenedesmus obliquus	Oscillation with glass beads and differential centrifugation	49–61	154, 182	52	—	117	—	31	*g*
Euglena gracilis	French press and differential centrifugation	146	237	12 117[d]		81[d]	39[d]	71[d]	*h*

[a] Activities are expressed as micromoles of oxidant reduced per milligram of chlorophyll per hour.

[b] Reaction mixture includes ferredoxin.

[c] Dichlorophenolindophenol/ascorbate as electron donor; water is the electron donor in the other cases.

[d] *Euglena* cytochrome *c* (552) included in reaction mixture.

[e] D. S. Gorman and R. P. Levine, *Proc. Nat. Acad. Sci. U.S.* **54,** 1665 (1965).

[f] A. L. Givan and R. P. Levine, *Plant Physiol.* **42,** 1264 (1967).

[g] L. H. Pratt and N. I. Bishop, *Biochim. Biophys. Acta* **153,** 664 (1968).

[h] S. Katoh and A. San Pietro, *Arch. Biochem. Biophys.* **118,** 488 (1967).

TABLE II
PHOTOPHOSPHORYLATION BY ALGAL CHLOROPLAST PREPARATIONS

Organism	Method of preparation of chloroplasts	Noncyclic photophosphorylation: Ferricyanide reduced[a]	Noncyclic photophosphorylation: ATP formed or P_i esterfied[b]	Noncyclic photophosphorylation: Coupling ratio P/2e	Cylic photophosphorylation: Phenazine methosulfate[b]	Cylic photophosphorylation: Pyocyanine[b]	Reference
Chlamydomonas reinhardi	Sand-ground and differential centrifugation	726	105	0.29	292 (+DCMU)	—	*e*
		685 ± 100[c]	37 ± 14[c]	0.11 ± 0.03[c]	178 ± 81[c] (+DCMU)	—	*f*
		596	192	0.64	440 (+DCMU)	—	*g*
Scenedesmus obliquus	Oscillation with glass beads and differential centrifugation	—	0[d]	—	8, 15	—	*h*
Euglena gracilis	Freeze–thaw, blending with polyethylene pellets and differential centrifugation	20–25	7.4–7.9	0.63–0.74	9.1	25.9	*i*
		—	9.5	—	—	51.4	*j*
		250–350	—	—	—	80–100	*k*

[a] Micromoles potassium ferricyanide reduced per milligram of chlorophyll per hour.
[b] Micromoles ATP formed or P_i esterfied per milligram of chlorophyll per hour.
[c] Standard deviation of mean.
[d] With either potassium ferricyanide or NADP as oxidant.
[e] D. S. Gorman and R. P. Levine, *Proc. Nat. Acad. Sci. U.S.* **54,** 1665 (1965).
[f] A. L. Givan and R. P. Levine, *Plant Physiol.* **42,** 1264 (1967).
[g] D. Graham and R. P. Levine, unpublished results (1969).
[h] L. H. Pratt and N. I. Bishop, *Biochim. Biophys. Acta* **153,** 664 (1968).
[i] J. S. Kahn, *Biochem. Biophys. Res. Commun.* **24,** 329 (1966).
[j] I. C. Chang and J. S. Kahn, *Arch. Biochem. Biophys.* **117,** 282 (1966).
[k] I. C. Chang and J. S. Kahn, personal communication (1970).

with *Euglena* chloroplasts isolated by the procedure of Eisenstadt and Brawerman[6] (Dr. J. A. Schiff, personal communication). However, high rates of incorporation (78 μmoles HCO_3^-/mg chlorophyll/mg) have been reported with preparations from *Acetabularia mediterrania*.[30]

[21] Algal Preparations with Photophosphorylation Activity

By PETER BÖGER

Most of our knowledge of the photophosphorylation processes *in vitro* is based on preparations from higher plants and purple bacteria. Chloroplasts isolated from algae are not usually active. The first active preparations, however, were obtained with the (prokaryotic) blue-green alga *Anabaena*[1,2] after disruption of the cells by mechanical means. These particles catalyzed only cyclic photophosphorylation [e.g., with *N*-methylphenazinium methyl sulfate (PMS)], although Hill reactions with artificial electron acceptors were consistently observed. A gentle rupture of the cell walls with enzymes was necessary to obtain particles active in both cyclic *and* noncyclic photophosphorylation.[3,4]

There are very few reports concerning the preparation of chloroplasts with photophosphorylation activity from *eukaryotic* algae. Preparations from the chrysomonad *Hymenomonas* yielded about 70 μmoles of ATP per milligram of chlorophyll per hour with PMS.[5] Rather high rates with this cofactor (approximately 180 μmoles ATP/mg Chl × hour) have been reported using chloroplasts from the green alga *Chlamydomonas reinhardi*.[6] This author has prepared chloroplasts with good yield and high phosphorylation activity from the heterokont *Bumilleriopsis filiformis*.[7] Active chloroplast preparations from other algae groups, e.g., red algae or diatoms, have not yet been reported.

In this article the methodology of photophosphorylation will be described for only those algae preparations which show reasonably high

[1] B. Petrack and F. Lipmann, *in* "Light and Life" (W. D. McElroy and B. Glass, eds.). Johns Hopkins Press, Baltimore, Maryland, 1961.
[2] W. A. Susor, W. C. Duane, and D. W. Krogmann, *Rec. Chem. Progr.* **25**, 197 (1964).
[3] B. Gerhardt and R. Santo, *Z. Naturforsch. B* **21**, 673 (1966).
[4] J. Biggins, *Plant Physiol.* **42**, 1447 (1967).
[5] S. W. Jeffrey, J. Ulrich, and M. B. Allen, *Biochim. Biophys. Acta* **112**, 35 (1966).
[6] A. Givan and R. P. Levine, *Plant Physiol.* **42**, 1264 (1967).
[7] P. Böger, *Z. Pflanzenphysiol.* **61**, 85 (1969).

photophosphorylation rates due to both cyclic *and* noncyclic electron transport.

Particles from *Anacystis nidulans* (Cyanophyceae, Blue-Green Algae)[3,8]

Preparation. Strain No. 1402-1 from the Algae Culture Collection, University of Göttingen, Germany, is grown in the medium of Kratz and Myers[9] at 30° with 3000 lux (fluorescent light) and gassed with CO_2-enriched air (1.5% v/v). The once washed algae are suspended in 5% sucrose containing 20 m*M* Tris buffer, pH 7.6, frozen at −50° to −60° and then lyophilized for 2–3 hours. The dry powder can be stored without loss over P_2O_5 up to 3 days at 5°. For use, it is resuspended in 40 m*M* $MgCl_2$ and lysozyme (Calbiochem, B grade, Los Angeles, California)—2 mg of enzyme per 0.2 mg of chlorophyll—is added. After standing for 3–5 hours at 15°, the particles are collected by centrifugation and resuspended in 5% sucrose–20 m*M* Tris, pH 7.6. They may be washed once with the same medium.

Photophosphorylation proceeds best when the particles are kept at room temperature. No data concerning the efficiency of cell wall lysis with this method were given by the authors.[3,8] However, lyophilization is necessary for the enzymatic action. Photophosphorylation with the same reaction mixture as described below can also be performed with lyophilized cells without lysozyme treatment. The rates are at best half of those from preparations treated with lysozyme.[10]

Reaction Conditions. Photophosphorylation is carried out in Warburg vessels with white light of 35,000 lux. The standard reaction mixture contains in a final volume of 3 ml (in μmoles): Tris buffer, pH 7.6, 60; $MgCl_2$, 40; ADP, 10; inorganic phosphate, 10 (with approximately 10^5 cpm ^{32}P); particles with 0.2 mg chlorophyll. The concentrations of some cofactors are indicated in Table I. Reaction time is 20–30 minutes; temperature, 15°[3] (or 25°[8]). Gas phase is prepurified nitrogen. ATP is determined by the method of Sugino and Miyoshi.[11] For ferredoxin-catalyzed cyclic phosphorylation *N*-2-hydroxyethylpiperazine-*N'*-2-ethanesulfonic acid (Hepes) buffer, pH 7, is used.[8]

Rates. Photophosphorylation rates with particles from *Anacystis* as reported by the Trebst group are comparable to those from spinach as far as the cyclic system with PMS and ferredoxin is concerned (Table I, part A). The rather high endogenous reaction of unwashed particles can

[8] H. Bothe, Thesis, Univ. of Göttingen, Germany, 1968; *Z. Naturforsch. B* **24,** 1574 (1969).
[9] W. A. Kratz and J. Myers, *Amer. J. Bot.* **42,** 282 (1955).
[10] B. Gerhardt and A. Trebst, *Z. Naturforsch. B* **20,** 879 (1965).
[11] Y. Sugino and Y. Miyoshi, *J. Biol. Chem.* **239,** 2360 (1964).

TABLE I
PHOTOPHOSPHORYLATION WITH PARTICLES FROM *Anacystis nidulans* (A) AND *Phormidium luridum* (B)[a]

Additions	Amount added (μmoles/3 ml reaction mixture)	Rates (μmoles ATP/mg Chl × hour)
Part A		
PMS	0.3	137
Ferricyanide	20.0	27
No cofactor; DCMU	— ; 3.0×10^{-3}	13.0
Fd; DCMU	0.10; 3×10^{-3}	61.5
NADP; (−)Fd	6.0; —	10.5
NADP; (+)Fd	6.0; 0.02	32.0
Part B		
No cofactor	— —	1–2
NADP; (−)Fd	6.0; —	132
NADP; Fd + Fd-NADP reductase	6.0; saturating amounts	125
Ferricyanide	10.0	181

[a] Chl, chlorophyll; PMS, *N*-methylphenazinium methyl sulfate; Fd, ferredoxin (prepared from spinach); DCMU, 3-(3,4-dichlorophenyl)-1,1-dimethylurea. The first two lines represent data from B. Gerhardt and R. Santo, *Z. Naturforsch. B* **21,** 673 (1966); the other data of part A are obtained with washed particles and are taken from R. Bothe, *Z. Naturforsch. B* **24,** 1574 (1969), Table 3. Data in part B are from J. Biggins, *Plant Physiol.* **42,** 1447 (1967). Ferredoxin is a crude preparation from *Phormidium* including the ferredoxin-NADP reductase. The amounts added are saturating for a Swiss chard chloroplast assay.

be stimulated by additional ferredoxin; (this was not observed with *Phormidium* particles,[4] Table I, part B). The ratio of ATP formed per two electrons transported in the noncyclic pathway (P:2*e* ratio) is about 0.3, and the phosphorylation rate is relatively low with ferricyanide and NADP. Photophosphorylation with the latter is not improved by adding ferredoxin to the reaction mixture. Repeated washings with dilute buffer, however, remove ferredoxin (and ferredoxin-NADP reductase, EC 1.6.1.1, to some extent), and the reduction of NADP can be partially restored with added ferredoxin.[8] It should be mentioned that the rates may be improved by using shorter reaction times (see below), e.g., in the minute range.

Particles from *Phormidium luridum* (Cyanophyceae)[4,12]

Preparation. The strain (*P. luridum* var. *olivaceae* Boresch) is provided by the Culture Collection of Algae, Indiana University, Blooming-

[12] J. Biggins, *Plant Physiol.* **42,** 1442 (1967).

ton, Indiana, and grown in Kratz and Myers' medium[9] with fluorescent light. The culture is gassed with air supplemented with 4% CO_2 (v/v). Cells are washed once with 0.5 *M* mannitol–30 m*M* K phosphate, pH 6.8, and resuspended in the same medium. Egg-white muramidase (Worthington Biochemical Corp., Freehold, New Jersey) is added in a final concentration of 0.05%, and incubation is at 35° for 2.5 hours. About 70% of the cells lose their cell wall. The cooled mixture is passed through a column (3 × 10 cm) of loosely packed glass wool, and the eluate contains the "protoplasts" (cells minus cell wall).[12] This is centrifuged (500 *g*, 4 minutes), and the preparation is washed once with mannitol–phosphate medium. The pellet is resuspended in 0.5 *M* mannitol–Tris buffer, pH 7.2–7.5.

Cell-free preparations are obtained by diluting the protoplast suspension 8-fold with the standard phosphorylation reaction mixture. Stronger dilution (40-fold) should be avoided since activity will decrease, particularly that associated with the noncyclic electron flow.

Reaction Conditions. Reaction is carried out in Warburg vessels with saturating white light. The final volume of 1.5 ml contains (in μmoles): Tris·maleate, pH 7.5, 40; $MgCl_2$, 30; ADP, 5; $K_2H^{32}PO_4$, 5; particles with 50 μg of chlorophyll and cofactors. Reaction time is 6 minutes at 25°. Thereafter the reaction mixture is inactivated with trichloroacetic acid and ATP formation measured by the method of Avron.[13] Conditions for cyclic photophosphorylation are slightly different.

Rates. Particles from this alga give the highest noncyclic phosphorylation rates so far reported for preparations from the blue-green algae (Table I, part B). No endogenous rate, i.e., without added cofactors, as described for *Anacystis* preparations, is observed, although addition of ferredoxin plus ferredoxin-NADP reductase has no stimulatory effect on NADP reduction (and photophosphorylation). P:2*e* ratios are about 0.8 and higher. Cyclic phosphorylation with high amounts of ferredoxin included in the reaction mixture was not investigated.

Particles from *Anabaena variabilis* (Cyanophyceae)

Krogmann and co-workers recently also succeeded in obtaining noncyclic phosphorylation with *Anabaena*.[14] The cells were grown according to the method of Kratz and Myers[9] and digested with lysozyme by a modified method of Biggins.[12] The photophosphorylation methods used are similar to those described above. The reaction mixture for NADP reduction and phosphorylation contains in a final volume of 3 ml (in

[13] M. Avron, *Biochim. Biophys. Acta* **40**, 257 (1960).

[14] S. S. Lee, A. M. Young, and D. W. Krogmann, *Biochim. Biophys. Acta* **180**, 130 (1969)

μmoles): Tricine–NaOH (*N*-tris-hydroxymethylglycine), pH 7.8, 50; $MgCl_2$, 20; ADP, 10; phosphate, 3; particles with 30 μg of Chl; NADP, 0.5; and ferredoxin plus ferredoxin-NADP reductase. Reaction time is 5 minutes at 25°; light intensity is ≈140,000 lux; gas phase, air. ATP formation is assayed by the use of ^{32}P.[15] A rate of 107 μmoles ATP/mg Chl × hour was obtained, whereas with PMS as cofactor this figure may be as high as approximately 500.

Chloroplasts from *Bumilleriopsis filiformis* Vischer (Xanthophyceae)[7]

Preparation. This alga is a member of the coccoid Xanthophyceae; the cells, however, stick together after division and form filaments consisting of several cells even in agitated cultures. They are grown autotrophically as described previously.[7,16] Each cell has several chloroplasts, which can be released by homogenizing the cells with glass beads (0.5 mm) in a rotary shaker for 1 minute at 4000 rpm at 0°–10° (Braun Comp., Melsungen, Germany). The algae are washed once with 0.9% NaCl and suspended in a freshly prepared ice-cold medium which contains (in millimolar concentration): sucrose, 350; Tricine–NaOH (*N*-tris-hydroxymethylglycine), pH 8, 65; sodium pyrophosphate, 3.5; $MgCl_2$ 1; 2-mercaptoethanol, 1.5; ascorbic acid, 0.4; EDTA, 0.45; NaOH, 6–8; and (in mg/ml) 7.7 mg of bovine serum albumin, and 35 mg of polyvinylpyrrolidone (molecular weight, approximately 24,000, FLUKA, Buchs, Switzerland). The cell number must be 1–3 × 10^7 per milliliter. The homogenate is freed from the beads by a suitable sintered-glass funnel, then centrifuged at 1100 *g* for 2 minutes at 3°. The supernatant solution is again centrifuged at 17,000 *g* for 5 minutes, and the pellet is suspended in a medium consisting of 0.44 *M* sucrose; 8 m*M* Tricine, pH 8; 0.16 m*M* $MgCl_2$, and 0.4 m*M* Na pyrophosphate. Chlorophyll content is 0.5 to 1 mg/ml. Chloroplasts retain full activity for at least 1 hour when kept at 0°C.

Under the light microscope, isolated chloroplasts appear as swollen vesicles with one or two strings of aggregated thylakoids traversing the vesicle.[7] Washing with dilute buffer has little effect on activity, the phosphorylation in the noncyclic system is irreversibly decreased to some extent.

Reaction Conditions. Reactions are performed in open cuvettes of approximately 1 cm light path with white light of 70,000 lux. The final volume of 1 ml contains generally (in μmoles): Mes–NaOH (morpholinoethanesulfonic acid), pH 8, 50; sucrose, 60; $MgCl_2$, 2; P_i plus ^{32}P (0.5–1

[15] W. C. Duane, M. C. Hohl, and D. W. Krogmann, *Biochim. Biophys. Acta* **109,** 108 (1965).

[16] P. Böger and A. San Pietro, *Z. Pflanzenphysiol.* **58,** 70 (1967).

TABLE II
PHOTOPHOSPHORYLATION WITH CHLOROPLASTS FROM *Bumilleriopsis filiformis* AND SPINACH UNDER THE SAME CONDITIONS[a]

Additions	Concentration (m*M*)	Rates in μmoles ATP/mg Chl × hours: *Bumilleriopsis filiformis* 2.3×10^{-5} μg Chl/cell	*Bumilleriopsis filiformis* 4.8×10^{-5} μg Chl/cell	Spinach
No cofactors	—	4	3	1.5
PMS	0.016	370	190	380
Ferricyanide	1.0	204	90	140
4-Ac-Py-I[b]	1.0	161	62	140
Fd	0.037	44	23	90
NADP; (−)Fd	0.5; —	5	4	3
NADP; (+)Fd	0.5; 2.2×10^{-3}	92	57	97

[a] Ferredoxins are from *Bumilleriopsis* and spinach, respectively, and prepared according to P. Böger, *Z. Pflanzenphysiol.* **61,** 447 (1969).

[b] *N*-Methyl-4-acetylpyridinium iodide [see P. Böger *et al., Biochemistry* **6,** 80 (1967) for this compound].

μCi), 1; ADP, 1.2; and algal chloroplasts with 8–20 μg of chlorophyll. The concentrations of some cofactors are listed in Table II. Reaction time is 3 minutes at 22°. Anaerobic experiments are run in Warburg vessels, flushed with nitrogen. After the phosphorylation reaction the chloroplasts are denatured by acidifying the reaction mixture with 0.4 *N* $HClO_4$; ATP is determined by a modification of the method of Avron.[13]

It should be mentioned that Mes buffer has an unfavorable pK and Hepes buffer may be substituted for it in case no ferredoxin is included in the reaction mixture.

Rates. The phosphorylation rates obtained are as high as with spinach chloroplasts under the same conditions (Table II). The endogenous rate without cofactors is low, ferredoxin has to be added to accomplish reduction of NADP and the phosphorylation accompanied by it. The P:2*e* ratio is 0.6–0.7. Anaerobiosis has no striking effect on the noncyclic electron flow or on PMS-mediated phosphorylation. There is ATP formation with added ferredoxin under air, but the rate is almost nil under nitrogen also when DCMU is present.[17]

Rates (expressed as ATP formed per chlorophyll of isolated chloroplasts) vary inversely with the chlorophyll content of the cell (Table II). Recent experiments (unpublished) indicate that this is due to different ratios of chlorophyll to (photosynthetic) chloroplast enzymes. Some of

[17] P. Böger, *Z. Pflanzenphysiol.* **61,** 447 (1969).

these quantitative relationships have been described for *Chlorella*[18] and *Euglena*.[16] In the latter, the amounts of ferredoxin and cytochrome *f* (552) are constant per cell, whereas the chlorophyll content increases during the cultivation time.

[18] A. Pirson and P. Böger, *Nature* (*London*) **205**, 1129 (1965); P. Böger, *Flora* **154**, 174 (1964).

[22] Chloroplasts (and Grana): Photosynthetic Electron Transport in Aldehyde-Fixed Material

By R. B. Park

Principle

Formaldehyde and glutaraldehyde treatment of isolated chloroplasts and various algae yields chemically fixed material capable of performing photosystem I and photosystem II reactions at reduced rates and quantum efficiencies.[1-3] The fixation procedure stabilizes the material toward subsequent treatments, such as detergent extraction.[4] Treated green, red, and blue-green algae become permeable to reaction mixture components without loss of accessory pigments in the latter two groups.[3,4] In proteins, glutaraldehyde reacts with ϵ-amino groups of lysine and to some extent with tyrosine, histidine, and sulfhydryl residues.[5] Both inter- and intramolecular cross-linkages occur.[6]

Reagents

Glutaraldehyde. Glutaraldehyde is available commercially as a technical grade aqueous solution or as purified material. Treatment of the technical grade is necessary to obtain satisfactory glutaraldehyde to preserve photosynthetic electron transport. About 20 mg of animal bone charcoal is added to 50 ml of a 20% glutaraldehyde solution.[7] This suspension is stirred at room temperature for 20 minutes. The charcoal is then removed by filtration through three layers of Whatman No. 1 filter

[1] R. B. Park, J. Kelly, S. Drury, and K. Sauer, *Proc. Nat. Acad. Sci. U.S.* **55**, 1056 (1966).
[2] U. W. Hallier and R. B. Park, *Plant Physiol.* **44**, 535 (1969).
[3] C. J. Ludlow and R. B. Park, *Plant Physiol.* **44**, 540 (1969).
[4] U. W. Hallier and R. B. Park, *Plant Physiol.* **44**, 544 (1969).
[5] A. F. S. A. Habeeb and R. Hiramoto, *Arch. Biochem. Biophys.* **126**, 16 (1968).
[6] F. M. Richards and J. R. Knowles, *J. Mol. Biol.* **37**, 231 (1968).
[7] P. J. Anderson, *J. Histochem. Cytochem.* **15**, 652 (1967).

paper. The resulting filtrate is washed with charcoal a second time as described above and is again clarified by filtration. This procedure removes from technical grade glutaraldehyde an inhibitory component that completely inactivates photosystem II reactions in isolated chloroplasts. The resulting glutaraldehyde solution should be close to neutral pH. Low pH indicates incompletely washed material. The final glutaraldehyde concentration is approximately 20% (w/v).

Formaldehyde. Formaldehyde is prepared from paraformaldehyde. Paraformaldehyde, 3.3 g, is placed in 100 ml of 30 mM (K) phosphate buffer, pH 7.4, and is heated to 75° with stirring for 5 minutes. The resulting solution is filtered through Whatman No. 1 filter paper and is titrated to pH 7.4 if necessary. The final formaldehyde concentration is approximately 3% (w/v).

Fixation Procedure

Leaves. Spinach leaves are cut with a razor blade into 1- to 2-cm squares and placed in a beaker with an equal volume of 6% glutaraldehyde at room temperature. The beaker is placed in a vacuum desiccator, evacuated for about 2 minutes, and returned to atmospheric pressure. The operation is repeated four times to fill the intercellular spaces of the leaf completely with fixative. The beaker containing the leaves in glutaraldehyde is then placed in the dark at room temperature for 1 hour. The leaves are then drained and rinsed with water. They are homogenized in water or dilute buffer either with a mortar and pestle or in a Waring Blendor and are poured through cheesecloth. A chloroplast fraction is collected by centrifuging the homogenate at 300 g for 2–3 minutes. The fraction is contaminated by attached mitochondria and nuclei. The crude preparation is then washed four times in distilled water or dilute buffer to remove an unidentified reductant present in the homogenate. The final precipitate is resuspended in buffer and is used in photochemical assays.

Chloroplasts. Isolated class II spinach chloroplasts in grinding buffer (50 mM Tricine, pH 7.4, 0.5 M sucrose, at 3°) are centrifugally washed three times to remove most of the stroma material. The final precipitate is resuspended in a small amount of grinding buffer. An equal volume of cold 10% glutaraldehyde solution containing 0.5 M sucrose is added to the suspension over a period of 2 minutes with stirring. The suspension is stirred in an ice bath for 20 minutes, followed by dilution with 10 volumes of cold grinding buffer. It is centrifuged for 8 minutes at 2500 g. The resulting precipitate is resuspended in grinding buffer containing 0.2 M methylamine, precipitated at 2500 g, and washed 2 times with grinding buffer. Methylamine is introduced to react with the excess aldehyde functions in the preparation. The final precipitate is resuspended in buffer and is used in photochemical assays.

Porphyridium cruentum, a unicellular red alga, is treated for 10 minutes at room temperature with 1.5% formaldehyde or for 4 minutes with 0.4% glutaraldehyde in 30 m*M* (K) phosphate buffer (pH 7.4). The cells are collected by centrifugation and are resuspended in 0.25 *M* mannitol, 30 m*M* (K) phosphate, pH 7.4. Glutaraldehyde-fixed cells are washed with 0.2 *M* methylamine before the final resuspension.

Chlorella pyrenoidosa may be subjected to the fixation procedure shown above though the glutaraldehyde concentration is not critical. Concentrations of glutaraldehyde from 0.1–6% yield active material.

Anacystis nidulans: Cells are preilluminated for 1 hour in growth medium and are then fixed in 1.5% formaldehyde for 4 minutes in 30 m*M* (K) phosphate, pH 7.4. Cells are resuspended as indicated for *Porphyridium.* Glutaraldehyde fixation is carried out in 10% glutaraldehyde for 30 minutes at room temperature in 30 m*M* (K) phosphate, pH 7.4. The cells are washed with methylamine and resuspended as described for *Porphyridium.*

Photochemical Properties

Chloroplasts from fixed leaves:

Photosystem II—Quantum conversion efficiency about 25% that of unfixed chloroplasts. Maximum rate of DCPIP reduction, 16 μM/hr per milligram of chlorophyll.

Photosystem I—Not investigated.

Fixed Chloroplasts:

Photosystem II—Maximum rate of DCPIP reduction, 80 μM/hr per milligram of chlorophyll.

Photosystem I—Maximum rate of methyl viologen reduction, 84 equivalents/hr per milligram of chlorophyll.

Algae:

Photosystems I and II present in fixed *Porphyridium* and *Chlorella* and *Anacystis.* Rates comparable to those for fixed chloroplasts.

[23] Chloroplast Preparations Deficient in Coupling Factor 1

By RICHARD E. MCCARTY

Photophosphorylation may be resolved into a soluble fraction and a particulate fraction.[1,2] The soluble fraction contains coupling factor 1 (CF_1), a latent, Ca^{2+}-dependent ATPase.[2] Photophosphorylation in the particulate fraction can be partially restored by the addition of CF_1 and Mg^{2+}.[1,2] Two methods have been used to deplete chloroplasts of their CF_1 content. One method involves the treatment of chloroplasts with dilute solutions of EDTA[1,3]; the other, sonic oscillation of chloroplasts followed by the extraction of the subchloroplast particles obtained with buffers of low ionic strength.[2]

Preparation of EDTA-Treated Chloroplasts[4]

Forty-five grams of washed, deveined spinach leaves were homogenized in a Waring Blendor for 10–15 seconds in 65 ml of a buffer which contained 20 m*M* Tricine–NaOH (pH 8.0), 10 m*M* NaCl, and 0.4 *M* sucrose. The chloroplasts were collected by centrifugation and resuspended in 30 ml of 10 m*M* NaCl. The resulting suspension was centrifuged at 3000 *g* for 10 minutes, and the pellets were resuspended in a small volume of 10 m*M* NaCl. The chlorophyll concentration of the suspension was determined and was adjusted by the addition of 10 m*M* NaCl to about 1.5 mg/ml. An aliquot of this suspension was diluted to a chlorophyll concentration of 0.1 mg/ml by the addition of 0.75 m*M* EDTA (pH 8.0). After 5 minutes, the suspension was centrifuged at 45,000 *g* for 20 minutes. The pellets were resuspended in a small volume of 10 m*M* NaCl. The supernatant fluid was clear and nearly colorless and was a convenient source of CF_1.

EDTA-treated chloroplasts catalyze pyocyanine-dependent cyclic phosphorylation at rates ranging from 1 to 25 μmoles P_i esterified per hour per milligram of chlorophyll. The stimulation of phosphorylation in EDTA-treated chloroplasts by saturating amounts of CF_1 ranged from 4- to 50-fold, depending on the activity of the preparation in the absence of added CF_1. Similar stimulations were observed in phosphorylation

[1] M. Avron, *Biochim. Biophys. Acta* **77**, 699 (1963).
[2] V. K. Vambutas and E. Racker, *J. Biol. Chem.* **240**, 2660 (1965).
[3] R. E. McCarty and E. Racker, *Brookhaven Symp. Biol.* **19**, 202 (1966).
[4] R. E. McCarty and E. Racker, *J. Biol. Chem.* **242**, 3435 (1967).

with other electron acceptors. On a few occasions, the EDTA-treatment, as described above, either failed to inhibit phosphorylation or inhibited it so severely that the addition of CF_1 did not restore phosphorylation. EDTA-treated chloroplasts lose activity rapidly under a variety of storage conditions and, therefore, should be used as soon after preparation as possible.

Resolved Subchloroplast Particles

A clarified suspension of azolectin[5] (Associated Concentrates Inc., Woodside, Long Island, New York) was prepared by homogenizing 5 g of azolectin with 50 ml of a solution containing 0.4 *M* sucrose, 10 m*M* Tricine–NaOH (pH 8), 0.5 m*M* EDTA, and 10 m*M* α-thioglycerol. The suspension was exposed to sonic oscillation for either 10 minutes with the Branson 20 kHz instrument at full output or for 30 minutes with the Raytheon 10 kHz instrument. The temperature during sonication was not allowed to rise above 10°.

The clarified suspension was dialyzed for 2–3 hours against 500 ml of the medium described above. The medium was then renewed and the suspension was further dialyzed for 12–14 hours. This dialyzed suspension was centrifuged at 104,000 *g* for 1 hour and the supernatant solution stored at 4° under an atmosphere of nitrogen.

To 14 ml of a chloroplast preparation (1.5–2 mg of chlorophyll per milliliter) suspended in a medium containing 0.4 *M* sucrose, 0.01 *M* NaCl, and 0.02 *M* Tricine–NaOH (pH 8) was added 6 ml of the clarified azolectin suspension. The resulting mixture was exposed to sonic oscillation for a total of 60 seconds using the Branson 20 Kc instrument equipped with a 0.5-inch solid step horn. The temperature of the mixture was kept below 8° during the sonication. The resulting subcloroplast particles were isolated by centrifugation as described in the preparation of phosphorylating subchloroplast particles.[6] The pellets were resuspended in 20 ml of a solution containing 2.5 m*M* Tricine–NaOH (pH 8) and 400 m*M* sucrose. The particles were again collected by centrifugation at 104,000 *g* for 45 minutes, and the pellets were resuspended in a small volume of a medium containing 20 m*M* Tricine–NaOH (pH 8), 10 m*M* NaCl, and 0.4 *M* sucrose. Aliquots of this suspension were stored at —80° under an atmosphere of nitrogen.

This method for the preparation of CF_1-deficient chloroplast fragments was not as reproducible as the EDTA-treatment method. Less than half of the preparations showed substantial stimulations of photophos-

[5] T. E. Conover, R. L. Prarie, and E. Racker, *J. Biol. Chem.* **238,** 2831 (1963).

[6] See this volume [30].

phorylation by CF_1. In many cases, preparations of resolved subchloroplast particles had only a very low rate of cyclic phosphorylation (less than 1 μmole of P_i esterified per hour per milligram of chlorophyll) and the addition of CF_1, had little effect. The major advantage of the resolved subchloroplast particles is that they may be stored for several weeks with little change in the activities they catalyze.

Assay for Coupling Factor Activity

EDTA-treated chloroplasts or resolved subchloroplast particles containing approximately 50 μg of chlorophyll were incubated in an ice bath with reagents added in the order given: 1 mg of defatted bovine serum albumin,[7] CF_1 (25 μg of protein if the purified protein was used, and 40–60 μg of protein if EDTA extracts were used), 0.5 μmole of EDTA (pH 7.9), and 5 μmoles of $MgCl_2$ in a final volume of 0.5 ml. After 15 minutes, a reaction mixture containing 50 μmoles of Tricine–NaOH (pH 8), 50 μmoles of NaCl, 3 μmoles of ADP, 2 μmoles of potassium phosphate buffer (pH 8) containing about 10^6 cpm of $^{32}P_i$ and 0.05 μmole pyocyanine or *N*-methyl phenazonium methosulfate or 1 μmole $K_3Fe(CN)_6$ was added in a final volume of 0.5 ml. The tubes were flushed with nitrogen for 15 seconds, tightly stoppered and illuminated for 2–5 minutes with about 2×10^6 erg/cm²/sec of white light. Esterified $^{32}P_i$ was determined in deproteinized aliquots of the reaction mixtures by the procedure of Lindberg and Ernster.[8]

[7] R. E. Chen, *J. Biol. Chem.* **242,** 173 (1967).

[8] O. Lindberg and L. Ernster, *Methods Biochem. Anal.* **3,** 1 (1956).

[24] Chloroplasts (and Grana): Heptane Treated

By CLANTON C. BLACK, JR.

Preparation

Principle. In the presence of a suitable electron acceptor isolated chloroplasts, when illuminated, readily synthesize ATP. Depending upon the exogenous acceptor added to illuminated chloroplasts, either a cyclic flow of electrons occurs, but without net production of a reductant, or noncyclic electron flow occurs and a reductant is produced. Concurrent with both types of electron flow, ATP is synthesized. However, a satisfactory procedure to distinguish whether or not ATP is synthesized at the same site or at a different site(s) during cyclic compared to noncyclic electron flow was not available until the discovery of a procedure for treating isolated chloroplasts with *n*-heptane. Heptane-treated chloroplasts synthesize ATP concurrent with cyclic electron flow. Noncyclic electron flow remains functional; however, the concurrent ATP synthesis is inhibited. The following procedure may be used to prepare spinach chloroplasts which synthesize ATP only during cyclic electron flow.

Reagents

Isolated spinach chloroplast or grana[1] suspension, approximately 1 mg of chlorophyll[2]/ml

n-Heptane[3]

Procedure. The routine procedure[4,5] employed for treatment of chloroplasts consists of layering 0.1 ml of *n*-heptane over 1 ml of chloroplasts in a test tube, 15×150 mm. The tube is placed in a Burrell "wrist action" shaker (130–150 strokes/minute) for 10 seconds. The tube is then stored in an ice bucket, and aliquots of chloroplasts are pipetted from beneath the heptane layer as needed. The chlorophyll concentration is not affected by this treatment.

Assays

Noncyclic electron transport and concurrent ATP synthesis with $NADP^+$ as the Hill oxidant is assayed in the presence of the following reagents, in 1-ml-volume cuvettes: 50 μmoles of Tricine[6]–KOH buffer, pH 7.9; 2 μmoles of $MgCl_2$; 0.3 μmole of NADP; 1 μmole of phosphate, containing about 0.5 μCi of ^{32}P; 1 μmole of ADP; about 1 mg of pure spinach ferredoxin[7]; and about 20 μg of chlorophyll as chloroplasts or grana. The optical density change at 340 nm of the suspension after 1 minute of illumination at 2000–3000 foot candles is taken as a measure of NADP reduction.[8] Addition of 0.1 ml of 20% trichloroacetic acid[9] will stop the reaction and chloroplasts, etc., can be removed by low speed centrifugation, 1000 *g* for 10 minutes, and ATP-^{32}P concentration

[1] This volume [18, 19].

[2] This volume [18].

[3] Routinely "Certified Spectranalyzed" *n*-heptane was purchased from the Fisher Scientific Company, Fair Lawn, New Jersey.

[4] C. C. Black, *Biochem. Biophys. Res. Commun.* **28**, 985 (1967).

[5] L. J. Laber and C. C. Black, *J. Biol. Chem.* **244**, 3463 (1969).

[6] Tris(hydroxymethyl)methylglycine purchased from the Calbiochem Company, Los Angeles, California.

[7] This volume [39].

[8] A. Trebst, this series, Volume XXIV.

[9] Purchased from the Fisher Scientific Company, Fair Lawn, New Jersey.

determined in the supernatant solution. An alternative assay is to omit NADP and ferredoxin from the assay mixture given above and to add 0.3–0.4 μmole of $K_3Fe(CN)_6$. Then the change in optical density at 420 nm upon illumination for 1–3 minutes is proportional to ferricyanide reduced.[8] ATP-^{32}P is determined after the addition of trichloroacetic acid.

Cyclic photophosphorylation is determined in the assay mixture given above omitting $NADP^+$ and ferredoxin and/or ferricyanide. PMS[10] is substituted at about 20 μ*M* final concentration in the assay mixture and following illumination ATP-^{32}P is determined after the addition of trichloroacetic acid.

Activities and Properties of Heptane-Treated Spinach Chloroplasts

Immediately[11] after heptane treatment, cyclic and noncyclic photophosphorylation and electron transport are not detectably affected.[4,5]

Cyclic photophosphorylation is inhibited in heptane-treated chloroplasts only about 50% even several hours after treatment. This level of inhibition is reached in about 10–40 minutes after treatment.[4]

Noncyclic photophosphorylation is inhibited about 90–95% within 10–40 minutes after heptane treatment.[4,5] Noncyclic electron flow, however, functions either at the same rate or is slightly stimulated if ADP and P_i are present in the assay mixtures. When ADP and P_i are omitted from the assay mixture, a more pronounced stimulation of electron flow occurs which is sometimes 50% above the control values.[4,5]

In addition to these effects, the treatment of chloroplasts with *n*-heptane results in changes in the activity of other chloroplast components.[12] A differential effect is observed upon the adeosine triphosphatase activities in chloroplasts in that the trypsin-activated ATPase is not affected by heptane treatment, whereas the Mg^{2+}-dependent ATPase activated by light in the presence of dithiothreitol is inhibited 80–90% within 10–30 minutes after treatment. The ability of heptane-treated chloroplasts to accumulate protons upon illumination is inhibited 40–60% and their ability to synthesize ATP when transferred from an acidic to alkaline environment is inhibited 50%.

The fast 520 nm change of spinach chloroplasts is completely inhibited within 40 minutes after heptane treatment, the 50-millisecond

[10] Phenazine methyl sulfate purchased from the Sigma Chemical Company, St. Louis, Missouri.

[11] Within about 2 minutes after shaking the heptane–chloroplast mixture.

[12] Methods for assaying these activities and more detailed data are given in references cited in footnotes 4 and 5.

and the 3.5-millisecond delayed light emission are 80–95% inhibited within 40 minutes, but the reversible P700 change is completely unaffected for periods up to 4 hours after heptane treatment.[13]

Comments

Note should be made that in a multiphase system such as water, chloroplasts, and a nonpolar solvent (*n*-heptane) deviations from the described procedures may give different results from those presented. For example, shaking the H_2O–chloroplasts–heptane mixture for 30 seconds or more will eliminate the stimulation of electron transport. Other nonpolar solvents, e.g., hexane, cyclohexane, carbon tetrachloride, benzene, and toluene also were tested as in the routine procedure, and different results were obtained.[5]

[13] C. C. Black and B. Mayne, unpublished observations.

[25] Bacterial Chromatophores

By Albert W. Frenkel and Rodger A. Nelson

The term "chromatophore" was originally applied to the relatively large particles isolated from photosynthetic bacteria containing all the photosynthetic pigments of the original cell.[1] The term has been widely accepted, but questions have been raised about its validity. It was recognized that, in contrast to chloroplasts and mitochondria, chromatophores are most likely formed as the result of the breakdown of a more complex membrane system existing within the organism,[2–4] and consequently, the term "chromatophore fragment" was proposed.[5] The latter term, however, has not been generally employed by investigators studying the photochemical and biochemical activities of chromatophore preparations; also such a term might lead to confusion with the term "subchromatophore fragment" (cf. this volume [31, 32]). More recently, the term "thylakoid" ("vesiclelike") has been introduced into the literature relating to the structure of photosynthetic lamellae.[6] This term appears

[1] H. K. Schachman, A. B. Pardee, and R. Y. Stanier, *Arch. Biochem. Biophys.* **38**, 245 (1952).
[2] A. L. Tuttle and H. Gest, *Proc. Nat. Acad. Sci. U.S.* **45**, 1261 (1959).
[3] D. D. Hickman and A. W. Frenkel, *J. Biophys. Biochem. Cytol.* **6**, 277 (1959).
[4] G. Cohen-Bazire and R. Kunisawa, *J. Cell Biol.* **16**, 401 (1963).
[5] M. D. Kamen, *in* "Bacterial Photosynthesis" (H. Gest, A. San Pietro, and L. P. Vernon, eds.), p. 448. Antioch Press, Yellow Springs, Ohio, 1963.
[6] W. Menke, *Annu. Rev. Plant Physiol.* **13**, 27 (1962).

to have found general acceptance in the cytological literature; however, there appears to be reluctance to apply the term to isolated, photochemically active particulates from photosynthetic bacteria, unless the vesicular structure of the isolated particles can actually be demonstrated.

One other shortcoming of the term "chromatophore" (meaning "bearing color") is that it is not very specific, as it has been used for pigmented structures in lower and higher plants and animals, and one probably should speak of "bacterial chromatophores" or "chromatophores derived from photosynthetic bacteria."

Definition of "Bacterial Chromatophore"

An operational definition can be formulated as follows: Chromatophores isolated from photosynthetic bacteria represent particulate structures which contain the photosynthetic pigments of the cells from which they were derived[1]; they also contain catalytic systems active in photochemical and dark reactions[7-11]; in size they are comparable to cytoplasmic vesicles or invaginations of the plasmalemma existing in cells from which they were isolated.[12-15]

It should be pointed out, however, that in some cases where the term "chromatophore" appears in the biochemical literature, no effort was made to examine the material for homogeneity, for nucleic acid content, or for the appearance of such preparations under the electron microscope. Consequently such preparations may have contained a considerable size range of particles including ribosomes, and possibly other cell components.

Photochemical Activities of Isolated Chromatophores

A number of photochemical reactions have been shown to be associated with suitably prepared chromatophores, such as the formation of at least one free radical, spectral changes in photosynthetic pigments

[7] E. C. Weaver, *Annu. Rev. Plant Physiol.* **19,** 283 (1969).

[8] L. P. Vernon, *Bacteriol. Rev.* **32,** 243 (1968).

[9] J. Lascelles, *in* "Advances in Microbial Physiology" (A. H. Rose and J. F. Wilkinson, eds.), Vol. 2, p. 1. Academic Press, New York, 1968.

[10] E. N. Kondrat'eva, "Photosynthetic Bacteria." Office of Technical Services, U.S. Department of Commerce, Washington, D.C. (1965).

[11] A. W. Frenkel, *Annu. Rev. Plant Physiol.* **10,** 53 (1959); *Biol. Rev.* **45,** 569 (1970).

[12] A. W. Frenkel and D. D. Hickman, *J. Biophys. Biochem. Cytol.* **6,** 285 (1959).

[13] J. A. Bergeron and R. C. Fuller, *in* "Macromolecular Complexes" (M. V. Edds, Jr., ed.), p. 179. Ronald, New York, 1961.

[14] D. D. Hickman and A. W. Frenkel, *J. Cell Biol.* **25,** 261 (1965).

[15] S. C. Holt and A. G. Marr, *J. Bacteriol.* **89,** 1402 (1965).

and cytochromes, photophosphorylation, photoreduction of pyridine nucleotides, photooxidations, pH changes, ion movements.[7-11]

Chloroplasts, when prepared under suitable conditions, will evolve oxygen in the light and simultaneously fix carbon dioxide into hexoses at rates comparable to those observed for overall photosynthesis of intact leaves.[16] Isolated chromatophores, however, when illuminated do not carry out active carbon dioxide fixation unless certain "soluble" components are added to such chromatophore preparations.[17] For the present, therefore, it does not appear proper to speak of isolated chromatophores as complete systems of bacterial photosynthesis. Instead, it appears preferable to refer to the catalytic activities of isolated chromatophores, involving both photochemical and thermochemical reactions. Evidence for the actual involvement of such partial reactions in the photosynthetic processes of the living cells is analyzed in recent reviews.[7-9]

Biochemical Uniqueness of Bacterial Chromatophores

Bacterial chromatophores exhibit certain biochemical properties which in part are similar to those of chloroplasts and in other respects are more comparable to those of plant or animal mitochondria. For instance, chromatophores carry out cyclic photophosphorylation similar to that of chloroplasts[18]; on the other hand, chromatophores like mitochondria catalyze a dark oxidation of certain substrates via an electron carrier system with oxygen as the final electron acceptor, and this oxidation may be accompanied by phosphorylation.[19] Thus, chromatophores present the investigator with an unusual combination of metabolic systems present in chloroplasts and mitochondria of higher organisms.

Biochemical Versatility of Bacterial Chromatophores

When a strain of photosynthetic bacteria is cultured under a variety of environmental conditions (some of which are listed below), one may detect changes in the yield of chromatophores, their pigment content, as well as other biochemical changes. Furthermore, one observes marked differences in the biochemical activities of chromatophores derived from different strains, species, and genera, examples of which are cited below. Important variations can be introduced by the use of mutant strains; for instance, recent advances in the study of reaction centers (described

[16] D. A. Walker, C. W. Baldry, and W. Cockburn, *Plant Physiol.* **43**, 1419 (1968).

[17] B. Buchanan, M. C. W. Evans, and D. I. Arnon, *Arch. Mikrobiol.* **59**, 32 (1967).

[18] M. Nozaki, K. Tagawa, and D. I. Arnon, *in* "Bacterial Photosynthesis" (H. Gest, A. San Pietro, and L. P. Vernon, eds.), p. 175. Antioch Press, Yellow Springs, Ohio, 1963.

[19] D. M. Geller, *J. Biol. Chem.* **237**, 2947 (1962).

in [67, 69]) have made effective use of certain carotenoid-deficient mutants of species of *Rhodospirillum* and *Rhodopseudomonas*.[20]

Relative Stability of Biochemical Reaction Systems of Chromatophores

A useful advantage of chromatophores over isolated chloroplasts or mitochondria is the relative stability of a number of their metabolic reaction systems at temperatures above freezing to about 3°. Storage of chromatophores in aqueous ethyleneglycol below —10° preserves photophosphorylation activity for several weeks[21]; certain other metabolic activities are also retained by lyophilization or during low temperature storage in aqueous media. Intact cells stored in aqueous media below —10°, when thawed and disintegrated, will yield active chromatophores which can catalyze photophosphorylation, the photoreduction of NAD, and numerous other metabolic reactions.

Comments on the Culture of Photosynthetic Bacteria as They Relate to the Isolation of Chromatophores

Culture conditions may not only affect the yield, structure, and chemical composition of the organism under study, but they also may condition the structure and biochemical activities of the chromatophores which can be isolated from them. Excellent reviews of culture conditions for these microorganisms are available.[22-24] For each detailed isolation procedure of chromatophores described below, the general culture conditions of the experimental organism will be described briefly.

Should the need arise for the isolation of chromatophores on a larger scale than has been reported in the literature thus far, it should be pointed out that several species of photosynthetic bacteria have been cultured successfully in the light on a relatively large-scale (20- to 25-liter bottles).[25,26] It should also be possible to culture certain species of the Athiorhodaceae on a large scale aerobically in the dark, or at reduced oxygen tensions if the formation of photochemically active chromatophores is desired.[27]

[20] D. W. Reed and R. K. Clayton, *Biochem. Biophys. Res. Commun.* **30**, 471 (1968).
[21] D. M. Geller and F. Lipmann, *J. Biol. Chem.* **235**, 2478 (1960).
[22] C. B. van Niel, *Bacteriol. Rev.* **8**, 1 (1944).
[23] S. K. Bose, *in* "Bacterial Photosynthesis" (H. Gest, A. San Pietro, and L. P. Vernon, eds.), p. 501. Antioch Press, Yellow Springs, Ohio, 1963.
[24] N. Pfennig, *Annu. Rev. Microbiol.* **21**, 285 (1967).
[25] P. Karrer and U. Solmssen, *Helv. Chim. Acta* **18**, 1306 (1935).
[26] T. E. Meyer, R. G. Bartsch, M. A. Cusanovich, and J. H. Mathewson, *Biochim. Biophys. Acta* **153**, 854 (1968).
[27] D. C. Pratt, A. W. Frenkel, and D. D. Hickman, *in* "Biological Structure and

Procedures Employed for the Isolation of Chromatophores from Photosynthetic Bacteria

The isolation procedures for chromatophores are organized here under the families[28] of the photosynthetic bacteria and are subdivided into those genera and species for which isolation procedures have been worked out in some detail. The grouping starts with the Athiorhodaceae, as certain species of this family have been studied in relatively greater detail with respect to fractionation procedures and the structure and function of isolated chromatophores.

Presently described isolation and fractionation procedures fall into the following groups:

I. Mechanical rupture of cells and dispersal of cell contents by sonic oscillation, or by the French pressure cell, or by grinding with alumina, etc.

II. Enzymatic disruption of cell walls formation of spheroplasts, followed by osmotic lysis or mild homogenization.

Procedures I and II may be followed by one or more of the following treatments:

1. DNase treatment
2. Differential centrifugation
3. Gradient centrifugation
4. Gradient electrophoresis
5. Other techniques possibly applicable to chromatophore isolation or purification, such as chromatography, treatment with enzymes other than DNase, or lysozyme, have not been reported on in sufficient detail to be included here.

Athiorhodaceae (Nonsulfur Purple Bacteria)

The species of this family generally are cultured photoorganotrophically under anaerobic or microaerophilic conditions. It is possible to culture certain species anaerobically in the light on a mineral medium, provided traces of certain growth factors are present.[22-24] Many species can be grown organotrophically in the dark at reduced oxygen tensions, while certain strains can be grown at normal atmospheric oxygen tension; in the latter case photosynthetic pigment production is completely or

Function" (T. W. Goodwin and O. Lindberg, eds.), Vol. II, p. 295. Academic Press, New York, 1961.

[28] C. B. van Niel, *in* "Bergey's Manual of Determinative Bacteriology" (R. S. Reed, E. G. D. Murray, and N. R. Smith, eds.), p. 35. Williams and Wilkins, Baltimore, Maryland, 1957.

almost completely suppressed.[27] Certain pigment mutants are available which can add greatly to the scope of experimental investigations.[9,20]

Rhodospirillum

A. Sonic Disruption Followed by Differential Centrifugation and Gradient Centrifugation; Photochemical Activities[12]

Step 1. Cultures of *R. rubrum* are grown photoorganotrophically at 30° for 18–24 hours in 1-liter bottles under anaerobic conditions, harvested, and washed twice with distilled water; finally they are suspended in cold 0.2 *M* potassium glycylglycine buffer, pH 7.5–7.6 (Gg-7.5), to yield a dense suspension. The cells are disrupted in a 10 kHz Raytheon magnetostrictive oscillator at 1°–3° for 1.5–2.5 minutes. The resulting suspension is centrifuged for 30 minutes at 25,000 *g* (g at R_{max}) at 1°–5°. (The sediment is discarded or pooled and used for isolation of pigments, enzymes, etc.).

Step 2. The supernatant from step 1 is centrifuged 50 minutes at 55,000–60,000 *g*. (The first high speed supernatant solution is discarded or saved for the isolation of smaller chromatophore fragments, ribosomes, soluble enzymes, nucleotides, etc.). The sediment is resuspended with a syringe in 0.1 *M* Gg-7.5 buffer in a ratio of about 1 part sediment to 10 parts of buffer. The resulting suspension is recentrifuged for 50 minutes at 50,000 *g*, and the resuspended pellet may be recentrifuged in the same manner. The supernatant solutions of the second and third high speed centrifugation, if clear, are discarded.

Step 3. The resuspended pellet from step 2 can be subjected to further fractionation by gradient centrifugation, details of which are given below in sections B and C for *Rhodospirillum*.

Results. A chromatophore preparation is obtained showing particles of relatively uniform size, with few apparent contaminating particles. From a 1-liter culture, it is possible to obtain 1–2 g (wet weight) of partially purified chromatophores within 4–5 hours starting with the harvest of the cells. A similar procedure has been described by Drews[29] for *R. rubrum* and for *R. molischianum* (cf. reference 14).

Preparations obtained in this manner carry out photophosphorylation,[30] NAD photoreduction,[27] a number of other light-induced reactions and several dark reactions catalyzed by enzymes bound more or less firmly to the chromatophores.[7–11]

[29] G. Drews, *Arch. Mikrobiol.* **48,** 122 (1964).

[30] A. W. Frenkel and K. Cost, *in* "Comprehensive Biochemistry" (M. Florkin and E. H. Stotz, eds.), Vol. 14, p. 397. Elsevier, Amsterdam, 1966.

B. Disruption by French Pressure Cell Followed by Differential Centrifugation and Centrifugation in Linear Ficoll Gradient; Removal of Ribosomes[31]

Step 1. Cultures of *R. rubrum* are grown photoorganotrophically in 50-ml bottles at 30°. Cells are harvested by centrifugation after 24 hours or when cultures are in exponential phase and are washed once with cold 20 m*M* Tris buffer, pH 7.8 (Tris-7.8). Cells are resuspended in 10 ml of 20 m*M* Tris-7.8 and passed twice through a precooled French pressure cell at 20,000 psi. The resulting homogenate is centrifuged at 4° for 20 minutes at 16,000 *g*. (It is assumed that the *g* values given here refer to *g* values at R_{max}.) The sediment is discarded.

Step 2. The low speed supernatant solution is centrifuged for 60 minutes at 177,000 *g*. The resulting pellet is resuspended in a minimal amount of 20 m*M* Tris-7.8 buffer

Step 3. Of the suspension prepared in step 2, 0.3 ml is placed on a linear gradient of Ficoll in 20 m*M* Tris-7.8 buffer, density range 1.02–1.06. The tubes are centrifuged for 2 hours at 83,000 *g* in a swinging-bucket rotor. Two pigment bands appear; they are recovered from the gradient by conventional methods.

Results. The lower band resulting from step 3 contains chromatophores. They appear as round particles with an average diameter of 73 nm in the electron microscope. The upper band contains particles with an average diameter of 36 nm which appear to be composed of cytoplasmic membrane fragments, etc. In extracts of dark aerobically grown cells (mutant strain M-46) which are free of pigments, the upper band is the only predominant component which appears in the gradient. The authors present data which indicate that Mg^{2+} causes the binding of 260 nm absorbing material to the isolated chromatophores. They suggest the use of 0.1% EDTA in the buffer solutions for the elimination of ribosomes and to prevent clumping of chromatophores. (Investigators using EDTA should check possible effects on photophosphorylation, etc.). No studies on photochemical or dark enzymatic reactions of the preparations are reported in this paper.

C. Disruption by French Pressure Cell Followed by DNase Treatment, Differential Centrifugation, Centrifugation in Continuous Sucrose Gradient and Gradient Electrophoresis; Removal of Ribosomes[32]

Step 1. Cultures of *R. rubrum* are grown photoorganotrophically in 56-ml test tubes or 1.25-liter Roux bottles sparged with 95% N_2 and

[31] J. Oelze, M. Biedermann, and G. Drews, *Biochim. Biophys. Acta* **173**, 436 (1969).
[32] S. C. Holt and A. G. Marr, *J. Bacteriol.* **89**, 1413 (1965).

5% CO_2 and illuminated with 375-W Sylvania movie lights, giving an incident intensity of 73 foot-candles. Cells in the exponential phase are harvested at 4°, washed by three successive centrifugations with cold deionized water at 3000 *g* for 20 minutes, and finally suspended in cold 20 m*M* phosphate buffer, pH 7.0, containing 10 m*M* $MgSO_4$ (P-Mg-7.0) to a concentration of 5 mg (dry weight) of cells per milliliter. This suspension is passed through a French pressure cell at 4° at 20,000 psi.

Step 2. The homogenate obtained from step 1 is treated with 0.3 μg of DNase per milliliter of suspension for 20 minutes at 25° and then is centrifuged at 104,000 *g* for 1 hour. The resulting pellet is resuspended in P-Mg-7.0 buffer.

Step 3. Two milliliters each of 2.0, 1.75, 1.50, 1.25, 1.0, 0.75, 0.50, and 0.25 *M* sucrose dissolved in P-Mg-7.0 buffer is layered into centrifuge tubes and held at 4° for 12–16 hours to form a linear gradient. A 3-mm band of the suspension from step 2 is layered on top of the gradient and the tubes are centrifuged for 90 minutes at 30,000 rpm (Spinco SW 39L rotor).

Results. Two bands form in the gradient. The lower band scatters light appreciably and is composed of intact cell envelopes with large fragments of membranous material; the upper band consists of a fairly homogeneous suspension of membranes shaped like disks and cups having a diameter of 70–90 nm as seen in the electron microscope. The latter band contains 16 nm particles assumed to be ribosomes which are removed by the following procedure.

Step 4. Gradient electrophoresis on a sucrose gradient (0.25–2.0 *M* sucrose in Tris, pH 7.0) which is prepared in each limb of an electrophoresis cell and allowed to equilibrate 8–10 hours at 4°. A 3 mm band of the sample is layered on the gradient, and a potential of 190 V at a current of 20 mA is applied to the cell. After electrophoresis the pigmented material is removed, centrifuged to a pellet, or left suspended in buffer. This procedure appears to remove ribosomes from the chromatophores as based on electron microscope observations.

Ketchum and Holt[32a] have further characterized certain chemical and physical properties of *R. rubrum* chromatophore fractions prepared by the above procedures. Their studies involved electron microscope observations, equilibrium density measurements on Ficoll, and sodium bromide gradients, as well as analyses of bacteriochlorophyll, hexoseamine, phospholipid, protein, and succinic dehydrogenase activity. Treatment of purified chromatophores by sonic oscillations or osmotic shock resulted in a differential solubilization of these chemical components.

[32a] P. A. Ketchum and S. C. Holt, *Biochim. Biophys. Acta* **196**, 141 (1970).

Chromatophores also have been freed of ribosomal contaminations by centrifugation in rubidium chloride[33] or cesium chloride[34] gradients.

Rhodopseudomonas spheroides

A. Disruption by French Pressure Cell Followed by Differential Centrifugation and Gradient Centrifugation.[35] *Enzyme and Lipid Analyses*[36,37]

This method is similar to those reported under Sections A and B for *Rhodospirillum rubrum*. Gradient centrifugation is carried out with a discontinuous sucrose gradient (centrifuged for 90 minutes at 53,000 *g*) yielding a deeply pigmented band centered in the 0.55 *M* sucrose zone with slight overlap into the 0.66 *M* sucrose layer (chromatophore fraction), the remainder of the pigmented material is found in the pellet. No photochemical activities are reported. Enzyme[36] and lipid[37] analyses of pigmented and nonpigmented cells and subcellular fractions are described.

B. Lysozyme-Versene Treatment Followed by Osmotic Lysis and Differential Centrifugation[38]

Step 1. A concentrated suspension of cells (0.5–1.0 ml, approximately 100 mg dry weight per milliliter) is added to 57 ml of sodium glycylglycine buffer pH 7.0 (Gg-7.0) with continuous stirring. Lysozyme (30 mg freshly dissolved in 1 ml distilled water) is added with continuous agitation, followed immediately by the addition of 1 ml of a Na-EDTA solution (96 mg/ml, pH 7.0). The suspension is incubated at 37° for 1.5 hours with frequent swirling to disperse any clumps that may form.

Step 2. The above preparation is centrifuged at low speed, washed once with Gg-7.0 buffer, and is resuspended in 60 ml distilled water by gentle passage through a hand-operated glass homogenizer. The suspension is left standing at room temperature for 15 minutes and then is centrifuged for 15 minutes at 20,000 *g*. The resulting pellet is homogenized again with distilled water and recentrifuged for 15 minutes at 20,000 *g*. This procedure is repeated three more times until the lysate is free of pig-

[33] P. B. Worden and W. R. Sistrom, *J. Cell Biol.* **23,** 135 (1964).
[34] K. D. Gibson, *Biochemistry* **4,** 2042 (1965).
[35] A. Gorchein, A. Neuberger, and G. H. Tait, *Proc. Roy. Soc. Ser. B* **170,** 229 (1968).
[36] A. Gorchein, A. Neuberger, and G. H. Tait, *Proc. Roy. Soc. Ser. B* **170,** 319 (1968).
[37] A. Gorchein, *Proc. Roy. Soc. Ser. B* **170,** 279 (1968).
[38] A. Gorchein, *Proc. Roy. Soc. Ser. B* **170,** 255 (1968).

ment. The pigment-containing supernatant solutions are combined and centrifuged at 45,000 *g* for 2 hours.

Step 3. The solubilized pellet resulting from step 2 is fractionated on a discontinuous sucrose gradient as in procedure "A" for *Rhodopseudomonas.*

Results. The pigmented particles isolated by this method have properties similar to those produced from cells disrupted by more vigorous methods, as judged by their appearance in the electron microscope, their position in the sucrose gradient employed, and their ultraviolet absorption. The milder method employed enables the visualization of previously undescribed cell fragments believed to be portions of the cytoplasmic membrane associated with the chromatophores. The authors consider this as evidence that the chromatophores are attached to the cytoplasmic membrane. Most consistent results are obtained with cell preparations stored for 4 weeks at —20° in media of high ionic strength (0.4 Gg; 0.2 *M* Tris).

Earlier Work. Use of lysozyme-EDTA plus polymixin B with sucrose as stabilizer leads to the rapid formation of spheroplasts from freshly harvested cells of *Rhodospirillum rubrum.* The spheroplasts thus formed do not release chromatophores upon osmotic lysis.[2]

Thiorhodaceae (Purple Sulfur Bacteria)

The Thiorhodaceae are considered generally to be photolithotrophs, but some species can be cultured photoorganotrophically under anaerobic conditions. They will not grow in the dark either anaerobically or at reduced oxygen tensions.

Chromatium sp. Strain D

A. Disrupted by Ribi Cell Fractionator Followed by Differential Centrifugation and Gradient Centrifugation in a Continuous Sucrose Gradient. Biochemical Characterization[39]

Step 1. Cells are cultured photoorganotrophically with succinate at 35° and incident light intensity of approximately 50 foot-candles. Cells are harvested and stored at —10° as an unwashed cell paste. The thawed cell paste is suspended in 4 volumes of 0.1 *M* potassium phosphate in 10% sucrose, pH 7.5 ("PS buffer" abbreviation used by Cusanovich and Kamen[39]). This preparation is treated with a Ribi cell fractionator at 20,000 psi, the temperature of the effluent not exceeding 20°. Alterna-

[39] M. A. Cusanovich and M. D. Kamen, *Biochim. Biophys. Acta* **153,** 376 (1968).

tively, preparations are ground with alumina (Alcoa A-305). The cell homogenate is centrifuged at 30,000 *g* for 1 hour. The resulting pellet is resuspended in 4 volumes of PS buffer and recentrifuged at 30,000 *g* for 1 hour. The supernatant solution from both centrifugations is saved and centrifuged at 144,000 *g* for 2 hours. The resulting pellet contains what the authors refer to as "classical chromatophores."

Step 2. The above pellet is resuspended in 4 volumes of PS buffer containing rubidium chloride (35% by weight). This suspension is centrifuged at 144,000 *g* for 2 hours. Ribosomes are found at the bottom of the tube, chromatophores float on top. The chromatophore fraction is removed by an aspirator tube and is diluted with 10 volumes of PS buffer and centrifuged again at 144,000 *g* for 2 hours.

Step 3. The pellet resulting from step 2 is resuspended in PS buffer, layered on a 10–50% sucrose gradient, and centrifuged at 25,000 rpm for 90 minutes (Spinco SW 25.1 rotor). The upper pigmented band with the center at 26% sucrose is referred to as "light" fraction and the lower band with the center at 48% sucrose as "heavy" fraction.

Step 4. The bands from step 3 are recentrifuged in 3 volumes of PS buffer at 144,000 *g* for 2 hours. The resulting pellets are resuspended and placed on a continuous sucrose gradient which is centrifuged at 25,000 rpm for 90 minutes. The pigmented bands resulting are separated and diluted with three parts of PS buffer, and these suspensions are centrifuged at 30,000 *g* for 10 minutes to remove aggregated material.

Results. The "light" particles can be prepared in yields of as much as 40% of the total photoactive pigment of the intact cells. The "heavy" particles contain only a small fraction of the total pigment and are derived probably from the cytoplasmic membrane.

The values of certain physical parameters reported here are consistent with earlier values reported for chromatophores from the same organism.[13]

Photochemical Activities. Detailed papers report on light-induced absorbancy changes involving cytochromes, bacteriochlorophyll, etc.,[40] and on photophosphorylation[41] carried out by the "light" particles isolated by the above procedure.

Chlorobacteriaceae (Green Sulfur Bacteria)

These are strictly anaerobic phototrophs, generally grown on mineral media; vitamin B_{12} may be required for optimal growth; certain low molecular weight organic compounds may have stimulating effects.[24]

[40] M. A. Cusanovich, R. G. Bartsch, and M. D. Kamen, *Biochim. Biophys. Acta* **153,** 397 (1968).

[41] M. A. Cusanovich and M. D. Kamen, *Biochim. Biophys. Acta* **153,** 418 (1968).

Chloropseudomonas ethylicum

A. *Disruption by French Pressure Cell Followed by DNase Treatment, Differential Centrifugation, Gradient Centrifugation, and Gradient Electrophoresis*[42]

Step 1. Chloropseudomonas ethylicum strain 2-K is cultured at 30° on a mineral medium containing 3 ml of 70% ethyl alcohol per liter. Greatest production of pigmented membranes occurs at a light intensity of 10 foot-candles. Cells harvested in exponential phase are washed three times with cold 20 m*M* phosphate buffer, pH 7.0, containing 1 m*M* $MgSO_4$ (P-Mg-7.0) by centrifuging at 5000 *g* for 20 minutes for each wash. The washed cells are resuspended in P-Mg-7.0 buffer to a concentration of 5 mg (dry weight) of cells per milliliter and are disintegrated in a French pressure cell at 20,000 psi.

Step 2. The preparation obtained in step 1 is treated with 0.5 μg of DNase per milliliter for 30 minutes (no temperature indicated) and then is centrifuged at 18,000 *g* for 30 minutes. The resulting supernatant is centrifuged at 144,000 *g* for 1 hour.

Step 3. The pellet from step 2 is resuspended in P-Mg-7.0 buffer and placed on a linear sucrose gradient ("C" under *Rhodospirillum*) and centrifuged at 25,000 rpm for 2 hours.

Results. The upper band obtained by gradient centrifugation contains ellipsoidal vesicles 100–150 nm long and 30–50 nm wide, comparable to those described earlier as "*Chlorobium* vesicles."[43] The lower band contains membrane and cell fragments. Photochemical activities were not tested.

Step 4. The upper band obtained in step 3 can be purified further by gradient electrophoresis according to the procedure described under "C," step 4 for *Rhodospirillum,* modified by the addition of 0.1 *M* L-cysteine to the Tris–sucrose buffer in the electrophoresis cell to prevent chlorophyll oxidation. This procedure separates the vesicles from certain contaminating cell fragments.

Analysis of Cell Rupture. Sonic treatment (Branson Sonifier) and ballistic disruption with glass beads (Mickle apparatus) are used to study the rate of release of material absorbing at 260 nm, the loss of sedimentable chlorophyll, the release of succinic dehydrogenase, and the apparent differential rupture of the cells. It is noted that succinic dehydrogenase can be separated from the pigment vesicles in contrast to *Rhodospirillum* chromatophores which exhibit tightly bound succinic dehydrogenase activity.

[42] S. C. Holt, S. F. Conti, and R. C. Fuller, *J Bacteriol.* **91,** 311 (1966).

[43] G. Cohen-Bazire, N. Pfennig, and R. Kunisawa, *J. Cell Biol.* **22,** 207 (1964).

Chlorobium thiosulfatophilum

A disruption procedure similar to "A" under *Rhodospirillum* yields particles approximately 15 nm in diameter which carry out active photophosphorylation.[44] The suggestion is made[42] that these particles may represent substructures of the "*Chlorobium* vesicles" described above. Such large vesicles can be obtained from *C. thiosulfatophilum* and can be partially purified by differential centrifugation.[43]

Conclusions

It appears likely that more efficient methods of chromatophore purification will be developed. There is increasing interest in the preparation of subchromatophore fragments (cf. [31, 32]) and in their chemical and physical characterization. Nevertheless, it appears advantageous to start with chromatophores of high purity for the preparation of such subchromatophore fragments. For the present, complex reactions such as light-induced phosphorylation and reverse electron transport generally are obtained only with chromatophores. Until these reactions can be demonstrated reproducibly with subchromatophore fragments, chromatophores remain the primary starting material for the investigation of such reactions.

Acknowledgments

The experimental work of this laboratory was supported by grants from the National Institute of Allergy and Infectious Diseases of the National Institutes of Health, U.S. Public Health Service (AI-02218) and the National Science Foundation.

[44] J. A. Bergeron and R. C. Fuller, *in* "Biological Structure and Function" (T. W. Goodwin and O. Lindberg, eds.), Vol. II, p. 307. Academic Press, New York, 1961.

[26] Subchloroplast Fragments: Digitonin Method

By N. K. Boardman

Detergents have been used for some years to disrupt chloroplasts into fragments of varying sizes, but in many instances the fragments have been devoid of photochemical activity. More recent studies have been directed toward the isolation of photochemically active subchloroplast fragments.[1] Digitonin, a nonionic detergent, was observed to disrupt chloroplasts into fragments with different pigment compositions and with

[1] N. K. Boardman, *Advan. Enzymol.* **30**, 1 (1968).

different photochemical activities. The larger fragments were found to be enriched in photosystem II, whereas the smaller fragments had the properties of photosystem I. Incubation of chloroplasts with digitonin therefore produces a physical separation of the photochemical systems. This article describes the preparative procedures and properties of the subchloroplast fragments.

Fractionation

Reagents

Sucrose, 0.3 *M*; KCl, 10 m*M*; phosphate buffer, 50 m*M*, pH 7.2
Phosphate buffer, 50 m*M*, pH 7.2 + KCl, 10 m*M*
Digitonin (British Drug Houses) analar grade, 4%, freshly made by dissolving in hot water and cooling rapidly

Procedure (Anderson and Boardman).[2] Chloroplasts are prepared from spinach leaves (*Spinacia oleracea* L.) and washed once with the sucrose–KCl–phosphate buffer. The chloroplasts from 8 g of leaves are resuspended in 8.5 ml of 50 m*M* phosphate buffer, pH 7.2, +10 m*M* KCl to give a suspension with a chlorophyll content of 0.3–0.4 mg/ml. Digitonin (4% solution in water) is added slowly to the chloroplast suspension to give a final digitonin concentration of 0.5%, and the mixture incubated for 30 minutes with gentle stirring. All operations are carried out at 0°–4°.

The subchloroplast fragments are separated by differential centrifugation. The first centrifugation is at 1000 *g* for 10 minutes in an SS-34 rotor of a Servall refrigerated centrifuge (RC-2). Subsequent centrifugations are at 10,000 *g* for 30 minutes in a Servall centrifuge, 50,000 *g* (23,000 rpm) for 30 minutes in a No. 40 rotor of a Spinco Model L centrifuge, and 144,000 *g* (40,000 rpm) for 60 minutes in a Spinco centrifuge. The pellets from each centrifugation are suspended in 50 m*M* phosphate buffer + 10 m*M* KCl and designated 1000 *g* (D-1), 10,000 *g* (D-10), 50,000 *g* (D-50), and 144,000 *g* (D-144) fractions. Brief sonication for 7 seconds in a sonic disintegrator (10-kHz, 250-W Raytheon) facilitates resuspension. Volumes of resuspension buffer for fragments prepared from 8 g of leaves are as follows: D-1, 2 ml; D-10, 6 ml; D-50, 1.5 ml; D-144, 1.5 ml.

Properties of Subchloroplast Fragments

The D-10 fraction accounts for about one-half of the chlorophyll of the chloroplast, while approximately one-third is divided about equally between D-50, D-144, and the 144,000 *g* supernatant solution (Table I).

[2] J. M. Anderson and N. K. Boardman, *Biochim. Biophys. Acta* **112**, 403 (1966).

TABLE I
DISTRIBUTION OF CHLOROPHYLL IN THE SUBCHLOROPLAST FRAGMENTS[a]

Fraction	Chl *a* + Chl *b* (%)	Ratio Chl *a*/Chl *b*
Chloroplasts	100	2.83
D-1	19.0	2.36
D-10	46.2	2.27
D-50	12.3	4.40
D-144	11.7	5.34
Supernatant	10.8	3.76

[a] From J. M. Anderson and N. K. Boardman, *Biochim. Biophys. Acta* **112,** 403 (1966).

The Chl *a*:Chl *b* ratios of D-1 and D-10 are lower than the ratio found with intact chloroplasts, whereas D-50, D-144, and the supernatant have higher ratios. D-10 shows enhanced absorption at 650 and 470 nm, due to its higher content of Chl *b* (Fig. 1). D-144 shows greater absorption at 705–710 nm, indicative of a higher content of a far-red absorbing form of chlorophyll.

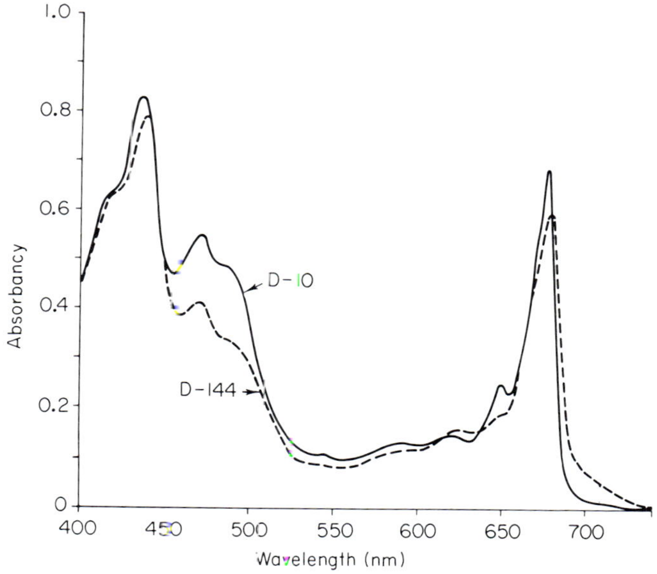

FIG. 1. Absorption spectra of subchloroplast fragments at 77°K. Solid line, D-10; broken line, D-144. From N. K. Boardman, S. W. Thorne, and J. M. Anderson, *Proc. Nat. Acad. Sci. U.S.* **56,** 586 (1966).

TABLE II
PHOTOCHEMICAL ACTIVITIES OF SUBCHLOROPLAST FRAGMENTS[a,b]

Fraction	Rates of reduction (μmoles/mg Chl/hr) Ferricyanide	TCIP	NADP	NADP (plus ascorbate–DCIP)
Chloroplasts	255	152	96	64
Chloroplasts after incubation with digitonin	138	81	33	18
D-1	209	139	24	14
D-10	160	61	17	17
D-50	43	0	0	70
D-144	0	0	0	123
Supernatant	0	0	0	103

[a] From J. M. Anderson and N. K. Boardman, *Biochim. Biophys. Acta* **112,** 403 (1966).

[b] Photochemical activities were determined essentially by the methods of A. T. Jagendorf and M. M. Margulies [*Arch. Biochem. Biophys.* **90,** 184 (1960)].

The photochemical activities of the fractions are shown in Table II. D-1 reduced ferricyanide and trichlorophenolindophenol (TCIP) at faster rates than did the digitonin-treated chloroplasts; D-10 was less active than D-1 and the other fractions were inactive or had low activity. Ferricyanide reduction was stoichiometrically coupled to oxygen evolution.[2] The D-1 and D-10 fractions reduced NADP at lower rates than did the digitonin-treated chloroplasts, and the other fractions were inactive. However, addition of the electron donor couple, ascorbate–dichlorophenolindophenol (ascorbate–DCIP) to the inactive fractions resulted in high rates of NADP reduction. The activity of D-144 for NADP reduction can vary markedly with different batches of digitonin. Preparations of D-144 giving low rates of NADP photoreduction are enhanced by providing additional amounts of plastocyanin.

Thus the small fragments (D-144) produced by incubating chloroplasts with digitonin have the photochemical properties of photosystem I, whereas the large fragments (D-1 and D-10) are enriched in photosytem II.[1]

The amount of Mn per mole of chlorophyll was 4.5 times higher in D-10 than in D-144.[3] Chloroplasts contained 1 Mn/73 Chl, compared with 1 Mn/52 Chl for D-10. Manganese is known to be an essential

[3] J. M. Anderson, N. K. Boardman, and D. J. David, *Biochem. Biophys. Res. Commun.* **17,** 685 (1964).

TABLE III
MOLAR RATIOS OF CHLOROPHYLL/CYTOCHROME[a] AND CHLOROPHYLL/P-700[b] FOR THE SUBCHLOROPLAST FRAGMENTS

Fraction	Molar ratios				
	Chl[c]/Cyt *b*	Chl/Cyt *f*	Chl/P700	Cyt *b*[d]/Cyt *f*	Cyt b_6/Cyt *f*
Chloroplasts	118	430	440	3.6	1.8
D-10	120	730	690	6.1	—
D-144	390	900	205	2.3	2.3
D-144 (dilution technique)	187	363	—	1.9	1.9

[a] From N. K. Boardman and J. M. Anderson, *Biochim. Biophys. Acta* **143,** 187 (1967).
[b] From J. M. Anderson, D. C. Fork, and J. Amesz, *Biochem. Biophys. Res. Commun.* **23,** 874 (1966).
[c] Chl = chl *a* + chl *b*.
[d] Cyt *b* = Cyt b_6 + Cyt *b*-559.

constituent of photosystem II. D-144 contained more Fe (1 Fe/29 Chl) and Cu (1 Cu/62 Chl) than did D-10 (1 Fe/48 Chl, 1 Cu/100 Chl).

Molar ratios of Chl/cytochrome and Chl/P700 are shown in Table III. Chloroplasts contain one *c*-type cytochrome (Cyt *f*) and at least 2 *b*-type cytochromes (Cyt b_6 and Cyt *b*-559).[1] Compared with the chloroplasts, D-10 was enriched in Cyt *b*-559, and it contained less Cyt *f* and Cyt b_6. Reduced *minus* oxidized difference spectra of D-144, determined at 77°K, indicated the presence of Cyt *f* and Cyt b_6, but Cyt *b*-559 was not detectable.[4]

The Cyt *f* contents of chloroplasts and D-10 were in good agreement with the P700 values.[5] D-144, however, contained considerably less Cyt *f* than P700, a result which suggested that some of the Cyt *f* was rendered soluble by the digitonin treatment. This loss of cytochromes (Cyt *f* and Cyt b_6) from D-144 may be minimized by diluting the chloroplast mixture 10-fold with the phosphate buffer immediately after incubation with digitonin and before differential centrifugation.

The molar ratios of Chl/P700, Chl/Cyt *f*, and Chl/Mn of D-10 indicate that this fraction contains about 70% photosystem II and 30% photosystem I, if it is assumed that the total chlorophyll is distributed equally between the two photosystems, and that P700 and Cyt *f* are localized in photosystem I and Mn in photosystem II.[4]

[4] N. K. Boardman and J. M. Anderson, *Biochim. Biophys. Acta* **143,** 187 (1967).
[5] J. M. Anderson, D. C. Fork, and J. Amesz, *Biochem. Biophys. Res. Commun.* **23,** 874 (1966).

TABLE IV
CAROTENOID COMPOSITION OF SUBCHLOROPLAST FRAGMENTS[a,b]

	D-10	D-144	Chloroplasts
Chl *a*	69	84	74
Chl *b*	31	16	26
β-Carotene	6	9	7
Lutein	11	7	11
Neoxanthin	4	2	2
Violaxanthin	4	5	4
Xanthophyll/carotene	3.8	1.7	2.6
Chl/carotenoid	4.0	4.3	4.0

[a] Calculated from data of N. K. Boardman and J. M. Anderson, *Biochim. Biophys. Acta* **143,** 187 (1967).
[b] Data are presented in moles and normalized to 100 moles of total chlorophyll.

Carotenoid analyses of D-10 and D-144 indicated the presence of the four major chloroplast carotenoids in both fractions (Table IV). D-10 showed a higher xanthophyll:β-carotene ratio, and D-144 a lower ratio as compared with chloroplasts.

Fluorescence Properties. The small fragments (D-144) fluoresced weakly at room temperature,[6] and the fluorescence yield was not influenced by redox condition, by the addition of CMU, an inhibitor of electron flow, or by ascorbate–DCIP and the enzymes required for the photoreduction of NADP by photosystem I. In contrast, D-10 was more fluorescent than chloroplasts (Fig. 2). The fluorescence of D-10 was quenched by ferricyanide and, at the low exciting intensity used, it was increased by dithionite or by CMU. D-10 showed the time-dependent rise in fluorescence observed previously with chloroplasts and attributed to the change in redox state of a quencher Q in photosystem II.

The fluorescence yields both of D-10 and D-144 were severalfold higher at 77°K than at room temperature. At 77°K, chloroplasts show a three-banded fluorescence emission spectrum, with maxima at about 683, 695, and 735 nm (F_{683}, F_{695}, F_{735}). The spectrum obtained with D-10 resembled that of the chloroplasts, but the relative emission at 735 nm was lower (60% of the total emission, compared with 75% for chloroplasts). D-144 showed only a small band at 683 nm at 77°K and 97% of the fluorescence came from the 735-nm band (Fig. 2). It was concluded that the fluorescence emitted at 735 nm by chloroplasts originates mainly from photosystem I and that emitted at 683 and 693 nm arises primarily from photosystem II.[6]

[6] N. K. Boardman, S. W. Thorne, and J. M. Anderson, *Proc. Nat. Acad. Sci. U.S.* **56,** 586 (1966).

Fluorescence excitation spectrum were obtained for F_{683} and F_{695} in D-10, and for F_{735} in D-144.[6] All spectra showed bands due to Chl *a*, Chl *b*, and carotenoids, indicating that quanta absorbed by Chl *b* and carotenoids, as well as those absorbed by Chl *a*, were active in promoting fluorescence in D-10 and D-144. Chlorophyll *b* and carotenoids, however, were relatively more active in D-10 (photosystem II). The excitation spectrum for F_{735} from D-144 showed an additional small band at 705 nm.

Electron Microscopy. Examination of chloroplast lamellae by the freeze-etch method reveals two types of subunits with average diameters of 175 and 110 Å.[7,8] The subchloroplast fragments produced by digitonin were studied by Arntzen, Dilley, and Crane.[9] The membrane fragments in D-1 and D-10 were found to have the 175-Å particles on most of their

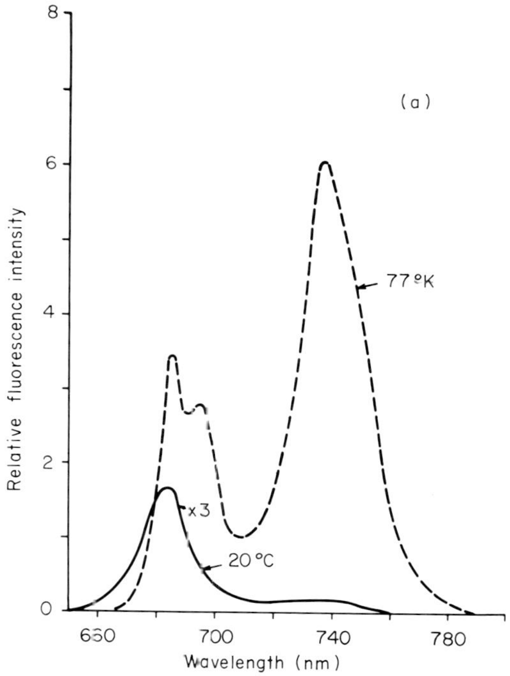

FIG. 2. Fluorescence emission spectra for subchloroplast fragments at 20°C (solid lines) and 77°K (broken lines). The fluorescence intensities at 20°C have been multiplied by a factor of 3. (a) Chloroplasts, (b) D-10, (c) D-144. From N. K. Boardman, S. W. Thorne, and J. M. Anderson, *Proc. Nat. Acad. Sci. U.S.* **56**, 586 (1966).

[7] K. Mühlethaler, H. Moor, and J. W. Szarkowski, *Planta* **67**, 305 (1965).
[8] D. Branton and R. B. Park, *J. Ultrastruct. Res.* **19**, 283 (1967).
[9] C. J. Arntzen, R. A. Dilley, and F. L. Crane, *J. Cell Biol.* **43**, 16 (1969).

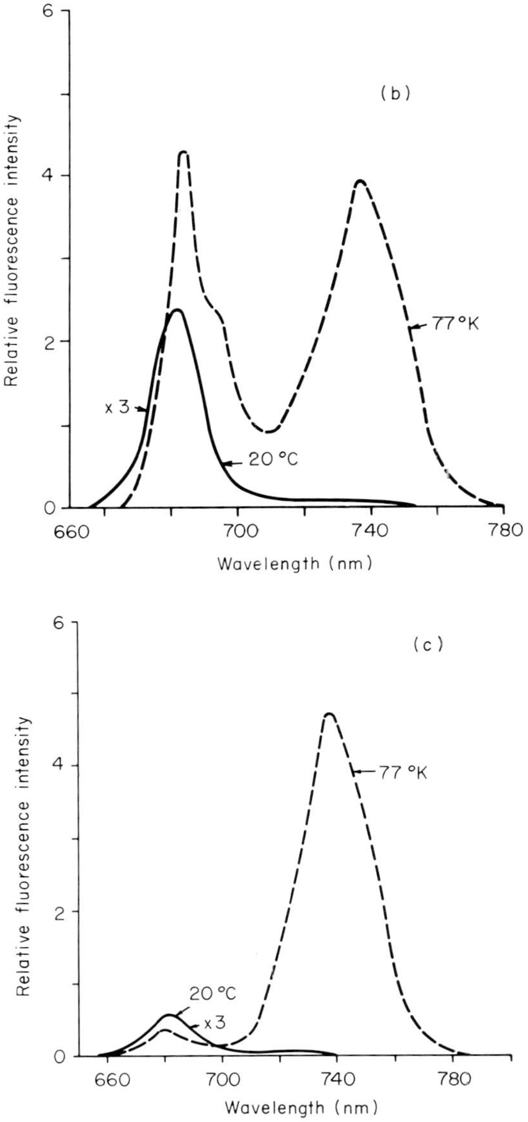

FIG. 2 (*continued*)

exposed faces. Freeze-etching of D-144 showed the presence of 110-Å particles. It is not known, however, whether the particles visualized by freeze-etching are the actual sites of the photochemical and electron transport reactions.

Thin sectioning of D-10 and D-144 indicated that both fractions contained membranes which were considerably thinner than chloroplast lamellae.[9] This result seems to support the view that digitonin splits the chloroplast membrane longitudinally into fragments which are characteristic of photosystem I and photosystem II, respectively. In their model of the chloroplast membrane, Arntzen, Dilley, and Crane place photosystem I and photosystem II on opposing sides of the membrane, photosystem I being located toward the outside of the thylakoid membrane and photosystem II towards the inside.

Purification of Subchloroplast Fragments by Density Gradient Centrifugation (Wessels)[10,11]

Spinach chloroplasts are disrupted in 1.3% digitonin for 30 minutes at 0° and the suspension centrifuged at 80,000 *g*. Samples (2.0 ml) of the supernatant solution are layered onto linear sucrose density gradients (10–30%) and centrifuged for 40 hours at 60,000 *g* in a swinging-bucket rotor (SW 25-1) of a Spinco Model L centrifuge.[10] The sucrose gradients contain 50 m*M* Tris buffer (pH 7.8), 5 m*M* $MgCl_2$, 2 m*M* EDTA, and 0.5% digitonin. Three colored bands are observed; the lowest band is blue-green (absorption maximum 678 nm), followed by a pink band containing cytochromes *f* and b_6, and a yellow-green band (absorption maximum 671 nm). The blue-green band has a Chl *a*/Chl *b* ratio of about 7 and it photoreduces NADP if supplied with ascorbate–DCIP, ferredoxin, ferredoxin–NADP reductase, and plastocyanin. It corresponds in its properties to the D-144 subchloroplast fragment. The blue-green fraction may be further purified by repeated density gradient centrifugation and chromatography on DEAE cellulose in the presence of 0.2% digitonin.[11] The Chl *a*:Chl *b* ratio of the purified material was 7.3, and it contained 75% protein and 25% lipid.

The yellow-green band is separated into two components either by chromatography on DEAE–cellulose, or by sucrose-density gradient centrifugation using a higher ratio of digitonin:Chl (10:1 w/w).[11] One component (fraction A) which is not adsorbed by DEAE is enriched in Chl *a* (*a*/*b* = 6.1); the other component (fraction B) has a Chl *a*: Chl *b* ratio of 1.8.

[10] J. S. C. Wessels, *Biochim. Biophys. Acta* **126**, 581 (1966).
[11] J. S. C. Wessels, *Biochim. Biophys. Acta* **153**, 497 (1968).

[27] Subchloroplast Fragments: Triton X-100 Method

By Leo P. Vernon and Elwood R. Shaw

The detergent Triton X-100 effectively fragments the membranes of photosynthetic systems and has been used extensively for this purpose.[1-4] This detergent behaves as do other nonionic detergents in that it removes from the membrane system a small particle which contains the components of photosystem I,[5] leaving a membrane residuum which contains photosystem II. Recent data indicate that Triton X-100 produces a cleaner fractionation, in terms of separation of photosystem I and photosystem II activities, than do other detergents.[6] The photosystem I particle so obtained, called the TSF-1 particle (Triton subchloroplast fraction 1) is enriched in chlorophyll *a* and β-carotene and has a lower content of the xanthophylls.

The ratio of Chl:P700 in the TSF-1 particle is approximately 100:1, and the use of higher concentrations of detergent does not improve this ratio. Application of the detergent to membrane systems which do not contain carotenoids, however, solubilizes a particle which is enriched in P700 and has a ratio of Chl:P700 of approximately 1:30. The preparation of both types of particles will be described below.

Chloroplast Preparation

Grind for a minimum time, 400–800 g of market spinach leaves in 400–800 ml of ice cold 50 m*M* Tris buffer, pH 7.8, containing 0.35 *M* NaCl. Press through four layers of cheesecloth and then pour through 12 layers. Centrifuge the filtrate at 1500 *g* for 10 minutes. Combine and wash the pellets by centrifugation in the same buffer. Decant the supernatant fluid and recentrifuge the pellets to remove most of the liquid.

Treatment with Detergent. Homogenize the chloroplasts with approximately 60 ml of an ice cold mixture of 5% Triton X-100, 0.5 *M* sucrose, and 50 m*M* Tris buffer, pH 7.5. Determine the total chlorophyll content of the suspension and then add sufficient sucrose and detergent from stock solutions of 2 *M* sucrose and 25% Triton X-100 so that the

[1] L. P. Vernon and E. R. Shaw, *Plant Physiol.* **40**, 1269 (1965).
[2] L. P. Vernon, E. R. Shaw, and B. Ke, *J. Biol. Chem.* **241**, 4101 (1966).
[3] L. P. Vernon, B. Ke, and E. R. Shaw, *Biochemistry* **6**, 2210 (1967).
[4] B. Ke and L. P. Vernon, *Biochemistry* **6**, 2221 (1967).
[5] L. P. Vernon, H. H. Mollenhauer, and E. R. Shaw, *in* "Regulatory Functions of Biological Membranes" (J. Järnefelt, ed.), p. 57. Elsevier, Amsterdam, 1968.
[6] L. P. Vernon and E. R. Shaw, *Biochem. Biophys. Res. Commun.* **36**, 878 (1969).

final concentrations are 0.4–0.5 *M* sucrose and 40 mg of chlorophyll per gram of detergent. Since the dilution affects both the detergent action and centrifugation characteristics, the final chlorophyll content of the suspension should be in the range 2.0–2.5 mg/ml. Stir the mixture for 1 hour at 0° (Fig. 1).

Photosystem II Particles (TSF-2). Centrifuge at 10,000 *g* for 10 minutes to sediment the large particles, which are discarded. Decant and centrifuge the supernatant solution at 144,000 *g* for 1 hour. Almost all photosystem II particles should sediment at this speed. Decant and centrifuge the supernatant solution for 5 or more hours at 144,000 *g*. The sediment is usually inactive and may be discarded.

Photosystem I Particles (TSF-1). Dilute the supernatant solution from above with an equal volume of water and centrifuge at 144,000 *g* for 10 hours. Discard the top one-third–one-half of the supernatant fluid from each tube and pour off the remainder. If the remainder of the

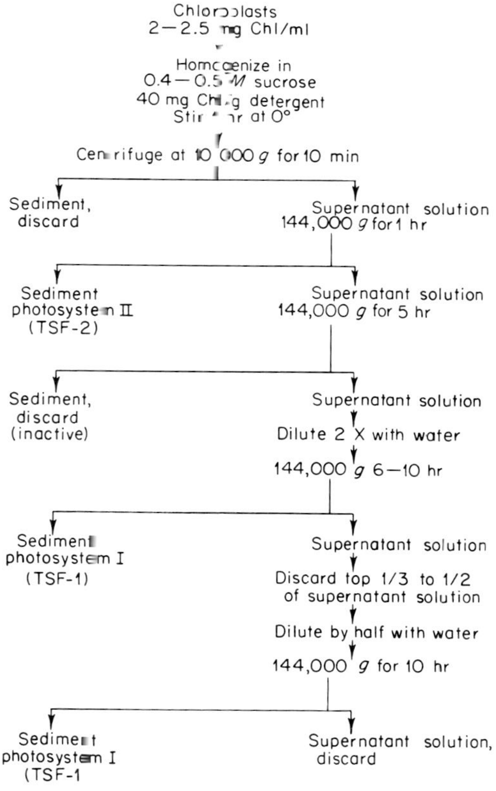

FIG. 1. Fractionation scheme for TSF-1 particles.

supernatant solution shows appreciable NADP photoreduction activity with the ascorbate–DPIP couple, it should be diluted by half and centrifuged again for 10 hours at 144,000 *g* to obtain another sediment.

Activities for the sediment from the first dilution range from 300 to 800 μmoles of NADP reduced per hour per milligram of chlorophyll, while those from the second dilution usually fall between 600 and 1200. Further dilutions and centrifugations have generally yielded particles of lower activity.

The pellets may be frozen as such for long storage or may be suspended in 0.5 *M* sucrose–50 m*M* Tricine, pH 7.5, and frozen for frequent thawing if required for daily use. The TSF-2 pellets retain their DPIP–Hill activity for several months and their DPIP–DPC activity (see below) for several years if deep frozen. TSF-1 pellets will retain their activity for a year, and the suspended particles are active after several months' storage in the frozen state.

Assays. NADP photoreduction is followed at 340 nm in 2.0 ml of reaction mixture containing 0.25 *M* sucrose, 50 m*M* phosphate buffer, pH 6.7, 0.4 m*M* NADP, 7.5 m*M* ascorbate, 50 μ*M* DPIP, 1 μ*M* plastocyanin, and saturating amounts of crude spinach ferredoxin or saturating amounts of purified spinach ferredoxin and of spinach ferredoxin–NADP reductase. At the above nonsaturating level of plastocyanin the chlorophyll content should be around 10–20 μg in the reaction mixture. The addition of small amounts of Triton X-100 (0.7 mg) to the reaction mixture generally increases the rate by 50–60%.

Illumination of the reaction mixture is with red light obtained with a tungsten microscope lamp and a Corning filter No. 2403. The light intensity at the reaction system should be approximately 2×10^5 ergs/cm^2 sec.

Photosystem II activity may be assayed in two ways. For the complete system, oxygen evolution (deriving from water oxidation) is coupled to the photoreduction of DPIP, which is measured at 590 nm. Two milliliters of reaction mixture contains 0.25 *M* sucrose, 7 m*M* phosphate buffer, pH 6.4, and 0.1 m*M* DPIP. The illumination conditions are the same as given above for photosystem I activity (NADP photoreduction).

Another assay for photosystem II involves the oxidation of an added electron donor, whose oxidation is coupled to the photoreduction of DPIP. The donor we have found best suited for this purpose is 1,5-diphenylcarbazide (DPC).[6,7] When this system is used, the oxidation of water is supplanted by the oxidation of the donor molecule, which is

[7] L. P. Vernon and E. R. Shaw, *Plant Physiol.* **44,** 1645 (1969).

present at a concentration of 0.5 m*M*. Typical rates for the DPIP–Hill reaction (water oxidation leading to oxygen evolution) observed with TSF-2 fragments are 40–80, while rates of DPIP photoreduction coupled to DPC photooxidation are 200–300 μmoles/hour per milligram of chlorophyll. In the case of the DPC-coupled reaction, the rates decrease with increasing pH and ionic strength of buffers.[7]

Properties of Fractions. The composition of the TSF-1 and TSF-2 fractions is given in Table I. There is an enrichment of TSF-1 in Chl *a*, β-carotene, and plastoquinones. Cytochromes b_6 and *f* are associated with

TABLE I
COMPOSITION OF SUBCHLOROPLAST PARTICLES OBTAINED THROUGH THE ACTION OF TRITON X-100

Component[a]	TSF-2	TSF-1	Untreated chloroplast lamellae
Chlorophyll *a*	67	85	70
Chlorophyll *b*	33	15	30
P700	0.5	1.4	0.4
Chl *a*/Chl *b*	2	5.7	2.4
β-Carotene	6	16	6.3
Lutene	16	6	9.6
Neoxanthin	2	1	2.5
Violaxanthin	3	1	2.8
Plastoquinone A	1.8	5.5	7.2
Plastoquinone B	0.2	4.5	4.2
Plastoquinone C	0	3.3	—
α-Tocopherolquinone	0.3	0.5	1.9
Cytochrome b_6 } [d]	0	2	—
Cytochrome *f* } [d]	0	2	—
Cytochrome 559	1	0	—
Monogalactosyl diglyceride[b]	53	25	115
Digalactosyl diglyceride	30	12	60
Phosphatidylglycerol	16	5	21
Sulfolipid	5	3	17
Lecithin	6	2	9
Protein[c]	0.5	1.0	0.4

[a] Data are presented in nanomoles and are normalized to 100 nmoles of total chlorophyll.

[b] The polar lipids were analyzed according to the procedure of C. F. Allen, P. Good, H. F. David, and P. Chisum [*J. Amer. Oil Chem. Soc.* **43,** 223 (1966)]. These data were made available by C. F. Allen.

[c] Protein in mg/100 nmoles of total chlorophyll for EDTA-washed lamellae.

[d] Some of this cytochrome is solubilized cytochrome which sediments with the TSF-1 particle. The particles were not purified by sucrose density gradient centrifugation prior to analysis.

the TSF-1 particle while cytochrome 559 is all found in the TSF-2 particle. Cytochromes b_6 and f were partly solubilized by the detergent, as indicated in Table I. Cytochrome 559, however, was tightly bound to the TSF-2 fragment.

The degree of separation of the two photosystems by means of treatment with the detergents Triton X-100 and digitonin is shown in Table II. In the absence of DPC, the rate of DPIP reduction represents the oxidation of water to produce oxygen. In all cases higher rates are

TABLE II
DISTRIBUTION OF PHOTOSYSTEM II AND PHOTOSYSTEM I ACTIVITIES IN FRACTIONS OBTAINED BY TREATMENT OF SPINACH CHLOROPLASTS WITH DETERGENTS[a]

	Photoactivities (μmoles/hr/mg Chl)				
	DPIP reduction photosystem II		NADP reduction photosystem I		
Method of preparation	−DPC	+DPC	No TX	0.035% TX	0.1% TX
Triton X-100					
125 mg Chl/gram detergent					
10,000 *g*, 10 min	74	104 (38)	—	—	—
10,000 *g*, 30 min	72	94 (46)	0	17	81 (46)
80,000 *g*, 30 min	34	36 (29)	0	33	180 (30)
105,000 *g*, 5 hr	0	3 (26)	67	128	260 (26)
40 mg Chl/gram detergent					
144,000 *g*, 1 hr (TSF-2)	81	262 (30)	0	0	58 (30)
144,000 *g*, 6 hr	0	35 (22)	0	132	79 (22)
144,000 *g*, 10 hr, after dilution of 1:1 (TSF-1)	0	0 (30)	435	735	470 (6)
144,000 *g*, 12 hr, after dilution of supernatant from above 2:3	0	0 (43)	504	935	480 (8)
Digitonin					
60 mg Chl/gram detergent					
1,000 *g*, 10 min	13	80 (51)	0	38	140 (30)
10,000 *g*, 30 min	75	142 (30)	13	39	104 (30)
50,000 *g*, 30 min	0	15 (35)	—	194	194 (18)
144,000 *g*, 60 min	0	8 (39)	75	435	169 (16)

[a] The assay systems and preparation methods are described in the text. The numbers in parentheses are the micrograms Chl in the assay systems used to determine photosystem I and photosystem II activities. In the absence of 1,5-diphenylcarbazide (DPC) the reduction of DPIP is coupled to oxygen evolution (Hill reaction). In the presence of DPC, this molecule serves as the electron donor for photosystem II. For NADP photoreduction, the amounts of Triton X-100 (TX) indicated were added to the assay system.

observed for photosystem II when DPC is added as an electron donor to photosystem II. The data for NADP photoreduction show that the addition of Triton X-100 to the assay system uniformly stimulates the reaction. The reason for this is not known, but is probably related to the ability of the detergent to keep the TSF-1 particles from aggregating. The data of Table 1 show that the cleanest separation of the two photosystems is obtained with the Triton X-100, using the detergent at a level of 40 mg Chl per gram of detergent. Under these conditions it is possible to obtain a TSF-1 particle preparation free of photosystem II activity, and to also obtain a TSF-2 fragment containing only a small residual amount of photosystem I activity.

Photosystem I Particles Enriched in P700

If the treatment with Triton X-100 is carried out on photosynthetic membrane material which is devoid of carotenoids, small particles are obtained which are enriched in P700 relative to total Chl, showing a ratio of Chl:P700 of 30:1.[8-11] This procedure can be carried out on normal photosynthetic membrane systems from which the carotenoids have been removed by treatment with organic solvents[8] or with the membrane fragments of *Anabaena variabilis* grown in the presence of diphenylamine (DPA *Anabaena*) to suppress the formation of carotenoids,[11] and with a mutant strain of *Scenedesmus* (6E) which lacks carotenoids.[11] This mutant strain was kindly made available to us by N. Bishop.

High P700 Containing (HP700) Particles from Spinach. All operations were performed at 0°. Chloroplasts are isolated from 800 g of deribbed spinach leaves by grinding in STN solution (0.25 *M* sucrose, 20 m*M* Tricine buffer, pH 8, 10 m*M* NaCl), filtering through 12 layers of gauze, and centrifuging at 2500 *g* for 10 minutes. The chloroplasts are washed successively in 1 liter each of 10 m*M* NaCl, 0.75 m*M* EDTA pH 8, and 2 m*M* Tricine and centrifuged at 25,000 *g* for 20, 30, and 60 minutes, respectively. After the three washings, the chloroplast material is resuspended in minimal volume and freeze-dried thoroughly. The bulk of the chlorophylls and carotenoids are extracted by resuspending the dried chloroplast materials in 240 ml of 15% acetone in hexane (precooled to −18°), and the suspension is centrifuged at 500 *g* for 5 minutes. This step is repeated seven times, using a glass homogenizer to

[8] L. P. Vernon, H. Y. Yamamoto, and T. Ogawa, *Proc. Nat. Acad. Sci. U.S.* **63,** 911 (1969).

[9] H. Y. Yamamoto and L. P. Vernon, *Biochemistry* **8,** 4131 (1969).

[10] T. Ogawa and L. P. Vernon, *Biochim. Biophys. Acta* **180,** 334 (1969).

[11] T. Ogawa and L. P. Vernon, *Biochim. Biophys. Acta* **197,** 292 (1970).

resuspend thoroughly the fragments on the first and fourth extractions. After the last extraction, the fragments are resuspended in cold hexane, divided into four equal portions, centrifuged at 5000 *g* for 2 minutes, and the pellets so obtained may be stored at $-70°$ until used.

The following describes the further treatment of one-fourth of the above preparation. The residual hexane is removed by evaporation under nitrogen, yielding a light-green powder which is homogenized in 60 ml of 0.05% Triton X-100 in 2 m*M* Tricine buffer to extract additional chlorophyll and cytochromes, and centrifuged at 39,000 *g* for 30 minutes. The resulting pellet is homogenized with 16 ml of 5% Triton X-100 in 2 m*M* Tricine buffer. This extracts the P700 particles which are separated from insoluble brown materials by centrifugation (39,000 *g* for 30 minutes). Portions (3 ml) of this extract are layered on discontinuous sucrose gradients (2–20% in 2% steps) and centrifuged at 131,000 *g* for 20 hours in a Spinco Model L2-65 ultracentrifuge with SW 27 rotor. The P700 particles are located in the 8% layer and the yield of P700 is approximately 60%; based on an assumed initial content of 1 P700 to 500 chlorophylls in chloroplasts.

HP700 Particle from Anabaena variabilis. Washed cells are suspended in 10 m*M* Tris·HCl buffer (pH 7.5) and sonicated in a 10 kHz Raytheon sonic oscillator for 15 minutes. The green fraction which sediments between 10,000 and 144,000 *g* contains membrane fragments and starch particles. This fraction is washed twice with the above buffer and lyophilized.

All steps are carried out at 0°, and 0.01 *M* Tris·HCl buffer (pH 7.5) is used where indicated. Lyophilized membrane fragments (200 mg) are extracted five times with hexane, and the pellet collected by low-speed centrifugation is extracted by 40% aqueous ethanol (v/v). A bluish-green pellet, obtained by centrifuging the aqueous ethanol suspension at 10,000 *g* for 10 minutes, is suspended in buffer. A solution of 1% Triton X-100 in buffer is added to an equal volume of this suspension (0.3 m*M* in terms of chlorophyll concentration) and the mixture kept at 0° for 30 minutes. Centrifugation of the suspension at 10,000 *g* for 15 minutes yields a pellet, which is resuspended in a solution of 5% Triton X-100 in buffer, and the mixture is kept at 0° for 30 minutes. The final concentration of chlorophyll in the suspension is usually 0.15 m*M*. Portions of the suspension (3 ml) are layered onto discontinuous sucrose density gradients of 10 layers formed from 3-ml portions of 30–3% sucrose solution, and centrifugation is performed at 131,000 *g* for 18 hours in the SW 27 rotor of the Spinco Model L2-65 ultracentifuge.

Cells of *Anabaena variabilis* grown in the presence of diphenylamine (DPA *Anabaena*) and cells of the *Scenedesmus* mutant 6E are treated directly with Triton X-100 to obtain the HP700 fragment. Membrane fragments of *Anabaena* are prepared by sonication as listed above, and *Scenedesmus* cells are broken in a like manner, except that 1 m*M* ascorbate is present to prevent chlorophyll photodestruction.

For the isolation of HP700 fragments from the membrane fragments of DPA *Anabaena* and *Scenedesmus* mutant 6E, all steps are performed at 0° in the presence of 10 m*M* Tris·HCl buffer, pH 7.5 (also containing 1 m*M* ascorbate for *Scenedesmus* mutant 6E). A solution of 10% Triton X-100 in buffer is added to an equal volume of the membrane fragment suspension (0.3 m*M* for DPA *Anabaena* and 0.2 m*M* for *Scenedesmus* mutant 6E in terms of chlorophyll concentration). A portion of the suspension (3 ml) is layered on a discontinuous sucrose gradient (2–20% in 2% steps) and centrifuged at 131,000 *g* for 20 hours in a Spinco Model L2-65 ultracentrifuge with SW 27 rotor. The HP700 fragments of both cells are located in the 8% layer.

Properties of the HP700 Particle. The absorption spectra of the TSF-1 and HP700 particles prepared from spinach are shown in Fig. 2. The absence of the carotenoids in the HP700 particle is the main difference observed in these spectra. Similar type preparations have been made from spinach, bean leaves, *Euglena* and *Anabaena variabilis*,[8,9] and all show the characteristic absorbance change in the 700-nm region asso-

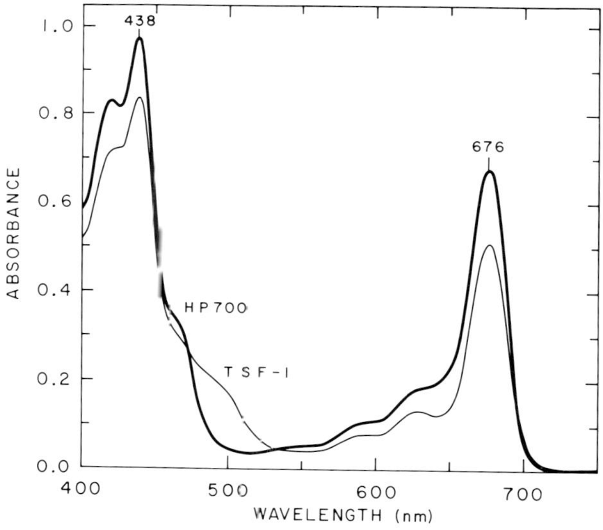

FIG. 2. Absorption spectra of HP700 particles (heavy line) and TSF-1 preparation (light line) from spinach chloroplasts. The spectra were obtained with a Cary 14 spectrophotometer, using particles suspended in 50 m*M* phosphate buffer, pH 7.2. [Reprinted from *Biochemistry* **8**, 4134 (1969). Copyright (1969) by The American Chemical Society. Reproduced by permission of the copyright owner.]

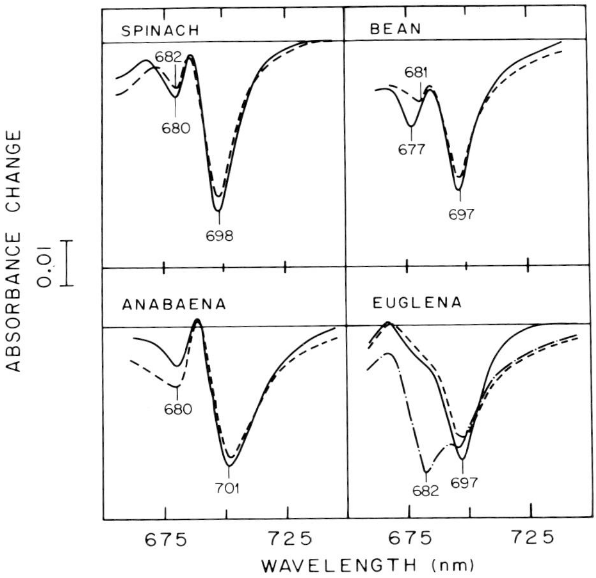

FIG. 3. Difference spectra for HP700 preparations from four plant species. The solid lines indicate the difference spectra obtained by oxidation with ferricyanide (oxidized minus reduced) while the dashed lines indicate the difference spectra obtained upon illumination (light minus dark). The particles were suspended in 5.0 m*M* phosphate buffer, pH 7.2, and the Chl concentrations (μg/ml) were: spinach, 17.5; bean, 17.5; *Anabaena*, 16.2; *Euglena*, 17.0. The dashed and dot–dashed lines for *Euglena* indicate the light-minus-dark spectra obtained after 1 and 8 minutes of illumination, respectively. [Reprinted from *Proc. Nat. Acad. Sci.* **63**, 913 (1969), by permission of the National Academy of Sciences.]

ciated with the oxidation of P700, as shown in Fig. 3. The location of the minima occur at slightly different wavelengths for each species, but in general the shapes of the difference spectra are similar. The minimum which occurs in the 680-nm region is attributed to another chlorophyll *a* molecule which is closely associated with the reaction center chlorophyll P700.[9,10]

The fluorescence properties of the HP700 particle prepared from spinach are shown in Fig. 4. One major feature is the marked decrease in the long wavelength fluorescence of the HP700 particle at the temperature of liquid nitrogen. Similar preparations obtained from the mutant *Scenedesmus* 6E (Fig. 5) show the almost complete absence of the 730-nm fluorescence in the HP700 particle which still shows P700 oxidation.[11] These data show that the long wavelength fluorescence observed at the temperature of liquid nitrogen, although it is characteristic of photosystem I, does not originate from the P700 itself.

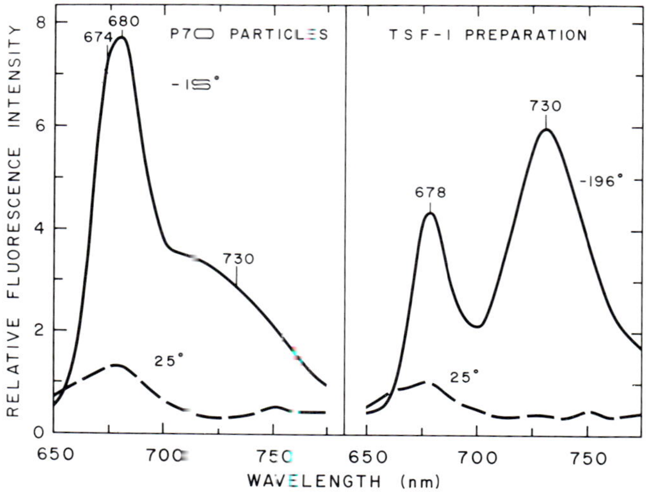

FIG. 4. Fluorescence spectra of HP700 particles and TSF-1 preparation at 25° (dashed line) and −196° (solid lines). The preparations were illuminated with 430-nm light from the front in a 0.15-mm path cell. The fluorescence emitted from the back of the cell was resolved with a Bausch and Lomb monochromater blazed at 1 μ and detected with a RCA 7102 photomultiplier cooled with dry ice. The preparations were suspended in 50 m*M* phosphate buffer, pH 7.2, and the Chl concentrations were 30 μg/ml. [Reprinted from *Biochemistry* **8**, 4135 (1969). Copyright (1969) by The American Chemical Society. Reproduced by permission of the copyright owner.]

The HP700 particle from spinach has been studied rather extensively.[8] The composition of such particles, showing the presence of both cytochrome *f* and cytochrome b_6 is shown in Table III. The ratio of Chl : P700 of 30 : 1 is characteristic of all such particles prepared to date,[8] which indicates a similar protein structure for the particles prepared by the use of Triton X-100. It is not known whether the Chl molecules

TABLE III
COMPOSITION OF HP700 PARTICLES FROM SPINACH CHLOROPLASTS[a]

Component	Amount
Chl *a*/Chl *b*	4.0
Chlorophyll	100 nmoles
P700	3.3 nmoles
P700/Chl	1:30
Cytochrome *f*	0.9 nmole
Cytochrome b_6	1.2 nmoles
Protein	3.2 mg

[a] The methods used for analysis are given by H. Y. Yamamoto and L. P. Vernon, *Biochemistry* **8**, 4131 (1969).

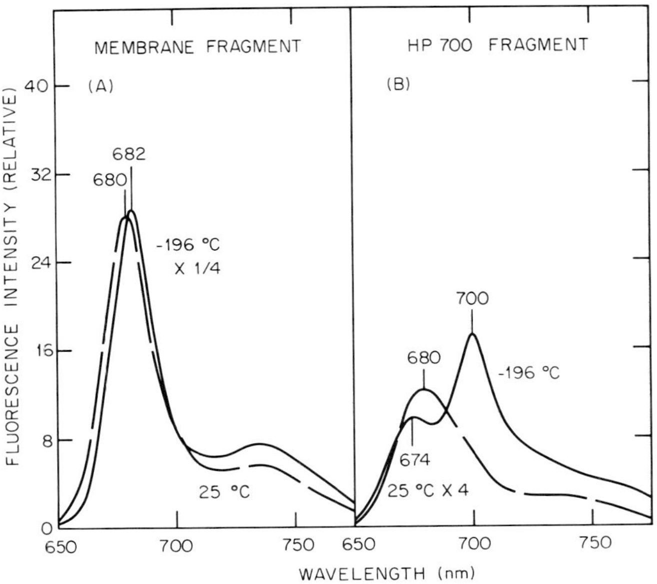

FIG. 5. Fluorescence spectra of the membrane (A) and HP700 (B) fragments of *Scenedesmus* mutant 6E at room (25°) and liquid nitrogen temperature (—196°). Each sample had the same chlorophyll content. [Reprinted from *Biochem. Biophys. Acta* **197**, 299 (1970) by permission of Elsevier Publishing Company.]

other than the P700 exist in their original conditions or have been influenced by the extraction procedures. The oxidation potential of the P700 in the particle prepared from spinach was determined to be +480 mV.[9]

The fractions prepared by the action of Triton X-100 on chloroplast membranes have characteristic shapes when examined with the electron microscope (Fig. 6). TSF-2 appears as a relatively smooth membrane, while TSF-1 preparations consist of small rod-shaped particles which show a marked tendency to aggregate. The individual rods are 70–80 Å in diameter. The HP700 preparations also contain small particles which are rod-to-ellipsoidal in shape (about 150 × 60 Å).

Whereas the HP700 particles prepared from normal membrane systems following extraction of carotenoids with organic solvents are only slightly active in NADP photoreduction,[9] the particles prepared from the diphenylamine-grown *Anabaena* or from *Scenedesmus* mutant 6E show good activity. This indicates that the low activity observed in the systems derived from solvent-extracted membranes reflects some modification caused by the extraction procedure. The systems from the DPA *Anabaena* and *Scenedesmus* thus allow further study of NADP photoreduction in this simplified system.

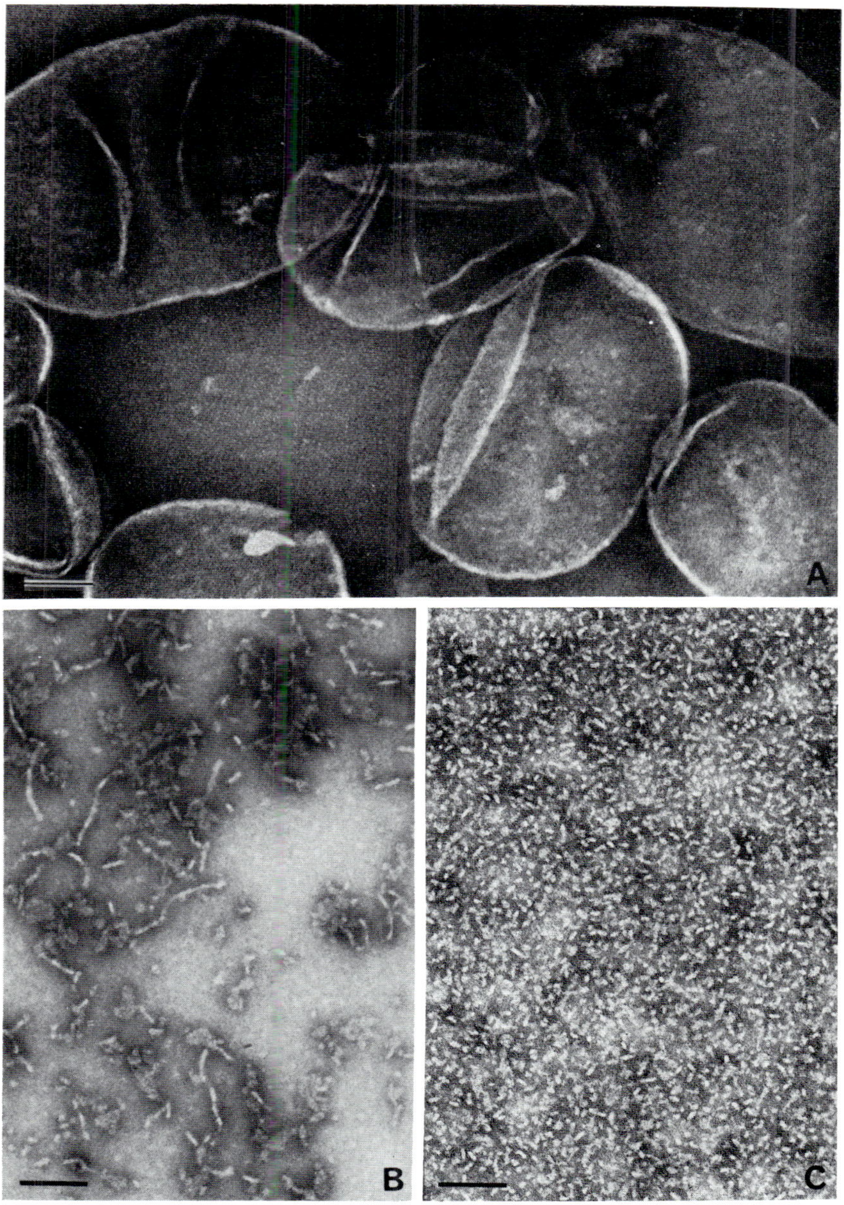

FIG. 6. Electron micrographs of TSF-2 preparation (A), TSF-1 preparation (B), and HP700 preparation (C) from spinach chloroplasts. The preparations were either dialyzed against 2 m*M* Tricine buffer, pH 8, or pelleted to remove sucrose and Triton X-100, negatively stained with phosphotungstic acid at pH 5.9 in 5 m*M* $MgCl_2$, and examined with a Philips EM 200 electron microscope. The bar represents 1000 Å.

The concentration of Triton X-100 used for the preparation of the TSF-1 particles described herein is selected to give the best possible separation of the two photosystems with the retention of activity in both. Briantais[12,13] has used a lower concentration of the detergent to remove a photosystem I particle which could be recombined with the photosystem II fragment. Kahn[14] had earlier used Triton X-100 to prepare a chlorophyll–protein complex which was purified by chromatography. This particle retained none of the activity normally associated with photosystems I and II, but carries out a slow photoreduction of ferricyanide in the absence of any added donor molecule.

Huzisige *et al.*[15] combined the use of digitonin, sonication, and Triton X-100 to purify a photosystem II particle that had no residual photosystem I activity. The reported oxygen evolution activity, however, was quite low as compared to the original activity of the chloroplast. The composition of the fractions obtained by this procedure is given by the authors.[15]

[12] J. M. Briantais, *Biochim. Biophys. Acta* **143,** 650 (1967).
[13] J. M. Briantais, *Bull. Soc. Fr. Physiol. Veg.* **14,** 227 (1968).
[14] J. S. Kahn, *Biochim. Biophys. Acta* **79,** 234 (1964).
[15] H. Huzisige, H. Usiyama, T. Kikuti, and T. Azi, *Plant Cell Physiol.* **10,** 441 (1969).

[28] Subchloroplast Fragments: Sonication Method

By G. Jacobi

Photochemical subunits are liberated from isolated chloroplasts by disintegration by ultrasonication or with detergents. The resulting particles are different in chemical composition and in their capacity to carry out photochemical reactions. Therefore, in some instances, a combination of sonication and detergents during the procedure of fragmentation will often give valuable information. However, even by the use of only one method, the isolated particles are not comparable because of different isolation procedure, pretreatment of the chloroplast suspensions, or even leaf material. Consequently, in order to obtain reproducible and comparable results, several factors have to be considered.

The use of sonication is generally preferred because it avoids the undefined influence of surface-active substances upon the membrane lipids resulting in irreversible inactivation of some processes, such as the Hill reaction. Conversely, the disadvantage of sonication is the effect of some physical parameters which remain unestimated. The frequency stated in kilohertz is by itself not the only factor affecting the disintegration of the biological structure. Another important factor influencing the effective cavitation is the amplitude. Unfortunately, no simple method for the estimation of the energy density given in watts per square centimeter is available yet. For the apparatus used by the author, Hübener *et al.*[1] calculated 40–50 W/cm^2 at 19.6 kHz and 0.6 A. Such a calculation is, however, only an approximation due to the irregular shape of the piston. References about the action of ultrasound upon biological material are given in the article of Hughes and Cunningham.[2]

Pretreatment of Chloroplasts and Sonication

Freshly harvested spinach leaves are ground with plexigum in 0.3 *M* sucrose containing 10 m*M* pyrophosphate, pH 7.8, and isolated as described.[3] In contrast to the preferred use of primary leaves leading to intact chloroplasts with a high protein content, for the disintegration experiments well expanded secondary leaves should be used because of the development of grana stacks. The disintegration is strongly dependent upon the medium employed for suspending the pellet of once-washed chloroplasts.

Medium I. Using slightly hypertonic salt solutions, including some additions, an excellent conservation of the thylakoid system as indicated by a close-packed grana area is achieved.[4] This medium should be used for isolation of grana stacks liberated by sonication. For this procedure a medium with the following composition was found to be convenient: 2% NaCl, 1 m*M* $MgCl_2$, and 1 m*M* Tricine buffer, pH 7.8. The smallest units obtained after centrifugation of the sonicated suspension in this medium is composed of particles exhibiting activities related to photosystem I only.

Medium II. In media of lower salinity, such as 15 m*M* NaCl,[5,6] 1 m*M* phosphate buffer,[7] or 30 m*M* Tricine buffer,[8] the contact between the grana stacks is loosened. During the disintegration, smaller units active in reactions related to photosystem II are liberated from the grana area.

The chloroplast material in either medium is adjusted to 0.5 mg of chlorophyll/ml and used immediately after dilution for sonication.

[1] H. J. Hübener, H. J. Gollnick, K. Tesser, W. Lippert, and L. Rossberg, *Biochem. Z.* **331**, 410 (1959).
[2] D. E. Hughes and V. R. Cunningham, *Biochem. Soc. Symp.* **23** (1963).
[3] G. Jacobi, *Z. Naturforsch. B* **18**, 312 (1963).
[4] G. Jacobi and H. Lehmann, *Z. Pflanzenphysiol.* **59**, 457 (1968).
[5] J. M. Becker, A. M. Shefner, and J. A. Gross, *Plant Physiol.* **40**, 243 (1965).
[6] G. Jacobi, *Z. Pflanzenphysiol.* **57**, 255 (1967).
[7] R. B. Park and N. G. Pon, *J. Mol. Biol.* **3**, 1 (1961).
[8] S. Izawa and N. E. Good, *Biochim. Biophys. Acta* **109**, 372 (1965).

Other media may influence the manner of disintegration and distribution of particle size after differential centrifugation. A higher yield of heavier particles was found with increased concentrations of phosphate.[9]

As noted before, no standardization applicable to all ultrasound apparatus available can be given. However, the comparison of the results published by several authors demonstrated that at least for one frequency the time of sonication is one of the most important factors responsible for the degree of disintegration. After sonication of only 15–20 seconds at 19.5 kHz and 0.4 A, the grana stacks are well preserved, but the intergrana area is disintegrated.[4] During longer sonication even in medium I, the grana stacks are also fragmented.

Another parameter very often neglected is the temperature of the suspension during sonication. The best conditions were found by using two vessels fitted into each other; the refrigerating solution is circulated continuously in the space between them. In short-time sonication, 15–20 seconds, a temperature of —4° is convenient to avoid inactivation at 19.5 kHz and 0.4 A. Sonication for a longer time requires lower temperatures. If the sonication is prolonged to 1 minute or longer, a temperature of —10° to —15° is necessary. The addition of serum albumin immediately after sonication was found to be favorable, perhaps by protecting the Hill activity against free fatty acids liberated during the fragmentation.[10]

Separation and Purification of Ultrasound Particles (UP)

The whole suspension of sonicated chloroplasts ($C_{1\,su}$) is fractionated by differential centrifugation. Some authors used five or six different centrifugal forces to separate fragments of varying size.[5-7] However, electron microscopic analysis demonstrated only three or four types present after sonication.[4,11] Thus, wider ranges of centrifugal forces will give the principal fractionation. Purifications are obtained by a density gradient centrifugation of the isolated fragments.

The separation procedure and the designation of the fraction are summarized in Fig. 1. The pellets after centrifugation are suspended in 0.1 M sucrose containing 0.05% bovine serum albumin and are designated as UP (ultrasound particles). The appended numbers indicate the centrifugal force ($g \times 10^3$), used to obtain the pellet.

The purification is achieved by centrifugation either in a continuous or discontinuous density gradient of sucrose containing 1 mM Tricine buffer, pH 7.8, and 0.05% bovine serum albumin. For the isolation of

[9] S. Katoh and A. San Pietro, *J. Biol. Chem.* **241**, 3575 (1966).

[10] R. E. McCarty and A. T. Jagendorf, *Plant Physiol.* **40**, 725 (1965).

[11] E. L. Gross and L. Packer, *Arch. Biochem. Biophys.* **121**, 779 (1967).

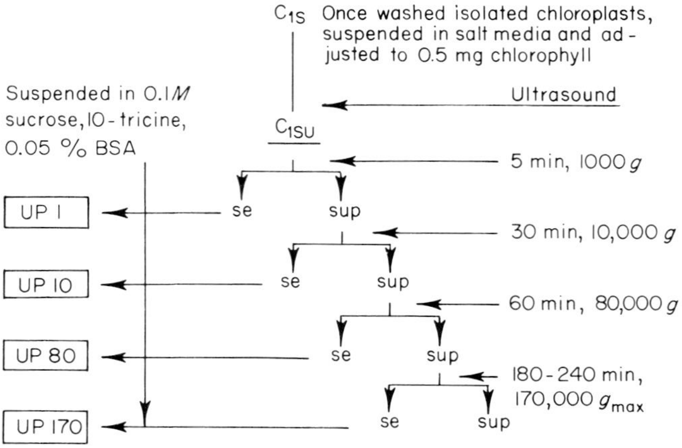

FIG. 1. Separation of sonicated isolated chloroplasts. BSA = bovine serum albumin; se = sediment, sup = supernatant, UP = ultrasonic particles.

grana stacks (UP 10) and UP 80, a gradient between 0.8 and 1.6 *M* sucrose is used. After 15 hours centrifugation at 100,000 *g*, a dark-green band is concentrated at 1.4–1.5 *M* sucrose. During this procedure the activity of reactions related to photosystem II decreases to about 50–60%. A slight stabilization is obtained after addition of 0.1 m*M* $MnCl_2$ and 0.1 μ*M* plastoquinone 45 to the gradient, but the inactivation is not avoided.

For the purification of UP 170 some prequisites must be considered. If the sonication is carried out in medium I for 15 seconds the yield of UP 170 is less than 1% of the total chlorophyll and thus not useful for further purification because of the low amount available. The proportion of particles present in UP 170 increases after longer sonication but this fraction now consists of at least two types of particles. Using a gradient between 0.5 and 1.6 *M* sucrose the light material is concentrated in the range 0.6–0.7 *M* sucrose and the heavier fragments are enriched in the same range as UP 80.[12]

Assay Methods

The assay methods used to measure Hill activity, NADP reduction and oxygen uptake are described in this series (Vol. XXIV). However, due to the changed properties induced by sonication, different conditions are required to obtain maximum rates. For the measurement of electron transport reactions correlated to either photosystem, the pH optimum is 6.3 instead of 8.4–8.7 as in intact chloroplasts (Fig. 2). Moreover,

[12] G. Jacobi, *Z. Pflanzenphysiol.* **61**, 203 (1969).

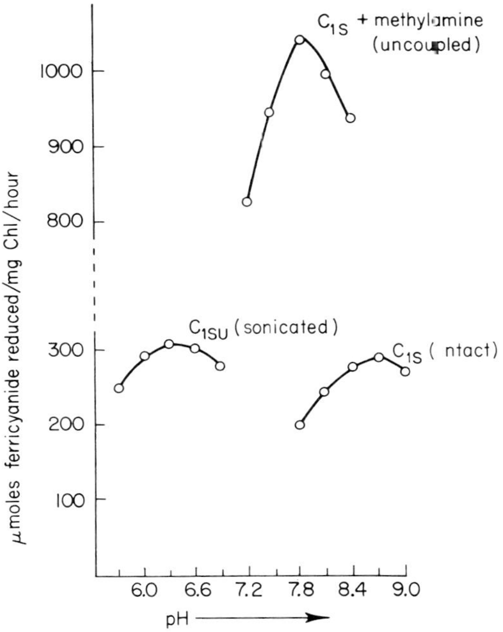

FIG. 2. pH optima of intact, uncoupled, and sonicated chloroplasts.

since UP 10 and sometimes UP 80 have some phosphorylating capacity,[13] the addition of methylamine is required to obtain maximal rates.[4,8]

Characteristics of the Fragments

Properties. After sonication a pronounced shift in the pH optima of the light-induced electron transport reactions is observed.[4,9] The difference of the pH optima for the Hill activity with ferricyanide for intact, uncoupled and sonicated chloroplasts is shown in Fig. 2. The same tendency is observed for the NADP-reducing systems but the maxima for the whole suspension (C_{1su}) and UP 10 are not so distinct as for the ferricyanide system.

Other properties which are changed by sonication are the concentration of ferricyanide necessary to obtain maximum velocity and the effect of light intensity.[9] In untreated chloroplasts the rate is essentially independent of oxidant concentration within the range of 5 μM to 1 mM. The velocity decreases unexpected in sonically treated chloroplasts when the concentration of ferricyanide is decreased below 40 mM. The relationship between light intensity and reaction rate of the Hill reaction

[13] Y. Chiba, K. Sugahara, and T. Oku, *Plant Cell Physiol.* **5**, 381 (1964).

with ferricyanide and with NADP is changed such that lower intensities are required for saturation with sonicated chloroplasts.[9] A double reciprocal plot indicated that both a light and a dark step were impaired.

Stability. Some authors reported a stability of oxygen evolution and electron transfer for special fractions during storage.[5] In contrast, other groups working under optimum conditions described a rapid loss of Hill activity.[9,14] Addition of 15% methanol or of 0.05% bovine serum albumin was found to be partially protective. In contrast to this lability of photosystem II, photosystem I is stable in all fractions during storage.

Distribution of Activities in the UP Fractions. Because of the very different pH optima, a direct comparison of the reaction rates between intact chloroplasts and the separated fragments is not possible. Therefore, only the distribution of the activities in the fragments of different size is given by taking the whole sonicated suspension ($C_{1\,su}$) as 100%.

The distribution of activities in the UP-fractions and the composition is strongly dependent upon the medium used for suspending the chloroplasts and on the time of sonication. After treatment for only 15 seconds in medium I, the capacity for the photoreduction of ferricyanide is concentrated in the fraction of grana stack (UP 10) but is absent in UP 170. On the basis of chlorophyll, the same values are estimated in $C_{1\,su}$ and in UP 10. Depending on the leaf material used, values between 200 and 400 μmoles of ferricyanide reduced per milligram of chlorophyll an hour are found. In Hill activity with NADP, the capacity is also retained in UP 10, but the values are more variable. The same rates found in the whole suspension and in the separated fraction of UP 10 are evident for the localization of the electron transport in the grana stacks.

For the activity associated with photosystem I, a reverse distribution is stated. The highest values between 700 and 1300 μmoles of NADP photoreduced are estimated in UP 170 provided higher amounts of plastocyanin (10–12 mμmoles/3.0 ml) are added. The particles liberated under these conditions exhibit approximately the same activity with and without DPIP.[12] In contrast photosystem I particles prepared from UP 10 by digitonin are different from UP 170 because of the requirement for DPIP and for lower concentrations of plastocyanin for the maximum activity. Upon addition of DPIP, the activity of NADP reduction increases by factors of 2.2–3.5 in the detergent particles, but only 1.0–1.4 in UP 170.

The oxygen uptake related to photosystem I, at higher concentrations of DPIP, in the heavier particles is independent of the addition of methyl viologen but is stimulated in UP 170. With the electron donor couple

[14] G. Jacobi, *Ber. Deut. Bot. Ges.* **79**, 72 (1966).

ascorbate and plastocyanin the requirement for methyl viologen becomes obligatory.[15,16]

Most of the experiments concerning the characterization of fragments obtained from sonicated chloroplasts are performed under drastic conditions of disintegration. The pattern of activity is different using longer time of sonication and in medium II or similar solutions.[4] Even the smallest particles isolated after prolonged sonication possess the capacity to reduce ferricyanide. However, under optimum conditions for measuring Hill activity, the values are lower with decreasing particle size.[4,8] UP 170 prepared by this method is composed of at least two types of particles, one derived from the grana stacks and active in the Hill reaction and the other identical with UP 170 as described earlier. A separation of the two types is achieved by density gradient centrifugation.

A part of the smallest units obtained after prolonged sonication was said to remain in the supernatant solution after 1 hour of centrifugation at 175,000 *g*.[9] In many respects these particles seem to be similar to UP 170 after short sonication in medium I and they sediment at 170,000 *g* only after several hours of centrifugation. The fragments from the supernatant solution show light-induced absorbance changes of P700 and electron spin resonance (ESR) signal I related to photosystem I.[17]

Chemical Composition. The chemical composition of fragments was undertaken to characterize the smallest units obtained from hypotonically suspended chloroplasts after prolonged sonication.[18,19] The same composition for chlorophylls, carotenoids, and quinones was found with intact chloroplasts and the subunits. This was generally confirmed for the particles present in the supernatant solution described by Vernon *et al.*[17] except for higher amounts of β-carotene and lower concentrations of plastoquinones. However, all those values are representative only of the smallest units obtained after drastic sonication. In all experiments resulting in small particles having the same composition as intact chloroplasts, the capacity for Hill activity is retained and thus photosystem II is present.

In contrast, under conditions protective for the grana area, the isolated heavy and light fractions differ in pigment composition. Corresponding values and electron micrographs were published for the heavier

[15] Y. Fujita and F. Murano, *Plant Cell Physiol.* **8**, 269 (1967).

[16] Y. Fujita and F. Murano, *in* "Comparative Biochemistry and Biophysics of Photosynthesis," p. 161. Univ. of Tokyo Press, 1968.

[17] L. P. Vernon, B. Ke, S. Katoh, A. San Pietro, and E. R. Shaw, *Brookhaven Symp. Biol.* **19**, 102 (1967).

[18] H. K. Lichtenthaler and R. B. Park, *Nature (London)* **198**, 1070 (1963).

[19] H. K. Lichtenthaler and M. Calvin, *Biochim. Biophys. Acta* **79**, 39 (1964).

fraction (UP 10) from two laboratories both using essentially the same procedure of sonication[4,11] Decreased ratios of chlorophyll *a/b*, between 1.8 and 2.4, were estimated in the fraction of grana stacks. Most of the total chlorophyll is concentrated in UP 80 (>80%). The same tendency is found in the heavy fraction obtained after treatment with low concentrations of digitonin.[20] This fraction was characterized as also composed of grana stacks,[21] having an ESR signal related to photosystem II[17] and active in performing the Hill reaction. Thus, the conclusion can be drawn that lower concentrations of digitonin cause the same type of disintegration as does short time sonication in medium I.

Under the same conditions of preparation, the ratio of chlorophyll *a/b* in UP 170 shows a tendency toward increase similar to the small particles from detergent-treated chloroplasts. Values between 3.8 and 6.2 were estimated. The yield of this fraction is quite low and seldom comes up to 1% of the total chlorophyll.

[20] J. M. Anderson and N. K. Boardman, *Biochim. Biophys. Acta* **112**, 403 (1966).
[21] W. Wehrmeyer, *Z. Naturforsch. B* **17**, 54 (1962).

[29] Subchloroplast Fragments: Sodium Dodecyl Sulfate Method

By KAZUO SHIBATA

One major obstacle in fractionating chloroplasts to obtain pigment–protein complexes or structural proteins is their insolubility in aqueous media. To overcome this difficulty, surface-active agents are used to disrupt the hydrophobic bonds between lipids and proteins or lipoproteins. Surface-active agents were first employed by Smith[1,2] and Smith and Pickels[3] for solubilization and fractionation of chloroplasts, and several natural surface-active agents, such as digitonin, bile salts, and sodium deoxycholate and a number of synthetic detergents, have since been applied to chloroplasts. Synthetic detergents include anionic, cationic, and nonionic ones; nonionic Triton X-100 (isooctylphenoxypolyethoxyethanol) and anionic SDS (sodium dodecyl sulfate) or DBS (sodium dodecylbenzene sulfonate) have been used most extensively. The applicability of SDS and DBS and some properties of the pigment–

[1] E. L. Smith, *J. Gen. Physiol.* **24**, 565 (1941).
[2] E. L. Smith, *J. Gen. Physiol.* **24**, 583 (1941).
[3] E. L. Smith and E. G. Pickels, *J. Gen Physiol.* **24**, 753 (1941).

protein complexes obtained with these detergents are reviewed in comparison with the data obtained with other surface-active agents.

Before the presence of two photochemical systems in chloroplasts was known, the experiments with detergents were carried out either to obtain a photosynthetic or structural units(s) or to follow the disintegration process as a way to study their structures. The sizes of the particles thus isolated ranged in sedimentation coefficient from 1.2 S or 1.7–2.6 S with DBS or SDS, respectively, to 13.5 S with digitonin, bile salts, or deoxycholate.[1-8] In general, anionic detergents as compared with other natural solubilizing agents disrupt chloroplasts into smaller components, although the Hill activity is completely abolished by treatment with SDS or DBS while most of the activity is preserved with digitonin. The effect of nonionic Triton X-100 seems to be somewhat between these effects of digitonin and anionic detergents, considering the fact that the chlorophyll–protein complex obtained with Triton X-100 could photoreduce ferricyanide at 10–20% of the rate with whole chloroplasts but did not reduce either DPIP (dichlorophenolindophenol) or NADP. This is seen more quantitatively in a comparison of the Hill activities[9] measured with these three solubilizing reagents for the same sample of chloroplasts fragments. The activity curve measured with methylamine as a function of reagent concentration showed a steep drop of activity at the same reagent concentration where the curve measured without methylamine showed a maximum. This concentration was approximately 0.05% for digitonin, 0.01% for Triton X-100, and 0.001–0.002% for SDS. Another example is found in the effects of synthetic detergents on the chromatophores of *Chromatium* D[10-12]; Triton X-100 and Sanizole-C (cationic detergent, alkylated dimethylbenzeneammonium chloride) cause reversible conversion of B850 to B810, while anionic detergents change absorption bands irreversibly. There are, however, some qualitative differences between the actions of SDS and Triton X-100. For example, SDS abolishes the Hill activity at appreciably lower concentrations than required to cause a pronounced fluorescence increase and a blue shift of the red

[4] M. Itoh, S. Izawa, and K. Shibata, *Biochim. Biophys. Acta* **69**, 130 (1963).
[5] J. J. Wolken, *J. Cell. Comp. Physiol.* **48**, 349 (1956).
[6] J. J. Wolken, *Brookhaven Symp. Biol.* **11**, 87 (1959).
[7] Y. Chiba, *Arch. Biochem. Biophys.* **90**, 294 (1960).
[8] J. S. Kahn, *Biochim. Biophys. Acta* **79**, 234 (1963).
[9] S. Izawa and N. E. Good, *Biochim. Biophys. Acta* **109**, 372 (1965).
[10] Y. Suzuki, S. Morita, and A. Takamiya, *Biochim. Biophys. Acta* **180**, 114 (1969).
[11] S. Izawa, M. Itoh, T. Ogawa, and K. Shibata, *in* "Studies on Microalgae and Photosynthetic Bacteria," p. 413, a special issue of *Plant Cell Physiol.,* Univ. of Tokyo Press, Tokyo, 1963.
[12] J. C. Goedheer, *Brookhaven Symp. Biol.* **11**, 325 (1959).

band whereas chloroplast fragments or quantasomes treated with Triton X-100 retain Hill activity at higher detergent concentrations than required to cause pronounced changes of absorption and fluorescence.[13,14] It is also to be noted that a 3:1 mixture of Dupanol C (mostly SDS) and Span 80 (sodium monooleate)[7,15] or a combination of low concentration of SDS with 0.5 *M* urea[16] was quite effective for solubilization of pigment–protein complexes or structural proteins.

Isolation of the Reaction Center from the Chromatophores of *Rhodopseudomonas viridis*

The strong solubilizing power of SDS was utilized by Thornber *et al.*[17] for isolation of photochemical reaction centers from the bulk light-harvesting chlorophylls in *Rhodopseudomonas viridis* cells containing a photobleachable pigment, B985 (B960 in dried films of chromatophores) and a pigment, B830 which undergoes a blue shift upon illumination. It is to be noted that the use of Triton X-100 did not enable a reaction center fraction to be obtained from this organism,[17,18] while photochemical reaction centers were successfully separated with this nonionic detergent from *Rhodopseudomonas spheroides*[19] and *Rhodospirillum rubrum*.[18]

The isolation from *Rhodopseudomonas viridis* was made in the following manner. Packed cells harvested either by centrifugation or by precipitation with alum were suspended in an equal volume of 50 m*M* Tris·HCl (pH 8.0), broken by sonication, and the sonicated material centrifuged at 15,000 *g* for 30 minutes. In the presence of alum added for the harvest, some chromatophores were present in the supernatant solution, but 90% or more of the bacteriochlorophyll *b* remained in the precipitate, whereas, in the absence of alum, high yields of chromatophores were obtained in the supernatant solution. The reaction center fraction was prepared in either case from the major chlorophyll-containing fraction by homogenizing that fraction with 1.0% SDS in 50 m*M* Tris (pH 8.0); the final ratio of SDS to bacteriochlorophyll *b* is 12 g/mmole. SDS in this ratio solubilized the light-harvesting bacteriochlorophyll as judged from the disappearance of the major peak of bacteriochlorophyll *b* *in vivo* at 1010 nm and the appearance of a peak

[13] K. Sauer and R. B. Park, *Biochim. Biophys. Acta* **79**, 476 (1964).
[14] S. Okayama, *Plant Cell Physiol.* **8**, 47 (1967).
[15] Y. Chiba and S. Okayama, *Plant Cell Physiol.* **3**, 379 (1962).
[16] R. S. Criddle and L. Park, *Biochem. Biophys. Res. Commun.* **17**, 74 (1964).
[17] J. P. Thornber, J. M. Olson, D. M. Williams, and M. L. Clayton, *Biochim. Biophys. Acta* **172**, 351 (1969).
[18] A. F. Garcia, L. P. Vernon, B. Ke, and H. Mallenhauer, *Biochemistry* **7**, 326 (1968).
[19] D. W. Reed and R. K. Clayton, *Biochem. Biophys. Res. Commun.* **30**, 471 (1968).

at 810 nm; the minor and broad band at 960 nm remained, but addition of excess SDS destroyed this 960-nm band. The SDS homogenate was then centrifuged at 15,000 *g* for 30 minutes, and the green supernatant solution was run into a column of hydroxyapatite[20,21] equilibrated in 10 m*M* sodium phosphate–0.2 *M* NaCl (pH 7.0). Elution with 0.2 *M* sodium phosphate–0.2 *M* NaCl (pH 7.0) removed a small fraction (approximately 8%) of the total bacteriochlorophyll *b* on the column. This fraction was rechromatographed on hydroxyapatite, after being precipitated with ammonium sulfate and redissolved in a minimum volume of 10 m*M* sodium phosphate–0.2 *M* NaCl (pH 7.0). The absorption spectrum of the subsequent eluate showed the 830- and 960-nm bands characteristic of the reaction center. The majority of the SDS-solubilized bacteriochlorophyll *b* (B810) was retained by the hydroxyapatite.

This reaction center preparation underwent expected spectral changes on illumination, and contained pigments and cytochromes C553 and C558 in a ratio of B960:lycopene:C553:C558 = 1:1:5:2. Ultracentrifugal analysis showed a colored boundary of 5 *S* in the presence of SDS.

Separation of the Two Photochemical Systems in Chloroplasts

Separation of different forms of pigment–protein complexes was suggested from spectral differences between the first and the second extracts with SDS from chloroplasts.[22] Since the success by Boardman and Anderson[23] in separating the two photochemical systems by differential centrifugation of digitonin-treated chloroplasts, a number of experiments have been reported on the separation with other surface-active agents including anionic detergents. The strong solubilizing power of anionic detergents, SDS[24,25] and DBS,[26,27] is suited for use in combination with the polyacrylamide gel electrophoresis, a powerful separation technique in protein chemistry. With digitonin or nonionic detergents[24,28] as the solubilizing agent, almost all the colored materials in chloroplasts remained

[20] H. W. Siegelman, G. A. Wieczorek, and B. C. Turner, *Anal. Biochem.* **13**, 402 (1965).

[21] J. P. Thornber, *Biochim. Biophys. Acta* **172**, 230 (1969).

[22] J. S. Brown and J. G. Duranton, *Biochim. Biophys. Acta* **79**, 209 (1964).

[23] N. K. Boardman and J. M. Anderson, *Nature (London)* **203**, 166 (1964).

[24] T. Ogawa, F. Obata, and K. Shibata, *Biochim. Biophys. Acta* **112**, 223 (1966).

[25] T. Ogawa, R. Kanai, and K. Shibata, *in* "Comparative Biochemistry and Biophysics of Photosynthesis" (K. Shibata, A. Takamiya, A. T. Jagendorf, and R. C. Fuller, eds.), p. 22. Univ. of Tokyo Press, Tokyo, 1968.

[26] J. P. Thornber, C. A. Smith, and J. L. Bailey, *Biochem. J.* **100**, 14P (1966).

[27] J. P. Thornber, R. P. F. Gregory, C. A. Smith, and J. L. Bailey, *Biochemistry* **6**, 391 (1967).

[28] J. L. Bailey, J. P. Thornber, and A. G. Whyborn, *in* "Biochemistry of Chloroplasts" (T. W. Goodwin, ed.), Vol. I, p. 243. Academic Press, New York, 1966.

at the origin or precipitated on the gel surface upon electrophoresis; in contrast, all the colored materials in SDS-solubilized chloroplasts entered the gel and were separated by electrophoresis. With cationic detergents,[24] the colored band diffused on electrophoresis without showing any trend of separation. DBS was preferred to SDS or Teepol 610 (secondary alkyl sulfonates) by Bailey *et al.*,[28] who reported that SDS and Teepol 610 interfered with the staining process. On the other hand, SDS was preferred to DBS by Ogawa *et al.*,[24] who found partial conversion of chlorophylls into pheophytins at low DBS concentrations before complete solubilization.

The following procedure was employed by Ogawa *et al.*[24] to obtain two pigment proteins from SDS-solubilized spinach chloroplasts. Chloroplasts were squeezed out of leaves through cotton cloth into 20 m*M* borate buffer of pH 10.3, the same pH as in electrophoresis, and sedimented at 200 *g*. The sediment was then resuspended in the same buffer, filtered through cotton cloth to remove cell debris, mixed with an equal volume of a solution of SDS in the same buffer, and shaken gently for 5 minutes in the dark at room temperature before being subjected to electrophoresis. The molar ratio, SDS/Chl (total chlorophylls in chloroplasts), rather than the absolute concentration of SDS in the mixture determines the degree of solubilization, so that the higher the chloroplast concentration, the higher the SDS concentration required for solubilization. In experiments with spinach chloroplasts, SDS/Chl = 125 was the lowest molar ratio for complete solubilization. Zone electrophoresis was conducted with a synthetic polyacrylamide gel as bed. Cyanogum 41 (5%) in 20 m*M* borate buffer (pH 10.3) was polymerized by addition of ammonium persulfate (10%) and DMAPN (American Cyanamid Co.) solutions; the volume ratios of these catalyst solutions to the cyanogum solution were 1:1000. Polymerization was made in a flat vessel for electrophoresis; air was excluded. A solution of solubilized chloroplasts was applied in a slot in the gel and electrophoresis was run for 2 hours at appropriate voltage and current; 2.3 V/cm and 10 mA.

Three colored components designated as components I, II, and III, in the order from the origin to the anode, were separated by the above procedure. Components I and II were pigment proteins and component III was a mixture of pigments released from the pigment proteins. Absorption spectra were measured directly on these components in the gel, and proteins were identified by staining with Amido Black 10B. These pigment proteins did not rapidly diffuse out of the gel when soaked in buffers or solvents, so that the following procedure was employed for elution. Each component area was cut out of the bed, and 12 such segments of a component were collected and arranged in the liquid medium of the above borate buffer in the flat vessel for electrophoresis. During a short period of electrophoresis, repeated three times with fresh medium, the pigment protein moved out of the gel into the liquid medium.

DBS was applied in the following manner by Thornber *et al.*[26,27] Washed chloroplasts were dispersed in 0.5% DBS–50 mM sodium borate, pH 8.0 (DBS/Chl = 2.5, w/w), and extracted several times with the same medium. Zone electrophoresis with polyacrylamide gel was carried out in 0.2% DBS–0.5 M sodium borate, pH 8.0, and the DBS extract mixed with propylene glycol (10%, v/v) was layered under the buffer on the top of a gel column. After electrophoresis, disks containing pigment proteins were cut into small pieces, which were disintegrated further by extrusion from a syringe through 100 μ nylon bolting cloth. The gel particles were then transferred to the top of a Sephadex G-200 column equilibrated in 0.02% DBS/50 mM sodium borate, pH 8.0, and the isolated components were eluted. By means of the above electrophoretic procedure, it was possible to obtain the two pigment–protein complexes separately from the DBS-solubilized chloroplasts, whereas the separation attempted by means of gel filtration with Sephadex G-200, chromatography with cellulose ion exchangers, and precipitation with ammonium sulfate was not satisfactory.[27] The two complexes, I and II, thus sepa-

PIGMENT DISTRIBUTION IN THE TWO COMPONENTS—LIGHT (L) AND HEAVY (H) PARTICLES—OBTAINED BY CENTRIFUGATION OF DIGITONIN-TREATED AND TRITON X-100-TREATED SPINACH CHLOROPLASTS, AND COMPONENTS I AND II SEPARATED BY ELECTROPHORESIS FROM SDS-TREATED SPINACH CHLOROPLASTS[a]

	Digitonin		Triton X-100		SDS		DBS	
Pigment	L	H	L	H	I	II	I	II
Chl *a*	85	67	85	67	87.5	65	92	55
Chl *b*	15	33	15	33	12.5	35	8	45
Chl *a*/Chl *b*	5.7	2	5.7	2	7.0	1.9	12	1.2
β-Carotene	14	7	16	6	14	5	—	—
Lutein	6	15	6	16	5	16	—	—
Neoxanthin	2	5	1	2	0	8	—	—
Violaxanthin	5	6	2	3	0	5	—	—

[a] Complexes I and II were separated by electrophoresis from DBS-treated spinach beet chloroplasts.[b]

[b] References: T. Ogawa, F. Obata, and K. Shibata, *Biochim. Biophys. Acta* **112,** 223 (1966); J. P. Thornber, C. A. Smith, and J. L. Bailey, *Biochem. J.* **100,** 14P (1966); J. P. Thornber, R. P. F. Gregory, C. A. Smith, and J. L. Bailey, *Biochemistry* **6,** 391 (1967); B. Ke, C. Seliskar, and R. Breeze, *Plant Physiol.* **41,** 1081 (1966); L. P. Vernon, B. Ke, S. Katoh, A. San Pietro, and E. R. Shaw, *Brookhaven Symp. Biol.* **19,** 102 (1967).

rated were proved[27] to be from photochemical systems I and II, respectively.

The pigment compositions of the two pigment proteins obtained from spinach chloroplasts with different solubilizing agents are listed in the table, where pigment fractions are normalized to 100 molecules of total chlorophyll. The light and heavy particles solubilized with digitonin[23,29] and with Triton X-100[29,30] were separated by differential or density gradient centrifugation. Components I and II of Ogawa *et al.* and complexes I and II of Thornber *et al.* solubilized with SDS and DBS, respectively, were separated by the above procedure of electrophoresis. High chlorophyll $a:b$ ratios and high β-carotene contents were found for the light particle, or component I or complex I, whereas low $a:b$ ratios and high xanthophyll contents were found for the heavy particle, or component II or complex II. The same trend of uneven distributions of carotene and xanthophylls was found for the two pigment proteins prepared similarly by electrophoresis from *Anabaena variabilis, Porphyra yezoensis,* and *Phaeodactylum tricornutum.*[25]

[29] B. Ke, C. Seliskar, and R. Breeze, *Plant Physiol.* **41**, 1081 (1966).
[30] L. P. Vernon, B. Ke, S. Katoh, A. San Pietro, and E. R. Shaw, *Brookhaven Symp. Biol.* **19**, 102 (1967).

[30] Preparation and Properties of Phosphorylating Subchloroplast Particles

By Richard E. McCarty

The rates of photophosphorylation and photoreduction catalyzed by chloroplasts vary considerably from preparation to preparation. Further, these activities are generally rapidly lost on storage of the chloroplasts. In view of these facts, it was of interest to develop a procedure for the preparation of subchloroplast particles capable of catalyzing photophosphorylation which could be stored for extended periods without loss of activity. Vambutas and Racker[1] reported the preparation of phosphorylating subchloroplast particles by exposure of chloroplasts to sonic oscillation in the presence of phosphatides. The following procedure is essentially a modification of that used by Vambutas and Racker.

Preparation

Chloroplasts were prepared from washed, market spinach leaves from which the midribs had been removed. All operations described below were

[1] V. K. Vambutas and E. Racker, *J. Biol. Chem.* **240**, 2660 (1965).

performed at 0°–4°. A total of 200 g of leaves were homogenized for 15 sec with 300 ml of a medium (STN) containing 0.4 *M* sucrose, 10 m*M* NaCl, and 20 m*M* tris(hydroxymethyl)methylglycine–NaOH (pH 8.0) in a Waring Blendor at top speed. The chloroplasts in the homogenate were isolated as described by Avron and Jagendorf[2] and were washed once with 150 ml of STN. The chloroplasts were resuspended by gentle homogenization in a minimal volume of STN, and the chlorophyll concentration of the suspension was adjusted to 2–2.5 mg/ml by the addition of STN.

Twenty milliliters of the chloroplast suspension was transferred to a 50-ml stainless-steel beaker which was immersed in an ice bath. A Branson 20-kHz Sonifier equipped with a 0.5-inch solid step horn was used to expose the chloroplasts to sonic irradiation. The probe was immersed about 1 cm into the chloroplast suspension and the Sonifier was run for 15 seconds at full output. The temperature rose to 8° during the sonication. After the temperature of the suspension fell to 4°, another 15-second period of sonication was given. This procedure was repeated once more, and the suspension was centrifuged at 6500 *g* for 10 minutes. The pellet which contained less than 10% of the total chlorophyll of the suspension was discarded. The supernatant fluid was centrifuged at 104,000 *g* for 45 minutes. The supernatant fluid was discarded and the pellet resuspended in 20 ml of STN by gentle homogenization. This suspension was centrifuged at 104,000 *g* for 45 minutes and the pellet resuspended in a small volume of STN. The chlorophyll concentration of the suspension was adjusted to 4–6 mg/ml by the addition of STN. Aliquots of 0.15 ml of the suspension were distributed to small tubes which were flushed for a few seconds with prepurified nitrogen and tightly stoppered. The tubes were stored at −80° in a Revco freezer.

Yield

From 200 g of leaves, subchloroplast particles equivalent to about 30 mg of chlorophyll were isolated. This recovery represents 60–70% of the chlorophyll contained in the original chloroplast suspension.

Properties

Under the electron microscope, subchloroplast particles negatively stained with phosphotungstic acid appeared to be vesicular. The diameter of the vesicles varied from about 0.08 to 0.5 μ, with an average diameter of 0.2 μ.

Subchloroplast particles catalyze noncyclic phosphorylation as well as cyclic phosphorylation (see the table). The rates of ferricyanide-dependent phosphorylation and pyocyanine-catalyzed cyclic phosphorylation

[2] M. Avron and A. T. Jagendorf, *J. Biol. Chem.* **234,** 967 (1959).

Some Activities Catalyzed by Subchloroplast Particles[a]

Activity measured	Reduction (μmoles reduced/hr/mg chlorophyll)	Phosphorylation (μmoles ATP formed/hr/mg chlorophyll)	P/e_2	H^+ uptake (μeq H^+ accumulated/mg chlorophyll)
Ferricyanide reduction and coupled phosphorylation	202–556	25–79	0.19–0.59	—
NADP reduction and coupled phosphorylation	32–98	32–48	0.3–0.7	—
Pyocyanine-dependent cyclic phosphorylation	—	125–490	—	—
Light-induced H^+ uptake	—	—	—	0.3–0.63
Acid-base induced ATP synthesis	—	35–60[b]	—	—
Postillumination ATP synthesis	—	25–40[b]	—	—

[a] Phosphorylation assays were carried out at room temperature in a reaction mixture which contained in a volume of 1.0 ml: 50 m*M* Tricine–NaOH (pH 8.0), 50 m*M* NaCl, 3 m*M* ADP, 2 m*M* potassium phosphate buffer (pH 8.0), 5 m*M* $MgCl_2$, 1 mg of defatted bovine serum albumin, subchloroplast particles equivalent to about 50 μg of chlorophyll and about 10^6, cpm $^{32}P_i$. For phosphorylation coupled to ferricyanide reduction, 1 m*M* $K_3Fe(CN)_6$ was present; for NADP reduction, 0.25 m*M* NADP was added; and for pyocyanine-dependent phosphorylation, 0.05 m*M* pyocyanine was used. The light intensity was 2×10^6 ergs/cm²/sec, and the gas phase was nitrogen. Ferricyanide and NADP reduction were determined spectrophotometrically, and $^{32}P_i$ esterification was assayed as previously described. Acid–base induced ATP formation and postillumination ATP synthesis were assayed as described elsewhere in this volume. Light-induced H^+ uptake was measured in the presence of 5 ml of a solution containing 50 m*M* NaCl, 50 μ*M* pyocyanine, 0.66 m*M* Tris·HCl, and subchloroplast particles equivalent to 100 μg of chlorophyll. The initial pH was 6.5 and the temperature, 5°.

[b] Nanomoles $^{32}P_i$ esterified per milligram of chlorophyll.

in subchloroplast particles averaged about half of those observed for these reactions in chloroplasts, whereas the rates of ferricyanide reduction were comparable to those in chloroplasts. In the presence of ferredoxin and ferredoxin–NADP reductase, NADP reduction with water as the electron donor and associated phosphorylation in subchloroplast particles also occurred. Maximal rates of phosphorylation were observed only in the presence of defatted bovine serum albumin (1 mg/ml) and of a nitrogen atmosphere.

The extent of the light-dependent H^+ uptake[3] in subchloroplast particles was about half that of chloroplasts. Further, ATP synthesis in the dark resulting from either a rapid change in the pH of a subchloroplast particle suspension[4] or from a prior illumination,[5] could be readily detected.

A major difference between chloroplasts and subchloroplast particles is the relative insensitivity of phosphorylation in the latter preparation to uncoupling by NH_4Cl and amines.[6] Whereas 2 m*M* NH_4Cl inhibited phosphorylation in chloroplasts by over 50%, it had only a slight effect on phosphorylation in subchloroplast particles.

The activities catalyzed by subchloroplast particles are quite stable to storage of the particles at −80°. For example, no decrease in either the rate of ferricyanide reduction or coupled ATP synthesis was observed after a year's storage.

[3] J. S. Neumann and A. T. Jagendorf, *Arch. Biochem. Biophys.* **107,** 109 (1964).
[4] A. T. Jagendorf and E. Uribe, *Proc. Nat. Acad. Sci. U.S.* **55,** 170 (1966).
[5] G. Hind and A. T. Jagendorf, *Proc. Nat. Acad. Sci. U.S.* **49,** 715 (1963).
[6] R. E. McCarty, *Biochem. Biophys. Res. Commun.* **32,** 37 (1968).

[31] Subchromatophore Fragments: *Chromatium, Rhodospirillum rubrum,* and *Rhodopseudomonas palustris*

By Augusto Garcia, J. Philip Thornber, and Leo P. Vernon

The photosynthetic membrane systems of some purple bacteria are readily cleaved by both nonionic and ionic detergents to yield distinct subchromatophore fragments which are characterized by the unique forms of bacteriochlorophyll contained in the particles. Other differences are also found, primarily in the cytochrome content and the presence of specific enzymes. The types of subchromatophore fragments obtained depend to a large extent upon the nature of the detergent, the presence or absence of carotenoids in the membrane, and the conditions used for growing the bacteria. In those bacteria examined to date, at least two fragments are produced, one containing the reaction center bacteriochlorophyll (BChl) and varying amounts of other associated BChl, while the other one contains only light-harvesting BChl (except for *R. rubrum*). In the case of *Chromatium,* fragmentation with sodium dodecyl sulfate (SDS) produces three fragments, but two of the three are quite similar in their general properties.

In vivo the BChl *a* (the form of BChl contained in the bacteria under consideration) exhibits an absorption band in the orange and a series of bands in the near infrared region. Although the exact locations of the absorption maxima of the three BChl *a* forms differ slightly in different bacteria, they are referred to as B800, B850, and B890. It is generally held that the different absorption properties of the BChl molecules in the near infrared region reflect different complexes of BChl with lipoproteins.

Fractions Obtained with Triton X-100

Chromatium

The photosynthetic apparatus of *Chromatium* cells is contained in internal structures which are derived from the cell membrane.[1] The early experiments of Komen with chromatophores of both *R. rubrum* and *Chromatium*[2] showed that SDS causes a lowering of the *in vivo* absorption bands of BChl and produces a new band at 760 nm. Bril[3,4] studied the effect of deoxycholate upon *Chromatium* chromatophores, and Clayton[5] showed that this detergent could be used to prepare a subchromatophore particle enriched in the B890 component. These early experiments indicated that the photosynthetic apparatus of *Chromatium* and other bacteria could be fragmented by the systematic application of detergents.

Chromatophore Preparation. Cells of *Chromatium* D, grown according to the directions of Hendley[6] are harvested after 3 days of growth, washed once by centrifugation, and suspended in 10 m*M* Tris buffer, pH 8.1, for a 3-minute sonication in a 10-kHz Raytheon sonic oscillator. The chromatophore fragment is obtained by centrifugation, collecting the material sedimenting between 20,000 and 110,000 *g* during 1 hour of centrifugation. The chromatophores are suspended in 10 m*M* Tris, pH 8.1, centrifuged again, and resuspended in fresh buffer.

Treatment with Detergent. Triton X-100 is added to the chromatophore preparation (2.6 mg of detergent/100 mg of protein), which is then allowed to stand in an ice bath, under argon, for 2 hours in the dark. The suspension is diluted to a concentration of 1.5 mg of protein per milliliter and applied to tubes containing a discontinuous sucrose gradient consisting of three different layers of sucrose (57, 24, and 14%) dissolved

[1] R. C. Fuller, S. F. Conti, and D. B. Mellin, *in* "Bacterial Photosynthesis" (H. Gest, A. San Pietro, and L. P. Vernon, eds.), p. 89. Antioch Press, Yellow Springs, Ohio, 1963.

[2] J. G. Komen, *Biochim. Biophys. Acta* **22,** 9 (1956).

[3] C. Bril, *Biochim. Biophys. Acta* **39,** 296 (1960).

[4] C. Bril, *Biochim. Biophys. Acta* **66,** 50 (1963).

[5] R. K. Clayton, *Photochem. Photobiol.* **1,** 201 (1962).

[6] D. D. Hendley, *J. Bacteriol.* **70,** 625 (1955).

in 10 mM Tris, pH 8.1. The tubes are centrifuged for 15 hours at 110,000 g in the 30 rotor of a Spinco Model L ultracentrifuge. The separated particles are then dialyzed overnight in the cold against 10 mM Tris buffer, pH 8.1.

For protein determination a small aliquot is extracted twice with a mixture of acetone and methanol (7:2, v/v); the precipitated proteins are dissolved in 0.5 M NaOH, and an aliquot is used to determine proteins by the method of Lowry *et al.*[7] The absorbance of the different preparations at 590 nm is used as a measure of the BChl content of the different fragments ($\epsilon_{590} = 20$ mM^{-1} cm^{-1}).

Two major particles are separated by density gradient centrifugation. A brownish band, designated the light (L) band is located in the upper portion of the tube and immediately below is a red band, called the heavy (H) band. The H band shows three BChl types, the B890 form being predominant. The L band shows the presence of both B800 and B850. Reextraction with Triton X-100 causes a marked decrease in the 800 and 850 components of H and yields a particle containing mainly the 890 component (Fig. 1). Similarly, reextraction of L partially removes or destroys the 850 component yielding a particle enriched in B800. It has

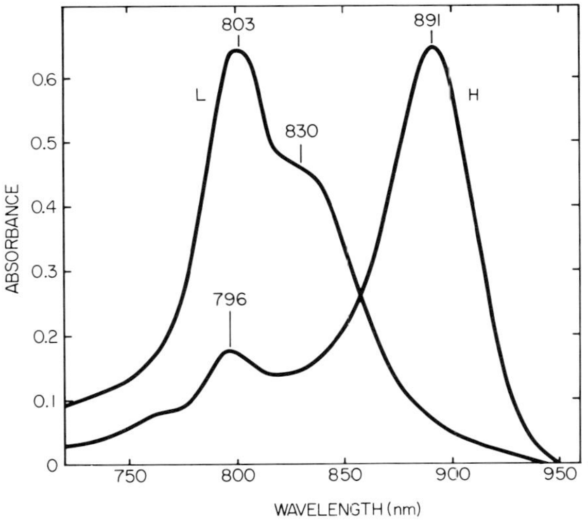

FIG. 1. Absorption spectra of the heavy (H) and the light (L) fractions after two treatments of *Chromatium* chromatophores with Triton X-100 and density gradient centrifugations. [Reprinted from *Biochemistry* **5**, 2401 (1966). Copyright (1966) by The American Chemical Society. Reproduced by permission of the copyright owner.]

[7] O. H. Lowry, N. J. Rosebrough, A. L. Farr, and R. J. Randall, *J. Biol. Chem.* **193**, 265 (1961).

not been possible to separate the B800 and the B850 components, which always appear together.

Properties. It is generally held that the physiologically significant photochemical reactions in bacterial systems are initiated at a reaction center which contains a BChl which is special in the sense that it is complexed with the necessary electron transfer agents, such as cytochromes. Although it has not been isolated from *Chromatium,* the reaction center BChl is readily detected by means of its light-induced absorbancy change. Fraction H shows this response, and the L fraction is devoid of this activity, as shown in Fig. 2. The heavy particle which is photochemically active shows an absorbance decrease (ΔA in the figure) at 890 nm (reaction center BChl). It has been shown by Ke *et al.*[8] that this fragment also shows a 275-nm change corresponding to the reduction of the endogenous ubiquinone.

Electron micrographs of *Chromatium* chromatophores show substructure that is clearly seen on the surface of the chromatophore as individual units of approximately 60–70 Å in diameter. The L particles, which are photochemically inactive, consist of the basic membrane matrix of the chromatophore along with some attached material. The H particles, which are photochemically competent, show none of the structure of the original chromatophore and apparently consist of small subunits released from the chromatophore.

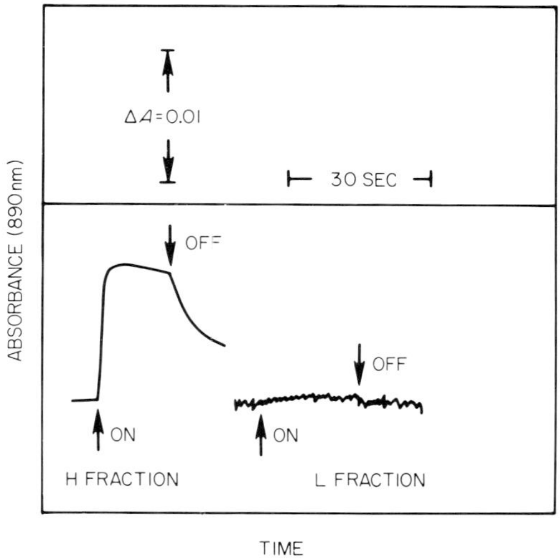

FIG. 2. Localization of the P890 reaction (reaction center BChl) in the heavy fraction from *Chromatium* chromatophores. [Reprinted from *Biochemistry* **5,** 2403 (1966). Copyright (1966) by The American Chemical Society. Reproduced by permission of the copyright owner.]

[8] B. Ke, L. P. Vernon, A. Garcia, and E. Ngo, *Biochemistry* **7,** 311 (1968).

High concentrations of Triton X-100 disrupt the chromatophore to yield two fragments and the BChl pigments segregate in the two particles. For additional information concerning the properties of the *Chromatium* subchromatophore particles, see the paper by Garcia *et al.*[9]

Rhodospirillum rubrum

Preparation. Rhodospirillum rubrum cells, strain S-1, were grown on malate medium according to the directions of Newton.[10] The experimental details concerning the preparation of chromatophores, apparatus, the procedure used for detergent treatment, density gradient centrifugation, dialysis of the separated particles, and protein determination are those described above for *Chromatium*. In the present case, a ratio of Triton X-100/BChl of 70/1 (mg/mg) is routinely used (this is equal to 2.5 mg of detergent per milligram of protein). Acetone–methanol is used to extract and measure the BChl of the particles, as described by Clayton.[11] Equilibrium centrifugation of *R. rubrum* chromatophores following treatment with Triton X-100 produces five different zones in a sucrose gradient. Practically all the pigmented material is localized in two sharp, red–purple bands designated the H (heavy) and L (light) bands. A diffuse upper brown band is designated the Pheo (pheophytin-containing) band. The zones in between the bands contain varying amounts of BChl. Most of the BChl is found in the H band, which is dark red because of the associated carotenoids.

Properties. The absorption spectra of the three pigment-containing bands are shown in Fig. 3. Both the L and H particles yield spectra that do not differ significantly from that of the intact chromatophore. There is a displacement of the 880-nm band to 882 nm for the H particle. There is no separation of the 880- and 800-nm component between the two fractions, however. The Pheo fraction shows an absorption spectrum similar to that of bacteriopheophytin. The Pheo fraction represented in Fig. 3 is obtained after dialysis of the original material and recentrifugation at 144,000 *g* for 2 hours.

One distinction between the pigment-containing fractions is apparent in their ability to catalyze the photoreduction of NAD, supported by either succinate or the ascorbate–DPIP couple as the electron donor system. Of the three pigmented fractions, only the L fraction has significant activity in this reaction. The rate of NAD reduction is low, however, when compared with rates observed for chromatophores (5 for L fraction vs. 20–40 μmoles/hr/mg of BChl for chromatophores). When

[9] A. Garcia, L. P. Vernon, and H. M. Mollenhauer, *Biochemistry* **5**, 2399 (1966).

[10] J. W. Newton, this series, Vol. 5, p. 70.

[11] R. K. Clayton, *in* "Bacterial Photosynthesis" (H. Gest, A. San Pietro, and L. P. Vernon, eds.), p. 495. Antioch Press, Yellow Springs, Ohio, 1963.

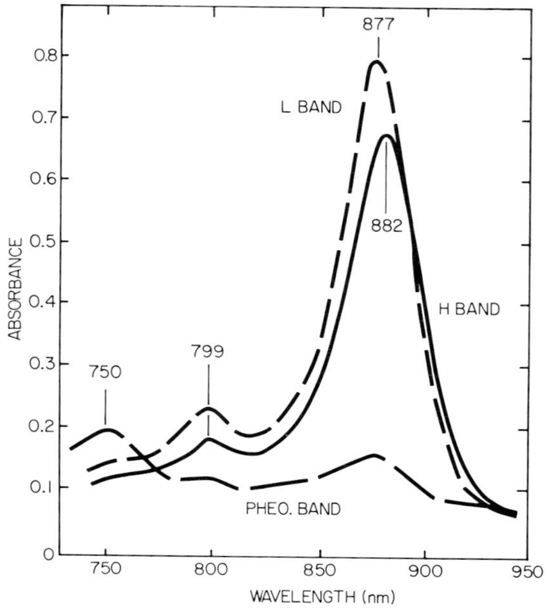

FIG. 3. Absorption spectra of the heavy (H), light (L), and pheophytin-containing (Pheo) fractions from *Rhodospirillum rubrum* chromatophores after separation by density gradient centrifugation. [Reprinted from *Biochemistry* **5**, 2409 (1966). Copyright (1966) by The American Chemical Society. Reproduced by permission of the copyright owner.]

the fractions are assayed for succinic dehydrogenase activity, only the light fraction and the Pheo fraction show this activity. Thus the activities which require the participation of both enzymes and BChl are restricted to the L particle.

Rhodospirillum rubrum chromatophores show a light-induced absorbance change at 890 nm, which is related to the reaction center BChl or P890. Both the L and H fractions show this light-induced change to about the same degree. A preparation of H particles (880-nm absorption = 1.84) gives an absorbance change of 0.043 at 890 nm upon illumination. A preparation of the L particles (880-nm absorption = 1.88) gives an absorption change of 0.034. It is apparent that in contrast to data for *Chromatium,* the P890 is partitioned between the two particles.

When observed under the electron microscope, it is apparent that the H fraction contain some unruptured chromatophores, but consists primarily of the residual chromatophore membrane and some attached material. The L fraction, on the other hand, consists of small subunits which have been removed by the detergent.[12]

[12] A. Garcia, L. P. Vernon, and H. H. Mollenhauer, *Biochemistry* **5**, 2408 (1966).

A method for dissolution of membranes from photosynthetic bacteria has been devised by Loach *et al.*[13] This method involves the use of high pH (11.0) coupled with exposure to 1.5% Triton X-100 and 6 *M* urea. This procedure has been applied to chromatophores of *Rhodopseudomonas spheroides*[13] as well as to *Rhodospirillum rubrum*.[14] In both cases a small particle is obtained which has a molecule weight of less than 200,000 and contains all the BChl and reaction center activity as well as the carotenoids. This treatment removes almost all the original lipid and replaces it with detergent.

Rhodopseudomonas palustris

Preparation. *Rhodopseudomonas palustris* cells are grown photosynthetically using the medium described by Cohen-Bazire *et al.*,[15] supplemented with 1% yeast extract and 1% peptone under a light intensity of 500 foot-candles. To prepare the chromatophore fraction, washed cells are suspended in 10 m*M* Tris buffer, pH 7.5, and sonicated in a Raytheon sonic oscillator for 2 minutes at full power. The fraction that sediments between 15,000 and 115,000 *g* is used as the chromatophore fraction. This preparation is resuspended in the same buffer and applied to the top of a 15-ml layer of a 60% (w/v) CsCl solution and centrifuged for 1 hour at 100,000 *g*. The pigmented fraction remains on the surface of the CsCl solution and a colorless protein residue can be seen at the bottom of the tube. This procedure greatly reduces the ratio of BChl to protein in the final chromatophore preparation.

Chromatophores are treated with 4% Triton X-100 for 1 hour at 0°. The suspension of treated chromatophores (1.5 mg of BChl/ml and 4% Triton X-100 final concentration) is then applied to the same gradient used to prepare the subchromotophore fragments from either *Chromatium* or *Rhodospirillum rubrum*.

Properties. The Triton treatment described here produces both a light (L) and a heavy (H) fraction. The absorption spectra of *Rhodopseudomonas palustris* chromatophores and derived fragments are shown in Fig. 4. The chromatophore fraction shows absorption peaks at 805 and 860 nm with a shoulder at approximately 880 nm. The H fragment obtained by Triton treatment shows a major peak at 870 nm, with a smaller band at 802 nm, showing clearly that the long wavelength BChl

[13] P. A. Loach, D. L. Sekura, R. M. Hadsell, and A. Stemer, *Biochemistry* **9**, 724 (1970).

[14] P. A. Loach, personal communication, 1969.

[15] G. Cohen-Bazire, W. R. Sistrom, and R. Y. Stanier, *J. Cell. Comp. Physiol.* **49**, 25 (1957).

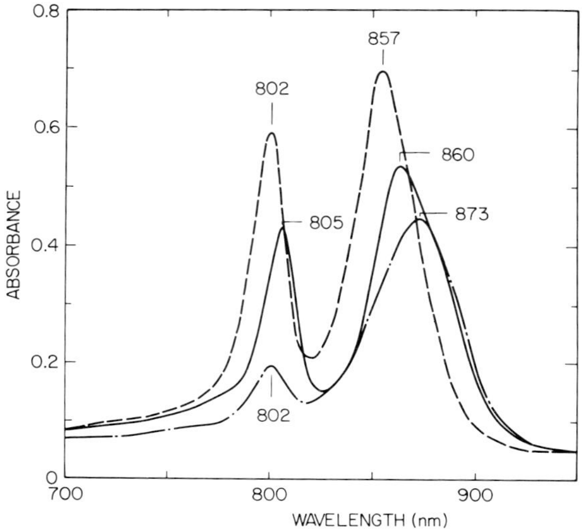

FIG. 4. Absorption spectra of *Rhodopseudomonas palustris* chromatophores (solid curve) and derived fragments (dashed and dot curve for H; dashed curve for L) in the near-infrared region. The H and L fragments were obtained by density gradient centrifugation and were diluted with 50 m*M* Tris (pH 7.5) prior to measurement of the absorption spectrum. [Reprinted from *Biochemistry* 7, 321 (1968). Copyright (1968) by The American Chemical Society. Reproduced by permission of the copyright owner.]

form is concentrated in this fragment. As usual, the two shorter wavelength forms are found in the L fraction, with peaks at 857 and 802 nm. The small amount of B802 appearing in the H fragment is presumably the special B800 material, called P800, that forms part of the reaction center.

For the reaction center distribution a situation similar to that of *Chromatium* (Fig. 5) is observed. This change is observed in the chromatophores and the H fraction, but not at all in the L fragment. For additional information on the properties of these paricles, see Garcia *et al.*[16]

Ke *et al.*[8] have studied in detail the photoreduction of endogenous quinone in *Chromatium* H particles, and showed that the photoreduction of ubiquinone is coupled to the photooxidation of the reaction center BChl. A similar situation occurs in *R. palustris*. The quinone reaction is followed at 275 nm, and the data show the presence of ubiquinone photoreduction in both the chromatophores and the H particle. The L particle is devoid of this activity (Fig. 5).

The kinetic characteristics of the 275-nm change are similar to the corresponding kinetic characteristics of the reaction center BChl. This would be consistent with the view that the endogenous ubiquinone func-

[16] A. Garcia, L. P. Vernon, and H. H. Mollenhauer, *Biochemistry* 7, 319 (1968).

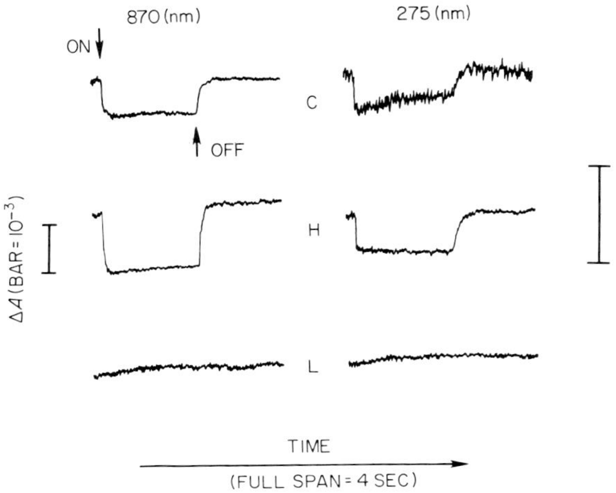

FIG. 5. Light-induced absorbance changes corresponding to P870 photooxidation and the coupled reduction of endogenous ubiquinone at 275 nm for the H and L fractions as well as chromatophores (C) from *Chromatium*. Actinic light at 875 nm (interference filter) was used for measuring the quinone change, and a broad band blue filter was used when P870 was determined. The intensities of both actinic lights were 10^5 ergs/cm^2/sec. [Reprinted from *Biochemistry* **7**, 322 (1968). Copyright (1968) by The American Chemical Society. Reproduced by permission of the copyright owner.]

tions as the initial electron acceptor in the primary photochemical act in these organisms. However, later data of Ke[17] demonstrate that in *Chromatium* H particles it is possible to abolish the 275-nm change without disturbing the 890-nm absorbance change of the reaction center BChl. This shows that ubiquinone cannot be the acceptor for P890 in the initial photochemical event, at least in *Chromatium*.

A comparison of the structural properties of the L and H fractions shows that the L fraction does not have any definite form and appears to consist of soluble material. The H fraction, on the other hand, is a "particulate fraction" in which the small particles are aggregated into linear arrays. It is difficult to see much detail in the substructure of these membranes in the original chromatophore to relate to the H fragment. However, the important point to emphasize is that the H and L fraction arise from two different environments with very different structural properties. The L fraction, whose function is postulated to be one of gathering light energy, does not reveal any regularity in its structure, as would be expected from its possible function. On the other hand, the H fraction containing the photochemically active part of the chromatophore vesicles has a more definite structure.

[17] B. Ke, *Biochim. Biophys. Acta* **172**, 583 (1969).

Fractions Obtained with Sodium Dodecyl Sulfate

This section describes what occurs when SDS is used for dissociation of the chromatophores, and alternative fractionation techniques are employed for the resolution of the solubilized material.[18] Using SDS as the dissociating agent, *Chromatium* chromatophores can be fractionated into three spectrally different carotenobacteriochlorophyll–protein complexes, compared to the two obtained with Triton X-100. The relative yield of each complex varies depending, among other things, upon the growth conditions of the cells; this variation is reflected in the diverse near infrared absorption spectra that are reported in the literature for *Chromatium* cells.

The three fractions that can be obtained are designated as follows: (1) Fraction A, which has its major near-infrared absorption maximum at 890 nm, and among other components, contain the known reaction center of the organism (P883).[19–21] (2) Fraction B, which has absorption maxima of approximately equal intensity at 800 nm and 850 nm; the precise location of the latter peak is variable (830–850 nm) depending upon growth conditions and age of the culture.[21] A photoactive pigment (P836) is present. (3) Fraction C, which has absorption maxima at 800 and 820 nm. This component contains no photoactive pigment.

Fractionation Procedure

The preparation is carried out at room temperature, preferably under green light (Sylvania F40 green fluorescent tubes covered by green Plexiglass 2092). The near-infrared spectrum of the solutions, obtained after various steps during the isolation procedure, is monitored, and used as an index of the degree of purity; the criterion of purity is that the derived fractions should have the spectra shown in Fig. 6. An additional criterion is that fractions B and C should contain no reversible light-induced bleaching at 883 nm; this property is a function of fraction A only. Some of the steps in the procedure can be repeated until the required criteria are obtained.

Step 1. *Chromatium* cells which have been grown photoautotrophically or photoheterotrophically[22] are harvested by centrifugation. Seven milli-

[18] Research carried out at Brookhaven National Laboratory under the auspices of the U.S. Atomic Energy Commission.

[19] L. N. M. Duysens, Ph.D. Thesis, State Univesity, Utrecht, The Netherlands, 1952.

[20] R. K. Clayton, *Photochem. Photobiol.* **1**, 201 (1962).

[21] M. A. Cusanovitch, R. G. Bartsch, and M. D. Kamen, *Biochim. Biophys. Acta* **153**, 397 (1968).

[22] C. Bril, Ph.D. Thesis, State University, Utrecht, The Netherlands, 1964.

liters of packed cells are suspended in 50 m*M* Tris·HCl, pH 8.0 (50 ml), and the pH is readjusted if necessary by the addition of solid Tris. The suspension is sonicated in a Raytheon 10-kHz oscillator at 1 A for 10 minutes, and the sonicate is centrifuged at 15,000 *g* for 15 minutes. Breakage in the French pressure cell is also satisfactory.

Step 2. The spectrum of the supernatant solution (crude chromatophores) is recorded, and the absorbance at 590 nm is used to estimate the bacteriochlorophyll (BChl) concentration ($\epsilon_{590} = 20/\text{m}M/\text{cm}$). A 10% SDS solution is then added to give a final SDS:BChl ratio of 12 g/mmole. The final concentration of SDS should be not less than 0.5% (w/v) if good yields are to be obtained from the ensuing chromatographic step.

Step 3. A column of hydroxyapatite (6.5 × 3.7 cm) is equilibrated with 10 m*M* sodium phosphate + 0.2 *M* NaCl, pH 7.0, and the SDS-solubilized chromatophores are run into the column. The column is washed with 100 ml of the equilibration buffer and then with the same volume of 100 m*M* sodium phosphate + 0.2 *M* NaCl, pH 7.0. Elution with 250 m*M* sodium phosphate + 0.2 *M* NaCl, pH 7.0, removes a large portion of the color which is absorbed to the column; this eluate contains a mixture of fractions A and B. Steps 4 and 5 describe the subsequent purification of each fraction. Elution with 350 m*M* sodium phosphate + 0.2 *M* NaCl, pH 7.0, removes another pink fraction which contains a mixture of fraction A, B, and C; this eluate is discarded. A spectrally pure form of fraction C is removed from the column by 700 m*M* sodium phosphate + 0.2 *M* NaCl. Addition of ammonium sulfate to the eluate precipitates fraction C, which is isolated by centrifugation, and then redissolved in 50 m*M* Tris, pH 8.0.

Step 4. Separation of Fractions A and B. A 50% (w/v) ammonium sulfate solution is added dropwise to the 250 m*M* sodium phosphate eluate from the hydroxyapatite column until a cloudiness develops, which occurs at a concentration of 7% (w/v) ammonium sulfate. The mixture is then centrifuged (15,000 *g*, 10 minutes), and the supernatant solution is separated from the precipitate. The precipitate (fraction A) is redissolved in 50 m*M* Tris, pH 8.0, and step 4 is repeated if necessary. The final solubilized precipitate is used for studies on fraction A.

Step 5. The supernatant solution of step 4 is made to 20% (w/v) ammonium sulfate and recentrifuged. The precipitate (mixture of further amounts of fraction A and some of fraction B) is discarded. The pH of the supernatant is lowered to pH 4.5 by addition of 1 *N* acetic acid, whereupon a cloudiness develops. The suspension is centrifuged and the precipitate (fraction B) is redissolved in 50 m*M* Tris, pH 8.0. If the spectrum of the solution does not meet the criteria of purity of fraction B, step 5 can be repeated.

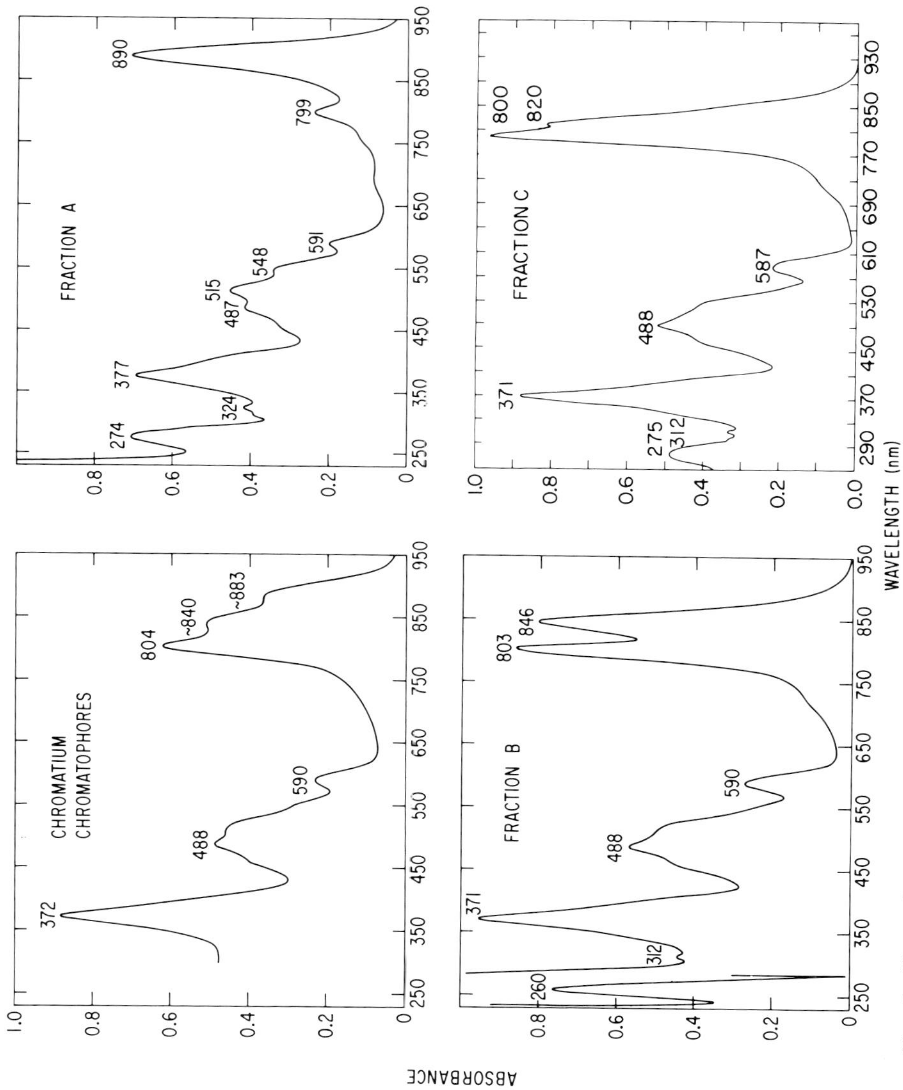

FIG. 6. Absorption spectra at room temperature for fractions A–C dissolved in 60 mM Tris·HCl, pH 8.0.

Properties

Spectral. a. ABSORPTION SPECTRA. The room temperature spectra of the fractions isolated from autotrophically grown cells are shown in Fig. 6. Cells grown in an acetate-containing media give fractions of identical spectra with one exception—the shorter infrared wavelength peak of fraction B is located at a lower wavelength (835 nm).

b. EMISSION SPECTRA. The emission spectra of fractions A and B at room temperature and at the temperature of liquid nitrogen are given in Fig. 7. The spectra were recorded for preparations which had their photo-oxidizable pigments in the reduced form.

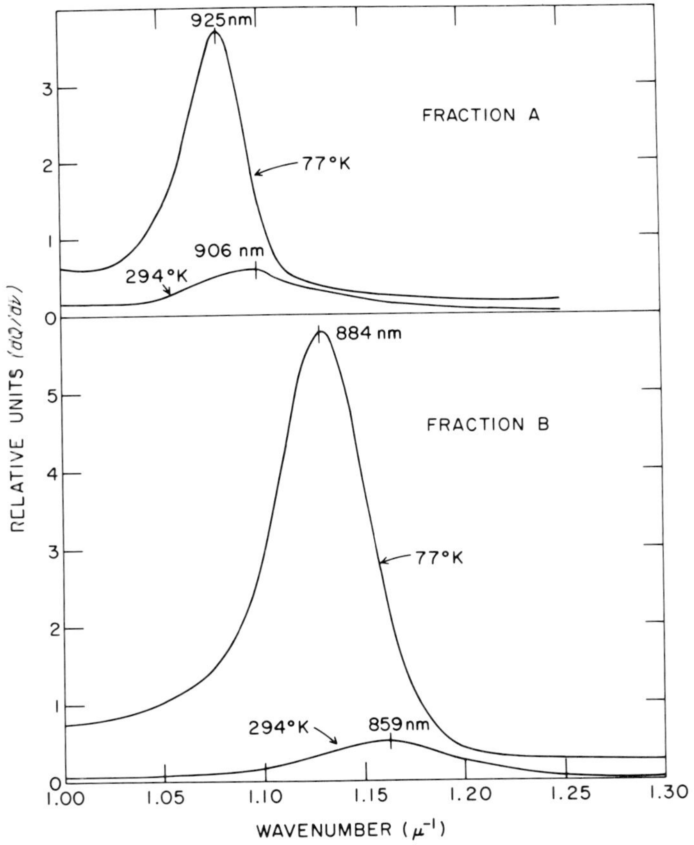

FIG. 7. Emission spectra of fractions A and B dissolved in 50 m*M* Tris·HCl, pH 8.0, at room temperature and at the temperature of liquid N_2. The spectra were recorded for samples having an absorbance at 590 nm of less than 0.2 cm^{-1}. The sample thickness was 1 mm. The spectra are corrected for the response of the monochromator and phototube. [Reprinted from *Biochemistry* **9**, 2692 (1970). Copyright (1970) by The American Chemical Society. Reproduced by permission of the copyright owner.]

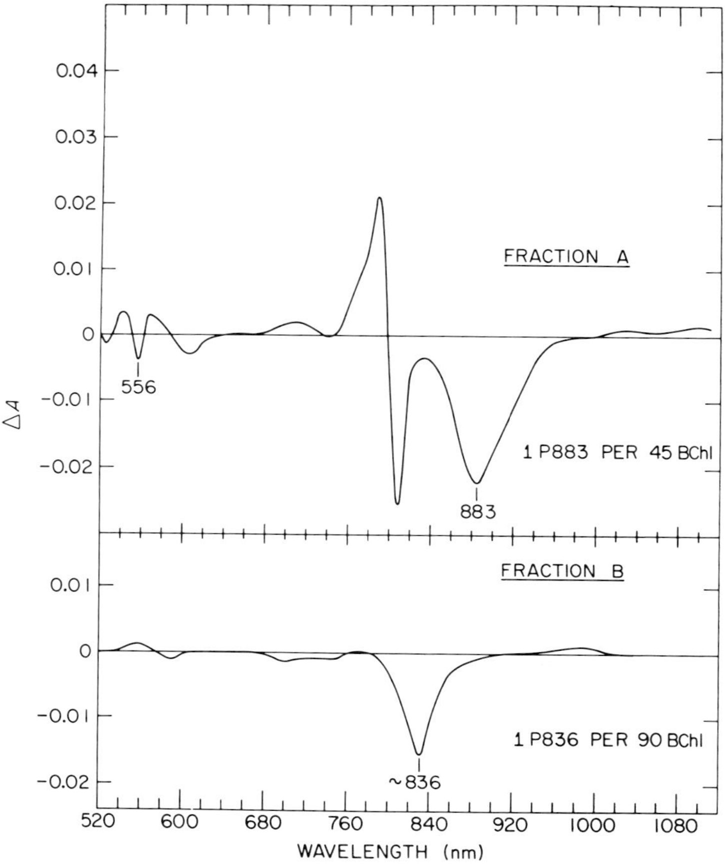

FIG. 8. Light-induced reversible absorbance changes in fractions A and B. The sample of fraction A contained 9 μM bacteriochlorophyll and 50 μM sodium ascorbate. The sample of fraction B contained 14 μM bacteriochlorophyll; no additions were made to fraction B. Light of 375 nm (6.2 $nE/cm^2/sec$) was used to oxidize both samples. [Reprinted from *Biochemistry* **9**, 2693 (1970). Copyright (1970) by The American Chemical Society. Reproduced by permission of the copyright owner.]

c. LIGHT-INDUCED SPECTRAL CHANGES. The reversible light vs. dark spectral changes that occur in fractions A and B are shown in Fig. 8; fraction C shows no reversible spectral changes. Fraction A demonstrates the changes associated with the known reaction center pigments (bleaching at 883 nm and an associated peak shift of P800) of the organism. A cytochrome (C556) also undergoes oxidation upon exposure to light (Fig. 9). The presence of phenazine methosulfate (5 μM) or sodium ascorbate (50 μM) permits a rapid recovery of the spectral changes following illumination; without these additions the recovery is very slow.

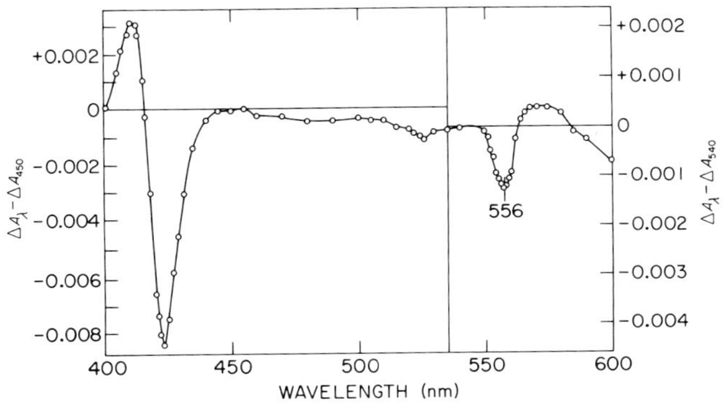

FIG. 9. Light-induced absorbance changes obtained by double-beam measurement of fraction A. The sample contained 2.5 μM phenazine methosulfate, and was illuminated by light of 825 nm (11.3 nE/cm²/sec) for 8 seconds with an intervening dark period of 30 seconds. [Reprinted from *Biochemistry* **9**, 2693 (1970). Copyright (1970) by The American Chemical Society. Reproduced by permission of the copyright owner.]

Fraction B shows light-induced bleaching at 835 nm; this spectral change is reversed upon returning the sample to darkness; cytochrome absorbance changes are not observed in fraction B. The reversible bleaching of P835 probably represents a second reaction center pigment in *Chromatium*. In both fractions A and B there are other reversible spectral changes, but the significance of most of them remains obscure.

If the spectral changes of fraction A are examined without addition of sodium ascorbate or phenazine methosulfate, then a reversible bleaching of a pigment whose absorption maximum is 890 nm is observed in addition to that of P883. This pigment is denoted as P′890, oxidation of which is not accompanied by a peak shift of P800.

The oxidation of the pigments in fractions A and B can be accomplished by chemical means as well as by light.

Composition. a. FRACTION A. This component contains the P883–P800 reaction center complex[19-21] of the organism, together with P′890 and light harvesting BChl (B890). Protein(s) and carotenoid (spirilloxanthin, 60%; lycoxanthin, 30%) are contained in the complex, together with two cytochromes (C552 and C556). C552 is present in the isolated complex in the oxidized state, whereas most of C556 is present in the reduced state; upon exposure of the complex to light, all the C556 (i.e., 2 moles/mole P883) becomes oxidized. The molar ratio of constituents so far identified is P883 (1) : P + B800 (4) : P′890 (~4) : B890 (40) : spirilloxanthin (12) : lycoxanthin (6) : C556 (2) : C552 (7).

b. FRACTION B. This is essentially a carotenochlorophyll–protein complex; it does not contain any cytochromes. The molar proportion of constituents so far identified is P835 (1) : B800 + B850 (90) : lycoxanthin

(14):spirilloxanthin (6). The proportion of P835 to light harvesting chlorophyll (B800 + B850) varies depending upon the growth conditions; on acetate medium the proportion is greater.

c. FRACTION C. Fraction C is also a cytochromeless carotenochlorophyll–protein complex, but it does not contain any reversibly photobleachable pigment. The carotenoid composition is identical to that of Fraction B.

Stability. The preparation can be stored in the dark at 4° for several weeks without any appreciable loss of photobleaching activity. Alternatively the fractions can be lyophilized and stored in a dry state for an indefinite period; addition of water to the dried powder results in a rapid dissolution of the complexes to give a clear pink solution; the light-induced spectral changes are still retained.

Molecular Size and Degree of Purity. Electrophoresis of the complexes in polyacrylamide gel containing SDS enables the size[23] and purity of the complexes to be estimated. Each isolated fraction electrophoreses as a single pink zone, and all the described constituents of each fraction are contained in that zone. Fraction A has a molecular weight of ~500,000; fractions B and C have the same molecular weight of 100,000 ± 10,000.

Function and Relationship to Other Chlorophyll–Protein Complexes. Fraction A is a photoelectron transport particle which is capable of carrying out a cyclic light-driven flow of electrons from P883 to a primary acceptor of unknown nature, and back to P883 via C556 and other as yet unidentified components.[24] In this respect it is identical to the *Rhodopseudomonas viridis* reaction center[25]; however, the major difference between the two is that fraction A contains a light-harvesting chlorophyll component (B890), the equivalent of which is absent in the *Rhodopseudomonas viridis* preparation. The light-harvesting chlorophyll may occur in fraction A as a BChl–protein complex similar to that described by Olson.[26] Fraction A possibly represents the minimum size intact photosynthetic unit.

Fraction B appears to represent a separate photosystem, the function of which is unknown. Fraction C is a light harvesting carotenochlorophyll–protein, which is closely allied to Fraction B in size and composition; in some batches of *Chromatium* cells its concentration is low.

[23] A. L. Shapiro, E. Viñuela, and J. V. Maizel, *Biochem. Biophys. Res. Commun.* **28**, 815 (1967).

[24] W. W. Parson, *Biochim. Biophys. Acta* **189**, 397 (1969).

[25] This volume [67].

[26] This volume [61].

[32] Subchromatophore Fragments: *Rhodopseudomonas spheroides*

By MICHAEL A. CUSANOVICH

In the search for the smallest membrane fragment capable of carrying on photosynthesis, that is, light-induced electron transfer and coupled energy conservation, some of the most successful preparations have been derived from the Athiorhodaceae, *Rhodopseudomonas spheroides*. The subchromatophore fractions isolated from *R. spheroides*, termed "reaction center preparations," are distinguished by their chemical and physical homogeneity, their small size, and the fact that they are devoid of "light harvesting" bacteriochlorophyll.

Reed and Clayton[1] have described the preparation of reaction centers from a carotenoidless mutant of *R. spheroides* (R-26) using the nonionic detergent Triton X-100. More recently Jolchine *et al.*[2] reported the isolation of reaction centers from a wild-type strain (*R. spheroides* strain Y) using the cationic detergent cetyltrimethyl ammonium bromide (CTAB). The two preparations although similar in bacteriochlorophyll composition have different properties.

Procedure

The growth conditions for the carotenoidless mutant (R-26) or wild-type *R spheroides* have been amply described.[3,4] However, when handling the R-26 mutant, care must be taken to protect the culture from strong light while oxygen is present in the medium. This is most easily accomplished by placing the culture in the dark for 12–24 hours immediately after inoculation and allowing the culture to metabolize dissolved oxygen prior to illumination. The cultures are then incubated in light (100–300 foot-candles) at 30°–37° for 24–48 hours and harvested by centrifugation. The harvested cells are washed by suspending in two volumes of 0.1 *M* Tris, pH 7.5, and centrifuging. Either fresh cells or cells stored at —20° can be used for the preparations. The washing and all subsequent manipulations should be carried out at 0°–4°, without prolonged exposure to either oxygen or room light.

[1] D. W. Reed and R. K. Clayton, *Biochem. Biophys. Res. Commun.* **30**, 471 (1968).
[2] G. Jolchine, F. Reiss-Husson, and M. D. Kamen, *Proc. Nat. Acad. Sci. U.S.* **64**, 650 (1969).
[3] W. R. Sistrom and R. K. Clayton, *Biochim. Biophys. Acta* **88**, 61 (1964).
[4] S. K. Bose, *in* "Bacterial Photosynthesis" (H. Gest, A. San Pietro, and L. P. Vernon, eds.), p. 501. Antioch Press, Yellow Springs, Ohio, 1963.

Triton Particles. The procedure described is essentially that of Reed and Clayton.[1] The washed cells are suspended in 10 m*M* Tris pH, 7.5 (20% wet weight cells/volume), and fractionated by passage through either a French pressure cell or preferably a Ribi Cell Fractionator at 20,000 psi. The temperature of the effluent should be kept at less than 20° and the pH checked immediately after breakage and adjusted to 7.5. Later centrifugations are facilitated if deoxyribonuclease (approximately 1 mg/100 ml of suspension) is added prior to breakage.

The broken cell suspension is then centrifuged for 30 minutes at 20,000–30,000 *g* to remove large pieces of the membrane. The supernatant is centrifuged at 240,000 *g* for 60 minutes to sediment chromatophores. The chromatophore pellet is suspended in 10 m*M* Tris, pH 7.5, to a final absorbance of approximately 50 at 870 nm. The chromatophore suspension is then titrated with a 10% solution of Triton X-100 (v/v) in the dark at 0°–4° with stirring. The final Triton concentration is critical since too little Triton will result in poor yields and too much in particles deficient in a number of components. A final concentration of 1% Triton X-100 will yield approximately 50% of the total P870 without extraction of cytochromes. Triton at concentrations greater than 2% will give a quantitative yield of P870, but cytochrome levels will be low.

The Triton-treated material is then layered on 0.6 *M* sucrose (1 volume particles to 2 volumes sucrose) and centrifuged for 90 minutes at 240,000 *g*. Two fractions are obtained, a pellet containing particles rich in light-harvesting bacteriochlorophyll and a supernatant solution which contains the reaction center particles, Triton X-100 and free bacteriochlorophyll. The supernatant is removed and dialyzed against 10 m*M* Tris, pH 7.5, for 40 hours (20 volumes buffer, 3 changes), or passed over Sephadex G-25 equilibrated with Tris buffer (10 ml of particle suspension per 100 ml of bed volume). The dialyzed solution is then concentrated, to a 870-nm absorbance of 2–3, by pressure dialysis (for example, using an Amicon ultrafiltration cell with a Um-10 membrane).

The final step in purification is accomplished by chromatography on Agarose A5m (1 ml of the reaction center suspension to 80–90 ml of bed volume) equilibrated with 10 m*M* Tris, pH 7.5, at a constant flow rate (approximately 40 ml/hour). Fractions are collected and those fractions having an A_{870}/A_{757} ratio of 1.4 or greater are pooled. If the fractions obtained have an A_{870}/A_{757} ratio less than 1.4, they contain excessive bacteriopheophytin and the preparation should be repeated.

CTAB Particles. This preparation from *R. spheroides* strain Y has been described by Jolchine *et al.*[2] Chromatophores are prepared as described in the Triton procedure, and further purified to "light particles."[5]

[5] This volume [31].

The light particles are suspended to an absorbance of 25–50 at 870 nm and incubated for 5 hours at 4° in the dark with 0.3% CTAB in 0.1 *M* potassium phosphate pH, 7.5. The CTAB-treated material is layered onto a discontinuous sucrose gradient (equal volumes of 0.6 and 1.0 *M* sucrose and 0.5 volume of particles) and centrifuged for 90 minutes at 240,000 *g*. Three fractions are obtained: a brown lower band containing light particles; a brown band in the middle containing aggregated material; and a yellow band on the top containing reaction centers. This upper band is removed and used as the reaction center particles.

Properties

Triton Particles. The final reaction center preparation should have absorbance maxima at approximately 1245, 865, 803, 757, 597, 532, 365, and 277 nm, although these maxima can vary slightly depending on the growth conditions of the organism. The ratio A_{277}/A_{803} should be between 1 and 1.2 with ratios larger than this indicating incomplete removal of Triton X-100.

With cross illumination (10,000–30,000 ergs/cm^2/sec) using white light or through interference filters (for example a 597- or 803-nm filter), the 865-nm band will completely bleach indicating the presence of only P870. For determination of the concentrations of P870 and P800, the extinction coefficients 127/m*M*/cm and 113/m*M*/cm, respectively, can be used.[6] These values should give two P800's per P870. For the light-induced bleaching at 865 nm, a ΔE_{mM} of 93 is used.[7]

By sizing on Agarose A5m and from dry weight determination, Reed[8] reports a particle weight of 6.5×10^5 with one P870 per particle and with the reaction centers 65% protein by weight. For particles prepared in 1% Triton X-100, the composition of electron carriers per mole of P870 are: cytochrome *c*-552, 1.0; cytochrome *b*-562, 1.8; iron, 16.0; copper, 9.0; ubiquinone, 13.5. With particles prepared from 2.3% Triton X-100, the particle size and bacteriochlorophyll content are unaffected, but the amounts of electron carriers mentioned above are reduced relative to P870. The Triton particles are homogeneous and show little tendency to aggregate.

CTAB Particles. The yield of these particles is approximately 30%, based on P870, and they contain approximately 30% of the protein present in the light particles, thus yielding approximately 5 μmoles of P870/g of protein in contrast to 2.4 μmoles of P870/g of protein for the Triton particles. Per mole of P870 the CTAB particles contain: P800, 1.9; ubiquinone, 4.5; heme *c*, 0.5.

[6] R. K. Clayton, *Photochem. Photobiol.* **5,** 699 (1966).
[7] J. Bolton, R. K. Clayton, and D. W. Reed, *Photochem. Photobiol.* **9,** 209 (1969).
[8] D. W. Reed, *J. Biol. Chem.* **244,** 4936 (1969).

CTAB particles contain carotenoids with absorption maxima at 498, 460, and 430 nm and bands at 855, 805, 589, 389, 362, and 784 nm due to bacteriochlorophyll and its degradation products. The light-induced bleaching is centered at 855 nm for the CTAB particles in contrast to 865 nm for Triton particles.

No estimate of the size of CTAB particles can be made, as they show a concentration-dependent tendency to aggregate.

Summary

As described, the Triton particles are the smallest, most homogeneous, and richest in electron carriers of the various subfractions of photosynthetic membranes isolated to date. Within the present understanding of the mechanism of bacterial photosynthesis these particles contain all the components necessary for cyclic electron transport.

Although the *R. spheroides* subchromatophore particles will not catalyze photophosphorylation, the presence of light-induced absorbance changes suggest that at least a portion of the electron transport chain is intact. Thus it appears that the *R. spheroides* particles are an ideal system for detailed studies on light-induced electron transport and reconstitution of the energy conservation system.

Section IV

Components

[33] Cytochrome Components in Chloroplasts of the Higher Plants

By D. S. BENDALL, H. E. DAVENPORT, and ROBERT HILL

Introduction

The first cytochrome to be described in the chloroplasts was cytochrome *f* which has been shown to be of the *c* type.[1-3] Evidence has been accumulated, however, which shows that cytochromes of both the *c* and the *b* types are associated with the photoinduced hydrogen or electron transport in chloroplasts.[4-6] So far only cytochrome *f* has been separated from the other chloroplast pigments in a soluble form. There is also experimental evidence which suggests the association of a particular cytochrome with one or other of the two definite photochemical systems of the chloroplast.[7] The determination of the chemical and physical characteristics of a cytochrome is an essential requisite for the examination of its behavior by application of spectroscopy to the cell and to photochemically active preparations pertaining to the chloroplasts.[8]

Preparation of the Chloroplasts for Studies of Cytochromes

It is recommended that in general the chloroplasts should be obtained in the intact state from the leaf. This is best accomplished by a very short grinding or blending of the leaf material; this and all subsequent processes are to be carried out below 5°. The leaf material is preferably freed from the main leaf veins and kept in a turgid condition. By using sucrose or other neutral carbohydrates to adjust the osmotic pressure, rather than salt, the chloroplast preparation can usually be obtained substantially free from other cytoplasmic organelles—this may be checked by suitable microscopic examination.[9]

For many studies it is desirable to obtain the chloroplast preparation without the intact envelopes or outer membranes.[10] This can be accomp-

[1] R. Scarisbrick, *Ann. Rep. Progr. Chem.* **44**, 226 (1947).
[2] R. Hill and R. Scarisbrick, *New Phytol.* **50**, 98 (1951).
[3] H. E. Davenport and R. Hill, *Proc. Roy. Soc. (London), Ser. B.* **139**, 327 (1952).
[4] L. N. M. Duysens and J. Amesz, *Biochim. Biophys. Acta* **64**, 243 (1962).
[5] W. A. Cramer and W. L. Butler, *Biochim. Biophys. Acta* **143**, 332 (1967).
[6] D. S. Bendall and R. Hill, *Ann. Rev. Plant Physiol.* **19**, 167 (1968).
[7] N. K. Boardman and J. M. Anderson, *Biochim. Biophys. Acta* **143**, 187 (1967).
[8] L. N. M. Duysens, *Progr. Biophys. Mol. Biol.* **14**, 1 (1964).
[9] This volume [18].

lished by suspending the freshly prepared chloroplast pellet in distilled water or in isotonic fluid diluted at least 10 times. Then by spinning this suspension, a pellet of chloroplast material relatively free from the stroma proteins can be obtained. If this pellet is now resuspended in isotonic sucrose, for example, instead of an hypertonic solution, it may be stored for longer periods in active condition. The preparation will now consist essentially of the membrane structures.

The membranous chloroplast preparations can be further broken down into smaller particles by a mechanical process, e.g., the French press,[11] or by sonication. Alternatively, treatment with a detergent can be used. The two types of disruption may be combined in one operation, or they may be applied successively.[12]

In certain cases it may be necessary to remove some or all of the chlorophyll from the chloroplast preparation. It has been found that acetone is most convenient for this purpose. Since 80% of acetone will extract all the chlorophyll, it is usual to add four volumes of acetone to an aqueous suspension of chloroplasts. Provided that the pH is above 7, the chlorophyll is removed but the haem remains in the precipitated chloroplast material. However, it happens that β-carotene is very slightly soluble in 80% acetone, and this may have to be removed by a further washing either with an excess of 90–95% acetone or with a non-polar solvent. Also, since the sugar alcohols mannitol and sorbitol are very slightly soluble in aqueous acetone, the presence of water soluble carbohydrates in the chloroplast preparation should be avoided.

Lyophilized preparations of chloroplasts may be extracted with solvents and yet retain certain photochemical activity.[13–15] Octane will remove carotenes and plastoquinone, leaving the xanthophylls and nearly all the chlorophyll. Aqueous acetone or octane containing a little acetone will remove xanthophylls together with chlorophyll.

The disadvantage of the use of organic solvents and detergents is that it may produce irreversible changes in the structure of the protein

---

[10] W. Cockburn, C. W. Baldry, and D. A. Walker, *Biochim. Biophys. Acta* **143**, 614 (1967).

[11] J.-M. Michel and M.-R. Michel-Wolwertz, *Carnegie Inst. Washington, Yearb.* **67**, 508 (1969).

[12] H. Huzisige, H. Usiyama, T. Kikuti, and T. Azi, *Plant Cell Physiol. Tokyo* **10**, 441 (1969).

[13] D. I. Arnon and A. A. Horton, *Acta Chem. Scand.* **17**, S135 (1963).

[14] N. I. Bishop, *in* "Quinones in Electron Transport" (G. E. W. Wolstenholme and C. M. O'Conner, eds.), p. 385, CIBA Foundation Symp. Churchill, London, 1961.

[15] P. M. Wood, H. N. Bhagavan, and F. L. Crane, *Plant Physiol.* **41**, 633 (1966).

moieties of certain cytochrome components and other components of the chloroplast.

Characterization of the Cytochromes

Absorption Spectra: In the reduced (ferrous) form, a cytochrome shows three well-defined regions of absorption usually termed α, β, and γ bands (Fig. 1). These are due to the haem group, and they appear in the yellow-green, green, and violet regions of the spectrum. There is also absorption in the ultraviolet due to the protein moiety. Cytochromes of the *c* type usually have an α band between 550 and 555 nm; those of the *b* type have an α band usually between 555 and 565 nm.

In the oxidized state the γ band is only slightly modified, but its position is altered to a definite extent. The α and β bands are replaced by a comparatively undefined region of absorption. Thus, the oxidation and reduction of cytochromes *in situ* may be determined by means of absorption spectra. The α bands of cytochromes of the *b* and *c* types occur in the region of relatively low absorption of the chlorophyll and other pigments present in chloroplasts.

For spectroscopic examination of chloroplast material at ordinary temperatures a suspension containing between 100 and 200 $\mu g/cm^2$ of chlorophyll may be used; the actual concentration in a given case depends upon the type of instrumentation. By having an equivalent thinner layer of chloroplast preparation and also by adjusting the refractive index of the sample with addition of glycerol or serum albumin, the transmission of light can be increased.

For examination of the preparations at low temperatures, as much as a 10-fold increase in intensity of absorption may be realized, due to ice crystals increasing the length of light path,[16,17] so that much less of the chloroplast preparation is required. The presence of sucrose in the suspending medium is a factor in determining the optimum size for the ice crystals.[18] By increasing the concentration of glycerol and allowing a sufficiently rapid cooling to a temperature below 100°K, the formation of ice can be delayed. For certain purposes it is advantageous to examine the preparation in this way; although there will be no intensification of the absorption bands the cytochrome spectra in the reduced form are markedly sharpened at low temperatures. In the cases so far examined, the α band of a cytochrome appears at a shorter wavelength at low temperatures. The displacement is of the order of 3 nm on cooling from

[16] D. Keilin and E. F. Hartree, *Nature* **164,** 254 (1949).
[17] D. Keilin and E. F. Hartree, *Nature* **165,** 504 (1950).
[18] D. F. Wilson, *Arch. Biochem. Biophys.* **121,** 757 (1967).

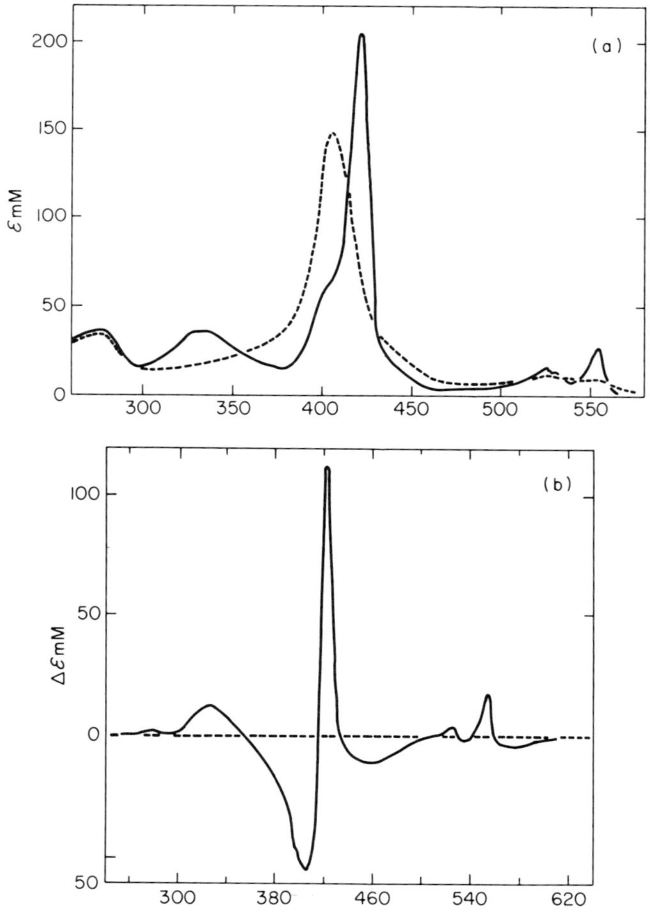

FIG. 1. Absorption spectra of cytochrome *f*: (a) oxidized (broken line) and reduced (solid line); (b) reduced–oxidized difference spectrum; (c) oxidized and reduced visible region; (d) difference spectrum visible region. (Abscissae, wavelength in nanometers.)

room temperature to 100°K, but the value of the shift depends upon the actual cytochrome component. Techniques for measurement at low temperature have been described by Chance[19,20] and by Bonner.[21] The absorption cell described by Bonner is shown in Fig. 2.

[19] B. Chance, this series Vol. IV, p. 273.

[20] B. Chance and B. Schoener, *J. Biol. Chem.* **241,** 4567 (1966).

[21] W. D. Bonner, *in* "Haematin Enzymes" (J. E. Falk, R. Lemberg, and R. K. Morton, eds.), p. 479, I.U.B. Symp. Pergamon Press, Oxford, 1961.

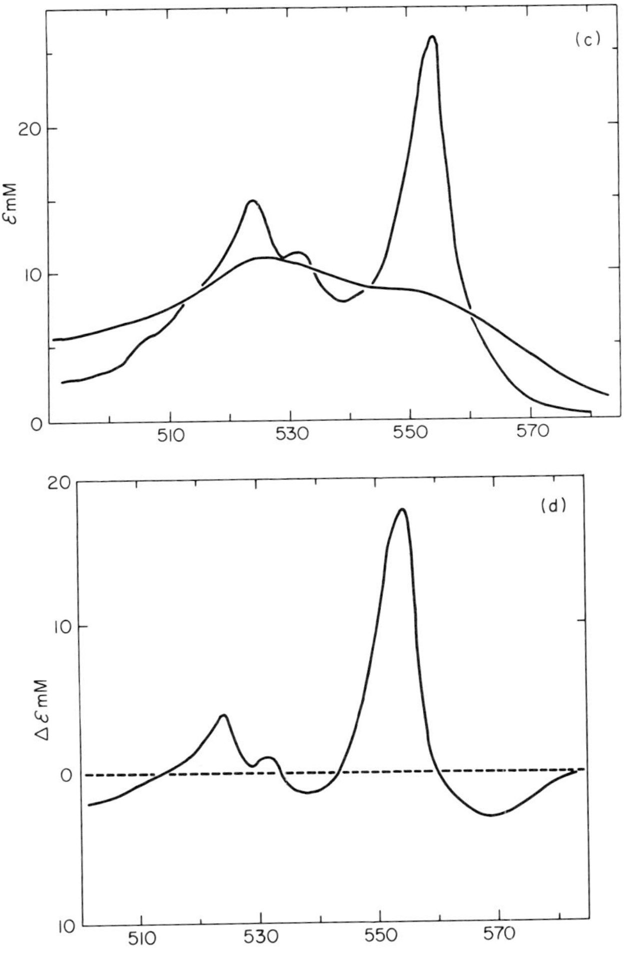

FIG. 1 (*continued*).

In the cytochrome structure the iron of the haem group is firmly associated with two ligands belonging to the protein part. In the ferrous state, in contrast to other simpler haem compounds, there is usually no reaction with carbon monoxide. Treatment of a cytochrome with carbon monoxide may be used as a guide in deciding whether a component has or has not been denatured. Relatively slight modification of the protein structure (for example by increasing pH of a solution) can render a ligand replacable by CO, giving a marked change in the absorption spectrum. Similarly treatment of a cytochrome with excess of pyridine in presence of a reducing agent will give a corresponding pyridine fer-

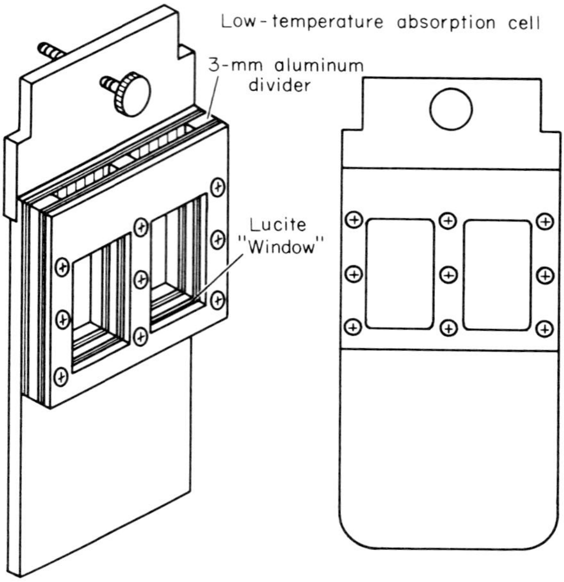

FIG. 2. An absorption cell described by Bonner[21] for low-temperature studies. During the optical measurements the lower part alone remains in the liquid nitrogen.

rohaemochrome (haemochromogen). Measurement of the change in absorption spectrum which results can give a guide towards determining the nature of the haem group.

Redox Properties: The various cytochrome components at present found to occur in the chloroplasts from a leaf of a plant show individual midpoint potentials at pH 7 which cover a range of nearly 0.6 V. For a given cytochrome (having a one-electron oxidation–reduction change) the span of potential at constant pH will be just over 0.2 V for a 99% conversion. Towards the acid side of neutrality the midpoint potential remains constant with pH in the majority of cases. This is because there is a loss of one positive charge when the cytochrome becomes reduced. Towards the alkaline side of neutrality a point may be reached where there is no change in the charge on reduction; this gives a change in potential towards less positive values of about 0.06 V per unit increase of pH. The value of the pH at which this change occurs (under standard conditions) is a characteristic property of a particular cytochrome. The relation of midpoint potential to pH for cytochrome *f* is shown in Fig. 3.

Prosthetic Groups: In the *c*-type cytochromes, the prosthetic group is attached to the protein by the addition of two SH groups in the protein across the two vinyl side chains of the porphyrin. The saturation of the vinyl groups of an iron porphyrin haemochrome will displace the α and

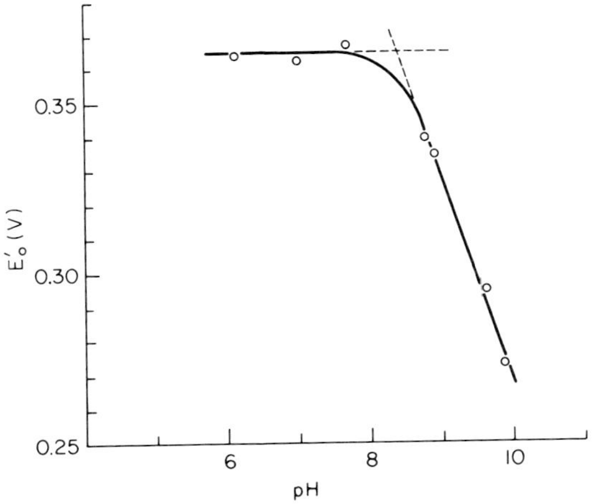

FIG. 3. Effect of pH on the midpoint potential of cytochrome *f*. (Taken from Davenport and Hill.[3])

β bands of the absorption spectrum by 10 nm towards shorter wavelengths. With a cytochrome component the actual positions of the absorption bands are also dependent on the nature of the proteins. The presence of the S linkages in a *c*-type cytochrome is inferred from certain tests:

1. The haem group is not removed by acid acetone.
2. The pyridine ferrohaemochrome (haemochromogen) produced by excess pyridine and a reducing agent has an α band at 550 nm.
3. On breaking the S bonds and removal of the iron a porphyrin corresponding to haematoporphyrin is liberated from the protein.[22]

The cytochrome components of the *b* type may be distinguished as follows:

1. The ferrohaemochrome formed by excess of pyridine has an α band at 556 nm (characteristic of protohaem).[23]
2. The haem group can be removed by acid acetone. In the measurement of the position of the α band of pyridine ferrohaemochrome, care has to be taken to avoid reaction with oxygen. Even in the presence of a reducing agent, the vinyl groups of the protohaem may be attacked leading to a shift in the position of the α band towards shorter wavelengths.

Reaction of Cytochromes with Oxygen and Carbon Monoxide: When

[22] H. E. Davenport, *Nature* **169,** 75 (1952).

[23] E. F. Hartree, *in* "Modern Methods of Plant Analysis" (K. Paech and M. V. Tracey, eds.), Vol. IV, p. 197. Springer, Berlin, 1955.

a cytochrome component has a midpoint potential of more than about +0.25 V, the reduced form reacts slowly or scarcely at all with atmospheric oxygen. When there has been any degradation of the protein, the reduced form will usually react rapidly with oxygen. Thus, no autoxidizability can sometimes be a reliable indication that a cytochrome component is in its native state. Cytochromes with more negative potentials, while being oxidized by atmospheric oxygen, do not necessarily combine with carbon monoxide. This is the case usually with cytochromes of the *b* type, so that while treatment with carbon monoxide may be used as a test in both *c* and *b* types of cytochrome, the effect of oxygen in this sense is only relevant for cytochrome of the high oxidation potential range.

Position of Cytochrome b_3: This cytochrome component,[2,24] obtained originally from leaves in a readily soluble form, still presents an uncertainty as to its origin, whether it is derived from the chloroplast or from the rest of the cell. The component b_3 was found not to be completely precipitated by saturated ammonium sulphate. It is destroyed by acetone, is reduced by ascorbate, and is oxidized by atmospheric oxygen. The α band is at 559 nm. While it was at one time thought to resemble one of the cytochrome components found in chloroplasts, there now seem to be definite differences, and also so far it has not been extracted directly from chloroplast preparations.

Extraction, Purification, and Properties of Cytochrome *f*

Cytochrome *f*, the characteristic *c*-type component of higher-plant chloroplasts, is not extracted from either leaves or isolated chloroplasts by aqueous solvents. It differs in this respect from analogous *c*-type components of some algae,[25,26] and in common with the other cytochrome components present in the chloroplast, it behaves as part of the insoluble "structural" protein. Hill and Scarisbrick,[2] however, by grinding leaves in the presence of ammoniacal ethanol, obtained a preparation of cytochrome *f* soluble in dilute salt solutions and susceptible to conventional methods of protein fractionation.

Extraction of Cytochrome f from Leaves of Parsley (Petroselinum sativum)

In spite of the universal occurrence of cytochrome *f* in higher-plant chloroplasts[2,3] the only highly purified soluble preparations to be de-

[24] H. Shichi, H. E. Kasinsky, and D. P. Hackett, *J. Biol. Chem.* **238**, 1162 (1963).
[25] S. Katoh, *J. Biochem. Tokyo* **46**, 629 (1959).
[26] J. J. Wolken and J. A. Gross, *J. Protozool.* **10**, 189 (1963).

scribed have been obtained from leaves of curled parsley.[3,27] The favorable characteristics of this material appear to be two-fold.

1. It is comparatively free of plant acids and tannins and other phenolic constituents capable of promoting protein denaturation, or, by the action of phenol oxidases, the production of colored oxidation products which can bind tenaciously to protein.
2. Optimal extraction of the component is attained when the concentration of ethanol in contact with the leaf at the moment of cell breakage is such as to give 50% ethanol in the leaf brei.[3] The tightly curled leaves of parsley retain relatively large volumes of the extraction medium, thus ensuring minimum departure from optimal conditions.

Nevertheless, the yield of cytochrome, at best about one-third of the total in the leaf, is highly variable and is greatly influenced by the state and age of the leaves and the method and scale of the grinding operation. Leaves should preferably have been ground rapidly. They should be made turgid if necessary by a preliminary soaking in tap water; wilted leaves give low yields. The highest yields relative to leaf chlorophyll are obtained by manual grinding of small amounts of leaf with ethanol in a mortar. Davenport and Hill[3] used a specially adapted mill for larger scale operations, and more recently Forti *et al.*[27] have shown that the addition of 1% of the nonionic detergent Triton X-100 to the ammoniacal ethanol allows the use of the more generally available Waring blendor.

Reagents:

Ethanol
Ammonium hydroxide (sp. gr., 0.88)
Triton X-100
Acetone
Ammonium sulphate
Dipotassium hydrogen phosphate

Maceration is carried out in a 1-gal (U.S.) Waring blendor at 0° ambient temperature. The blades are covered with 500 ml of a mixture of equal parts of water and ethanol containing 1.5% of ammonium hydroxide and 1% Triton X-100 (both v/v). Parsley leaves (500 g) freed from petioles are packed lightly into the container, and 500 ml of the ethanol/ammonia/Triton mixture (at −20°) are poured over them The blendor is run at maximum speed for 1 minute, and the brei is squeezed through nylon mesh and centrifuged at 40,000 *g* for 10 minutes

[27] G. Forti, M. L. Bertolè, and G. Zanetti, *Biochim. Biophys. Acta* **109, 33 (1965).**

at −10°. To the yellow supernatant fluid 1.1 volumes of acetone (at −20°) are added with vigorous stirring, and the yellow precipitate containing the cytochrome is filtered off on a large (20-cm diameter) Buchner funnel using celite as filter aid. The filter pad is washed with a mixture of acetone/ethanol/water in the proportions 2:1:1 containing 0.5% ammonium hydroxide (all v/v), and washing continued until the runnings are colorless. The filter pad is then washed with 66% saturated ammonium sulphate solution, again continued until the washings are colorless. The filter pad is suspended in 50% saturated ammonium sulphate and transferred to a smaller sintered glass filter of such a diameter that the filter pad is about 4 cm deep. Protein containing cytochrome is dissolved from the filter in a minimum volume of 0.05 *M* K_2HPO_4, and the brownish pink solution is dialyzed against a large volume of 1 m*M* phosphate buffer, pH 7.2. Yield is estimated in this crude material by the absorbancy at the α-band maximum (554.5 nm) of the reduced cytochrome relative to the absorbancy at 580 nm using a millimolar extinction coefficient of 26.12,[27] and it is usually 0.45–0.6 μmole/kg of leaves. The preparation may be stored frozen until a sufficient amount has been accumulated for further purification.

Extraction of Cytochrome f from Isolated Chloroplast Material

Since cytochrome *f* is confined to chloroplasts, it would appear advantageous that the starting material should be isolated chloroplasts or their fragments freed by washing from cytoplasmic proteins, soluble chloroplast proteins, vacuolar acids, and phenolics. Ammoniacal ethanol, however, extracts insignificant amounts of the cytochrome from such material.[3] Hotta *et al.*[28] have supplemented this extraction mixture with Triton X-100 and have obtained soluble cytochrome *f* from tobacco chloroplasts. The following method[29] has been successfully applied to chloroplasts derived from a variety of higher plants, including spinach (*Spinacia oleracea*), New Zealand spinach (*Tetragonia expansa*), Chicory (*Chicorum intybus*), garden pea (*Pisum sativum*), various *Brassica oleracea* cultivars, and beet (*Beta vulgaris*), but the following refers specifically to spinach.

Additional Materials:

Ethyl acetate

NaCl (0.4 *M*) containing phosphate (0.05 *M*), pH 7.2, and potassium metabisulphite (4 m*M*)

Diethyl aminoethyl cellulose

[28] R. Hotta, S. Shimizu, and E. Tamaki, *Bot. Mag. Tokyo* **80,** 23 (1967).
[29] H. E. Davenport, manuscript in preparation.

All operations are carried out at 0° ambient. Spinach leaves (3 kg), freed from petioles, are blended in a large Waring blendor in batches of 500 g with 1 liter of the NaCl solution for 1 minute. The homogenate is centrifuged at 2,000 *g* for 5 minutes, to remove debris, and the supernatant solution is centrifuged at 20,000 *g* for 20 minutes to sediment chloroplast material. This (from 3 kg of leaf) is resuspended in 500 ml of the NaCl medium by maceration in a smaller blender. The suspension is transferred to the large blendor, and a freshly prepared mixture of 500 ml of ethanol:1.5 liters of ethyl acetate:7.5 ml of ammonium hydroxide (sp. gr., 0.88) held at —20° is added. The blendor is immediately run at full speed for 30 seconds to form an unstable emulsion which is decanted and allowed to settle for 10 minutes at —20°. As much as possible of the dark green upper phase is poured off and discarded, and the remainder is centrifuged at 20,000 *g* for 10 minutes at —5°. The centrifuge bottles now contain two liquid phases separated by a layer of swollen, partly decolorized chloroplast debris, and the lower predominantly aqueous phase containing the cytochrome is collected by suction. To this (500 ml) add 1.1 volumes of acetone (—20°); the precipitated protein is collected by centrifugation (20,000 *g*, 10 minutes, —5°). The precipitate is extracted twice with a total of 50 ml of 0.05 *M* K_2HPO_4, and the supernatant solution, after centrifuging at 20,000 *g* for 10 minutes at 0°, is dialyzed against 3 liters of 0.001 *M* phosphate buffer, pH 7.5. The dialyzed material containing soluble cytochrome is then applied to a DEAE column (2 cm diam × 5 cm depth) equilibrated against the dilute phosphate buffer. The cytochrome absorbed on the column is eluted in 0.3 *M* phosphate, pH 7.5, leaving a broad band of black material at the top of the column. The eluted material, usually containing 0.7 μmole of cytochrome haem, is transferred to a dialysis sac. This is immersed in sufficient water to cover the sac and that amount of solid ammonium sulphate which will bring the total liquid volume to 70% saturation is added. Dialysis at 0° is continued with stirring until equilibrium is reached. This procedure serves to diminish the volume of the sac contents so that the finely divided precipitate may be collected by high speed centrifugation (e.g., 100,000 *g* for 2 hours at 0°). The precipitate is dissolved in 0.05 *M* K_2HPO_4 and dialyzed against 0.001 *M* phosphate buffer pH 7.5; then it is stored frozen pending the processing of sufficient material for further purification.

Purification of the Extracted Cytochrome

Forti *et al.*[27] have improved upon the earlier method[3] of purification by using gel filtration on Sephadex G100 and G200, and they have obtained preparations having absorbancy ratios A_{422}/A_{278} of 2.8 indicative of a high haem/protein ratio. This they regarded as the pure cytochrome. It has been found, however, that the following modification of the method of Davenport and Hill[3] yields fractions having A_{422}/A_{278} ratios up to 5.8 in preparations either from leaves of parsley or from isolated chloroplasts of spinach.

The accumulated frozen preparations derived by either extraction procedure are thawed, combined, and taken by dialysis to 50% saturation with respect to ammonium sulphate. The precipitate, collected by centrifugation (20,000 g, 30 minutes, 0°), is dissolved in a minimum amount of 0.05 M phosphate, pH 7.5, and equilibrated with a solution of 0.002 M NH_4HCO_3 either by dialysis or on Sephadex G25. To the cytochrome solution, successive additions of calcium phosphate gel, each followed by a short centrifugation, are made until absorbancy measurements at 554.5 nm (α-band maximum) or 422 nm (γ-band maximum) indicate that about 90% of the cytochrome is absorbed (see Fig. 1).

The cytochrome is then eluted from the gel by suspending the gel pellet in increasing concentrations of K_2HPO_4 in steps of 0.05, 0.1, 0.2, and 0.5 M. After each addition of the eluting medium, the gel suspension is centrifuged at $\approx$5,000 g for 1 minute, and the ratio A_{422}/A_{278} in the supernatant fluid is measured. Those fractions having the higher ratio are pooled, reprecipitated with 50% saturated ammonium sulphate, dialyzed against 0.002 M NH_4HCO_3, and the absorption on calcium phosphate gel is repeated. The selection of fractions for further treatment is arbitrary, but the highest A_{422}/A_{278} ratios occur in fractions obtained in 0.1–0.2 M K_2HPO_4. The progress of purification in a typical preparation from parsley leaves is indicated in Table I. In other more extensively purified samples, the absorbancy ratio has been raised to 5.8 with a recovery of less than 1% of the total cytochrome in the crude material. It has been found, however, that little further improvement is achieved

TABLE I
PURIFICATION OF CYTOCHROME FROM PARSLEY (12 KG)

Purification stage	Haem (μmoles)	Ratio (A_{422}/A_{278})	Haem (% recovery)
Initial extract	6.22	0.59	—
Adsorptions on calcium phosphate gel eluted in;			
1st (0.1–0.5 M K_2HPO_4)	3.0	2.44	48
2nd (0.08–0.3 M K_2HPO_4)	1.23	3.96	19.8
3rd (0.08–0.2 M K_2HPO_4)	0.65	4.5	10.4
4th (0.08–0.2 M K_2HPO_4)	0.2	4.77	3.2

in attempts to purify those fractions rejected at each step because of their low A_{422}/A_{278} ratios.

Properties of the Purified Cytochrome

Oxidation–Reduction Potential: The potential ($E'_0 = +0.365$ V) recorded for relatively impure preparations[3] is unchanged by further purification, but these preparations show an increased tendency to slow autooxidation on storage. The value of E'_0 is identical in cytochrome from parsley, spinach, New Zealand spinach, and pea.

Spectroscopic Properties: Iron estimation on two samples from spinach and parsley having A_{422}/A_{278} ratios of 4.5 and 5.6, respectively, gave identical extinction coefficients in the range 300–600 nm.[29] The data for the parsley preparation is summarized in Fig. 1(a), and the difference spectrum (reduced–oxidized) is plotted in Fig. 1(b). Although the largest differences occur in the γ-band region of the spectrum (422 nm), the smaller changes associated with the α band are those principally observed in spectroscopic observations of redox changes of the cytochrome in intact chloroplasts (see below). For this reason, extinction values for the visible region are plotted on an expanded scale [Fig. 1(c)] together with the difference spectrum in this region [Fig. 1(d)]. A summary of the spectral properties is given in Table II.

Molecular Weight: The earliest estimate of the molecular weight of parsley cytochrome *f* was 110,000 for a molecule containing two haem groups.[3] This has been modified by Forti *et al.*[27] who deduced by gel filtration, a molecular weight of 250,000, compatible with $S_{20,w} = 6.5 \times 10^{-13}$ and containing four haem groups. In more highly purified preparations[29] from spinach and parsley ($A_{422}/A_{278} = 4.37$) values of $S_{20,w}$ were

TABLE II
SPECTROSCOPIC PROPERTIES OF REDUCED AND OXIDIZED CYTOCHROME *f* INCLUDING PRINCIPAL ISOSBESTIC POINTS (SEE FIG. 1)

Wavelength (nm)	ε (m*M*) Reduced	ε (m*M*) Oxidized	Δε (m*M*) Red − ox
560	7.1	7.1	isosbestic
554.5	26.0 (α)	8.3	+17.7
543.5	9.1	9.1	isosbestic
526	—	11.0 (max)	—
524	14.9 (β)	11.0	+3.9
432.5	35.0	35.0	isosbestic
422	204.0 (Soret)	89	+115.0
415.5	115.0	115.0	isosbestic
410	92.0	147.0	−55.0

8.69×10^{-13} and 8.5×10^{-13}, respectively. The parsley preparation gave $D_{20} = 3.88 \times 10^{-7}$ indicating a probable molecular weight of 206,000 and possessing one haem group in 46,000 g dry weight. All more highly purified samples showed a single Schlieren peak, and the light absorption gradient due to cytochrome corresponded with this peak. Recent experimental evidence,[29] however, suggests that cytochrome *f* is a small protein of ≈12,000 molecular weight and readily forming dissociable aggregates with other fragments of chloroplast membrane protein.

Estimation *in situ* of Individual Cytochromes of the Chloroplasts

Principles: The absorption spectra of spinach chloroplast preparations in the α-band region of the cytochromes were described in detail by Boardman and Anderson.[7] These authors recognized three components, namely cytochrome *f* and two *b* components with α-peaks at 559 and 563 nm, respectively. More recent studies[30] of the oxidation–reduction potentials and the kinetics of reduction by dithionite distinguish two *b* components with peaks at 559 nm; these differ in potential by about 300 mV. The relevant properties are summarized in Table III and characteristic difference spectra of spinach chloroplasts at room temperature are shown in Fig. 4.

A rigorous method of estimation is available at the present time only for cytochrome *f*, for which extinction coefficients have been measured on the partially purified protein.[27,29] For the three *b* components, which have not been obtained in soluble form, extinction coefficients must be assumed. A millimolar extinction coefficient of 20 cm^2/mmole is frequently taken as characteristic of the height of the α peak of a *b* cytochrome in a difference spectrum.

TABLE III
PROPERTIES OF CHLOROPLAST CYTOCHROMES

Component	α Peak (nm)	E'_0, pH 7 (V)	Sensitivity to organic solvents, detergents, heating to 50°	References
f	554	+0.36	Resistant	3, 30
b-559_{HP}	559	+0.37	Sensitive	30
b-559_{LP}	559	+0.065[a]	Resistant	*b*
b-563	563	−0.18	Resistant	*c*

[a] Etiolated barley plastids.
[b] D. S. Bendall, unpublished observations.
[c] H. N. Fan and W. A. Cramer, *Biochim. Biophys. Acta* **216**, 200 (1970).

[30] D. S. Bendall, *Biochem. J.* **109**, 46P (1968).

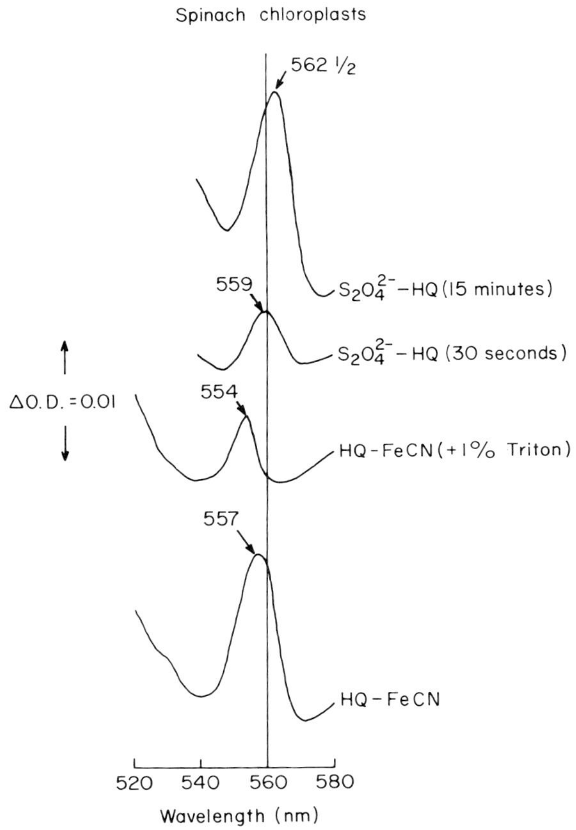

FIG. 4. Characteristic difference spectra (reduced minus oxidized) in the α-band region of the cytochromes of spinach chloroplasts at room temperature. Abbreviations: HQ, hydroquinone; FeCN, potassium ferricyanide. The conditions of observation are referred to in the text.

Measurements are best taken with a sensitive split-beam recording spectrophotometer that has a photomultiplier arranged to subtend the largest possible solid angle at the sample. A dual-wavelength instrument may also be used. Suitable instruments have been described by Chance,[19] and some commercial instruments are available from Cary, Phoenix, and Aminco.

The methods described below depend on the measurement of absorption difference spectra of chloroplast preparations between 580 and 520 nm under a variety of conditions. This is a region of the spectrum in which chlorophyll absorbs relatively weakly. Nevertheless, some interference from changes in chlorophyll absorption is inevitable. For example, ferricyanide, especially at high concentration, causes a slow bleaching of chlorophyll, and mild reducing agents such as hydroquinone or ascorbate cause a rapid bleaching to a limited extent.

Some of the reagents used do not readily penetrate the chloroplast envelope, and measurements are best made on hypotonically shocked chloroplasts. Chloroplast cytochromes fall into two groups according to their oxidation–reduction potential. A high-potential group, which comprises cytochrome *f* and cytochrome *b*-559_{HP}, can be distinguished by the difference spectrum: reduced with hydroquinone minus oxidized with ferricyanide. The spectrum obtained (Fig. 4) represents the fused peaks of the two cytochromes, but after treatment of the suspension with the neutral detergent Triton X-100, cytochrome *f* alone is seen because the cytochrome *b*-559 has been converted into a compound of lower potential that is no longer reducible by hydroquinone. A low-potential group is distinguished by the difference spectrum: reduced with dithionite minus reduced with hydroquinone. This group comprises cytochrome *b*-559_{LP}, which is rapidly reduced by dithionite, and cytochrome *b*-563, which is slowly reduced.[30] Reduction by dithionite is of the first order with respect to cytochrome when catalase is present to remove the hydrogen peroxide which is formed as the product of the reaction between dithionite and oxygen.

Ascorbate, which has frequently been used as a mild reducing agent for chloroplast preparations, causes a slow, incomplete reduction of cytochrome *b*-559_{LP}. Hydroquinone discriminates more successfully between the two *b*-559 components, but it is best used on the acid side of neutrality because of its autoxidizability at alkaline pH.

The methods given below are described for use with the type of split-beam spectrophotometer described by Chance.

Cytochrome f[31]*:* The chloroplast preparation is diluted in a medium containing 0.33 *M* mannitol, 1 m*M* $MgCl_2$ 1 m*M* $MnCl_2$, 2 m*M* EDTA, 50 m*M* potassium phosphate buffer (pH 6.5), and 1% (w/v) Triton X-100 to give a concentration of about 170 μg of chlorophyll/ml. The sample is divided equally between two 1-cm cuvettes, and the base line is drawn with the spectrophotometer. Sufficient 0.5 *M* hydroquinone is added to one cuvette and 0.5 *M* $K_3Fe(CN)_6$ to the other to give a final concentration of each reagent of 1.25 m*M*. The difference spectrum is then determined between 580 and 520 nm. The concentration of cytochrome *f* is calculated from the height of the α peak at 554 nm above a line drawn through the isosbestic points at 543.5 and 560 nm (corrected

[31] M. Plesničar and D. S. Bendall, *Biochim. Biophys. Acta* **216**, 192 (1970).

for any nonlinearity in the base line) for which a millimolar extinction coefficient of 17.7 cm^2/mmole is used.[29]

As an alternative to treatment with Triton X-100 an acetone powder of the chloroplasts may be prepared.[7]

The concentration of photoreactive cytochrome *f* may be determined by measurement of the difference spectrum between a sample illuminated with red actinic light in the presence of 1 μM DCMU and a dark sample. Both samples should contain a low concentration of ascorbate or hydroquinone to ensure dark reduction of cytochrome *f*. The photomultiplier is protected from the actinic light with a complementary filter.

Cytochrome b-559$_{HP}$: The chloroplast preparation is diluted about 10 times in distilled water, and then an equal volume is added of a medium containing 0.66 *M* mannitol, 2 m*M* $MgCl_2$, 2 m*M* $MnCl_2$, 4 m*M* EDTA, and 100 m*M* potassium phosphate buffer (pH 6.5) so that the final chlorophyll concentration is the same as that used for the determination of cytochrome *f*. The difference spectrum (hydroquinone reduced minus ferricyanide oxidized) is determined as for cytochrome *f*. A measurement is taken of the height of the peak at 559 nm above a line drawn through the minima at either side of the peak. This optical density change is corrected for the contribution of cytochrome *f* ($0.29 \times \Delta$O.D. at 554 nm, measured as described above), and the concentration of cytochrome

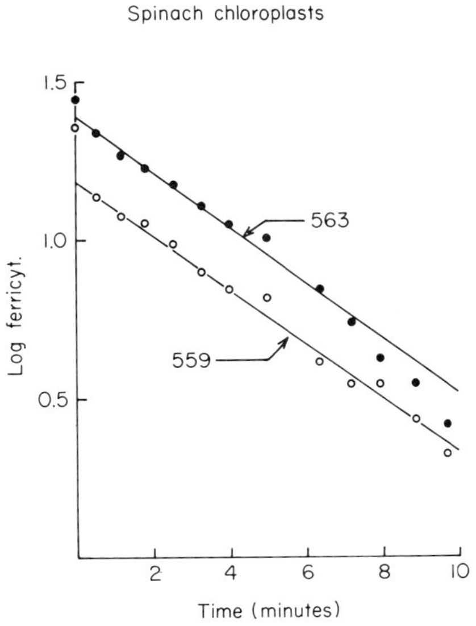

FIG. 5. Time course of reduction of cytochrome *b*-563 and *b*-559$_{LP}$ in spinach chloroplasts by dithionite. The conditions of measurement are given in the text.

b-559$_{HP}$ is calculated on the assumption of a value of 20 cm^2/mmole for the millimolar extinction coefficient.

Cytochrome b-559$_{LP}$ and b-563: An osmotically shocked dilution of the chloroplast preparation is made as described for cytochrome b-559$_{HP}$, with the further addition of 1.25 mM hydroquinone and catalase (400 U/ml of suspension, i.e., 0.5 μl crystalline suspension from Boehringer to each milliliter of suspension) to the suspension before it is divided between two 1-cm cuvettes. The base line is drawn with the spectrophotometer. A freshly prepared solution of sodium dithionite (for each milliliter of suspension, 10 μl of a solution containing 50 mg/ml) is added with rapid mixing to one cuvette. The spectrum is scanned rapidly and repeatedly between 580 and about 530 nm, and the time after the addition of dithionite is noted each time the peak position is reached. Measurements are continued until no further increase is detectable (about 10 minutes, but the rate of reduction can be controlled by the amount of dithionite added). For each time point measurements are taken on the heights of the peak at 559 and 563 nm above the line drawn through the minima at either side of the peak. For each wavelength the log of the difference between the final peak height and the height at time t is plotted against time (Fig. 5). From the intercept the peak heights of the rapidly reduced b-559$_{LP}$ and the slowly reduced b-563 are calculated. A millimolar extinction coefficient of 20 cm^2/mmole is again assumed for each component.

[34] Cytochromes: Bacterial

By ROBERT G. BARTSCH

As is indicated by the tabulation of presently known soluble cytochromes of photosynthetic bacteria (see Table I), this group of organisms is a prolific source of diverse cytochromes. Many of the cytochromes have distinctive properties which make them interesting examples of the variety of environments the apoproteins can impose on hemes. The information summarized in the first six columns of Table I will aid in identifying those bacterial cytochromes presently known. With the aid of information presented in the last four columns of Table I, it should be possible to devise suitable chromatographic procedures to purify the cytochromes listed. In general, pI values less than 7 indicate that the cytochrome can be chromatographed on DEAE–cellulose; values greater than 7 indicate basic cytochromes which are chromatographed on CM–cellulose.

Assay Methods

Absorption spectra measurements are routinely used to assay the cytochromes. Because of the high concentration of strongly absorbing carotenoids and chlorophylls in wild strains of the photosynthetic bacteria, it is difficult to detect cytochrome absorption bands with a hand spectroscope. In addition, the effect of light scattering makes it difficult to measure cytochrome spectra of whole cell suspensions even with many sensitive spectrophotometers. Reduced-minus-oxidized difference spectra between two initially equivalent buffered samples must be resorted to for an indication of the cytochrome content of a suspension of cells or of membrane fragments. With a split-beam spectrophotometer such as a Cary 14 (or 15) instrument, one of the pair of samples may be left unaltered, chemically oxidized with a small amount of potassium ferricyanide or even sodium hypochlorite, or exposed to photoactive light to test for a light-induced reaction. The other sample may be reduced by endogenous reductants, or by the requisite amount of chemical reductants such as succinate, ascorbate, NADH, sodium dithionite, or thiol compounds such as 2-mercaptoethanol or dithiothreitol. Alternatively a dual wavelength spectrophotometer may be used in the manner described elsewhere in this series.[1] The same techniques are applicable to cell-free extracts and purified cytochromes.

Because of interference by chlorophyll and especially carotenoid pigments, because cytochromes such as cytochrome *cc′* lack distinctive absorption peaks, or because the cytochrome absorption peaks in a complex mixture overlap, it is essentially impractical to estimate accurately the amounts of the individual cytochromes contained in whole cells, or even in most crude extracts. Relatively simple cytochrome mixtures such as that in *R. rubrum* extracts are an exception (see Table III). A further complication is the failure of bound cytochromes *cc′* and certain others to react with carbon monoxide,[2] whereas the unbound forms react readily, with the consequence that the carbon monoxide derivatives of these cytochromes are not reliable indexes of their concentration in the cells or chromatophores.

The description of cytochrome analysis procedures as applied to mitochondrial systems by Rieske[3] is a useful example of techniques applicable to the bacteria.

[1] This series, Vol. XXIV.

[2] M. A. Cusanovich and M. D. Kamen, *Biochim. Biophys. Acta* **153**, 376 (1968).

[3] J. S. Rieske, this series, Volume 10 [76].

TABLE I
Properties of Cytochromes of Photosynthetic Bacteria

Cytochrome	Organism	Extinction coefficients[a]: α peak λ_{nm}	α peak ϵ_{mM}	γ peak λ_{nm}	γ peak ϵ_{mM}	Molecular weight $\times 10^{-3}$ [b]	Number of hemes	$E_{m,7}$[c]	pI	Salt concn. to elute[d]	Purity index	Approx. yield[e]
c_2	*Rhodospirillum rubrum* strain 1	550	32	415.5	156[f]	12.8[g]	1	0.32[g]	6.2[h]	30 m*M* T, pH 8.0 (D)[h]	$\frac{A_{272,\ red}}{A_{415,\ red}} = 0.24$[h]	1.4[h]
	Rhodopseudomonas spheroides (ATCC 11167)	550	27.5	416	129[i]	~13[j]	1	0.346[i]	5.5[j]	40 m*M* NaCl, 20 m*M* T, pH 7.3 (D)[j]	$\frac{A_{275,\ ox}}{A_{416,\ red}} = 0.19$[i]	—
	Rhodospirillum molischianum (ATCC 14031)	549.5	—	415[j]	—	~13[k]	1	0.29[k]	10[k]	10 m*M* T, pH 7.3, (C)[j]	—	—
		549.5	—	415[j]	—	~10[k]	1	0.38[k]	10.5[k]	20 m*M* T, pH 7.3 (C)[j]	$\frac{A_{280,\ ox}}{A_{415,\ red}} \leq 0.3$[j] Not pure	—
	Rhodopseudomonas palustris (van Niel No. 2.1.37)[l]	551.5	26.6	418	168	13	1	0.33	9.7	20 m*M* P, pH 6.0 (C)	$\frac{A_{273,\ red}}{A_{419,\ red}} = 0.14$	—
	Rhodopseudomonas capsulata (H. Gest, St. Louis strain)[j]	550	—	416	—	—	—	—	—	10 m*M* NaCl, 10 m*M* T, pH 8.0 (D)	—	—
cc'[m]	*R. rubrum*	550	25.3	423	210[f]	29.8[g]	2	−0.008[g]	5.6[h]	50 m*M* T, pH 8.0 (D)[h]	$\frac{A_{282,\ ox}}{A_{390,\ ox}} = 0.277$[h]	0.6[h]
	R. spheroides[j]	545	10.4	423	99[a]	~25	—	—	4.9	70 m*M* NaCl, 20 m*M* T, pH 7.3 (D)	$\frac{A_{280,\ ox}}{A_{400,\ ox}} = 0.31$	—
	R. spheroides[j]	548	13.4	430[n]	119	~13	1	—	4.4	20 m*M* NaCl, 20 m*M* T, pH 7.3 (D)	$\frac{A_{280,\ ox}}{A_{400,\ ox}} \leq 0.2$ Not pure	—
	R. molischianum[j]	555	—	426	—	~36	—	—	7.2	20 m*M* T, pH 7.3, (D)	$\frac{A_{280,\ ox}}{A_{398,\ ox}} \leq 0.44$ Not pure	—
	R. palustris[l]	552	10.3	426	99	15	1	0.10	9.4	5 m*M* P, pH 6.0 (C)	$\frac{A_{284,\ ox}}{A_{397,\ ox}} = 0.21$	—
	Rhodopseudomonas gelatinosa (ATCC 11169)	550	—	425	—	~36	—	—	9.6	20 m*M* P, pH 7.0 (C)	—	—
	R. capsulata[j]	550	—	425	—	—	—	—	—	60 m*M* NaCl 10 m*M* T, pH 8.0 (D)	$\frac{A_{280,\ ox}}{A_{400,\ ox}} \leq 0.38$	—
	Chromatium strain D	547	21	426	190[o]	28[p]	2	−0.005[p]	4.6[j]	0.1 *M* NaCl, 20 m*M* T, pH 8.0 (D)[p]	$\frac{A_{280,\ ox}}{A_{400,\ ox}} = 0.33$[p]	—

c_3	*R. spheroides*[j]	551.5	29.5	419	216	~21	—	−0.22	4.1	0.12 *M* NaCl, 20 m*M* T, pH 7.3 (D)	$\frac{A_{280,\ ox}}{A_{419,\ red}} = 0.08$	—
	R. palustris[l]	551.5	—	418.5	—	~16	—	−0.15	6.1	20 m*M* T, pH 8.0 (D)	$\frac{A_{270,\ ox}}{A_{418,\ red}} \leq 0.2$ Not pure	—
	Chloropseudomonas ethylicum strain 2k[j]	551.5	30.5	418	196	~11	—	−0.16	4.1	0.2 *M* NaCl, 10 m*M* P, pH 7.8 (C)[q]	$\frac{A_{200,\ ox}}{A_{418,\ red}} = 0.05$	—
c_{554}	*R. spheroides*[j]	554[r]	—	419	—	~44	—	0.12[r]	4.1	0.16 *M* NaCl, 20 m*M* T, pH 7.3 (D)	$\frac{A_{278,\ ox}}{A_{419,\ red}} \leq 0.37$ Not pure	—
	R. palustris[l]	554	27	418.5	163	~40	—	−0.006	—	50 m*M* P, pH 6.0 (C)	$\frac{A_{280,\ ox}}{A_{418.5,\ red}} = 0.29$	—
f-type (double α peak)	*Chromatium*[s]	553 (550 sh)	16.3	417.5	115	13–50	1 (min)	0.32	4.4[j]	20 m*M* T, pH 7.3, (D)	$\frac{A_{275,\ red}}{A_{417.5,\ red}} = 0.12$	0.2
	R. gelatinosa[j]	554 (550 sh)	—	418	—	~27	—	—	9.6	Acetone powder extract, 10 m*M* P, pH 6.0 (C)	Not purified	—
	C. ethylicum[j]	555 (549 sh)	—	418[q]	—	—	—	—	5.0	0.1 *M* NaCl, 10 m*M* P, pH 7.8 (C)[q]	$\frac{A_{275,\ red}}{A_{418,\ red}}$ 0≅.18[q]	—
	Chlorobium thiosulphatophilum (NCIB 8346)[t]	555 (551 sh)	26.2	418.5	184	~10	1	0.145	10.5	20 m*M* NaCl, 20 m*M* T, pH 7.8 (C)	$\frac{A_{275}}{A_{418,\ red}} = 0.16$	—
Flavin-containing cytochromes	*Chromatium*	552	54	416.5	310[o]	72[u]	2 (1 FMN)	0.01[p]	4.5[j]	0.16 *M* NaCl, 20 m*M* T, pH 8.0 (D)[p]	$\frac{A_{278,\ ox}}{A_{410,\ ox}} = 0.55$[p]	0.8[p]
	C. thiosulphatophilum[t]	553.5	30	416.7	156	~50	1 (1 flavin)	0.098	6.7	20 m*M* NaCl, 20 m*M* T, pH 7.8 (D)	$\frac{A_{280,\ ox}}{A_{410,\ ox}} = 0.94$	—
Misc. cytochromes *c*	*R. palustris*[l]	555	19.9	420	136	~12	1	0.23	—	2 m*M* P, pH 6.0 (C)	$\frac{A_{275,\ red}}{A_{415,\ ox}} = 0.18$	—
	R. palustris[l]	551.5	—	420	—	~13	1	—	—	5 m*M* P, pH 6.0 (C)	Not pure	—
	R. gelatinosa[j]	552.5	—	418.5	—	~50	—	—	—	Poorly adsorbed by DEAE, fractionate with ammonium sulfate		
	R. gelatinosa[j]	551	—	417	—	~12	—	—	10.5	40 m*M* P, pH 7.0 (C)	—	—
	C. thiosulphatophilum[t]	551	17.6	416	114	45	2	0.135	6.0[j]	40 m*M* NaCl, 20 m*M* T, pH 7.8 (D)	$\frac{A_{280,\ ox}}{A_{416,\ red}} = 0.36$	—
b	*R. rubrum*[h]	557.5	—	425	—	450	—	−0.21	4.6	10 m*M* P, pH 7.0 (D)	$\frac{A_{275,\ ox}}{A_{417,\ ox}} = 1.3$	—
	R. spheroides[v]	559	—	426	—	—	—	—	—	Chromat. on Amberlite CG-50, pH 5	—	—

(*Continued*)

TABLE I (*Continued*)

Cytochrome	Organism	Extinction coefficients[a] α peak λ_{nm}	Extinction coefficients[a] α peak ϵ_{mM}	γ peak λ_{nm}	γ peak ϵ_{mM}	Molecular weight $\times 10^{-3}$ [b]	Number of hemes	$E_{m,7}$[c]	pI	Salt concn. to elute[d]	Purity index	Approx. yield[e]
	R. palustris[w]	559	—	425.5		—	—	—	4.0	0.14 *M* NaCl, 20 m*M* T, pH 7.3 (D)	$\frac{A_{280,\,ox}}{A_{419,\,ox}} = 1.3$	—
a	*R. spheroides* (undesignated strain, G. Kikuchi)[x]	587	—	445		—	—	—	—	Extracted from membrane fragments with Triton X-100, contains also Cyt *c* and Cyt *b*	—	—

[a] Most of the listed extinction coefficient values are determined as $\epsilon_{mM} = A\lambda_{max}$/mmole heme (determined by alkaline pyridine ferrohemochrome assay).

[b] The molecular weight values marked with ~ are approximate values obtained by the gel filtration technique, see text.

[c] Most of the $E_{m,7}$ values cannot be critically evaluated. The standard $E_{m,7}$ values of the redox buffers used in the titrations are very sensitive to ionic strength and specific salts used (W. M. Clark, "Oxidation Reduction Potentials of Organic Systems," p. 455. Williams and Wilkins, Baltimore, 1960). Adequate control of these factors is seldom indicated together with the reported $E_{m,7}$ values of the cytochromes.

[d] The symbol T means Tris·HCl buffer, P means potassium phosphate buffer, D means DEAE-cellulose, and C means CM-cellulose.

[e] μmoles/100 g wet weight cells.

[f] K. Sletten and M. D. Kamen, *in* "Structure and Function of Cytochromes" (K. Okunuki, M. D. Kamen, and I. Sekuzu, eds.), p. 472. Univ. of Tokyo Press, Tokyo, 1968.

[g] T. Horio and M. D. Kamen, *Biochim. Biophys. Acta* **48,** 266 (1961).

[h] R. G. Bartsch, T. Kakuno, T. Horio, and M. D. Kamen, submitted to *J. Biol. Chem.*

[i] S. Morita, *Bot. Mag.* (*Tokyo*) **79,** 630 (1966).

[j] T. E. Meyer, Ph.D. Thesis, Univ. of California, San Diego, 1970.

[k] M. D. Kamen, K. Dus, T. Flatmark, and H. De Klerk, *in* "Treatise on Electron and Coupled Energy Transfer in Biological Systems" (T. E. King and M. Klingenberg, eds.), Chapter 5. Dekker, New York, 1971.

[l] R. G. Bartsch, T. Horio, and M. D. Kamen, submitted to *Biochim. Biophys. Acta.*

[m] The so-called α peak of cytochromes *cc′* is broad and poorly defined, unlike the α-ferrohemochrome peak of typical cytochromes *c*.

[n] The γ peak of this cytochrome decreases one-third in amplitude upon reduction.

[o] R. G. Bartsch, *in* "Bacterial Photosynthesis" (H. Gest, A. San Pietro, and L. P. Vernon, eds.), p. 475. Antioch Press, Yellow Springs, Ohio, 1963.

[p] R. G. Bartsch and M. D. Kamen, *J. Biol. Chem.* **235,** 825 (1960).

[q] J. M. Olson and E. K. Shaw, *Photosynthetica* **3,** 288 (1969).

[r] J. A. Orlando, *Biochim. Biophys. Acta* **57,** 373 (1962). The preparation described by Orlando probably contained cytochrome $c_{551.5}$ (c_3) which skewed the peak of the mixture to 553 nm rather than 555 nm.

[s] M. A. Cusanovich and R. G. Bartsch, *Biochim. Biophys. Acta* **189,** 245 (1969).

[t] T. E. Meyer, R. G. Bartsch, M. A. Cusanovich, and J. H. Mathewson, *Biochim. Biophys. Acta* **153,** 854 (1968).

[u] R. G. Bartsch, T. E. Meyer, and A. B. Robinson, *in* "Structure and Function of Cytochromes" (K. Okunuki, M. D. Kamen, and I. Sekuzu, eds.), p. 443. Univ. of Tokyo Press, Tokyo, 1968.

[v] J. A. Orlando and T. Horio, *Biochim. Biophys. Acta* **50,** 367 (1961).

[w] R. G. Bartsch and T. E. Meyer, private communication, 1970.

[x] Y. Motokawa and G. Kikuchi, *Biochim. Biophys. Acta* **120,** 274 (1966).

TABLE II
ALKALINE PYRIDINE FERROHEMOCHROME EXTINCTION COEFFICIENTS

Cytochrome	λ_{nm}	ϵ_{mM}	$\Delta\epsilon_{mM,\ red\text{-}ox}$
c	550 ± 1	31.18[a]	19.1[b]
b	556 ± 1	34.7[c]	30[d]
a	587 ± 2	24[e]	—

[a] T. Flatmark, unpublished value for pure ox heart cytochrome *c*, quoted by K. Dus, H. De Klerk, R. G. Bartsch, T. Horio, and M. D. Kamen, *Proc. Nat. Acad. Sci. U.S.* **57,** 367 (1967).
[b] H. Theorell, *Biochem. Z.* **285,** 207 (1936).
[c] K. G. Paul, H. Theorell, and A. Åkeson, *Acta. Chem. Scand.* **7,** 1284 (1953).
[d] See J. S. Rieske, this series, Volume X [76].
[e] W. A. Rawlinson and J. H. Hale, *Biochem. J.* **45,** 247 (1949).

In Table I are listed the available extinction coefficients for the cytochromes. Most of the values are determined as absorbance per mole of heme. The heme content of the proteins is estimated by measuring the alkaline pyridine ferrohemochrome spectra. For this purpose the protein is dissolved in a 3:1 mixture (v/v) of 0.2 *M* NaOH solution and pyridine (reagent grade), the absorption spectrum of the ferrihemochrome is measured, then several crystals of sodium dithionite are added to reduce the sample, and the ferrohemochrome spectrum is measured. The extinction coefficients used for the several types of cytochrome are listed in Table II. Alternatively, the reduced-minus-oxidized difference spectrum may be useful with samples such as whole cells which scatter light, although ordinarily most cells and proteins dissolve completely in the strongly alkaline solution.

It is necessary to remove interfering carotenoid and chlorophyll pigments from cells or membrane preparations before preparing the alkaline pyridine hemochrome solutions. For this purpose the material is extracted and centrifuged several times in a centrifuge tube with acetone–methanol (7:2, v/v)[2] until the residue is essentially free of the soluble pigments.

The alkaline pyridine ferrohemochrome spectra give qualitative identification of the type of cytochrome in a preparation. The minimum requirement suggested[4] for classifying a cytochrome as to type is to measure the alkaline pyridine ferrohemochrome spectrum of the cytochrome as well as of the heme derived from the protein. To extract dissociable heme from a cytochrome, the ice cold preparation is made pH 2 with HCl and is extracted with 2–3 vol of cold acetone, where-

[4] M. Florkin and E. Stotz, "Comprehensive Biochemistry," Vol. 13, p. 18. Elsevier, Amsterdam, 1963.

upon the protein is precipitated, or with 1–2 vol of ice-cold methylethylketone, which is immiscible with the aqueous phase.[5] The protoheme of cytochrome *b*, and the formyl heme of cytochrome *a*, are extracted into the organic phase, whereas the covalently bound mesoheme of cytochrome *c* remains with the protein in the aqueous phase. The organic phase may be evaporated and the residue examined directly in the alkaline pyridine solvent, or better, the heme is transferred to ethyl ether, and then is extracted into 0.1 *M* NaOH. Finally the alkaline pyridine ferrohemochrome spectrum is measured. The protein or aqueous residue is also tested for the alkaline pyridine ferrohemochrome spectrum of cytochrome *c*. The diagnostically useful absorption maxima for identifying the hemes are indicated in Table II. Once the heme group is identified, the cytochrome can generally be given its proper designation.

Methods for Determining Useful Physical Properties

Electrophoresis of proteins using the isoelectric focusing technique of Vesterberg and Svensson[6] provides useful information about the charge properties, or isoelectric pH, as well as purity of the cytochromes. The technique is also useful as an analytical method for separating differently charged members of a group of cytochromes. For example, some seven or more cytochromes *cc′* ranging between pI 4 and pI 9, and eight or more cytochromes c_2 can be separated from a crude extract of *R. rubrum* cells by the isoelectric focusing method.[7,8] The technique has been used on a small scale preparative basis to separate mixtures for amino acid analysis. The Ampholine (LKB) electrolytes required seem to bind tightly to the separated proteins, but these can be separated by precipitation of the protein by ammonium sulfate, which presumably displaces the electrolyte.

The measurement of oxidation–reduction potentials of small cytochrome samples is most conveniently done by the method of mixtures as described by Davenport and Hill[9] and by Velick and Strittmatter,[10] wherein the relative concentration of oxidized and reduced cytochrome is determined spectrophotometrically in the presence of different ratios of oxidized and reduced components of a redox buffer. The redox buffer

[5] F. W. J. Teale, *Biochim. Biophys. Acta* **35**, 543 (1957).

[6] O. Vesterberg and H. Svensson, *Acta Chem. Scand.* **20**, 820 (1966).

[7] K. Sletten and M. D. Kamen, *in* "Structure and Function of Cytochromes" (K. Okunuki, M. D. Kamen, and I. Sekuzu, eds.), p. 472. Univ. of Tokyo Press, Tokyo, 1968.

[8] R. G. Bartsch, T. Kakuno, T. Horio, and M. D. Kamen, submitted to *J. Biol. Chem.*

[9] H. E. Davenport and R. Hill, *Proc. Roy. Soc. Ser. B* **139**, 327 (1952).

[10] S. F. Velick and P. Strittmatter, *J. Biol. Chem.* **221**, 265 (1956).

is at least 100-fold more concentrated than the cytochrome and may consist of reagents such as ferri–ferrocyanide ($E_{m,7} \simeq 0.4$ V),[9] ferri–ferro-oxalate ($E_{m,7} = 0$ V),[11] or ferri–ferro-EDTA ($E_{m,7} = 0.1$ V).[12] Alternatively the initially oxidized cytochrome plus any of the above redox mediators, or various redox dyes of suitable potential range, may be titrated with 10 mM sodium dithionite in 0.1 M Tris·HCl, pH 7.5–8.0, or other alkaline buffer. The E_h of the mixture may be measured vs. a Pt electrode, and the extent of reduction of the cytochrome is measured spectrophotometrically. In general, anaerobic conditions must be used to exclude interference by atmospheric oxygen in any of these titrations, although high redox potential systems such as cytochrome c_2 or f may be titrated in open cuvettes with a ferri–ferrocyanide buffer with no noticeable effect by atmospheric O_2.

Molecular weights or, more accurately, relative cross-sectional sizes of the cytochromes are generally approximated by use of the Andrews[13] technique of comparing the relative rates of migration of the protein and protein standards through Sephadex G-75 or G-100 columns. Such approximate values can be used to determine the scale factor required to convert accurate amino acid analysis values to an accurate formula weight of the heme protein.

If the heme content and formula weight are determined with aliquots of the same sample, the number of hemes per molecule can be derived,[14] or at least the number of hemes per minimum integral number of amino acid residues can be determined.

General Purification Procedures

The purification of most of the cytochromes listed in Table I is carried out in a standardized manner as illustrated in the schematic flow sheet (Fig. 1). For desalting extracts or cytochrome solutions, it is convenient to use a Sephadex G-25 column equilibrated vs. distilled water or a low concentration of buffer suitable for starting chromatographs on DEAE– or CM–cellulose. The conductivity of the effluent from Sephadex G-25 desalting columns is conveniently monitored with a flow cell and conductivity meter such as the Radiometer CDM. Customarily, the average conductivity of the desalted solution is kept below one-third that of the initial column equilibration buffer, and for this purpose the sample size is kept somewhat less than one-third the Sephadex G-25

[11] R. Hill, *Nature (London)* **174,** 501 (1954).
[12] G. Schwarzenbach and J. Heller, *Helv. Chim. Acta* **34,** 576 (1951).
[13] P. Andrews, *Biochem. J.* **91,** 222 (1964).
[14] K. Dus, H. De Klerk, R. G. Bartsch, T. Horio, and M. D. Kamen, *Proc. Nat. Acad. Sci. U.S.* **57,** 367 (1967).

Cell suspension (100 g wet wt in 400 ml of 0.2 *M* Tris·HCl, pH 7.3–8.0)
- Break cells with Sorvall-Ribi cell fractionator, 20,000 psi, 20°
- Centrifuge 10–30 minutes at 30,000 *g*

Precipitate cellular debris | Cell-free extract

Cell-free extract
- Centrifuge 2–3 hours at 100,000 *g*

Precipitate chromatophores | Supernatant solutions

Supernatant solutions
- DEAE–cellulose, Selectacel Standard, 75-ml bed volume, +0.2 *M* buffer
- Wash with 400 ml of application buffer

Unadsorbed proteins | Ferredoxin (Elute with 0.3–0.4 *M* NaCl in 20 m*M* Tris·HCl, pH 7.3)

Unadsorbed proteins
- Desalt, Sephadex G-25
- DEAE–cellulose, type 20, 100-ml bed volume, +1–10 m*M* Tris·HCl, pH 8.0
- Wash with 200 ml of application buffer

Acidic cytochromes:
- Elute cytochrome zones with increasing concentration of NaCl in buffer
- Desalt each zone, Sephadex G-25
- Concentrate on DEAE–cellulose, Type 20, 3-ml bed volume, elute with 0.5 *M* NaCl in 20 m*M* buffer
- Fractionate with ammonium sulfate[a]
- Desalt and chromatograph on DEAE–cellulose columns
- Chromatograph on Sephadex G-100
- Concentrate
- Crystallize from ammonium sulfate solution

Acidic cytochromes

Unadsorbed basic proteins
- CM-cellulose, Selectacel standard, 30–50 ml bed volume, +1 m*M* phosphate, pH 5.0–6.0
- Wash with 200 ml application buffer
- Elute colored bands with increasing concentrations of buffer
- Desalt each zone, Sephadex G-25 or Bio-Gel P-2
- Concentrate on CM–cellulose, standard, 3-ml bed volume, elute with 0.1 or 0.2 *M* P_i, pH 7.0
- Fractionate with ammonium sulfate[a]
- Chromatograph on Sephadex G-100
- Concentrate
- Crystallize

Basic cytochromes

FIG. 1. General purification scheme used for bacterial cytochromes. See text for details.

[a] A. A. Green and W. L. Hughes, this series, Volume 1 [10].

column bed volume. Some crude extracts stain the columns with colored membrane fragments which can generally be removed by eluting the column with Triton X-100 in 10 mM NaOH (1%, v/v).

The general procedure for use of DEAE– and CM–cellulose chromatography columns described by Peterson and Sober[15] is followed. The CM–cellulose and DEAE–cellulose Selectacel-Standard, a rapid flow grade, and the DEAE–cellulose Selectacel Type 20, a moderate flow grade, are obtained from the Brown Co., Berlin, New Hampshire. To regenerate DEAE–cellulose columns it is useful to include Triton X-100 in 0.2 M NaOH (1%, v/v) plus 0.5 M NaCl to elute colored membrane fragments which otherwise adhere irreversibly to the adsorbent.

Several nearly standardized preparations are given here to illustrate the general scheme. For this purpose the descriptions are scaled for 100 g wet weight of cells.

Rhodospirillum rubrum Cytochromes

The simplest cytochrome pattern encountered is that of *Rhodospirillum rubrum,* from which only cytochrome c_2 and cytochrome cc' with numerous isocytochrome forms[7,8] and cytochrome b_{558},[8] have been isolated. Cytochrome c_2 (pI = 6.2) and cytochrome cc' (pI = 5.6) predominate, to the extent of approximately one-half of the total content of the two cytochromes in crude extracts. A small part of the same predominant cytochromes have been tentatively identified as the tightly bound chromatophore cytochromes which can be shown to undergo light-induced oxidation–reduction reactions.[16] The three cytochromes are acidic and consequently are chromatographed on DEAE–cellulose. The purification of the two cytochromes c is summarized in Table III. For none of the cytochromes discussed in this article is there available a complete set of yield and purity values per purification step to make such a table complete

Cytochrome b_{558}. To separate conveniently cytochrome b_{558}, the crude eluate from the first DEAE–cellulose column used to remove ferredoxin is fractionated with ammonium sulfate. Cytochrome b_{558} is precipitated in the range 15–30% ammonium sulfate saturation, together with an FMN protein, the bulk of which remains in solution. The precipitate is dissolved in 40 ml of 0.1 M Tris·HCl, pH 8.0, and centrifuged for 10–12 hours at 100,000 g. The precipitate of cytochrome b is dissolved in 10 ml of 0.1 M Tris·HCl, pH 8.0, and centrifuged at 30,000 g for 1 hour to remove insoluble material. The supernatant solution is adjusted to pH 4.6 by addition of 1 M acetic acid at 0° and immediately centrifuged

[15] E. A. Peterson and H. A. Sober, this series, Volume V [1].

[16] T. Kakuno, R. G. Bartsch, K. Nishikawa, and T. Horio, submitted to *J. Biochem.* (Japan).

TABLE III
PURIFICATION OF *Rhodospirillum rubrum* CYTOCHROMES

	Cytochrome c_2[a] yield (μmoles)	Cytochrome cc'[b] yield (μmoles)	Cytochrome b_{558}[c] yield (ΔA_{425})
Crude extract	4.7	2.2	77
1st DEAE–cellulose	2.3	0.86	
1st Crystallization	1.4	0.6	

[a] $\Delta A_{550\ \mathrm{nm}}$ corrected for $\Delta A_{550\ \mathrm{nm}}$ due to cytochrome cc'. $\Delta\epsilon_{\mathrm{m}M,\ 550(c_2)} = 22$, $\Delta\epsilon_{\mathrm{m}M,\ 550(cc')} = 25$.

[b] $\Delta A_{638\ \mathrm{nm}}$, $\Delta\epsilon_{\mathrm{m}M,\ 638\ \mathrm{nm}} = 4$.

[c] $\Delta A_{425\ \mathrm{nm},\ Na_2S_2O_4} - \Delta A_{425\ \mathrm{nm},\ \mathrm{NADH}}$. At pH 7, 1 m$M$ NADH completely reduces cytochromes c_2 and cc' in the crude extract.

at 30,000 g for 10 minutes to remove unwanted colorless protein. The supernatant solution is immediately neutralized with 1 M sodium hydroxide. The clear solution is chromatographed on a Sephadex G-100 column (1 liter bed volume per 20–30 ml of solution) equilibrated with 1 M NaCl in 0.1 M phosphate buffer, pH 7.0, in the cold room. The cytochrome is eluted in the void volume, followed by the remaining FMN protein. The best cytochrome b_{558} fractions are combined, desalted with the aid of a Sephadex G-25 column, equilibrated with 1 mM phosphate buffer, pH 7.0, and then chromatographed on a DEAE–cellulose column (type 20, 30 ml bed volume) equilibrated with the same buffer. The cytochrome is eluted with 10 mM phosphate buffer, pH 7.0. Finally, the best fractions are pooled and precipitated with ammonium sulfate, over the range 15–20% saturation.

Cytochromes c_2 and cc'. Cytochromes c_2 and cc can conveniently be precipitated from the ammonium sulfate supernatant solution remaining from the cytochrome b preparation by saturating the solution with ammonium sulfate. Residual cytochrome c_2 remaining in solution can be concentrated by filtering the ammonium sulfate solution through a column of DEAE–cellulose (6-ml bed volume); this serves as a convenient filter aid which is superior to the more conventional types tried. The accumulated protein can be eluted off the column with water or 0.1 M Tris·HCl, pH 7.3. The precipitated proteins are dissolved in sufficient 0.1 M Tris·HCl, pH 8.0, to reduce the residual ammonium sulfate concentration to less than 20% saturation. The solution is desalted with Sephadex G-25 equilibrated with 1 mM Tris·HCl, pH 8.0. Alternatively, the crude ferredoxin-free extract may be desalted directly if no cytochrome b is to be recovered. The protein solution is then chromatographed at room temperature on a DEAE–cellulose column (type 20,

100-ml bed volume) equilibrated with 1 mM Tris·HCl, pH 8.0. After charging with the protein solution, the column is washed with 300 ml of 1 mM Tris·HCl, pH 8.0, and then with 300 ml of 10 mM Tris·HCl, pH 8.0, to distribute the protein in the column. Cytochrome c_2 is then eluted with 400 ml of 30 mM Tris·HCl, pH 8.0, and the cytochrome cc' is eluted with 500 ml of 50 mM Tris·HCl, pH 8.0. There remains on the column as much as one-third of the total cytochrome which consists of the more acidic isocytochromes c_2 and cc' found in *R. rubrum* extracts. These isocytochromes may be eluted with 0.5 M NaCl in 20 mM Tris·HCl, pH 8.0; after desalting with Sephadex G-25, they are best separated by electrophoresis with the isoelectric focusing technique.[8] The best fractions of cytochrome c_2 in the 30 mM Tris·HCl eluate are pooled, desalted with Sephadex G-25, concentrated on a DEAE–cellulose column (type 20, 2-ml bed volume) equilibrated with 1 mM Tris·HCl, pH 8.0, and eluted in concentrated solution with 0.5 M NaCl in 20 mM Tris·HCl, pH 8.0.

The best cytochrome cc' fractions are combined, desalted, concentrated on a DEAE–cellulose column (type 20, 2-ml bed volume) and eluted with the strong salt solution. Primarily to remove slowly migrating residual cytochrome c_2, the cytochrome cc' is next chromatographed on a Sephadex G-100 column, equilibrated with 0.2 M NaCl in 20 mM Tris·HCl, pH 7.3, in the cold. Again, the best fractions of cytochrome cc' are combined, desalted, and concentrated.

Both concentrated cytochromes are next precipitated with ammonium sulfate, between 70 and 100% saturation. Finally the two cytochromes are induced to crystallize at room temperature from ~60% saturated ammonium sulfate in 0.1 M Tris·HCl, pH 8.0. By three successive crystallizations of the main crops of each cytochrome, the predominant forms, isocytochrome c, pI 6.2, and isocytochrome cc', pI 5.6, can be isolated in pure form.

Properties of R. rubrum Cytochromes. The characterization of cytochrome b_{558} is incomplete. One peculiarity to note is the extreme slowness with which the cytochrome becomes reduced, even with sufficient excess of sodium dithionite as reducing agent to make the reaction mixture completely anaerobic. Several hours may be required for complete reduction to occur in a spectrophotometer cuvette. The cytochromes c_2 and cc' serve as the type examples for these two classes of cytochrome, and they have been extensively characterized (see Table I and indicated references). The amino acid sequence of cytochrome c_2 has been reported.[17]

[17] K. Dus, K. Sletten, and M. D. Kamen, *J. Biol. Chem.* **243**, 5507 (1968).

Chromatium Cytochromes

At least three cytochromes can be isolated from extracts of *Chromatium* strain D. The extract free from ferredoxin is desalted on a Sephadex G-25 column, made 2 m*M* with respect to Tris·HCl, pH 8.0, and chromatographed at 4° on a DEAE–cellulose column (Type 20, 100-ml bed volume) equilibrated with the same buffer. After preliminary washing with 400 ml of the application buffer to distribute the proteins through the column, the small amount of soluble cytochrome $c_{553(550)}$ (*Chromatium* cytochrome *f*) is eluted with 2 m*M* Tris·HCl, pH 8.0. Some more of the cytochrome may be eluted together with oxidized high potential iron protein (HiPIP) with 400 ml of 20 m*M* NaCl in 20 m*M* Tris·HCl, pH 8.0. Reduced HiPIP is eluted with 400 ml of 40 m*M* NaCl in 20 m*M* Tris·HCl, pH 8.0. Next, an as yet uncharacterized cytochrome c_{551} may be eluted with 400 ml of 60–80 m*M* NaCl in the buffer. Cytochrome *cc'* is eluted with 80–100 m*M* NaCl in the buffer, followed by cytochrome c_{552} eluted with 400 of ml 0.14–0.18 *M* NaCl plus buffer.

A severalfold greater yield of cytochrome $c_{553(550)}$ can be extracted from cells, or the crude chromatophores from the 100,000 *g* centrifugation, by extracting twice with 50% acetone plus cell or particle suspension in 10 m*M* Tris·HCl, pH 7.3, in the cold.[18] The chilled acetone (−10°) is rapidly stirred into the buffered chromatophore suspension in the cold, and then left for 10 minutes. The suspension is centrifuged for 10 minutes at 10,000 *g* at −20°. The residue is resuspended in 10 m*M* Tris·HCl, pH 7.3, and again extracted with 50% cold acetone. The supernatant solutions are decanted and poured immediately through a DEAE–cellulose column (type 20, 50-ml bed volume) equilibrated with 10 m*M* Tris·HCl, pH 7.3. A small red-colored zone of the cytochrome accumulates ahead of a broad brown-colored zone consisting of HiPIP plus chromatophore fragments. If whole cells are extracted, ferredoxin also accumulates at the very top of the column. The cytochrome is eluted from the column with 20 m*M* Tris·HCl, pH 7.3.

The cytochrome $c_{553(550)}$ fractions are desalted with Sephadex G-25, adjusted to 1 m*M* Tris·HCl, pH 8.0, and chromatographed again on a DEAE–cellulose column (type 20, 20 ml bed volume) in the manner described. The best fractions ($A_{275}/A_{417,\text{ red}} \simeq 0.12$) are pooled, desalted, and concentrated with the aid of a small DEAE–cellulose column.

Both the cytochrome *cc'* and cytochrome c_{552} fractions are desalted on a Sephadex G-25 column, adjusted to pH 8.0, adsorbed on DEAE–cellulose columns and eluted in concentrated solution with 0.5 *M* NaCl

[18] M. A. Cusanovich and R. G. Bartsch, *Biochim. Biophys. Acta* **189**, 245 (1969).

plus 20 mM Tris·HCl, pH 7.3. The concentrated solutions are fractionated with ammonium sulfate. The precipitate formed between 50 and 100% saturated ammonium sulfate is dissolved in 0.1 M Tris·HCl, pH 7.3.

The cytochrome *cc'* fraction is desalted with a Sephadex G-25 column, the solution adjusted to pH 8.0 and adsorbed on a DEAE–cellulose column (type 20, 30-ml bed volume) equilibrated with 20 mM Tris·HCl, pH 8.0. After preliminary washing with 400 ml of 20 mM Tris·HCl, pH 8.0, followed by 300 ml of 60 mM NaCl in 20 mM Tris·HCl, pH 8.0, the cytochrome zone is eluted with 500 ml of 80–100 mM NaCl in the buffer. The best fractions (purity index $A_{284}/A_{390} = 0.4–0.5$) are pooled, desalted, and concentrated with the aid of a small DEAE–cellulose column. Finally, the concentrated solution is fractionated with ammonium sulfate, the main fraction of the cytochrome precipitates between 70 and 100% saturation. The precipitate is dissolved in 0.1 M Tris·HCl, pH 7.3. The sample is essentially homogeneous if $A_{280}/A_{390,\mathrm{ox}} \simeq 0.33$ is attained, repetition of the fractionation with ammonium sulfate may be needed to reach this purity level. The cytochrome has never been crystallized.

The cytochrome c_{552} fraction precipitated by ammonium sulfate is dissolved in 0.1 M Tris·HCl, pH 7.3, and chromatographed on a Sephadex G-100 column (1-liter bed volume per 20–30 ml solution) equilibrated with 0.2 M NaCl in 20 mM Tris·HCl, pH 7.3. The first colored fraction consists of membrane fragments in the void volume followed by a flavoprotein with dye reductase capabilities.[19] Cytochrome c_{552} is eluted next, followed by residual cytochrome *cc'*. The best fractions of cytochrome c_{552} are pooled, desalted with a Sephadex G-25 column, and concentrated with the aid of a DEAE–cellulose column. Dependent on the purity achieved, the sample may be chromatographed again on DEAE–cellulose (type 20, 30-ml bed volume) in the manner described for cytochrome *cc'*, with the difference that the column is first eluted with 300 ml of 0.12 M NaCl in 20 mM Tris·HCl, pH 8.0, and then with 0.16 M NaCl in the same buffer to elute the cytochrome. The best fractions are pooled, desalted and concentrated. The cytochrome prepared this way, or possibly the best fraction from the Sephadex G-100 chromatogram, is subjected to stepwise ammonium sulfate fractionation over the range 60–90% saturation in 10% saturation increments. The best fraction may crystallize upon standing in ammonium sulfate solution of the same or slightly lower concentration from which it was precipitated, generally 60–70%

[19] M. A. Cusanovich, R. G. Bartsch, and M. D. Kamen, *Biochim. Biophys. Acta* **153**, 397 (1968).

saturation. For the essentially homogeneous cytochrome the purity index, $A_{280}/A_{410,\,ox} = 0.55$.

The flavin of this cytochrome slowly dissociates throughout the purification. Consequently, in part, the purification procedure eliminates denatured flavin-free heme protein. The absorption ratio, A_{480}/A_{520}, for the oxidized protein is used as a measure of the relative amounts of flavin and heme of the preparation, for the best samples the ratio $\simeq 1.3$. The flavin can be dissociated from the cytochrome by incubation at pH > 9, with saturated urea or 4 *M* guanadine hydrochloride, or with a mercurial such as *p*-chloromercuribenzoate.[20]

Properties of the Chromatium Cytochromes. Cytochrome $c_{553(550)}$ is a nonautoxidizable cytochrome, ($E_{m,7} = +0.32$ V) which is isolated predominantly in the reduced state. The purified protein appears to polymerize upon storage, inasmuch as molecular weights ranging from 13,000 to 50,000 daltons are found by sedimentation-equilibrium analysis. The cytochrome resembles algal cytochromes *f* in spectroscopic and physical properties, and may be the soluble form of the high-potential chromatophore-bound cytochrome c_{555} which undergoes light-induced oxidation in chromatophores and whole cells.

Cytochromes *cc'* and c_{552} are autoxidizable and are isolated in the oxidized state. On the basis of similar absorption spectra and redox potentials, cytochrome c_{552} has been tentatively identified with cytochrome $c_{423.5}$ which undergoes light-induced oxidation in anaerobic whole cells[21] or in chromatophores adjusted to $E_h \leq 0$ V.[22] The same cytochrome in starved cells responds to a reducing substrate such as sodium thiosulfate, and therefore has been implicated in a noncyclic electron transfer system.[23]

Both of the isolated autoxidizable cytochromes combine with carbon monoxide, but the bound forms in the chromatophore fail to react with the ligand.[2] The bound cytochromes can be solubilized from an acetone powder of the particles or by extracting the particles with 2% Triton X-100 in 0.1 *M* Tris·HCl, pH 7.3. Upon purification by the above procedure, the cytochromes have properties identical with the proteins isolated from the aqueous extract.

[20] R. G. Bartsch, T. E. Meyer, and A. B. Robinson, *in* "Structure and Function of Cytochromes" (K. Okunuki, M. D. Kamen, and I. Sekuzu, eds.), p. 472. Univ. of Tokyo Press, Tokyo, 1968.

[21] J. M. Olson and B. Chance, *Arch. Biochem. Biophys.* **88**, 26 (1960).

[22] M. A. Cusanovich, R. G. Bartsch, and J. M. Olson, *in* "Comparative Biochemistry and Biophysics of Photosynthesis" (K. Shibata, A. Takamiya, A. T. Jagendorf, and R. C. Fuller, eds.), p. 186. Univ. of Tokyo Press, Tokyo, 1968.

[23] S. Morita, M. Edwards, and J. Gibson, *Biochim. Biophys. Acta* **109**, 45 (1965).

Rhodopseudomonas palustris Cytochromes

Rhodopseudomonas palustris (van Niel strain No. 2.1.37) provides the most complex assortment of cytochromes encountered among the photosynthetic bacteria so far examined. Two acidic cytochromes plus five cytochromes with basic isoelectric pH values can be isolated from the organism.[24,25] Except for cytochromes c_2 and c', the following directions are not optimized for recovery of these 7 cytochromes. The extraction procedure and ferredoxin removal method already described are used. The ferredoxin-free extract is desalted on a Sephadex G-25 column equilibrated with 1 m*M* Tris·HCl, pH 8.0. The solution is next passed into a DEAE–cellulose column (Selectacel-Standard, 270-ml bed volume), equilibrated with 1 m*M* Tris·HCl, pH 8.0. The column is washed with 500 ml of the application buffer to stabilize the protein zones and to wash off unadsorbed proteins (eluate 1). The column is next washed with 500 ml of 0.1 *M* Tris·HCl, pH 8.0, to remove a mixture of cytochromes and dehydrogenases (eluate 2). The column is then eluted with 500 ml of 0.2 *M* NaCl in 20 m*M* Tris·HCl, pH 8.0, to remove a mixture of a cytochrome *b*, an iron protein with an absorption spectrum like that of adrenal redoxin and a possible NADPH dehydrogenase[26] (eluate 3). Eluates 2 and 3 are thereby considerably concentrated relative to the starting extract and the basic proteins are separated in eluate 1.

Eluate 1 is adjusted to pH 5.5, either by careful titration with 1 *M* acetic acid, or by addition of small increments of Dowex 50-H^+ ion exchange resin, which is filtered off once the desired pH is reached. With the latter method, the ionic strength of the solution is kept at a minimum. The solution is made 100 μM in 2-mercaptoethanol to reduce high potential cytochromes and is then chromatographed on a CM–cellulose (Selectacel-Standard, 30-ml packed volume) equilibrated with 1 m*M* phosphate buffer, pH 6.0. The column is washed with 400 ml of the application buffer to develop the clearly separated colored bands and then the following cytochromes are eluted with 400–500 ml portions of the buffer indicated: cytochrome c_{555} (2 m*M* phosphate buffer, pH 6.0), cytochrome $c_{551.5}$ plus cytochrome c' (5 m*M* phosphate buffer, pH 6.0), cytochrome $c_{2\,551.5}$ (20 m*M* phosphate buffer, pH 6.0), and cytochrome c_{554} (50 m*M* phosphate buffer, pH 6.0). Each cytochrome zone is separately desalted by passage through a Sephadex G-25 column equilibrated with 1 m*M* phosphate buffer, pH 5.5, and concentrated with the aid of small (3-ml packed volume) CM–cellulose columns. With the exception of the cyto-

[24] R. G. Bartsch, T. Horio, and M. D. Kamen, submitted to *Biochim. Biophys. Acta.*
[25] R. G. Bartsch and T. E. Meyer, private communication, 1970.
[26] T. Yamanaka and M. D. Kamen, *Biochim. Biophys. Acta* **131,** 317 (1967).

chrome c' fraction, the proteins are eluted with 0.1 M phosphate buffer, pH 7.0. The cytochrome c' solution is made 100 μM with 2-mercaptoethanol and adsorbed on a concentration column. The contaminating cytochromes c_{555} and $c_{551.5}$ are reduced by the 2-mercaptoethanol and therefore are easily eluted with 2 mM phosphate buffer, pH 5.5, leaving the more positively charged, oxidized cytochrome c' adsorbed. The latter cytochrome is eluted with 5 mM phosphate buffer, pH 5.5. The now separated fractions are concentrated in the above manner.

A recurrent problem with highly basic cytochromes is the obvious loss of a portion of the colored proteins by adsorption on the Sephadex G-25 column. This loss is not completely eliminated by the use of the low concentration of buffer required for subsequent adsorption on CM–cellulose columns. Freshly hydrated Bio-Gel P-2 desalting columns do not initially adsorb the cytochromes, but slowly deteriorate with use and eventually behave like Sephadex G-25.

The concentrated cytochrome c' is precipitated over the range 70–90% saturated ammonium sulfate. The precipitate is dissolved in 3 ml of 0.1 M Tris·HCl, pH 8.0, solid ammonium sulfate is added to ~70% saturation and the cytochrome crystallizes after several days standing at room temperature.

Cytochrome c_2 is passed through a Dowex-2 anion exchange column (2-ml bed volume) charged to two-thirds capacity with ferricyanide. The solution of completely oxidized cytochrome is then desalted by passage through a Bio-Gel P-2 column equilibrated with 1 mM phosphate buffer, pH 6.0. The desalted solution is then chromatographed on a CM–cellulose column (Selectacel-Standard, 40-ml bed volume) equilibrated with 1 mM phosphate buffer, pH 6.0. The main zone of cytochrome c_2 is eluted with 500 ml of 25 mM phosphate buffer, pH 6.0, desalted with the Bio-Gel P-2 column, concentrated on a small CM–cellulose column and eluted with 1 M NaCl in 0.1 M Tris·HCl, pH 8.0. The cytochrome is precipitated between 65 and 90% saturated ammonium sulfate, then dissolved in 3 ml of 0.1 M Tris·HCl, pH 8.0, and reprecipitated in the same manner. The precipitate is dissolved in 0.1 M Tris·HCl, pH 8.0, containing 20 mM 2-mercaptoethanol to reduce all the cytochrome. Sufficient solid ammonium sulfate is added to the solution at room temperature until slight turbidity appears (at ~70% saturation). After 10 minutes standing the solution is centrifuged at 20° for 5 minutes at 30,000 g to remove the precipitate. Upon further standing fine pink crystals develop somewhat more rapidly than does a gelatinous precipitate. The early crop of crystals is collected before much gelatinous material accumulates. Recrystallization three or four times in succession in the same manner gives ~20% recovery of cytochrome c_2, free of the amorphous material. Thus far no other technique for removing the latter material has been devised.

The three remaining concentrated basic cytochromes are precipitated with ammonium sulfate: cytochrome c_{554} (50–70% saturation), cytochrome $c_{551.5}$ (60–80% saturation) to cytochrome c_{555} (80–100% saturation). Each precipitate is dissolved in 3 ml of 1 *M* phosphate buffer, pH 7.0, and chromatographed on a Sephadex G-75 column equilibrated with 0.1 *M* NaCl in 10 m*M* phosphate buffer, pH 7.0, to remove colorless impurities. The best cytochrome fractions are pooled, desalted and concentrated with small CM–cellulose columns. Small quantities of nearly homogeneous cytochromes are obtained.

Eluate 2 is cooled to 0° and adjusted to pH 5 with 2 *M* acetic acid. The suspension is centrifuged at 30,000 *g* for 10 minutes to remove the bulky precipitate. Occasionally a portion of the cytochrome c_{552} is precipitated and may be extracted from the bulk of denatured precipitate with 0.1 *M* Tris·HCl, pH 8.0. The supernatant solution is immediately adjusted to pH 6.0. Any precipitate which forms is removed by centrifugation. The solution is then desalted with Sephadex G-25 equilibrated with 1 m*M* Tris·HCl, pH 8.0, and chromatographed at room temperature on a DEAE–cellulose column (type 20, 55-ml bed volume) equilibrated with the same buffer. Cytochrome c_{552} (c_3) is eluted with 450 ml of 2 m*M* Tris·HCl, pH 8.0. Two NADH-dehydrogenases contaminated with cytochrome c_{552} may subsequently be eluted.[24] As an alternative procedure the eluate 2 is precipitated with saturated ammonium sulfate, the precipitate is dissolved in 0.1 *M* Tris·HCl, pH 7.3 and the solution is chromatographed on a Sephadex G-100 column (1-liter bed volume for each 20–30 ml of concentrated solution) equilibrated with 0.2 *M* NaCl in 20 m*M* Tris·HCl, pH 7.3, in the cold. Presumed membrane fragments are eluted in the void volume, followed by the flavin-containing dehydrogenases and finally cytochrome c_{552}.

The cytochrome c_{552} fraction is desalted with Sephadex G-25 and concentrated with a small DEAE–cellulose column. The concentrated solution is fractionated with 80–100% saturated ammonium sulfate to precipitate the cytochrome. The precipitate is dissolved in 0.1 *M* phosphate buffer, 7.0, and chromatographed on a Sephadex G-75 column to eliminate colorless impurities. The best fractions are pooled, desalted, and concentrated with a small DEAE–cellulose column. Even the best preparation made to date may not be homogeneous.

Although certainly present in eluate 3, cytochrome b_{558} has not been successfully purified from extracts of *R. palustris* strain 2.1.37. However, an undesignated strain of *R. palustris* isolated as a contaminant from a culture of *Chlorobium thiosulfatophilum* yields the cytochrome.[25] Eluate 3 from this organism is chilled in an ice bath, and 2 *M* acetic acid is added to adjust the solution to pH 5. The precipitate formed is removed by centrifugation, and the supernatant solution is adjusted to pH 8.0 by cautious addition of 2 *M* NaOH and then is desalted with a Sephadex G-25 column. The solution is next chromatographed on a DEAE–cellulose column (type 20, 20-ml bed volume) equilibrated with 0.1 *M* Tris·HCl, pH 7.3. After a preliminary wash with 200 ml of 0.1 *M* Tris·HCl, pH 7.3, to remove flavoproteins, an iron protein with the plant ferredoxin spectrum followed by cytochrome b_{558} are eluted with 0.14 *M* NaCl in 20 m*M* Tris·HCl, pH 7.3. The cytochrome is desalted with a Sephadex G-25 column and concentrated with the aid of a small DEAE–cellulose column. The concentrated solution is then chromatographed on a Sephadex G-100 column (1-liter bed volume per 20–30 ml of solution) equilibrated with 0.2 *M* NaCl in 20 m*M* Tris·HCl, pH 7.3. The cytochrome is eluted slightly behind the void volume, followed immediately by the remaining iron protein. Again the best fractions of cytochrome b_{558} are pooled, desalted, and concentrated with a small DEAE–cellulose column. Solid ammonium sulfate is added to the concentrated cytochrome solution to ~30% saturation. After standing in the cold for 5–10 hours, deep-red crystals form and continue to grow over a period of several days. However, the bulk of the cytochrome crystallizes within 24 hours and is collected by centrifugation before much amorphous colorless material accumulates. Upon recrystallization the purity index, $A_{280}/A_{419,\,ox} = 1.3$, is attained. The cytochrome is slowly reduced with sodium dithionite, several hours are required for a sample in an open cuvette to become completely reduced after addition of excess reductant.

Properties of the R. palustris Cytochromes. The available information about the properties of the *R. palustris* cytochromes is summarized in Table I. As yet no studies on the biochemical role of any of the cytochromes has been reported. Cytochromes c_2 and cc' are the most abundant cytochromes *c* isolated, the others are recovered in no more than one-tenth the yield of those two.

Chloropseudomonas ethylicum Cytochromes

An exception to the general methodology already described is offered by *Chloropseudomonas ethylicum*. An unknown component of the extracts prevents complete adsorption from a crude extract of cytochrome $c_{551.5}$ to DEAE–cellulose, although the cytochrome c_{555} can be successfully chromatographed (see Table I). Fractionation with ammonium sulfate, either by eluting a total precipitate with an inverse ammonium sulfate gradient,[27] or by precipitating the crude extract between 40 and 100% saturated ammonium sulfate before chromatographing on a DEAE–cellulose column[28] removes the interfering factor. After desalting with a Sephadex G-25 column, cytochrome $c_{551.5}$ is chromatographed on a DEAE–cellulose column (type 20, 20-ml bed volume) equilibrated with 20 mM Tris·HCl, pH 7.3. Elution of the cytochrome is accomplished with 0.2 M NaCl in the buffer, after preliminary elution of the column with 0.1 M NaCl in the buffer. Both cytochromes are further purified by chromatography on a Sephadex G-100 column in the manner described for *R. palustris* cytochrome *b*. Finally the cytochromes are desalted and concentrated on DEAE–cellulose columns (type 20, 3-ml bed volume) and eluted with 0.5 M NaCl in 20 mM Tris·HCl, pH 7.3.

Cytochromes from Dark-Aerobic Bacteria

From *R. rubrum* grown aerobically in the dark, cytochrome c_2 has been prepared essentially indistinguishable from that of light-anaerobic cells. The cytochrome *cc'* content of the aerobic cells was negligible. A membrane-bound cytochrome *b* with properties like those of cytochrome *o* was detected,[29] but was not solubilized. A different pattern was found in one strain of *R. spheroides*.[30,31] In aerobic cells there is produced a membrane-bound cytochrome c_{551} different from the anaerobic cytochrome c_2, and the content of cytochrome c_{553} (probably equal to cytochrome c_{555} in Table I) is greatly increased.[30] A short-term (18 hours) aerobic culture of the *R. spheroides* strain develops a cytochrome *a*, which can be solubilized and shown to function as an oxidase.[31] The cytochrome c_{551}, as well as a cytochrome *b*, are components of the solubilized preparation. Upon longer incubation, the cytochrome *a* disappears and a cytochrome *o* appears to act as an oxidase.[31]

[27] J. M. Olson and E. K. Shaw, *Photosynthetica* **3**, 288 (1969).

[28] T. E. Meyer, Ph.D. Thesis, Univ. of California, San Diego, 1970.

[29] S. Taniguchi and M. D. Kamen, *Biochim. Biophys. Acta* **96**, 395 (1965).

[30] Y. Motokawa and G. Kikuchi, *Biochim. Biophys. Acta* **120**, 274 (1966).

[31] G. Kikuchi and Y. Motokawa, *in* "Structure and Function of Cytochromes" (K. Okunuki, M. D. Kamen, and I. Sekuzu, eds.), p. 174. Univ. of Tokyo Press, Tokyo, 1968.

[35] Cytochromes: Algal

By EIJIRO YAKUSHIJI

High potential *c*-type cytochromes from algae, which are believed to be involved in the photosynthetic electron flow, are described.

Petalonia fascia Cytochrome c_{553}[1]

Assay

The purity index is defined as the ratio of absorbancy for the reduced form at the absorption maximum at 553 nm to that at 273 nm. The amount of cytochrome can be calculated by the specific absorbancy at 553 nm using the extinction coefficient of 2.71 liter/g/cm.

Preparation

Step 1. About 2.5 kg of fresh thalli, which contain about 88% water are immersed in 10 liters of ammonia water making the final concentration 20 m*M* NH_4OH and left at room temperature (about 15°) for about 50 hours with occasional stirring. The extract is then filtered through absorbent cotton.

Step 2. One liter of 3% acrinol solution is then added to the resulting 10 liter filtrate, and it is filtered through cotton to remove yellow flocculent material. The yellowish-pink filtrate is then half saturated with ammonium sulfate and filtered on a Büchner funnel through a thin layer of talc. The precipitate is discarded. The brownish-pink filtrate is saturated with ammonium sulfate. The pink precipitate is collected on a Büchner funnel and dissolved in 200 ml of 5 m*M* phosphate buffer, pH 7.0. The crude preparation thus obtained contains about 40 mg of cytochrome. The purity index in this stage is about 0.42.

Step 3. The crude preparation is dialyzed against 5 m*M* phosphate buffer (pH 7.0) overnight and then placed on a DEAE–cellulose column (diameter, 2 cm; length, 15 cm) which has been equilibrated with the same buffer before use. It is preferable to reduce the cytochrome solution completely by the addition of a minute amount of solid ascorbic acid before the column chromatography to ensure uniformity. The flow rate is adjusted to 3 ml/minute. The cytochrome is adsorbed on the top of the column contaminated with a brown substance(s). The column is washed with 500 ml of 5 m*M* phosphate buffer and the cytochrome is developed and eluted with about 2 liters of 10 m*M* phosphate buffer,

[1] Y. Sugimura and E. Yakushiji, *J. Biochem.* **63,** 261 (1968).

leaving the brown substance(s) on the top of the column. The eluate is diluted by adding an equal volume of deionized water, and the cytochrome is adsorbed on a second column of DEAE–cellulose (2 cm × 10 cm). The column is developed with 10 m*M* phosphate buffer until just before the cytochrome reaches the bottom of the column. The upper part of DEAE–cellulose in the column, which contains no more cytochrome, is removed from the column and discarded. The cytochrome is then eluated with 100 ml of 0.1 *M* phosphate buffer. The eluate contains about 30 mg of ferrocytochrome with a purity index of 0.95.

Step 4. Crystallization. The eluate is saturated with ammonium sulfate and centrifuged at 5000 rpm for 15 minutes. The pink precipitate is dissolved in 5 ml of deionized water and centrifuged in the same way. Finely powdered ammonium sulfate is added to the supernatant solution until a slight turbidity appears. The solution is stored in the cold with the addition of a trace of solid sodium dithionite. The crystallization of the cytochrome is completed within a few days. After repeating the recrystallization three times under the same conditions, about 20 mg of crystalline cytochrome is obtained.

Properties

Spectral Features. The ferrocytochrome shows absorption maxima at 273.0, 293.0, 317.5, 415.5, 521.5 and 553.0 nm. The α band at 553.0 nm is characteristic with a shoulder at 549.0 nm forming an asymmetrical band. The ratio of absorbances, A_γ/A_α, is 6.9. The ferricytochrome has main absorption peaks at 360.0 and 409.8 nm. The isosbestic points in the difference spectrum (reduced minus oxidized) are located at 336.0, 410.0, 433.5, 506.0, 529.5, 541.5, and 560.5 nm. The maxima in the difference spectrum are found at 317.5, 387.0, 417.0, 521.0, and 553.0 nm and the minima at 364.0, 370.0, 400.0, 453.0, and 536.5 nm. The pyridinehemochrome has main absorption peaks at 413.5, 520.0, and 549.5 nm.

Iron Content, Molecular Weight. The cytochrome contains 0.530% iron. Assuming that one molecule of the cytochrome contains one heme, as was established by Katoh[2] for *Porphyra tenera* cytochrome $c_{553.0}$ by ultracentrifugal analysis, the molecular weight of the *Petalonia* cytochrome is calculated to be 10,500.

Specific Absorbancy. The gram specific absorbancy at 553.0 nm is 2.71 liter/g/cm and the molar extinction coefficient is 28.5/m*M*/cm.

Normal Redox Potential. The $E_{m,7}$ value of +0.360 V was determined by the method of Davenport and Hill[3] where Clark's value for ferri-ferrocyanide of +0.43 V was used.

[2] S. Katoh, *Plant Cell Physiol.* **1,** 91 (1960).
[3] H. E. Davenport and R. Hill, *Proc. Roy. Soc. Ser. B* **139,** 327 (1952).

TABLE I
ABSORPTION PEAKS OF REDUCED ALGAL CYTOCHROMES[a]

Source	λ_{max} (nm)			A_γ/A_α
	α Peak	β Peak	γ Peak	
Red algae				
* *Bangia fusco-purpurea*[b]	553.5	521.5	415.5	6.7
* *Porphyra tenera*	553.0	522.0	416.0	7.0
* *Porphyra yezoensis*	553.0	522.0	416.0	7.7
* *Porphyra pseudolinealis*	553.0	522.0	416.0	—
Nemalion vermiculare	553.0	522.0	416.0	—
Scinaia japonica	553.0	521.5	415.5	6.7
Gloiophloea okamurai	553.0	521.0	416.0	5.7
Pterocladia tenuis	552.5	521.5	415.5	5.6
Grateloupia filicina	552.5	521.8	416.0	6.6
Pachymeniopsis lanceolata	552.5	521.8	415.3	7.2
* *Gloiopeltis complanata*	553.0	521.0	415.0	—
Gracilaria verrucosa	552.8	521.5	415.5	6.4
Gracilaria textorii	553.0	522.0	416.0	6.8
Chondrus giganteus	553.0	521.5	415.5	6.3
Rhodoglossum pulcherum	553.5	521.5	415.5	5.9
* *Polysiphonia urceolata*	551.5	522.5	416.3	5.7
Chondria crassicaulis	553.0	521.5	415.5	5.9
Brown algae				
Scytosiphon lomentaria	553.0	521.5	415.5	—
Ishige okamurai	553.0	521.5	415.5	6.7
* *Endarachne binghamiae*	553.0	521.5	415.5	6.6
* *Petalonia fascia*	553.0	521.5	415.5	6.9
Yellow-green alga				
Vaucheria sp.	553.0	521.5	415.3	5.7
Green algae				
Ulva pertusa	553.0	522.3	416.3	6.3
* *Enteromorpha prolifera*	552.5	522.3	416.3	6.7
Cladophora sp.	553.0	522.3	416.0	6.0
Chaetomorpha spiralis	553.0	521.5	416.5	5.9
Chaetomorpha crassa	552.8	522.5	416.0	6.5
Bryopsis maxima	554.0	523.5	417.3	—
Bryopsis sp.	553.0	522.5	—	—
Caulerpa brachypus	552.5	523.0	416.0	6.6
Codium latum	554.5	523.0	417.0	7.0
Codium fragile	554.5	523.0	417.0	—

[a] Y. Sugimura *et al.*, Ref. 4.
[b] Nine cytochromes marked with asterisks were obtained in crystalline form.

Other Algal Cytochromes[4]

Preparation

In the case of red and green algae, the method of preparation is fundamentally the same as described above, except that tap water or deionized water is used for the extraction instead of ammonia water. Sometimes it is convenient to plasmolyze the fresh material with solid ammonium sulfate which corresponds to about a quarter of the weight of the material. A considerable amount of cytochrome is brought into solution by this method and the remainder in the thalli can be easily extracted by adding sufficient water.

Properties

Spectral Features. The results obtained with cytochromes of 32 species of algae are listed in Table I. The location and the shape of α band and the ratio A_γ/A_α are appropriate characteristics for comparison of these cytochromes. Algal cytochromes predominantly have their

TABLE II
$E_{m,7}$ OF ALGAL CYTOCHROMES[a]

Source	$E_{m,7}$ (V)
Red algae	
Porphyra yezoensis	0.342
Porphyra pseudolinealis	0.340
Nemalion vermiculare	0.363
Gloiopeltis complanata	0.340
Brown algae	
Endarachne binghamiae	0.361
Petalonia fascia	0.360
Green algae	
Enteromorpha prolifera	0.364
Cladophora sp.	0.378
Chaetomorpha spiralis	0.382
Chaetomorpha crassa	0.381
Bryopsis sp.	0.385
Caulerpa brachypus	0.385
Codium latum	0.390

[a] Y. Sugimura *et al.*, Ref. 4.

[4] Y. Sugimura, F. Toda, T. Murata, and E. Yakushiji, *in* "Structure and Function of Cytochromes" (K. Okunuki, M. D. Kamen, and I. Sekuzu, eds.), p. **452**. Univ. of Tokyo Press, Tokyo, and Univ. Park Press, London, 1968.

α peak at about 553 nm accompanied by a shoulder at 549 nm. Exceptions are *Polysiphonia* cytochrome $c_{551.5}$ and *Caulerpa* cytochrome $c_{552.5}$ which have symmetrical α bands.

Normal Redox Potential. The values of $E_{m,7}$ of 13 cytochromes obtained from red, brown, and green algae are listed in Table II.

[36] *Euglena* Cytochromes

By Akira Mitsui

Several cytochromes can be isolated from *Euglena* species, e.g., c_{552},[1-7] *c*-type 556,[3-6] *b*-type 561 (558 and 563),[5,6,8,9] and *a*-type 605.[6] *Euglena* cytochrome c_{552} and *b*-type 561 (558 and 563) are functionally localized in the chloroplasts,[2,4-6,8,9] while cytochrome *c*-type 556 and *a*-type 605 are associated with the respiratory system.[3-6] Cytochrome c_{552} and cytochrome *c*-type 556 are soluble proteins, while cytochrome *b*-type 561 (558 and 563) and cytochrome *a*-type 605 are particle bound. Only cytochrome c_{552} has been highly purified,[7] and fully characterized.[4-7]

Purification of Cytochrome c_{552}

Wild-type strains of *Euglena gracilis* are grown in an illuminated vessel at room temperature (25°),[10,11] and harvested after maximum growth is achieved. The cells are washed once with 0.9% NaCl and after centrifugation the cell paste is stored at −15° until use. Mutant strains may contain reduced amounts of cytochrome c_{552} when grown under

[1] H. E. Davenport and R. Hill, *Proc. Roy. Soc. Ser. B* **139,** 327 (1950).
[2] M. Nishimura, *J. Biochem.* **46,** 219 (1959).
[3] J. A. Gross and J. J. Wolken, *Science* **132,** 357 (1960).
[4] J. J. Wolken and J. A. Gross, *J. Protozool.* **10,** 189 (1963).
[5] R. M. Smillie, *Can. J. Bot.* **41,** 123 (1963).
[6] F. Perini, M. D. Kamen, and J. A. Shiff, *Biochim. Biophys. Acta* **88,** 74 and 91 (1964).
[7] A. Mitsui and K. Tsushima, *in* "Structure and Function of Cytochromes" (K. Okunuki, M. D. Kamen, and I. Sekuzu, eds.), p. 459. Univ. of Tokyo Press, Tokyo, and Univ. Park Press, Baltimore, 1968.
[8] J. M. Olson and R. M. Smillie, *in* "Photosynthetic Mechanism of Green Plants" (B. Kok and A. T. Jagendorf, eds.), p. 56. Nat. Acad. Sci.—Nat. Res. Council Publ. 1145 (1963).
[9] I. Ikegami, S. Katoh, and A. Takamiya, *Biochim. Biophys. Acta* **162,** 604 (1968).
[10] M. Cramer and J. Myers, *Arch. Mikrobiol.* **17,** 384 (1952).
[11] C. L. Greenblatt and J. A. Schiff, *J. Photozool.* **6,** 23 (1959).

these conditions.[6] Dark-grown *Euglena* strains do not contain this cytochrome.[2,4-6]

Step 1. Preparation of the Crude Extract. The frozen cell paste is thawed in the cold room and homogenized into a heavy cell suspension using a minimal amount of water. The cell suspension is mixed with enough cold acetone (—10°) to give a final concentration of 95% acetone. The slurry is stirred for 1 hour; occasionally a small piece of dry ice is added to maintain the temperature at about —10°. The acetone powder is collected by filtration on a Büchner funnel, washed several times with 100% cold acetone (—10°), and dried in a vacuum desiccator over P_2O_5. Two hundred grams of the acetone powder is suspended in 2 liters of 10 m*M* phosphate buffer, pH 7.3, and ground in a mortar for 15 minutes. The suspension is left standing for 1 hour at room temperature and then centrifuged at 100,000 *g* for 60 minutes. The clear supernatant solution is collected, and the pellet is resuspended in the same buffer, the extraction procedure is repeated. This second extraction yields about 15% of the amount obtained in the initial extraction. Both supernatant solutions are combined. However, if the solution is turbid the flow rates in the subsequent column chromatographic procedures will be slow.

Step 2. DEAE–Cellulose Chromatography. The solution is passed through a column of DEAE–cellulose (5 × 15 cm). The coarse grade of DEAE–cellulose should be used in order to maintain a high flow rate. Cytochrome c_{552} and ferredoxin are adsorbed at the top of the column. The column is washed with 0.1 *M* phosphate buffer, pH 7.4, until the edge of the pink cytochrome *c* band reaches the bottom of the column. The cytochrome c_{552} is then eluted with 0.1 *M* phosphate buffer, pH 7.4, containing 0.1 *M* NaCl. The ferredoxin which remains adsorbed on the column can be further purified and crystallized.[12]

Step 3. Second Chromatography on DEAE–Cellulose. The pink fraction containing cytochrome c_{552} is diluted 10 times with distilled water and added to a fresh column of DEAE–cellulose (4 × 45 cm). The adsorbed c_{552} is moved down the column with 30 m*M* phosphate buffer, pH 7.4, as before and then eluted with 60 m*M* phosphate buffer, pH 7.4.

Step 4. Hydroxyapatite–Cellulose Treatment. The pink cytochrome c_{552} fraction is immediately passed through a hydroxyapatite–cellulose column (5 × 30 cm). The majority of the cytochrome passes through this column whereas contaminating material which absorbs at 260 nm is retained. Cytochrome c_{552} remaining on the column is recovered by washing with small volumes of 60 m*M* phosphate buffer, pH 7.4. Both pink cytochrome solutions are combined. Compared with the eluate of

[12] A. Mitsui, unpublished data (1964).

step 3, this procedure increases the ratio, $A_{552\,\mathrm{nm}}/A_{276\,\mathrm{nm}}$, from 0.76 to 0.88.

Step 5. DEAE–Cellulose Chromatography. The pink cytochrome solution is diluted 4 times with distilled water and placed immediately on a fresh column of DEAE–cellulose (4 × 50 cm). The column is washed with 20 m*M* phosphate buffer and the cytochrome fraction eluted with 30 m*M* phosphate buffer, pH 7.4. The pink cytochrome moves slowly on this column. Again the pink fractions are collected, diluted with distilled water (1:3) and added to a small column of DEAE–cellulose (3 × 8 cm). The adsorbed cytochrome is eluted with a minimal volume of 0.15 *M* phosphate buffer, pH 7.4, using a reduced flow rate, i.e., 1 ml/minute, to obtain a concentrated cytochrome solution.

Step 6. Ammonium Sulfate Fractionation. Ammonium sulfate is added to the concentrated cytochrome solution, and the precipitate obtained between 70 and 80% saturation is collected by centrifugation. After each addition of ammonium sulfate, the solution should stand for 1 hour before centrifugation at 20,000 *g*, for 30 minutes. The absorbance at 552 nm of the cytochrome solution before ammonium sulfate fractionation should be higher than 2.0. If it is less than 2.0, the concentration procedure should be repeated using a smaller column and slower flow rates.

Step 7. Crystallization. The ammonium sulfate precipitate is dissolved in a minimal amount of 0.1 *M* phosphate buffer containing small amounts of reducing agents, i.e., ascorbate or dithionite. The solution is centrifuged at 4000 *g* for 10 minutes and any precipitate is discarded. Finely powdered ammonium sulfate is then added carefully to the pink supernatant solution until it becomes slightly turbid. The pH is adjusted to 5.4 with a small amount of 1 *M* sulfuric acid and the solution is placed in the cold (4°) overnight. Any precipitate is removed by centrifugation as before. A slight amount of finely powdered ammonium sulfate is again added to the supernatant solution. Small red crystals appear within 1 hour. After standing for a few days in the cold, the crystals grow, but the mother solution remains slightly pink indicating that not all the cytochrome has come out of the solution.

Crystals are collected by centrifugation at 4000 *g* for 10 minutes. Finely powdered ammonium sulfate is added to the supernatant solution as before to give additional crystals.

Crystals are dissolved in a minimal amount of 0.1 *M* phosphate buffer, pH 7.4, containing either ascorbate or dithionite and centrifuged. Ammonium sulfate powder is added to the supernatant solution, and cytochrome c_{552} is recrystallized as before.

Crystals are large, red, square, or rectangular and can be seen without magnification. The suspension of crystals at this stage shows no observable change in spectral properties or stability characteristics of the cytochrome, even after storage for two years at 0°.

Notes

1. All procedures except the preparation of the acetone powder can be performed at room temperature. There is no difference in recovery of cytochrome prepared at 5° or 25°. However, dilute solutions of cytochrome can be partially denatured if left standing for a couple of days in dilute salt solution at room temperature. Dilute solutions should be adsorbed on DEAE–cellulose immediately because the cytochrome is stable in this state.

2. Several methods for the extraction of the cytochrome c_{552} have been described, e.g., with sulfuric acid,[2] with digitonin,[3,4] freezing and thawing,[6] sonication,[6] as well as acetone treatment.[6,7] The acetone treatment described above gives the highest yield of cytochrome c_{552}. This method also provides a high yield of ferredoxin. The freezing and thawing method is also recommended for the extraction process, although the recovery is lower and the ferredoxin is decomposed.

Properties

Spectral Properties. The absorption peaks of the reduced form of crystalline cytochrome c_{552} solution are at 552 nm (α band), 523 nm (β band), 416 nm (Soret), 319 nm (δ band), and 276 nm, whereas those of the oxidized form are at 530, 412, and 355–362 nm.[7] The α band of the reduced form is symmetrical in shape without any shoulder. In the ultraviolet region there is one peak at 276 nm and two shoulders at 282 and 290 nm. Absorption ratios are: 5.3 ($A_{416\,\text{nm red.}}/A_{552\,\text{nm red.}}$); 1.6 ($A_{552\,\text{nm red.}}/A_{522\,\text{nm red.}}$); 1.01 ($A_{552\,\text{nm red.}}/A_{280\,\text{nm}}$); and 1.21 ($A_{280\,\text{nm}}/A_{260\,\text{nm}}$). The extinction coefficients (ϵ, mM) of the reduced form of this protein at 552, 523, and 416 nm are 29.7, 18.6, and 157.4, respectively.

Physicochemical and Physiological Properties. The molecular weight of *Euglena* cytochrome c_{552} is estimated to be 11,000[4] or 13,500[6] and contains one heme per molecule.[6] The isoelectric point is 5.5.[4,6] The oxidation–reduction potential (E_0' at pH 7.0) is estimated to be +0.37[6] or +0.38 V.[4] The physiological role of *Euglena* cytochrome c_{552} may be as an electron carrier between photosystems I and II.[8,13]

[13] S. Katoh and A. San Pietro, *Arch. Biochem. Biophys.* **118,** 488 (1967).

[37] Quinones in Algae and Higher Plants

By RITA BARR and F. L. CRANE

I. Preparation of Spinach Chloroplasts

A. Procedure[1]

Spinach leaves deveined, washed, and cooled to 4°
Grinding and suspension medium, 2 liters [0.5 *M* sucrose, dissolved in 2 m*M* aqueous phosphate buffer, (pH 7.5)], cooled to 4°
Four 1-liter centrifuge bottles, packed in ice
Cheesecloth (45 × 45 cm), four layers
Waring blendor jar, 1-gallon size, cooled to 4°
Five-liter beaker, surrounded by ice
Pair rubber gloves
Plastic tubing, 1.5 × 100 cm
Plastic funnel, 17 cm diameter
Cheesecloth, 20 × 20 cm, two layers
Potter-Elvehjem homogenizer tube (50 ml) cooled to 4°
"Foamkil" (Nutritional Biochemicals Corporation)

Preparation of chloroplasts is carried out in dim light (7-W night lights). Spinach leaves in 200-g quantities are ground in 1.5 liters of grinding medium for 1 minute at low speed (15,500 rpm without a load). The resulting slurry is filtered through four layers of cheesecloth into a chilled 5-liter beaker. Manual squeezing may be necessary to drain the pellet which is discarded. The solution containing chloroplasts is transferred to two 1-liter centrifuge bottles though plastic tubing. If draining is done from the bottom of the container by the insertion of weighted tubing to make it sink, much useless foam is left behind. Alternatively, a drop of antifoam silicone spray can be added to the grinding solution to avoid the formation of foam. The filled bottles are centrifuged first at 600 *g* for 15 minutes. The supernatant solution is poured through two layers of cheesecloth suspended over a funnel into another pair of clean centrifuge bottles. The second centrifugation, at 1500 *g* for 20 minutes, collects chloroplasts as a pellet. The supernatant solution is discarded. The chloroplast pellet is suspended in 50 ml of grinding solution and homogenized with a Potter-Elvehjem homogenizer. Two 1-ml aliquots are then withdrawn for a chlorophyll determination according to Arnon's

[1] R. Barr, M. D. Henninger, and F. L. Crane, *Plant Physiol.* **42**, 1246 (1967).

method.[2] Fifty milliliters of the chloroplast suspension are frozen immediately for future use in quinone extraction.

B. Chlorophyll Determination

Chloroplasts, 1 ml
Acetone, 10 ml
Glass centrifuge tubes, 15 ml, with stoppers
Water bath at 60°
Vortex mixer
Clinical centrifuge
Cuvettes, 1 ml
Spectrophotometer

The chlorophyll content of spinach chloroplasts is determined essentially by Arnon's method.[2] Acetone, 9 ml, is added to 1 ml of chloroplast suspension in a 15-ml stoppered glass centrifuge tube which is shaken on a Vortex mixer for 1 minute before immersion into a water bath for 2–3 minutes with gentle shaking. A stopper on the centrifuge tube, held down manually, is an absolute necessity, because acetone boils at 56.1°C and is lost quickly unless prevented from doing so. As soon as the acetone extract starts bubbling, white protein flakes can be seen settling toward the bottom of the tube. They are removed by centrifugation in a clinical centrifuge at 3000 rpm for 5 minutes.

The green supernatant is used for a spectrophotometric assay for chlorophyll a (663 nm) and b (645 nm) read against an acetone blank. The absorbance at 663 nm multiplied by a factor of 0.00802 gives the amount of chlorophyll a in milligrams present per milliliter of solution; the reading at 645 nm multiplied by 0.0202 gives the milligrams of chlorophyll b per milliliter of solution. The combined values multiplied by a dilution factor, if one is used, represent total chlorophyll per milliliter of solution.

Two hundred grams of spinach chloroplasts prepared by the above method yield between 1–2 mg of chlorophyll/ml.

II. Extraction of Quinones[1]

A. Extraction of Spinach Chloroplasts

Spinach chloroplasts, 100 ml, suspended in 0.5 *M* sucrose with 2 m*M* phosphate buffer, pH 7.5
Deionized water, 400 ml
Isopropanol, 500 ml

[2] D. I. Arnon, *Plant Physiol.* **24**, 1 (1949).

Heptane, 500 ml (Phillips, redistilled)
Dark glass container, 2 liter

The above ingredients are mixed in the dark glass container and shaken on a reciprocal shaker for 1 hour at low speed and at room temperature.

B. Extraction of Whole Leaves

Fresh leaf material, 100–200 g
Phosphate buffer, 0.2 *M* (pH 7.5), 100 ml
Deionized water, 400 ml
Isopropanol, 500 ml
Heptane, 500 ml
Glass funnel, 17 cm diameter
Cheesecloth, 45 × 45 cm, two layers
Miracloth (Chicopee Mills, Inc., 1450 Broadway, New York, N.Y. 10018), (30 × 30 cm)
Waring blendor, gallon size
Dark glass container, 2 liter

Leaves are ground in water, buffer, and isopropanol in a Waring blendor at low speed for 2 minutes. The slurry is filtered through two layers of cheesecloth and Miracloth over a funnel into a dark glass container, with gentle squeezing if necessary. After the addition of heptane, the extract is shaken on a reciprocal shaker for 1 hour at low speed and room temperature.

III. Separation of Aqueous and Lipid Fractions of the Extract[1]

Separatory funnel, 2000 ml
Heptane, 1000 ml
Acetone, 500 ml
Benzene, 100 ml
Methanol, 1000 ml
Deionized water, 1000 ml
Glass funnel, 17 cm diameter
Sodium sulfate, 50 g

The extract from chloroplasts or whole leaves containing plastoquinones and other quinones is transferred to a separatory funnel. After separation of the initial two phases, the green lipid epiphase is set aside while the aqueous hypophase is rewashed with 500 ml of heptane, which is later added to the previously collected heptane epiphase. The remaining brownish-green hypophase is washed once more with a mixture of 500 ml of heptane, 100 ml of benzene, 500 ml of acetone, and 500 ml of methanol. If the hypophase contains no more chlorophyll by visual inspection, it is discarded. When all aliquots of the green lipophilic heptane epiphase have been pooled, it is washed once or twice with a 1:1 mixture of methanol and water to remove glycolipids from chlorophyll and quinones. Finally, the epiphase is dried with sodium sulfate and evaporated to dryness in a rotary evaporator.

IV. Alternative Quinone Extraction Procedures (modified Bucke and Hallaway[3])

A. Extraction of Chloroplasts with 90% Acetone

Spinach chloroplasts, 100 ml, in 0.5 *M* sucrose and 0.002 *M* phosphate buffer (pH 7.5) containing 150–200 mg of chlorophyll
Acetone, 1500 ml
Petroleum ether, 250 ml, bp 40–60°
Chloroform–methanol (3:1), 400 ml
Beaker, 2000 ml
Glass stirring rod
Sintered glass funnel, 250 ml, with coarse pores (Pyrex)
Buchner flask, 1000 ml, connected to an aspirator
Round bottom flask, 2000 ml
Round bottom flask, 100 ml
Evaporator

Spinach chloroplasts prepared in 0.5 *M* sucrose and phosphate buffer, pH 7.5, are stirred with acetone to a 90% concentration (1:9 chloroplasts–acetone). The suspension is immediately filtered through a sintered glass funnel connected to an aspirator to obtain a chloroplast-free acetone extract which is evaporated in a 2000-ml round bottom flask. The chloroplast residue which collects in the filter is washed twice with 300-ml aliquots of acetone to remove as much chlorophyll and quinones as possible. The final acetone-washed chloroplast residue is suspended in chloroform–methanol (3:1) and allowed to soak overnight as a precaution against incomplete extraction of quinones by acetone.

After filtration the acetone extracts are combined and evaporated to dryness. It may be necessary to pour additional acetone or absolute ethanol into a partially dry evaporating flask to dry out the contents completely.

[3] C. Bucke and M. Hallaway, *Biochem. Chloroplasts* **1**, 153 (1966).

TABLE I

THE LIPOPHILIC QUINONE CONTENT OF SPINACH CHLOROPLASTS DETERMINED BY THREE DIFFERENT METHODS

Method	PQ A	K_1	PQ B	PQ C_{1-4}	PQ $C_{5,6}$	α-TQ
	(μmole of quinone/mg of chlorophyll)					
I,[a] replication 1	0.061	0.007	0.007	0.018	0.006	0.019
I, replication 2	0.071	0.005	0.007	0.019	0.008	0.032
I, average	0.066	0.006	0.007	0.019	0.007	0.026
II,[b] replication 1	0.051	0.006	0.007	0.022	0.010	0.024
II, replication 2	0.069	0.006	0.007	0.025	0.007	0.036
II, average	0.060	0.006	0.007	0.024	0.009	0.030
III,[c] replication 1	0.060	0.005	0.006	0.017	0.010	0.018
III, replication 2	0.079	0.005	0.005	0.030	0.007	0.023
III, replication 3	0.070	0.005	0.006	0.024	0.008	0.021

[a] Heptane–isopropanol–water extraction (1:1:1).[1]
[b] Acetone–water extraction (9:1).[3]
[c] Acetone extraction of lyophilized chloroplasts.[4]

When the acetone extract in the 2000-ml round bottom flask is dry, chloroplast pigments and quinones are transferred to petroleum ether by washing the flask several times with 50-ml portions of petroleum ether. A large evaporating flask is superior to a small one for the initial evaporation, because a thin layer of pigments and quinones spread over a large surface is more accessible to petroleum ether and, therefore, is extracted more easily from the gummy residue. The separate petroleum ether extracts are combined and evaporated to dryness in a small evaporating flask. Resuspension of the contents in a known volume of petroleum ether or heptane (5–10 ml) yields a chloroplast extract which is ready for column or thin layer chromatography to separate the various quinones.

If the results of a quantitative assay for PQ A (Section VII) show that less than 0.05 μmole of PQ A/mg of chlorophyll is present in the extract, chances are that the acetone extraction was incomplete and that a portion of quinones is still present in the final chloroform–methanol overnight soaking solution. This extract can be filtered through a sintered glass funnel and evaporated to dryness. The residual quinones are again transferred to petroleum ether and evaporated to dryness, and after resuspension they are added to the general chloroplast extract. Our data from Table I is obtained in this manner. Rewashing is a necessary precaution to take with every 90% acetone extraction of quinones from chloroplasts.

[4] J. P. Williams, *J. Chromatogr.* **36**, 504 (1968).

B. Extraction of Lyophilized Chloroplasts with Acetone (Modified Williams[4])

- Spinach chloroplasts, 100 ml, made in 0.5 *M* sucrose with 0.002 *M* phosphate buffer (pH 7.5), centrifuged once to remove sucrose and suspended in as little water as necessary to transfer to a 1000-ml round bottom flask for lyophilization
- Lyophilizer
- Acetone, 1500 ml
- Petroleum ether, 250 ml, bp 40°–60°
- Sintered glass funnel, 250 ml, with course pores (Pyrex)
- Buchner filter flask, 1000 ml, connected to an aspirator
- Evaporating flask, 2000 ml
- Evaporating flask, 100 ml
- Evaporator

Since chloroplasts cannot be lyophilized to a fine, dry powder if sucrose is present, the chloroplasts which are prepared as usual in sucrose are centrifuged and suspended in enough water to transfer to a lyophilizer flask.

When lyophilization is complete, the powdery chloroplasts are washed three times with 500 ml of acetone by a swirling motion. The extract is filtered through a course sintered glass funnel connected to an aspirator. After the final wash, all acetone extracts are combined and evaporated to dryness in a 2000-ml round bottom flask. Chloroplast pigments and quinones are transferred to petroleum ether by several 50-ml washes of the dry flask containing total acetone-extractable materials. All petroleum ether extracts are combined, evaporated to dryness in a small evaporating flask, and suspended in a known volume of heptane or petroleum ether for column or thin layer separation of the various quinones.

We have found (Table I) that acetone alone completely removes chloroplast pigments and quinones from lyophilized spinach chloroplasts so that soaking of lyophilized chloroplast residue in chloroform–methanol as for 90% acetone extraction is unnecessary.

V. Fractionation of Lipid-Quinone Extract[1]

A. Column Chromatography on an Alumina Column

- Chloroplast or whole leaf extract suspended in 5 ml of petroleum ether
- Merck's acid-washed alumina, 100 g
- Distilled water, 6 ml
- Chromatography column, 45 × 2.5 cm, plugged with glass wool

Twelve Erlenmyer flasks, 250 ml
Glass stirring rod, 0.5 × 25 cm
Petroleum ether, 1000 ml
Diethyl ether, 300 ml
Pipette, 10 ml
Funnel, 7-cm diameter
Round bottom flask, 250 ml

All steps in this procedure are carried out at room temperature, but in dim light (7-W night lights).

Before building a column, alumina and water are stirred vigorously with a glass rod in a dry, clean Erlenmyer flask until all traces of water disappear and the flask feels warm to the touch from the outside. Next, 100 ml of petroleum ether are added. The column is made from the resulting slurry with difficulty, with more or less petroleum ether needed to get the alumina out of the flask and into the column. No special equilibration period is required. As soon as the solvent line is down to the top of the alumina proper, the sample, suspended in petroleum ether, is added to the top of the column with a pipette. It can carefully be stirred into the top layer of the alumina with a glass rod. After this, fractions are collected in the following order: fraction 1, petroleum ether; fractions 2–10, 0.2, 2, 4, 8, 12, 16, 20, 24, and 28% diethyl ether in petroleum ether; fraction 11, diethyl ether. If chlorophyll derivatives still remain on the column after fraction 11, the column can be flushed with 100 ml of absolute ethanol.

The sequence in which quinones are eluted from an alumina column is reported in Table II. The collected fractions are evaporated to

TABLE II
THE ELUTION PATTERN OF CHLOROPLAST COMPONENTS FROM AN ALUMINA COLUMN

Fraction	Composition (% diethyl ether in petroleum ether)	Component eluted
2	0.2	β-Carotene and vitamin K_1
3	2	PQ A and B
4	4	PQ A, B, and Q
5	8	α-Tocopherol and plastochromanol
6	12	The tocopherols
7	16	PQ C_{1-4}
8	20	PQ $C_{5,6}$ and α-TQ
9	24	Tocopherylquinones and chlorophyll
10	28	Tocopherylquinones, mainly chlorophyll
11	100	Green components and xanthophylls

dryness and suspended in a known volume of heptane for thin layer chromatography.

An alumina column generally gives better recoveries of the plastoquinone C series than other types of columns.

B. Column Chromatography on a Silicic Acid–Super-Cel Column

Chloroplast or whole leaf lipid extract (100–200 mg of chlorophyll) suspended in 5 ml of heptane
Silicic acid, 10 g, meta precipitated ($H_2SiO_2 \cdot nH_2O$) from Fisher Scientific Co.
Hyflo Super-Cel, 30 g (Fisher)
Glass chromatography column, 65 × 2.5 cm, plugged with glass wool
Pipette, 10 ml
Twelve Erlenmyer flasks, 250 ml
Glass stirring rod, 0.5 × 25 cm
Tank of nitrogen
Heptane, 500 ml (Phillips, redistilled)
Chloroform, 100 ml

All steps in this procedure are carried out at room temperature in dim light (7-W night lights). Dry silicic acid and Super-Cel are mixed in an Erlenmyer flask with a glass stirring rod before the addition of 100–200 ml of heptane. The column is packed stepwise in 1 hour under nitrogen (3–4 lbs pressure/inch^2). Chloroplast extract is added to the top of the column with a pipette when the solvent line is even with the top of the packing material. It is then allowed to soak in evenly before more solvent is added to the column. Chlorophylls and xanthophylls are absorbed and held by the column while plastoquinones are eluted with increasing concentrations of chloroform in heptane as follows: fraction 1, 200 ml of heptane to elute β-carotene; fraction 2, 100 ml of 2% chloroform in heptane to elute vitamin K_1; fraction 3, 100 ml of 4% chloroform in heptane to elute PQ A and PQ B; fraction 4, 100 ml of 10% chloroform in heptane to elute PQ C_{1-4}; fraction 5, 100 ml of 25% chloroform in heptane to elute PQ $C_{5,6}$ and α-TQ; fraction 6, 100 ml of 30% chloroform in heptane to elute α-TQ and other TQ's; fraction 7, 100 ml of absolute ethanol to elute chlorophylls and xanthrophylls.

If lipid extracts are not entirely free from glycolipids, the chlorophylls and other green components may elute sooner than fraction 7. In such a case, fractions 5 and 6 have to be purified more thoroughly by thin layer chromatography to obtain fractions which can be used for a spectrophotometric quinone assay.

When all fractions are collected, they are evaporated to dryness and suspended in a known volume of heptane for purification by thin layer chromatography if purified quinone fractions are desired.

C. Thin Layer Chromatography of Extract

1. Preparation of Thin Layer Plates:

Five thin layer plates, 20 × 20 cm, or two (20 × 40 cm) plus one (20 × 20 cm)
Silica gel GHR, 30 g (Brinkman's)
Distilled water, 60 ml
Applicator
Oven set to 110°
Erlenmyer flask, 500 ml, with a stopper

Silica gel GHR is combined with water in an Erlenmyer flask and shaken vigorously for 1 minute. The slurry is applied to plates with an applicator to a thickness of 250 μ. After drying in air for 5 minutes, the plates are oven-dried at 110° for 30 minutes. After cooling these plates are stored in a dessicator. Only freshly made plates are used for streaking chloroplast or whole leaf extracts to reduce tailing.

2. Preparation of Paraffin Oil-Impregnated Thin Layer Plates for Reverse-Phase Chromatography

Thin layer plate (20 × 20 cm) coated with silica gel GHR after activation in the oven
Paraffin oil, 5.5 ml (U.S.P., white, heavy; viscosity 335–350 from Fisher Scientific)
Petroleum ether, 95 ml
Acetone, 95 ml
Distilled water, 5 ml
Graduated cylinder, 100 ml
Glass container, 25 × 25 cm or larger

Paraffin oil (5 ml) and petroleum ether (95 ml) are mixed in a glass container. Coated silica gel plates are dipped into the mixture, face down, until the whole surface is thoroughly wet. Gentle swirling of the container by the edges may be necessary. The impregnated plate is then removed from the container, set right side up, and air-dried for 5 minutes under the hood. Only freshly impregnated plates are used for streaking or spotting.

D. Application of Extract to Thin Layer Plates

For quantitative measurement of quinones, a known volume of chloroplast or whole leaf lipid extract is applied to a plate. To get a sharp separation of bands, the applied extract should contain no more than 3 mg of chlorophyll. Extract suspended in a fast-drying solvent, such as ether or heptane, spreads less than ethanol and reduces tailing. A fan to facilitate the drying of the applied extract also helps.

The extract can be applied to a plate in various ways: with a special applicator, a Pasteur pipette, or an ordinary long-tip pipette. It is applied 1.5 cm above the bottom of the plate without injury to the silica gel. If the volume of extract applied is 1 ml, for example, an additional 0.1 ml of extract is applied as a spot at the origin on the left or right of the main streak for the purpose of spraying to identify various quinone bands.

E. Development of Plates

No single solvent separates all chloroplast quinones equally well. Chloroform or benzene tends to crowd most components into the upper half of a thin layer plate, while heptane, pentane, and isooctane are too nonpolar to accomplish much separation and keep all bands near the origin. Mixtures of chloroform with heptane (such as 80:20) or heptane–benzene (15:85) are good general solvents for preliminary separation of the various quinones. In these two solvent systems, chlorophyll serves as a divider between plastoquinones, the coenzyme Q family and vitamin K_1 which are located above chlorophyll and the various tocopherylquinones which run below the highest chlorophyll band. The main advantages of using chloroform–heptane (80:20) also include a clean separation of six plastoquinone C types from chlorophyll and the tocopherylquinones. A disadvantage encountered at times, particularly if a plate is overloaded, is the tailing of β-carotene over the PQ A region and the poor separation between vitamin K_1 and various PQ B types.

The other general pair of solvents, heptane with benzene (15:85) takes care of this problem, giving a finer separation of all components from the upper half of a plate. Thus, β-carotene can be separated from PQ A and K_1, PQ A_{20} and PQ A_{10} from PQ A_{45}, and PQ B from coenzyme Q and all other lipophilic quinones. The separation of PQ C from chlorophyll, on the other hand, is poor in this solvent system. Tocopherylquinones stay at the origin and are completely masked by green components.

Other solvent systems for quinone separation, with an R_f value for individual quinones, are presented in Table III.

TABLE III

R_f VALUES OF PLASTOQUINONES A, B, C_{1-6}, VITAMIN K_1, COENZYMES Q_{10} AND Q_6, AND α-TOCOPHERYLQUINONE IN VARIOUS SOLVENT SYSTEMS

No.	Solvent system	Dielectric constant	R_F values vit. PQ A	PQ B	PQ C_{1-4}	PQ $C_{5,6}$	K_1	Co. Q_{10}	Co. Q_6	α-TQ	PQ A 20[a]
1.	Heptane–benzene (15:85)	—	0.77	0.68	0.19	0.15	0.72	0.35	0.27	0.05	0.70
2.	Chloroform–heptane (97:3)	—	0.92	0.94	0.57	0.48	0.83	0.69	0.67	0.29	—
3.	Chloroform–isooctane (80:20)	—	0.91	0.83	0.41	0.35	0.88	0.57	0.54	0.23	0.89
4.	Diisopropylether–benzene (20:80)	—	0.95	0.97	0.81	0.71	0.88	0.85	0.79	0.34	—
5.	Pentane–diethylether–acetic acid (70:30:2)	2.4	0.98	0.98	0.78	0.72	0.95	0.85	0.78	0.45	—
6.	Pentane–ethanol (100:0.5)	—	0.20	0.08	0.33	0.00	0.29	0.33	0.00	0.00	—
7.	Pentane–ethanol (100:0.8)	—	0.57	0.58	0.47	0.00	0.73	0.24	0.17	0.20	—
8.	Pentane–ethanol (100:2)	—	0.71	0.81	0.36	0.27	0.79	0.55	0.47	0.23	—
9.	Chloroform–cyclohexane–methanol (90:10:5)	—	0.92	0.91	0.87	0.74	0.87	0.92	0.89	0.73	—
10.	Chloroform–cyclohexane (80:20)	—	0.77	0.77	0.35	0.27	0.71	0.52	0.45	0.25	—
11.	Methanol	32.6	0.61	0.64	0.77	—	0.71	0.73	0.78	0.77	—
12.	Methanol–pentane (90:10)	—	0.69	0.60	0.82	0.73	0.67	0.63	0.82	0.81	—
13.	Methanol–cyclohexane (60:40)	—	0.62	0.53	0.67	—	0.60	0.61	0.63	0.69	—
14.	*N*-amyl acetate	—	0.85	0.92	0.89	—	0.84	0.88	0.83	0.75	—
15.	Diisopropyl ether	—	0.87	—	0.79	0.73	0.81	0.77	0.69	0.48	—

16. Pentane–diisopropylether (90:10)	—	0.85	0.87	0.21	0.13	0.77	0.44	0.33	0.15	—
17. Petroleum ether–benzene–ethanol (80:30:7.2)	—	0.93	0.89	0.63	0.57	0.85	0.76	0.69	0.48	
18. Methanol–pentane–acetic acid (80:20:1)	—	0.59	0.54	0.63	0.71	0.57	0.59	0.72	0.73	—
19. Pentane	1.8	0.07	0.47	0.27	0.00	0.09	0.00	0.00	0.00	—
20. Ethyl acetate–chloroform (90:10)	—	0.89	0.92	0.89	—	0.85	0.89	0.85	0.82	—
21. Chloroform–ethyl acetate (90:10)	—	0.90	0.95	0.87	0.82	0.85	0.89	0.83	0.55	—
22. Tertiary butyl alcohol	11.7	0.73	0.71	—	—	0.65	0.70	0.67	0.68	—
23. Isopropanol	18.1	0.65	0.63	0.64	—	0.57	0.63	0.61	0.57	—
24. Ethanol	24.3	0.80	0.81	0.81	—	0.74	0.79	0.78	0.79	—
25. Heptane–ethanol (90:10)	—	0.95	0.95	0.95	—	0.91	0.95	0.93	0.89	—
26. Pentane–tertiary amyl alcohol (85:15)	3	0.99	0.98	0.97	—	0.97	0.99	0.98	-0.98	—
27. Benzene–methanol (95:5)	3	0.95	0.93	0.87	—	0.66	0.92	0.89	0.73	—
28. Methanol–dichloroethane–water (100:18:20)	—	0.78	0.77	0.66	—	0.79	0.73	0.69	0.41	—
29. Pentane–ethanol (50:50)	—	0.99	0.99	0.96	—	0.97	0.99	0.97	0.97	—
30. Toluene–ethyl formate–formic acid (50:40:10)	10	0.92	0.91	0.87	—	0.81	0.91	0.83	0.74	—
31. Pentane–ethanol (100:1)	—	0.78	0.70	0.08	0.47	0.85	0.21	0.17	0.53	—
32. Cyclohexane	2	0.07	0.00	0.00	0.00	0.47	0.00	0.00	0.00	—
33. Cyclohexane–chloroform (60:40)	—	0.37	0.21	0.10	0.07	0.31	0.15	0.13	0.40	—
34. Chloroform–cyclohexane (60:40)	—	0.57	0.50	0.18	0.15	0.51	0.28	0.25	0.08	—
35. Benzene	—	0.68	0.51	0.14	0.10	0.62	0.30	0.23	0.06	0.58

[a] Synthetic PQ A 20 standard.

Thin layer plates (20 × 20 cm) are allowed to develop to a height of 15 cm above the origin in chromatography tanks lined with filter paper, and they are also allowed to equilibrate half an hour before use. Development to 15 cm above origin takes 20–30 minutes. After removal from the chromatography tank, plates are allowed to dry 5 minutes for solvent mixtures containing chloroform, 10 minutes for mixtures containing benzene before spray is applied.

F. Detection of Lipophilic Chloroplast Quinones on Thin Layer Plates by the Use of Sprays

1. Reduced Methylene Blue Spray:

Aqueous methylene blue chloride (Matheson Coleman and Bell), 0.1%, 50 ml
Zinc dust, 2–5 g
Concentrated sulfuric acid, 2–3 ml
Erlenmyer flask, 100 ml
Funnel (7 cm diameter)
Glass wool
Chromatogram sprayer

To make reduced methylene blue spray, 50 ml of methylene blue chloride solution is mixed with zinc dust and concentrated sulfuric acid. The mixture is shaken lightly for 1 minute and allowed to sit under the hood undisturbed for 5–10 minutes before use. Such aging of the spray avoids coloring the background of the plate. Immediately before use, the spray is filtered through glass wool and sprayed over a developed spot on the right or left side of a chromatography plate while the main streak on which extract was applied quantitatively is protected from spray with a glass plate. Plastoquinones, coenzyme Q, and tocopherylquinone give a deep blue spot almost instantaneously when sprayed, while vitamin K_1 becomes visible in 10 minutes–2 hours after spraying.

The mechanism of the spray reaction with methylene blue is one of redox reactions: Quinones as mild oxidizing agents are able to oxidize the leuco form of methylene blue back to its blue oxidized form.

2. Reduced Nile Blue Spray:

Aqueous solution of Nile blue A (Matheson Coleman and Bell), 0.1%, 50 ml
Zinc dust, 2–5 g
Concentrated sulfuric acid, 2–3 ml
Erlenmyer flask, 100 ml

Funnel, 7 cm diameter
Glass wool
Sprayer for chromatograms

Nile blue solution, zinc dust, and concentrated sulfuric acid are mixed and treated as described for reduced methylene blue spray. The only difference between these two sprays is that vitamin K_1 and other K-like compounds give a bluish-black spot immediately after spraying, as do other quinones with reduced methylene blue spray, but Nile blue identifies K_1 better than methylene blue because its redox potential is lower.

G. Detection of Tocopherols, Plastochromanol, and Reduced Quinones on Thin Layer Chromatograms by the Use of Sprays

1. Emmerie–Engel Reagent[5]*:*

Ferric chloride in absolute ethanol, 0.2%, 5 ml
2,2′-Bipyridine in absolute ethanol (from Eastman Organic Chemicals), 0.5%, 5 ml
Sprayer for chromatograms

Tocopherols, reduced quinones, and plastochromanol give a pink spot on thin layer chromatograms with the Emmerie–Engel reagent. The two reagent solutions are stored under refrigeration and are mixed in equal proportions immediately before use.

The test is based on the reduction of Fe^{3+} ions to Fe^{2+} ions by the tocopherols or other reducing substances which then form an intensely red complex with the bipyridine, allowing pink spots to appear on the chromatogram.

2. Diazotized o-Dianisidine Reagent[6]*:*

Spray I: 5% (w/v) aqueous sodium carbonate, freshly made, 10 ml
Spray II: 0.5 g of *o*-dianisidine hydrochloride dissolved in 60 ml of distilled water (Eastman Organic Chemicals)
Concentrated hydrochloric acid, 6 ml
Aqueous sodium nitrite, 5% (w/v), 12 ml
Aqueous urea, 5% (w/v), 12 ml
Erlenmeyer flask, 250 ml

Diazotized dianisidine spray is used mainly to identify plasto-

[5] A. Emmerie and Ch. Engel, *Rec. Trav. Chim. Pays Bas* **57,** 1351 (1938).
[6] P. J. Dunphy, K. J. Whittle, and J. F. Pennock, *Biochem. Chloroplasts* **1,** 165 (1966).

chromanol on thin layer chromatograms. The first step in the procedure is to make fresh diazotized dianisidine for spray II. Sixty milliliters of the aqueous dye solution (preferably in the hydrochloride form which is more soluble) and 6 ml of concentrated hydrochloric acid are combined in an Erlenmyer flask which is embedded in ice to reduce heat formation. Next, 12 ml of the sodium nitrite solution are added with gentle shaking. After 5 minutes, 12 ml of urea are added. If diazotization proceeds normally, the end product is a deep red solution.

To use the spray, a developed thin layer plate is sprayed with spray I and allowed to dry 3–5 minutes before spraying with spray II. Plastochromanol gives a slate-blue spot.[6]

H. Recovery of Quinones and Other Compounds from Thin Layer Chromatograms

1. The Regular Procedure

Twelve centrifuge tubes, 12 or 15 ml
Absolute ethanol
Spatula or razor blade
Weighing paper, 6 $\times$ 6 inches
Clinical centrifuge
Vortex mixer

After spraying, areas of the unsprayed main streak corresponding to blue spots given by reduced methylene blue or other sprays are scraped off and collected into numbered centrifuge tubes. A known amount of absolute ethanol is then added to each tube, and the contents are mixed thoroughly by hand with a spatula or mechanically on a Vortex mixer. The resulting slurry is centrifuged in a clinical centrifuge for 5–10 minutes at high speed (3400 rpm) to get rid of silica gel from the plate. The supernatant solution containing quinones or other compounds is saved. The silica gel can be rewashed with additional ethanol, centrifuged, and added to the corresponding supernatant fractions, although if mixing has been complete, recoveries of quinones in the second supernatant solution amount to less than 5%.

When all corresponding supernatant solutions have been pooled and reduced in volume by evaporation if necessary, the samples are now ready for a spectrophotometric assay.

2. An Alternate Procedure:

Rhodamine 6 G solution, 0.1%, in absolute ethanol (Allied Chemical Co.), 10 ml
Diethylether, 50 ml

After taking a developed plate from a chromatography tank and drying it, the whole plate can be sprayed with a solution of rhodamine. Alternatively, rhodamine can be incorporated into thin layer plates when they are made. All quinone-type and other compounds become visible as deep reddish bands over a lighter background. The contrast is even more striking if the plate is viewed under uv light. Bands of interest can then be marked with confidence, knowing that no part will be missed, as is possible when using reduced methylene blue spray, especially if bands did not run straight as anticipated. However, rhodamine is not as specific for quinones as methylene blue, so misidentification, particularly in a new solvent system, is likely to occur.

If rhodamine spray is used, elution of quinones from scraped bands is accomplished with diethyl ether, because quinones are soluble in this solvent while rhodamine is not. Thus, it is possible to centrifuge out rhodamine with the silica gel, obtaining colorless or light yellow solutions of quinones. A quantitative spectrophotometric assay of quinones in ether is impossible, however, because quinones can't be reduced with borohydride in the presence of ether. Therefore, ether fractions are evaporated first and suspended in a known volume of ethanol for a spectrophotometric assay.

If quinones are to be recovered from a large number of thin layer plates, as with mass preparations, the fastest way to recover various compounds from the silica gel is to make a slurry of material from recovered bands with ethanol and then pour it through a sintered glass filter (Pyrex, 4-cm diameter) fitted into a side-arm vacuum flask (50 ml) attached to an aspirator. This method also allows fast rewashing of the silica gel with additional solvent in the funnel while it is disconnected from the aspirator.

When all corresponding eluates have been combined, they can be evaporated to dryness and suspended in a small volume of solvent for further purification if necessary.

VI. Extraction of Quinones from Algae and Bacteria

Plastoquinones of the A, B, and C types which occur in all taxonomic divisions of higher plants[7] have also been found in all classes of algae, including representatives of greens, yellow-greens, blue-greens, reds, browns, and the flagellate, *Euglena*.[8]

Quinones from algae can be extracted by the heptane–isopropanol–water (1:1:1) method described in Sections II and III, or by the alter-

[7] R. Barr and F. L. Crane, *Plant Physiol.* **42,** 1255 (1967).

[8] E. Sun, R. Barr, and F. L. Crane, *Plant Physiol.* **43,** 1935 (1968).

native quinone extraction procedures with acetone discussed in Section IV. However, since algal cells and chromatophores are hard to break up, sonication of the material in phosphate buffer (0.02 *M* at pH 7.5) before extraction is recommended. A 100-ml suspension of some blue-greens, such as *Anacystis* or *Anabaena,* or the green alga, *Chlorella,* may require as many as two or three sonication periods, one at the beginning, the second after a preliminary separation of the lipid–aqueous phases of the extract, and a third one if the aqueous phase still appears green after an extraction in the separatory funnel and if organic solvents like acetone and methanol don't dissolve all of the pigments adhering to incompletely broken-up chromatophores.

If all of the above treatments are only partially successful in releasing the green pigments and associated quinones from cell fragments, soaking in chloroform–methanol (3:1) overnight can be tried as a last resort. Finally, the combined heptane–benzene epiphases from extraction in the separatory funnel, and the chloroform–methanol extracts are pooled and evaporated to dryness to give a single lipid extract which can be used for column or thin layer chromatography.

Since algae contain many other lipids besides quinones, it is often necessary to chromatograph the lipid extract on a thin layer plate once in chloroform–heptane (80:20) and scrape the upper and lower halves of the plate separately using chlorophyll as the dividing line, without quinone identification. Extract from the upper and lower halves of the plate can then be rechromatographed in the same solvent system. Individual quinone bands can now be identified with reduced methylene blue spray (Section V, F, 1) and assayed spectrophotometrically (Section VII).

Analysis of quinones from algae[8] shows that in addition to PQ A, several types of B and the usual six types of PQ C, the reds (*Gigartina, Rhydomela*) contain an additional PQ C.

Some algae also contain different types of vitamin K (Table IV). On the basis of the chromatographic properties of these naphthoquinones on thin layers of silica gel GHR, we arbitrarily designate K_1 and related forms as "high K" and more polar types as "low K."[8] Henninger[9] has also found a "low K" type in *Chlorella* cells grown on high glucose concentration in the culture medium. Henninger *et al.*[10] find a hydroxy naphthoquinone and a more polar type in *Anacystis nidulans.*

[9] M. D. Henninger, *Biochem. Biophys. Res. Commun.* **19,** 233 (1965).

[10] M. D. Henninger, H. N. Bhagavan, and F. L. Crane, *Arch. Biochem. Biophys.* **110,** 69 (1965).

TABLE IV
CHROMATOGRAPHIC DIFFERENCES BETWEEN "HIGH" AND "LOW" FORMS OF VITAMIN K FOUND IN ALGAE

Compound	$R_f{}^a$	$R_f{}^b$
K_1—*cis* form	0.57	—
K_1—*trans* form	0.48	—
Chlorella "low" K	0.05	—
Nostoc "high" K	0.50	—
Anabaena "high" K	0.49	—
Chlorella K_1	—	0.71
Chlorella "low" K	—	0.48
Anacystis naphthoquinone	—	0.55
Anacystis naphthoquinone (polar)	—	0.17

[a] On silica gel GHR plates developed in hexane–butyl ether (90:10).[8]
[b] On silica gel G plates developed in benzene.[9,10]

Bacteria, in contrast to higher plants and algae, contain no plastoquinones, but they do contain homologs of the coenzyme Q series and various menaquinone isoprenologs. Methods for the extraction and purification of coenzyme Q and menaquinones are described in a previous volume of this series.[11]

VII. A Spectrophotometric Assay of Plastoquinones and Tocopherylquinones

Ethanolic solution of plastoquinone A, 1 ml, containing 0.2–1 μmole of PQ A
KBH_4, potassium borohydride (Sigma)
Cuvettes, 1 ml
Recording spectrophotometer
Absolute ethanol, 1 ml plus extra for cleaning cuvettes
Deionized water for cleaning cuvettes

All plastoquinones, PQ A, B, and C, give the same absorption spectrum in ethanol (Fig. 1) with maximum absorption at 255 nm. The tocopherylquinone spectra (Fig. 2) show their maximum absorption between 250 and 270 nm; γ-hydroxy plastoquinone A, the oxidation product of plastochromanol, gives a spectrum like γ-tocopherylquinone.

Upon reduction with potassium borohydride, all plastoquinones develop a new absorption maximum at 290 nm, tocopherylquinones between 285 and 290 nm.

[11] F. L. Crane and R. Barr, this series, Vol. XVIIIC [220].

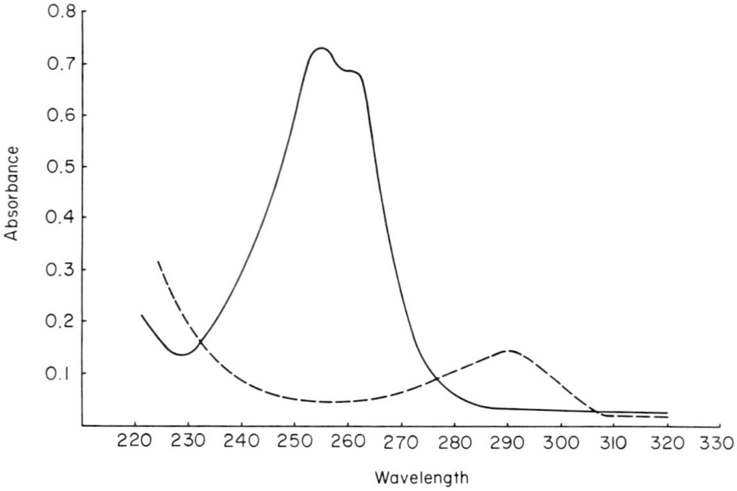

FIG. 1. The uv spectrum of plastoquinone A in absolute ethanol taken against an ethanol blank. The solid line denotes oxidized form; the dashed line denotes reduced form after the addition of potassium borohydride. The uv spectra of PQ B and C are similar.

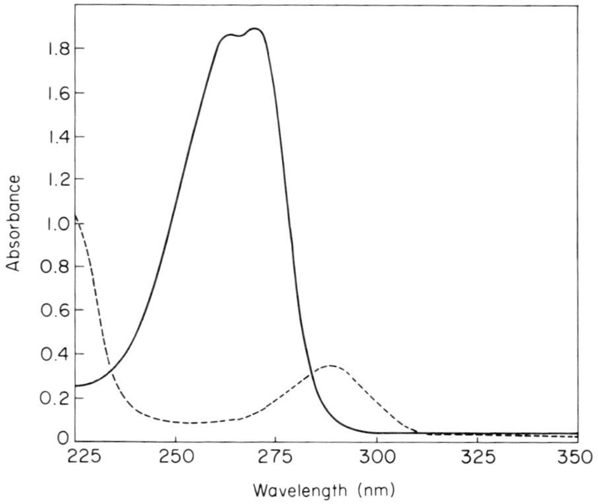

FIG. 2. The uv spectrum of α-tocopherylquinone in absolute ethanol taken against an ethanol blank. The solid line denotes oxidized form; the dashed line denotes reduced form after the addition of potassium borohydride.

A single representative assay to determine how much plastoquinone A is present in an extract is performed as follows: The spectrum on a 1-ml ethanolic solution from a PQ A band is taken from 220–340 nm against an absolute ethanol blank. Dilutions of sample with ethanol may be necessary if the absorbance is too high at 255 nm. After the first spectrum is completed, a pinch of potassium borohydride is added to the sample, the suspension is gently shaken for 5 seconds, and the spectrum is retaken in the 220- to 340-nm region. At this time it can be seen that absorption decreases at 255 nm but increases at 290 nm, which indicates that reduction of PQ A is taking place. To be sure that reduction is complete, the spectrum from 220–340 nm is retaken once more. If the third line coincides or runs close to the second line, reduction is considered complete.

The changes in absorbancy between oxidized and reduced spectra at 255 nm in case of PQ A and other plastoquinones after reduction with borohydride can be used as a quantitative test to calculate how much PQ is present in the extract. The difference is absorbancy between oxidized and reduced forms at 255 nm is divided by the millimolar extinction coefficient (e) to obtain micromoles of quinone per ml sample in the cuvette (or mmoles per liter):

$$\mu\text{moles of PQ} = \Delta A \text{ oxidized} - \text{reduced at 255 nm}/\Delta e$$

The Δe (255 nm, oxidized minus reduced) is 15 for any plastoquinone.

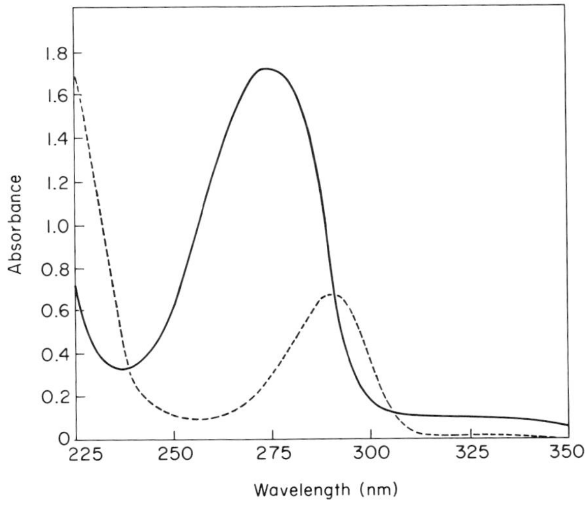

FIG. 3. The uv spectrum of coenzyme Q_{10} in absolute ethanol taken against an ethanol blank. The solid line denotes oxidized form; the dashed line denotes reduced form after the addition of potassium borohydride.

TABLE V

EXTINCTION COEFFICIENTS OF VARIOUS QUINONES IN ETHANOL

Quinone or other chloroplast component	Oxidized λ_{max} (nm)	Oxidized $E^{1\%}_{1cm}$	Reduced λ_{max} (nm)	Reduced $E^{1\%}_{1cm}$	$\Delta E^{1\%}_{1cm}$ (oxidized minus reduced)	Molar, millimolar extinction coefficient	Reference
PQ A_{45}	255					15	Bucke and Hallaway[3]
PQ A_{45}	255				200		Bucke *et al.*[a]
PQ A_{45}	255	210	290	46	198		Crane[b]
PQ A_{45}	255	210			198	15,200	Crane and Dilley[c]
PQ A_{45}	255	200					Eck and Trebst[d]
PQ A_{45}	255	198					Gaunt and Stowe[e]
PQ A_{45}	255	246					Henninger and Crane[f]
PQ A_{45}	255					15,000	Hindberg and Dam[g]
PQ A_{45}			290	48			Trenner *et al.*[h]
PQ A_{45}	255					15,000	Wellburn and Hemming[i]
PQ A_{45}	255					15,000	Williams[4]
PQ A_{20}	255	342					Eck and Trebst[e]
PQ B	255	202					Henninger and Crane[f]
PQ B	255	218					Henninger and Crane[j]
PQ B	255	227					Henninger *et al.*[k]
PQ B	255	202					Misiti *et al.*[l]
PQ C_{1-4}	255	96					Henninger and Crane[f]
PQ $C_{5,6}$	255	55					Henninger and Crane[f]
PQ C_{1-4}	255	203					Henninger *et al.*[j]
PQ $C_{5,6}$	255	129					Henninger *et al.*[j]
PQ C_{1-4}	255	66					Misiti *et al.*[l]

PQ $C_{5,6}$	255	75					Misiti *et al.*[l]
PQ C_{1-4}	255	194	291	42	179		Threlfall *et al.*[m]
PQ $C_{5,6}$	255	165	291	38	150		Threlfall *et al.*[m]
Plastochromanol	294 and 300.5[n]	55					Dunphy *et al.*[6]
Plastochromanolquinone	257.5	153	290		107		Dunphy *et al.*[6]
K_1	270	330					Gaunt and Stowe[e]
K_1	249	300					Hall and Laidman[o]
Q_{10}	275	165			142	14,000	Crane and Dilley[c]
Q_9	275	172				13,700	Crane and Dilley[c]
Q_8	275	190				13,800	Crane and Dilley[c]
Q_7	275	202				13,900	Crane and Dilley[c]
Q_6	275	240			198	15,200	Crane and Dilley[c]
Q_{10}	275				142		Olson *et al.*[p]
Q_9	275				158		Olson *et al.*[p]
Q_{10}	275				142		Williams[4]
α-TQ						17.8	Bucke and Hallaway[3]
α-TQ	262				400		Bucke *et al.*[a]
α-TQ	262				375	15,100	Crane and Dilley[c]
α-TQ	262	412			397	15,100	Dilley and Crane[13]
α-TQ	262	414					Gaunt and Stowe[e]
α-TQ		414					Henninger and Crane[f]
α-TQ	262				390		Hall and Laidman[o]
α-TQ	262				397		Williams[4]
β-TQ	262				395		Hall and Laidman[o]
β-TQ	261	430					Henninger and Crane[f]
γ-TQ	258	430					Henninger and Crane[f]
δ-TQ	253					17.8	Barr and Arntzen[q]

(*Footnotes appear on p. 394.*)

(*Continued*)

TABLE V (*Continued*)

[a] C. Bucke, R. M. Leech, M. Hallaway, and R. A. Morton, *Biochim. Biophys. Acta* **112,** 19 (1966).
[b] F. L. Crane, *Plant Physiol.* **34,** 546 (1959).
[c] F. L. Crane and R. A. Dilley, *Methods Biochem. Anal.* **11,** 279 (1963).
[d] H. Eck and A. Trebst, *Z. Naturforsch.* **18B,** 446 (1963).
[e] J. K. Gaunt and B. B. Stowe, *Plant Physiol.* **42,** 851 (1967).
[f] M. D. Henninger and F. L. Crane, *Biochemistry* **2,** 1168 (1963).
[g] I. Hindberg and H. Dam, *Physiol. Plant.* **18,** 838 (1965).
[h] N. R. Trenner, B. H. Arison, and R. E. Erickson, *J. Amer. Chem. Soc.* **81,** 2026 (1959).
[i] A. R. Wellburn and F. W. Hemming, *Biochem. Chloroplasts* **1,** 173 (1966).
[j] M. D. Henninger and F. L. Crane, *Plant Physiol.* **39,** 598 (1964).
[k] M. D. Henninger, R. Barr, and F. L. Crane, *Plant Physiol.* **41,** 696 (1966).
[l] D. Misiti, H. W. Moore, and K. Folkers, *J. Amer. Chem. Soc.* **87,** 1402 (1965).
[m] D. R. Threlfall, W. T. Griffiths, and T. W. Goodwin, *Biochim. Biophys. Acta* **102,** 614 (1965).
[n] Determined in cyclohexane.
[o] G. S. Hall and D. L. Laidman, *Biochem. J.* **108,** 465 (1968).
[p] R. E. Olson, G. H. Dialameh, R. Bentley, C. M. Springer, and V. G. Ramsey, *J. Biol. Chem.* **240,** 514 (1965).
[q] R. Barr and C. J. Arntzen, *Plant Physiol.* **44,** 591 (1969).

TABLE VI
PROPERTIES OF CHLOROPLAST QUINONES

Quinone	Absorbance in ethanol maximum (nm)	Absorbance in ethanol shoulder (nm)	Isosbestic points[a] (nm)
PQ A_{45}	255	262	276,233
PQ A_{20}	255–256	262	276,232
PQ B_2	255	262	276,233
PQ C_{1-4}	255	262	276,232
PQ $C_{5,6}$	255	262	276,232
α-TQ	262–263	269	282,232
β-TQ	261	—	280,229
γ-TQ	258	—	279,230
δ-TQ	253	—	278,230
Q_{10}	275	—	294,231
K_1 oxidized	268–270	261	—
K_1 reduced	246–248	—	280,252

[a] Determined after reduction with potassium borohydride.

The millimolar extinction coefficients to calculate μmoles of other quinones present in a chloroplast or whole leaf lipid extract in ethanol are as follows: for coenzyme Q, 14 at 275 nm (see Fig. 3); α-tocopherylquinone, 17.8 at 263 nm; vitamin K_1, 19 at 248 nm (Table V).

The isosbestic points, i.e., points where the reduced line crosses the oxidized line on a quinone spectrum, give an indication of the purity of the compound. In case of PQ A they occur at 232 and 276 nm. Other chloroplast quinone isosbestic points with absorption maxima are listed in Table VI.

VIII. Purification and Identification of Individual Quinones

A. Plastoquinone A_{45}

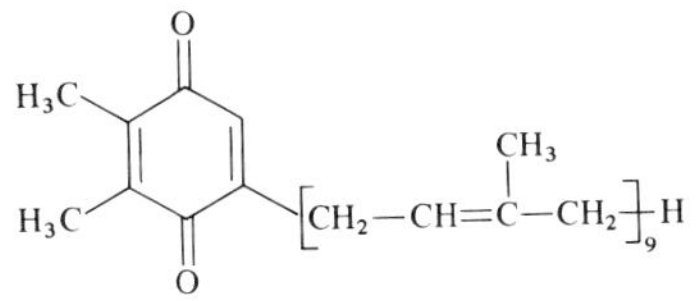

(I)

This quinone can almost always be obtained pure enough for a spectrophotometric assay by developing a thin layer chromatogram streaked with chloroplast or whole leaf extract once in chloroform–heptane (80:20). However, if there is interference from β-carotene, the impure fraction can be restreaked on another plate which is developed in chloroform–heptane (50:50) or heptane–benzene (15:85). If the impurity found with PQ A is vitamin K_1, the impure PQ A can be crystallized at —20° from absolute ethanol and removed from K_1 in the supernatant solution by centrifugation while cold.

B. Reduced Plastoquinone A_{45}

1. Reduction of Plastoquinone A[12]*:*

Ethanolic solution of PQ A containing 1–5 μmoles PQ A, 0.5 ml
HCl, 1 *N*, 2 ml
Distilled water, 10 ml
Isooctane, 10 ml
Potassium borohydride, a pinch
Stoppered 25-ml separatory funnel
Round bottom flask, 25 ml
Evaporator
Cuvettes, 1 ml
Spectrophotometer

Reduced plastoquinones A can be made easily from oxidized PQ A_{45} by reduction with potassium borohydride in a small separatory funnel with gentle swirling for 1 minute. After reduction is complete, 5 ml of water with 2–3 drops of 1 *N* HCl are added, followed by 10 ml of isooctane. Since the reduced quinone partitions into the isooctane phase, the aqueous hypophase can be discarded. The isooctane layer is rewashed once more with 5 ml of water acidified with 2–3 drops of 1 *N* HCl and evaporated to dryness as fast as possible to prevent reoxidation by oxygen from the air. The dry sample is suspended in a known volume of ethanol or methanol.

To check if all of the quinone is in the reduced form, a uv spectrum can be taken from 220–340 nm against an absolute ethanol blank. A major absorption peak at 290 nm and little or no absorption at 255 nm indicates that the quinone has been successfully reduced.

2. Determination of the Redox State of Plastoquinone A in Spinach Chloroplasts[12]*:*

Aqueous chloroplast suspension with 2–3 mg of chlorophyll, 7 ml
Absolute ethanol, 75 ml
Round bottom flask, 250 ml
Evaporator
Heptane, chloroform, and methanol, 3 ml total

[12] P. M. Wood, PhD thesis, Purdue Univ., 1968.

Five thin layer plates made from 5 g of Kieselgel GHR UV 254, 15 g of Kieselgel GHR, and 10 g of silica gel G
Chromatography tank containing chloroform–heptane (40:60)
Chromatography tank containing chloroform–heptane (25:75)
Absolute ethanol or methanol, 10 ml
Cuvettes, 1 ml
Spectrophotometer

An aqueous chloroplast suspension representing 2–3 mg of chlorophyll is added to or injected into 75 ml of absolute ethanol to make an azeotropic mixture of chloroplasts and ethanol, with final ethanol concentration of 92%. This mixture is rapidly evaporated to dryness and extracted three times with a mixture of chloroform, heptane, and methanol (1:1:1). The extraction is performed by allowing the solvent mixture to drip freely from a pipette onto the walls of the round bottom flask while rotating the flask. The final combined volume of the extracts (3 ml) is immediately streaked onto a specially prepared thin layer plate allowing no more than 3 minutes of drying time before it is placed into a chromatography tank. The plate is developed to a height of 15 cm above the origin in a solvent consisting of heptane–chloroform (60:40). The quinone form of PQ A is located as a dark, absorbing spot under uv light against a fluorescent background. It is scraped off and eluted from the silica gel with a known amount of absolute ethanol. To further purify this PQ A fraction, it can be restreaked on another special thin layer plate and developed in chloroform–heptane (25:75) before a spectrophotometric assay.

The quinol fraction is found by a visual scanning of the landmarks of the developed plate. It is generally located between a carotenoid and a pheophytin band, as previously ascertained by the use of a standard. In case of doubt, the quinol can also be located as one of a number of pink spots given by spraying a portion of the plate with the Emmerie–Engel reagent.[5] After scraping a broad band in the quinol region, the quinol is eluted from the silica gel with methanol.

The amount of quinone in the oxidized form present in this particular sample of chloroplasts is determined by a spectrophotometric assay as described (Section VII).

The amount of quinol can be measured spectrophotometrically at 290 nm. Alternatively, it can be oxidized to the quinone form first and then assayed as above.

3. Oxidation of Reduced Plastoquinone A[12]*:*

Reduced PQ A solution in methanol, 4 ml
Potassium ferricyanide, 0.005 *M*, 1 ml

Petroleum ether, 10 ml
Stoppered separatory funnel, 25 ml
Round bottom flask, 25 ml
Evaporator
Special chromatography plate, as in Section VIII, B, 2
Chromatography tank containing chloroform–heptane (65:35)
Cuvettes, 1 ml
Absolute ethanol, 10 ml

The reduced form of plastoquinone A or any other reduced quinone can be oxidized with potassium ferricyanide as follows: To the reduced quinone suspended in 4 ml of methanol in a separatory funnel, 1 ml of 0.005 *M* potassium ferricyanide is added. After gentle swirling for 1 minute, 5 ml of petroleum ether are added. Both phases are saved. The methanolic hypophase is washed once more with 5 ml of petroleum ether before being discarded. The combined petroleum ether extracts containing the quinone in the oxidized form are evaporated to dryness, suspended in a volatile solvent such as heptane or ether, and restreaked on a special thin layer plate developed in heptane–chloroform (35:65). The PQ A band is scraped, eluted with ethanol and used for a spectrophotometric assay as before (Section VII).

When the relative amounts of plastoquinone from the oxidized and reduced fractions are calculated after a spectrophotometric assay, the proportion of the two forms of PQ A in spinach chloroplasts can be determined.

C. Plastoquinone A_{20}

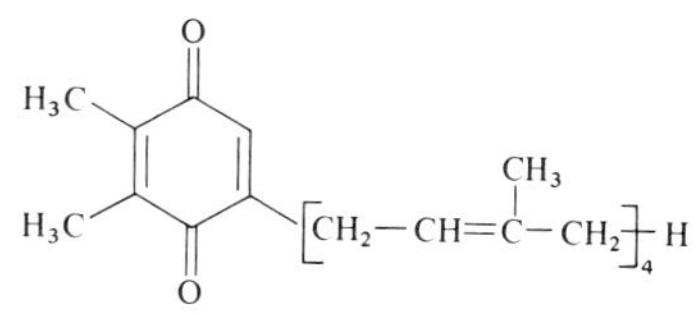

(II)

A natural form of this plastoquinone with a shorter side chain than PQ A_{45} occurs in horse-chestnut leaves.[13] It can be separated from other quinones on thin layer chromatography in heptane–benzene (15:85) where it runs below PQ A_{45}. It can be located on the plate with reduced methylene blue spray. It gives a spectrophotometric assay as PQ A_{45} (Section VII). After reduction with potassium borohydride, the amount of PQ A_{20} can be calculated using a coefficient of 15 or $E^{1\%}_{1\,cm}$ of 342.[13]

[13] H. Eck and A. Trebst, *Z. Naturforsch.* **18B**, 446 (1963).

D. Plastochromanol

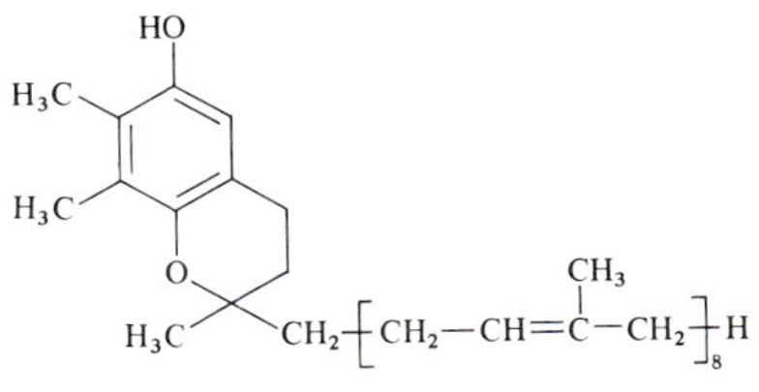

(III)

1. Detection of Plastochromanol

Plastochromanol, the chromanol form related to plastoquinone A_{45}, is found in chloroplast or whole leaf lipid extracts in small amounts (< 0.01 μmole/mg of chlorophyll) except in the milky juice of the rubber tree (*Hevea brasiliensis*) where it occurs in abundance (318 μg/g wet wt).[6] On thin layer chromatograms made with silica gel GHR and developed in chloroform–heptane (80:20), it is found closely associated with the α-tocopherol fraction, mixed with or running slightly below. The two can be separated more easily by running a plate in heptane–benzene (15:85) or from a paraffin-oil impregnated plate by reverse phase thin layer chromatography (Section V, C, 2).

Both compounds can be located on a chromatogram by spraying with the Emmerie–Engel reagent (Section V, G, 1). In addition, purified plastochromanol gives a slate-blue spot with diazotized dianisidine spray (Section V, G, 2). Purified plastochromanol has an absorption maximum at 294 nm in cyclohexane and the $E^{1\%}_{1\,\text{cm}}$ is about 55.[6]

2. Oxidation of Plastochromanol to γ-Hydroxy Plastoquinone by the Gold Chloride Method[14]

Ethanolic solution of plastochromanol (0.1–0.5 mg), 1 ml
Aqueous gold chloride (wt./v; from Baker Chemical Co.), 20%, 0.2 ml
Heptane, 10 ml
Separatory funnel, 125 ml
Distilled water, 30 ml
Round bottom flask, 50 ml
Evaporator
Absolute ethanol, 5 ml

An ethanolic solution of plastochromanol and 0.2 ml of gold chloride

[14] R. A. Dilley and F. L. Crane, *Anal. Biochem.* **5,** 531 (1963).

solution are mixed in a separatory funnel with gentle swirling and are allowed to sit undisturbed in dim light for 20–30 minutes, but no longer. Then 10 ml of heptane are added to the reaction mixture, again with gentle swirling, but taking care not to disturb the layers in order to prevent gold chloride from mixing with heptane, if possible. As soon as layers are clearly separated, the gold chloride–ethanol hypophase is discarded while the heptane phase is washed three times with 10-ml portions of distilled water.

After three washes, the heptane–quinone solution is evaporated to dryness, and the residue is suspended in a known volume of absolute ethanol, such as 5 ml, for a spectrophotometric assay as described for plastoquinone A (Section VII).

The quinone obtained from gold chloride oxidation of plastochromanol has an absorbance maximum at 258 nm. It can be reduced with potassium borohydride, giving rise to a new reduced peak at 290 nm in ethanol.

E. Plastoquinone B[1]

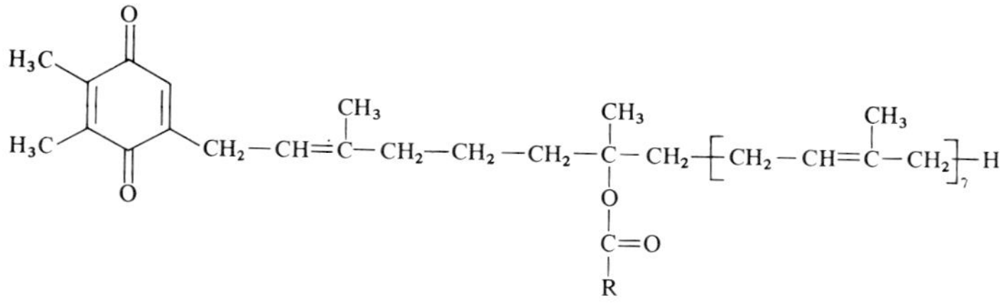

(IV)

A purified form of this quinone can be obtained by restreaking the single diffuse band obtained from the original separation of chloroplast or whole leaf lipid extract in chloroform–heptane (80:20). The solvent system for the development of the second thin layer plate is heptane–benzene (40:60). PQ B runs below PQ A_{45} and PQ A_{20} in this system and can be identified as a paler blue spot with reduced methylene blue spray. If this band is scraped and eluted with ethanol, and the concentrated eluate is applied to a third thin layer plate, which is developed in heptane–benzene (15:85), three different PQ B bands can be observed running between PQ A and PQ C standards.

To demonstrate the existence of six different types of PQ B,[1,15] each of the three previously obtained PQ B bands is spotted on another thin layer plate which is developed in diisopropyl ether–petroleum ether

[15] W. T. Griffiths, J. C. Wallwork, and J. F. Pennock, *Nature* **211**, 1037 (1966).

(12:88). It can be seen after spraying with reduced methylene blue spray that each of the three spots is now double, providing evidence for the existence of six different types of PQ B.

Plastoquinone B can unequivocally be distinguished from PQ A or any other chloroplast quinone by reverse-phase thin layer chromatography (Section V, C, 2). It is the only quinone or series of quinones which run below PQ A on a 5% paraffin oil-impregnated plate developed in acetone–water (95:5 or 93:7), saturated with an additional 0.5 ml of paraffin oil.[1]

F. Plastoquinone C

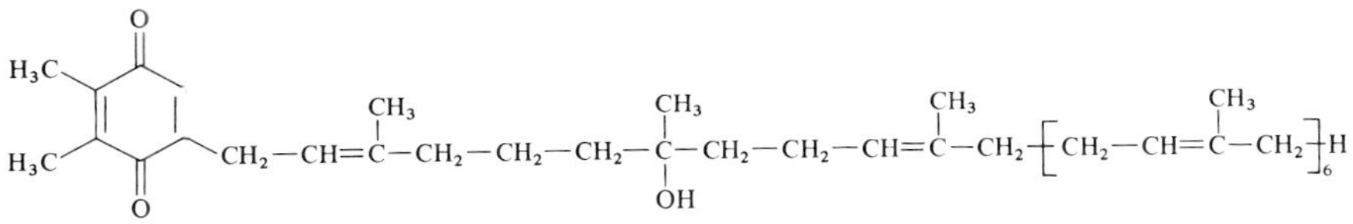

(V)

Sometimes six different bands of PQ C can be seen from chloroplast or whole leaf extracts after development of a thin layer plate in chloroform–heptane (80:20) which lie directly above the topmost chlorophyll band. However, a diagonal array of PQ C spots can be obtained from a

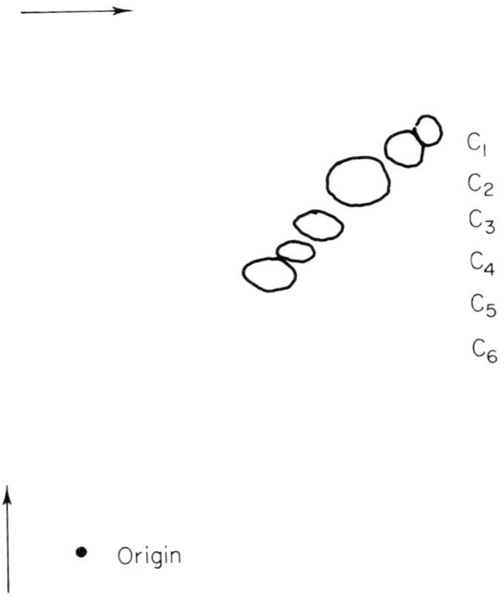

FIG. 4. Six forms of tomato plastoquinone C from high pigment (hp) mutant separated in diisopropylether-benzene (15:85) in both directions.

TABLE VII
RELATIVE AMOUNTS OF THE VARIOUS TYPES OF PLASTOQUINONE C FOUND IN SPINACH AND A TOMATO MUTANT (HP)

Plastoquinone C	Spinach: μmoles of PQ C per mg of chlorophyll	Spinach: Percent of total PQ C	Tomato (hp): μmoles of PQ C per mg of chlorophyll	Tomato (hp): Percent of total PQ C
Type				
PQ C_1	0.001	5	0.009	30
PQ C_2	0.004	20	0.0045	15
PQ C_3	0.008	40	0.003	10
PQ C_4	0.002	10	0.0045	15
PQ C_5	0.001	5	0.0015	5
PQ C_6	0.004	20	0.0075	25

two-dimensional chromatogram[15] where all six types of PQ C are applied as a single spot at the origin. The plate is developed in diisopropyl ether–benzene (15:85) in both directions. After spraying with leucomethylene blue spray, six different PQ C types are visible as blue spots (Fig. 4). The relative amount of each of the six types of PQ C in spinach and a tomato mutant (hp) is shown in Table VII.

G. Vitamin K_1

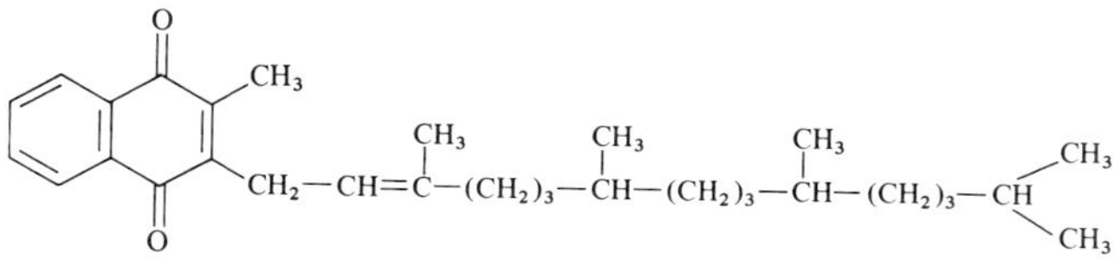

(VI)

Vitamin K_1 can be best identified from its characteristic uv absorption spectrum (Fig. 5) after reduction with potassium borohydride. Complete reduction, followed with a recording spectrophotometer, takes 5–10 minutes in absolute ethanol. This can also be used for a quantitative assay. The millimolar extinction coefficient at 248 nm between oxidized minus reduced forms is 19.

A vitamin K_1 band or spot on a thin layer plate can be located by means of reduced Nile blue spray (Section V, F, 2). This reaction is fast, giving a bluish-black spot immediately upon spraying, whereas reduced

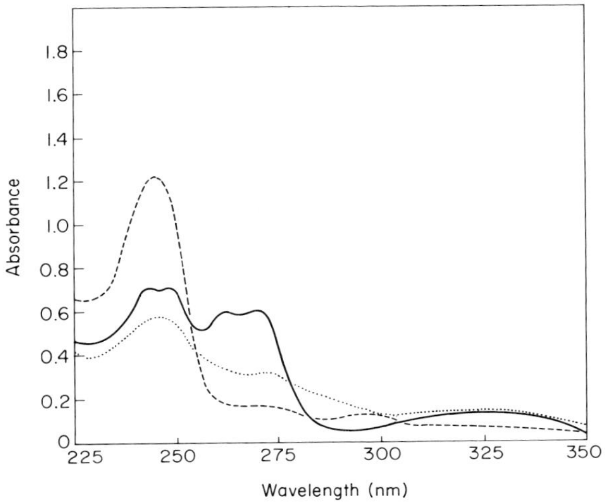

FIG. 5. The uv spectrum of vitamin K_1 in absolute ethanol taken against an ethanol blank. The solid line denotes oxidized form; the dashed line denotes reduced form 6–8 minutes after the addition of potassium borohydride.

methylene blue spray is slow, allowing vitamin K_1 spots to become bluish gradually over 10 minutes–2 hours after spraying.

The *cis* and *trans* forms of vitamin K_1 are separable upon thin layer chromatography in butyl ether–hexane (1:9).[16]

H. Coenzyme Q

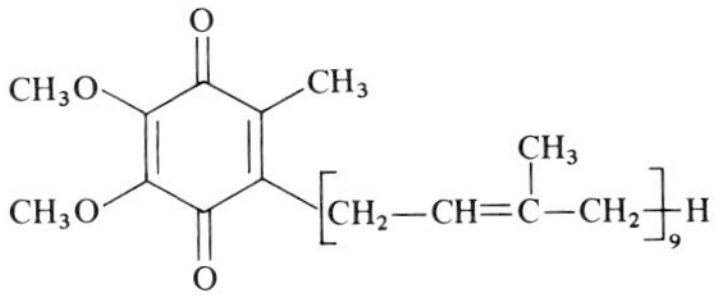

(VII)

The various coenzyme Q homologs found in whole leaf lipid extracts, Q_6–Q_{10}, are located in the PQ B region between PQ A and PQ C on thin layer chromatograms developed in chloroform, chloroform–heptane (80:20), benzene, or heptane–benzene (15:85). They can best be separated from each other on 5% paraffin oil-impregnated reverse phase thin

[16] D. L. Gutnick, P. J. Dunphy, H. Sakamoto, P. G. Phillips, and A. F. Brodie, *Science* **158**, 1469 (1967).

layer plates developed in acetone–water (90:10 or 95:5), saturated with an additional 0.5 ml of paraffin oil or by reverse phase chromatography on silicone-impregnated filter paper according to the method of Lester and Ramasarma.[17] The isolation of various ubiquinones from bacteria and yeasts is described by Crane and Barr.[11]

Coenzyme Q can be located on thin layer chromatograms by spraying with reduced methylene blue spray (Section F, 1). It gives an immediate blue spot with concentrations as low as 1–5 μg.

A quantitative test for conclusive coenzyme Q identification is the spectrophotometric assay with potassium borohydride as for PQ A (Section VII). The oxidized form of coenzyme Q gives maximum absorption at 275 nm, the reduced form at 290 nm in ethanol. The millimolar extinction coefficient for oxidized minus reduced forms at 275 nm is 14 per ml of solution.

Reverse Phase Chromatography of Coenzyme Q Homologs on Silicone-Impregnated Paper[17]

Whatman's 3 mm filter paper, 40 × 56 cm
Dow Corning 550 Silicone Fluid (Dow Corning Corporation, Midland, Michigan), 5% by weight per volume
Chloroform, 95 ml
n-Propanol, 90 ml
Distilled water, 10 ml
Container for dipping paper
Round jar with lid for chromatography

A sheet of filter paper is dipped in a mixture of Dow Corning 550 silicone fluid and chloroform until the paper is uniformly impregnated. It is ready to use as soon as it is dry, but such impregnated papers can also be stored for 6 months or longer.

The samples are spotted near the origin as usual until a yellow spot is visible. Pencil marks on the paper with sample designations don't interfere with separation. After spotting, the paper is rolled into a cylindrical form and stapled without overlap of the outer edges. The cylinder is then taped to the inside of the lid of the chromatography jar with freezer tape and inverted to hang suspended into a previously equilibrated chromatography jar which contains the solvent and a sheet of extra filter paper to keep the atmosphere inside the tank saturated at all times. As soon as the paper cylinder with the spotted samples has saturated with solvent, it falls to the bottom of the jar into the solvent mix-

[17] R. L. Lester and T. Ramasarma, *J. Biol. Chem.* **234,** 672 (1959).

ture, and development of the chromatogram begins. It takes 6–12 hours to develop a single chromatogram to a height of 30 cm by this method.

When development is complete, the paper chromatogram is removed from the jar and dried at least 15 minutes before spraying with reduced methylene blue spray to visualize quinone spots.

I. Tocopherylquinones

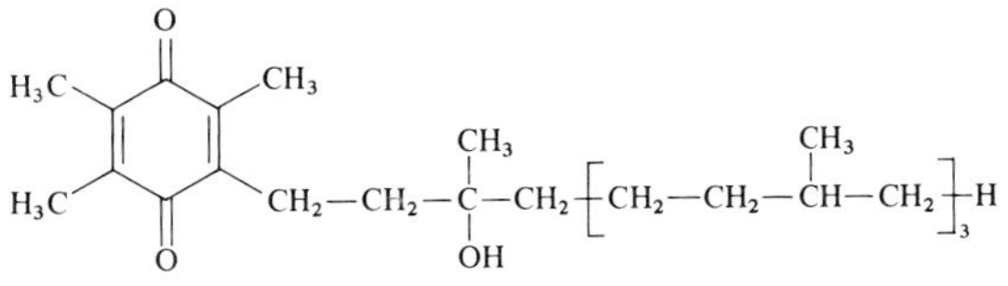

α-Tocopherylquinone

(VIII)

α-Tocopherylquinone (VIII)

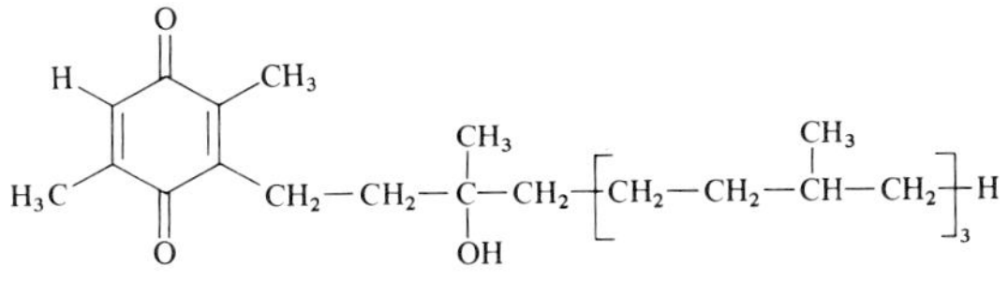

β-Tocopherylquinone

(IX)

β-Tocopherylquinone (IX)

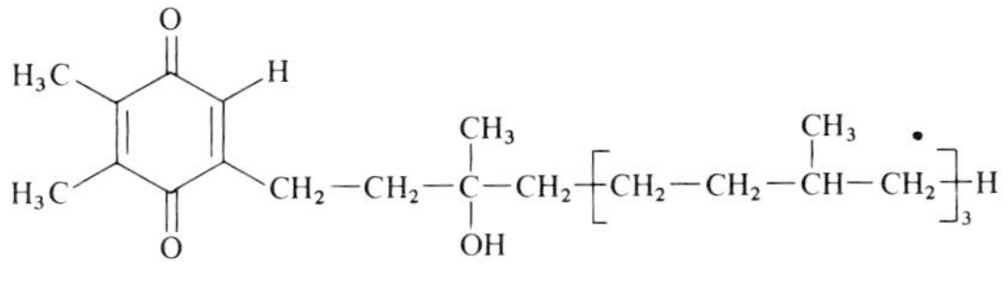

γ-Tocopherylquinone

(X)

γ-Tocopherylquinone (X)

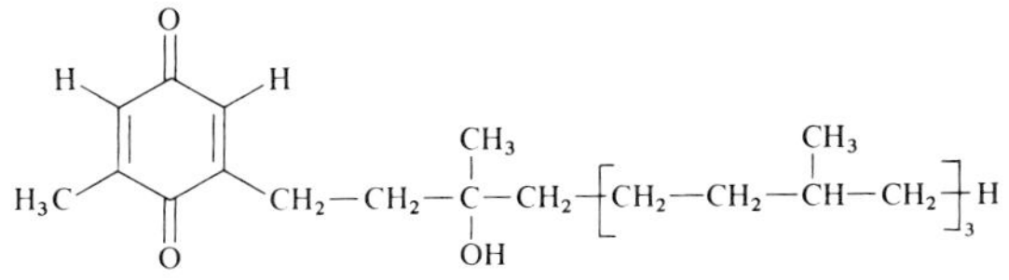

δ-Tocopherylquinone

(XI)

δ-Tocopherylquinone (XI)

α-, β-, γ-, and δ-tocopherylquinones are the most polar group of quinones found in chloroplast or whole leaf lipid extracts. In a general solvent for thin layer chromatography, such as mixtures of chloroform–heptane (80:20), they may be masked by green components of the extracts or yellow xanthophylls unless long thin layer plates (40 × 20 cm) are used. A spectrophotometric assay as for plastoquinone A (Section VII) using a millimolar extinction coefficient of 17.8 between oxidized minus reduced forms at 263 nm for α-TQ, with potassium borohydride as reducing agent, can be performed on partially purified fractions. However, it is best to repurify the lower bands which originally gave a blue spot with reduced methylene blue spray in a more polar solvent system, such as heptane–ethanol (90:10) or benzene–acetone–heptane (96:4:2).[7]

A conclusive test to distinguish between purified plastoquinones and tocopherylquinones is as follows: A thin layer plate made with silica gel GHR is divided into two halves with a vertical line. On the left side, a streak of PQ A or some other plastoquinone is applied, and on the right, a streak of α-TQ or some other tocopherylquinone. The plate is developed in a general solvent system such as chloroform–heptane (80:20). The developed plate is not sprayed, but it is placed into another chromatography tank saturated with iodine vapor for 10 minutes. The quinone bands show up as faint brownish-yellow outlines on a white background. The respective bands can be scraped, eluted with absolute ethanol and used for a spectrophotometric assay as previously described. It can be seen that PQ A or any other plastoquinone type quinone has lost its characteristic absorption spectrum whereas a tocopheryl–quinone gives a characteristic spectrophotometric assay as before and can be reduced with potassium borohydride.

J. The Tocopherols

1. Detection on Thin Layer Plates

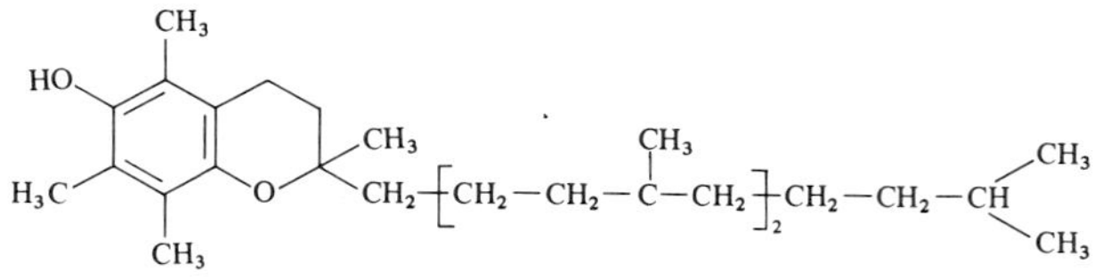

α-Tocopherol

(XII)

α-Tocopherol (XII)

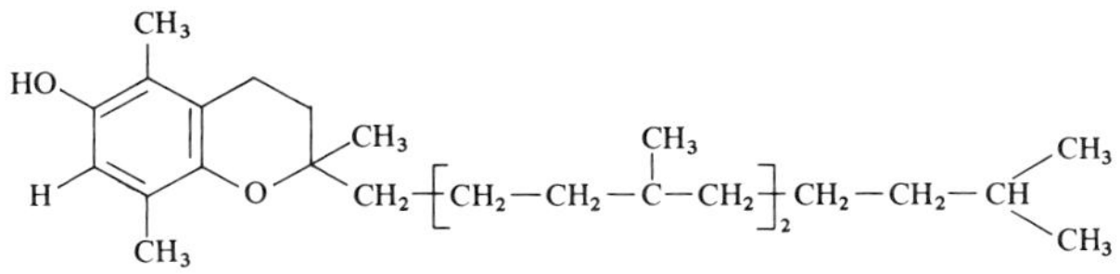

β-Tocopherol

(XIII)

β-Tocopherol (XIII)

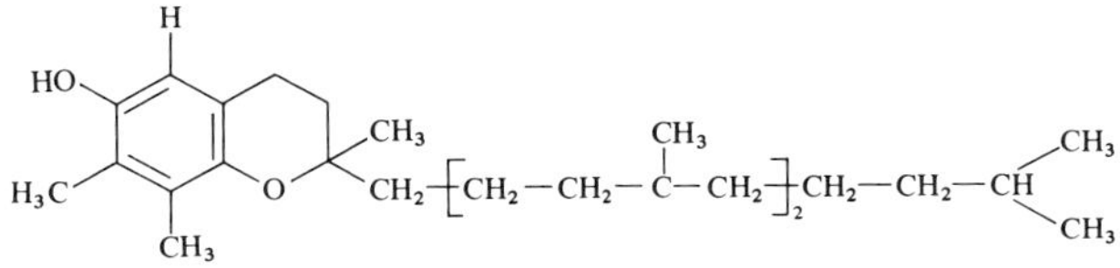

γ-Tocopherol

(XIV)

γ-Tocopherol (XIV)

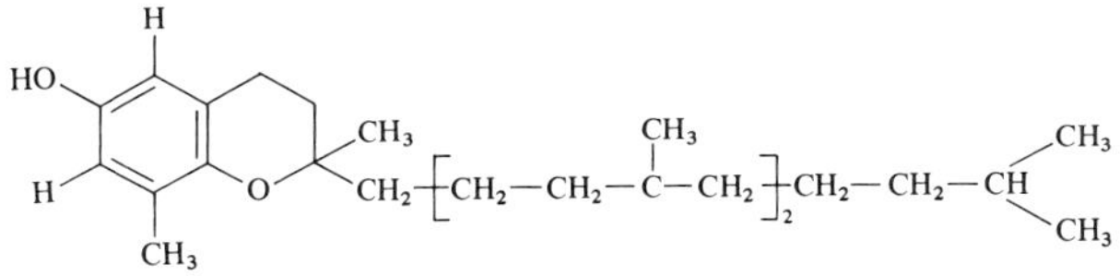

δ-Tocopherol

(XV)

δ-Tocopherol (XV)

α-, β-, γ-, and δ-tocopherols may be found in various chloroplast or whole leaf extracts. In a general solvent system, such as chloroform, benzene, chloroform–heptane (80:20), or heptane–benzene (15:85), they are found in the middle regions of a thin layer chromatogram between coenzyme Q and plastoquinone C. The various tocopherol bands can be visualized by spraying with the Emmerie–Engel reagent (Section V, G, 1) as pink spots on a white background. A spectrophotometric test with this reagent shows that a quantitative conversion occurs with each of the tocopherols which can be measured at 520 nm. However, this reaction is not specific for tocopherols, so that reduced quinones, if present as impurities in the same fractions as the tocopherols, can also give a positive test.

A specific identification of each of the tocopherols is possible when they are oxidized by gold chloride (Section VIII, D, 2) to yield their corresponding tocopherylquinones which can be separated by thin layer chromatography and identified by a spectrophotometric assay as described for α-TQ (Section VII).

All of the tocopherols also give a positive Furter–Meyer test for cyclization.[18]

2. Procedure

Ethanolic solution containing 1–5 mg α-tocopherol (not less than 0.3 mg), 5 ml
Concentrated nitric acid, 1 ml
Two or three glass beads or boiling chips
Reflux condenser which fits into a 50-ml round bottom flask
Boiling water bath
Two 1-ml cuvettes

An ethanolic solution of α-tocopherol, 1 ml of concentrated nitric acid, and boiling chips are combined in a round bottom flask with continuous gentle swirling. Then a reflux condenser is connected to the round bottom flask. It is secured with clips from a ring stand. The whole apparatus is set over a boiling water bath and is allowed to reflux 3–5 minutes until a red color develops. When the solution cools in 1–2 minutes a spectrophotometric test is performed taking a reading of the sample against a blank (83.5% absolute ethanol with 16.5% nitric acid) at 460–470 nm. The results are quantitative and correspond to the amount of tocopherol present in the original sample.

This test is based on the formation of tocopherol red from the individual tocopherols employed in the test. All tocopherols form this oxidation product in nitric acid but of widely differing color intensities.

[18] M. Furter and R. E. Meyer, *Helv. Chim. Acta* **22**, 240 (1939).

[38] Plastocyanin

By Sakae Katoh

Plastocyanin is a copper protein which occurs in a wide variety of photosynthetic organisms including higher plants and algae, but not photosynthetic bacteria.[1-3] It was shown in spinach leaves that plasto-

[1] S. Katoh, *Nature (London)* **186**, 533 (1960).
[2] S. Katoh, I. Suga, I. Shiratori, and A. Takamiya, *Arch. Biochem. Biophys.* **84**, 136 (1961).
[3] J. J. Lightbody and D. W. Krogmann, *Biochim. Biophys. Acta* **131**, 508 (1967).

cyanin is associated with the chloroplasts in the cells.[2] The copper protein is solubilized from the chloroplasts by various treatments of the plastids including acetone treatment, detergent treatment, or sonic oscillation.[4] The purification procedures described below are designed for the preparation of a relatively large amount of highly purified plastocyanin from spinach leaves.[5]

Assay

Principle. The concentration of plastocyanin in an appropriate buffer of pH near neutrality is estimated spectrophotometrically from the absorbancy at 597 nm. The protein must be solubilized from the chloroplasts prior to the assay.

Procedure. The absorbance at 597 nm (optical path, 1 cm) is determined: (1) after the blue color of the oxidized protein is fully developed by the addition of a few crystals of potassium ferricyanide (A_{ox}); and (2) then after the reduction of the protein with a few milligrams of potassium ascorbate (A_{red}). The difference between these readings ($A_{ox} - A_{red}$) corresponds to the amount of plastocyanin. The concentration of plastocyanin is determined[5] using ϵ at 597 nm $= 9.8/\text{m}M/\text{cm}$.

The absorption index, $E_{278/597}$, defined as the ratio of absorbance at 278 nm to that at 597 nm, is a convenient estimate of the approximate purity of the protein.[5] The absorbance at 278 nm has to be determined before addition of any redox reagents, and that at 597 nm after oxidation of the protein with potassium ferricyanide. The absorption index $E_{278/597}$ for the highly purified sample of spinach plastocyanin ranges from 0.80 to 1.00.

Purification Procedures[5]

Unless otherwise specified, the procedures are carried out at room temperature during the early stages of purification. All the chromatographic manipulations are, however, performed at 4°.

Step 1. Acetone Treatment of Spinach Leaves. Washed spinach leaves are freed from stems, cut into pieces, and weighed. Four or five 1.5-kg lots of leaves are usually handled at one time. Each lot is minced and ground for 40 minutes in a mechanical mortar with 750 g of washed quartz sand and 200 g of solid Na_2HPO_4. After grinding, the homogenate is chilled to 0°, and 4.5 liters of cold acetone (—15°) is added with stirring. After standing for 10 minutes, the suspension is centrifuged at 3000 *g* for 10 minutes. The supernatant solution is decanted, and the

[4] S. Katoh and A. San Pietro, *in* "Biochemistry of Copper" (J. Peisach, P. Aisen, and W. E. Blumberg, eds.), p. 407. Academic Press, New York, 1966.

[5] S. Katoh, I. Shiratori, and A. Takamiya, *J. Biochem.* **51,** 32 (1962).

pale green precipitate is spread on plates and dried with a fan to remove acetone completely.

Step 2. Extraction of Plastocyanin. The greenish powder prepared from 1.5 kg of leaves is suspended in 3 liters of deionized water and allowed to stand for several hours with occasional stirring. The suspension is centrifuged at 3000 *g* for 15 minutes, and the sediment is discarded.

Step 3. Fractionation with Ammonium Sulfate. Three hundred and eighty grams of solid ammonium sulfate is dissolved in 1 liter of the dark brown supernatant liquid from step 2, and the precipitate formed is removed by centrifugation. An equal amount of ammonium sulfate is further added to the supernatant fraction and the heavy precipitate formed is collected by centrifugation and dissolved in 1 liter of 10 m*M* phosphate buffer, pH 7.0. The fractional precipitation of the protein with ammonium sulfate is repeated. The final precipitate obtained is dissolved in 100 ml of 10 m*M* phosphate, pH 7.0, and dialyzed against a large volume of the same buffer for 24 hours.

Step 4. First Chromatography on DEAE–Cellulose Column. The dialyzate is placed onto a column of DEAE–cellulose (3×10 cm) which has been previously equilibrated with 10 m*M* phosphate buffer, pH 7.0. In order to oxidize completely the protein, a few milligrams of potassium ferricyanide is added to the dialyzate prior to application to the column. The charged column is washed with 2 liters of 10 m*M* phosphate buffer, pH 7.0, and the protein is then eluted with 50 m*M* phosphate buffer, pH 7. The location of the oxidized protein in the column can be visually followed by its blue color (usually appearing greenish in tone on the brown background), although the color fades gradually during the progress of elution, probably because of a slow reduction of the protein by reducing contaminants in the preparation. The presence of plastocyanin in the eluates is, therefore, confirmed by adding a drop of ferricyanide solution (1 m*M*). This is especially important for testing the eluate fractions immediately following the blue zone, since these fractions may contain the reduced form of the protein which descends the column more slowly than the oxidized form.

Step 5. Fractionation with Ammonium Sulfate. To 100 ml of the combined fractions containing plastocyanin is added 50 g of solid ammonium sulfate. The supernatant solution obtained by centrifugation is brought to saturation with ammonium sulfate by adding 25 g of the salt. After standing for 1 hour the precipitate is collected by centrifugation, dissolved in a small amount of 10 m*M* of phosphate buffer, pH 7.0, and dialyzed against a large volume of the buffer.

Step 6. Second Chromatography with DEAE–Cellulose Column. The dialyzate is absorbed on a DEAE–cellulose column (2 × 20 cm). The column is washed with 2 mM phosphate buffer, pH 7.0, and the protein is eluted with 50 mM buffer as described above.

Further purification of plastocyanin, if necessary, is achieved by repeating the chromatographic procedure. In this case, the ionic strength of the eluate is lowered by adding 2 volumes of deionized water since prolonged dialysis of the purified protein causes a gradual release of copper from the protein. The diluted sample is then adsorbed on a short column of DEAE–cellulose (1 cm in height), and the plastocyanin-charged adsorbent is transferred to the top of a larger column, followed by elution with 50 mM of phosphate buffer, pH 7.0. Concentration of the protein is carried out by a similar procedure; dilution with deionized water, adsorption on a short column of DEAE–cellulose and elution with 0.2 M phosphate buffer, pH 7.0, since it is difficult to precipitate the purified protein from a dilute solution even with a saturating amount of ammonium sulfate.

A summary of the purification and yields of plastocyanin from spinach leaves (6 kg fresh weight) is shown in the table. The concentrations and the absorption indices of plastocyanin in the original extract and in the precipitates from the first and second fractionation with ammonium sulfate are not given in the table since the colored contaminants, present in large quantities in these fractions, are apt to cause significant uncertainty in the spectrophotometric determination. In the experiments shown in the table, the eluates from the second and third columns were collected in 10-ml aliquots; the extent of variations of absorbances at 278 and 597 nm and of the absorption index among the aliquots is indicated in the table. Only those fractions with relatively high levels of

Yields and Purity at Each Step of Purification of Spinach Plastocyanin

Fractions	Volume (ml)	Absorbance at 597 nm	Absorbance at 278 nm	Absorption index, $E_{278/597}$	Yield (μmoles plastocyanin)
Eluate, 1st chromatography	350	0.135	0.861	7.02	4.8
Precipitate between 66 and 100% saturation with ammonium sulfate	210	0.205	0.960	4.68	4.4
Eluates, 2nd chromatography	120	0.126–0.981	0.212–2.15	1.7–2.8	3.6
Eluates, 3rd chromatography	199	0.054–0.135	0.046–0.143	0.80–1.12	2.45

absorption indices were combined and subjected to subsequent purification. This is the reason for the apparent decrease in chromatographic recovery as seen in the last column of the table. The actual recovery of the protein in the chromatographic procedures is reasonably good and amounts to more than 90% of the original material applied.

Modified Procedures

Depending upon the laboratory equipment available, and the quantity of the copper protein required, the procedures for extraction and purification may be modified as follows.

Modified Procedure for Acetone Treatment of Leaves. When a large mechanical mortar is not available, spinach leaves, freed from stems and cut into pieces, are homogenized in a Waring Blendor with cold acetone (—15°) for 3 minutes. Usually 100 g of spinach leaves are treated with 600 ml of acetone. After the homogenate has been allowed to stand at 0° for 15 minutes, the supernatant solution is carefully decanted and the loosely sedimented material is centrifuged at 3000 *g* for 5 minutes. The precipitate formed is collected, dried, and subjected to the extraction of the protein as described above in steps 1 and 2.

Simplified Procedure for the Preparation of Small Amount of Plastocyanin. Chloroplasts or large fragments of the chloroplasts are prepared from spinach leaves according to the method described by Arnon. To 100 ml of the chloroplast suspension containing 1–5 mg of chlorophyll per milliliter, 900 ml of cold acetone (—15°) is added slowly and with continuous stirring. The precipitate is collected rapidly by centrifugation, dried under reduced pressure, and then suspended in 100 ml of 10 m*M* phosphate buffer, pH 7.0. After standing for 1 hour with occasional stirring, the suspension is centrifuged at 6000 *g* for 10 minutes. The chloroplast extract thus obtained contains much less impurities, as compared to the extract prepared from the leaves, and can be directly applied to the DEAE–cellulose column without preliminary fractionation with ammonium sulfate. Plastocyanin extracted from the acetone-treated chloroplasts usually amounts to 1 μmole of protein per 600–1000 μmoles of chlorophyll.[2] Because of a limited yield of the chloroplasts, however, the yield of the protein on a basis of the amount of the starting material (spinach leaves) is considerably lower in this simplified procedure than in the method described above.

Properties

Stability. Spinach plastocyanin is stable and can be stored at —20° for 1 year without any appreciable change in the spectral properties and the solubility characteristics of the protein. Incubation of the protein for

5 minutes at temperatures up to 60° or at pH's between 5.0 and 9.5 causes no change in its absorption spectrum.[6] However, dialysis of plastocyanin, in a highly purified state, against deionized water may lead to a gradual loss of copper from the protein.

Spectral Properties.[5] The oxidized protein appears blue and exhibits absorption bands at 460, 597 and 770 nm. The reduced protein has no absorption in the visible and far-red region. The absorption spectrum in the ultraviolet region of the protein shows, besides the main absorption peak at 278 nm, vibrational fine structure bands at 253, 259, 265, and 284 nm.

Physicochemical Properties.[5] The molecular weight of spinach plastocyanin is estimated to be about 21,000. The protein is considerably acidic, with an isoelectric point of less than pH 4.0. The oxidized protein is readily reduced by ascorbate, hydroquinone, and other mild or strong reductants. The reduced protein is oxidized by ferricyanide, but not by molecular oxygen. The normal oxidation reduction potential is 370 mV between pH 5.4 and 9.9 and increases at a rate of 60 mV/pH at pH's lower than pH 5.4.

Chemical Composition. The protein contains, per molecule, about 200 amino acid residues, a small amount of carbohydrate and two atoms of copper.[5] Copper is released reversibly on incubating the protein at pH 2.0.[6]

Enzymatic Activity. The protein exhibits no enzymatic activity. The role of plastocyanin is considered to be that of an electron carrier between the two photosystems of the photosynthetic electron transfer chain, the reaction site being localized at the electron donating side of photosystem I.[4-7]

[6] S. Katoh and A. Takamiya, *J. Biochem.* **55,** 378 (1964).
[7] D. S. Gorman and R. P. Levine, *Plant Physiol.* **41,** 1648 (1966).

[39] Ferredoxins from Photosynthetic Bacteria, Algae, and Higher Plants

By Bob B. Buchanan and Daniel I. Arnon

The name ferredoxin was introduced in 1962 by Mortenson, Valentine, and Carnahan[1] to denote a nonheme, iron-containing protein which they

[1] L. E. Mortenson, R. C. Valentine, and J. E. Carnahan, *Biochem. Biophys. Res. Commun.* **7,** 448 (1962).

isolated from *Clostridium pasteurianum*. In *C. pasteurianum* and in other nonphotosynthetic anaerobic bacteria where ferredoxin was later found,[2-4] it appeared to function as an electron carrier either between molecular hydrogen (activated by hydrogenase) and different electron acceptors or in the breakdown of compounds which, like pyruvate, generate strong reducing power.

The isolation of ferredoxin from *C. pasteurianum*—an anaerobic bacterium devoid of chlorophyll and normally living in the soil at a depth to which sunlight does not penetrate—did not at first concern photosynthesis. A connection between bacterial ferredoxin and photosynthesis was established, also in 1962, when Tagawa and Arnon[5] crystallized *C. pasteurianum* ferredoxin and found it to mediate the photoreduction of NADP by spinach chloroplasts. In this reaction *Clostridium* ferredoxin replaced a native chloroplast protein that was earlier known as methemoglobin-reducing factor,[6] TPN-reducing factor,[7] or photosynthetic pyridine nucleotide reductase.[8] The elucidation of the mechanism of NADP reduction by chloroplasts[9,10] and the newly found similarity in chemical properties between bacterial ferredoxin and the chloroplast protein prompted the renaming of the chloroplast protein ferredoxin.[5] Ferredoxins are now known to be present in every type of photosynthetic cell including photosynthetic bacteria and blue-green algae, which do not contain chloroplasts.

The role of ferredoxin in photosynthetic cells is not limited to NADP reduction. The photoreduction of ferredoxin in chloroplasts was found to be stoichiometrically coupled to oxygen evolution and ATP formation in noncyclic photophosphorylation.[11] Ferredoxin was also found to catalyze cyclic photophosphorylation in a manner which points to its being the

[2] R. C. Valentine, R. L. Jackson, and R. S. Wolfe, *Biochem. Biophys. Res. Commun.* **7**, 453 (1962).

[3] H. R. Whiteley and C. A. Woolfolk, *Biochem. Biophys. Res. Commun.* **9**, 517 (1962).

[4] W. Lovenberg, B. B. Buchanan, and J. C. Rabinowitz, *J. Biol. Chem.* **238**, 3899 (1963).

[5] K. Tagawa and D. I. Arnon, *Nature (London)* **195**, 537 (1962).

[6] H. E. Davenport, R. Hill, and F. R. Whatley, *Proc. Roy. Soc. Ser. B* **139**, 346 (1952).

[7] D. I. Arnon, F. R. Whatley, and M. B. Allen, *Nature (London)* **180**, 182, 1325 (1957).

[8] A San Pietro and H. M. Lang, *J. Biol. Chem.* **231**, 211 (1958).

[9] D. I. Arnon, *Science* **149**, 1460 (1965).

[10] B. B. Buchanan and D. I. Arnon, *Advan. Enzymol.* **33**, 120 (1970).

[11] D. I. Arnon, H. Y. Tsujimoto, and B. D. McSwain, *Proc. Nat. Acad. Sci. U.S.* **51**, 1274 (1964).

physiological catalyst of that process in chloroplasts.[12-14] Other functions of ferredoxin in photosynthesis for which there is now evidence include: (a) activation of the chloroplast fructose diphosphatase[15,16] (a key enzyme in carbohydrate synthesis); (b) serving as a direct reductant in carbon assimilation via the reductive carboxylic acid cycle in photosynthetic bacteria[17,18]; and (c) serving as an electron carrier in the photoreduction of nicotinamide adenine dinucleotides by subcellular particles of photosynthetic bacteria.[19]

Ferredoxins are iron–sulfur proteins. Their definitive characterization and classification as a distinct group of cellular electron carriers must await the isolation of a common prosthetic group in ferredoxins of different species. Pending such isolation of a common prosthetic group, which might or might not distinguish ferredoxins from other iron–sulfur proteins, it seems useful to consider them provisionally as a distinct group of electron carriers that function on the hydrogen side of nicotinamide adenine dinucleotides. The properties which now distinguish ferredoxins from other iron–sulfur proteins include a characteristic composition and sequence of amino acids, a low molecular weight, an oxidation–reduction potential close to that of hydrogen gas, and the fact that they are, at least in part, functionally interchangeable in the photoreduction of NADP by isolated chloroplasts. The ability to catalyze photoreduction of NADP by washed chloroplasts is provisionally a useful test in identifying ferredoxins. All ferredoxins isolated so far, including the newly identified *Azotobacter* ferredoxin,[20] exhibit this property. By contrast, the replaceability of different ferredoxins in other enzymatic reactions is less consistent.

Ferredoxins were divided earlier[9] into a bacterial and a plant (chloroplast) type. The recent recognition of a third, *Azotobacter* type, led us to introduce a flexible system of nomenclature for classifying members

[12] K. Tagawa, H. Y. Tsujimoto, and D. I. Arnon, *Proc. Nat. Acad. Sci. U.S.* **49,** 567 (1963).

[13] D. I. Arnon, H. Y. Tsujimoto, and B. D. McSwain, *Nature (London)* **214,** 562 (1967).

[14] D. I. Arnon, *Naturwissenschaften* **56,** 295 (1969).

[15] B. B. Buchanan, P. P. Kalberer, and D. I. Arnon, *Biochem. Biophys. Res. Commun.* **29,** 74 (1967).

[16] B. B. Buchanan, P. P. Kalberer, and D. I. Arnon, *Fed. Proc., Fed. Amer. Soc. Exp. Biol.* **27,** 344 (1968).

[17] M. C. W. Evans, B. B. Buchanan, and D. I. Arnon, *Proc. Nat. Acad. Sci. U.S.* **55,** 928 (1966).

[18] B. B. Buchanan, M. C. W. Evans, and D. I. Arnon, *Arch. Mikrobiol.* **59,** 32 (1967).

[19] B. B. Buchanan and M. C. W. Evans, *Biochim. Biophys. Acta* **180,** 123 (1969).

[20] D. C. Yoch, J. R. Benemann, R. C. Valentine, and D. I. Arnon, *Proc. Nat. Acad. Sci. U.S.* **64,** 1404 (1969).

A System of Nomenclature for the Ferredoxins

Class	Occurrence	Representative type	Characteristics
Ferredoxin *a*	Fermentative (nonphotosynthetic) and photosynthetic green anaerobic bacteria	*Clostridium pasteurianum*	Absorption maxima at 390, 280 nm, shoulder at 300 nm; molecular weight, 6000; contains more than two iron and sulfide groups per molecule
Ferredoxin a_1	Photosynthetic purple sulfur anaerobic bacteria	*Chromatium* strain D	Absorption maxima at 385, 280 nm; shoulder at 300 nm; molecular weight, 10,000; contains seven to eight iron and sulfide groups per molecule
Ferredoxin *b*	Algae and higher plants	Spinach chloroplast	Absorption maxima at 465, 425, 325, 280 nm; molecular weight, 12,000; contains two iron and sulfide groups per molecule
Ferredoxin *c*	Aerobic nitrogen-fixing bacteria	*Azotobacter vinelandii*	Absorption maximum at 400 nm; provisional molecular weight, 20,000; six iron and sulfide groups per molecule

of this growing family of proteins. We have proposed, by analogy with cytochromes, to designate each type of ferredoxin (regardless of the kind of cells in which it occurs) by a letter, modified by subscript number to denote distinct properties within a given type.[10] We have designated clostridial-type ferredoxins as ferredoxin *a*; *Chromatium* ferredoxin as a variant within this type becomes ferredoxin a_1. Chloroplast or plant-type ferredoxin is designated ferredoxin *b*; and the newest addition, the *Azotobacter* type, becomes ferredoxin *c* since, though not yet fully characterized, it is distinctly different from ferredoxins *a* and *b*. The respective properties of ferredoxins *a*, *b*, and *c* are summarized in the table.

In this paper we describe procedures for isolation in pure form of ferredoxin *a* from the photosynthetic green bacterium *Chlorobium thiosulfatophilum;* ferredoxin a_1 from the photosynthetic purple sulfur bacterium *Chromatium;* and ferredoxin *b* from higher plants (spinach and alfalfa leaves) and algae (blue-green, green, red, and brown types). A method recently devised for isolation of ferredoxin from the photosynthetic purple nonsulfur bacterium, *Rhodospirillum rubrum* also is described.

Several reviews describing the role of these diverse types of ferredoxins in photosynthesis have been published.[9,10,21,22]

Ferredoxin Assay

Assay Method

Principle. The method is based on the requirement of ferredoxin for the photochemical reduction of NADP by isolated spinach chloroplasts using reduced 2,6-dichlorophenolindophenol (DPIP) as the electron donor [Eqs. (1) and (2)].

$$2\ \mathrm{Fd_{ox}} + \mathrm{DPIP_{red}} \xrightarrow[\text{chloroplasts}]{\text{light}} 2\ \mathrm{Fd_{red}} + \mathrm{DPIP_{ox}} \tag{1}$$

$$2\ \mathrm{Fd_{red}} + \mathrm{NADP} \xrightarrow[\text{reductase}]{\text{Fd:NADP}} 2\ \mathrm{Fd_{ox}} + \mathrm{NADPH_2} \tag{2}$$

Fd:NADP reductase, catalyzing reaction (2) independently of light, is bound to chloroplast fragments (prepared as described below) and need not be added separately.

Under the given experimental conditions, the reduction of NADP is linear with time and proportional to the concentration of ferredoxin. As already mentioned, all ferredoxins isolated so far are active in this assay.

[21] A. San Pietro and C. C. Black, *Annu. Rev. Plant Physiol.* **16,** 155 (1965).
[22] D. I. Arnon, *Experientia* **22,** 273 (1966).

Reagents

Tris· or Tricine·HCl buffer, 1 *M*, pH 7.8
2,6-Dichlorophenolindophenol, 2 m*M*
Sodium ascorbate, 0.2 *M*
Washed spinach chloroplasts (equivalent to 0.5 mg of chlorophyll/ml)
Ferredoxin (up to 50 μg for the plant ferredoxins and 100 μg for the bacterial ferredoxins)
NADP, 10 m*M*

Procedure. The reaction is carried out in a cuvette (3-ml capacity, 1-cm light path) containing 2.4 ml of H_2O and 0.1 ml each of buffer, DPIP, sodium ascorbate, chloroplasts, ferredoxin, and NADP. Since the reduction of ferredoxin by chloroplasts is light-dependent, strong illumination is to be avoided during the preliminary steps of the assay.

Cuvettes may be illuminated at room temperature with incandescent light filtered through a water filter (about 5 cm thick) to protect from excess heat. Reduced NADP is determined by the increase in absorbance at 340 nm and is measured every 3 minutes in a conventional spectrophotometer. A cuvette containing the complete reaction mixture but kept in the dark serves as a blank. More elaborate assay methods involving continuous illumination and a recording spectrophotometer have been described,[23] but offer no advantage other than convenience.

Preparation of Ferredoxins

Preparative Materials

Tris(hydroxymethyl)aminomethane(Tris)·HCl buffer, pH 7.3. A stock 1 *M* solution of Tris·HCl is adjusted to pH 7.3 with HCl and is diluted to the desired concentration as needed. (A 0.15 *M* Tris·HCl buffer solution at pH 7.3 contains 0.12 *M* Cl from the HCl used in adjusting the pH.) For the alfalfa and *Scenedesmus* ferredoxin procedures, the stock Tris·HCl buffer solution is adjusted to pH 7.5 and 8.0, respectively. For the *Chromatium* and *R. rubrum* ferredoxin procedures, a stock 1 *M* potassium phosphate solution, pH 7.3, is prepared. For the *Chlorobium thiosulfatophilum* ferredoxin procedure a stock 1 *M* potassium phosphate solution, pH 6.5, is prepared in addition to the above Tris·HCl buffer at pH 7.3.

DEAE–cellulose (diethylaminoethyl cellulose; Schleicher and Schuell, Keene, New Hampshire). The resin is soaked overnight in 0.5 *M* NaCl,

[23] B. D. McSwain and D. I. Arnon, *Proc. Nat. Acad. Sci.* **61**, 989 (1968).

poured into a column, and charged as follows. The column is washed first with an acid–salt solution (0.01 N HCl, 1 M NaCl) until a heavy brown band is eluted (this step is completed within an hour to avoid hydrolysis of the cellulose). The column is then washed with a base–salt solution (0.5 M NaOH, 0.5 M NaCl) to remove additional brown material (this step is completed within 2 hours). The column is finally washed with distilled water until a neutral or slightly acidic pH is achieved. (The same procedure is satisfactory for regenerating used DEAE–cellulose.) The washed resin may be stored indefinitely.

Unless indicated otherwise, all steps of the ferredoxin preparation procedures described below are carried out at 4°. All dialysis steps are carried out with constant stirring.

Preparation of Chloroplasts

Washed spinach chloroplast particles (P_1S_2 or C_1S_2) are prepared in isotonic sodium chloride according to Whatley and Arnon[24] or in isotonic sorbitol media according to Kalberer *et al.*[25] The chloroplast preparation retains sufficient activity for the ferredoxin assay after 1 week at 4°.

Growth of Microorganisms

Nostoc sp. is cultured with or without combined nitrogen as described previously for *Anabaena cylindrica*.[26] *Scenedesmus* sp. may also be cultured in this medium in the presence of 10 mM KNO_3 as nitrogen source. *Porphyridium cruentum* and *Botrydiopsis alpina* are cultured as described by Jones *et al.*[27] *Chromatium* strain D is grown on the malate medium as described by Arnon *et al.*[28]; *Chlorobium thiosulfatophilum* (strain Tassajara) on Pfennig's medium as previously described[29]; and *Rhodospirillum rubrum* on medium S as given by Lascelles.[30]

Ferredoxin from Spinach Leaves

Spinach ferredoxin is purified and crystallized by a procedure developed in this laboratory by K. Tagawa and R. K. Chain, similar to that described originally for *Clostridium* ferredoxin by Tagawa and Arnon.[5] The procedure includes: (1) extraction of an acetone precipitate of a

[24] F. R. Whatley and D. I. Arnon, this series, Vol. VI, p. 308.

[25] P. P. Kalberer, B. B. Buchanan, and D. I. Arnon, *Proc. Nat. Acad. Sci. U.S.* **57,** 1542 (1967).

[26] M. B. Allen and D. I. Arnon, *Plant Physiol.* **30,** 366 (1955).

[27] R. F. Jones, H. L. Speer, and W. Kury, *Physiol. Plant.* **16,** 636 (1963).

[28] D. I. Arnon, V. S. R. Das, and J. D. Anderson, *Studies on Microalgae and Photosynthetic Bacteria* (special issue, *Plant Cell Physiol.*) p. 529 (1963).

[29] M. C. W. Evans and B. B. Buchanan, *Proc. Nat. Acad. Sci. U.S.* **53,** 1420 (1965).

[30] J. Lascelles, *Biochem. J.* (*London*) **62,** 78 (1956).

leaf homogenate; (2) adsorption of crude ferredoxin on a DEAE–cellulose bed and elution with Tris·NaCl buffer; (3) chromatography on a DEAE–cellulose column; (4) concentration of the main effluent to give a concentration of ferredoxin not less than 1%; and, if desired (5) crystallization with ammonium sulfate. This procedure may be used to process a minimum of 1 kg of spinach leaves, but batches of 10–50 kg of leaves usually are used in our laboratory. About 20 mg of pure ferredoxin is obtained per kilogram of spinach leaves. This procedure has been used also for preparing pure ferredoxins from *Amaranthus edulis* ("pigweed") and *Polystichum munitum* ("sword fern") (P. Schürmann and B. B. Buchanan, unpublished results). A similar procedure giving about the same yield of ferredoxin has been recently reported.[31]

Step 1. Preparation of Acetone Extract. This step is a modification of the procedure described by San Pietro and Lang.[32] Spinach leaves are freed of stems, washed, packed in plastic bags (1 kg/bag), and stored overnight at —20°. The frozen leaves, crushed by hand, are blended for 3 minutes (high-power setting) with cold distilled water (1500 ml/1 kg) in a 1-gallon Waring Blendor (Model CB-5). The resulting slurry is adjusted to pH 7.3–7.5 with 1 *M* Tris·HCl buffer and freed of solid residue either by centrifugation (3 minutes, 1000 *g*) or passage through cheesecloth bags of double thickness. Acetone (precooled to —20°) is added to the supernatant fraction to give a final concentration of 75%. The 75% acetone solution is left to stand 1 hour at —20°. The green supernatant layer is then decanted and discarded, and the acetone precipitate is concentrated by centrifugation (first for 2 minutes at 1000 *g* to collect the precipitate and then for 10 minutes at 25,000 *g* to remove excess acetone from the combined precipitate fraction). The gray precipitate is immediately dried under a stream of cool air (supplied by a hair dryer) until excess acetone is removed. While drying, the precipitate is continuously kneaded with a spatula, and after 20–30 minutes, the precipitate darkens and assumes a powdery consistency. (To ensure maximum recovery of ferredoxin, it is important to begin drying within 30 minutes after collecting the 75% acetone precipitate.) The gray powder, free of acetone, is suspended in 0.15 *M* Tris·HCl, pH 7.3 (50 ml of buffer per kilogram of leaves) and is stirred mechanically until all the lumps have disappeared. The turbid protein suspension is dialyzed overnight against at least 10 volumes of 1 m*M* Tris·HCl buffer, pH 7.3. The following day, the precipitated material is removed by centrifugation (30 minutes, 35000 *g*) and discarded. The clear, dark-red supernatant solution contains the ferredoxin.

[31] M. T. Borchert and J. S. C. Wessels, *Biochim. Biophys. Acta* **197**, 78 (1970).
[32] A. San Pietro and H. M. Lang, *J. Biol. Chem.* **231**, 211 (1958).

Step 2. Adsorption of Ferredoxin on DEAE–Cellulose. Solid NaCl is added to the supernatant solution to give a final concentration of 0.2 *M*. The solution is then passed through a DEAE–cellulose bed (3 cm diameter × 2 cm high per kilogram of leaves) equilibrated beforehand with a solution containing 0.15 *M* Tris·HCl buffer, pH 7.3, and 80 m*M* NaCl [(Cl^-) = 0.2 *M*]. The reddish ferredoxin accumulates just below the dark-brown impurities adsorbed at the top of the bed. (The passed solution is yellowish brown and contains the flavoprotein enzyme ferredoxin–NADP reductase[33] and plastocyanin,[34] which may be further purified.) The charged DEAE–cellulose bed is successively washed with chloride solutions of increasing chloride concentration [0.15 *M* Tris·HCl buffer, pH 7.3, containing 0.08, 0.11, and 0.14 *M* NaCl; corresponding (Cl^-) = 0.2 *M*, 0.23 *M*, and 0.26 *M*] until the reddish ferredoxin band descends to within 0.5 cm from the bottom of the bed. Ferredoxin is eluted from the bed with a solution containing 0.55 *M* NaCl and 0.30 *M* Tris·HCl buffer, pH 7.3 [(Cl^-) = 0.8 *M*]. About 5 ml of eluate is obtained per kilogram of spinach leaves.

Step 3. Chromatography of Ferredoxin on DEAE–Cellulose. The eluate is diluted with 2.5 volumes of distilled water and placed on a DEAE–cellulose column (4 cm diameter × 50 cm high per 5 kg of leaves) which has been equilibrated beforehand and later developed with a chloride·buffer mixture of 0.15 *M* Tris·HCl, pH 7.3, and 0.25 *M* NaCl; [(Cl^-) = 0.37 *M*]. The reddish ferredoxin band will be displaced by passing through the column a volume of Tris·chloride buffer about 3 times greater than the volume of the column. The 420/276 nm absorbance ratio and the pure fractions (showing a 420/276 nm ratio of 0.45 or greater) are combined and treated as described below. In certain cases it may be necessary to repeat the DEAE–cellulose chromatography step for a portion or for the entire preparation.

Step 4. Concentration of Ferredoxin on DEAE–Cellulose. After recovery from the column, the ferredoxin effluent is diluted with 4 volumes of distilled water and is then adsorbed on a DEAE–cellulose bed (3 cm diameter × 2 cm high per 5 kg of leaves) equilibrated beforehand with 0.1 *M* Tris·HCl buffer [(Cl^-) = 80 m*M*]. Ferredoxin is eluted from the bed with a solution containing 0.3 *M* Tris·HCl buffer and 0.55 *M* NaCl, pH 7.3 [(Cl^-) = 0.8 *M*]. Ferredoxin concentration in the main effluent fraction should be not less than 1% at this step (20–30 mg/ml is possible). The concentrated pure ferredoxin, when frozen and stored under nitrogen, is stable for several months. Alternatively, crystallization from the concentrated eluate may be carried out as described below.

[33] M. Shin, K. Tagawa, and D. I. Arnon, *Biochem. Z.* **338**, 84 (1963). See also this volume [40] and [41].

[34] S. Katoh, I. Shiratori, and A. Takamiya, *J. Biochem. (Tokyo)* **51**, 32 (1962).

Step 5. Crystallization of Ferredoxin. Ammonium sulfate, 0.6 g, is added to 1 ml of the concentrated ferredoxin solution. The resulting precipitate is removed by centrifugation and solid ammonium sulfate is added to the clear supernatant solution until the solution becomes slightly turbid. Crystals appear after the solution is left standing several days in the refrigerator.

Properties. The absorption spectrum of crystalline spinach ferredoxin is shown in Fig. 1. The oxidized protein has peaks in the visible region at 465 and 420 nm, in the near ultraviolet at 330 nm, and in the ultraviolet at 276 nm. Pure spinach ferredoxin has an absorption ratio 420/276 nm = 0.49; this absorption ratio serves as a purity index and, as with other plant ferredoxins, may be used to monitor ferredoxin purity throughout the preparation procedure. The extinction coefficient, $\epsilon_{420\text{ nm}}$, for spinach ferredoxin[35] is 9.7 mM^{-1} cm^{-1}.

Spinach ferredoxin may be reduced photochemically with chloroplasts,[5,36] with H_2–hydrogenase independent of light,[5] or with dithionite[5]; on reduction, spinach ferredoxin accepts one electron per molecule.[35–38] Its standard oxidation–reduction potential, E'_0 is —0.42 V[35]; on reduction, the absorption peaks in the visible region disappear.[5]

Crystalline spinach ferredoxin is homogeneous in the ultracentrifuge

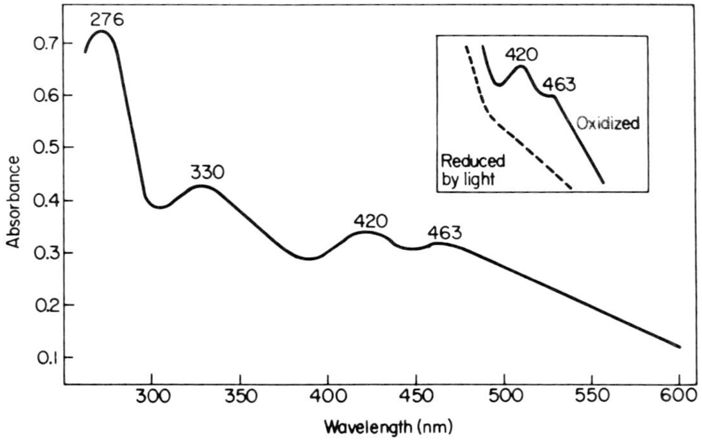

FIG. 1. Absorption spectrum of spinach ferredoxin.

[35] K. Tagawa and D. I. Arnon, *Biochim. Biophys. Acta* **153,** 602 (1968).
[36] F. R. Whatley, K. Tagawa, and D. I. Arnon, *Proc. Nat. Acad. Sci. U.S.* **49,** 266 (1963).
[37] T. Horio and A. San Pietro, *Proc Nat. Acad. Sci. U.S.* **51,** 1226 (1964).
[38] M. C. W. Evans, D. O. Hall, H. Bothe, and F. R. Whatley, *Biochem. J. (London)* **110,** 485 (1968).

and migrates as a single band in polyacrylamide gel electrophoresis. It contains two atoms each of iron and labile sulfide and has five SH groups, contributed by its five cysteine residues.[35,39] The molecular weight is 11,500 based on amino acid composition[39] and iron–sulfur content.[35] The complete amino acid sequence of spinach ferredoxin is known.[39]

Ferredoxin from Alfalfa Leaves

Alfalfa ferredoxin is purified by a modification of the procedure given above for spinach ferredoxin, which was developed by Keresztes-Nagy and Margoliash.[40] The major change in the procedure for alfalfa ferredoxin is elimination of the acetone precipitation step. The procedure consists of (1) preparation of a leaf extract; (2) adsorption of ferredoxin on successive beds of DEAE–cellulose and elution with Tris·NaCl buffer; (3) Sephadex G-25 gel filtration; (4) ammonium sulfate fractionation and concentration on DEAE–cellulose; (5) a series of three DEAE–cellulose column chromatography steps. The procedure below is adapted to 10 kg of alfalfa leaves and yields 200–250 mg of pure ferredoxin. If at any step the preparation is discontinued, the solution containing ferredoxin is deaerated and stored under nitrogen.

Step 1. Preparation of Leaf Homogenate. Fresh alfalfa, cut with short stems, is packed in plastic bags (0.5 kg per bag) and stored at —15° for periods up to 5 months without appreciable loss of ferredoxin. Frozen leaves, 0.5 kg, are crushed in the bag by hand and added to 1.3 liters distilled water containing 1 g solid Tris·HCl buffer in a 1-gallon Waring Blendor (Model CB-5). The suspension is blended 0.5-1 minute (power setting, high); the green homogenate is freed of debris by filtering through a funnel covered with a double layer of cheesecloth and fitted with a glass-wool plug. The precipitate is discarded and the filtrate is applied directly to DEAE–cellulose. To process 10 kg of the original frozen leaf material, this step must be repeated 20 times; the resultant yield is 23–24 liters of leaf extract.

Step 2. Adsorption of Ferredoxin on DEAE–Cellulose and Concentration by Successive DEAE Column Steps. A sufficient amount of NaCl is added to the filtrate to increase the ionic strength to 0.15 and the solution is passed through a sintered-glass funnel containing a bed of DEAE–cellulose (16 cm diameter × 3 cm high) equilibrated beforehand with a solution containing 10 m*M* Tris·HCl buffer, pH 7.5, and 0.15 *M* NaCl. Sufficient suction is applied to the funnel to permit a flow rate of 20–30 liters/hour. During the adsorption step, green particulate material is

[39] H. Matsubara, R. M. Sasaki, and R. K. Chain, *Proc. Nat. Acad. Sci. U.S.* **57,** 439 (1967)

[40] S. Keresztes-Nagy and E. Margoliash, *J. Biol. Chem.* **241,** 5955 (1966).

trapped in the resin; this may be dispersed by repeatedly suspending the resin in 3 liters of the above solution and washing under moderate suction. After the washing step, ferredoxin is eluted in a volume of 1–1.5 liters with a solution containing 10 m*M* Tris·HCl buffer, pH 7.5, and 0.8 *M* NaCl. The dark brown eluate still contains some suspended green particulate material, which is removed by centrifugation (30 minutes, 10,000 *g*) and discarded. The clear solution is concentrated to about 100 ml by adsorption and elution from three successively narrower but longer DEAE–cellulose columns (9.5 × 4 cm, 6.5 × 8 cm, and 4.5 × 10 cm). Ferredoxin is eluted in each case with a solution containing 10 m*M* Tris·HCl buffer, pH 7.5, and 0.8 *M* NaCl and is diluted 1:5 with water before adsorption on the next column. Yellow-brown impurities remain on the columns.

Step 3. Sephadex G-25 Gel Filtration. When the ferredoxin volume is reduced to about 100 ml by the successive DEAE–cellulose column steps, the solution is passed through a Sephadex G-25 column (4.5 cm diameter × 45 cm high; fine bead form from Pharmacia) equilibrated beforehand with 20 m*M* Tris·HCl buffer, pH 7.5. The excluded brownish-red ferredoxin fraction (120–140 ml) separates from a slower moving orange fraction and from a yellow fraction tightly bound to the Sephadex. The visible spectrum of ferredoxin is evident in the excluded fraction and the amount of ferredoxin estimated from the absorption at 422 nm is 400–500 mg. The maximum absorption in the ultraviolet region at this step is at 258 nm and shows an intensity 60 times that at 422 nm.

Step 4. Ammonium Sulfate Fractionation and Concentration on DEAE–Cellulose. Solid ammonium sulfate, 0.6 g, is added per milliliter of ferredoxin solution obtained in the Sephadex G-25 column step. The precipitate is removed by centrifugation (10 minutes 36000 *g*) and discarded. The supernatant solution containing the bulk of the ferredoxin is diluted 50-fold with water and is then adsorbed on a DEAE–cellulose column (3 cm diameter × 5 cm high) equilibrated beforehand with 20 m*M* Tris·HCl buffer, pH 7.5, and eluted with a solution containing 10 m*M* Tris·HCl buffer, pH 7.5, and 0.8 *M* NaCl. The yield of ferredoxin at this stage is about 350 mg, and the 422/258 nm absorbance ratio is 0.1. (It is important to carry out the preparation up to this stage in 2 days or less.)

Step 5. DEAE·Cellulose Chromatography Steps. Final purification of alfalfa ferredoxin is achieved by three successive DEAE–cellulose column chromatography steps (twice on columns 2 cm diameter × 50 cm high and once on a column 0.9 cm × 50 cm). (Each column is equilibrated beforehand with 10 m*M* Tris·HCl buffer, pH 7.5.) Ferredoxin is eluted from each of the three columns by using a linear gradient of 0.2–0.4 *M* NaCl (in 10 m*M* Tris·HCl buffer, pH 7.5).

The central fractions of the ferredoxin peak (containing 80% of the ferredoxin applied initially to the column) are pooled and diluted 1:1 with water. (At this step, the absorption maximum of ferredoxin in the ultraviolet region shifts from 258 to 274 nm and the 422/274 absorbance ratio is increased to 0.4.)

Ferredoxin from the pooled fractions is absorbed on the second DEAE–cellulose column prepared and developed as just described above. The central ferredoxin fractions are again pooled, diluted 1:1 with water, and applied to a third DEAE–cellulose column (0.9 × 50 cm), otherwise prepared and developed as described above.

Following the third chromatography, ferredoxin is again diluted 1:1 with water, adsorbed on a DEAE–cellulose column (1.5 × 3 cm) and eluted with a solution of 10 m*M* Tris·HCl buffer, pH 7.5, and 0.8 *M* NaCl in a volume of 15–20 ml. The solution, containing 200–250 mg of pure ferredoxin, is deaerated under reduced pressure and stored under N_2 at −15°. Ferredoxin stored under these conditions is stable indefinitely.

Properties.[40,41] The absorption spectrum of highly purified alfalfa ferredoxin is similar to that of spinach ferredoxin shown in Fig. 1. Spectral absorption maxima for alfalfa ferredoxin occur at 465, 422, 331, and 277 nm. The absorption ratio, 422/277 nm = 0.48, compares to 0.49 for spinach ferredoxin. The extinction coefficient ϵ_{422} nm for alfalfa ferredoxin is 9.1 mM^{-1} cm^{-1}.

Pure alfalfa ferredoxin contains two atoms each of iron and sulfide and has six SH groups contributed by its six cysteine residues. The molecular weight is 11,500. The complete amino acid sequence of alfalfa ferredoxin is known.[41]

Ferredoxin from the Blue-Green Alga, *Nostoc*

Nostoc ferredoxin is purified and recrystallized by a modification of the procedure described above for spinach ferredoxin. This method, which is used also for *Porphyridium* and *Botrydiopsis* ferredoxins, is based on the recently described detailed procedure by Mitsui and Arnon.[41a] The procedure includes the following steps: (1) preparation of an acetone powder of *Nostoc* cells; (2) extraction of acetone powder and adsorption of crude ferredoxin on a DEAE–cellulose bed and elution with Tris·NaCl buffer; (3) Sephadex G-25 gel filtration; (4) chromatography on DEAE–cellulose and concentration of the main effluent to give a concentration not less than 1%; and (5), if desired, crystallization with ammonium sulfate.

Step 1. Preparation of Acetone Powder. Cells are thawed, suspended in water (1:2 w/v), and acetone precooled to −20° is added to give a final

[41] S. Keresztes-Nagy, F. Perini, and E. Margoliash, *J. Biol. Chem.* **244**, 981 (1969).
[41a] A. Mitsui and D. I. Arnon, *Physiol. Plant.* **24** (1971), in press.

concentration of 80%. After standing 1 hour at 4°, the mixture is filtered through a Büchner funnel, and the residue is washed with cold acetone until the filtrate is colorless. The filtrate is discarded and the blue residue is dried overnight at 25° in a vacuum desiccator over P_2O_5.

Step 2. Extraction of Acetone Powder and Adsorption of Ferredoxin on DEAE–Cellulose. The dried acetone powder (15 g) is pulverized with a mortar and pestle and suspended in 200 ml of 0.15 *M* Tris·HCl buffer, pH 7.3. After 2 hours at room temperature, the mixture is centrifuged (90 min, 36000 *g*). The precipitate is discarded and the blue supernatant fluid is applied to a DEAE–cellulose bed (4 cm diameter × 10 cm high) equilibrated with 0.15 *M* Tris·HCl buffer, pH 7.3. The column is washed with 3 liters of a solution containing 0.15 *M* Tris·HCl buffer, pH 7.3, and 0.1 *M* NaCl [(Cl^-) = 0.22 *M*]; ferredoxin is then eluted with a solution containing 0.3 *M* Tris·HCl buffer, pH 7.3, and 0.55 *M* NaCl [(Cl^-) = 0.8 *M*]. The brownish-red ferredoxin eluate is diluted with 10 volumes of water and applied to a second DEAE–cellulose column (2.5 × 20 cm) equilibrated beforehand with a solution containing 0.15 *M* Tris·HCl buffer, pH 7.3, and 0.2 *M* NaCl ([Cl^-] = 0.32 *M*). The column is then washed with 3 liters of the same buffer solution, and ferredoxin is eluted in 7 ml of a solution containing 0.3 *M* Tris·HCl buffer, pH 7.3, and 0.55 *M* NaCl ([Cl^-] = 0.8 *M*).

Step 3. Sephadex G-25 Filtration. Eluted ferredoxin is applied to a Sephadex G-25 column (2.5 cm diameter × 30 cm high), equilibrated with a solution containing 0.15 *M* Tris·HCl buffer, pH 7.3, and 0.2 *M* NaCl [(Cl^-) = 0.32 *M*]. The ferredoxin component, which is readily recognized by its brownish-red color, is collected as a single fraction.

Step 4. DEAE–Cellulose Chromatography. The ferredoxin fraction is chromatographed on a third DEAE–cellulose column (2.5 cm diameter × 30 cm high), equilibrated beforehand with a solution containing 0.15 *M* Tris·HCl buffer, pH 7.3, and 0.2 *M* NaCl [(Cl^-) = 0.32 *M*]; the column is developed with the same buffer solution. Ferredoxin is eluted in the 720- to 960-ml fraction; this fraction is diluted 1:5 with water and collected on a fourth DEAE–cellulose column (1 × 4 cm), equilibrated with 0.15 *M* Tris·HCl buffer, pH 7.3, and eluted in 2 ml of a solution containing 0.3 *M* Tris·HCl buffer, pH 7.3, and 0.55 *M* NaCl [(Cl^-) = 0.8 *M*].

Step 5. Crystallization. Crystalline ferredoxin may be obtained from the above concentrated brownish-red solution by adding powdered $(NH_4)_2SO_4$ until a slight turbidity is observed. After several days at 4°, crystals are observed.

Properties. The absorption spectrum of crystalline *Nostoc* ferredoxin is shown in Fig. 2. The oxidized protein has peaks at 470, 423, 331, and

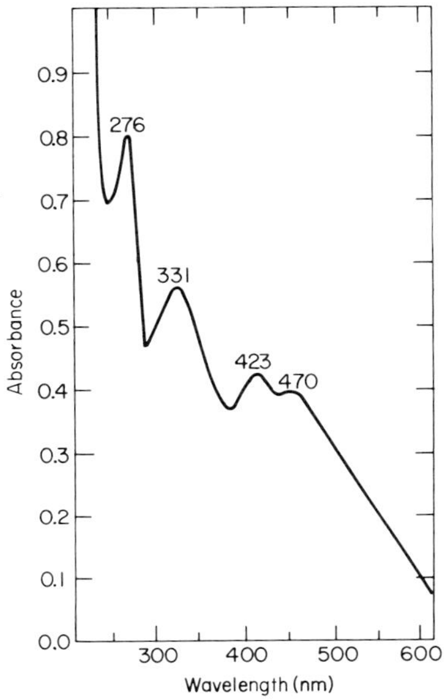

FIG. 2. Absorption spectrum of *Nostoc* ferredoxin.

276 nm. The purest preparations obtained have a 423/276 nm absorption ratio of 0.57 (compared to 0.49 for spinach ferredoxin).

Crystalline *Nostoc* ferredoxin contains two atoms each of iron and labile sulfide per molecule (based on an assumed molecular weight of 12,000). The standard oxidation–reduction potential of *Nostoc* ferredoxin, obtained by the H_2–hydrogenase method,[35] is −0.41 V. The ferredoxin isolated from another blue-green alga, *Anacystis nidulans,* is reported to have similar properties.[38,42]

Ferredoxin from the Red Alga *Porphyridium cruentum* and the Brown Alga *Botrydiopsis alpina*

In several explorative experiments, ferredoxin has been partially purified from *P. cruentum* and *B. alpina* by a slight modification of the procedure developed for *Nostoc* ferredoxin. For this procedure an acetone powder of *Prophyridium* and *Botrydiopsis* cells is prepared and extracted as described above for *Nostoc.* However, unlike *Nostoc,* the acetone powder extract of both *Porphyridium* and *Botrydiopsis* cells is highly viscous when suspended in aqueous solutions and cannot be passed

[42] T. Yamanaka, S. Takenani, K. Wada, and K. Okunuki, *Biochim. Biophys. Acta* **180,** 196 (1969).

through a DEAE–cellulose bed. DEAE–cellulose (equilibrated with 0.15 *M* Tris·HCl buffer, pH 7.3) therefore is directly added to the acetone powder extract of these cells after the extract is diluted 1:10 with 0.15 *M* Tris·HCl buffer, pH 7.3 (0.5 g of wet DEAE–cellulose is added per milliliter of diluted acetone powder extract). The preparation is stirred for 10 minutes at 25°, then the DEAE–cellulose is collected by centrifugation (5 minutes, 36,000 *g*). The supernatant fluid is discarded and the adsorbed ferredoxin is eluted by washing the DEAE–cellulose pellet twice with a solution containing 0.3 *M* Tris·HCl buffer, pH 7.3, and 0.55 *M* NaCl [(Cl^-) = 0.8 *M*]. The remaining steps of the purification of *Porphyridium* and *Botrydiopsis* ferredoxins are as described for *Nostoc* ferredoxin (through step 4). The ferredoxins so far obtained are free of other colored materials but, as reflected by high ultraviolet absorption, are not pure. Their crystallization has not been attempted.

Properties. The absorption spectrum of partially purified *Porphyridium* ferredoxin is similar to that of spinach ferredoxin shown in Fig. 1. Spectral absorption maxima for *Porphyridium* ferredoxin occur at 465, 418, 320, and 276 nm. The absorption ratio 422/276 nm of the purest preparation obtained is 0.32. *Botrydiopsis* ferredoxin shows similar peaks in the visible region (465, 418 nm) and a shoulder at 330 nm; however, the protein peak has not been identified because of contaminating ultraviolet absorbing material present in the preparations obtained so far.

Ferredoxin from the Green Alga *Scenedesmus*

Scenedesmus ferredoxin is purified and crystallized by the procedure of Matsubara,[43] which is similar to the procedure for alfalfa ferredoxin described above. The procedure includes the following steps: (1) preparation of a cell extract; (2) adsorption of crude ferredoxin on a DEAE–cellulose bed and elution with Tris·HCl buffer; (3) dialysis, concentration by ammonium sulfate precipitation, and adsorption on DEAE–cellulose; (4) Sephadex G-75 column chromatography and concentration of the main effluent on DEAE–cellulose; (5) hydroxyapatite column chromatography; and if desired, (6) crystallization with ammonium sulfate. The procedure is adapted for 1800 g of cells and yields about 120 mg of pure ferredoxin.

Step 1. Preparation of Leaf Extract. Four packages, each containing 450 g of frozen *Scenedesmus* cells, are thawed overnight in a cold room. To each 450 g of the cells are added 400 ml of cold water and 300 g of ice cubes. The mixture is homogenized in a 1-gallon Waring Blendor (Model CB-5) at low speed for 30 seconds; 75 ml of 2 *M* Tris·HCl

[43] H. Matsubara, *J. Biol. Chem.* **243,** 370 (1968).

buffer, pH 8, and 900 g of previously chilled glass beads are added to the cell suspension; and the mixture is blended at low speed for an additional 30 seconds and at high speed for 4 minutes. The homogenates from the four 450-g packages of cells are combined and left to stand until the glass beads settle. The cell extract is decanted, the glass beads are washed twice with 4 liters of distilled water, and the washings are added to the original cell extract. The combined preparation is filtered through Hyflo Super-Cel (300 g of Super-Cel per 2 liters of cell extract) on a large Büchner funnel. The residue is washed with 0.5 to 1 liter of water until the filtrate becomes light green (0.5–1 liter of water per 1800 g of cells). The initial filtrate and the washings are combined; the residue is discarded.

Step 2. Adsorption of Ferredoxin on DEAE–Cellulose. The combined filtrate is passed through a DEAE–cellulose column (4.8 cm diameter × 6 cm high) equilibrated beforehand with 2 *M* Tris·HCl buffer, pH 8, and then washed with water until free of buffer. A fall of about 50 cm between the column and a reservoir is maintained to keep the flow rate sufficient to complete this step in 2–3 hours; occasionally the surface of the cellulose column is stirred to prevent clogging. Dark, reddish-black material is adsorbed at the top of the column; green material is apparent near the end of the column.

The column is washed with a volume of 0.02 *M* Tris·HCl buffer, pH 8, equivalent to the holdup volume of the column; the DEAE–cellulose is then transferred to a beaker and suspended in 2 volumes of this buffer. After thorough mixing, the DEAE–cellulose is allowed to settle; the green, turbid supernatant solution is decanted and discarded. This process is repeated twice, and the DEAE–cellulose is then repacked in the original column (after adding a 1-cm layer of new DEAE–cellulose at the bottom of the column). The repacked column is washed with 1 liter 0.15 *M* Tris·HCl buffer, pH 7.5, followed by 1 liter of 0.2 *M* Tris-HCl buffer, pH 7.5, to remove contaminating yellow and light green material. The reddish-brown band of ferredoxin moves down the column about 0.5 cm during the washing procedure. Ferredoxin is eluted with a solution containing 0.02 *M* Tris·HCl buffer, pH 8, and 0.8 *M* NaCl.

Step 3. Dialysis, Ammonium Sulfate Precipitation, and Concentration of Ferredoxin on DEAE–Cellulose. The DEAE–cellulose eluate is dialyzed against 5 m*M* Tris·HCl buffer, pH 8, for 30 minutes. Sufficient solid ammonium sulfate to give a final concentration of 85% saturation is added, with stirring, to the dialyzed solution. (The pH of the solution is maintained at approximately 8 with concentrated NH_4OH during ammonium sulfate addition.) The solution is centrifuged (10 minutes, 39,000 *g*) and the green precipitate is discarded. The supernatant solution is diluted 50-fold with water and passed through a DEAE–cellulose column (4.5 cm diameter × 4 cm high) to adsorb the ferredoxin. The column is washed with 500 ml 0.2 *M* Tris·HCl buffer, pH 7.5, and the portion of the DEAE–cellulose containing the red ferredoxin band is then transferred to a smaller column (0.9 cm diameter). Ferredoxin is eluted from the newly packed column with a solution containing 0.02 *M* Tris·HCl buffer, pH 8, and 0.8 *M* NaCl.

Step 4. Sephadex G-75 Column Chromatography and Concentration of Ferredoxin on DEAE–Cellulose. The ferredoxin solution is passed through a Sephadex G-75 column (2 cm diameter × 30 cm high) equilibrated beforehand with 20 m*M* Tris·HCl buffer, pH 8. The ferredoxin fraction (which is followed by a slower moving yellow impurity) is collected, adsorbed on a DEAE–cellulose bed (2 × 3 cm) and eluted with 0.02 *M* Tris·HCl buffer, pH 8, containing 0.8 *M* NaCl. Ferredoxin is then dialyzed for 3–4 hours against distilled water and an additional 16 hours (against 10 m*M* potassium phosphate buffer, pH 7.5).

Step 5. Chromatography of Ferredoxin on Hydroxyapatite and Concentration on DEAE–Cellulose. The dialyzed ferredoxin solution is adsorbed on a hydroxyapatite column (2 cm diameter × 20 cm high) prepared according to Levin (this series, Vol. V, p. 27) and equilibrated beforehand with 5 m*M* potassium phosphate buffer, pH 7.5. The column is washed successively with 10 m*M*, 40 m*M*, and 70 m*M* solutions of potassium phosphate buffer, pH 7.5. The ferredoxin present in the fractions at the beginning and at the end of this elution procedure represents a minor fraction of the total ferredoxin and is of low purity; the middle fractions contain most of the ferredoxin, and at this stage it is nearly pure. If further purification is needed, the ferredoxin in the fractions may be chromatographed again on hydroxyapatite just as described above. (The original hydroxyapatite column may be used repeatedly after washing with 0.2 *M* potassium phosphate buffer, pH 7.5, and equilibrating with 0.5 m*M* potassium phosphate buffer, pH 7.5.) The ferredoxin fractions are pooled, applied to a DEAE–cellulose bed (2 × 3 cm) equilibrated beforehand with 20 m*M* Tris·HCl buffer, pH 8; ferredoxin is eluted with a solution containing 20 m*M* Tris·HCl buffer, pH 8, and 0.8 *M* NaCl, and is dialyzed for 1 hour against distilled water. The concentrated ferredoxin, which is pure at this stage, may be frozen and is stable for several months when stored under nitrogen. If desired, crystallization may be carried out as described below. About 120 mg of pure ferredoxin is obtained from 1800 g of frozen cells.

Step 6. Crystallization of Ferredoxin. The solution is brought to 85% saturation by addition of solid ammonium sulfate (as before, pH is maintained near 8 with concentrated NH_4OH during this procedure), and the precipitate is removed by centrifugation (10 minutes, 39,000 *g*) and discarded. A small additional amount of solid ammonium sulfate is added to the supernatant solution, which is kept in an ice bath for one to 2 hours. The solution becomes viscous, and then fine, needlelike crystals are observed. Crystallization is almost complete in one step. The crystals are collected by centrifugation as above and dissolved in a small amount of 0.01 *M* Tris·HCl buffer, pH 8. Recrystallization may be carried out by the gradual addition of ammonium sulfate. The crystalline solution is dialyzed against 0.01 *M* Tris·HCl buffer, pH 8, and the solution is stored in a freezer under nitrogen.

Properties.[43,44] As observed with other plant-type ferredoxins, *Scenedesmus* ferredoxin shows the characteristic absorption maxima at 464, 421, 330, and 276 nm. The absorption ratio 421/276 nm is 0.6, which is the highest value observed for any of the plant ferredoxins. This high ratio is a reflection of the relatively lower ultraviolet absorption shown by *Scenedesmus* ferredoxin which is probably due to the absence of tryptophan in this ferredoxin.

Crystalline *Scenedesmus* ferredoxin is homogeneous in the ultracentrifuge and migrates as a single band in polyacrylamide gel electrophoresis. Like other plant ferredoxins, *Scenedesmus* ferredoxin contains two atoms each of iron and labile sulfide. Its molecular weight determined from amino acid composition and ultracentrifugation data is 11,500. The complete amino acid sequence of *Scenedesmus* ferredoxin has been reported.[44]

Ferredoxin from the Photosynthetic Purple Sulfur Bacterium, *Chromatium*

Chromatium ferredoxin is purified and crystallized by an unpublished procedure from this laboratory (K. T. Shanmugam, R. Bachofen, B. B. Buchanan, and D. I. Arnon) similar to that described originally for *Clostridium pasteurianum* ferredoxin by Mortenson.[45] The procedure includes the following steps: (1) preparation of a cell-free extract and treatment with Triton and acetone; (2) adsorption of crude ferredoxin on a DEAE–cellulose bed and elution with phosphate–NaCl buffer; (3) concentration of ferredoxin on a DEAE–cellulose column; (4) chromatography on a DEAE–cellulose column; (5) concentration of the effluent to give a concentration of ferredoxin not less than 1%; and, if desired, (6) crystallization with ammonium sulfate. The procedure below is adapted to 1 kg of cells and yields about 40 mg of pure ferredoxin.

[44] K. Sugeno and H. Matsubara, *Biochem. Biophys. Res. Commun.* **32,** 951 (1968).
[45] L. E. Mortenson, *Biochim. Biophys. Acta* **81,** 71 (1964).

Step 1. Preparation of Cell-Free Extract. One kilogram of frozen cell paste is thawed and suspended in 1 liter of 50 mM potassium phosphate buffer, pH 7.3. The cell suspension is sonicated in an ice bath for 5 minutes (Branson sonifier, power setting 8). Triton X-100 is added to give a final concentration of 5%, and the cell suspension is incubated 30 minutes at 4° with stirring. (Triton is added to solubilize bound ferredoxin; pure ferredoxin may be isolated without Triton treatment but with variable and reduced yield.) Acetone, precooled to —20°, is added to give a concentration of 50%; and the cell suspension is incubated an additional 30 minutes at 4° with stirring. The cell suspension is centrifuged (20 minutes, 13,000 g), and the heavy red precipitate is discarded.

Step 2. Adsorption of Ferredoxin on DEAE–Cellulose. The red 50% acetone supernatant fraction is passed through a DEAE–cellulose bed (3 cm diameter × 15 cm high) equilibrated beforehand with 20 mM potassium phosphate buffer, pH 7.3. The column is washed with 500 ml 20 mM potassium phosphate buffer, pH 7.3, and then with 500 ml of a solution containing 10 mM potassium phosphate buffer, pH 7.3, and 0.26 M NaCl. Ferredoxin is eluted with a solution containing 10 mM potassium phosphate buffer, pH 7.3, and 0.8 M NaCl.

Step 3. Concentration of Ferredoxin on DEAE–Cellulose. The dark brown DEAE–cellulose eluate is diluted 1:4 with distilled water and is passed through a second DEAE–cellulose column prepared and equilibrated as described for step 2. Ferredoxin appears as a dark brown band at the top of the column. The column is washed with 500 ml of a solution containing 10 mM potassium phosphate buffer, pH 7.3, and 0.26 M NaCl and then with 500 ml of a solution containing 10 mM potassium phosphate buffer, pH 7.3, and 0.30 M NaCl. Ferredoxin is eluted with a solution containing 10 mM potassium phosphate buffer, pH 7.3, and 0.8 M NaCl.

Step 4. Chromatography on DEAE–Cellulose. The DEAE–cellulose eluate is diluted 1:4 with distilled water and applied to a DEAE–cellulose chromatography column (2.5 cm diameter × 40 cm high) equilibrated beforehand with 0.02 M potassium phosphate buffer, pH 7.3, and 0.4 M NaCl. Ferredoxin is eluted slowly with this salt concentration. The ferredoxin fractions, recognized by their characteristic brown color, are monitored for purity by determining the 385/280 nm absorption ratio. The ferredoxin fractions are pooled and diluted 1:4 with distilled water.

The diluted ferredoxin solution is applied to a second DEAE–cellulose chromatography column equilibrated and developed as described above. Several ferredoxin fractions are collected at this step and their absorption spectra are measured. (The absorption ratio 385/280 nm for the fractions generally ranges from 0.65 to 0.75.)

Step 5. Concentration of Ferredoxin on DEAE–Cellulose. The ferredoxin fractions from above are pooled, diluted 1:5 with distilled water, and applied to a DEAE–cellulose column (2 cm diameter × 3 cm high) equilibrated beforehand with 20 m*M* potassium phosphate buffer, pH 7.3. The dark-brown ferredoxin band is eluted with a solution containing 10 m*M* potassium phosphate buffer, pH 7.3, and 0.8 *M* NaCl. Ferredoxin concentration in the main effluent fraction should be not less than 1% at this step.

Step 6. Crystallization of Ferredoxin. Ferredoxin is dialyzed against 50 m*M* Tris·HCl buffer, pH 7.3, to remove excess salt. Ammonium sulfate, 0.5 g, is added to 1 ml of the ferredoxin solution, and the resulting precipitate is removed by centrifugation. Solid ammonium sulfate is added to the clear supernatant solution until the solution becomes turbid. Long, needlelike crystals appear after the solution is left standing several hours in the refrigerator.

Properties. The absorption spectra for oxidized and reduced *Chromatium* ferredoxin are shown in Fig. 3. The oxidized protein resembles ferredoxins from nonphotosynthetic bacteria[1-4] and shows absorption peaks at 385 and 280 nm, with a shoulder at 300 nm. Crystalline *Chromatium* ferredoxin has a 385/280 nm absorption ratio equal to 0.75.[46]

When *Chromatium* ferredoxin is reduced with illuminated chloroplasts, the peak at 385 nm readily disappears (Fig. 3). Other agents which

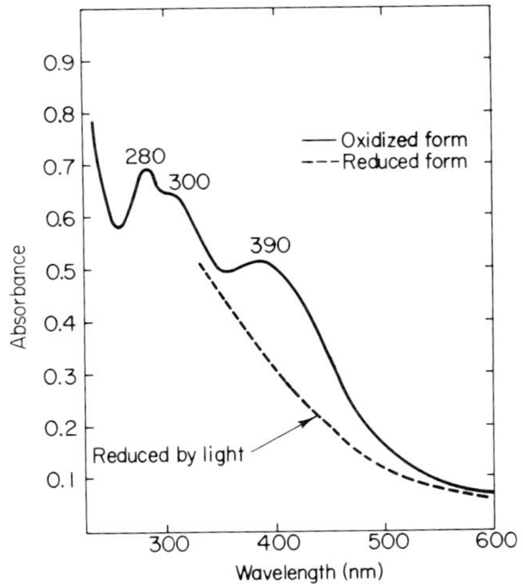

FIG. 3. Absorption spectrum of *Chromatium* ferredoxin.

[46] R. Bachofen and D. I. Arnon, *Biochim. Biophys. Acta* **120,** 259 (1966).

reduce ferredoxins from plants and nonphotosynthetic bacteria[5,35] [e.g., H_2–hydrogenase or sodium dithionite (Fig. 1)] only partly reduce *Chromatium* ferredoxin[46]—an observation that has been interpreted[46] as evidence for a more negative potential for *Chromatium* ferredoxin. On the basis of H_2–hydrogenase reduction data, Bachofen and Arnon[46] calculated a provisional standard oxidation–reduction potential of —0.49 V.

The extent of ferredoxin reduction by dithionite (but not by H_2) is greatly increased by adding a catalytic amount of methylviologen (0.1 μM), as was first observed by Dr. M. C. W. Evans, King's College, University of London.

Although potentiometric measurements are incomplete, *Chromatium* ferredoxin has been reported to accept two electrons per molecule on reduction.[38] Reasoning by analogy with the more complete data available for the clostridial ferredoxins,[35,47,48] it is possible that *Chromatium* ferredoxin can accept either one or two electrons per molecule. (See Buchanan and Arnon[10] for a more complete discussion of stoichiometry of electron transfer by ferredoxins.)

Crystalline *Chromatium* ferredoxin migrates as a single band in polyacrylamide gel electrophoresis. Based on amino acid composition and gel filtration data, *Chromatium* ferredoxin is larger than other bacterial ferredoxins examined and has a molecular weight of about 10,000.[49,50] Its extinction coefficient is ϵ_{385} nm = 31 mM^{-1} cm^{-1}.[49] On the basis of our most recent analyses, *Chromatium* ferredoxin, like clostridial ferredoxins,[51] contains eight atoms each of iron and labile sulfide per molecule (K. T. Shanmugam and B. B. Buchanan, unpublished data). This finding suggests a basic similarity between *Chromatium* and clostridial ferredoxins despite their different molecular weights. The complete amino acid sequence of *Chromatium* has been reported recently.[52]

Ferredoxin from the Photosynthetic Purple Nonsulfur Bacterium *Rhodospirillum rubrum*

Rhodospirillum rubrum ferredoxin is purified by a procedure (K. T. Shanmugam, B. B. Buchanan, and D. I. Arnon, unpublished) similar to that described for *Chromatium* ferredoxin. The procedure includes: (1)

[47] K. K. Eisenstein and J. H. Wang, *J. Biol. Chem.* **244,** 1720 (1969).
[48] B. E. Sobel and W. Lovenberg, *Biochemistry* **5,** 6 (1966).
[49] R. M. Sasaki and H. Matsubara, *Biochem. Biophys. Res. Commun.* **28,** 467 (1967).
[50] T. H. Moss, A. J. Bearden, R. G. Bartsch, M. A. Cusanovich, and A. San Pietro, *Biochemistry* **7,** 1591 (1968).
[51] J.-S. Hong and J. C. Rabinowitz, *J. Biol. Chem.* **245,** 4982 (1970).
[52] H. Matsubara, R. M. Sasaki, D. I. Tsuchiya, and M. C. W. Evans, *J. Biol. Chem.* **245,** 2121 (1970).

preparation of a cell-free extract and treatment with Triton and acetone; (2) adsorption of crude ferredoxin on a DEAE–cellulose bed and elution with phosphate–NaCl buffer; (3) concentration of ferredoxin on DEAE–cellulose; (4) chromatography on DEAE–cellulose columns; and (5) concentration of the effluent on DEAE–cellulose. The procedure below is adapted to 1 kg of cells and yields 12–15 mg of pure ferredoxin.

Step 1. Preparation of Cell-Free Extract and Extraction with Triton and Acetone. One kilogram of cells is suspended in 1200 ml of 50 m*M* potassium phosphate buffer, pH 7.3. The cell suspension is sonicated for 5 minutes (Branson sonifier, power setting 8) at ice temperature. A volume of Triton X-100 sufficient to give a final concentration of 5% is added, and the suspension is incubated 30 minutes at 4° with constant stirring. (As for *Chromatium*, the added Triton solubilizes bound ferredoxin so far, we have obtained only marginal amounts of ferredoxin from *Rhodospirillum rubrum* in the absence of Triton treatment.)

Acetone, precooled to —20°, is added to give a concentration of 30% and the cell suspension is incubated an additional 30 minutes at 4° with stirring. The cell suspension is centrifuged (20 minutes, 13000 *g*) and the heavy red precipitate is discarded. The supernatant fraction, containing most of the ferredoxin solubilized, is centrifuged a second time (15 minutes, 37,000 *g*) to remove small particles which interfere in the DEAE–cellulose step.

Step 2. Adsorption of Ferredoxin on DEAE–Cellulose. The greenish blue 30% acetone supernatant fraction is passed through a DEAE–cellulose bed (3 cm diameter × 15 cm high) equilibrated beforehand with 20 m*M* potassium phosphate buffer, pH 7.3. The column is washed with 500 ml of a solution containing 10 m*M* potassium phosphate buffer, pH 7.3, and 0.2 *M* NaCl, and ferredoxin is eluted with a solution containing 10 m*M* potassium phosphate buffer, pH 7.3, and 0.8 *M* NaCl.

Step 3. Concentration of Ferredoxin on DEAE–Cellulose. The ferredoxin fraction is diluted 1:4 with distilled water and is applied to a DEAE–cellulose column (3 cm diameter × 10 cm high) equilibrated beforehand with 20 m*M* potassium phosphate buffer, pH 7.3. Ferredoxin appears as a brown band at the top of the column at this stage and is eluted in dilute solution with a solution containing 10 m*M* potassium phosphate buffer, pH 7.3, and 0.3 *M* NaCl. The ferredoxin fraction, which is readily recognized by its brown color, shows an absorption peak at about 400 nm and a shoulder at 350 nm and contains a large amount of ultraviolet-absorbing material.

Step 4. Chromatography on DEAE–Cellulose. The DEAE–cellulose eluate is adsorbed on a DEAE–cellulose chromatography column (2.5 cm diameter × 40 cm high) equilibrated beforehand with 20 m*M* potassium phosphate buffer, pH 7.3. The column is developed with a solution containing 10 m*M* potassium phosphate buffer, pH 7.3, and 0.3 *M* NaCl. The ferredoxin fractions are monitored for purity by determining their 385/280 nm absorption ratio.

Step 5. Concentration of Ferredoxin on DEAE–Cellulose. The ferredoxin fraction is diluted 1:2 with distilled water and applied to a DEAE–cellulose collecting column (2 cm diameter × 3 cm high) equilibrated beforehand with 20 m*M* potassium phosphate buffer, pH 7.3. Ferredoxin appears as a sharp, dark brown band at the top of the column and is eluted in a volume not exceeding 3 ml with a solution containing 10 m*M* potassium phosphate buffer, pH 7.3, and 0.8 *M* NaCl.

Step 6. Sephadex G-50 Column Chromatography and Concentration of Ferredoxin on DEAE–Cellulose. The concentrated DEAE–cellulose effluent is applied to a Sephadex G-50 column (2 cm diameter × 40 cm high), medium grade from Pharmacia, equilibrated beforehand and developed with 50 m*M* potassium phosphate buffer, pH 7.3. Ferredoxin, recognized by its brown color, separates on the column into two fractions (the first is designated ferredoxin II, and the second fraction because of its similarity to other bacterial ferredoxins is designated ferredoxin I). Both ferredoxins are active in the chloroplast-NADP assay described above.

Properties.[53] *Rhodospirillum rubrum* ferredoxin I (the slower moving Sephadex fraction) migrates as a single band in polyacrylamide gel electrophoresis; ferredoxin I shows absorption peaks at 385 and 280 nm (Fig. 4). Typical of other bacterial ferredoxins, *R. rubrum* ferredoxin I shows a 385/280 nm absorption ratio of 0.76. The minimum molecular weight of *R. rubrum* ferredoxin I is 8700; our best analyses show that ferredoxin II contains six atoms each of iron and labile sulfide per molecule (based on an $\epsilon_{385\,nm}$ of 24 mM^{-1} cm^{-1}).

Rhodospirillum rubrum ferredoxin II (the faster moving Sephadex fraction) also shows absorption peaks at 385 and 280 nm (Fig. 4). However, in contrast to ferredoxin I, the 385/280 nm absorption ratio of *R. rubrum* ferredoxin II is 0.48. Ferredoxin II migrates as a single band in polyacrylamide gel electrophoresis. *Rhodospirillum rubrum* ferredoxin II has a molecular weight of 7500 (minimum value, based on amino acid composition) and contains two iron and sulfide groups per mole (based on an $\epsilon_{385\,nm}$ of 8.8 mM^{-1} cm^{-1}).

Rhodospirillum rubrum provides the first evidence for two different ferredoxins in the same cell. The two ferredoxins are distinguished by absorption spectra and amino acid composition but show a similar iron–

[53] K. T. Shanmugam, B. B. Buchanan, and D. I. Arnon, manuscript in preparation.

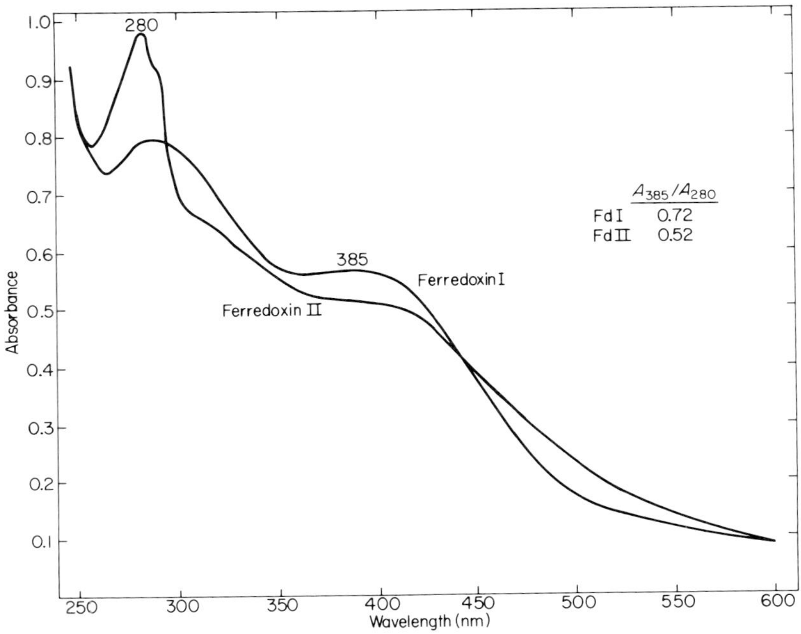

FIG. 4. Absorption spectrum of *Rhodospirillum rubrum* ferredoxin.

sulfur content and equivalent activity in the chloroplast–NADP assay. Both ferredoxins are also active in promoting the synthesis of α-ketoglutarate by *Chlorobium* α-ketoglutarate synthase. The relative amount of the two ferredoxins is not constant in different lots of *R. rubrum* cells, but ferredoxin II is consistently a major constitutent of cells grown anaerobically in the light and may account for up to 50% of the total ferredoxin content.

Ferredoxin from the Photosynthetic Green Bacterium *Chlorobium thiosulfatophilum*

Chlorobium thiosulfatophilum ferredoxin is not stable at any level of purity, but it may be obtained in pure form when the preparation procedure is carried out rapidly, the total time not exceeding 3 days.[29] The purification procedure is similar to that described originally for clostridial ferredoxins by Buchanan, Lovenberg, and Rabinowitz.[54] Evans[55] has reported the purification of ferredoxin from another photosynthetic green

[54] B. B. Buchanan, W. Lovenberg, and J. C. Rabinowitz, *Proc. Nat. Acad. Sci. U.S.* **49,** 345 (1963).

[55] M. C. W. Evans, *Biochem. Biophys. Res. Commun.* **33,** 146 (1968).

bacterium, *Chloropseudomonas ethylicum,* by a procedure described earlier for *Chromatium* ferredoxin.[38]

The procedure for *Chlorobium thiosulfatophilum* ferredoxin includes: (1) preparation of a cell-free extract; (2) adsorption of crude ferredoxin on DEAE–cellulose and elution with Tris·HCl buffer; (3) concentration on DEAE–cellulose collecting columns; (4) chromatography and concentration on DEAE–cellulose columns. The procedure is adapted to 400 g of cells and yields up to 40 mg of pure ferredoxin.

Step 1. Preparation of Cell-Free Extract. Frozen cell paste, 400 g, is thawed and suspended with a Waring Blendor in 1 liter 20 m*M* potassium phosphate buffer, pH 6.5. A small amount of DNase is added, and the cell suspension is sonicated for 5 minutes (Branson sonifier, power setting 8) at ice temperature. The cell suspension is centrifuged (20 minutes, 13000 *g*) and the heavy green precipitate is discarded. The green supernatant fraction is used below.

Step 2. Adsorption of Ferredoxin on DEAE–Cellulose. DEAE–cellulose from a column (5 cm diameter × 15 cm high), equilibrated beforehand with 20 m*M* potassium phosphate buffer, pH 6.5, is added to the supernatant fraction and is mixed thoroughly. The slurry is poured into the original column, and the resulting DEAE–cellulose column is washed with 20 m*M* potassium phosphate buffer, pH 6.5, until the effluent is colorless. Ferredoxin is eluted with a solution containing 0.3 *M* Tris·HCl buffer and 0.55 *M* NaCl, pH 7.3 [(Cl^-) = 0.8 *M*].

Step 3. Concentration of Ferredoxin on DEAE–Cellulose. The brownish eluate is diluted 1:10 with water, and since ferredoxin is poorly adsorbed at this step, it is collected on a series of DEAE–cellulose columns (3 cm diameter × 15 cm high) equilibrated beforehand with 20 m*M* potassium phosphate buffer, pH 6.5. Usually, most of the ferredoxin is adsorbed on the first two collecting columns, but with certain preparations up to four collecting columns have been necessary. Ferredoxin from each of the collecting columns is eluted with a solution containing 0.3 *M* Tris·HCl buffer, pH 7.3, and 0.55 *M* NaCl [(Cl^-) = 0.8 *M*]. The column eluates are combined, diluted 1:10 with water, and passed through another DEAE–cellulose collecting column (3 × 15 cm) equilibrated beforehand with 20 m*M* potassium phosphate buffer, pH 6.5. Ferredoxin adsorbed as a dark brown band at the top of the column is eluted with a solution containing 0.3 *M* Tris·HCl buffer, pH 7.3, and 0.55 *M* NaCl [(Cl^-) = 0.8 *M*].

Step 4. Chromatography of Ferredoxin on DEAE–Cellulose. The ferredoxin eluate is diluted 1:4 with water and applied to a DEAE–cellulose chromatography column (5 cm diameter × 30 cm high) equilibrated beforehand with 20 m*M* potassium phosphate buffer, pH 6.5. The ferredoxin fraction, readily recognized by its characteristic brown color and absorbance peak at about 400 nm, is eluted with a solution containing 0.15 *M* Tris·HCl buffer, pH 7.3, and 0.16 *M* NaCl [(Cl^-) = 0.28 *M*]. The eluted ferredoxin is diluted 1:4 with water and collected on a DEAE–cellulose column (5 × 18 cm) equilibrated beforehand with 20 m*M* potassium phosphate buffer, pH 6.5. The ferredoxin band is eluted from the column with a solution containing 0.3 *M* Tris·HCl buffer, pH 7.3, and 0.55 *M* NaCl [(Cl^-) = 0.8 *M*].

Chlorobium thiosulfatophilum ferredoxin is usually pure after this step and has a 385/280 nm absorbance ratio of about 0.7. If further purification is desired, the DEAE–cellulose chromatography step may be repeated. About 40 mg of pure ferredoxin is recovered from 400 g of cells, but the yield may be considerably reduced if the total preparation time takes more than 2 days.

Properties.[56] The instability of *Chlorobium thiosulfatophilum* ferredoxin mentioned above has prevented a detailed characterization of its physical and chemical properties. However, the properties described below have been observed with several pure preparations of the protein.

C. thiosulfatophilum ferredoxin, like other bacterial ferredoxins, shows absorption maxima at 385 and 280 nm, with a shoulder at 300 nm (Fig. 5). The purest preparations have an absorption ratio 385/280 nm equal to 0.71. The instability of *C. thiosulfatophilum* was seen in one

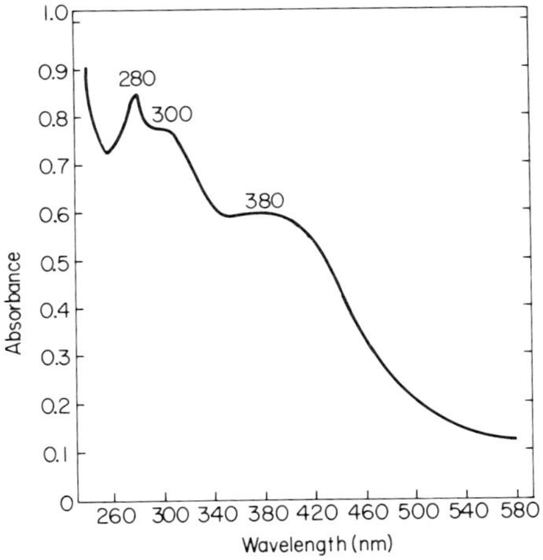

FIG. 5. Absorption spectrum of *Chlorobium thiosulfatophilum* ferredoxin.

[56] B. B. Buchanan, H. Matsubara, and M. C. W. Evans, *Biochim. Biophys. Acta* **189**, 46 (1969).

preparation in which the 385/280 nm ratio dropped from 0.71 to 0.60 after the preparation had stood for 5 hours at 4° in air.

Like ferredoxins from nonphotosynthetic bacteria, *C. thiosulfatophilum* ferredoxin has a molecular weight of 6000 as determined by amino acid composition analyses. It contains four to five atoms of iron and sulfide per molecule. (Based on an assumed extinction coefficient of 30 mM^{-1} cm^{-1}, *Chlorobium thiosulfatophilum,* like other typical bacterial ferredoxins, contains eight atoms each of iron and sulfide per molecule.) Like *Chromatium* ferredoxin, the reduction of *Chlorobium thiosulfatophilum* ferredoxin, by dithionite is most marked in the presence of a catalytic amount of methylviologen (1×10^{-7} *M*).

The carboxyl and amino terminal groups of *C. thiosulfatophilum* ferredoxin[56] (and ferredoxins from two other photosynthetic green bacteria[57]) are similar to those of *Chromatium* ferredoxin, while the molecular weights of the ferredoxins from the green bacteria resemble those of the clostridial ferredoxins. These observations led Buchanan *et al.*[56] to suggest that, in the evolutionary development of photosynthesis, the green bacteria may represent a link between nonphotosynthetic anaerobes (like *Clostridium pasteurianum*) and the photosynthetic purple sulfur bacteria (like *Chromatium*).

[57] K. K. Rao, H. Matsubara, B. B. Buchanan, and M. C. W. Evans, *J. Bacteriol.* **100,** 1411 (1969).

[40] Ferredoxin-NADP Reductase from Spinach

By Masateru Shin

2-Reduced ferredoxin + NADP $\rightleftharpoons$ 2-Oxidized ferredoxin + NADPH

Ferredoxin–NADP reductase is a chloroplast flavoprotein and catalyzes the reduction of NADP with reduced ferredoxin. The enzyme was first crystallized by Shin *et al.*[1] The crystalline enzyme showed not only ferredoxin–NADP reductase activity, but a number of other enzymatic activities, such as NADPH diaphorase, transhydrogenase, and hemoprotein reductase activity. On the basis of comparative studies,[2] ferredoxin–NADP reductase is the same flavoprotein enzyme as the NADPH

[1] M. Shin, K. Tagawa, and D. I. Arnon, *Biochem. Z.* **338,** 84 (1963).
[2] G. Forti and G. Zanetti, *Biochem. Chloroplasts* **2,** 523 (1966).

diaphorase of Avron and Jagendorf,[3] the transhydrogenase of Keister *et al.*,[4] and the NADPH–cytochrome *f* reductase of Forti *et al.*[5]

Assay Method

Principle. Ferredoxin–NADP reductase catalyzes reversible electron transport between ferredoxin and NADP. The usual assay method is based on the reduction of NADP by reduced ferredoxin in the presence of a regenerating system, either illuminated chloroplasts or the H_2–hydrogenase system in the dark. The assay method for the reverse reaction involves the reduction of cytochrome *c*, as the terminal electron acceptor, in the presence of ferredoxin and NADPH.

Treated Chloroplast Method

Reagents

Treated chloroplast suspension,[6] 150 μg of chlorophyll per milliliter. Treated chloroplasts are prepared by extraction of the bound enzyme from the broken spinach chloroplasts (defined as P_{1s1} by Whatley and Arnon[7]). The broken chloroplast suspension, containing 150 μg of chlorophyll per milliliter in 50 m*M* Tris·HCl buffer, pH 7.6, is incubated overnight at room temperature, and then washed twice with the same buffer solution by centrifugation at 18,000 *g* for 10 minutes.

Ferredoxin, 1 mg/ml in 5 m*M* Tris·HCl buffer, pH 7.3

NADP, 5 m*M*

2,6-Dichlorophenolindophenol, 2 m*M*

Ascorbate, 0.1 *M*

Tris·HCl buffer, pH 7.8, 0.1 *M*

Procedure. To a cuvette of 1 cm light path are added 1.0 ml of Tris·HCl buffer, 0.1 ml each of ascorbate, 2,6-dichlorophenolindophenol and

[3] M. Avron and A. T. Jagendorf, *Arch. Biochem. Biophys.* **65**, 475 (1956).

[4] D. L. Keister, A. San Pietro, and F. E. Stolzenbach, *J. Biol. Chem.* **235**, 2989 (1960). See also Vol. VI [61].

[5] G. Forti, M. L. Bertolé, and B. Parisi, *in* "Photosynthetic Mechanisms of Green Plants," p. 284, Nat. Acad. Sci.—Nat. Res. Council Publ. No. 1145, Washington, D.C., 1963. See also this volume [41].

[6] The treatment used to remove the bound enzyme from spinach chloroplasts also results in the loss of the oxygen-evolving capacity. The ascorbate–indophenol dye system is, therefore, used as the substitutive electron donor for NADP photoreduction. When pea chloroplasts are treated, the oxygen-evolving capacity remains after removal of the bound enzyme.[24]

[7] See Vol. VI [37].

ferredoxin, 0.2 ml of treated chloroplasts, and enzyme in a final volume of 2.5 ml. The reaction mixture should be prepared in the dark. The reaction is initiated with red actinic light. A tungsten filament 500-W lamp is the source of the actinic light; a red plus a $CuSO_4$ filter (0.45% solution, 2 cm thick) are placed between the cuvette and lamp. Under continuous illumination the increase in absorbance at 340 nm is measured against a blank which contains everything except enzyme. An interference filter for 340 nm is placed between the photodetector and the cuvette to prevent actinic light from reaching the photodetector.

Hydrogenase Method

Reagents

Hydrogenase, 200 units/ml. It is prepared from *Desulfovibrio desulfuricans* cells cultured by the method described by Sanada and Jagennathan.[8] The acetone powder of the cells (1 g) is extracted with 10 ml of 0.1 *M* Tris·HCl buffer, pH 8.4. The extract is applied to a DEAE–cellulose column equilibrated with 0.15 *M* Tris·HCl buffer, pH 7.3. Hydrogenase is adsorbed on the column at almost the same position as the red band of cytochrome c_3. The red band containing active hydrogenase is eluted from the column with 0.2 *M* Tris·HCl buffer, pH 7.3, and is used as the hydrogenase preparation.

Ferredoxin, 1 mg/ml in 50 m*M* Tris·HCl buffer, pH 7.3

NADP, 5 m*M*

$FeSO_4$, 20 m*M*

Cysteine, 0.1 *M*

Benzyl viologen, 0.5 m*M*

Tris·HCl buffer, pH 7.8, 0.1 *M*

Procedure. To a Thunberg-type cuvette are added 1.0 ml of Tris·HCl buffer, 0.2 ml of hydrogenase, 0.1 ml each of NADP, $FeSO_4$, cysteine, and benzyl viologen, and to the side arm 0.1 ml of ferredoxin and enzyme in a final volume of 2.5 ml. The cuvette is flushed with hydrogen, and the gas pressure in the cuvette is maintained at 1 atmosphere. Wait about 10 min for the complete reduction of benzyl viologen to occur. The reaction is then initiated by tipping the contents of the side arm into the main compartment and the increase in absorbance at 340 nm is measured.

It should be noted that the reduction of ferredoxin by the *Disulfo-*

[8] J. C. Sanada and V. Jagennathan, *Biochim. Biophys. Acta* **19**, 440 (1956).

vibrio hydrogenase is quite slow, but it is increased strikingly by adding a catalytic amount of benzyl viologen (20 μM). This amount of benzyl viologen gives negligible reduction of NADP without added ferredoxin.

Cytochrome c Reduction Method

Reagents

Cytochrome *c*, 0.5 m*M*
Ferredoxin, 1 mg/ml in 50 m*M* Tris·HCl buffer, pH 7.3
NADPH, 20 m*M*
Tris·HCl buffer, 0.1 *M*, pH 7.8

Procedure. To a cuvette are added 0.1 ml each of Tris·HCl buffer, cytochrome *c*, ferredoxin, and enzyme in a final volume of 1.0 ml. The reaction is initiated by the addition of 10 μl of NADPH, and the increase in absorbance at 550 nm is measured.

Purification Procedure

Preparation of Tris·HCl Buffer. The stock solution of Tris·HCl buffer used for DEAE–cellulose chromatography contains 0.3 *N* HCl and 0.45 *M* Tris. For convenience the concentration of Tris·HCl buffer is expressed by the normality of HCl in the buffer solution; for example, 0.3 *N* HCl·Tris buffer for the stock solution. Another stock solution of Tris·HCl buffer (for ferredoxin) is made by mixing 0.3 *N* HCl and 0.35 *M* Tris. The Tris·HCl buffer is only for the procedure in Step 3.

During purification of the enzyme, the NADPH–diaphorase activity, measured as described by Jagendorf,[9] serves as a routine assay.

Step 1. Preparation of Crude Homogenate. Three kilograms of spinach leaves, previously deveined and washed with water, are ground for 2 minutes in 1.5 volumes of ice water in a Waring Blendor. The homogenate is filtered through a double layer of cheesecloth on a large Büchner funnel to remove debris. The filtrate is adjusted to pH 7.5–7.8 by adding about 5 ml of a 1 *M* Tris solution per liter.

Step 2. Acetone Fractionation of Crude Homogenate. Acetone, previously cooled to −15°, is added to give 35% by volume with mechanical stirring. The precipitate is removed by centrifugation at 5000 *g* for 10 minutes. Acetone is added to the supernatant fluid to a final concentration of 75% in the same manner. The precipitate settles within 30 minutes. After the greater part of the supernatant fluid is decanted, the precipitate is collected by centrifugation at 3000 *g* for 10 minutes and

[9] See Vol. VI [60].

washed twice with cold acetone. The residual acetone is vaporized in air by spreading the precipitate on large filter paper.

Step 3. Removal of Ferredoxin. The dried acetone powder is dissolved in 1 liter of 0.05 *N* HCl·Tris buffer (for ferredoxin) and centrifuged to remove undissolved substances. The supernatant fluid is passed through a DEAE–cellulose column (5 × 25 cm), equilibrated with 0.15 *N* HCl·Tris buffer (for ferredoxin). Ferredoxin and the bulk of colored substances are adsorbed on the upper part of the column. Ferredoxin on the column can be purified by the method described by Tagawa and Arnon.[10] The column is washed with 400 ml of the same buffer.

Step 4. Ammonium Sulfate Fractionation. The yellow eluate and washings are fractionated with ammonium sulfate, 40–65% saturation, in the presence of 50 m*M* sodium pyrophosphate, which protects the enzyme against denaturation.

Step 5. First Chromatography on DEAE–Cellulose Column. The precipitate from ammonium sulfate fractionation is dissolved in 150 ml of 0.09 *N* HCl·Tris buffer and passed through a Sephadex G-25 column (6 × 45 cm), equilibrated with the same buffer. The eluate is subjected to chromatography on a DEAE–cellulose column (5 × 45 cm), equilibrated with the same buffer. The enzyme comes off the column after one-half column volume of effluent has passed through.

Step 6. Second Chromatography on DEAE–Cellulose Column. The enzyme fraction from the first chromatography is diluted three times with water and passed through a DEAE–cellulose column (4 × 10 cm), equilibrated with 0.03 *N* HCl·Tris buffer. The yellow color of the enzyme is seen in the upper portion of the column. The yellow fraction is eluted with 0.3 *N* HCl·Tris buffer. The eluate is diluted five times with water and subjected again to chromatography on a DEAE–cellulose column (4 × 45 cm), equilibrated with 0.067 *N* HCl·Tris buffer. The main enzyme fraction is eluted from the column after the first two column volumes of effluent have passed through.

Step 7. Crystallization. The main enzyme fraction from the second chromatography is diluted 2.3 times with water and the enzyme is collected on a DEAE–cellulose column (3 × 10 cm), equilibrated with 0.03 *N* HCl·Tris buffer. The enzyme adsorbed on the column is eluted with 0.3 *N* HCl·Tris buffer. The enzyme solution is fractionated with ammonium sulfate at 50–60% saturation. The yellow precipitate is dissolved in a minimum amount of water. A small amount of powdered ammonium sulfate is added until the enzyme solution becomes turbid. The solution is placed in a refrigerator and crystals appear in about a week.

[10] K. Tagawa and D. I. Arnon, *Nature* (*London*) **195,** 537 (1962). See also this volume [39].

Properties[11]

The ferredoxin–NADP reductase has a molecular weight of about 40,000, contains 1 FAD per mole[9,12] and shows a typical flavoprotein absorption spectrum with absorption maxima at 275, 385, and 456 nm and minima at 321 and 410 nm.[1] Shoulders are seen at 283, 290, 437, and 485 nm. The molar extinction coefficients are summarized in the table.

EXTINCTION COEFFICIENTS

	Molar extinction coefficient[a] M^{-1} cm^{-1}			
Enzyme	456	385	275	456/275
Oxidized	10,740	8,850	85,000	0.127[b]
Reduced	2,160	—	—	—

[a] The molar extinction coefficients are based on the value of 10,740 at 456 nm reported by Forti [*Brookhaven Symp. Biol.* **19,** 195 (1967)]. More recently, Foust *et al.* [G. P. Foust, S. G. Mayhew, and V. Massey, *J. Biol. Chem.* **244,** 964 (1969)] have reported a value of 10,300.

[b] The ratio of absorbance at 456 nm to 275 nm varies from 0.110 to 0.127 depending on the crystalline preparation.

The fully reduced enzyme is obtained by reduction with $Na_2S_2O_4$ or the H_2–hydrogenase–ferredoxin system.[13] The reduction by $Na_2S_2O_4$ is fairly slow.[14] Two electrons are required per mole of FAD in the full reduction of the enzyme.[15] When the enzyme is reduced anaerobically with either NADPH or NADH, full reduction is not achieved, suggesting that its oxidation–reduction potential is quite low, approximately −370 mV.[13] The partially reduced enzyme shows a spectrum typical of a flavin semiquinone.[13,15] An electron spin resonance (ESR) signal of the enzyme partially reduced by NADPH is observed at $g = 2.0042$ in the frozen state.[16] However, semiquinone formation has not been observed during reduction by $Na_2S_2O_4$ or by the H_2–hydrogenase–ferredoxin system. Circular dichroic spectrum of the enzyme has positive peaks at 270 and 376 nm and negative peaks at 286 and 475 nm.[17] Fluorescence of

[11] See also this series, Vol. VI [60] and [61].

[12] G. Zanetti and G. Forti, *J. Biol. Chem.* **241,** 279 (1966).

[13] M. Shin and D. I. Arnon, *J. Biol. Chem.* **240,** 1405 (1965).

[14] M. Shin *in* "Flavins and Flavoproteins" (K. Yagi, ed.), p. 1. Univ. of Tokyo Press, Tokyo, and Univ. Park Press, London, 1968.

[15] G. Forti B. A. Melandri, A. San Pietro, and B. Ke, *Arch. Biochem. Biophys.* **140,** 107 (1970). See also this volume [41].

[16] K. Huang, S. I. Tu, and J. H. Wang, *Biochem. Biophys. Res. Commun.* **34,** 48 (1969).

[17] M. Shin, unpublished data.

FAD in the enzyme is quenched considerably and is increased upon denaturation of the enzyme by urea.[18]

Specificity. PYRIDINE NUCLEOTIDE.[13,14] The crystalline enzyme reduces both NAD and NADP. The reduction of both pyridine nucleotides are inhibited by 2′-AMP in a competitive manner. When both NAD and NADP are present in the reaction mixture, there is mutual inhibition of the reduction of each other. This evidence indicates that NAD and NADP are attached to the same site on the enzyme and reduced by the same mechanism. The reported Michaelis constant for NAD is 3.77 m*M* and that for NADP is 7.72 μM. However, for NAD reduction a four-times greater quantity of enzyme is required to give approximately the same maximum velocity. Michaelis constants for the pyridine nucleotides can also be estimated from the transhydrogenase action of the enzyme. Thereby, 2.56 m*M* for NAD and 2.78 μM for NADPH were determined. The enzyme is specific for NADP reduction under the physiological conditions.

FERREDOXIN. The Michaelis constant of the enzyme for spinach ferredoxin is 0.33 μM.[19] Bacterial ferredoxin[9] and rubredoxin[20,21] are also capable of reacting with the enzyme, but adrenodoxin is inert.[22]

HEMOPROTEIN AND PLASTCYANIN. Cytochrome *f*,[5] b_2, and b_5 are reduced directly by the enzyme with NADPH. On the other hand, cytochrome *c*,[23] methemoglobin,[24] and peroxidase[1] are reduced only in the presence of ferredoxin or substitutive electron carriers, such as FAD, FMN, and some other dyes.

Plastcyanin is directly reduced by the enzyme with NADPH.

Complex Formation with Ferredoxin and with NADP.[25-28] The enzyme reacts with oxidized spinach ferredoxin to form a protein complex in the molecular ratio of 1:1. The dissociation constant for the protein complex is 50 n*M*. Bacterial ferredoxin, rubredoxin, and flavodoxin also form complexes with the enzyme. Complex formation changes the absorption spectrum of the enzyme and also the circular dichroic spectrum.[17]

[18] G. Forti, *Brookhaven Symp. Biol.* **19,** 195 (1967).
[19] K. Tagawa, H. Y. Tsujimoto, and D. I. Arnon, *Nature (London)* **199,** 1247 (1963).
[20] J. A. Peterson, D. Basu, and M. J. Coon, *J. Biol. Chem.* **241,** 5162 (1966).
[21] M. Kusunose, K. Ichihara, E. Kusunose, J. Nozaka, and J. Matsumoto, *Agr. Biol. Chem.* **31,** 990 (1967).
[22] Y. Ichikawa and T. Yamano, *Biochim. Biophys. Acta* **153,** 753 (1968).
[23] R. A. Lazzarini and A. San Pietro, *Biochim. Biophys. Acta* **62,** 417 (1962).
[24] H. E. Davenport, *Nature (London)* **199,** 151 (1963).
[25] M. Shin and A. San Pietro, *Biochim. Biophys. Res. Commun.* **33,** 38 (1968).
[26] G. P. Foust, S. G. Mayhew, and V. Massey, *J. Biol. Chem.* **244,** 964 (1969).
[27] N. Nelson and J. S. Neumann, *J. Biol. Chem.* **244,** 1926 (1969).
[28] N. Nelson and J. S. Neumann, *J. Biol. Chem.* **244,** 1932 (1969).

The absorption maxima due to the interaction of the proteins are at 395 and 465 nm. When the SH groups of the enzyme are masked by PCMB, complex formation is inhibited. The complex is completely dissociated at high ionic strength, suggesting that the forces responsible for binding are primarily electrostatic in nature. The effect of salt on the protein complex and the inhibition of NADP photoreduction by chloroplasts in the presence of high concentrations of salt suggest that the protein complex is important physiologically.

NADP also causes perturbations in the visible spectrum and in the circular dichroic spectrum of the enzyme.

[41] NADPH-Cytochrome *f* Reductase from Spinach

By GIORGIO FORTI

This enzyme, an FAD-containing flavoprotein, is identical with the ferredoxin–NADP reductase of chloroplasts.[1-4] It has been shown that the enzyme catalyzes the following reactions:

(a) NADPH to ferricyanide or to indophenol dyes[2,3,5]
(b) NADPH to NAD (transhydrogenase reaction)[6]
(c) NADPH to cytochrome *f* or cytochrome 552 of *Euglena gracilis*[2,3,7]
(d) Reduced ferredoxin to NADP[8,9]

Occurrence of the Enzyme

The enzyme is present in rather large amounts in all green leaves so far tested, as well as in algae. Its activity as ferredoxin–NADP reductase is an essential step of photosynthetic electron transport. It has

[1] G. Forti and G. Zanetti, *Biochem. Chloroplasts* **2** (1966).
[2] G. Zanetti and G. Forti, *J. Biol. Chem.* **241,** 274 (1966).
[3] G. Forti, M. L. Bertolè, and B. Parisi, Nat. Acad. Sci. Nat. Res. Counc. Publ. **1145,** 284 (1963).
[4] See [40] this volume.
[5] M. Avron and A. T. Jagendorf, *Arch. Biochem. Biophys.* **65,** 475 (1956).
[6] D. L. Keister, A. San Pietro, and F. E. Stolzenbach, *J. Biol. Chem.* **235,** 2898 (1960).
[7] G. Forti and E. Sturani, *Eur. J. Biochem.* **3,** 461 (1968).
[8] M. Shin, K. Tagawa, and D. I. Arnon, *Biochem. Z.* **338,** 84 (1963).
[9] M. Shin and D. I. Arnon, *J. Biol. Chem.* **240,** 1405 (1965).

also been shown that this flavoprotein is an essential catalyst of cyclic photophosphorylation in intact chloroplasts.[10]

Activity Measurement: The most convenient method of estimating the activity of the enzyme is by measuring its NADPH–ferricyanide diaphorase activity. This also can be done in crude extracts, because the enzyme accounts for most if not all of the NADPH-specific diaphorase of green leaves. No information is available on the validity of this statement in the case of algae.

To measure the activity, the reduction of ferri- to ferrocyanide is followed spectrophotometrically at 420 nm in the presence of NADPH, buffer, ferricyanide, and enzyme. The cuvette, in a final volume of 1 ml, contains the following: 0.05 *M* Tris buffer, pH 8.0; 0.7 m*M* ferricyanide; 0.5 m*M* NADP; 5 m*M* G-6-P; and Glucose 6-P dehydrogenase in large excess. The enzyme is added last, and the decrease of absorbance at 420 nm is immediately recorded after mixing the contents of the cuvette. The reaction velocity is linear until more than 50% of the ferricyanide has been reduced and is proportional to enzyme concentration. One unit of activity is defined as that amount which catalyzes the reduction of 1 μmole of ferricyanide per minute, at 30°. The millimolar extinction coefficient of ferricyanide at 420 nm is 1.02.

Cytochrome f Reductase Activity: This activity is measured as above, with ferricytochrome *f* (50 μM) from higher plants instead of ferricyanide, as the *increase* of absorbancy at 555 nm due to cytochrome reduction.[2,3] The millimolar extinction coefficient[11] of ferro- minus ferricytochrome *f* at 555 nm is 19.6. The oxidation of cytochrome *f* is performed by dialysis against 0.1 m*M* ferricyanide in 0.05 *M* Tris buffer, pH 8.0, followed by dialysis against the buffer to remove the excess ferricyanide. The same procedure applies to *Euglena* cytochrome 552. In the case of this latter cytochrome, the measuring wavelength is 552 nm.

Purification of the Enzyme

The method described here applies to spinach leaves and is essentially the same as described by Forti and Sturani.[7] The leaves, depetiolated and rinsed with distilled water, are ground in a Waring blendor with 130 ml of double-distilled water per 100 g of leaves. The blendor is operated at full speed for 2–3 minutes in a cold room. All operations from this point on are carried on at 2–4°. The homogenate is filtered through two layers of cheesecloth, and Tris buffer, pH 8.0, is added to the filtrate to give a final concentration of 0.05 *M*.

[10] G. Forti and G. Zanetti, *in* "Progress in Photosynthesis Research" (H. Metzner, ed.), Vol. III, p. 1213. Metzner, Tubingen, 1969.

[11] G. Forti, M. L. Bertolè, and G. Zanetti, *Biochim. Biophys. Acta* **109**, 33 (1965).

Step 1—Acetone Fractionation: Cold acetone (—20°) is added to the filtered homogenate to give a final concentration of 35% (v/v), and the precipitate formed is removed by centrifugation in the cold (—10° to —15°). Acetone is added to the supernatant fluid to a final concentration of 75%. The large amount of precipitate formed, which contains both the flavoprotein and ferredoxin, is collected by centrifugation at —20° and redissolved in 0.05 *M* Tris buffer, pH 7.4. The enzyme solution is then dialyzed for 12–15 hours against 5 m*M* Tris pH 7.4 and clarified by centrifugation. NaCl is then added to a final concentration of 0.22 *M*. The enzyme solution is passed through a DEAE–cellulose column (2.5 × 65 cm for a preparation from 8–10 kg of leaves) equilibrated with 0.1 *M* Tris buffer, pH 7.4, containing 0.114 *M* NaCl. The flavoprotein is washed out of the column, while ferredoxin is strongly retained on it. The column is washed with the same buffer until the activity in the eluate becomes negligible. The ferredoxin, which can easily be seen as a red band in the upper part of the column, can be eluted with 0.5 *M* Tris, pH 7.4, and further purified as described elsewhere in this volume.

Step 2—Ammonium Sulfate Fractionation: Solid ammonium sulfate is added slowly to give 40% saturation. During the addition of ammonium sulfate the pH must be maintained above 7.0 by the addition of a few drops of 1 *M* Tris (unbuffered). After 10–15 minutes in the cold, the precipitate formed is separated by centrifugation and discarded. The supernatant solution is then brought to 75% saturation with ammonium sulfate, any acidification being avoided as above. After allowing several hours for complete precipitation, the suspension is centrifuged, the precipitate redissolved in 0.03 *M* Tris, pH 7.4, and dialyzed against the same buffer. After the dialysis, the enzyme solution is clarified by centrifugation, if necessary.

Step 3—DEAE–Cellulose Chromatography: The dialyzed ammonium sulfate fraction is applied on a DEAE–cellulose column (5 × 60 cm), equilibrated with 0.035 *M* Tris·HCl, pH 7.4. The column is then washed with the same buffer until the effluent is almost protein-free (the absorbance at 280 nm is below 0.05). The enzyme is eluted with a linear gradient of Tris·HCl, pH 7.4, from 0.035–0.35 *M*. The best fractions (containing more than 100 units of activity per unit of absorbance at 280 nm) are pooled and further purified by one of two alternative procedures.

Step 4A—Chromatography on Hydroxyapatite: The enzyme is dialyzed against 0.01 *M* phosphate buffer, pH 7.15, and applied on a column of hydroxyapatite (2.5 × 20 cm), prepared according to Jenkins,[12] and

[12] J. Jenkins, *Biochem. Prep.* **9**, 83 (1962).

equilibrated against the same buffer. The column is then washed with two volumes of the equilibrating buffer. Elution is started with 0.1 *M* phosphate, pH 7.15. One column volume of this buffer will spread the enzyme down the column. The elution is then achieved by increasing the phosphate concentration to 0.22 *M*. The best fractions are pooled, and if necessary, they can be concentrated by precipitation with ammonium sulfate (70% saturation) with essentially complete recovery.

Step 4B—Chromatography on Calcium Phosphate Gel: A column is prepared by mixing calcium phosphate gel (prepared according to Swingle and Tiselius)[13] with cellulose powder in a ratio of 1:9 (w/w). The mixture is deaerated before being poured into the column. Except for the slightly different ratio of gel to cellulose, the preparation of the column is identical to that described by Massey.[14] The enzyme from step 3 is dialyzed against 5 m*M* Tris·HCl, pH 7.4, and is applied on the column washed with the same buffer. The flavoprotein is adsorbed on top of the column which is then washed with the 5 m*M* Tris buffer.

The enzyme yellow band is then spread down the column by 0.05 *M* phosphate buffer, pH 7.6. This buffer is allowed to flow until the yellow band has traveled down about two-thirds the length of the column. Elution is then achieved with 0.1 *M* phosphate, pH 7.6, containing 3%

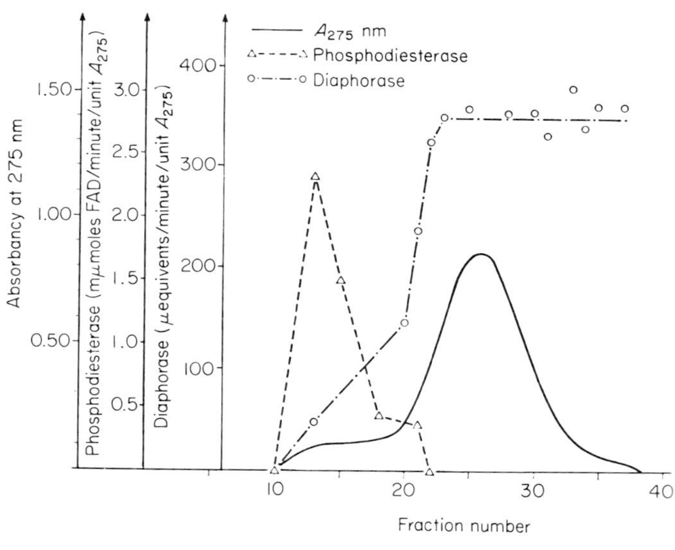

FIG. 1. Elution pattern obtained by chromatography on Sephadex G-150 dextran gel.

[13] S. M. Swingle and A. Tiselius, *Biochem. J.* **48**, 171 (1951).
[14] V. Massey, *Biochim. Biophys. Acta* **37**, 310 (1960).

(w/v) ammonium sulfate. The best fractions are pooled and they can be concentrated by ammonium sulfate precipitation as above.

Step 5—Chromatography on Sephadex G-150 Dextran Gel: The enzyme from step 4A or step 4B can be purified finally by Sephadex G-150 chromatography. This last step removes traces of phosphodiesterase still present in the enzyme of step 4B, sometimes also in 4A. The enzyme is applied to a column of Sephadex G-150, equilibrated with 0.05 *M* Tris·HCl, pH 7.5, to 8.0. Phosphodiesterase is eluted first, then the pure flavoprotein. The elution pattern is shown in Fig. 1.

Properties of the Enzyme

The flavoprotein is a FAD-containing enzyme of molecular weight of 40,000.[2,7] It exhibits a typical flavoprotein spectrum, with peaks (in the oxidized state) at 458, 384, and 276 nm, and a shoulder at 482 nm. The molar extinction coefficients are 10,740 and 85,600 at 458 and 276 nm, respectively, per mole of FAD.[2,7,8] The enzyme-bound FAD is not available to phosphodiesterase in the native enzyme, but it is hydrolyzed by the diesterase if the enzyme is treated with urea or with mercurials.[7]

Oxidation–Reduction: The enzyme can be rapidly reduced by NADPH or by $Na_2S_2O_4$ to the semiquinone level, Fp.FADH•, and more slowly to the fully reduced form.[2,15] The semiquinone exhibits a long wavelength absorption band, from 530 to beyond 600 nm, with a molar extinction coefficient of 1810 at 600 nm.[15] This absorption is bleached upon complete reduction to the level of Fp.$FADH_2$.

Stability: The enzyme can be stored at —20° in 0.05–0.1 *M* Tris·HCl, pH 7.5–8.0, for over one year without loss of activity.

Inhibitors: The enzyme is inhibited by mercurials,[7,16] the most effective being *p*-chloromercuriphenylsulphonate (PCMS). However, the sensitivity to these inhibitors as well as to *N*-ethylmaleimide is low: 0.1 m*M* PCMS inhibits 12% when added to the complete reaction mixture.[7] Preincubation with the mercurials produces inactivation[7] accompanied by the appearance of FAD fluorescence, which is absent in the native enzyme.[7] This process is greatly accelerated if the enzyme is reduced by NADPH in the presence of SH reagents. Under these conditions, 1 μ*M* PCMS inactivates almost completely the enzyme in less than 5 min.[7] The diaphorase activity of the enzyme, but not the cytochrome *f* reductase, is inhibited by preincubation with NADPH.[2,7]

Inorganic pyrophosphate, as well as 2′-AMP and 3′-AMP,[5,6] are inhibitors competitive with NADPH and, in the case of pyrophosphate, with reduced ferredoxin.

[15] G. Forti, B. A. Melandri, A. San Pietro, and B. Ke, *Arch. Biochem. Biophys.* **140,** 107 (1970).

[16] G. Zanetti and G. Forti, *J. Biol. Chem.* **244,** 4757 (1969).

[42] Analytical Procedures for the Isolation, Identification, Estimation, and Investigation of the Chlorophylls[1]

By Harold H. Strain, Benjamin T. Cope, and Walter A. Svec

I. Requirements for Analytical Studies

Analytical studies of the chlorophylls require intimate knowledge of the function, reactions, occurrence and distribution of these green pigments in plants. Most of these studies and techniques must be based upon the physical and chemical properties of the individual pigments. Many of these properties can be determined only with the pure substances isolated from plant material.

Various properties and functions of the chlorophylls have been summarized in reviews[1a-10] and comprehensive books,[11-15] which provide a key to the extensive literature. In the limited space available here, emphasis has been placed upon recent reports most pertinent to useful analytical procedures.

II. Significance of the Chlorophylls

Chlorophylls are the preponderant photosynthetic pigments of the verdant tissues of vascular plants, liverworts, and various algae. They

[1] Work performed under the auspices of the U.S. Atomic Energy Commission.

[1a] S. Aronoff and R. K. Ellsworth, *Photosynthetica* **2**, 288 (1968).

[2] K. Egle, *in* "Handbuch der Pflanzenphysiologie" (W. Ruhland, ed.), Vol. V, Part 1, p. 323. Springer, Berlin, 1960.

[3] J. H. C. Smith and A. Benitez, *in* "Modern Methods of Plant Analysis" (K. Paech and M. V. Tracey, eds.), p. 142. Springer, Berlin, 1955.

[4] Z. Sestak, *Photosynthetica* **1**, 269 (1967).

[5] C. S. French, *in* "Handbuch der Pflanzenphysiologie" (W. Ruhland, ed.), Vol. V, Part 1, p. 252. Springer, Berlin, 1960.

[6] S. Aronoff, *in* "Handbuch der Pflanzenphysiologie" (W. Ruhland, ed.), Vol. V, Part 1, p. 234. Springer, Berlin, 1960.

[7] H. H. Inhoffen, *Naturwissenschaften* **55**, 457 (1968).

[8] N. K. Boardman, *Advan. Enzymol.* **30**, 1 (1968).

[9] D. C. Fork and J. Amesz, *Annu. Rev. Plant Physiol.* **20**, 305 (1969).

[10] P. Daniel, *Wirtschaftseigene Futter* **12**, 346 (1966).

[11] G. P. Gurinovich, A. N. Sevchenko, and K. N. Solov'ev, "Spektroskopiya Khlorofilla i Rodstevennykh Soedinenii." Nauk i Tekhnika, Minsk, 1968.

[12] A. San Pietro (ed.), "Harvesting the Sun: Photosynthesis in Plant Life." Academic Press, New York, 1967.

[13] M. B. Allen (ed.), "Comparative Biochemistry of Photoreactive Systems," Academic Press, New York, 1960.

[14] L. P. Vernon and G. R. Seely (eds.), "The Chlorophylls," Academic Press, New York, 1966. Especially Chapters II, IV, VI, VII, IX.

[15] E. Rabinowitch and Govindjee, "Photosynthesis." Wiley, New York, 1969.

also occur in certain chemoautotrophic bacteria. In most of these organisms, they impart the characteristic green color that usually obscures the yellow and red carotenoid pigments always found with them.[14,16,17]

Only plants containing a particular green pigment, chlorophyll *a*, are capable of photosynthesis with the evolution of oxygen. In some plants, as the yellow-green algae (Heterokontae), the *a* is the only green pigment. In others, it is accompanied by one or two additional green pigments The action spectrum of the green pigments in these plants, indicated by the level of photosynthesis in monochromatic light, at various wavelengths, parallels the absorption spectrum of the chlorophylls. Conversely, organisms lacking chlorophyll, as certain mutants and colorless and yellow chimeras of familiar plants, are incapable of photosynthesis and an autotrophic existence.[15,17]

Chlorophylls do not occur randomly or solitarily within the living cells. They are associated with one or two carotenes (usually β-carotene $\pm$ α-carotene) and several xanthophylls (usually in combinations characteristic of the organisms of each major taxonomic class as may be seen in Table I). All these pigments occur in minute, highly organized, subcellular organs, the chloroplasts. In the organisms of one taxonomic group, the blue-green algae, the diffuse chloroplasts contain proteinaceous pigments, blue phycocyanin in all species plus red C-phycoerythrin in others. In the organisms of another group, the red algae, the discrete chloroplasts also contain proteinaceous pigments, red phycoerythrin in all species plus blue R-phycocyanin in others. In all plants, as shown by microscopy and ultramicroscopy, the chloroplasts have a very labile, lamellar structure with discrete layers of pigments and structural material. Chemical analysis has revealed that, in addition to the pigments, the chloroplasts are also rich in proteins, fats, and other fatty substances.[18-20]

The chlorophyll content of plants varies with their age or stage of development, with the seasons, with the mineral nutrition, and with the intensity and duration of the illumination. Vernal greening, autumnal coloration, and the ripening of many fruits illustrate extreme variation of the pigment content. In green algae and chloroplasts, the pigments may amount to nearly 2% of the fresh weight and nearly 8% of the dry weight.

[16] H. H. Strain, *Biochem. Chloroplasts, Proc. Aberystwyth, Engl. 1965* **1,** 387 (1966).

[17] H. H. Strain, *Annu. Priestley Lect.* **32** (1958).

[18] G. A. Lipskaya, *Usp. Sovrem. Biol.* **65,** 362 (1968).

[19] A. Ongun, W. W. Thomson, and J. B. Mudd, *J. Lipid Res.* **9,** 416 (1968).

[20] R. A. Olson, W. H. Jennings, and J. M. Olson, *Arch. Biochem. Biophys.* **129,** 30 (1969).

TABLE I
PRINCIPAL, FAT-SOLUBLE CHLOROPLAST PIGMENTS OF VARIOUS AUTOTROPHIC PLANTS AND THEIR CHROMATOGRAPHIC SEQUENCE IN COLUMNS OF POWDERED SUGAR WASHED WITH PETROLEUM ETHER PLUS 0.5–2.0% *n*-PROPANOL

Plants	Pigments (most adsorbed at left)
Vascular plants and liverworts[a,b]	Neoxanthin, violaxanthin, Chl *b*, lutein ± zeaxanthin, Chl *a*, β ± α-carotene
Green algae, except Siphonales[c–e]	Neoxanthin, ± loroxanthin, violaxanthin, Chl *b*, lutein ± zeaxanthin, Chl *a*, β ± α-carotene
Siphonalean green algae[f]	±Siphonaxanthin, neoxanthin, violaxanthin, ±siphonein, Chl *b*, lutein ± zeaxanthin, Chl *a*, α + β-carotene
Euglena[g]	Neoxanthin, Chl *b* + diadinoxanthin, Chl *a*, β-carotene
Yellow-green algae (*Vaucheria*)[h,i]	Heteroxanthin, partial ester of vaucheriaxanthin, diatoxanthin, diadinoxanthin, Chl *a*, β-carotene
Brown algae[a,b]	Chls *c*, neofucoxanthins A and B, fucoxanthin, violaxanthin, Chl *a*, β ± α-carotene
Diatoms[a,b,g,j]	Chls *c*, neofucoxanthins A and B, fucoxanthin, diadinoxanthin, diatoxanthin, Chl *a*, β ± α-carotene
Dinoflagellates[b,k]	Chls *c*, neoperidinin, peridinin, neodinoxanthin + neodiadinoxanthin, dinoxanthin + diadinoxanthin, Chl *a*, β-carotene
Red algae[b]	±Chl *d*, zeaxanthin or lutein + zeaxanthin, Chl *a*, β ± α-carotene or α-carotene
Blue-green algae[b,l]	Myxoxanthophyll, zeaxanthin ± lutein, Chl *a*, myxoxanthin, β-carotene

[a] H. H. Strain, *Biochem. Chloroplasts, Proc. Aberystwyth, Engl. 1965* **1,** 387 (1966).
[b] H. H. Strain, *Annu. Priestley Lect.* **32** (1958).
[c] K. Aitzetmüller, H. H. Strain, W. A. Svec, M. Grandolfo, and J. J. Katz, *Phytochemistry* **8,** 1761 (1969).
[d] J. Sherma and G. Zweig, *J. Chromatogr.* **31,** 589 (1967).
[e] H. H. Strain and J. Sherma, *J. Chem. Educ.* **46,** 476 (1969).
[f] H. H. Strain, *Biol. Bull.* **129,** 366 (1965).
[g] K. Aitzetmüller, W. A. Svec, J. J. Katz, and H. H. Strain, *Chem. Commun.* p. 32 (1968).
[h] H. H. Strain, W. A. Svec, K. Aitzetmüller, M. Grandolfo, and J. J. Katz, *Phytochemistry* **7,** 1417 (1968).
[i] K. Egger, H. Nitsche, and H. Kleinig, *Phytochemistry* **8,** 1583 (1969).
[j] A. K. Mallams, E. S. Waight, B. C. L. Weedon, D. J. Chapman, F. T. Haxo, T. W. Goodwin, and D. M. Thomas, *Chem. Commun.* p. 301 (1967).
[k] J. P. Riley and T. R. S. Wilson, *J. Mar. Biol. Ass. U.K.* **47,** 351 (1967).
[l] S. Hertzberg and S. L. Jensen, *Phytochemistry* **5,** 557, 565 (1966).

The synthesis of chlorophylls within the chloroplasts is influenced genetically, as in the colorless, yellow, and pale green mutants frequently observed among seedings.[21,22] In most plants, the synthesis of the chlorophyll is coupled with the action of light, because it does not occur in yellow, etiolated seedlings grown in the dark. These seedlings usually contain very small quantities of a unique green pigment, protochlorophyll or its chlorophyllide, that disappears as chlorophyll *a* and *b* are formed upon exposure of the seedlings to visible light.[23-28] Irradiation of plant material with γ rays inhibits chlorophyll formation.[29-31]

Thus far, a chemical participation of chlorophyll in the process of photosynthesis has not been established. Although the chlorophyll serves as the primary absorber of the radiant energy of the sunlight utilized in photosynthesis, the mechanism for the utilization of this absorbed energy remains a matter of controversial speculation. According to one proposal, for example, each absorbed quantum may split a C–O or O–H bond. According to another, the energy may be conveyed to special reactive sites.[8,15,32-34]

Because of their function as the primary agents for the absorption and utilization of sunlight through photosynthesis, the chlorophylls are indispensable to the production of virtually all organic matter. In this way, they are responsible for the nutrition of all the other organisms. Through their activity in past ages, they have provided the prodigious quantities of the organic fuels that now power man's mechanized industrial activities.

[21] F. Lang, L. M. Vorob'eva, and A. A. Krasnovskii, *Dokl. Akad. Nauk SSSR* **183**, 711 (1968).
[22] C. R. Benedict and R. J. Kohel, *Plant Physiol.* **43**, 1611 (1968).
[23] G. Akoyunoglou and J. H. Argyroudi-Akoyunoglou, *Physiol. Plant.* **22**, 288 (1969).
[24] L. M. Vorob'eva and A. A. Krasnovskii, *Biofizika* **13**, 456 (1968).
[25] L. Bogorad, L. J. Laber, and M. Gassman, *in* "Comp. Biochem. Biophys. of Photosynthesis" (K. Shibata, A. Takamiya, A. T. Jagendorf and R. C. Fuller, eds.), p. 299. Univ. of Tokyo Press, Tokyo, Japan, 1968.
[26] A. B. Rudoi, A. A. Shlyk, and A. Y. Vezitskii, *Dokl. Akad. Nauk SSSR* **183**, 215 (1968).
[27] C. Sironval, M. R. Michel-Wolwertz, and A. Madsen, *Biochim. Biophys. Acta* **94**, 344 (1965).
[28] J. B. Harris and A. W. Naylor, *Tobacco Sci.* **12**, 171 (1968).
[29] R. Ziegler, S. H. Schanderl, and P. Markakis, *J. Food Sci.* **33**, 533 (1968).
[30] C. Sironval, R. Kirchman, R. Bronchart, and J. M. Michel, *Photosynthetica* **2**, 57 (1968).
[31] W. Gottschalk and F. Müller, *Planta* **61**, 259 (1964).
[32] E. Roux, *C. R. Acad. Agr. Fr.* **54**, 329 (1968).
[33] W. Arnold and J. R. Azzi, *Proc. Nat. Acad. Sci. U.S.* **61**, 29 (1968).
[34] A. Joliot, *Physiol. Veg.* **6**, 235 (1968).

Analytical determinations of the chlorophylls are of increasing interest in various fields as plant physiology, biochemistry, plant nutrition, the organic production of land and sea, medicine, paleontology, and food technology. In the preservation and marketing of food and fodder, chlorophyll content is constantly employed, even if subconsciously, as an index of quality.[35,36]

III. Occurrence of Chlorophylls

Natural Chlorophylls

The word chlorophyll was derived from Greek roots indicating the green of leaves. In the early investigations, this green matter was presumed to be a single substance, but classical partition and spectroscopic investigations indicated that two green chlorophylls occur in leaves. The highly definitive chromatographic experiments by Tswett and by many subsequent workers have shown that only these two green pigments are present in leaves.[37] The major green component is chlorophyll *a*; the minor component is chlorophyll *b*. In certain algae, by contrast, one or two minor green pigments, as *d* or modifications of *c*, accompany the *a* (Table I). Chlorophyll and chlorophylls have, therefore, become generic terms for the several green constituents of photosynthetic organisms.

Both the *a* and the *b*, in the ratios (*a* to *b*) of about 2:1 to 3:1, occur in vascular plants, liverworts, green algae, and Euglenophyceae. The *b* does not occur elsewhere in the vegetable kingdom. It plays a functional role, but not an indispensable role, because certain barley mutants without the *b* are truly autotrophic.[38,39] In the yellow-green and blue-green algae, the *a* is unaccompanied by other green pigments; hence it is capable of its photosynthetic function without the aid of auxiliary green pigments.

Tswett's early chromatographic experiments established the presence of a minor chlorophyll in brown algae (Phaeophyceae) in addition to *a*. This chlorophyll *c*, also found in diatoms and dinoflagellates, has been isolated as infusible crystals. As obtained from diatoms and brown algae, these crystals are comprised of two molecular species that differ by two hydrogen atoms and that are readily detectable by mass spectrometry of derivatives.[40] These two species, c_1 and c_2,[41] are readily separable by

[35] M. S. Eheart, *Food Technol.* (*Chicago*) **23,** 238 (1969).

[36] F. M. Clydesdale and F. J. Francis, *Food Technol.* (*Champaign, Ill.*) **22,** 793 (1968).

[37] H. H. Strain and J. Sherma, *J. Chem. Educ.* **44,** 235, 238 (1967).

[38] N. K. Boardman and S. W. Thorne, *Biochim. Biophys. Acta* **153,** 448 (1968).

[39] A. Meister and T. G. Maslova, *Photosynthetica* **2,** 261 (1968).

[40] R. C. Dougherty, H. H. Strain, W. A. Svec, R. A. Uphaus, and J. J. Katz, *J. Amer. Chem. Soc.* **88,** 5037 (1966); **92,** 2826 (1970).

chromatography upon a special preparation of polyethylene.[41,42] The less sorbed component is c_1; the other is c_2.

Chlorophyll *d* is a minor component of some red algae (Rhodophyceae), but it is absent in others.[17] Chlorophyll *e*, reported as a minor chlorophyll in an old, natural stand of yellow-green algae,[17] has not been found in well-nourished, pure cultures of these organisms.

As noted already, photochlorophyll occurs in minute quantities in yellow, etiolated seedlings, obtained by germinating seeds in the dark. In these seedlings, the green pigment may occur in two forms, the normal ester with phytol or the free acid, protochlorophyllide.[23-26]

Bacteriochlorophyll is the principal chlorophyll of the chemoautotrophic purple sulfur bacteria. Another green pigment, first called bacterioviridin but later chlorobium chlorophyll, is characteristic of the green, chemoautotrophic sulfur bacteria, wherein it is sometimes accompanied by small quantities of bacteriochlorophyll. Different isolates of these bacteria contain chlorobium chlorophylls that differ slightly, especially in the wavelengths of their absorption maxima; hence they are known as chlorobium chlorophylls 650 and 660.[14]

Isotopically Modified Chlorophylls

The usual or normal chlorophylls have been obtained in various, isotopically labeled modifications by supplying the growing plants with isotopic compounds of carbon and hydrogen. This method with $^{13}CO_2$ and $^{14}CO_2$ has provided chlorophylls randomly labeled with ^{13}C and ^{14}C.[42,43] With culture media containing D_2O and T_2O, the chlorophylls are randomly labeled with D and T. With pure D_2O, in which some algae are capable of autotrophic development, the chlorophylls are fully deuterated, all the H being replaced by D.[14] With isotopes of magnesium, the chlorophylls are correspondingly labeled. With media containing mostly D, ^{13}C, ^{15}N, and ^{18}O, autotrophic organisms composed of these isotopes have been obtained.[44]

Contaminants of Chlorophylls

Many fatty or fatlike substances are extracted from plant material with the solvents required for the extraction of the fat-soluble chlorophylls.[18,19] These natural contaminants must, therefore, be considered in

[41] S. W. Jeffrey, *Biochem. Biophys. Acta* **177**, 456 (1969).

[42] R. F. McFeeters and S. H. Schanderl, *J. Food Sci.* **33**, 547 (1968).

[43] R. Brown *J. Fish. Res. Board Can.* **25**, 523 (1968).

[44] J. J. Katz and H. L. Crespi, *in* "Recent Advances in Phytochemistry" (M. K. Seikel and V. C. Runeckles, eds.), Vol. II, p. 1. Appleton-Century-Crofts, New York, 1969.

the selection of methods for investigation of the green pigments. The most conspicuous contaminants are the yellow and orange carotenoid pigments. More abundant, but less conspicuous, colorless contaminants include hydrocarbons and ketones of large molecular weight, sterols, fats, and various additional fatty substances. Alteration products of chlorophylls, formed upon injury of the plants or in solutions of the pigments, often appear as contaminants of the normal green pigments.[14]

IV. Physical Properties of Chlorophylls

Properties Common to Chlorophylls

Chlorophylls have many striking physical properties in common. They are green to gray-green and purple pigments with pronounced absorption bands in the blue-green and red spectral regions. They are strongly fluorescent, the wavelength of the emitted light usually corresponding to that of the principal absorption band in the red spectral region.

Chlorophylls have tetrapyrrolic molecular structures, usually with an additional isocyclic ring. The four pyrrole rings are linked by >CH (methine) bridges in a larger, cyclic, nearly planar structure. In this way, the nitrogen atom of each pyrrole ring is held near the center. A magnesium atom is fixed among these four nitrogen atoms thus forming the "hub" of the multicyclic molecule.

The structure for chlorophyll *a* is shown in Fig. 1. The structures

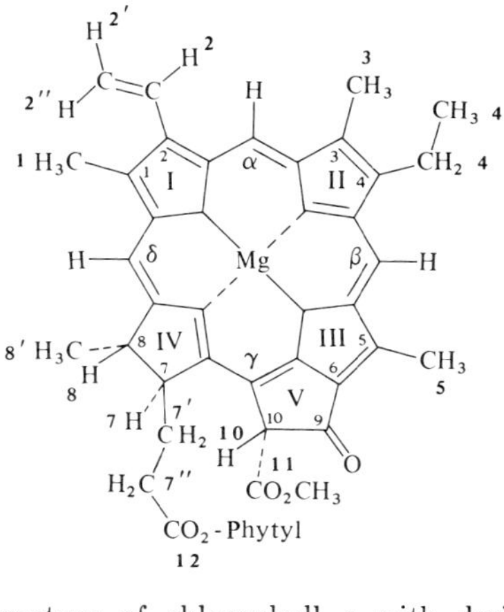

FIG. 1. Molecular structure of chlorophyll *a* with designations of the carbon atoms (light face numerals and Greek letters), protons (bold face numerals), and rings (Roman numerals).

TABLE II
PROPERTIES OF CHLOROPHYLLS[a]

Chlorophyll a
Formula: $C_{55}H_{72}O_5N_4Mg$. Mol. Wt.: 893.48. % Mg: 2.722
Structure: see Fig. 1
Abs. max. ether: 428.5 nm; *e*, 125,000. 660 nm; *e*, 96,600. Ratio: 1.294; see Fig. 2
Fluorescence max. ether: 668 (723) nm. NMR. IR

Deuteriochlorophyll a
Formula: $C_{55}D_{72}O_5N_4Mg$. Mol. Wt.: 965.48. % Mg: 2.519
Structure: H— replaced by D— in Fig. 1
Abs. max. ether: 428.0 nm; *e*, 116,100. 659.0 nm; *e*, 88,600. Ratio: 1.310; see Fig. 2

Chlorophyll b
Formula: $C_{55}H_{70}O_6N_4Mg$. Mol. Wt.: 907.46. % Mg: 2.680
Structure: —CHO in place of —CH_3 at C-3 in Fig. 1
Abs. max. ether: 452.5 nm; *e*, 175,300. 642.0 nm; *e*, 61.8. Ratio: 2.837
Fluorescence max. ether: 649 (708) nm. NMR. IR

Deuteriochlorophyll b
Formula: $C_{55}D_{70}O_6N_4Mg$. Mol. Wt.: 977.46. % Mg: 2.488
Structure: H— replaced by D— in chlorophyll *b*
Abs. max. ether: 451.0 nm; *e*, 165,700. 640.5 nm; *e*, 57,900. Ratio: 2.862

Chlorophyll c_1[b,c]
Formula: $C_{35}H_{30}O_5N_4Mg$. Mol. Wt.: 610.54. % Mg: 3.980
Structure: —CH=CHCOOH in place of the propionyl phytyl ester at C-7, double bond at C-7 to C-8 in Fig. 1
Abs. max. ether: av. 444.4 nm, 628.2 nm. Ratio: 10.0

Chlorophyll c_2[b,c]
Formula: $C_{35}H_{28}O_5N_4Mg$. Mol. Wt.: 608.53. % Mg: 3.997
Structure: as c_1 but with —CH=CH_2 in place of ethyl at C-4
Abs. max. ether: av. 448.4 nm. 628.8 nm. Ratio: 14.5

Chlorophyll d
Formula: $C_{54}H_{70}O_6N_4Mg$. Mol. Wt.: 895.45. % Mg: 2.716
Structure: —CHO in place of CH_2=CH— at C-2 in Fig. 1
Abs. max. ether: 447 nm; *e*, 97,800. 688 nm; *e*, 110,400. Ratio: 0.889
Fluorescence max. ether: 693 ($\approx$705) nm

Protochlorophyll
Formula: $C_{55}H_{70}O_5N_4Mg$. Mol. Wt.: 891.46. % Mg: 2.728
Structure: double bond between C-7 and C-8 in Fig. 1
Abs. max. ether: 432 nm; *e*, 325,500. 623 nm; *e*, 39,900. Ratio: 8.158
Fluorescence max. ether: 629 nm

Bacteriochlorophyll
Formula: $C_{55}H_{74}O_6N_4Mg$. Mol. Wt.: 911.49. % Mg: 2.668
Structure: CH_3C(=O)— in place of CH_2=CH— at C-2 in Fig. 1
Abs. max. ether: 357.7 nm; *e*, 87,600. 770 nm; *e*, 102,000. Ratio: 0.859
Fluorescence max. ether: 695 (805) nm

(Continued)

TABLE II (*Continued*)

Deuteriobacteriochlorophyll

Formula: $C_{55}D_{74}O_6N_4Mg$. Mol. Wt.: 985.49. % Mg: 2.468
Structure: H— replaced by D— in bacteriochlorophyll
Abs. max. ether: as for bacteriochlorophyll

Chlorobium chlorophyll 660

Formula: uncertain due to some six components. Mol. Wt.: 781.31 to 837.42
Structure: ethyl, *n*-propyl or isobutyl at C-4; methyl or ethyl at C-5; methyl or ethyl at C-δ; farnesol in place of phytol; see Fig. 1
Abs. max. ether: (mixture) 430 nm; *e*, 151,100. 660 nm; *e*, 98.600. Ratio: 1.532
Fluorescence max. ether: 667

Chlorobium chlorophyll 650

Formula: uncertain due to some six components. Mol. Wt.: 767.29 to 809.36
Structure: ethyl, *n*-propyl or isobutyl at C-4; methyl or ethyl at C-5 in different combinations than in the 660, but H— remaining at C-δ; farnesol in place of phytol; see Fig. 1
Abs. max. ether: (mixture) 425 nm; *e*, 146,000. 650 nm; *e*, 113,400. Ratio: 1.286

[a] L. P. Vernon and G. R. Seely (eds.), "The Chlorophylls." Academic Press, New York, 1966. Especially Chapters II, IV, VI, VII, IX.
[b] R. C. Dougherty, H. H. Strain, W. A. Svec, R. A. Uphaus, and J. J. Katz, *J. Amer. Chem. Soc.* **88,** 5037 (1966); **92,** 2826 (1970).
[c] S. Jeffrey, *Biochim. Biophys. Acta* **177,** 456 (1969).

of the other chlorophylls are easily related to this structure for *a* as summarized in Table II.

Solubility properties of the chlorophylls are of great significance for the development and application of analytical techniques. All the chlorophylls, like the yellow carotenoids, are soluble in fats and fat solvents, but the phycobilins of the blue-green and red algae are insoluble. Chlorophylls and carotenoids are insoluble in water, but with emulsifying and stabilizing agents, they form colloidal suspensions from which they are extractable with great difficulty by the water-insoluble, fat solvents as ether and petroleum ether.[45,46]

Properties of Individual Chlorophylls

The development and application of analytical methods for studies of the chlorophylls are based upon the properties of the individual pigments and of the chlorophyll alteration products. Preparative procedures for isolation of the individual chlorophylls have been developed into a highly specialized art. Further reference to them will be made in the sections

[45] J. P. Thornber, *Biochim. Biophys. Acta* **172,** 230 (1969).
[46] C. N. Cederstrand, E. Rabinowitch, and Govindjee, *Biochim. Biophys. Acta* **120,** 247 (1966).

TABLE III

CHROMATOGRAPHIC SEQUENCE OF VARIOUS CHLOROPHYLLS AND SOME OF THEIR ISOMERS IN COLUMNS OF POWDERED SUGAR WITH PETROLEUM ETHER PLUS 0.5–2.0% *n*-PROPANOL AS THE WASH LIQUID[a]

Various chlorophylls	Color of zones
Chlorophylls *c* (most adsorbed)[b]	Yellow green
[Chlorobium chlorophyll 650	Green
Chlorobium chlorophyll 660]	Green
[Deuteriochlorophyll *b*	Yellow green
Chlorophyll *b*]	Yellow green
Chlorophyll *d*	Yellow green
[Bacteriochlorophyll	Purplish gray
Deuteriochlorophyll *b′*	Yellow green
Chlorophyll *b′*]	Yellow green
Chlorophyll *d′*	Yellow green
Isochlorophyll *d*	Green
[Deuteriochlorophyll *a*	Green
Chlorophyll *a*]	Green
Protochlorophyll	Green
Isochlorophyll *d′*	Green
[Deuteriochlorophyll *a′*	Green
Chlorophyll *a′* (least adsorbed)]	Green

[a] L. P. Vernon and G. R. Seely (eds.), "The Chlorophylls." Academic Press, New York, 1966. Especially Chapters II, IV, VI, VII, IX.

[b] Brackets indicate pigments incompletely separated under the adsorption conditions.

dealing with chromatography and with the separation of the chlorophylls in plant extracts (see also Tables I, III, and IV).

Once determined, the constants for the chlorophylls may be employed without the necessity of reisolating the pure pigments. In this brief summary, it is impossible to present all the properties of each chlorophyll. It has been necessary, therefore, to select those properties of primary significance in analytical procedures.

Fluorescence. The fluorescence, produced by ultraviolet light, of chlorophylls in plants is relatively weak compared to their fluorescence in solution. Moreover, the fluorescence of the several chlorophylls and their alteration products vary with conditions as temperature and salt concentration.[47–49] They also overlap so that fluorescence alone does not serve as a precise basis for identification and estimation of the individual

[47] N. Murata, *Biochim. Biophys. Acta* **172,** 242 (1969).

[48] G. Hind, *Biochim. Biophys. Acta* **172,** 290 (1969).

[49] F. A. J. Armstrong, C. R. Stearns, and J. D. H. Strickland, *Deep-Sea Res. Oceanogr. Abstr.* **14,** 381 (1967).

TABLE IV
CHROMATOGRAPHIC SEQUENCE AND ABSORPTION MAXIMA OF VARIOUS ALTERATION PRODUCTS OF CHLOROPHYLLS *a* AND *b* IN COLUMNS OF POWDERED SUGAR WITH PETROLEUM ETHER PLUS 0.5–2% *n*-PROPANOL[a]

Pigments	Formation	Abs. maxima (ether) Blue (nm)	Abs. maxima (ether) Red (nm)	Ratio blue/red
Chlorophyllide *b*[b]	Enzym.	451	641.5	—
Chlorophyllide *a*	Enzym.	428.5	660.5	—
[Oxidized chlorophyll *b*	Enzym.	450.5	641	2.81
Methyl chlorophyllide *b*]	Enzym.	451	641.5	2.84
Pyrochlorophyll *b*	Heat, pyridine	—	—	—
Chlorophyll *b*	Natural	452.5	642	2.84
[Chlorophyll *b′*	Heat	≈451	641	—
Methyl pyrochlorophyllide *a*]	Enzym. + heat, pyr.	428	659	1.52
Oxidized chlorophyll *a*	Enzym.	428	660.5	1.29
Methyl chlorophyllide *a*	Enzym.	428.5	660.5	1.29
Methyl chlorophyllide *a′*	Enzym. + heat	—	—	—
Methyl pheophorbide *a*	HCl + CH_3OH	408.5	667	2.07
Methyl pyropheophorbide *a*	Heat and acid	408.5	667	2.09
Pyrochlorophyll *a*	Heat, pyridine	429	659.5	1.49
Chlorophyll *a*	Natural	428.5	660.5	1.295
Chlorophyll *a′*	Heat	≈428	660	—
Pheophytin *a*	HCl	409	667	2.09
Pyropheophytin *a* (least adsorbed)	HCl	409	667	2.09

[a] L. P. Vernon and G. R. Seely (eds.), "The Chlorophylls." Academic Press, New York, 1966. Especially Chapters II, IV, VI, VII, IX.

[b] Brackets indicate pigments incompletely separated under the adsorption conditions.

pigments within the chloroplasts[50,51] or in the extracts.[52–54] For the location of chlorophylls in plants and for the detection of trace amounts of chlorophylls and of their alteration products in chromatograms, however, fluorescence is a remarkably sensitive property.

Spectral Absorption of Visible Light. The electronic absorption has been the most widely employed method for the description, identification, and estimation of the individual chlorophylls and many of their colored

[50] M. Brody, B. Nathanson, and W. S. Cohen, *Biochim. Biophys. Acta* **172,** 340 (1969).

[51] W. A. Cramer and W. L. Butler, *Biochim. Biophys. Acta* **172,** 503 (1969).

[52] V. L. Kaler and G. M. Podchufarova, *Fiziol-Biokhim. Issled. Rast., Inst. Eksp. Bot. Mikrobiol. Akad. Nauk Beloruss. SSR.* 20 (1965).

[53] L. O. Bjorn, *Physiol. Plant.* **22,** 1 (1969).

[54] Y. Saijo and S. Nishizawa, *Mar. Biol.* **2,** 135 (1969).

alteration products. Absorption has been employed colorimetrically in plants[5,55,56] and in extracts,[57,58] and spectrophotometrically, as summarized in several reviews.[3,5,10,14,59-61] The absorption spectra have been determined with pigments under many different conditions, as in living plant cells,[62] in suspensions of chloroplasts, in extracts of plant material, and in suspensions and solutions of the individual pigments.

The absorption spectra have also been utilized in various ways. Wavelengths of the absorption maxima have served for the description and identification of the so-called "forms" of chlorophyll in living cells and of the individual pigments, as chlorobium chlorophylls 650 and 660. The qualitative absorption spectra serve to indicate the complexity of many mixtures of pigments, especially the presence of colored contaminants in pigment preparations. For quantitative estimations, the specific extinction coefficients, *e*, of the pure pigments have provided the basis of the widely employed, and convenient, spectrophotometric methods.[3,5,10,14]

Within the chloroplasts, the chlorophylls exhibit great variation of their spectral properties. The spectral absorption maxima, not nearly so pronounced as in solutions of the pigments, vary with the age and condition of the plant. These effects result, in part, from the unusually high concentration of the pigments, from the presence of much colorless, structural material that scatters the incident light, and from the combination of the chlorophyll with itself and with various constituents of the chloroplasts.[62,63]

In solution in organic liquids, the chlorophylls exhibit well-defined and readily reproducible absorption properties. Regardless of the association or combinations of the chlorophyll in solid preparations or in the chloroplasts,[14,27,64-66] only the nonagglomerated pigment is found in the

[55] W. F. McClure, *Tobacco Sci.* **168,** 22 (1969).

[56] H. I. Virgin and L. G. Arvidsson, *Physiol. Plant.* **21,** 1177 (1968).

[57] L. A. Appelquist and S. A. Johansson, *Sver. Utsaedesfoeren. Tidskr.* **78,** 415 (1968).

[58] H. Borriss and K. H. Koehler, *Jena Rev.* (*Engl. Transl.*) **13,** 232 (1968).

[59] V. B. Evstigneev and L. I. Prokhorova, *Biokhimiya* **33,** 286 (1968).

[60] A. A. Shlyk, *Biokhimiya* **33,** 275 (1968).

[61] A. Ballester, *Invest. Pesquera* **30,** 613 (1966).

[62] I. L. Pyrina and N. P. Mokeeva, *Tr. Inst. Biol. Vnutr. Vod, Akad. Nauk SSSR* **11,** 198 (1966).

[63] Y. Mukohata, *in* "Comp. Biochem. Biophys. of Photosynthesis" (K. Shibata, A. Takamiya, A. T. Jagendorf, and R. C. Fuller, eds.), p. 89. Univ. of Tokyo Press, Tokyo, Japan, 1968.

[64] Y. E. Giller, G. V. Krasichkova, and D. I. Sapozhnikov, *Dokl. Akad. Nauk SSSR* **182,** 1230 (1968).

[65] A. A. Krasnovskii and M. I. Bystrova, *Dokl. Akad. Nauk SSSR* **182,** 211 (1968).

[66] K. Ballschmiter, T. M. Cotton, H. H. Strain, and J. J. Katz, *Biochim. Biophys. Acta* **180,** 347 (1969).

solutions and extracts prepared with polar or weakly polar liquids at room temperature.[67] Spectra of the extracted or dissolved chlorophylls vary with the proportion of water in the extracts, as in acetone extracts, and with the presence of colloidal chlorophylls and colloidal, colorless substances.[14] Colloidal substances scatter the incident light so that their presence is indicated by apparent absorption in the red spectral region beyond that absorbed by the chlorophyll itself. In the solutions, however, the light scattering is much greater in the blue spectral region than in the red. Consequently, a scattering value determined in the red region is not precisely applicable in the blue and green regions.[68]

The interference of colloidal substances with the spectral absorption may be avoided by *quantitative* transfer of the pigments from the extracts to diethyl ether, as by the addition of this liquid and an excess of aqueous salt solution to acetone or methanol extracts. (With the colloidal pigments, however, quantitative transfers are often very tedious, because several extractions may be required, and complete transfer is difficult to achieve and to establish.) If extraction is incomplete, the analysis must be discarded. With any valuable or irreplaceable material, the solution may be diluted with alcohol or acetone until the colloid is dissociated and the pigment is then transferred to ether as just described.

Interpretation of absorption measurements of chlorophyll solutions is most precise when only one natural pigment is present and when chlorophyll alteration products are absent. If two or three chlorophylls with absorption maxima at different wavelengths are present, determination of the absorbance at the absorption maximum of each pigment in the red spectral region provides a basis for estimation of each pigment as *a* and *b*[3,5,10,14,59-61,69,70] and *a*, *b*, and *c*.[71-78] Such measurements do not, however, provide significant information concerning the number of alteration products of each chlorophyll that might be present. It is a desirable precaution, therefore, to make a chromatographic examination of the

[67] S. S. Brody and S. B. Broyde, *Biophys. J.* **8,** 1511 (1968).
[68] J. C. Madgwick, *Deep-Sea Res.* **13,** 459 (1966).
[69] R. Ziegler and K. Egle, *Beitr. Biol. Pflanz.* **41,** 11 (1965).
[70] J. Billot, *Physiol. Veg.* **5,** 341 (1967).
[71] K. Banse and G. C. Anderson, *Limnol. Oceanogr.* **12,** 696 (1967).
[72] E. V. Belogorskaya, *Tr. Sevastopol. Biol. Sta. Akad. Nauk Ukr. SSR.* **17,** 221 (1964).
[73] R. Boje, *Limnologica* **4,** 397 (1966).
[74] Z. Z. Finenko and L. A. Lanskaya, *Okeanologiya* **8,** 839 (1968).
[75] T. R. Ricketts, *Phytochemistry* **4,** 725 (1965).
[76] T. R. Ricketts, *Phytochemistry* **6,** 1353 (1967).
[77] B. Wauthy and J. Le Bourhis, *Oceanogr.* **5,** 59 (1967).
[78] B. Wauthy and J. Le Bourhis, *Oceanogr.* **4,** 3 (1966).

extract. With extracts in which a portion of the chlorophyll is present as the magnesium-free pheophytin, it is desirable to convert all the remaining chlorophyll to the pheophytins with acid. These are then estimated by spectral absorption either before or after transfer to ether.[14,79-81]

Spectral absorption curves may be determined in any solvent that will dissolve the chlorophylls without decomposition. In practice, however, acetone, which may be employed for direct extraction of fresh plant material, and ether, into which the pigments may be transferred after extraction from the plant material with acetone or methanol plus ether or petroleum ether, have proved most useful and convenient.

Typical quantitative, spectral absorption curves of chlorophyll *a* and of fully deuterated chlorophyll *a* dissolved in ether are illustrated in Fig. 2. Corresponding curves for the other chlorophylls are readily available.[14]

For a given pigment in a given solvent, the spectral absorption curves plotted as log *e* versus the wavelength are superposable regardless of the concentration of the pigment. Variation of the shape of the curves indicates contamination by other pigments. In practice, this comparison is commonly restricted to determination of the ratio of the absorption at

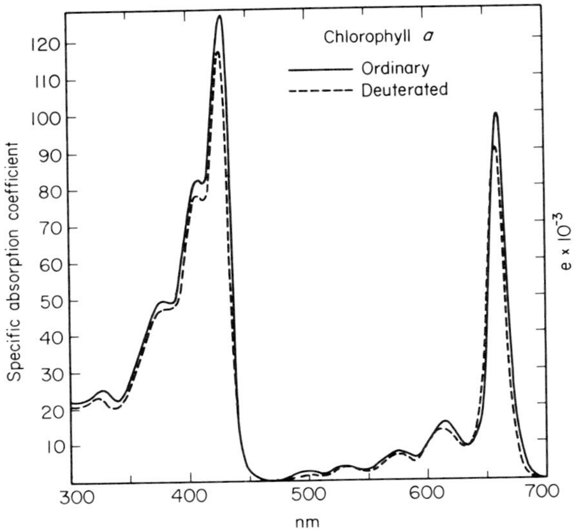

FIG. 2. Spectral absorption curves for ordinary chlorophyll *a* (solid line) and fully deuterated chlorophyll *a* (dashed line) in ether.

[79] C. J. Lorenzen, *Limnol. Oceanogr.* **12,** 343 (1967).
[80] B. Moss, *Limnol. Oceanogr.* **12,** 335, 340 (1967).
[81] S. Jeffrey and K. Shibata, *Biol. Bull.* **136,** 54 (1969).

the maximum in the blue region to that in the red (see Tables II and IV).

Spectral absorption curves of some chlorophyll alteration products, as chlorophyllide *a* and methyl chlorophyllide *a* (Table IV), are identical to those of the parent chlorophyll *a* with respect to the shape of the curves but not with respect to the magnitude of the coefficients. If the *a* were converted to these two products, through the action of chlorophyllase during the extraction, use of the *a* coefficients would give the correct value for the unchanged *a* in the plant, but would not provide any information about the presence of the altered pigments.

Absorption in the Infrared. The IR absorption by chlorophylls is useful for determination of their molecular structure and for ascertaining the degree and nature of their mild association, commonly referred to as agglomeration or aggregation. The IR must be employed with the individual pigments, which may be in the form of thin layers or films, melts on KBr plates, KBr pellets, or solutions in various solvents that are transparent to the IR.[14] This technique is not readily applicable to the precise estimation of chlorophylls.

Nuclear Magnetic Resonance. Owing to the development of extremely sensitive and reliable NMR spectrometers, the resonance frequencies of the protons in the chlorophyll molecules may be determined with great precision, and the characteristic frequencies have been assigned to par-

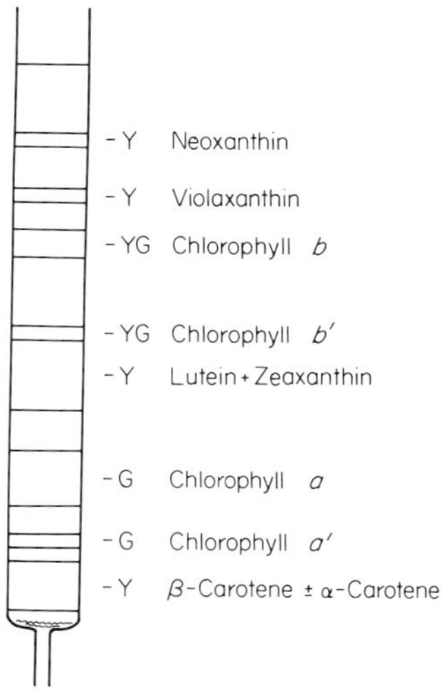

FIG. 3. Chlorophylls *a* and *b*, chlorophyll isomers *a′* and *b′*, and carotenoids of leaves heated in boiling water. Separation with powdered sugar and petroleum ether plus 0.5% *n*-propanol. Y, yellow; YG, yellow green; G, green.

ticular protons, as indicated by the boldface numbers in Fig. 1. This technique facilitates the determination of the molecular structure of isomerized chlorophylls, as a' and b',[82] of natural chlorophylls as c,[40] of various deuterium exchange products, and of chemical alteration products as pyrochlorophyll and oxidized chlorophyll.[83] It also aids in establishing the molecular interactions in agglomerated chlorophylls and the interactions between the solvent and the dissolved chlorophyll.[19]

Chromatography. Solution adsorption chromatography is the simplest and most effective method for the isolation of the chlorophylls from plant extracts. This method, in its various geometric forms as columns, thin layers, and paper with one-way and two-way or transverse development, serves for the separation of the chlorophylls from most, but not necessarily all, of the carotenoids and colorless substances extracted with the pigments (see Table I and Figs. 3 and 4). It also serves for the separation of the chlorophylls and their alteration products from one another (see Fig. 3 and Tables III and IV).[84-86]

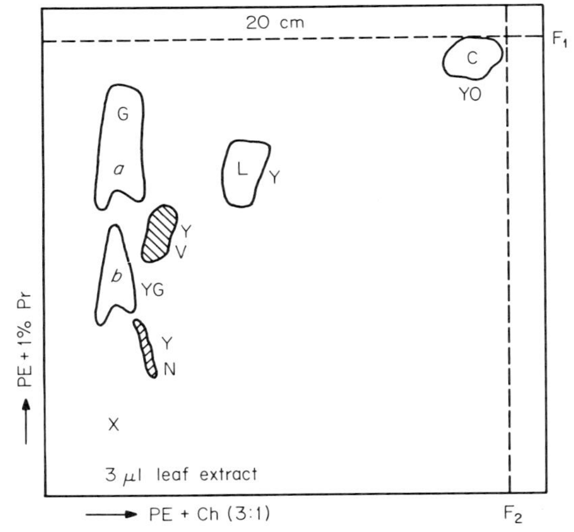

FIG. 4. Chlorophylls a (G, green) and b (YG, yellow green), carotenes (C; YO, yellow orange), and xanthophylls (N, neoxanthin; V, violaxanthin; L, lutein; Y, yellow) separated by two-way chromatography on paper washed first with petroleum ether (PE) plus 1% n-propanol (Pr), then transversely with petroleum ether plus chloroform (Ch) (3:1).

[82] J. J. Katz, G. D. Norman, W. A. Svec, and H. H. Strain, *J. Amer. Chem. Soc.* **90,** 6841 (1968).

[83] F. C. Pennington, H. H. Strain, W. A. Svec, and J. J. Katz, *J. Amer. Chem. Soc.* **89,** 3875 (1967).

Much attention has been devoted to the selection of adsorbents and solvents for the chromatography of the chlorophylls.[86,87] Many inorganic adsorbents alter the pigments so that they do not form discrete zones and so that their elution is incomplete. Mild inorganic adsorbents, as calcium carbonate and certain silica gel preparations, have, however, provided useful separations.[86,87] Many organic adsorbents produce a minimum of alteration. Of these, the saccharides have found extensive use. Powdered glucose[68] and sucrose,[17,85,86] both in the form of columns and thin layers, have effected excellent separations and have provided quantitative recovery of the pigments. Cellulose also provides approximately equivalent separations when employed as columns, as thin layers, and as paper.[4,88,89] With thin layers of cellulose or with paper exposed to light and air, oxidative changes of the adsorbed pigments may occur with the formation of secondary green substances.[90] The chromatographic separation of all these compounds depends upon the properties of the solvent or wash liquid as well as upon the sorbent.[84] For the separations on sugar, an effective solvent or wash liquid is petroleum ether plus 0.25 to about 1% or 2% *n*-propanol.

The separation of the mixtures of pigments varies with the loading as well as with the sorbent and the solvent. The lighter the loading, the more effective the separations. At the lower loading, however, the minor pigments may be very difficult to detect. For preparative purposes, it is economical to overload the sorbent and to purify each pigment further by readsorption.[86]

Even after preparation by chromatography and rechromatography, the chlorophylls may be contaminated with colorless substances. Consequently, pure chlorophylls will not be obtained if the solutions of the eluted pigments are evaporated to dryness. It is better to transfer the eluted pigments to petroleum ether (b.p. 20°–40°) containing a little methanol. When these solutions are washed with water (to remove the methanol), the chlorophylls separate, especially after cooling the solutions, and are collected by centrifugation and dried in vacuum.[14,85]

Chlorophylls prepared in this way always contain varying amounts of water. Some of this water may be entrapped in the solid, but much of it is combined in the solid, agglomerated chlorophyll. Chlorophyll *a*, for

[84] H. H. Strain and J. Sherma, *J. Chem. Educ.* **46,** 476 (1969).
[85] H. H. Strain and W. A. Svec, *Advan. Chromatog.* **8,** 119 (1969).
[86] H. H. Strain, J. Sherma, and M. Grandolfo, *Anal. Biochem.* **24,** 54 (1968).
[87] J. Sherma and G. S. Lippstone, *J. Chromatogr.* **41,** 220 (1969).
[88] H. H. Strain, J. Sherma, F. L. Benton, and J. J. Katz, *Biochim. Biophys. Acta* **109,** 1, 16, 23 (1965).
[89] H. A. W. Schneider, *J. Chromatogr.* **21,** 48 (1966).
[90] M. F. Bacon, *Biochem. J.* **101,** 340 (1966).

example, may contain as much as one mole of water per mole of pigment when freshly prepared (i.e., nearly 2% water). Upon prolonged drying of the chlorophyll at low pressure, most of this water may be removed.[66] Use of this chlorophyll for the determination of the specific extinction coefficients introduces a small but variable error.

Chromatography provides a convenient test for the homogeneity of chlorophyll preparations. It also serves as the most sensitive method for ascertaining the number of green pigments extracted from plant material treated in various ways. It should be employed for the supplemental examination of the pigments in extracts in which the chlorophyll is estimated by spectrophotometric methods.

Although chromatography is the most effective method for the separation of the chlorophylls and for the determination of their homogeneity, it does not in itself provide a technique for the estimation of the green pigments. With suitable precautions, spectrophotometry, or even colorimetry, may be employed for the estimation of each chlorophyll separated by chromatography. The spectrophotometric curves also provide supplemental identification of the separated pigments. In spite of the extra time and inconvenience involved in the chromatography and elution of the pigments, this procedure is the most reliable and specific method for the estimation and identification of the chlorophylls.

The chromatographic sequences of the chlorophylls and the carotenoids, reported in Table I, provide a convenient basis for the identification and description of the green pigments. The sequences of the chlorophylls and their alteration products are also useful for the description and identification of these substances as indicated in Tables III and IV. For example, chlorophyll *a′*, which is difficult to detect spectrophotometrically in the presence of *a*, appears below the *a* in columns of powdered sugar; solvent, petroleum ether plus 0.25% *n*-propanol. Similarly, methyl chlorophyllide *a*, which cannot be detected spectroscopically in the presence of *a*, appears above the *a* in the sugar columns. This methyl derivative separates above the lutein of leaves whereas the *a* is well below the lutein.

With some of the pigments as shown in Tables I, III, and IV, chromatography fails to effect a complete separation. The two components of chlorophyll *c* are not readily separable with thin layers of cellulose or with columns of powdered sugar. They have been separated with special preparations of polyethylene, which are readily available.[40] The chlorobium chlorophylls 650 and 660 are incompletely separated in long columns of powdered sugar. The half-dozen components presumed to be present in each of these pigments, an inference based upon the gas chromatography of the oxidation products, have never been separated by any method.[14]

Partition between Immiscible Liquids. Partition of the pigments between immiscible liquids serves several useful functions in analytical and preparative studies of the chlorophylls. The more important of these is the transfer of the pigments from aqueous acetone or methanol plus petroleum ether extracts to a liquid, such as diethyl ether, in which the chlorophylls are readily soluble. By this transfer, which is facilitated by dilution of the acetone or methanol extracts with aqueous salt solution, water-soluble substances are separated from the pigments. Moreover, the solution of the pigments in the low-boiling liquid may be utilized directly for chromatographic and spectrophotometric examination. The liquid in these solutions may also be evaporated leaving a residue that can be preserved in vacuum or redissolved in another liquid for further studies.

The partition phenomenon may also be made the basis of chromatographic[91-93] and countercurrent[94] procedures. For example, the green and yellow leaf pigments are readily separable with dimethyl formamide fixed in paper when petroleum ether is the wash liquid.[84]

V. Chemical Reaction Products of Chlorophylls

Reactions Common to Chlorophylls

Many chlorophylls undergo analogous chemical reactions, usually with the formation of similarly colored reaction products. Many, but not all, chlorophylls form interconvertible isomers, especially when the plants or the solutions are heated. Thus $a \rightleftharpoons a'$, $b \rightleftharpoons b'$, and $d \rightleftharpoons d'$. Chlorophylls *c* do not yield unique isomers (as would be expected from their truly planar, symmetrical structure).

All chlorophylls yield pheophytins when the magnesium is removed with acids. All chlorophylls except the chlorobium chlorophylls and chlorophylls *c* are phytyl esters. Replacement of the phytyl group by hydrogen, as by hydrolysis with chlorophyllase contained within the green tissues, provides the acidic chlorophyllides.[14] Replacement of the phytyl by methyl, as by methanolysis with chlorophyllase, provides the methyl chlorophyllides. Enzymatic or induced enzymatic oxidation provides the 10-hydroxychlorophylls.[83] All the altered chlorophylls containing magnesium are converted into their magnesium-free pheophytin equivalents by reaction with acids. Chlorophylls are converted to green or green-brown, water-soluble products with alkalies without loss of magnesium.[95]

[91] A. Hager and T. Meyer-Bertenrath, *Planta* **69,** 198 (1966).
[92] J. Schenk and H. G. Daessler, *Pharmazie* **24,** 116 (1969).
[93] W. S. Kim, *Biochim. Biophys. Acta* **112,** 392 (1966).
[94] H. Arn, E. C. Grob, and R. Signer, *Helv. Chim. Acta* **49,** 851 (1966).
[95] G. Oster, S. B. Broyde, and J. S. Bellin, *J. Amer. Chem. Soc.* **86,** 1309 (1964).

Some of the abundant and clearly defined alteration products of the chlorophylls are listed in Tables III and IV. Table IV also includes the conditions for their formation as well as some spectral properties.

For studies of the alteration products of the chlorophylls, it is sometimes possible to select plant material and conditions so that one or two products predominate. Examples are the preparation of the 10-hydroxychlorophylls and the methyl chlorophyllides. Even under the most favorable conditions, however, secondary products are often formed in such abundance that they must be separated by chromatography.

Conditions That Lead to Alteration of the Chlorophylls

As indicated above, the chlorophylls in healthy, green, vegetative tissues are those characteristic of the major group to which the plant belongs. Colored alteration products are absent (see Table I).

If the pigments are extracted quickly from the vegetative tissues under carefully controlled conditions, the chlorophylls may be obtained with a minimum of alteration and contamination by colored reaction products. If, however, the plant cells are injured, allowed to deteriorate, treated with mild anesthetics, frozen and thawed, heated, or allowed to wilt and dry, especially in bright light, the chlorophylls undergo extensive deterioration with the formation of various colored products. These reactions not only reduce the quantity of the chlorophyll itself, but they also yield various colored products that interfere with the spectrophotometric determination of the chlorophyll.[96] Colored alteration products are also formed if extracts containing the chlorophylls are heated, acidified, exposed to acidic solids as ion exchangers[97] and certain adsorbents, exposed to alkalies, to very bright light, or if the pigments are dissolved in anhydrous methanol exposed to air or oxygen (allomerization).[14,93]

Alteration Products and the Estimation of Chlorophylls

When present in plant material or in the extracts of plant material, chlorophyll alteration products must be converted to substances whose properties can be related to the natural chlorophylls. For example, a mixture of substances as chlorophyll *a*, pheophytin *a*, and methyl chlorophyllide *a* may be converted into their magnesium-free equivalents with acid. The spectral absorption then provides a basis for calculation of the *a* initially present. With much stored and preserved plant material, however, the chlorophylls are so extensively and variously altered that they

[96] M. F. Bacon and M. Holden, *Phytochemistry* **61**, 193 (1967).

[97] S. A. Chernomorskii, A. I. Fragina, V. T. Kurnygina, and F. T. Solodkii, *Rast. Resur.* **4**, 395 (1968).

cannot be converted to a common pigment or to pigments with the same absorption spectra; hence, the spectrophotometric methods of analysis are inapplicable. Moreover, the properties and the nature of many of these products have not been established; consequently, there may be no precise way to estimate the chlorophylls with respect to the alteration products.

Chemical Alteration for Estimation of Chlorophylls

Few chemical reactions have been employed in the estimation of the chlorophylls. Removal of the magnesium from chlorophylls with acid provides pheophytins and magnesium ions. The former may be estimated spectrophotometrically, the latter are estimated in various ways. If only one chlorophyll of known magnesium content is present, the quantity of pigment may be calculated from the amount of magnesium ions.[3] But if several chlorophylls or their alteration products, with unknown or different magnesium content, are present, this method is inapplicable. Determination of the pheophytins spectrophotometrically is limited to use with mixtures of only one or two pigments, as noted above. Chlorophyll *b* is convertible to the oxime with hydroxylamine and may be estimated from the accompanying spectral shift.[98]

VI. Typical Analytical Procedures

Plant Material

Determinations of the nature and the quantity of the chlorophylls must usually be related to the quantity of the chlorophyllous plant material. The plant sample may be specified on a fresh weight or dry weight basis. With leaves, portions of the soft tissue minus major veins and midribs may be employed. With unicellular and some filamentous algae, the sample may be specified as weight of centrifuged or dried cells. It may also be related to single cells. With wilted, frozen, brined, canned, and partially dehydrated plant material, it may be necessary to reduce the samples to a dry weight basis, a rather variable basis at best.[14] For preparation of the individual pigments, the precise specification of sample size is usually not important.[85]

Estimation of Chlorophylls Directly in Plant Material

There is no dependable method for the estimation and identification of chlorophylls within the plant itself.[55,56,58] Approximate values may be obtained with instruments that collect all the scattered light as in an

[98] T. Ogawa and K. Shibata, *Photochem. Photobiol.* **4**, 193 (1965).

integrating sphere or by using scattered light from opal glass to examine the absorption by unicellular algae or dispersed chloroplasts.[5]

Extraction of the Chlorophylls

The extraction procedure employed for the quantitative estimation of the chlorophylls depends primarily upon the condition of the plant material, secondarily upon the analytical procedures. With fresh material that disintegrates in a blender, samples of about 1–5 g are agitated with about 100 ml of the extractant. Acetone usually provides rapid and complete extraction of the pigments. If it does not, the extractions may often be made with methanol. Certain unicellular algae and some large species, even though composed of a thallus only one or two cells thick, as with *Ulva* and *Enteromorpha*, are not easily disintegrated in a blender and do not readily yield their pigments to acetone or even to methanol. Such material liberates its pigments much more readily if it is first placed in boiling water for about 1 minute, then cooled, collected, and extracted. Large, tough red algae and resilient, mucilaginous brown algae, which are not easily disintegrated in a blender, may be triturated with sand and the extractant in a large glass mortar.

Exposure of the plant material to hot water always causes some isomerization of most chlorophylls. Because the isomers are so similar to the parent chlorophylls, only a small error is introduced into the analytical determinations based upon spectrophotometry.

With most fresh plant material, the volume of the extractant should be kept large relative to the size of the sample (about 30 to 1). This reduces alteration of the pigments by secondary reactions, especially those catalyzed by enzymes, as oxidation and hydrolysis or alcoholysis.

Plant material that is acidic (begonia, cactus, mesembryanthemum, sorrel, rhubarb, etc.) yields pheophytins when placed in boiling water or even treated with the extraction solvents. This acidity must be neutralized with buffers (such as ammonia, Na_2CO_3, Na_2HPO_4) added, in excess, to the boiling water or to the extraction liquids. Contrary to many reports in the literature, slightly soluble buffers (such as $CaCO_3$ and $MgCO_3$) do not neutralize the acidity before it alters the chlorophylls.

Dried plant materials (such as dehydrated peas, broccoli, spinach, and alfalfa) encountered in hay, meal, and pelletized fodder retain their green pigments when treated with the common chlorophyll extractants. If these plant materials are reconstituted with water, subsequent extraction of the pigments is usually improved. Frequently, however, the chlorophylls have been altered not only to a great degree, but also to a great variety of products.

For isolation of the solid chlorophylls in a state of high purity, some

50–200 g of fresh, chlorophyll-rich algae or 1–2 kg of fresh green leaves may be required. Leaves, such as cocklebur, mallow, or spinach, may be placed in boiling water for about 1 minute, filtered on a sieve, cooled with cold water or ice, pressed between paper towels, and extracted with methanol or methanol plus ether or petroleum ether. Unicellular algae, as *Chlorella* and diatoms, are handled by centrifugation as described above by using a large centrifuge for collection of the cells.

Spectrophotometric Determinations. If the chlorophylls are to be estimated spectrophotometrically in the extracts, acetone is the preferred extractant. The acetone concentration should be maintained near 80 or 90%. Before the measurements, the extracts may be clarified by centrifugation or by filtration through paper or a short column of Celite 545. If there is much residual turbidity, indicated by absorption beyond 700 nm, the pigments should be transferred to ether, as described in the next section.

Equations for the estimation of chlorophylls *a* and *b* in methanol, ether, and 80% and 90% acetone extracts have been derived from the specific extinction coefficients of the pure pigments and the observed absorption, *A*.[99,100] Those for the 80% extracts at 649 and 665 nm are[100]:

$$\text{Chl } a\ (\mu\text{g/ml}) = 11.63\ (A_{665}) - 2.39\ (A_{649})$$
$$\text{Chl } b\ (\mu\text{g/ml}) = 20.11\ (A_{649}) - 5.18\ (A_{665})$$
$$\text{Total Chl}\ (\mu\text{g/ml}) = 6.45\ (A_{665}) + 17.72\ (A_{649})$$

Corresponding equations have also been developed for the estimation of the chlorophylls after conversion to pheophytins with oxalic acid in acetone extracts.[100,101]

Several spectrophotometric methods have been developed for the estimation of chlorophyll *c* in the presence of *a* and *b*.[71-78] Now, however, in view of the complexity of *c*,[40,41] it seems unlikely that a precise method can be perfected until the specific spectral extinction coefficients for the components of *c* have been determined.

Nomograms facilitate reading the *a* and *b* concentrations from the absorption, A, measurements with 80% acetone.[102,103] A nomogram for *a*, *b*, and *c*[104] is subject to the limitations of the coefficients for *c*.

Transfer of Extracted Pigments to Ether and to Petroleum Ether

For many spectrophotometric estimations of the chlorophylls, the pigments in the acetone, or methanol, or methanol plus petroleum ether

[99] J. Lenz and B. Zeitzschel, *Kiel. Meeresforsch.* **24,** 41 (1968).
[100] L. P. Vernon, *Anal. Chem.* **32,** 1144 (1960).
[101] M. D. Nutting and R. Becker, *J. Food Sci.* **31,** 210 (1966).
[102] J. T. O. Kirk, *Planta* **78,** 200 (1968).
[103] Z. Sestak, *Biol. Plant., Acad. Sci. Bohemoslov.* **8,** 97 (1966).
[104] G. A. W. Battin, *J. Mar. Biol. Ass. U. K.* **47,** 407 (1967).

or ether extracts are transferred quantitatively to ether. The extracts of about 1–5 g of green tissues are treated with this solvent and are then diluted with an excess of aqueous salt solution. Usually most of the pigment appears in the ether layer with little or no pigment remaining in the aqueous layer. A second or third extraction may be necessary to ensure a quantitative transfer from the aqueous layer. The combined ether layers are then brought to a suitable concentration, and the chlorophyll is determined spectrophotometrically. As colloidal matter may be present and as the ether will be saturated with water, it is preferable to evaporate the solution to dryness with an aspirator and a rotary evaporator, maintaining the temperature below about 30°. The residue is then dissolved in pure anhydrous ether.

The residue from the evaporation of the ether solution may be dissolved in other solvents for various other tests. For chromatographic examination in columns, for example, the residue may be dissolved in petroleum ether or in a little ether which is then diluted with petroleum ether.

Spectrophotometric Determinations. Equations for the estimation of chlorophylls *a* and *b* in the ether solutions are[5]:

$$\text{Chl } a\ (\mu\text{g/ml}) = 9.93\ (A_{660}) - 0.777\ (A_{642.5})$$
$$\text{Chl } b\ (\mu\text{g/ml}) = 17.6\ (A_{642.5}) - 2.81\ (A_{660})$$
$$\text{Total Chl}\ (\mu\text{g/ml}) = 7.12\ (A_{660}) + 16.8\ (A_{642.5})$$

Determination of Magnesium. If the green pigment in the ether solutions is a single chlorophyll of known molecular weight, the quantity of pigment present may be estimated from the amount of magnesium extracted with aqueous mineral acids. To this end, the ether solution of the pigment is shaken with aqueous HCl; the extracted magnesium is then estimated by any standardized method as gravimetry, colorimetry, spectrophotometry, fluorimetry, or atomic absorption.

Separation and Identification of Chlorophylls in Plant Extracts by Chromatography. A few chromatographic procedures for the isolation, identification and estimation of the chlorophylls are presented here. Many other modifications of these techniques have been reported and are readily available.[85–89,91–93]

The leaf pigments have been utilized to demonstrate applications of various geometric modifications of solution adsorption chromatography. Conversely, these methods may be utilized for separation and examination of the chlorophylls in plant extracts.[84]

Columns of powdered sugar are convenient for examination of the chlorophylls. Columns 0.5–1 cm in diameter and about 20–25 cm in length are used with a petroleum ether solution of the pigments prepared as described above. Formation of the chromatogram is effected with petroleum ether plus 0.5% *n*-propanol. With extracts of leaves, the sequence of pigments is: neoxanthin (most sorbed) violaxanthin, chlorophyll *b*, lutein plus traces of zeaxanthin, chlorophyll *a*, and carotenes, usually β-carotene $\pm$ α-carotene (see Table I). For the best separations, the columns should be lightly loaded with the pigment extract. Usually this loading is established empirically by running several columns.

Columns of powdered sugar are also very effective for the isolation of solid chlorophylls in about 0.1-g quantities. A 1- to 2-kg portion of deep green leaves is placed in boiling water for about 2 minutes, cooled, squeezed in a towel, and extracted with methanol plus ether. The pigments are transferred to ether, as described above; the ether is evaporated; and the residue is dissolved in a small quantity of ether (about 100 ml), which is then diluted with petroleum ether (about 400 ml). Aliquot portions of this solution are chromatographed in 10 large sugar columns (8 $\times$ 36 cm) (packed mechanically). The individual chlorophyll zones from each column are pressed into chromatographic tubes, and the pigments are eluted with ethanol plus petroleum ether. The elutriates from the zones of each pigment are combined; each pigment is transferred to ether and recovered by evaporation of the solvent. Each residue is again dissolved in ether, diluted with petroleum ether, and readsorbed as just described using only 3–7 large sugar columns, depending upon the amount of the pigment obtained from the first adsorption. The readsorbed pigments are eluted with ethanol plus petroleum ether, transferred to petroleum ether, and induced to separate by washing the ethanol from the petroleum ether with water. Detailed directions for the isolation of the principal chlorophylls by this procedure are now available.[41,85]

Thin layers of powdered sugar are useful for the separation and quantitative recovery of chlorophylls on an analytical scale. To this end, the sugar is spread on glass plates, and the plant extract is added as a number of small initial spots varying in concentration. The chromatogram is formed by upward development with the solvent. The separations are essentially the same as those observed in columns of powdered sugar. Each green zone is then removed, and the pigments are eluted for spectrophotometric measurements.

Commercially prepared, thin layers of cellulose and sheets of filter paper are also useful for the isolation of the chlorophylls. For one-way development with the same solvent, the results are similar to those observed with powdered sugar.[87] Two-way development also provides an effective separation of the pigments as illustrated in Fig. 4.

[43] Biological Forms of Chlorophyll *a*

By J. S. BROWN

Definition

Even relatively inaccurate absorption spectra of leaf chloroplasts and algae indicate that the red absorption band is complex as well as shifted to longer wavelengths as compared to the absorption of chlorophyll *a* in solution. The shift has been attributed to aggregation of the chlorophyll and/or the formation of chlorophyll–protein complexes. Since chlorophyll in solution will not mediate photosynthesis, an understanding of the complexity of the absorption *in vivo* might also explain how chlorophyll functions. The individual overlapping absorption bands that comprise the complex spectrum observed *in vivo* have been called the biological forms of chlorophyll *a*. The number and wavelength peak positions of these forms are similar in chloroplasts prepared from different higher plants and in samples of an alga grown in the same way but vary considerably among different genera of algae. There is as yet no testable hypothesis to explain either the chemical nature or the variability of the biological forms of chlorophyll *a*. The following is a description of our current knowledge of these forms in a variety of plants.

Measurement

Since the components of chlorophyll *a* occur only in particles and absorb within 15 nm of each other, their measurement requires special spectrophotometry. The demonstration of the biological forms of chlorophyll in a wide variety of plants was carried out largely with a derivative spectrophotometer constructed by French.[1,2] Because derivative spectrophotometers are not widely available commercially and their spectra are difficult to interpret, a preferable way to measure the chlorophyll forms is at low temperature (—196°) with a conventional spectrophotometer that collects a large angle of the scattered light. This has two advantages: the individual absorption bands are sharpened at lower temperatures, and the ice forms a highly scattering medium with an increased optical path length giving intensification of the apparent absorption. Since the same absorbing forms can be detected in derivative spectra measured at 20° and in conventional spectra at —196°, the

[1] M. B. Allen, C. S. French, and J. S. Brown, *in* "Comparative Biochemistry of Photoreactive Systems" (M. B. Allen, ed.), p. 33. Academic Press, New York, 1960.
[2] J. S. Brown, *Photochem. Photobiol.* **2**, 159 (1963).

freezing does not in itself change the structural arrangement of chlorophyll *in vivo*. Butler[3] has presented an excellent review of possible absorption artifacts due to light scattering and suggested ways to overcome them in part (Low temperature and derivative spectrophotometry are discussed in Vol. XXIV by Butler).

Another source of error in measurements of the absorption of intact algae or chloroplasts can be caused by a too high concentration of chlorophyll within the cells. In this case a greater proportion of the light is transmitted at the peak of an absorption band than at the sides; this results in an artificial flattening of the maximum. Breaking the cells into submicroscopic particles largely overcomes this error. However, artificial enhancement of the absorption band on its long wavelength side due to anomalous dispersion may still cause inaccuracy. In a few algae, including *Ochromonas* and *Phaeodactylum,* some of the chlorophyll forms are destroyed by breaking the cells.

Description of Examples

Absorption. Spectra of representative plants that contain chlorophyll *b* and also illustrate the variability of chlorophyll *a* absorption are shown in Fig. 1. Spectra of fine homogenates of the cells or chloroplasts were measured at −196°. No variability in the Soret band of chlorophyll *a in vivo* has been observed that could be correlated with the different red-absorbing forms. The maximum at 650 nm is from chlorophyll *b*. The small maximum near 620 nm is believed to be the satellite band of chlorophyll *a* that is also observed in solution. The spectra of all the plants have a doublet structure between 670 and 680 nm, indicating at least two major forms of chlorophyll *a*. We refer to these as "C*a*670" and "C*a*680" in quotation marks to emphasize the possibility that each of these major forms may actually be composed of more than one chlorophyll type absorbing near 670 and 680 nm.

The spectrum of spinach chloroplast particles is similar to spectra of all the other plant chloroplasts that have been examined. The *Chlorella* spectrum is representative of several species of this genus, including *C. pyrenoidosa, C. sorokinii, C. vulgaris,* and *C. protothecoides* as well as *Chlamydomonas rheinhardii. Scenedesmus obliquus* D_3 has a higher proportion of "C*a*680" than *Chlorella* and also a small amount of a pigment type absorbing near 700 nm, C*a*700. This latter band is missing from Bishop's *Scenedesmus* mutant No. 8. *Stichococcus bacillaris* has more chlorophyll *b* than most other green algae and relatively less "C*a*670." When *Euglena gracilis* has been grown slowly with dim light,

[3] W. L. Butler, *Annu. Rev. Plant Physiol.* **15,** 451 (1964).

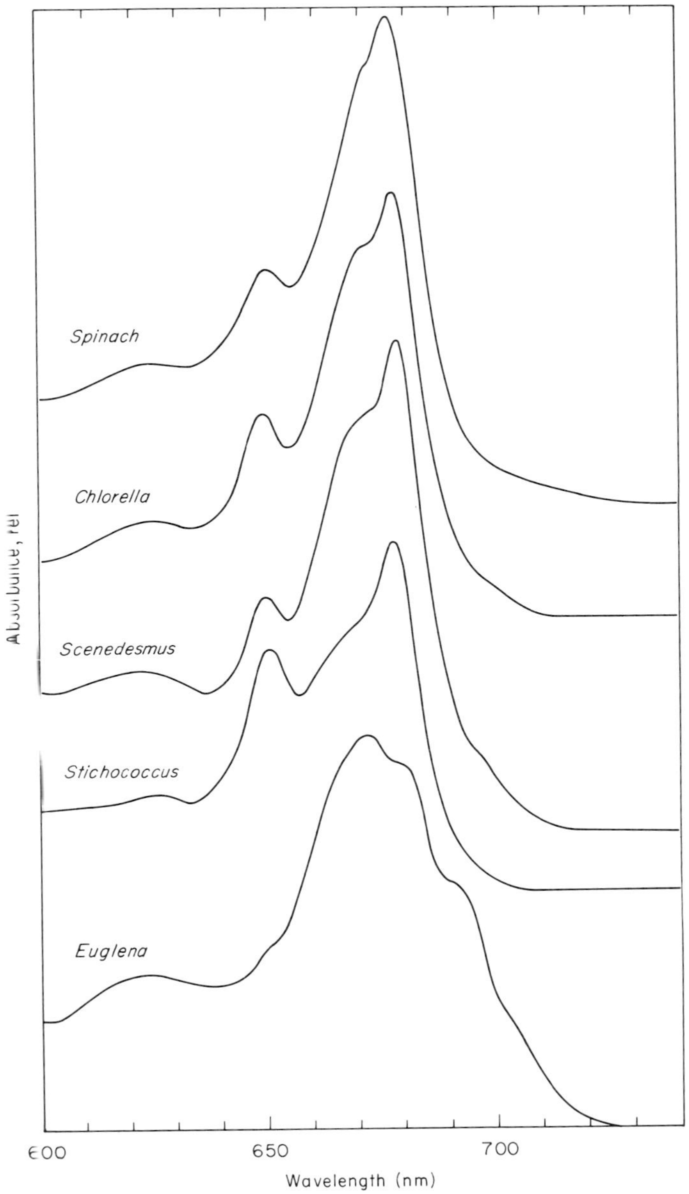

FIG. 1. Absorption spectra measured at −196° of supernatant chloroplast particles from spinach, *Chlorella pyrenoidosa, Scenedesmus obliquus* D-3, *Stichococcus bacillaris,* and *Euglena gracilis* prepared by breaking the cells in the French press and centrifuging at 3000 *g* for 10 minutes.

its chloroplasts have less chlorophyll *b* and more of the chlorophyll *a* form at 695 nm, as shown in Fig. 1. Rapidly dividing *Euglena* have more chlorophyll *b* and often no detectable C*a*695.

Ochromonas danica has three distinct forms of chlorophyll *a* (Fig. **2**) similar to *Euglena,* but their proportions are not influenced by growth conditions. The spectrum of the diatom *Phaeodactylum tricornutum* shows

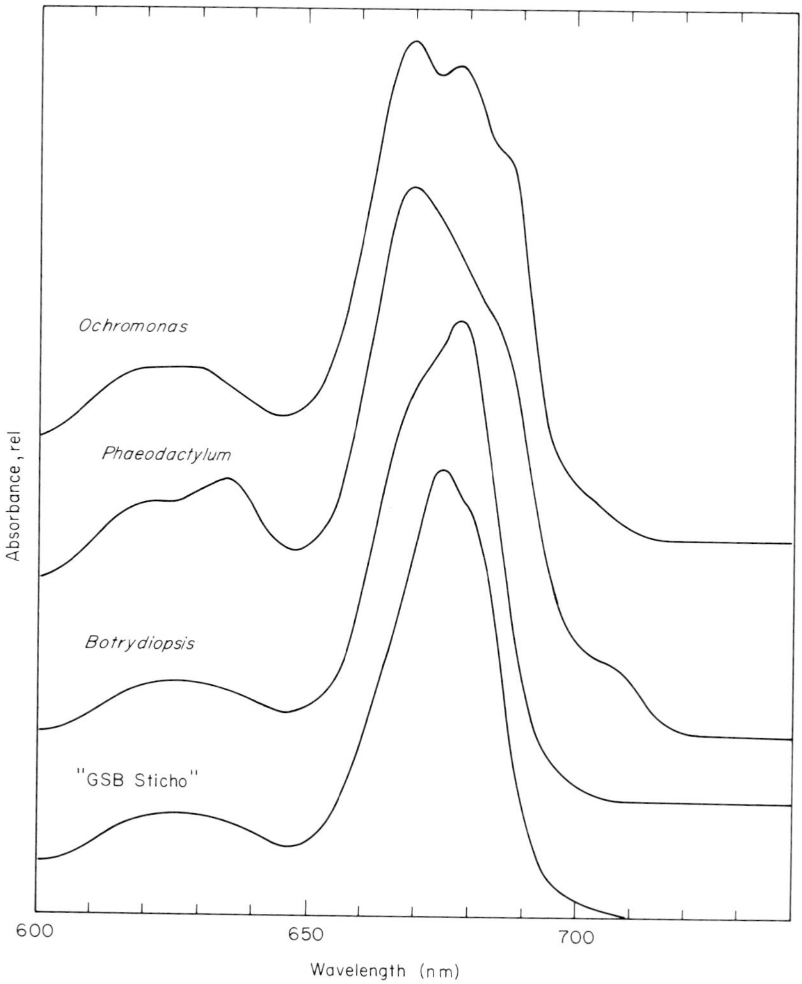

FIG. 2. Absorption spectra measured at −196° of *Ochromonas danica* and *Phaeodactylum tricornutum* cells and of *Botrydiopsis alpina* and "GSB Sticho" particles prepared as for Fig. 1.

chlorophyll *c* absorbing near 635 nm and three chlorophyll *a* bands. The long wavelength band near 710 nm increases to the maximum proportion shown in Fig. 2 when the cells are grown slowly in dim light, as does C*a*695 in *Euglena*.

The spectra of *Botrydiopsis* and a "GSB Sticho" isolate of *Stichococcus cylindricus* from Woods Hole Oceanographic Institute are typical of algae containing only chlorophyll *a*. *Tribonema* shows a spectrum similar to *Botrydiopsis*. If a homogenate of *Botrydiopsis* ages for a day at 4°, the absorption peak shifts to shorter wavelengths and the spectrum resembles that of "GSB Sticho."

The spectra in Fig. 3 of the red alga *Porphyridium cruentum* and

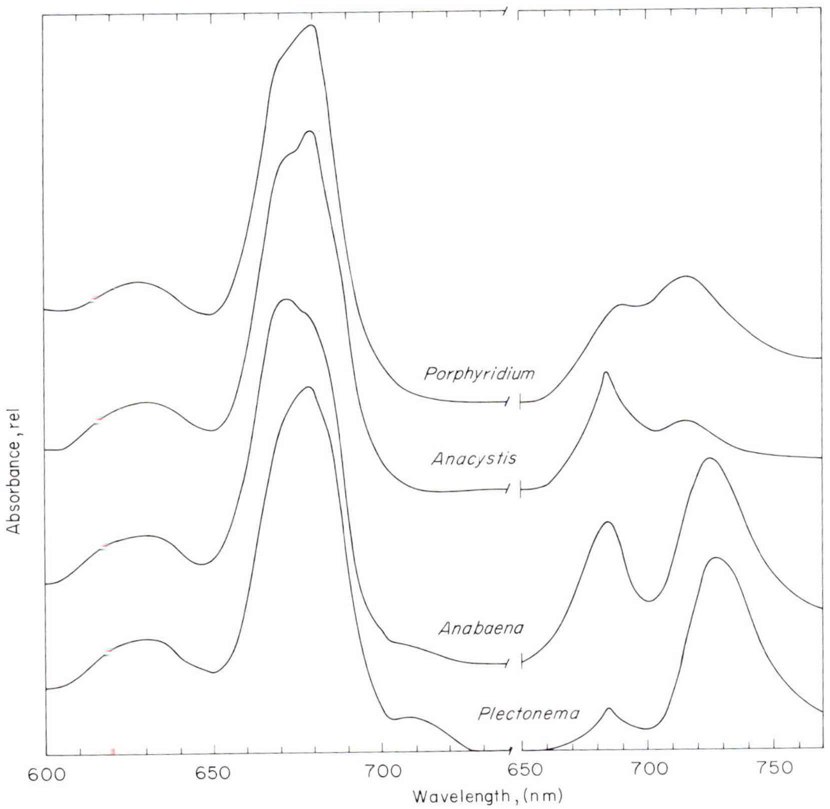

FIG. 3. Absorption and fluorescence spectra measured at −196° of fraction 1 particles prepared from *Porphyridium cruentum, Anacystis nidulans, Anabaena cylindrica,* and *Plectonema boryanum*. Excitation at 435 nm.

three species of blue-green algae were measured on particles that had been largely freed from the phycobilin pigments by centrifuging broken cells in a sucrose gradient. The soluble phycobilins remained at the top of the tube and were thus separated from the chlorophyll-containing particles below. The spectra of *Porphyridium* and *Anacystis* particles are very similar, but the differences in absorption between *Anacystis, Anabaena,* and *Plectonema* are striking. Since these four spectra were measured in the same way on particles of similar size and density, the absorption differences cannot be caused by light-scattering artifacts, but must be attributed to different arrangements of chlorophyll in the lipoprotein complexes.

Fluorescence. In theory it may be possible to detect certain of the forms of chlorophyll *a* by their separate fluorescence emission bands. The absence of emission from an absorption band indicates that the pigment in that form may be aggregated. Action spectra for separate emission bands may provide evidence for energy transfer from pigments absorbing at shorter wavelengths or for additional absorption bands of the same chlorophyll form. In spite of a great deal of study by Butler,[4] Goedheer,[5] Govindjee *et al.*,[6] and many others, no completely consistent hypothesis has yet emerged that correlates absorption and emission of the biological forms of chlorophyll *a*.

Fluorescence spectra are more difficult to interpret than absorption spectra, because the shorter wavelength part of the emission may be reabsorbed by the long wavelength tail of the emitting pigment or adjacent pigments. Reabsorption is particularly troublesome when trying to interpret chlorophyll spectra measured *in vivo,* because apparently the longer wavelength absorbing chlorophyll in pigment system I is only weakly fluorescent or nonfluorescent, but it may reabsorb the fluorescence emitted at shorter wavelengths.

It is generally but not entirely agreed that the main fluorescence band between 680 and 685 nm observed at room temperature in all plants is emitted by "C*a*670." At low temperature an additional long wavelength emission maximum that may be as high or higher than the 680-nm peak appears in spectra of nearly all broken chloroplasts and algae. The wavelength peak position of this low-temperature band varies between 715 and 740 nm in different kinds of plants. The reason for this variation is not at all understood. One can find examples such as the

[4] W. L. Butler, *in* "The Chlorophylls" (L. P. Vernon and G. R. Seely, eds.), p. 343. Academic Press, New York, 1966.

[5] J. C. Goedheer, *Biochim. Biophys. Acta* **88,** 304 (1964).

[6] Govindjee, G. Papageorgiou, and E. Rabinowitch, *in* "Fluorescence" (G. G. Guilbault, ed.), p. 511. Dekker, New York, 1967.

three species of blue-green algae shown in Fig. 3 where the long wavelength absorption and emission bands are correlated both in wavelength position and relative heights. These lead to the conclusion that the long wavelength emission comes from "C*a*700." On the other hand, numerous other examples can be observed where "C*a*680" appears to be the logical source of long wavelength emission. The absorption and emission spectra of the spinach fractions in Fig. 4 illustrate this type of correlation in which no separate small bands are detectable on the long wavelength side of the absorption band of fraction 1. Several other higher plants and green algae have been fractionated in the same way as spinach, described below, to form a fraction 1 enriched in both "C*a*680" and long wavelength fluorescence.

Another fluorescence band near 700 nm has been observed in some plant material measured at —196° or below (see fraction 2, Fig. 4), but it is impossible to identify the source of this emission from absorption spectra.

Fractionation

Obviously, if the several biological forms could be physically separated from each other, their study would be greatly facilitated. To date, several methods have been found that will fractionate chloroplasts into particles with different absorption spectra indicating a partial separation of the chlorophyll *a* forms and chlorophyll *b*. (The procedures that use detergents such as digitonin or Triton are discussed elsewhere in this volume, see [26, 27].)

Since detergents can bind irreversibly to the chlorophyll proteins and complicate further analysis, a mechanical fractionation procedure devised by Michel and Michel[7] may be preferable. (Jacobi describes a method using sonication in a high salt buffer in [28].) In their procedure, chloroplasts or algae are suspended in a high salt buffer and disintegrated by the French needle-valve press. The homogenate, layered on a gradient of 10–50% (w/v) sucrose in the same high salt buffer, is centrifuged at 60,000 *g* for 30–60 minutes in a swinging-bucket rotor. The resulting separated layers of pigmented particles in the centrifuge tube can be removed with a syringe and further analyzed immediately or after dialysis to remove the sucrose.

Spectra of spinach fractions measured at —196° are shown in Fig. 4. These fractions were prepared by passing spinach chloroplasts, suspended in 0.3 *M* KCl and 50 m*M* Tricine adjusted to pH 8 with KOH, through

[7] J.-M. Michel and M.-R. Michel-Wolwertz, *Carnegie Inst. Wash. Yearb.* **67**, 508 (1968).

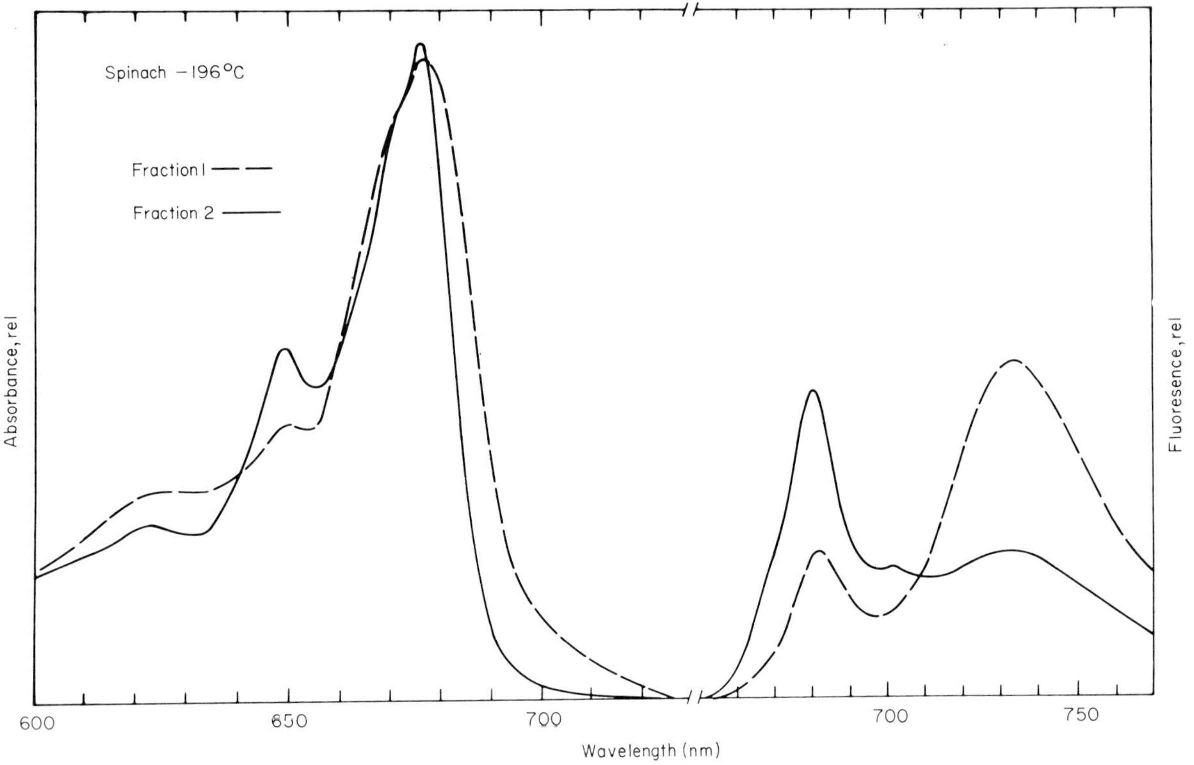

FIG. 4. Absorption and fluorescence spectra of fraction **1** (-----) and fraction **2** (——) prepared from spinach. Excitation at **435** nm.

the French press three times. The homogenate was layered on a sucrose step gradient [10–30–50% (w/v), top to bottom], and the tube was centrifuged for 30 minutes. The green layer above the 30% sucrose is called fraction 1 and that just above the 50% sucrose is fraction 2. The absorption and fluorescence spectra of these fractions correspond to the spectra of photosystem I and II particles prepared with detergents. Fraction 2 has proportionately more chlorophyll *b* and a sharp chlorophyll *a* peak at 677 nm. The "C*a*680" in fraction 1 apparently has a much broader bandwidth and therefore considerably greater absorption between 680 and 710 nm than does the "C*a*680" in fraction 2. Because of this difference in shape between the bands with maxima near 680 nm in each fraction, we may consider that there are in fact two kinds of C*a*680. The curve analysis results of French, discussed below, also indicate that both "C*a*670" and "C*a*680" could contain more than one chlorophyll form.

A large variety of algae have been fractionated by the Michel procedure. Since some algae are hard to break with the needle-valve, the Braun Mechanical Cell Homogenizer which shakes the cells rapidly with glass beads is more practical. If a mixture of 10 g of wet algal paste, 10 ml of KCl–Tricine buffer, and 50 g of 0.25- to 0.30-mm glass beads is shaken at 4000 cpm for 2 minutes, breakage is sufficient for further separation on the gradient. Chlorophyll fractions that differ in absorption and fluorescence have been obtained from several higher plants, a liverwort, several green algae, and the diatom *Phaeodactylum*. The red alga *Porphyridium* and three species of blue-green algae yielded only one kind of chlorophyll-containing particle each (Fig. 3).

The degree of separation of the chlorophyll forms shown by the spinach fractions in Fig. 4 has not been improved thus far. It is perhaps significant that the photochemically active fractions produced by any of the detergent fractionation procedures have the same spectroscopic differences as the mechanically separated fractions. All these fractionation results indicate that at least two distinct chloroplast particles with different proportions or kinds of chlorophyll *a* can be separated from most plant material. The question has not been resolved whether the lack of a more complete separation of the absorbing forms is caused by contamination of each fraction by the other or by the presence of different proportions of each form in both particle fractions.

Stability

In general, the longer wavelength forms of chlorophyll are destroyed first when exposed to such agents as intense light, heat, high or low pH, detergents, and organic solvents. In *Ochromonas* mild deleterious treatments, such as freezing and thawing or heating to 40° for 10 minutes, cause a conversion of the chlorophyll to a pheophytin-like pigment having a single red absorption peak at 671 nm within the cells. *Ca*695 in *Euglena* particles can be enzymatically converted to a shorter wavelength absorbing form.[8] The shift to shorter wavelengths of the absorption maximum of *Botrydiopsis* particles during aging was noted above. Frequently partial destruction of a long wavelength absorption band coincides with an increase in absorption at shorter wavelengths. Possibly this results from a conversion of the chlorophyll in protein complexes to a lipid soluble form.

Reversible Bleaching.[9] Preliminary results[10] indicate that "*Ca*680" in both spinach and *Chlorella* particles can be reversibly bleached by heating to 40° and cooling again.

It should be realized that although maxima observed in spectra of light-induced absorbancy changes may be at the same wavelength position as the various biological forms of chlorophyll *a*, their size is at least one hundred times smaller. Either only a small portion of a chlorophyll form can be reversibly bleached, or the pigments that bleach reversibly such as P700 are completely different from the biological forms discussed here.

Curve Analysis

In order to define the peak position and shape of each of the forms of chlorophyll, attempts have been made to resolve the complex spectra into their individual components. An analysis of derivative spectra of *Chlorella* particles indicated that two major forms at 672 and 683 nm were present in about equal proportions but with different bandwidths.[11] Rabinowitch and co-workers[6] have also done similar curve fitting with Gaussian components. Recently recorded spectra of chlorophyll particle fractions cast doubt on the validity of some of the earlier analyses. Now French is attempting to resolve spectra with the aid of a digital computer. The Carnegie Institution Year Books 67 through 69 have extensive discussions of the program and its modifications. The results so far are hopeful but not definitive.

Theoretical

Although no experimentally verifiable explanation for the absorbing forms of chlorophyll *a* has been proposed, there is no lack of speculation. One hypothesis, based upon spectroscopic studies of chlorophyll in vari-

[8] M.-R. Michel-Wolwertz, *Carnegie Inst. Wash. Yearb.* **67,** 505 (1968).

[9] See this volume [47], and Ke and Gorman, Vol. XXIV.

[10] J. S. Brown, *Carnegie Inst. Wash. Yearb.* **67,** 530 (1968).

[11] J. S. Brown and C. S. French, *Plant Physiol.* **34,** 305 (1959).

ous solvents, concentrations, and physical states, suggests that the absorption bands *in vivo* are from separate aggregates or oligomeres of chlorophyll *a*. Evidence for this is amply reviewed by Goedheer, Katz *et al.*, and Ke.[12]

Another hypothesis, derived from studies of the effects of enzymes, detergents, heat, and other agents on plant material, explains the several absorption bands by assuming that chlorophyll is attached to different lipoproteins or oriented in different ways in the lamellar matrix. The ability to prepare subchloroplast particles that differ in absorption supports this second concept.

One promising line of investigation is optical rotatory dispersion[13] that can be used to distinguish between states of chlorophyll in purified, natural chlorophyll–protein complexes such as those prepared by Thornber.[14] Kreutz[15] presented a model of the possible arrangements of the absorbing forms of chlorophyll *a* between protein and lipid layers based partly upon measurements of X-ray diffraction, dichroism, and fluorescence polarization. Eventually these kinds of measurements performed on more purified preparations of chlorophyll fractions may lead to a consistent description of the photochemically active states of chlorophyll *in vivo*.

[12] See L. P. Vernon and G. R. Seely (Eds.), "The Chlorophylls." Academic Press, New York, 1966.
[13] See Sauer, Vol. XXIV.
[14] See this volume [66].
[15] W. Kreutz, *Z. Naturforsch. B* **23**, 520 (1968).

[44] Nitrite Reductase

By MANUEL LOSADA and ANTONIO PANEQUE

$$NO_2^- + 6e^- + 8H^+ \rightarrow NH_4^+ + 2H_2O$$

Assay Method

Principle. The usual assay involves sodium dithionite as reductant and either ferredoxin or its artificial substitute, methyl viologen, as electron carrier.[1] Enzymatic activity can be best followed by measuring colorimetrically the rate of disappearance of nitrite.[2]

[1] J. M. Ramirez, F. F. del Campo, A. Paneque, and M. Losada, *Biochim. Biophys. Acta* **118**, 58 (1966).

Reagents

Tris·HCl buffer, pH 8.0, 0.5 *M*
Sodium nitrite, 20 m*M*
Ferredoxin,[3] 1 m*M* (For routine assay of the enzyme, 5 m*M* methyl viologen can be used instead.)
Sodium dithionite solution. Just prior to use, dissolve 25 mg of sodium dithionite in 1 ml of 0.29 *M* $NaHCO_3$
Diazo-coupling reagents. 1% sulfanilamide in 3 *M* HCl, and 0.02% *N*-(1-naphthyl)ethylenediamine hydrochloride[2]

Procedure. The reaction is carried out aerobically in a 10-ml test tube. Mix 0.3 ml of Tris buffer, 0.2 ml of sodium nitrite, either 0.2 ml of ferredoxin or 0.3 ml of methyl viologen, 0.1–0.2 unit of enzyme (see definition below) and 0.3 ml of fresh dithionite solution, to give a final volume of 2 ml. After incubation for 10 minutes at 30°, the reaction is stopped by vigorous shaking in a cyclomixer until the dithionite is completely oxidized and the dye becomes colorless. One milliliter of each of the diazo-coupling reagents is then added to a 2-ml aliquot of a 100-fold dilution of the reaction mixture, and the volume is made up to 5 ml with water. After 10 minutes, the optical density of the solution is determined at 540 nm, and the nitrite content is calculated from a standard curve. Minor variants of this procedure have been described.[4,5]

Definition of Unit and Specific Activity. One unit of enzyme is defined as that amount which catalyzes the reduction of 1 μmole of nitrite per minute under the above conditions. Specific activity is expressed as units per milligram of protein. The protein content of the enzyme preparation is determined by the ultraviolet absorption procedure of Warburg and Christian.[6]

Purification Procedure

The purification procedure presented below has been described for the enzyme from spinach leaves.[1] Similar procedures have been applied in other laboratories for the purification of the enzyme from maize[4] and

[2] F. D. Snell and C. T. Snell, "Colorimetric Methods of Analysis," 3rd ed., Vol. 2. Van Nostrand, New York, 1949. See also this series, Vol. II [59].

[3] For the preparation of ferredoxin (photosynthetic pyridine nucleotide reductase), see this series, Vol. VI [62].

[4] K. W. Joy and R. H. Hageman, *Biochem. J.* **100,** 263 (1966).

[5] A. Hattori and I. Uesugi, *Plant Cell Physiol.* **9,** 689 (1968).

[6] O. Warburg and W. Christian, *Biochem. Z.* **310,** 384 (1941–1942). See also this series, Vol. III [73].

Anabaena.[5] All steps in the purification are performed in the cold at 0° to 4°.

Steps 1 and 2. Preparation of Crude Homogenate; Extract of Acetone Precipitate. The preparation of the homogenate is carried out as described by San Pietro[3] for the purification of ferredoxin. Precipitation with acetone, extraction of the precipitate, and dialysis of the extract are also carried out according to San Pietro,[3] except that 10 m*M* Tris, pH 8.0, is used.

Step 3. Removal of Ferredoxin. The dialyzed enzyme solution is supplemented with NaCl to give a final concentration of 0.2 *M* Cl^- and then passed through a DEAE–cellulose column (50 mm × 32 mm) previously equilibrated with 10 m*M* Tris·0.2 *M* Cl^-, pH 8.0, in order to adsorb the ferredoxin.

Step 4. Adsorption on DEAE–Cellulose. The ferredoxin-free eluate is diluted 10 times with water and put on a DEAE–cellulose column (60 mm × 32 mm), equilibrated beforehand with 10 m*M* Tris, pH 8.0. Unwanted protein is removed with 10 m*M* Tris·0.1 *M* Cl^-, pH 8.0, and the enzyme is eluted with a solution of 10 m*M* Tris·0.3 *M* Cl^-, pH 8.0.

Step 5. Chromatography on DEAE–Cellulose. The eluate is diluted 1.8 times with water and subjected to chromatography on a DEAE–cellulose column (400 mm high and 27 mm in diameter), previously equilibrated with 10 m*M* Tris·0.17 *M* Cl^-, pH 8.0. Fractions (15 ml) of the effluent are analyzed for protein content and for nitrite reductase activity. The major part of the enzyme is present in that portion of the effluent (fractions 13–21) which is obtained after one and a half column volumes has passed through.

By this procedure, nitrite reductase can be purified about 500-fold with 30% recovery. A summary of the yields and purifications ob-

Purification Procedure for Nitrite Reductase from Spinach Leaves

Fraction	Total volume (ml)	Total protein (mg)	Total units	Specific activity (units/mg protein)	Yield (%)
1. Crude homogenate	1,100	48,400	395	0.008	100
2. Extract of acetone precipitate	120	1,320	290	0.215	73
3. Ferredoxin-free eluate	140	1,120	310	0.272	77.5
4. Eluate of DEAE–cellulose	17	560	225	0.400	57
5. Pool of fractions (13–21) after chromatography on DEAE–cellulose	135	38	121	3.250	31

tained at each step in the preparation of the enzyme is presented in the table.

Properties

Stability. The purified enzyme can be stored in the deep-freeze for a month with no loss in activity. By heating at 60° for 10 minutes, the activity is totally lost.[1,4,5]

Specificity. There is a marked specificity for reduced ferredoxin as the natural electron donor. The purified enzyme is completely inactive with either reduced pyridine nucleotides or reduced flavin nucleotides as electron donor. Among the artificial substitutes for ferredoxin, methyl viologen is the most effective one examined.[1,4,5]

The enzyme specifically catalyzes the stoichiometric reduction of nitrite to ammonia (hydroxylamine is not a free intermediate in the reaction) and has been physically separated from other ferredoxin-dependent photosynthetic enzymes, i.e., NADP reductase, sulfite reductase, and hydroxylamine reductase.[1,4,5,7,8]

Physiologically, the following electron donor systems are effective as the source of reducing power for the reaction catalyzed by different ferredoxin–nitrite reductases: illuminated grana, NADPH ferredoxin–NADP reductase and H_2 hydrogenase.[1,4,5,9] In the presence of fresh grana, ferredoxin, and nitrite reductase, the light dependent reduction of nitrite to ammonia is coupled to the evolution of oxygen and to the formation of ATP.[9-13] The molar ratio between nitrite reduced, ammonia formed, orthophosphate esterified, and oxygen evolved is 1:1:3:3.[9]

Physical Properties. The molecular weight of the enzyme estimated by gel filtration with calibrated Sephadex G-200 columns is about 60,000 for higher plants as well as for algae.[5,7,8]

According to its absorption spectrum, the enzyme does not appear to be a flavoprotein.[1,8,12] However, the absorption spectrum of the most puri-

[7] E. J. Hewitt and D. P. Hucklesby, *Biochem. Biophys. Res. Commun.* **25**, 689 (1966).

[8] W. G. Zumft, A. Paneque, P. J. Aparicio, and M. Losada, *Biochem. Biophys. Res. Commun.* **36**, 980 (1969).

[9] A. Paneque, J. M. Ramirez, F. F. del Campo, and M. Losada, *J. Biol. Chem.* **239**, 1737 (1964).

[10] M. Losada, A. Paneque, J. M. Ramirez, and F. F. del Campo, *Biochem. Biophys. Res. Commun.* **10**, 298 (1963).

[11] H. Huzisige, K. Satoh, K. Tanaka, and T. Hayasida, *Plant Cell Physiol.* **4**, 307 (1963).

[12] M. Shin and Y. Oda, *Plant Cell Physiol.* **7**, 643 (1966).

[13] G. F. Betts and E. J. Hewitt, *Nature (London)* **210**, 1327 (1966).

fied preparations from spinach and *Chlorella*, which contain iron as constituent of the enzyme, shows two peaks at 400 and 570 nm.[14]

The K_m's for ferredoxin and nitrite are approximately 10 μM and 0.1 mM, respectively.[1,4,5] The K_m for methyl viologen is about 0.1 mM.[1,5]

pH Optimum. The enzyme exhibits maximal activity in the pH range from 7.1 to 7.8.[1,4,5]

Inhibitors. Potassium cyanide, but not sodium azide, causes a total inhibition at a concentration of 1 mM.[1,9,15]

p-Chloromercuribenzoate at a concentration of 0.1 mM is without effect on the activity of the enzyme itself using methyl viologen as electron carrier but is inhibitory in the ferredoxin system.[1]

Intracellular Localization. Isolated chloroplasts contain nitrite reductase.[1,10,13,16] After breaking the chloroplasts, most of the enzyme is recovered in the chloroplast extract, and only a small part remains bound to the grana.[1]

[14] P. J. Aparicio, J. Cárdenas, W. G. Zumft, J. M. Vega, J. Herrera, A. Paneque, and M. Losada, *Phytochemistry*, in press.

[15] C. F. Cresswell, R. H. Hageman, E. J. Hewitt, and D. P. Hucklesby, *Biochem. J.* **94**, 40 (1965).

[16] G. L. Ritenour, K. W. Joy, J. Bunning, and R. H. Hageman, *Plant Physiol.* **42**, 233 (1967).

[45] Nitrate Reductase from Higher Plants[1]

By R. H. Hageman and D. P. Hucklesby

$$NO_3^- + NADH + H^+ \rightarrow NO_2^- + NAD^+ + H_2O$$

Preparation

Material

Nitrate reductase has been extracted from such diverse tissue as leaves, petioles, stems, shoots, roots, barley aleurone layers, corn scutella, cotyledons, glumes from seed of pod corn, corn husks, and cultured (e.g., tobacco pith) cells. However, the amount of enzyme that can be extracted from these tissues varies from traces to 60 μmoles of NO_2^- produced per equivalent grams fresh weight per hour as measured by crude homogenate activities. Higher activities are obtained from chlorophyllous

[1] EC 1.6.6.1, NADH:nitrate oxidoreductase. The NADH-dependent enzyme is most prevalent in plants, however, as will be detailed, some tissues also possess NADPH activity.

than from nonchlorophyllous tissue. The source material for the enzyme is most important because the amount of extractable enzyme varies drastically with: (a) plant species, (b) varieties within a species, (c) plant age, and (d) cultural techniques.

Because the enzyme is substrate inducible and exhibits a diurnal variation, highest activities are obtained with cotyledons and leaf tissue from vigorous young seedlings grown on a high (at least 15 mM) nitrate medium, and with adequate illumination (at least a 2500-fc, 12-hour photoperiod). Since minimal activity would be encountered at the termination of the dark period, a 3- or 4-hour period of illumination prior to harvest is suggested.

Tissue and seedling age has a drastic effect on extractable activity with some species, but not with others. In cotyledonous tissue from radish, cucumber, or marrow plants, highest activities are concurrent with attainment of maximum fresh weight. In general, the same statement can be made for individual leaves. With respect to plant age, maize is most unusual in that a peak of maximum activity occurs within 7–10 days after planting. In contrast, leaf tissues from wheat and foxtail exhibit maximal activity at about the same age as maize, but retain this high activity for a much longer period.

While flat- or tray-grown seedlings are convenient for most laboratories, field-grown crop plants or weeds provide an abundant source of material of high activity. However, the presence of the enzyme is a function of plant age (stage of development), nitrate supply and environment. Specific details of the level of nitrate reductase activity throughout the life cycle of maize and wheat have been published.[2,3] With pigweed and lambsquarter, the leaf tissue exhibits high activity through their life cycle.

Although Evans and Nason[4] demonstrated that the nitrate reductase from soybean cotyledons was able to utilize either NADH or NADPH as electron donor, subsequent work[5] demonstrated that the leaf tissue from 15 plant species had a specific or preferential requirement for NADH as cofactor. Recently, it has been shown[6] that when duckweed plants were supplied with exogenous sucrose, an increase in NADPH-dependent nitrate reductase occurred. It may be inferred that the initial level of

[2] J. F. Zieserl, W. L. Rivenbark, and R. H. Hageman, *Crop Sci.* **3,** 27 (1963).

[3] G. L. Eilrich, Ph.D. Thesis, Univ. of Illinois, Urbana, Illinois, 1968.

[4] H. J. Evans and A. Nason, *Plant Physiol.* **28,** 233 (1953).

[5] L. Beevers, D. Flesher, and R. H. Hageman, *Biochim. Biophys. Acta* **89,** 453 (1964).

[6] A. P. Sims, B. F. Folkes, and A. H. Bussey, *in* "Recent Aspects of Nitrogen Metabolism in Plants" (E. J. Hewitt and C. V. Cutting, eds.), Academic Press, New York, 1968.

NADPH-dependent activity was very low. Thus, the relative proportions of NADH- and NADPH-dependent nitrate reductases may depend not only upon the species[4] but upon the kind of metabolic activities occurring in the plant prior to sampling.

Based upon published results[5] and unpublished data of this laboratory, the following two groups of plant species are given as good sources of nitrate reductase:

Species with NADH nitrate reductase

Lambsquarter (*Chenopodium album* L.)	(leaves)
Pigweed (*Amaranthus hybridus* L.)	(leaves)
Broccoli (*Brassica oleracea italica*)	(leaves)
Radish (*Raphanus sativus* L.)	(cotyledons)
Cucumber (*Cucumis sativus* L.)	(cotyledons and leaves)
Marrow (*Cucurbita pepo* L.)	(cotyledons and leaves)
Spinach (*Spinacia oleracea* L.)	(leaves)
Wheat (*Triticum aestivum* L.)	(leaves)
Green Pepper (*Capsicum frutescens*)	(leaves)
Tomato (*Lycopersicon esculentum* Mill.)	(leaves)[7]
Maize (*Zea mays* L. var. Hy2 × Oh7, B14 × Oh43, and WF9 × C103)	(leaves)
Duckweed (*Lemna minor* L.)	(whole plant)[8]

Species with NADH and NADPH nitrate reductases

Maize (variety C103 × B37)	(leaves)
Maize (all varieties tested)	(scutellum)
Soybean (*Glycine max.* L.)	(leaves)[4,5]
Foxtail (*Setaria faberii* Herrm.)	(leaves)

In clarified (30,000 *g*) homogenates of maize (variety C103 × B37) and foxtail, the NADPH activity is approximately 30% of the NADH activity, while in the maize scutellum (5 days after seeding) the NADPH activity varies from 40 to 160% of the NADH activity, depending on genotype.

Extraction

The tissue is washed, blotted dry, and cut into small (1 × 1 cm) pieces when appropriate, prior to homogenization. With leaf tissue the large veins, midribs, or petioles are removed and discarded. For small samples, the mortar and pestle or Ten Broeck homogenizer, with or with-

[7] G. W. Sanderson and E. C. Cocking, *Plant Physiol.* **39,** 416 (1964).

[8] K. W. Joy, *Plant Physiol.* **44,** 849 (1969).

out sand or alumina powder as grinding aids, are most effective methods of homogenization.[9] For larger samples, the Waring Blendor has been a convenient and effective method of grinding. Foaming encountered during homogenization can be reduced by addition of a small (1-second spray) amount of Dow Antifoam, (Dow Chemical Co., Midland, Michigan). With some tissues (e.g., leaves from young soybean plants) homogenization fails to disrupt all the cells. To achieve a more complete recovery of the enzyme the extract can be sonicated or passed through a French press.

A standard procedure for preparation of crude extracts from green leaf tissue is as follows: The material is homogenized for 90 seconds in a medium of 1 m*M* EDTA, 10 m*M* cysteine, and 25 m*M* potassium phosphate, adjusted to a final pH of 8.8 with KOH. Six milliliters of grinding medium is added for each gram of fresh weight of tissue. The homogenate is pressed through four layers of cheesecloth or Mirracloth (Chicopee Mills Inc., 1450 Broadway, New York, New York), and the filtrate centrifuged for 15 minutes at 30,000 *g*. The supernatant fluid is then decanted through glass wool. The homogenates and extracts are kept cold (2°–3°) throughout the extraction process. Tests with tissue from corn, wheat, and marrow show that the enzyme is completely solubilized by this procedure.

Cysteine. The concentration of cysteine in the extraction medium is most crucial, and preliminary tests to establish the optimum concentration for each tissue and plant species are suggested. Glutathione and dithiothreitol are slightly more effective than cysteine as a protectant of the enzyme during extraction. With marrow tissue, 5 m*M* cysteine was the optimum level, and higher concentrations were as detrimental as omission.[10] With mature corn leaf tissue, 10 m*M* cysteine is mandatory for extraction of the enzyme, and while higher concentrations are not detrimental, they are not beneficial. With corn scutellum 2.0 m*M* cysteine is optimal, and higher concentrations produce extracts with lower activities. With tomato tissue, ground with mortar and pestle, 1 m*M* cysteine was optimal, and essentially no activity was observed with 10 m*M* cysteine was used.[7] In general, lower concentrations of cysteine are required with etiolated or young tissue than with mature tissue and with mortar and pestle grinding than with mechanical homogenization.[10]

Buffer. Although 100 m*M* Tris buffer has been extensively used for

[9] E. J. Hewitt and D. J. D. Nicholas, *in* "Modern Methods of Plant Analysis." Springer, Berlin, 1964.

[10] R. H. Hageman, C. F. Cresswell, and E. J. Hewitt. *Nature* (*London*) **193,** 4812 (1962) and unpublished work.

the extraction of nitrate reductase,[8,10,11] recent work (unpublished) has shown that the use of 25 m*M* phosphate buffer produces extracts of higher activity. Other buffers tested (e.g., *N*-hydroxyethylpiperazine-*N*′-2-ethanesulfonic acid, HEPES) have not proved to be superior to phosphate. The pH of the extraction medium is altered for each kind of tissue used so that the homogenate will have a pH of 7 after grinding.

Special Protectants. The introduction of the use of insoluble polyvinylpyrrolidone[12] (Polyclar AT from General Aniline and Film Corp., Dyestuffs and Chem. Div., 436 Hudson St., New York) as a protectant of enzymes from plant species with high phenolic content has made it possible to extract nitrate reductase from such species.[13] Details of these procedures are given in these two publications.

Assay Method (*in Vitro*)

$$NO_3^- + AH_2 \rightarrow NO_2^- + A + H_2O$$

Principle. Nitrate reductase is capable of utilizing reduced pyridine-nucleotides, flavin,[14] or benzyl viologen[9] as electron donors for the reduction of nitrate to nitrite. These reductants can be added in the reduced form or generated enzymatically in the reaction mixture. Since the NADH-dependent nitrate reductase is most prevalent in plants, NADH has been the most commonly employed reductant. Enzyme activity is usually measured by the colorimetric determination[15] of the nitrite formed during a timed incubation period at a fixed temperature, regardless of electron donor used. The activity can also be measured by following the oxidation of the pyridine nucleotides at 340 nm. The relative merits and limitations of these methods have been presented.[9] The stoichiometric relationship between nitrite produced and pyridine nucleotide or flavin oxidized has been established.[4,16]

Reagents for Reaction

Potassium phosphate buffer, 0.1 *M*, pH 7.5
Potassium nitrate, 0.1 *M*
NADH, 2 m*M*

[11] R. H. Hageman and Donna Flesher, *Plant Physiol.* **35**, 700 (1960).
[12] W. D. Loomis and J. Battaile, *Phytochemistry* **5**, 423 (1966).
[13] L. Klepper and R. H. Hageman, *Plant Physiol.* **44**, 110 (1969).
[14] A. Paneque, F. F. del Campo, J. M. Rameriz, and M. Losada, *Biochim. Biophys. Acta* **109**, 79 (1965).
[15] F. D. Snell and C. T. Snell, "Colorimetric Methods of Analysis." Van Nostrand, Princeton, New Jersey, 1949.
[16] L. E. Schrader, G. L. Ritenour, G. L. Eilrich, and R. H. Hageman, *Plant Physiol.* **43**, 930 (1968).

Reagents for Nitrite Assay

Sulfanilamide, 1% w/v in 1.5 *N* HCl
N-(1-naphthyl) ethylenediamine hydrochloride, 0.02% w/v

Procedure (NADH). The basic assay mixture contains, in micromoles, potassium phosphate, 50; KNO_3, 20; NADH, 0.4, and enzyme (0.1–0.2 ml of the crude extract equivalent to 0.2–1.0 mg of trichloroacetic acid-precipitable protein), in a final volume of 2.0 ml. The reaction is usually initiated by addition of the enzyme. A zero time or minus NADH reaction mix is used as control. After incubation at 30° for 15 minutes the reaction is terminated by the rapid addition of 1 ml of sulfanilamide reagent followed by 1 ml of the *N*-(1-naphthyl)ethylenediamine reagent. The color is allowed to develop for 30 minutes prior to reading at 540 nm. The reaction rate is linear over a 30-minute period.

The same proportions of reaction reagents are used when NADH oxidation is followed spectrophotometrically (340 nm; 1- or 3-ml volume; 1-cm light path). The linear initial reaction rate is used to compute the activity.

Activity is expressed as micromoles of nitrite produced per equivalent gram fresh weight per hour or per milligram of protein per minute. The protein precipitated by 5% trichloroacetic acid is determined by any of the several standard protein methods.

Procedure (NADPH). With extracts from maize scutella, the assay is identical as described for NADH, except for the cofactor added. With leaf extracts, maximum activity is obtained when the pH of the reaction mixture is 6.5.[5]

Procedure ($FMNH_2$). With the exception of substitution of $FMNH_2$ for the reduced pyridine nucleotides, the reagents and procedures are the same as previously described. The following modifications are concerned only with the methods of introducing $FMNH_2$ into the reaction mixture.

Chemical Reduction. The FMN (1.6 μmoles) is added to the reaction mixture, and the reaction is initiated by adding sodium hydrosulfite (8.0 μmoles $Na_2S_2O_4$ in 0.3 ml of 25 m*M* potassium phosphate buffer, pH 7.5). After the incubation period the reaction is terminated by vigorous agitation for 20 seconds [a Vortex mixer (Scientific Industries Inc., Queens Village, New York) is suggested]. Zinc acetate with or without ethanol is then added and nitrite determined as described (see precautions).

Enzymatic Reduction. The FMN can also be reduced with NADPH as the prime electron donor. The FMN (1.6 μmoles) and NADP: reductase (0.2 mg protein as prepared by Schrader *et al.*[16]) are added to the reaction mixture held in a Thunburg tube. The NADPH (2.0 μmoles) is added to the side arm and the tube evacuated prior to tipping in the NADPH. The reaction is stopped by addition of zinc acetate, with or without ethanol.

Prereduced FMN. In this procedure reduced FMN is added directly to the reaction mixture. After adding phosphate buffer, nitrate, and enzyme to a test tube, the tube is stoppered with a serum stopper. The tube is then evacuated. (A hypodermic syringe needle, coupled by tubing to a vacuum pump is a most convenient method.) The $FMNH_2$ (1.6 μmoles) is added with a hypodermic syringe. The FMN can be reduced in the dark by use of palladized asbestos and hydrogen gas. The reaction is stopped by removal of the serum stopper and addition of zinc acetate with or without ethanol.

Coupled Assays. The most convenient enzyme system for the generation of NADH in the reaction mix is glyceraldehyde-3-phosphate dehydrogenase with 3-phosphoglyceraldehyde as the electron donor.[17]

Light mediated by chloroplast grana can also be used as the prime energy source for the reduction of pyridine nucleotides,[4,18] flavins,[19] or benzyl viologen.[20] These reductants in turn can be coupled to nitrate by nitrate reductases.

Precautions. In the colorimetric procedure, reduced pyridine nucleotides, especially NADH present at the end of the reaction period, will interfere with color development. The excess of reduced pyridine nucleotides can be removed by precipitation with zinc acetate, zinc acetate and ethanol,[9] or by enzymatic oxidation.[9] A standard procedure is as follows: The reaction is stopped by the addition of 0.2 ml of 1 *M* zinc acetate. The resultant precipitate is removed by centrifugation (5000 *g* for 5 minutes) and an aliquot of the supernatant solution used for assay of nitrite.

The spectrophotometric assay, is not sensitive enough for use with many crude preparations and its use is suggested only with very active or partially purified preparations. Crude preparations from many plant species contain interfering pyridine nucleotide oxidases.

The enzyme for some plant species is very unstable even at 0°. Therefore, minimum time should elapse between extraction and assay. Alternatively, enzyme inactivation rate for the particular tissue or species should be determined.

[17] D. Spencer, *Aust. J. Biol. Sci.* **12,** 181 (1959).
[18] M. Losada and A. Paneque, *Biochim. Biophys. Acta* **126,** 578 (1966).
[19] M. Losada, J. M. Ramirez, A. Paneque, and F. F. del Campo, *Biochim. Biophys. Acta* **109,** 89 (1965).
[20] F. F. del Campo, J. M. Ramirez, A. Paneque, and M. Losada, *Biochim. Biophys. Acta* **66,** 450 (1963).

Assay Method (*in Vivo*)

Principle. Plant tissue containing or supplied with adequate amounts of carbohydrate when submerged in a solution containing nitrate and placed in the dark is capable of accumulating and exuding nitrite into the medium.[21]

Procedure. Freshly harvested leaves are washed and blotted dry and cut into 10-mm diameter disks or squares. Thirty leaf sections (0.3-g fresh weight) are then placed into a 50-ml Erlenmeyer flask containing cold (3°) infiltration medium. The infiltration medium is composed of 200 m*M* KNO_3 and 1 m*M* potassium phosphate, pH 7.5. The flask is then evacuated (6 mm Hg for 30 seconds), the vacuum released and the process repeated. (The tissue should be visibly wetted and sink below the surface of the medium.) After releasing the vacuum, the flask is wrapped in aluminum foil to exclude light and incubated in a water bath at 33° with gentle shaking. At timed intervals, 0.2-ml aliquots of the medium are removed for determination of nitrite.[15] Nitrite production is linear for most plant tissues unless carbohydrate supply is limited. At the end of the incubation period (1–4 hours), the tissue and remaining medium is transferred to a Ten Broeck homogenizer with 5 ml of 0.1 *N* HCl. The resultant homogenate is centrifuged (10,000 *g* for 5 minutes) and an aliquot removed for estimation of total nitrite produced.

When tissue is harvested from dark-pretreated (8–12 hours) plants or from "low carbohydrate" plants, marked increases in nitrite production are obtained by addition of 30 μmoles of 3-phosphoglyceraldehyde or fructose-1,6-diphosphate to the infiltration medium (cf. Note added in proof).

Comments. The estimated nitrate reductase activities are minimal values as some nitrite is assimilated under the dark anaerobic conditions. Estimation of the amount of nitrite lost can be made by substitution of 0.6 m*M* KNO_2 for the KNO_3 in the infiltration medium. All other reagents and procedures are the same; tissue must be nitrate free.

The *in vivo* nitrate reductase assay is ideally suited for demonstration of the occurrence of high amounts of the enzyme (e.g., giant ragweed, *Ambrosia trifida* L.; smartweed, *Polygonum pennsylvanicum* L.). Even with the use of Polyclar AT, it has not been possible to obtain extractable nitrate reductase from such species.

The amount of KNO_3 required in the infiltration medium for optimum production of nitrite is a function of plant species and tissue, and should be determined.

[21] L. Klepper, Ph.D. Thesis. Univ. of Illinois, Urbana, Illinois, 1969.

Purification

No method for the definitive purification of the enzyme from higher plants has been devised. In all purification methods so far developed, chromatography on calcium phosphate gel features prominently, used either in batches[4] or in columns.[22] Nitrate reductase precipitates with exceptional ease on addition of ammonium sulfate, and this step may be worth using more than once in a purification method. The following procedure was used by Schrader *et al.*[23] for marrow cotyledons and with slight modifications for corn and spinach leaves.

Extraction. A clarified extract was prepared as previously described. Where it was particularly important to remove ferredoxin from the preparation,[14,23] the extract was passed through a column of DEAE–cellulose which had been equilibrated with the extraction medium, adjusted to pH 7.5. This step is probably unnecessary for most purposes since the later purification steps, especially ammonium sulfate precipitation, should also remove ferredoxin.

Ammonium Sulfate Precipitation and Chromatography on Calcium Phosphate. Solid ammonium sulfate was added to the clarified homogenate to 25% saturation while the pH was maintained at 7.5 by the addition of KOH (50 m*M*). After 15 minutes standing time, the precipitate was removed by centrifugation (15 minutes, 15,000 g) and the supernatant solution brought to 42% saturation with ammonium sulfate in the same way. The precipitate was dissolved in a minimal amount of potassium phosphate buffer (25 m*M*, pH 7.5) and mixed with an equal volume of calcium phosphate gel (pH 7.5, 180-mg dry wt/ml). After 10 minutes of standing and occasional stirring, the mixture was centrifuged (5 minutes, 1000 *g*). The precipitate was washed with the potassium phosphate buffer (25 m*M*, pH 7.5), adding a volume one-half of that of the discarded supernatant solution. The enzyme was then eluted with sodium pyrophosphate (100 m*M*, pH 7.5), and the mixture was centrifuged. Ammonium sulfate precipitation (25–42% saturation) was then repeated as described above using the supernatant fluid from the pyrophosphate elution. The precipitate obtained at the 42% level was dissolved in a minimal amount of potassium phosphate (25 m*M*, pH 7.5); the solution was clarified by centrifugation and then placed on a column (1.5 cm diameter, 1.5 cm long) of calcium phosphate gel. The column was washed successively with potassium phosphate buffers (7.5 ml of 25 m*M* and 6.0 ml of 50 m*M*) at pH 7.5 and eluted with 6.5 ml of sodium

[22] W. F. Anacker and V. Stoy, *Biochem. Z.* **330**, 141 (1958).
[23] L. E. Schrader, G. L. Ritenour, G. L. Eilrich, and R. H. Hageman, *Plant Physiol.* **43**, 930 (1968).

pyrophosphate (100 mM, pH 7.5). All buffers used throughout the above steps contained cysteine hydrochloride (1 mM) and the pH given refers to the final pH after the addition of this compound.

A slightly different sequence of steps was used for the enzyme from corn and spinach leaves.

Alumina Gel. A further increment of purification of nitrate reductase from spinach leaves was obtained by chromatography on alumina C_γ gel after a procedure which was basically similar to the one described above.[24] The enzyme was adsorbed onto the alumina gel (1 mg of gel/mg of protein) which was then washed with sodium pyrophosphate (20 mM, pH 7.0); the enzyme was eluted with the same buffer at 100 mM.

Column Chromatography on DEAE–Cellulose. Schrader *et al.*[23] used a DEAE–cellulose column (4×12 cm) equilibrated with Tris·HCl (25 mM, pH 7.5) during the purification of nitrate reductase from marrow cotyledons. After adsorption of the enzyme, the column was eluted with a linear gradient of Tris·HCl, passing from 25 mM to 150 mM in an elution volume of 600 ml. Repetition of the DEAE–cellulose step gave only a minimal increase in purification. DEAE–cellulose chromatography may be inserted at any point in the purification procedure given above, provided that the ionic concentration of the enzyme solution is low enough to permit adsorption.

Other Methods. Since higher plant nitrate reductases, in contrast to the enzyme from *Neurospora*,[25] have not been highly purified, the following comments may be useful.

Solvent precipitation and resin chromatography usually inactivate the enzyme. Molecular exclusion chromatography has not been much used, presumably because of the generally high molecular weights of nitrate reductase enzymes, but agarose gels may prove useful. Purification by isoelectric precipitation and sucrose density gradient centrifugation show promise.[26] Some nitrate reductase activity from radish cotyledons was recovered after gel acrylamide electrophoresis,[27] and this may become a useful tool for purification.

Characteristics

The properties of nitrate reductases from higher plants have been recently reviewed.[28]

Molecular Weight. The estimates of molecular weight so far made are

[24] A. Paneque and M. Losada, *Biochim. Biophys. Acta* **128,** 202 (1968).
[25] R. H. Garrett and A. Nason, *Proc. Nat. Acad. Sci. U.S.* **58,** 1603 (1967).
[26] Unpublished results of this laboratory.
[27] J. Ingle, *Biochem. J.* **108,** 715 (1968).
[28] L. Beevers and R. H. Hageman, *Annu. Rev. Plant Physiol.* **20,** 495 (1969).

500,000–600,000 (wheat leaves[22]), greater than 100,000 (*Neurospora*[25]) 25,000 (*Perilla* leaves[29]) and 160,000 (maize tissues[26]).

Substrate Affinity. The Michaelis constants (K_m) for nitrate are of the order of 0.2 mM.[4,5,23,30]

Pyridine Nucleotide Specificity. Nitrate reductase is predominantly NADH-specific. The K_m values for NADH are of the order of 2.5 μM.[5] Preliminary work with corn scutellum and foxtail leaves, which have unusually high activity with NADPH as electron donor, suggests the presence of two enzymes. In each of these species, one of the enzymes is able to use both pyridine nucleotides; in addition corn scutellum contains an NADH-specific nitrate reductase. A single enzyme nonspecific for the pyridine nucleotide was reported to occur in soybean leaves. Plant nitrate reductases may therefore classify in two enzyme categories, EC 1.6.6.1, and EC 1.6.6.2 (cf. Note added in proof).

Flavin Requirement. FAD is a constituent of the soybean enzyme; half-maximal activities were obtained at concentrations of 0.1 μM and 3.7 μM of FAD and FMN, respectively[4] when a pyridine nucleotide was the electron donor. With dithionite-FMN as reductant, the K_m values for flavins are generally much higher, i.e., of the order of 0.5 mM for FMN.[23] An exogenous supply of flavin is necessary for the assay with dithionite.

Involvement of Metals. There is indirect, but good evidence that plant nitrate reductases contain molybdenum.[22,31] Presumably this participates in electron transfer in the manner described for *Neurospora*.[32,33] The enzyme from this latter source appears to have cytochrome b_{557} as a prosthetic group,[25] but no cytochrome involvement has been demonstrated for higher plant enzymes.

Phosphate Requirement. Nitrate reductase from *Neurospora* has a requirement for orthophosphate.[34,35] Its probable function is to facilitate the reduction of molybdenum, by forming complexes with it. No extensive studies have been made of the phosphate requirements of the plant enzymes. However, addition of phosphate to extracts made with Tris buffer enhances activity.[26]

Sequence of Electron Transfer. The sequence of electron transfer for the *Neurospora* nitrate reductase seems to be[32]:

$$\text{Pyridine nucleotide} \rightarrow \text{FAD} \rightarrow \text{Mo} \rightarrow \text{nitrate}$$

[29] C. G. Kannangara and A. W. Woolhouse, *New Phytol.* **67,** 533 (1968).
[30] D. Spencer, *Aust. J. Biol. Sci.* **12,** 181 (1959).
[31] H. J. Evans and H. S. Hall, *Science* **122,** 922 (1955).
[32] D. J. D. Nicholas and A. Nason, *J. Biol. Chem.* **211,** 183 (1954).
[33] D. J. D. Nicholas and H. M. Stevens, *Nature (London)* **176,** 1066 (1955).
[34] D. J. D. Nicholas and J. H. Scawin, *Nature (London)* **178,** 1474 (1956).
[35] S. C. Kinsky and W. D. McElroy, *Arch. Biochem. Biophys.* **73,** 466 (1958).

with cytochrome b_{557} tentatively assigned to a position between the flavin and molybdenum.[25] A similar sequence is probable with higher plant enzymes, but bearing in mind the reservation concerning the presence of a cytochrome. Partial purifications of nitrate reductases from spinach, corn and marrow[23,24] have so far failed to separate the system into an NADH-diaphorase reducing FAD, and a nitrate reductase transferring electrons from reduced flavin to nitrate. NADH–nitrate reductase from leaves can be readily separated from the photosynthetic ferredoxin–NADP reductase, which is not a part of NADH–nitrate reductase.[23]

Inhibitors. Nitrate reductase is sensitive to inhibition by pCMB (1 m*M* to 1 μ*M*) especially when a pyridine nucleotide is the electron donor; cysteine or glutathione in 5- to 100-fold excess protects the enzyme against inhibition.[4,23] The enzyme is sensitive to reagents that react with metals. Cyanide and azide are particularly effective, while organic-chelating reagents give various degrees of inhibition.[4] Atabrin inhibits marrow nitrate reductase at a concentration of 5 m*M*. Neither carbon monoxide nor fluoride (1 m*M*) inhibit soybean nitrate reductase.[4]

Cellular Location. Evidence that nitrate reductase is a cytoplasmic enzyme has been obtained by the use of differential centrifugation and marker enzymes on aqueous and nonaqueous leaf extracts[36] and by studies with inhibitors of chloroplastic protein synthesis.[37] In preparations of whole chloroplasts, which were made by the method of Jensen and Bassham[38] and shown by microscopic examination to contain a high proportion of intact plastids, nitrate reductase activity was associated predominantly with the supernatant rather than with the chloroplast fraction.[26] Coupé,[39] however, using density gradient centrifugation techniques, concluded that the enzyme is located in the chloroplasts. In this work, activity was determined solely in relation to chlorophyll and protein distribution. This question of subcellular location has important bearings upon another controversy, namely the nature of the immediate energy source for nitrate reduction in the leaf.[28]

Induction. Nitrate reductase is adaptive in that activity is inducible by nitrate in intact plants or tissue[40–42] and by molybdenum in tissue from molybdenum-deficient plants.[43] Optimum induction is often ob-

[36] G. L. Ritenour, K. W. Joy, J. Bunning, and R. H. Hageman, *Plant Physiol.* **42**, 233 (1966).

[37] L. E. Schrader, L. Beevers, and R. H. Hageman, *Biochem. Biophys. Res. Commun.* **26**, 14 (1967).

[38] R. G. Jensen and J. A. Bassham, *Proc. Nat. Acad. Sci. U.S.* **56**, 1095 (1966).

[39] M. Coupé, M. L. Champigny, and A. Moyse, *Physiol. Veg.* **5**, 271 (1967).

[40] P. S. Tang and H. Y. Wu, *Nature (London)* **179**, 1355 (1957).

[41] E. J. Hewitt, *Nature (London)* **180**, 1020 (1957).

[42] L. Beevers, L. E. Schrader, D. Flesher, and R. H. Hageman, *Plant Physiol.* **40**, 691 (1965).

tained in tissue when the induction medium is pH 4–5 and contains 10–100 m*M* nitrate. Both pH and nitrate concentration of the induction medium, as well as light,[42] influence the absorption of nitrate by the tissue. The importance of these three factors varies according to species and tissue and must be carefully determined. The use of inhibitors of protein and nucleic acid synthesis suggests that *de novo* synthesis of the enzyme occurs during induction.

An absolute and direct requirement of light and CO_2 fixation for the induction of nitrate reductase in leaves of *Perilla* has been indicated.[44] However, because nitrate reductase occurs in nonchlorophyllous tissue, including roots, and since induction of nitrate reductase in radish cotyledons in the dark has been demonstrated, it appears that light is not an obligatory requirement for synthesis of the enzyme. A more complete discussion of the role of light in induction is available.[28]

Stability. Nitrate reductase is unstable both *in vivo* and *in vitro*. In excised whole maize seedlings at 30°, the half-life of the enzyme was estimated to be 4 hours.[16] With intact maize plants placed in the dark, at 25°, a half-life of 12–14 hours is indicated.[45] The stability *in vitro* and *in vivo* varies greatly with species, plant age, and tissue. Increases in temperature decrease the stability. Partially purified preparations can be stored at —15° to 20°, for 2–3 months with little loss of activity; however, this stability is also dependent upon the species and tissue.

Note added in proof: With respect to the *in vivo* assay (p. 498) subsequent work has shown that: (a) The phosphate concentration of the infiltration medium should be 0.1–0.15 *M* for maximum activity; (b) glucose is the best source of energy for nitrate reduction for "carbohydrate deficient" tissue; and (c) excretion of nitrite into the infiltration medium is not always linear for some tissues, especially during the initial (15–30 minute) phase of incubation. Since addition of organic solvents (e.g., 5% acetone) increases the appearance of nitrite in the medium, at least with soybean leaves, the permeability of the tissue to both external nitrate and internal nitrite is considered to be the cause of nonlinear excretion.

With respect to the pyridine nucleotide specificity (p. 501), it recently has been shown[46] that the nitrate reductase extracted from leaves of soybean, maize, and foxtail is NADH-specific. The apparent ability of the crude or partially purified enzyme, obtained from certain plant species, to utilize NADPH reflects the presence of a phosphatase which rapidly converts NADPH to NADH.

[43] M. M. R. K. Afridi and E. J. Hewitt, *J. Exp. Bot.* **15**, 251 (1964).
[44] C. G. Kannangara and H. W. Woolhouse, *New Phytol.* **66**, 553 (1967).
[45] R. H. Hageman and D. Flesher, *Plant Physiol.* **35**, 700 (1960).
[46] G. N. Wells and R. H. Hageman, *Plant Physiol. Suppl.* **46**, 45 (1970).

[46] Phytoflavin

By ROBERT M. SMILLIE and BARRIE ENTSCH

The discovery of *Anacystis* phytoflavin[1-4] has led to recognition of a new class of FMN-containing electron transfer proteins which have been found in blue-green algae and nitrogen-fixing bacteria. The highest concentrations of these proteins are found in iron-deficient cells. Phytoflavin will substitute for ferredoxin in photosynthetic electron transfer reactions, and this may be its physiological role in photosynthetic cells growing under conditions of iron insufficiency.

Assay of Phytoflavin

The reduction of NADP by illuminated chloroplast fragments is measured spectrophotometrically.[3,5] The reduction is dependent upon the presence of either phytoflavin or ferredoxin.

Reagents

Tricine–NaOH buffer, 200 m*M*, pH 7.8
$MgCl_2$, 25 m*M*
NADP, 20 m*M*
Chloroplast fragments (120 μg of chlorophyll/ml). Washed chloroplast fragments are prepared from the leaves of 9- to 12-day-old pea plants (*Pisum sativum* L.) following the procedure used for spinach by San Pietro and Lang.[6] The fragments are stored in ice and should be used as soon as possible.

Procedure. The first three reagents are mixed (0.1 ml of each) and phytoflavin and water are added to give a final volume of 0.9 ml. Chloroplast fragments (0.1 ml) are added, and the reaction mixture is illuminated with red or white light. The increase in absorbancy at 340 nm is measured for 1–3 minutes, preferably using a recording spectrcphotometer.[5] An illuminated control in which phytoflavin has been omitted is included, and corrections should be made for any absorbancy changes occurring in nonilluminated samples. These are usually negligible. The

[1] R. M. Smillie, *Plant Physiol. Suppl.* **38,** xxviii (1963).
[2] R. M. Smillie, *Biochem. Biophys. Res. Commun.* **20,** 621 (1965).
[3] R. M. Smillie, *Plant Physiol.* **40,** 1124 (1965).
[4] R. M. Smillie, *Aust. J. Sci.* **28,** 79 (1965).
[5] R. M. Smillie, *Plant Physiol.* **37,** 716 (1962).
[6] A. San Pietro and H. M. Lang, *J. Biol. Chem.* **231,** 211 (1958).

reaction is carried out at 23°, and care should be taken to ensure that the temperature of the sample does not rise significantly during illumination. The initial rate of reduction is proportional to the amount of phytoflavin in the reaction mixture. The use of the assay is largely confined to comparative measurements using a single batch of chloroplasts as it is difficult to obtain the same activities from chloroplasts isolated from different batches of leaves.

The assay procedure does not distinguish between phytoflavin and ferredoxin in crude extracts of *Anacystis nidulans*.

Preparation of Phytoflavin from *Anacystis nidulans*

Growth of the Cells

Growth medium (modified from Kratz and Myers[7])

1. $NaNO_3$, 2.0 g
2. $MgSO_4 \cdot 7\ H_2O$, 0.25 g
3. $Ca(NO_3)_2 \cdot 4\ H_2O$, 0.02 g
4. EDTA, disodium salt, 0.065 g
5. K_2HPO_4, 1.0 g
6. Iron solution. A stock solution of $(NH_4)_2SO_4 \cdot Fe_2(SO_4)_3 \cdot 24\ H_2O$, 2.5 m*M* with respect to Fe, is prepared.
7. Micronutrients. Into 1 liter of demineralized water is dissolved $ZnSO_4 \cdot 7\ H_2O$, 8.82 g; $MnCl_2 \cdot 4\ H_2O$, 1.44 g; $(NH_4)_6Mo_7O_{24} \cdot 4\ H_2O$, 0.74 g; $CuSO_4 \cdot 5\ H_2O$, 1.57 g; $Co(NO_3)_2 \cdot 6\ H_2O$, 0.49 g; and H_3BO_3, 2.86 g.

For 1 liter of medium, components 1 to 4 are dissolved in 850 ml of demineralized water; 1.0 ml of iron solution and 1.0 ml of micronutrient solution are added, and the solution is diluted to 900 ml. Component 5 is dissolved in 100 ml of water. The two solutions are autoclaved separately and mixed when the solutions have cooled.

Procedure. Cells of *Anacystis nidulans* are grown in 20-liter bottles, each containing 6 liters of medium. The bottles are placed on their sides on the platform of a reciprocal shaker and illuminated by an overhead bank of white fluorescent lights to give an intensity at the surface of the culture of between 6000 and 10,000 lux. The temperature of the culture is maintained at 35°–37°, although satisfactory growth can also be obtained at 25°. The cultures are gassed with 1.0–1.5% CO_2 in air and are shaken continuously. Each culture is started with a 5% inoculum

[7] W. A. Kratz and J. Myers, *Am. J. Bot.* **42,** 282 (1955).

and is allowed to grow until the culture approaches the stationary phase of growth (5–7 days).

The cells are harvested by centrifugation and washed twice with 50 m*M* Tris·HCl buffer, pH 7.8.

Extraction of the Cells

Reagents

50 m*M* Tris·HCl, pH 7.8
Acetone
Diethyl ether

Procedure. Washed cells (100-g fresh weight) are suspended in an equal volume of 50 m*M* Tris·HCl buffer, pH 7.8, and cooled to 0°–5°; 10–12 vol of acetone at —20° are slowly added with stirring. The mixture is allowed to stand for a few minutes until the suspended matter settles, and the clear green supernatant fluid is decanted and discarded. The suspension is rapidly filtered through filter paper using suction and washed with 5 vol of the cold acetone, followed by 5 vol of diethyl ether, also at —20°. The cake of extracted cells is spread thinly on brown paper and allowed to dry at room temperature for 5 minutes. The blue powder is mixed with 50 m*M* Tris·HCl buffer, pH 7.8 (4–5 vol per original wet weight of cells) and extracted for 1–2 hours at 0°–4°. The suspended material is recovered by centrifugation at 34,000 *g* for 15 minutes and reextracted as before. The extracts are combined and fractionated as described below.

Alternatively, the cells can be broken by passage through a French pressure cell at 20,000 psi as described previously.[2,4] The above method is preferred since use of the pressure cell results in the extraction of large amounts of pteridine, and phycocyanin and other proteins which interfere in the subsequent purification steps.

Purification of Phytoflavin

Reagents

Ammonium sulfate
Tris·HCl buffer, 5 m*M*, pH 7.8
Whatman DE-32 (DEAE–cellulose)
NaCl, 0.1 *M*, Tris·HCl buffer, 5 m*M*, pH 7.8
NaCl, 0.6 *M*, Tris·HCl buffer, 5 m*M*, pH 7.8
NaCl, 0.5 *M*, Tris·HCl buffer, 5 m*M*, pH 7.8

Fractionation with Ammonium Sulfate. Solid ammonium sulfate is added to the cell-free extract at room temperature to a final concentration of 60% saturation. After it has stood for 1 hour, the mixture is centrifuged at 10,000 *g* for 15 minutes. The green to yellow-brown supernatant fluid is retained, and ammonium sulfate is added to 100% saturation. The mixture is left standing for 1 hour (12 hours if ferredoxin is also to be recovered) and is then centrifuged at 10,000 *g* for 15 minutes; the precipitate is retained.

Chromatography on DEAE–Cellulose. The precipitate obtained between 60 and 100% saturated ammonium sulfate is dissolved in and diluted with 5 m*M* Tris·HCl buffer, pH 7.8 to give a final salt concentration of less than 0.1 *M*. The solution is placed on a column (16 × 2.2 cm diameter) of Whatman DE-32, and the absorbed protein is eluted with a linear gradient of 0.1–0.6 *M* NaCl containing 5 m*M* Tris·HCl buffer, pH 7.8. This column is usually run at 0°–4°, but phytoflavin is stable enough for the column to be run at room temperature. Phytoflavin is a bright yellow protein which is eluted at about 0.35 *M* salt. As the salt concentration is increased, ferredoxin is eluted as a well-separated fraction.

The behavior of these two proteins on a similar column is shown in Fig. 1 In this experiment Whatman DE-11 and a nonlinear NaCl gradient of 0–0.6 *M* were used[2] and two *c*-type cytochromes were obtained as by-products. Two fractions containing chloroplast ferredoxin activity were obtained—one corresponding to phytoflavin, the other to *Anacystis* ferredoxin. The tailing which is especially evident in the case of ferredoxin is largely eliminated by substituting DE-32 for DE-11.

Rechromatography on DEAE–Cellulose. Additional purification is obtained by rechromatographing the phytoflavin on a DE-32 column using a linear gradient of 0.1 *M* to 0.5 *M* NaCl containing 5 m*M* Tris·HCl, pH 7.8. Phytoflavin can be concentrated by precipitation with 80% saturated ammonium sulfate.

Yield. The yield is about 2.0 μmoles of protein (40 mg) per 100-g wet weight of cells.

Purity. The protein moves as a single band in disk electrophoresis in 12.5% polyacrylamide gels at either pH 5.3 or pH 8.9. The purity is also indicated by the ratio of the absorption at 275 nm to that at 465 nm (see below). Constant spectral ratios were obtained in fractions taken across the phytoflavin peak eluted from the second DE-32 column. In a similar way, constant ratios were obtained when the phytoflavin was passed through a column of Sephadex G-75.

Stability. Phytoflavin is stable for some months when stored at −20° and pH 7.8. Phytoflavin is not sensitive to light.

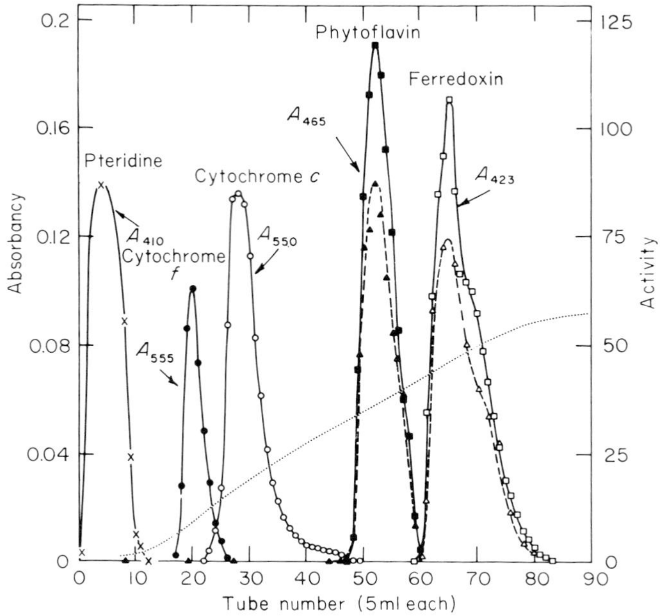

FIG. 1. Chromatography of proteins from *Anacystis nidulans* on DEAE-cellulose. A 65–100% saturated ammonium sulfate fraction of an extract from cells of *A. nidulans* was absorbed onto a DEAE-cellulose column. Protein was eluted by an increasing nonlinear gradient of NaCl [R. M. Smillie, *Biochem. Biophys. Res. Commun.* **20,** 621 (1965)]. The continuous line in the figure shows the absorbancy of individual fractions at the wavelengths indicated. The dashed line and triangles show chloroplast ferredoxin activities in millimicromoles of NADP reduced per minute per milliliter of fraction. The dotted line indicates the salt gradient (conductivity measurements). The activity and absorbancy values shown for fractions containing ferredoxin are one-fifth the actual values obtained.

Properties of Phytoflavin

Prosthetic Group. The yellow color of phytoflavin is due to FMN.[3] FAD is absent. Iron is also absent as indicated by the absorption spectrum and by the absence of radioactivity in phytoflavin isolated from cells grown in the presence of ^{55}Fe.[3] Acid-labile sulfur of the type found in chloroplast ferredoxin[8] is also absent.[3]

Absorption Spectrum. The absorption spectrum of oxidized phytoflavin is shown in Fig. 2. There are peaks at 275, 377, and 465 nm, with shoulders at 284, 292, and 492 nm, and troughs at 248, 315, and 408 nm.

[8] K. T. Fry and A. San Pietro, *in* "Photosynthetic Mechanisms of Green Plants," p. 252, *Nat. Acad. Sci.—Nat. Res. Counc., Misc. Publ. No.* **1145** (1963).

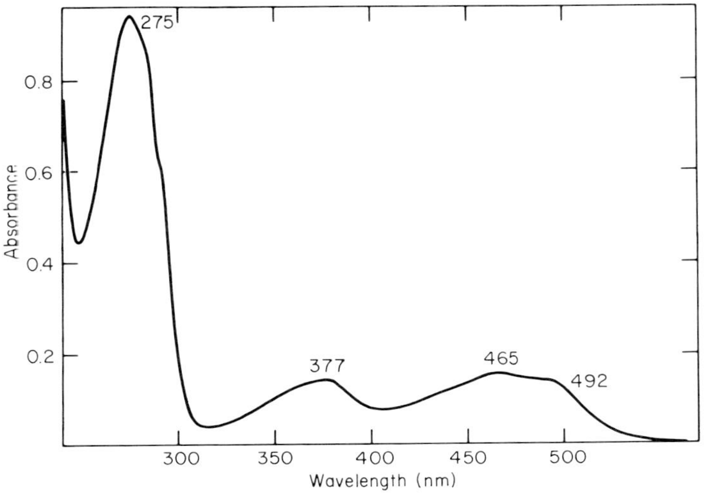

FIG. 2 Absorption spectrum of oxidized phytoflavin in 0.35 *M* NaCl, 5 m*M* Tris·HCl (pH 7.8).

The ratio of $A_{275\,\mathrm{nm}}$ to $A_{465\,\mathrm{nm}}$ is 6.0, and that of $A_{377\,\mathrm{nm}}$ to $A_{465\,\mathrm{nm}}$ is 0.91. The visible absorption bands are 5–20 nm higher than those of the flavoprotein from *Clostridium pasteurianum* (flavodoxin), which can also substitute for ferredoxin in a number of enzymatic reactions[9,10] and those of a FMN-protein of unknown function isolated from *Azotobacter vinelandii*.[11–14]

The millimolar extinction coefficient of phytoflavin at 465 nm is 9.2 ± 0.2. This value was determined from amino acid analysis and the FMN content.

Reduction of Phytoflavin. Phytoflavin is slowly reduced by sodium dithionite under anaerobic conditions (Fig. 3). The reduction, which takes about 2.5 hours, probably proceeds to the hydroquinone via the semiquinone, as indicated by the appearance of a small absorption band

[9] E. Knight, Jr., A. J. D'Eustachio, and R. W. F. Hardy, *Biochim. Biophys. Acta* **113,** 625 (1966).

[10] E. Knight, Jr. and R. W. F. Hardy, *J. Biol. Chem.* **242,** 1370 (1967).

[11] J. W. Hinkson and W. A. Bulen, *J. Biol. Chem.* **242,** 3345 (1967).

[12] J. W. Hinkson, *Biochemistry* **7,** 2666 (1968).

[13] Y. I. Shethna, P. W. Wilson, and H. Beinert, *Biochim. Biophys. Acta* **113,** 225 (1966).

[14] P. Hemmerich, C. Veeger, and H. C. S. Wood, *Angew. Chem. Int. Ed. Engl.* **4,** 671 (1965).

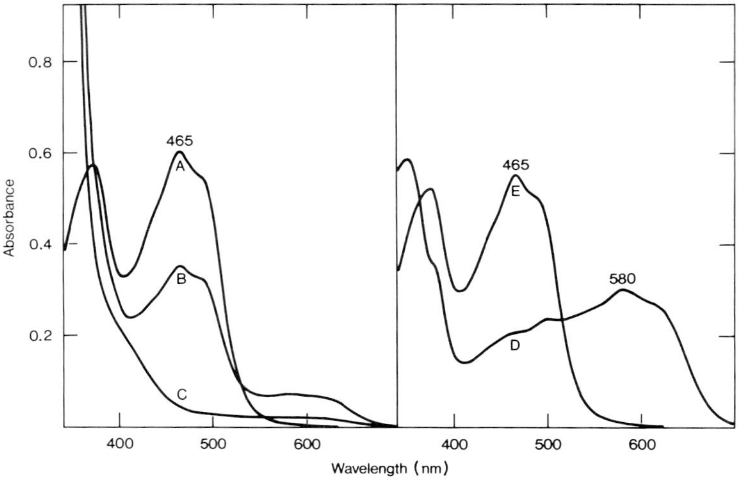

FIG. 3. Reduction and oxidation of phytoflavin. Reduction of phytoflavin (66 μM solution in 25 mM potassium phosphate buffer, pH 7.5) was carried out at 23° and in an atmosphere of N_2 using a 50–100 M excess of sodium dithionite. Curve A, oxidized phytoflavin; curve B, approximately 50% reduced; curve C, fully reduced. Air was then admitted, and the spectrum was immediately determined (curve D). After the further addition of potassium ferricyanide, curve E was obtained.

around 600 nm during the course of the reduction (Fig. 3, curve B). However, the intermediate could also be formed by reaction between the fully oxidized and fully reduced species. Reaction rates are dependent upon pH.

Reduced phytoflavin is rapidly oxidized by oxygen to a sky-blue intermediate (Fig. 3, curve C) which has the spectral characteristics of the neutral species of semiquinone of FMN.[15] Oxidation of the semiquinone form by oxygen is by comparison very slow, and oxidation of a 18.0 μM solution of the semiquinone in air takes about 2.5 hours. Potassium ferricyanide rapidly converts the semiquinone form of phytoflavin to the oxidized form.

In the presence of ferredoxin-NADP reductase[16,17] and NADPH, phytoflavin is reduced to the semiquinone form under aerobic conditions.

Oxidation–Reduction Potential. Bothe[18] estimated that the redox

[15] V. Massey and G. Palmer, *Biochemistry* **5,** 3181 (1966).

[16] D. L. Keister, A. San Pietro, and F. E. Stolzenbach, *J. Biol. Chem.* **235,** 2989 (1960).

[17] M. Shin, K. Tagawa, and D. I. Arnon, *Biochem. Z.* **338,** 84 (1963).

[18] H. Bothe, "Ferredoxin und Phytoflavin in photosynthetischen Reaktionen einer Präparation aus der Blaualge *Anacystis nidulans.*" Doctoral Dissertation, University of Göttingen, 1968.

potential of the couple, oxidized phytoflavin/fully reduced phytoflavin, was more positive than that of the couple, NADP/NADPH. In this estimation no allowance was made for the stable semiquinone form of phytoflavin.

Using hydrogenase from *Clostridium pasteurianum*, we found the E_0 at pH 7.0 of the couple phytoflavin semiquinone/fully reduced phytoflavin to be -0.450 ± 0.005 V. This value is independent of pH over the range 6.5–8.5. Thus at pH 7.0, phytoflavin is a stronger reducing agent than *Anacystis* ferredoxin (E_0 at pH 7.0 = -0.38 V). This couple (semiquinone/fully reduced phytoflavin) is probably the way phytoflavin acts as an electron carrier in the photoreduction of NADP.

Molecular Weight and Amino Acid Composition. The molecular weight of phytoflavin is 20,600 $\pm$ 500 as estimated by the method of Andrews[19] using Sephadex G-75 and 20,300 by amino acid composition (see the table).

The amino acid composition of phytoflavin is shown in the table. Corresponding values for flavodoxin and *Anacystis* ferredoxin are included. All are small, negatively charged proteins which can act as electron transfer carriers in the same reactions, but there is probably no close evolutionary relationship between them. According to the Difference Index of Metzger *et al.*,[20] phytoflavin and flavodoxin are the most closely related, with a value of 16.4, followed by phytoflavin and *Anacystis* ferredoxin. Proteins with a close compositional relationship usually have a value of less than 10.

Biological Activity. PHOTOREDUCTION OF NADP. Phytoflavin will replace the requirement for ferredoxin in the photoreduction of NADP by isolated chloroplast particles (Fig. 4). The rates attained with phytoflavin are comparable to those with ferredoxin (mole FMN in phytoflavin to mole ferredoxin).[2] Phytoflavin substitutes for ferredoxin in this reaction using chloroplasts isolated from pea, spinach, wheat, lettuce, and *Euglena gracilis*, or using chlorophyll-containing particles isolated from *Anacystis nidulans*.[18,21,22] Phytoflavin can also replace ferredoxin in the photoreduction of NADP by spinach chloroplasts or particles from *A. nidulans*, ascorbate being used as the electron donor.[18,21]

Phytoflavin mediates the photoreduction of the 3-acetyl pyridine analog of NADP as well as NADP. NAD is also reduced, but at a rate slower than NADP, at a pyridine nucleotide concentration of 0.22 m*M*

[19] P. Andrews, *Biochem. J.* **91,** 222 (1964).
[20] H. Metzger, M. B. Shapiro, J. E. Mosimann, and J. E. Vinton, *Nature (London)* **219,** 1166 (1968).
[21] A. Trebst and H. Bothe, *Ber. Deut. Bot. Ges.* **79,** 44 (1966).
[22] B. Entsch and R. M. Smillie, unpublished results (1968).

COMPOSITION OF PHYTOFLAVIN, FLAVODOXIN, AND *Anacystis* FERREDOXIN

Components	Phytoflavin from *Anacystis nidulans*[b]	Flavodoxin from *Clostridium pasteurianum*[c]	Ferredoxin from *A. nidulans*[b]	Ferredoxin from *A. nidulans*[d]
Amino acids[a]				
Lysine	8	10	3	3
Histidine	Absent	Absent	1	1
(Ammonia)	19	—	7	4
Arginine	2	2	1	1
Aspartic acid	28	18	15	15
Threonine	8	4	10	12
Serine	13	14	6	7
Glutamic acid	26	19	11	11
Proline	4	4	2	3
Glycine	18	17	6	6
Alanine	13	14	12	12
Half-cystine	1	1	4	6
Valine	11	16	9	9
Methionine	1	4	Absent	Absent
Isoleucine	13	5	5	5
Leucine	13	12	7	7
Tyrosine	8	2	5	5
Phenylalanine	8	3	2	2
Tryptophan	5	4	Absent	Absent
Others[a]				
FMN	1	1	0	0
Iron	0	0	2	2
Acid-labile sulfur	0	0	1	1
Molecular weight (from amino acid analysis)	20,300	14,600	10,700	11,000

[a] Number per molecule protein.
[b] B. Entsch and R. M. Smillie, unpublished results (1969).
[c] E. Knight, Jr. and R. W. F. Hardy, *J. Biol. Chem.* **242,** 1370 (1967).
[d] T. Yamanaka, S. Takenami, K. Wada, and K. Okunuki, *Biochim. Biophys. Acta* **180,** 196 (1969).

(Fig. 4). At higher concentrations (2–5 m*M*), the rate with NAD approaches that of NADP. The specificity of the phytoflavin-mediated system is thus similar to that of the ferredoxin-mediated system.[15] These results, together with the observation that the phytoflavin-mediated photoreduction of NADP by "aged" spinach chloroplasts is stimulated by ferredoxin-NADP reductase[21] and the spectral demonstration of phytoflavin reduction via the reductase (see above), all point to phyto-

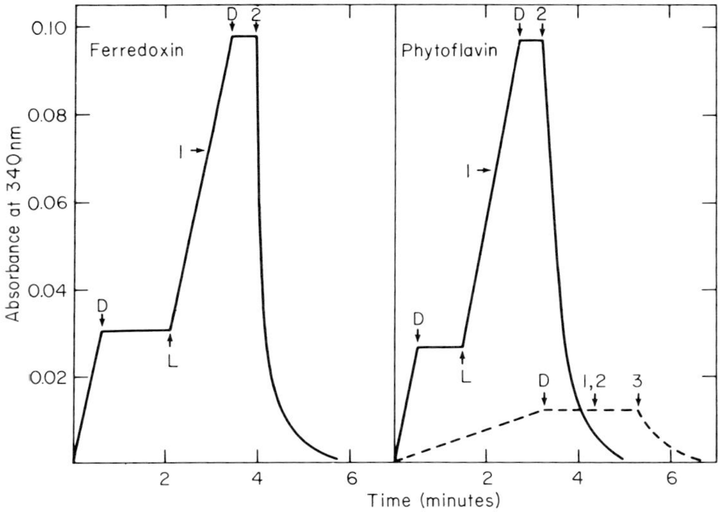

FIG. 4. Photoreduction of NADP mediated by chloroplast-type ferredoxin from *Anacystis nidulans* (left side) and phytoflavin (right side). The reaction mixture contained pea leaf chloroplasts (12 μg of chlorophyll), 20 m*M* Tris·HCl buffer, pH 7.8, 2.5 m*M* $MgCl_2$, 0.22 m*M* NADP or NAD, and 0.88 μ*M* ferredoxin or 0.81 μ*M* phytoflavin. Solid line, NADP; broken line, NAD. *1,* glutathione reductase added; *2,* 2.5 m*M* oxidized glutathione added; *3,* malate dehydrogenase and 2.5 m*M* oxaloacetate added. D, dark; L, light.

flavin acting in the chloroplast system through ferredoxin-NADP reductase. Attempts to demonstrate a phytoflavin-specific reductase using chloroplasts reacted with antibody to ferredoxin NADP reductase were unsuccessful.[18]

PHOTOREDUCTION OF NITRITE AND NITRATE. In the presence of illuminated chloroplasts and the enzyme nitrite reductase (EC 1.6.6.4), ferredoxin mediates the reduction of nitrite.[23] Phytoflavin can replace ferredoxin in this reaction.[18]

Ferredoxin will not act as an electron donor for nitrate reductase (EC 1.6.6.2).[24] Instead it has been proposed that free flavin nucleotides are the natural electron carriers for nitrate reduction in spinach leaves.[25-27] While it is known that flavin nucleotides are reduced by

[23] A. Paneque, J. M. Ramírez, F. F. del Campo, and M. Losada, *J. Biol. Chem.* **239,** 1737 (1964).

[24] A. Paneque, F. F. del Campo, J. M. Ramírez, and M. Losada, *Biochim. Biophys. Acta* **109,** 79 (1965).

[25] F. F. del Campo, A. Paneque, J. M. Ramírez, and M. Losada, *Nature* (*London*) **205,** 387 (1965).

illuminated chloroplasts under anaerobic conditions,[28] reoxidation takes place under aerobic conditions. The semiquinone form of phytoflavin is not easily autoxidizable, and if phytoflavin is shown to be present in chloroplasts, it may be a more suitable electron carrier than free flavin nucleotide for coupling between the site of reduction in the lamella and nitrate reductase. This would be especially true if, as the studies of Ritenour *et al.*[29] indicate, the nitrate reductase is located outside of the chloroplasts. We have compared the photoreduction of nitrate mediated by phytoflavin or FMN in a system containing purified soybean nitrate reductase[30] and pea chloroplasts. While both compounds supported nitrate reduction, the concentrations required were in excess of 100 times the concentration of phytoflavin required for comparable rates of NADP reduction.

Other reactions involving phytoflavin. Phytoflavin catalyzes cyclic photophosphorylation (in the presence of DCMU) by spinach chloroplasts or particles from *A. nidulans*.[18] It also acts as an electron carrier in the reduction of cytochrome *c* by either illuminated chloroplasts or NADPH and NADP-ferredoxin reductase. Phytoflavin mediates the photooxidation of *Euglena* cytochrome *c* (552) coupled to NADP reduction catalyzed by isolated *Euglena* chloroplasts.[4]

Bothe[18] has shown that phytoflavin will substitute for ferredoxin in the reductive formation of pyruvate from acetylphosphate and CO_2 catalyzed by extracts of *Clostridium aceticium*.

Phytoflavin shows neither NAD(P)H-diaphorase or transhydrogenase activity.[3]

Distribution. Phytoflavin has been isolated from the blue-green algae *Anacystis nidulans*[1–4,18,21] and *Anabaena cylindrica*.[3] Cells growing under conditions of partial iron deficiency produce the highest levels of phytoflavin.[18,21,22] It will be interesting to see whether phytoflavin is also synthesized in iron-deficient plants and photosynthetic bacteria.

[26] J. M. Ramírez, F. F. del Campo, A. Paneque, and M. Losada, *Biochem. Biophys. Res. Commun.* **15**, 297 (1964).

[27] M. Losada, J. M. Ramírez, A. Paneque, and F. F. del Campo, *Biochim. Biophys. Acta* **109**, 86 (1965).

[28] L. P. Vernon and W. S. Zaugg, *J. Biol. Chem.* **235**, 2728 (1960).

[29] G. L. Ritenour, K. W. Joy, J. Bunning, and R. H. Hageman, *Plant Physiol.* **42**, 233 (1967).

[30] H. J. Evans and A. Nason, *Plant Physiol.* **28**, 233 (1953).

[47] Detection and Isolation of P700

By T. V. Marsho and B. Kok

The pigment designated P700[1,2] has been detected in all normal photosynthetic algae and green plants studied. It occurs in small concentrations (1/200–1/1000) relative to the "bulk" chlorophyll(s) and is characterized by absorption bands around 430 and 700 nm. The location of these bands at slightly longer wavelengths than those of the light-harvesting chlorophyll *a*, and the similar solubility properties of the two pigments suggest that P700 is a form of chlorophyll *a* segregated as a result of environmental influences. P700 behaves as a one-electron redox component ($E_0' \simeq 430$ mV) and is thought to function as the reaction center of photosystem I in which light energy is converted into chemical energy.[3,4] Quanta absorbed by the harvesting pigment of photosystem I are transferred to P700, causing transfer of an electron to an unidentified acceptor (X) of low potential [Eq. (1)].

$$P700^- + X^+ \underset{\text{dark cycle}}{\overset{K_L \text{ system I}}{\rightleftarrows}} P700^+ + X^- \quad (1)$$

This photoevent, which has a quantum efficiency approaching 1, leaves P700 in the oxidized state and is accompanied by a (partial?) loss of the pigment's absorption bands at 430 and 700 nm. Reduced P700 can also be oxidized with chemical agents such as ferricyanide. Reduction of P700 occurs in one of several possible dark reactions [Eq. (2)].

$$P700^+ + \text{donor}^- \left\{ \begin{matrix} \text{Cyt } f\text{, Pc,} \\ \text{exogenous} \\ \text{donors} \end{matrix} \right\} \xrightarrow[k_D]{\text{dark}} P700^- + \text{donor}^+ \quad (2)$$

In vivo cytochrome *f* (cyt *f*) and plastocyanin (Pc) are probably the major electron donors, which in turn are reduced by photosystem II. In addition, reduction may occur via a cyclic back flow of electrons from (X^-) as shown in Eq. (1). Exogenous reducing agents such as ascorbate or ferrocyanide can also restore photooxidized P700.

[1] B. Kok, *Biochim. Biophys. Acta* **22,** 399 (1956).
[2] B. Kok, *Acta Bot. Neer.* **6,** 316 (1957).
[3] R. K. Clayton, *in* "The Chlorophylls" (L. P. Vernon and G. R. Seely, eds.), p. 609. Academic Press, New York, 1966.
[4] B. Kok, *in* "Plant Biochemistry" (J. Bonner and J. Varner, eds.), p. 903. Academic Press, New York, 1966.

Assay Methods

The detection of P700 has generally involved the measurement of absorption changes at 700 nm (or 430 nm) induced by light or chemical oxidation–reduction. Alternate methods, such as the measurement of a specific electron spin resonance (ESR) signal (see below), NADP or methylviologen reduction rates, are not discussed. Although these techniques avoid the inherent difficulties in optically detecting P700, they are less direct and in the case of rate measurements provide only a relative index of P700.

Light-Induced P700 Measurements

Principle. To measure P700 by means of light–dark spectroscopy, a monochromatic (700 or 430 nm) "detecting light" (I_{det}) is passed through a sample and monitored by a photocell. The change of transmission of I_{det} is observed upon illumination of the sample by a second "actinic light" (I_{act}) which is blocked from the photocell. Since this change is small ($\Delta I/I_0 < 1\%$), the detecting light must be stable and the noise in the photocurrent minimal. The signal-to-noise ratio is proportional to the square root of the number of quanta ($I_{det} \times t$) which are "seen" by the photocathode and used for the determination. It is desirable to use a strong detecting beam and to collect efficiently the light transmitted and scattered by the sample on a sensitive photocathode (S20 response for 700 nm). However, a strong 700-nm light perturbs the system since it sensitizes the photooxidation of P700 [Eq. (1)]. If I_{det} provokes a rate of oxidation $k_{L(det)}$, and the dark reduction rate is k_D, the system will attain a steady state in which $[P_{700}^+]/[P_{700}^-] = k_{L(det)}/k_D$. To avoid photooxidation by the detecting beam we thus desire $k_{L(det)} \ll k_D$.

The rate of P700 reduction (k_D) can be increased by exogenous electron donors like ascorbate plus catalytic amounts of phenazine methosulfate (PMS) or 2,6-dichlorophenolindophenol (DPIP) to override the effects of the detecting beam. On the other hand, a high value of k_D (such as in the presence of high concentrations of reduced DPIP or PMS) requires a strong I_{act} to convert P700 into the oxidized form, $[P_{700}^+]/[P_{700}^-] = k_{L(act+det)}/k_D$. Although one should be aware of this aspect, it poses no serious problem since it is not difficult to select $I_{act} \gg I_{det}$ and a suitable k_D value.

Procedure. In complete photosynthetic systems the oxidized state of P700 can be induced rapidly by strong, rate-saturating light or by weak light which excites predominantly photosystem I ($\lambda > 700$ nm). The rate of P700 reduction (k_D) generally is rapid in whole cells or fresh chloroplasts due to the simultaneous excitation of photosystem II by I_{act} and/or cyclic back flow of electrons from $[X^-]$. The latter effect can be minimized in chloroplasts by the addition of a low potential, autoxidizable electron acceptor such as methyl viologen (10–100 μM) which removes electrons from $[X^-]$. The uncertain rate k_D due to partial photosystem II excitation, however, may present a problem and, depending upon conditions such as instrumental response time, choice of wavelength, or intensity of I_{act}, a complete photooxidation of P700 may not be observed. In addition, the P700 measurements at 430 nm are complicated by other light-induced absorption changes.

The observation of P700 can be greatly facilitated after photosystem II is inactivated or uncoupled from photosystem I and its action replaced by that of an exogenous electron donor. The rate k_D can now be controlled by varying the concentration of donor (ascorbate) and catalyst (DPIP or PMS).[5] The proper donor concentrations will depend upon the type and response time of the instrument used.

For example, chloroplast or cell preparations may be assayed in a medium containing 1–2 mM ascorbate and ~10 μM PMS provided instruments with response times ≤ 1 msec (single-beam instruments) and saturating flashes or very strong I_{act} are used. Using weaker actinic light or in measurements where I_{act} is separated from I_{det} (phosphoroscope arrangement), the reaction mixture should contain ascorbate alone or ascorbate plus lesser amount of PMS (0.1–1 μM) or DPIP (1 μM).

The P700 changes are relatively stable and survive a variety of treatments which effectively remove photosystem II activity from photosystem I, and we briefly enumerate the procedures which have been used successfully. In all cases, the chloroplast or cell preparations may be assayed in a medium containing a suitable buffer such as Tris·HCl or phosphate (~50 mM) pH 7–8, methyl viologen (10–100 μM), and an appropriate electron donor as discussed above.

(a) Addition of 1–10 μM DCMU to isolated chloroplasts or whole cells.[5]

(b) Aging of chloroplasts suspended in neutral buffer for about 1 week at 0°–4°.[6,7]

(c) Suspension of chloroplasts in buffer containing digitonin for 15–30 minutes. Digitonin/chlorophyll $\simeq$ 12–80/1 (w/w).[6,8]

(d) Sonication or other disruption of chloroplasts or whole cells. Time and method required for breakage varies with the plant material

[5] B. Rumberg, P. Schmidt-Mende, J. Weikard, and H. T. Witt, *in* "Photosynthetic Mechanisms of Green Plants," p. 18. Nat. Acad. Sci.—Nat. Res. Counc., Publ. No. 1145, Washington, D.C., 1963.

[6] B. Rumberg and H. T. Witt, *Z. Naturforsch. B* **19,** 693 (1964).

[7] B. Ke, *Biochim. Biophys. Acta* **88,** 289 (1964).

[8] B. Ke, *Biochim. Biophys. Acta* **88,** 297 (1964).

(5–15 minutes). Fragments sedimenting between 10,000 and 144,000 g are collected and suspended in the reaction mixture.[9]

(e) Extraction with petroleum ether.[6] Lyophilized chloroplasts ($\approx$100 mg) are extracted with cold (4°) petroleum ether ($\approx$100 ml) in a tube containing a fine fritted-glass filter. Excess petroleum ether is evaporated under a stream of N_2.[10]

(f) Measurement at low temperature, e.g., in liquid nitrogen. Under these conditions $k_D = 0$ and only "single shot" photooxidation experiments are possible. The preparation is cooled with P700 in the reduced state.[6]

The various treatments may not exclude cytochrome interference around 430 nm dependent upon the plant material or treatment used, and observation at 700 nm is generally preferred.

Chemically Induced P700 Measurements

Chemically induced P700 spectra are most conveniently measured in preparations enriched in P700 but can be obtained using sonicated chloroplast or cell preparations (see above). We normally measure oxidized minus reduced spectra with a specially constructed split-beam spectrophotometer.[11] With sonicated preparations of blue-green algae (*Anabaena*), in which the ratio of P700/chlorophyll is higher than in green cells, a Cary Model 14 recording spectrophotometer has been used.[12]

Particles are suspended in Tris·HCl or phosphate buffer (~50 m*M*) at pH 7–8. Equal samples containing about 25 μg of chlorophyll/ml (giving an absorbance of about 2 at 680 nm) are placed in identical cuvettes and used to record a baseline. One of the two samples is then treated with ferricyanide (~1 m*M*), the other with ascorbate (~2 m*M*) and allowed to equilibrate prior to recording a difference spectrum.

Enrichment of P700

A 2 to 7-fold purification of P700 (e.g., a larger absorption change per unit chlorophyll) can be obtained either by extraction of chloroplasts with acetone or by isolation of a small-particle fraction prepared by detergent treatment of chloroplasts or broken algae.

Acetone Extraction.[11] Fresh chloroplasts are suspended in phosphate buffer at pH 7–8 (final concentration ~1 mg of chlorophyll/ml). Cold

[9] B. Ke, S. Katoh, and A. San Pietro, *Biochim. Biophys. Acta* **131,** 538 (1967).

[10] N. I. Bishop, *Proc. Nat. Acad. Sci. U.S.* **44,** 501 (1958).

[11] B. Kok, *Biochim. Biophys. Acta* **48,** 527 (1961).

[12] T. Ogawa, L. P. Vernon, and H. H. Mollenhauer, *Biochim. Biophys. Acta* **172,** 216 (1969).

acetone (—10°) is added during constant stirring to give a final concentration of 70–72% (v/v). The light-green precipitate is immediately collected by centrifugation, washed, and suspended in phosphate buffer. Although rather variable, in good preparations up to 85% of the chlorophyll can be removed without a loss of P700 signal, thus yielding about a 6-fold enrichment with respect to Chl_{total}. To measure P700, either by light or chemically induced absorption changes, a homogeneous suspension is required. Accordingly, the preparation is first sonicated and then treated with detergent [1% (v/v) Triton X-100]. Exogenous reductant (ascorbate–PMS) is added prior to measuring the light induced absorption changes.

Particle Preparation following Detergent Treatment. A 2- to 4-fold purification of P700 can be obtained by isolation of photosystem I chloroplast particles following treatment with digitonin[13] or Triton X-100.[14,15] Particles about 4-fold enriched in P700 have also been isolated from blue-green algae after carotenoid extraction with hexane and ethanol, treatment with Triton X-100, and sucrose density gradient centrifugation.[16]

To prepare digitonin-treated particles, chloroplasts are first diluted in cold (4°) 50 m*M* phosphate or Tris·HCl buffer (pH 7.4) containing 0.4 *M* sucrose to a final concentration of 1 mg of chlorophyll/ml. Digitonin (5 mg/ml) is added and the preparation kept for 30 minutes at 4° previous to centrifugation. The supernatant solution obtained after successive centrifugations at 10,000 *g* for 30 minutes and 50,000 *g* for 30 minutes is centrifuged at 144,000 *g* for 1 hour. The resulting pellet is suspended in buffer and used to measure P700. Light-induced measurements again require the addition of a suitable reductant.

Instrumentation

Several types of sensitive difference spectrophotometers have been especially built to measure light-induced absorption changes in photosynthetic tissues. The design principles of the various types of apparatus with the advantages and limitations of each have previously been discussed.[17,18] In addition to these instruments, commercially available dual-wavelength spectrophotometers have been used to measure P700.[16]

[13] J. M. Anderson, D. C. Fork, and J. Amesz, *Biochem. Biophys. Res. Commun.* **23,** 874 (1966).
[14] L. P. Vernon, E. R. Shaw, and B. Ke, *J. Biol. Chem.* **241,** 4101 (1966).
[15] L. P. Vernon, B. Ke, and E. R. Shaw, *Biochemistry* **6,** 2210 (1967).
[16] T. Ogawa and L. P. Vernon, *Biochim. Biophys. Acta* **180,** 334 (1969).
[17] L. N. M. Duysens, "Progress in Biophysics," Vol. 14. Pergamon, New York, 1964.
[18] R. K. Clayton, "Molecular Physics in Photosynthesis." Blaisdell, New York, 1965.

The type of instrument used is generally dictated by the type of observation one wishes to make.

Under conditions where the P700 changes occur rapidy (<1 sec), one can use a single-beam instrument, the absorption changes induced by a brief saturating flash being amplified and recorded directly. The (photo)oxidized state can be induced rapidly by a light flash of sufficient quantum content. Such can be obtained using a laser (~ nsec), a xenon discharge lamp (~ μsec) or a rotating disk cutting through the image of bright continuous sources such as xenon arcs or incandescent lamps (~ msec). In reaction systems where the rate of P700 reduction (k_D) is rapid, the redox state of P700 can be alternated repetitively by giving a sequence of (N) properly spaced flashes. As long as the reaction system is stable so that the recurrent event remains constant, one can average the observations and so greatly improve the signal to noise ratio ($S/N \simeq \sqrt{N}$).[2,19,20]

In the measurement of slower events (>1 sec), interferences such as settling of the particles or light-induced swelling or shrinking, generally prohibit single-beam measurements. Such interfering variations of I_{det} can be balanced out by using a time-sharing, second light beam (I_{ref}) having either: (1) a different wavelength (e.g., 720 nm) which passes through the same sample and varies like I_{det} with the interfering effects, but is not affected by the actinic light (dual-wavelength apparatus),[21] or (2) the same wavelength which passes through a parallel sample which is not illuminated with the actinic light (split-beam apparatus).[22]

Preferably, one observes the absorption changes at about 700 nm. In most cases the 430-nm band is less suitable since the strong absorption by bulk chlorophyll and carotenoids and absorption changes of other pigments such as cytochromes might interfere. In measuring P700 at 700 nm several problems may be encountered:

(1) The actinic beam and the chlorophyll fluorescence which it excites (some 5% of I_{act}, at wavelengths between 680 and 750 nm) should not be seen by the photocathode. This problem can be solved with either time separation or color separation. In time separation, a phosphoroscope arrangement is used. The photocell is darkened during the time the actinic beam hits the sample. Single-beam, split-beam, or dual-wavelength spectrophotometers have been built along this principle.[2,23] Time separa-

[19] B. Ke, R. W. Treharne, and C. McKibben, *Rev. Sci. Instr.* **35**, 296 (1964).

[20] H. T. Witt, *in* "Nobel Symposium 5" (S. Claesson, ed.), p. 81. Wiley (Interscience), New York, 1967.

[21] See Chance, Vol. XXIV.

[22] L. N. M. Duysens, *in* "Research in Photosynthesis" (H. Gaffron, ed.), p. 59. Wiley (Interscience), New York, 1957.

[23] B. Kok, *Plant Physiol.* **34**, 184 (1959).

tion avoids interference of I_{act} and its fluorescence and allows actinic illumination with any wavelength. There is, however, an inherent loss of time during the switching from I_{act} to I_{det}. Delayed light emission induced by I_{act} ($\sim$0.1% of it) can interfere with the subsequent sampling of I_{det} This interference may be checked for by observing the Δ signal in the absence of I_{det}.

The alternate solution is color separation: The actinic light contains only wavelengths longer ($> \cong 720$ nm) or shorter ($< \cong 690$ nm) than the detection wavelength (700 nm), and an appropriate color filter transmits only the latter to the photocathode. Fluorescence and delayed light emission excited by I_{act} still interfere and pose major problems in single-beam instruments. Again these artifacts can be checked by decreasing or removing I_{det} and recording the apparent transmission changes using the normal settings of I_{act} and instrument sensitivity. Since the intensity of fluorescence decreases with the square of the distance, its interference can be selectively minimized by placing the photomultiplier some distance back from the sample.[13] Fluorescence and luminescence occur in a wide spectral range and their interference can be diminished further with a narrow band 700-nm filter in front of the photocell. In a split-beam or dual-wavelength apparatus, where the measuring and reference beams are alternated at a certain frequency and their signals subtracted, the residual transmission and the fluorescence from a continuous exciting beam should not, in principle, affect the measurements. In practice one cannot take this for granted and checks as mentioned above must be made.

(2) A potential fluorescence problem inherent in measurements at 700 nm with photosystem II operative is a change of the yield of the fluorescence excited by I_{det} caused by I_{act}. This effect is generally assumed to be small and neglected. Again, it can be checked for and minimized by moving the photomultiplier back from the sample and/or inserting a 700-nm interference filter between sample and photocell.

Properties

Absorption and Band Shape. The absorption minima in the oxidized minus reduced spectrum of P700 occur between 698 and 709 nm in the red and between 430 and 435 nm in the blue. The red absorption band is often asymmetric, being skewed toward the far-red. Possibly the net change is due to both a decrease of absorption and a band shift toward shorter wavelengths.[1]

Extinction Coefficient and Concentration of P700. It has been assumed that P700 at $\sim$700 nm, has a molar extinction similar to that of chlorophyll *a* ($\simeq$80/cm/mM) and that upon oxidation the extinction becomes zero. The blue difference band of P700 is about one-half the magnitude of red band and a molar extinction of $\sim$40/cm/mM has been ascribed to the blue difference band. Based on these assumptions the amount of P700 relative to total chlorophyll ranges between about 1/200 to 1/1000 in algae or chloroplasts with the higher ratios occurring in blue-green algae. Recent measurements[24,25] in which the amount of ferricyanide or DPIP reduced in a short flash was correlated with the simultaneous absorption change at 705 nm, suggested a difference molar extinction for P700 of 36–42/cm/m*M*, a number which correlates better with certain ESR data (see below).

Redox Potential. The oxidation–reduction potential (E_0') of P700 has been measured in acetone-extracted[11] or aged spinach chloroplasts.[26] Titration with ferro/ferricyanide mixtures showed a midpoint potential of +430 to +460 mV. The oxidation of P700 appears to involve a single electron transfer which is pH independent.

Stability. The light-induced oxidation of P700 is unaffected by pH between 5 and 11 and remains observable in chloroplasts heated at 55° for 5 minutes.[6] At 0°–5° it is stable for over a week, and at lower temperatures it can be stored indefinitely.

ESR Signal. One of the light-induced ESR signals observed in algae and chloroplasts (the R signal) peaks at $g = 2.0025 \pm 0.005$, is 7–9 gauss wide, and displays no hyperfine structure.[27] The occurrence and kinetics of this signal correlate with those of the oxidized form of P700 suggesting that $P700^+$ is responsible for the R signal.[14,28,29] Quantitative comparison of the number of unpaired spins ($P700^+$ molecules) produced by light and chemical oxidation varied for different preparations and conditions, but, on the average, a ratio of about 1 would correspond to a molar Δ extinction of ~40/m*M*/cm for P700 at 700 nm.

Acknowledgment

Preparation of this chapter was supported in part by the Atomic Energy Commission, Contract No. AT(30-1)-3706, the National Aeronautics and Space Administration Contract ASW-1592, and the National Institutes of Health, 5 FO2 GM 24,449-02.

[24] P. Schmidt-Mende and B. Rumberg, *Z. Naturforsch. B* **23**, 225 (1968).
[25] W. Schliephake, W. Junge, and H. T. Witt, *Z. Naturforsch. B* **23**, 1571 (1968).
[26] B. Rumberg, *Z. Naturforsch. B.* **19**, 707 (1964).
[27] B. Commoner, *in* "Light and Life" (W. D. McElroy and B. Glass, eds.), p. 356. Johns Hopkins Press, Baltimore, Maryland, 1961.
[28] H. Beinert and B. Kok, *Biochim. Biophys. Acta* **88**, 278 (1964).
[29] See this series, Vol. XXIV.

[48] Acyl Lipids in Photosynthetic Systems

By C. Freeman Allen and Pearl Good[1]

Lipid Composition

The lamellar membranes of chloroplasts and blue-green algae contain only four lipids in appreciable amounts: the two neutral glycolipids, monogalactosyl diglyceride (I) and digalactosyl diglyceride (II), and the two anionic lipids, phosphatidylglycerol (III) and plant sulfolipid (IV).

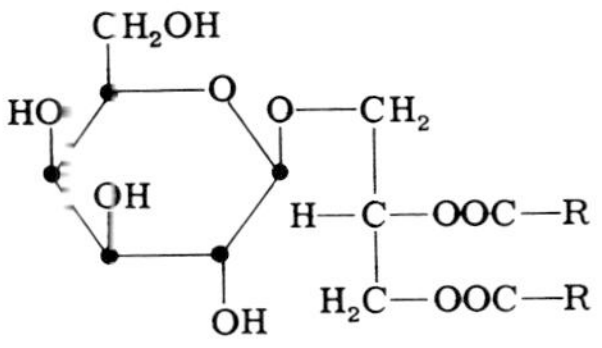

Monogalactosyl diglyceride
(I)

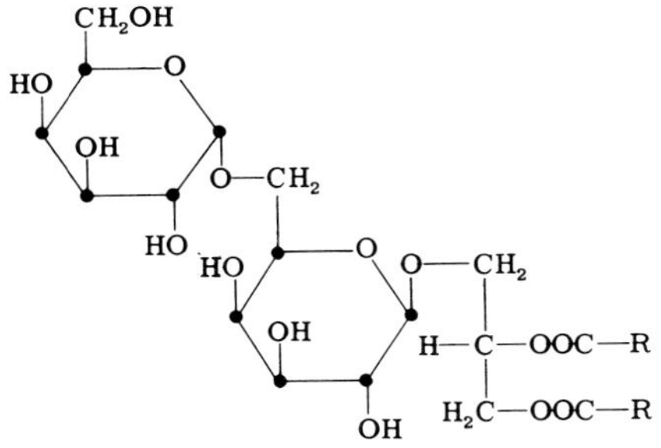

Digalactosyl diglyceride
(II)

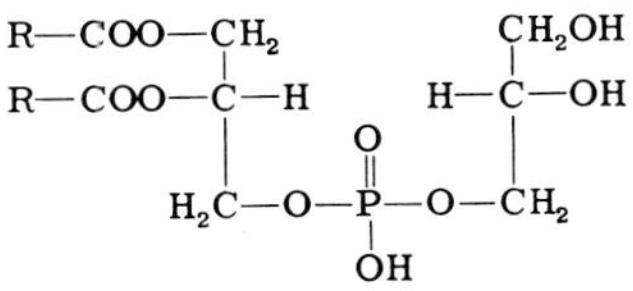

Phosphatidylglycerol
(III)

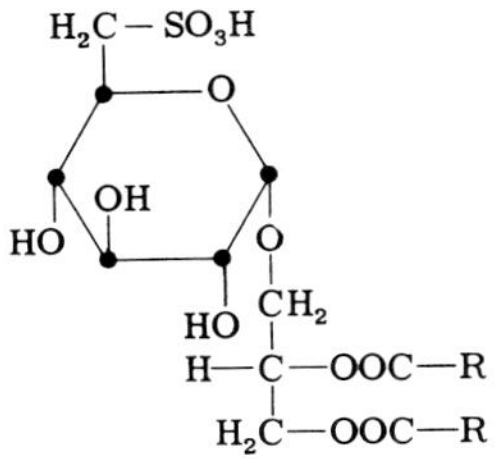

Plant sulfolipid
(IV)

Whole chloroplasts may contain lecithin in addition, although it is found in decreasing amounts as the purity of the chloroplast preparation increases. As usually isolated by differential centrifugation of aqueous plant homogenates, chloroplasts are contaminated with other organelles

[1] This work was supported by Grant AI-04788 from the National Institutes of Health, U.S. Public Health Service.

and tissue fragments which may introduce lipids such as phosphatidylethanolamine, phosphatidylinositol, cardiolipin, phosphatidylserine, cerebroside, sterol esters, and sterol glycoside. Fatty acid is also frequently found in subchloroplast fragments, but usually it results from enzymatic degradation, particularly of the galactolipids and plant sulfolipid.[2-4]

Phosphatidylglycerol is also rapidly degraded by some plant extracts. A good discussion of the acyl lipids and fatty acids of photosynthetic plant tissue has appeared in recent reviews.[2]

Photosynthetic bacteria have a less consistent gross lipid composition.[5,6] Galactosyl glycerides and sulfolipid are present in some species but absent in others. *O*-Ornithylphosphatidylglycerol (V) is frequently a major component of such bacteria and is found in the chromatophores.[7,8] Phosphatidylglycerol seems to be the one lipid common to photosynthetic plants, algae, and bacteria.[9]

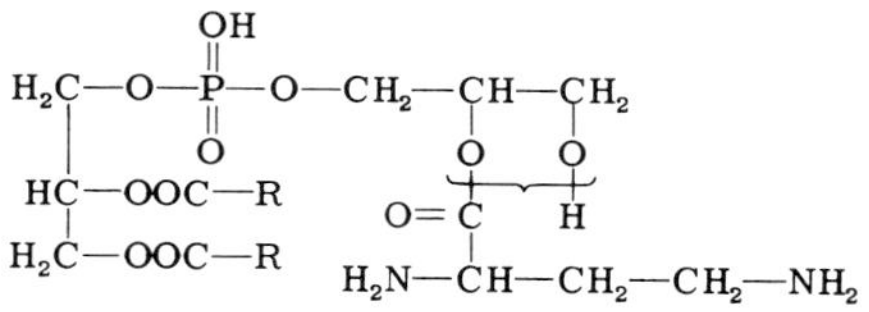

O-Ornithylphosphatidylglycerol[9]

(V)

The fatty acids derived from plant lipids vary widely with the species and tissue.[2] The major components of higher plant chloroplasts are palmitic, linoleic, and linolenic acids. Relatively little is known of the fatty acid composition of algal chloroplast lipids although the gross composition of a number of species has been reported.[2] All contain

[2] B. W. Nichols and A. T. James, *Progr. Phytochem.* **1,** 1 (1968); also a shorter version in "Plant Cell Organelles" (J. B. Dridham, ed.), p. 163. Academic Press, New York, 1968.

[3] P. S. Sastry and M. Kates, *Biochemistry* **3,** 1280 (1964).

[4] T. Yagi and A. A. Benson, *Biochim. Biophys. Acta* **57,** 601 (1962).

[5] C.-E. Park and L. R. Berger, *J. Bacteriol.* **93,** 221 (1967). This paper briefly reviews literature on lipids of photosynthetic bacteria. C. Constantopoules and K. Bloch, *J. Bacteriol.* **93,** 1788 (1967), discuss glycolipids.

[6] M. Kates, *Advan. Lipid Res.* **2,** 17 (1964). A review on biosynthesis of bacterial lipids.

[7] A. Gorchein, *Biochim. Biophys. Acta* **84,** 356 (1964).

[8] B. J. B. Wood, B. W. Nichols, and A. T. James, *Biochim. Biophys. Acta* **106,** 261 (1965).

[9] U. M. T. Houtsmuller and L. L. M. Van Deenen, *Biochim. Biophys. Acta* **70,** 211 (1963).

TABLE I
MAJOR FATTY ACIDS OF LAMELLAR FRAGMENT LIPIDS ISOLATED FROM SPINACH CHLOROPLASTS (PERCENT COMPOSITION)

Fatty acid	14:0	16:0	Δ^3-*trans* 16:1	16:2	16:3	18:0	18:1	18:2	18:3
Monogalactosyl diglyceride	—	—	—	—	25	—	—	2	72
Digalactosyl diglyceride	—	3	—	—	5	—	2	2	87
Phosphatidylglycerol	1	11	32	—	2	—	2	4	47
Sulfolipid	—	39	—	—	—	—	—	6	52
Lecithin	—	12	—	—	4	—	9	16	58
Phosphatidylinositol	4	34	6	—	3	2	7	15	27

polyunsaturated acids, with the exception of a few species of blue-green algae, such as *Anacystis nidulans*. Lipids of photosynthetic bacteria contain only saturated and monounsaturated acids.[10]

In plants, the galactosyl lipids are rich in polyunsaturated acids, the sulfolipid and phosphatidylglycerol contain a relatively high proportion of palmitic acid, and *trans*-3-hexadecenoic acid is found almost exclusively in phosphatidylglycerol. This is true in the whole tissue[2] as well as in spinach lamellar fragments.[11] In *Anacystis nidulans* and the photosynthetic bacteria, which lack polyunsaturated acids, the acyl group composition also varies among the lipid classes (see Tables I–III).

TABLE II
MAJOR FATTY ACIDS OF *Anacystis nidulans* LIPIDS[a] (PERCENT COMPOSITION)

Fatty acid	14:1	15:1	16:0	16:1	17:1	18:0	18:1
Unsaturation at	Δ^9	Δ^9	—	Δ^9	Δ^9	—	Δ^9(40%) Δ^{11}(60%)
Monogalactosyl diglyceride	2	2	36	28	2	—	13
Digalactosyl diglyceride	2	—	43	58	1	1	10
Phosphatidylglycerol	1	—	42	32	2	1	22
Sulfolipid	2	—	49	40	1	1	6
Total lipid extract	3	1	40	45	1	—	9

[a] C. F. Allen, O. Hirayama, and P. Good, *Biochem. Chloroplasts* **1**, 195 (1966).

[10] A. T. James and B. W. Nichols, *Nature (London)* **210**, 372 (1966).
[11] C. F. Allen, P. Good, H. Davis, P. Chisum, and S. Fowler, *J. Amer. Oil Chem. Soc.* **43**, 223 (1966).

TABLE III

LIPIDS AND FATTY ACIDS (%) OF SOME PHOTOSYNTHETIC BACTERIA (ANAEROBIC GROWTH)[a]

Fatty acid	*Rhodopseudomonas gelatinosa*[b]	*Rhodopseudomonas spheroides* b	*Rhodopseudomonas spheroides* c	*Rhodopseudomonas capsulata*[b]	*Rhodopseudomonas palustris*[b]	*Chromatium*[c]	*Rhodospirillum rubrum* b	*Rhodospirillum rubrum* c
Sulfolipid		+	+					
Phosphatidylglycerol	+	+	+	+	+	+	+	+
Lecithin		+	+	+	+			+
Phosphatidylethanolamine	+	+	+	+	+	+	+	+
Cardiolipin		+	+		+	tr	+	+
O-Ornithylphosphatidylglycerol (probable identity)	+	+	(−?)	+	tr		+	
14:0	3		1			tr	2	3
14:1			1					
16:0	33	2	9	2	13	29	16	19
16:1	51	1	7	2	3	33	38	26
18:0	tr	5	11	4	6	+	1	tr
18:1	6	91	70	90	72	39	37	52

[a] Glycosyl glycerides are present in appreciable amounts in *Chloropseudomonas ethylicum*,[d] but are absent in *Rhodopseudomonas palustris*,[d] *Rhodospirillum rubrum*,[b] *Rhodopseudomonas spheroides*.[b] Aerobic grown organisms are similar.

[b] B. J. B. Wood, B. W. Nichols, and A. T. James, *Biochim. Biophys. Acta* **106,** 261 (1965).

[c] F. Haverkate, F. A. G. Teuling, and L. L. M. Van Deenen, *Proc. Kon. Ned. Akad. Wetenschap. Ser. B*, **68,** 154 (1965).

[d] C. Constantopoulos and K. Bloch, *J. Bacteriol.* **93,** 1788 (1967).

Lipid Methodology

The techniques of lipid chemistry described below have been used successfully in the authors' laboratory with lipids of higher plant tissue, green algae, blue-green algae, and photosynthetic bacteria. As such, they reflect current personal preferences and experience. An excellent book on lipid techniques with contributions from several laboratories is available[12] which contains a wealth of useful information of a more general nature.

Lipid Stability and Other Problems

Enzyme-catalyzed degradation and oxygen-initiated oxidation are problems which must not be neglected in working with lipids.

Plant tissue is often a rich source of enzymes[2] which alter lipids during storage of the tissue, isolation of subcellular fractions, or extraction of the lipids. In prolonged preparative procedures there is probably little that can be done other than to work rapidly in the cold. Some lipolytic enzymes remain active or become activated during extraction of lipids with organic solvents. We have not commonly found this to be a problem if operations are carried through quickly in the cold. At the extraction stage hot isopropanol has been found quite effective in minimizing the problem.[13] Phosphatidylglycerol, sulfolipid, and perhaps most frequently, galactolipids are lost with formation of fatty acids, phosphatidic acid, or lysosulfolipid. Lysogalactolipids, if formed at all, are apparently deacylated very rapidly, as they are not found even in instances where there has been a heavy loss of monogalactosyl diglyceride. If an alcohol has been used in the lipid extraction, its phosphatidyl ester may be found in the lipid extracts as a consequence of enzymatic transesterification of phospholipid.

The quantity of free fatty acid found in lipid extracts of plant tissue is a good measure of the extent of enzymatic deacylation. Sometimes the composition of the derived fatty acids will indicate the identity of the lipid which has been lost.

Once extracted, lipids can be destroyed by oxidation or polymerization, especially if the acyl residues are highly unsaturated. These processes are autocatalytic; once peroxide content reaches a threshold level, decomposition becomes rapid and difficult to control. Wherever possible, operations should be carried out in an atmosphere of inert gas, such as nitrogen or carbon dioxide. This is of greatest importance whenever

[12] "Lipid Chromatographic Analysis" (G. V. Marinetti, ed.), Vol. I. Dekker, New York, 1967.

[13] M. Kates, *Can. J. Bot.* **35,** 895 (1957).

highly unsaturated lipids are highly concentrated, have a large surface area, and are freed of natural antioxidants. Thus thin films of purified lipids, as might be formed by evaporation of isolated fractions, are most susceptible and must be treated with great care or avoided.

Lipids in solution are relatively stable, especially if the solvent has been deaerated with bubbling nitrogen for a few minutes and the solution is kept under a nitrogen atmosphere in the cold. Such solutions may survive for months or years without decomposition. In sealed glass containers under vacuum even sensitive lipids can be stored for years with little or no loss. Vials with Teflon liners in tight sealing caps are much more convenient for shorter periods of storage.

In general the more unsaturated a lipid, and the more it is to be handled, the more care must be taken to minimize contact with oxygen. Lipids lacking polyunsaturated acids, such as those from bacteria or *Anacystis nidulans* are quite stable to oxidation, and deoxygenation is rarely necessary. In the procedures to follow, common sense tempered by experience should be a guide as to the caution required.

Oxidation becomes most readily evident in the fatty acid composition of lipids. The percentage of polyunsaturated species falls off, and can be followed by gas chromatography. A common indication is streaking of lipid spots on thin-layer chromatograms when oxidation is in advanced stages. Contamination of lipid preparations with stopcock grease, soluble plastics, plasticizers, extracts of rubber, and skin lipids can be a nuisance. These materials should not be allowed to come into contact with lipid extracts, especially where quantitative work is in progress.

Extraction of Lipids[14]

Extraction is initiated with a methanol-rich chloroform–methanol mixture (1:2 by volume) in sufficient amount to completely dissolve tissue water and thus to form one liquid phase. This greatly facilitates extraction and filtration while removing most of the lipid. After an additional wash with this solvent the residue is further extracted with chloroform, to complete extraction. Water added to the combined extracts causes the separation of a second liquid phase. Only the lower phase, which is mostly chloroform, contains appreciable amounts of lipid.

Reagents. All solvents should be redistilled to remove nonvolatile impurities, and deaerated for 30 minutes with inert gas from a gas dispersion tube if sensitive lipids are to be handled.

[14] E. G. Bligh and W. J. Dyer, *Can. J. Biochem. Physiol.* **37,** 911 (1959), provide the basis for this procedure.

Chloroform, ACS reagent, stabilized with alcohol
Methanol, ACS reagent
Nitrogen, not oil pumped (commercial nitrogen distilled from liquid nitrogen is supplied with less than two parts per million of oxygen)

Extractions of Larger Samples of Tissue. Five hundred grams of tissue (e.g., spinach leaf) is blended with 3 liters of ice cold chloroform–methanol (1:2 by volume) for about 1 minute or until finely divided. A gallon-size blender is most useful, but a quart size used on smaller batches also works well. If lipids are highly unsaturated it is best to replace air above the mixture with inert gas (nitrogen or carbon dioxide). The blender can be tightly covered with oil-free aluminium foil and should not be more than two-thirds full to avoid splashing. There will be some volatilization of solvent during the first few seconds, and blending should therefore be commenced cautiously.

The homogenate is filtered under vacuum through a medium- or coarse-porosity sintered-glass or Büchner funnel (13.5 cm is a useful diameter for this quantity of extract). An inert atmosphere can be maintained over the filter with an inverted conical funnel through which nitrogen is passed. An additional 300 ml of chloroform–methanol (1:2) is allowed to pass through the filter cake and is followed by 1100 ml of chloroform in several portions.

The residue will normally be essentially free of extractable lipids at this point. Cold water (1500 ml) mixed with the filtrate causes two liquid phases to form. The lower (chloroform) layer is drawn off from a separatory funnel with a grease-free stopcock (Teflon plug). The aqueous methanol (upper) phase should be washed in two or three portions with a total of about 500 ml of chloroform which is combined with the main chloroform extract. In some instances it will be most convenient to use centrifugation to separate the phases quickly.

Residues of lipids in the aqueous phase and in the extracted residue will usually be negligible at this point.

In some tissues complete removal of carotenoids requires further extraction with hexane or other similar solvent.

Certain highly polar lipids such as phytoglycolipid or highly phosphorylated lipids require other techniques for their quantitative isolation.[12] However, this is not likely to be a problem with photosynthetic tissue.

Most of the soluble nonlipid material will remain in the methanol–water phase, which is discarded.

The chloroform-rich phase, which contains a few percent methanol and only a trace of water is evaporated under vacuum in a rotary evaporator. The receiver should be cooled with ice or dry ice, and the vacuum high enough to permit distillation of solvent at close to room temperature. An inert atmosphere can be maintained by breaking the vacuum with nitrogen.

The lipid residue is dissolved in a small volume of chloroform or chloroform–methanol (9:1). If the solution is to be used for spotting a thin-layer plate, the methanol containing solvent is preferred.

A residue of insoluble nonlipid will almost certainly remain and can be further extracted with chloroform-methanol mixtures such as 2:1 or 2:1 with a trace of water. Lipid content of such extracts is easily checked by thin-layer chromatography on a portion of the sample. In our experience such extracts have rarely contained more than a trace of lipid. Concentration of lipid in the bulk solution is best determined by evaporation of a small aliquot in a weighed vial with the aid of a stream of inert gas. An electrically heated metal block (Temp-Blok Module Heater, Lab-Line Instruments, Inc., Melrose Park, Illinois) at 40° is particularly convenient for this operation. The vial is cooled and final traces of solvent are removed in a vacuum desiccator before weighing. This procedure avoids difficulties inherent in thorough drying of a large waxy mass of lipid and the danger of further decomposition. In a typical preparation, 500 g of spinach leaf yields about 4.4 g of lipid of which two-thirds is glycerolipids.

Extraction of Smaller Batches. The same solvents used for larger samples can be scaled down easily. Most cells and subcellular fractions do not require blending but can be extracted directly. Extraction of some cells with impervious cell walls is facilitated by prior freezing. Leaf or other tissue can be frozen with liquid nitrogen in a porcelain mortar and ground with the extraction solvent.

In certain unusual cases (e.g., *Cyanidium caldarium*) lipids are not extracted by chloroform–methanol, but a mixture of equal volumes of glacial acetic acid, methanol, and chloroform added to the tissue in that order is effective. Extraction can then be completed with the usual chloroform–methanol mixtures. Resistant cells can also be broken by homogenization with glass beads or alumina before extraction of lipids with chloroform–methanol. Risk of enzymatic lipid degradation is increased with this procedure.

For extraction and washing any convenient working volume of chloroform–methanol (1:2) can be used, but 10 ml is a reasonable volume for a gram or less of tissue. This is followed by a wash with chloroform equal to one-third the volume of the first extract. Enough water is added to the combined extracts to give a final proportion of chloroform–

methanol–water of 1:1:0.9, which will result in formation of the two liquid phases which are separated. The upper phase is washed with two or three small portions of chloroform.

The combined chloroform rich extracts can be conveniently evaporated under a stream of nitrogen in a vial warmed slightly above room temperature. If an accurate weight of lipid extract is desired, the vial is brought to constant weight in a vacuum desiccator, the lipids taken up in chloroform, and the dried vial plus insoluble residue is reweighed. For most purposes, a solution of 10 mg of lipid per milliliter is convenient.

Analysis of Lipids

The procedure described here has been developed in the author's laboratory for analysis of plant lipids, for which it has been used repeatedly since 1965. The general topic of quantitative analysis of lipids by combined TLC and GPC has been reviewed.[15]

Qualitative analysis of lipid composition is currently most commonly carried out by thin-layer chromatography (TLC) on silicic acid. With the techniques described here a single spot containing 2–3 μg of acyl lipid can be reliably quantitated, and smaller amounts can be detected. For quantitation, lipid acyl groups are converted to methyl esters: The silica gel containing the lipid in question is removed from the plate and treated with warm methanolic sulfuric acid to effect transesterification. Heptadecanoic acid (or another acid not present in the lipids) is added to serve as a quantitative internal standard and is simultaneously esterified, the combined methyl ester mixture is extracted with hexane and quantitatively analyzed by gas chromatography (GPC). With a nonpolar column (SE-30) esters of the same carbon number are eluted together and give a relatively simple chromatogram; or a polar column (Reoplex 400) can be used to determine relative amounts of each of the individual fatty acids. The accuracy of quantitation of a 1-μg lipid spot is approximately 10% of the value determined.

With 10 μg or more, accuracy of 5% can be expected.

Thin-Layer Chromatography and Transesterification

Materials. Silicic acid thin-layer plates 20 × 20 cm, 0.25 mm layer, without fluorescent indicator, are used. These plates may be purchased ready-made (Merck 5763 distributed by Brinkman Instruments, Inc., Cantiague Road, Westbury, New York 11590) or spread from silicic acid with calcium sulfate binder (Silica Gel G, Merck). The readymade plates spread on glass backing have been of uniform and satisfactory quality.

[15] A. Kuksis, *Chromatogr. Rev.* **8**, 172 (1966).

Aluminum-backed plates of the same manufacturer have been badly contaminated with oil or other lipophilic impurities. For quantitative analysis of lipids even the glass backed plates should be washed by allowing the developing solvent (chloroform–methanol–water) to run to the top followed by drying half an hour in the open air, or for a few minutes in a 100° oven. In either case the plate should then be stored overnight in a desiccated cabinet over indicating silica gel. The plate should be handled carefully after washing to avoid fingerprints of skin lipids.

Chloroform–methanol–water (65:25:4 by volume)
Chloroform–methanol–saturated ammonium hydroxide (65:35:5 by volume)
Chloroform–methanol–isopropylamine–saturated ammonium hydroxide (65:35:0.5:5 by volume)
Sulfuric acid, 5% (by volume) in redistilled absolute methanol. (We routinely prepare this fresh for each batch of analyses.)
Saturated aqueous sodium bicarbonate extracted three times with hexane
Rhodamine 6G, 0.006%, in water prepared from a 1% stock solution (Rhodamine 6G, Matheson Coleman and Bell, RX90)
Aqueous sodium hydroxide, 2 *N*
Heptadecanoic acid, 99% + purity (Applied Science Laboratories, State College, Pennsylvania), 0.1000 mg/ml in absolute methyl alcohol
Hexane, commercial grade

All organic solvents are redistilled.

Procedure. Silicic acid is removed from a chromatography plate along lines 1 inch from each of two adjacent edges to serve as solvent front boundaries. The front to which impurities have been washed should be along one of these strips (Fig. 1).

The lipid mixture is applied at a spot on the opposite corner 1 inch from each edge and allowed to spread to a round spot 7–10 mm in diameter. For typical lipid mixtures, 400 μg of acyl lipid is a convenient amount permitting determination of lipids which make up even less than 1% of the mixture. Chloroplast extracts contain about one-third carotenoids and chlorophyll; thus a 500- to 700-μg sample is about right. For a mixture containing 10 mg/ml, a 100-μl syringe or repeated applications from a 1–1.5-mm-i.d. capillary tube is convenient for this operation. In any case, a little practice may be needed.

Solvent evaporation can be hastened with a gentle jet of inert gas, taking care not to grossly distort the spot shape. If lipid molar ratios only are to be determined, the quantity of lipid spotted need not be accurately known. Determination of absolute amounts will, of course, require precise knowledge of the sample size.

Reference lipids to aid in identification can be applied on the strips at the edge of the plate; 5–10 μg of an individual lipid is applied as a bar 7–10 mm long. The chromatogram should be developed without delay once spotting is accomplished if the lipids are highly unsaturated.

Chromatography tanks lined with filter paper (Whatman 3 MM) are filled to a depth of about 0.5 inch with developing solution, covered with a glass plate and tipped to thoroughly moisten the filter paper liner and saturate the atmosphere.

For the first direction, the chloroform–methanol–water mixture is allowed to run to the scribed front. The plate is removed to a vacuum oven and dried at about 35° for 5 minutes at 100 torr or less. Or the plate may be allowed to dry under a slow stream of inert gas if a vacuum oven is not available.

The sample is then chromatographed in the chloroform–methanol–isopropylamine–ammonia system in the second direction. (The solvent mixture without the isopropylamine gives a poorer separation of sulfolipid and lecithin with some lots of silicic acid plates).

R_f values for lipids will be consistent only if care is used to reproduce conditions of moisture content in the plate, chamber saturation, chamber size, and solvent composition. Two chromatograms can be run together

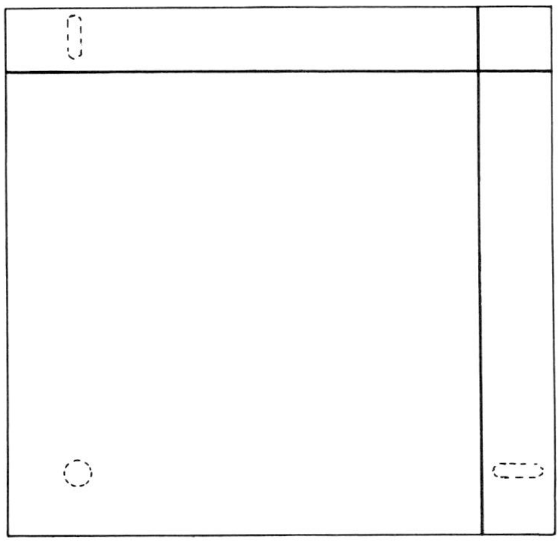

FIG. 1. Appearance of a thin-layer plate prepared for chromatography. Lipids are applied in the dotted areas.

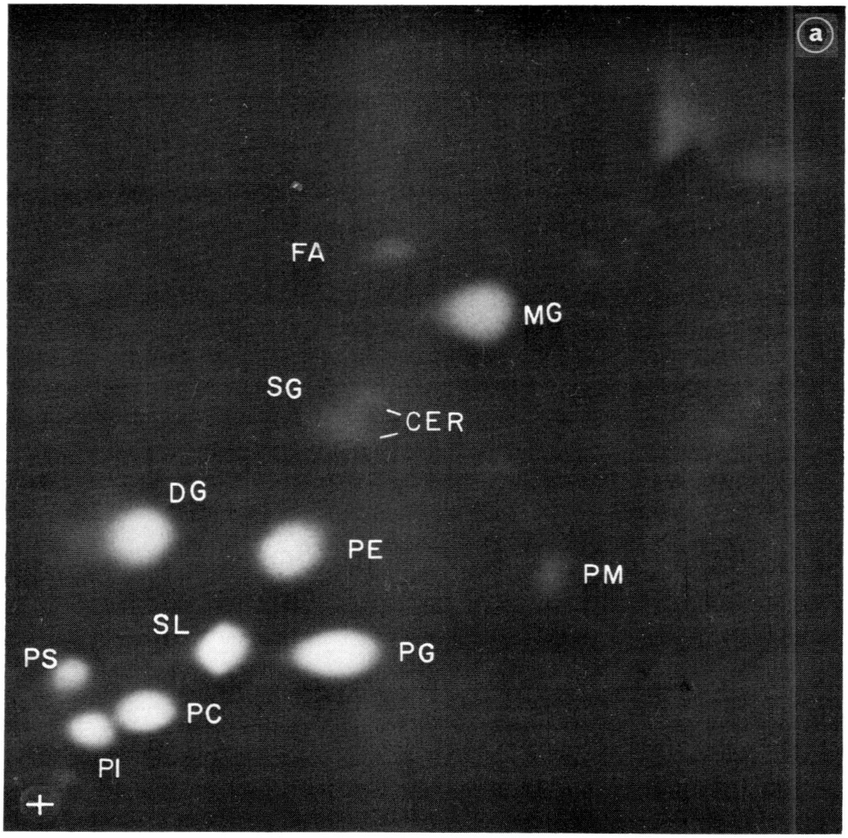

FIG. 2. Thin-layer chromatograms run as described in the text. Vertical development in chloroform:methanol:water (65:25:4); horizontal development in chloroform:methanol:isopropylamine:concentrated ammonium hydroxide (65:35:0.5:5). (a) Lipid extract of total spinach leaf. (b) Lipid extract of *Rhodospirillum rubrum*. (c) Lipid extract of *Chromatium*. Lipids are: MG, monogalactosyl diglyceride; DG, digalactosyl diglyceride; CER, cerebroside; SG, sterol glycoside; SL, plant sulfolipid; PG, phosphatidylglycerol; PE, phosphatidylethanolamine; PS, phosphatidylserine; PC, phosphatidylcholine; PI, phosphatidylinositol; OPG, *O*-ornithylphosphatidylglycerol; FA, free fatty acid. The spot PM is phosphatidylmethanol, an artifact sometimes formed in small amounts, and rarely in substantial quantities. The phospholipid reported by H. Debuch and E. Rotsch [*Hoppe Seylers Z. Physiol. Chem.* **347**, 79 (1966)] is believed to be such an artifact (H. Debuch, private communication). (d) Lipid extract of whole spinach leaf prepared as described in the text except that the leaf homogenate was kept at room temperature for 30 minutes before lipid extraction. Numerous lipid degradation products have appeared [compare with (a)].

with the adsorbant layers facing the filter papers. The plates can be held together at the top in the center of the chamber with a small metal clip such as is used in paper chromatography.

The developing solvent should be changed frequently. When this is done the filter paper liner should also be dried out; failure to do this can lead to extraordinary changes in the R_f values and separations.

The photographs of chromatograms in Fig. 2 show the lipid pattern of total spinach leaf, *Rhodospirillum rubrum* and *Chromatium*.

Excellent separations of plant lipids have also been achieved with chloroform–methanol–7*N* ammonium hydroxide (65:25:4 by volume) in the first direction, and chloroform–methanol–acetic acid–water (170:

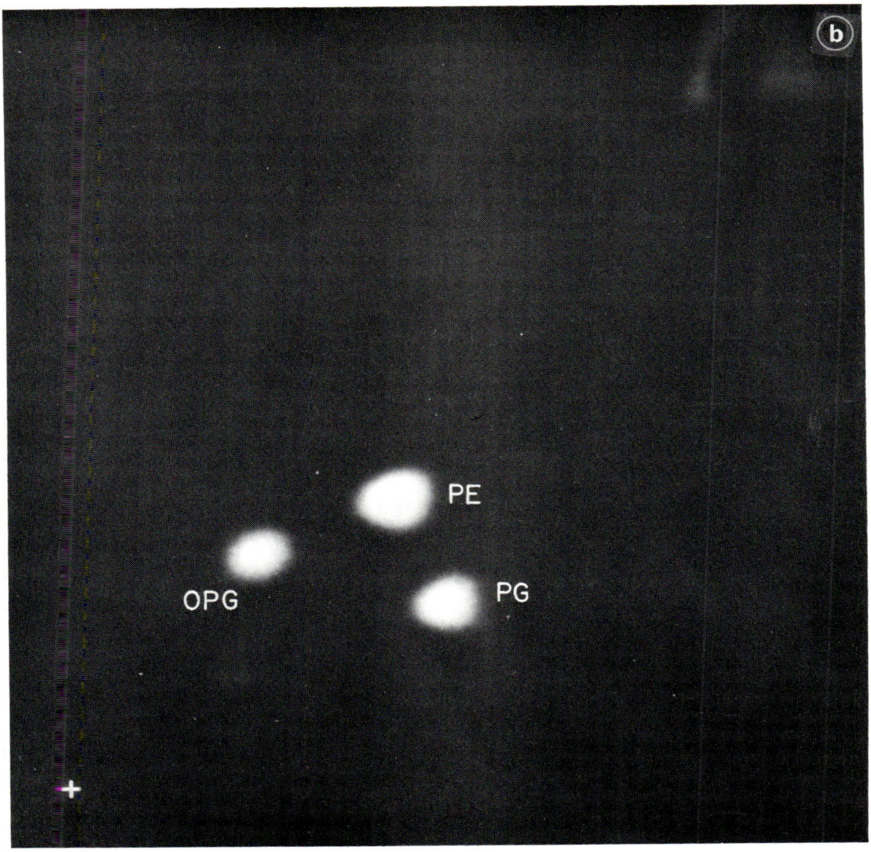

FIG. 2 (*continued*)

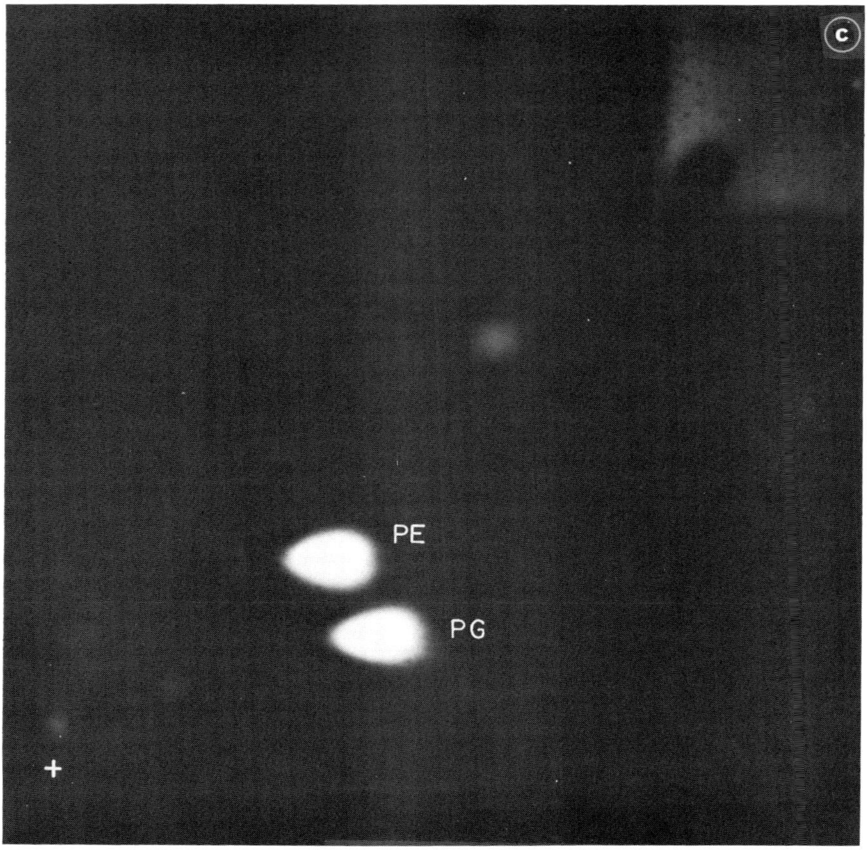

FIG. 2 (*continued*)

25:25:4 by volume) in the second.[16] The acetic acid is somewhat more difficult to remove, which is a disadvantage when the Rhodamine 6G–sodium hydroxide spray is to be used. To bring out the lipid spots, the plate, dried in air for a few moments, is evenly and thoroughly moistened with a spray of Rhodamine 6G solution prepared just before use by mixing 10 ml of 0.006% aqueous Rhodamine 6G with an equal volume of 2 N sodium hydroxide. Under short wavelength ultraviolet illumination (Model C-81 lamp, Ultraviolet Products, Inc., San Gabriel, California) lipid areas fluoresce with a yellow-orange hue on a dark background. Freshly sprayed plates will usually have an undesirable yellow fluorescence in the background, which is eliminated by evaporating excess water

[16] B. W. Nichols and A. T. James, *Fette, Seifen Anstrichm.* **66,** 1003 (1964).

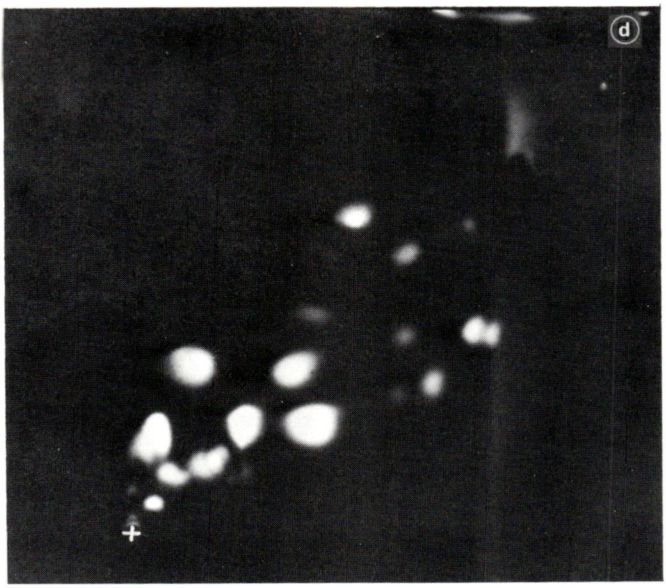

FIG. 2 (*continued*).

with a hair dryer. Drying should not be overdone or a bright yellow background fluorescence will return. A photograph of the plate made at this point will provide a convenient record of the separation attained and can be used for a rough quantitation of lipid composition by comparison of spot size and density. For this we use a Polaroid MP-3 camera with Kodak Royal Pan 4 × 5 inch sheet film exposed for 30 seconds at *f*4.5 through an orange filter (Harrison YL-6). For even illumination the ultraviolet lamp is supported on a bracket to the front of the camera lens 1 foot above the plate. The exposed film is developed in DK-60 for 6 minutes at 20°.

Fluorescent lipid areas of interest are outlined with a scribe (not a pencil), scraped from the plate, and transferred to a 15-ml screw-capped vial containing 5 ml of the 5% sulfuric acid in methanol transesterification mixture and an accurately measured quantity of the internal standard. For most mixtures, 100 μl of the methanolic solution containing 10 μg of heptadecanoic acid will suffice. For minor lipid spots a fifth or a tenth of this amount will simplify quantitation.

The vials and all other glassware used in the transesterification and subsequent extraction should be carefully washed, before use, with chloroform to remove traces of lipid contamination.

With particularly sensitive lipids it may be necessary to deoxygenate the transesterification solution (before standard and lipid are added) and maintain an inert atmosphere during the heating period, but this precaution is not necessary with lipids as unsaturated as those of spinach.

The vials are loosely closed with a Teflon-lined cap and placed in an oven at 68°–70° for 10 minutes before the caps are securely tightened. New caps are less likely to leak than reused ones. The Teflon liner should be rinsed in chloroform before use.

After 2 hours in the oven the vials are cooled, 5 ml of distilled water is added, the phases are mixed by shaking, and methyl esters are extracted with three successive 2-ml portions of distilled hexane. Pasteur pipettes are used to separate the hexane extracts which are pooled in a 6-inch test tube. The hexane solution of methyl esters is washed and neutralized by shaking with 5 ml of saturated aqueous sodium bicarbonate. The hexane solutions are transferred to 1-dram vials and evaporated to about 1 ml in a Temp-Blok heater at 35° with the aid of a jet of nitrogen. The vials are capped (chloroform-washed Teflon liner) and stored in a freezer until needed for gas chromatography, which should be done as soon as possible to minimize danger of loss of unsaturated esters.

Gas Chromatography and Quantitation

Equipment and Reagents

Analytical gas chromatograph with hydrogen flame detector (e.g., Varian-Aerograph Model 204B). Polar and nonpolar columns which will provide adequate resolution. Data presented here were determined with the following pair:
- Reoplex 400 column, ⅛ inch × 6 feet 5% on DCMS treated Chromosorb G, 100–120 mesh
- SE-30 column, ⅛ inch × 5 feet 20% on HMDS treated Chromosorb W, 60–80 mesh

Carbon disulfide, analytical reagent, redistilled.

Procedure. The remaining hexane is evaporated from the solution of methyl esters with a jet of nitrogen. The residue is immediately taken up in 5 drops of carbon disulfide (this solvent gives a much smaller solvent peak and a better baseline with a hydrogen flame detector than does hexane).

For sample injection, 2 μl of carbon disulfide are taken up in a 10-μl fixed-needle syringe (Hamilton Company, Inc., Whittier, California) followed by about 5 μl of the ester solution. This technique helps ensure introduction of a representative sample of the ester mixture on the column.

The SE-30 column is used at about 230° with the helium flow rate adjusted to elute methyl palmitate in about 3 minutes. On this column esters with the same number of carbons will be eluted together as a group of unresolved peaks in the order indicated (Table IV). The internal standard must be eluted by itself and be well resolved from other peaks.

For more detailed acyl group composition studies, the Reoplex 400 column used at about 185° will provide resolution of the individual species. Retention times for some common esters are indicated in the Table IV relative to methyl heptadecanoate. Note that there will be interference of methyl hexadecadienoate with the standard; if this ester is to be determined it will be necessary to choose another standard, such as pentadecanoic acid or a branched-chain 17-carbon acid, or to use another column where interference does not occur.

Retention times, particularly for polar columns, will vary with the position and degree of unsaturation and branching among acids with the same number of carbons. Homologous series of esters will give a straight line if the log of relative retention time is plotted against the number of carbons in the acid for a given column at a given temperature. Reference acids are available from several suppliers (Applied Science Laboratories, Inc., State College, Pennsylvania; Supelco, Inc., Bellefonte,

TABLE IV

UNBRANCHED FATTY ACID METHYL ESTERS: RETENTION TIMES RELATIVE TO METHYL HEPTADECANOATE = 10.00

	SE-30 column			Reoplex column			
Carbons in acid	Saturated	Mono-, di-, and triunsaturated	Saturated α-hydroxy	Saturated	Monounsaturated	Diunsaturated	Triunsaturated
12	2.11	2.10	3.58	2.08	2.31	—	—
14	3.90	3.80	6.40	3.76	4.36	—	—
16	7.21	6.93	11.5	7.20	8.16[a]	9.93	12.7
18	13.5	12.5	20.8	13.8	15.5	19.1	24.5[b]
20	25.0	23.3	37.1	26.6	29.4	36.2	46.4
22	46.5	42.3	67	49.8	56.5	—	—
24	88	78.0	—	97.5	107	—	—

[a] Palmitoleic acid. The *trans*-3-hexadecanoic acid of spinach has a retention time of 8.74.

[b] 9,12,15-Octadecatrienoic acid.

Pennsylvania), or mixed esters from tissue of known composition can be used. In cases of uncertain identification retention times should be checked on more than one column. Several pertinent reviews[17] and articles[18] have more detailed discussions of GPC of methyl esters.

For the quantitation of methyl esters, and thus their parent lipids, the ester peak areas are compared to that of the internal standard. With a hydrogen flame ionization detector, response is very nearly proportional to the number of carbons in the acid. However, there may be selective loss of certain esters on the column or elsewhere in the chromatograph.

The actual response factors should be determined by use of standard ester mixtures of accurately known composition which include the range of esters of interest in the analysis. Standard mixtures are available from the above firms specializing in lipids.

Such factors determined with the author's chromatograph are shown in Table V.

For an individual lipid spot isolated from a thin-layer chromatogram:

$$\text{nanomoles lipid} = \text{nanomoles standard}\left(\frac{\text{sum of adjusted ester peak areas}}{\text{peak area of standard}}\right)\frac{1}{n}$$

where n is the number of acyl residues in the lipid molecule.

TABLE V
RESPONSE FACTORS FOR FLAME IONIZATION DETECTOR[a]

	Relative to methyl heptadecanoate = 1.00	
Methyl ester	By weight[a]	Molar[b]
12:0	1.00	1.33
14:0	1.00	1.17
16:0	1.00	1.05
17:0	1.00	1.00
18:0	1.00	0.953
18:1	1.00	0.959
18:2	1.00	0.966
18:3	1.00	0.972
20:0	1.00	0.871

[a] Ratios are averages of numerous runs of standard ester mixtures on Reoplex-400 and SE-30 columns. These averages have been reproducible to $\pm 1\%$.

[b] Measured peak areas are multiplied by these factors to convert to molar ratios relative to heptadecanoic acid.

[17] A. T. James, *Methods Biochem. Anal.* **8**, 1 (1960); W. R. Supina, *in* "Biomedical Applications of Gas Chromatography" (H. A. Szymanski, ed.), p. 271. Plenum Press, New York, 1964.

[18] R. G. Ackman and R. D. Burgher, *J. Chromatogr.* **11**, 185 (1963).

Peak area can be determined conveniently with the aid of an electronic digital integrator which will print out areas during the run (the Hewlett Packard integrator has given good results; the cost is in the 5000-dollar range). The cut and weigh technique has been slightly more accurate in our experience, but is much more time consuming, especially if a polar column is being used to separate the unsaturated esters. The original chromatographic record can be preserved if Xerox copies of the peaks are used for the quantitation.

The adjusted peak area is obtained by multiplying the measured peak area by the proportionality factor (Table V) for conversion to a molar basis.

Determination of *total* acyl lipid content of a sample without separation of the individual lipids can be done much more quickly since the thin-layer chromatography and most of the gas chromatography are unnecessary. In such an analysis there may occasionally be substances in the hexane extracts which interfere with the gas chromatographic analysis by overlap with the methyl ester peaks. This can be checked by comparing results on both the polar and nonpolar columns,[18] or by removing the interfering substances by a procedure such as the following:

Hexane is evaporated from the extract, and the residue is taken up in 4 ml of methanol and 0.5 ml of 6 *M* aqueous sodium hydroxide. This solution is heated at 68° in a tightly capped vial for 1.5 hours and extracted with hexane as before. The hexane extracts are discarded. The methanolic solution is acidified cautiously with 0.2 ml of concentrated sulfuric acid (sodium sulfate will precipitate) and heated as before for about 2 hours at 68°. Methyl esters are extracted and analyzed as before.

General Comments

The success of the quantitation is dependent on careful attention to avoid contamination from reagents and equipment which may contain traces of fatty acid or skin lipids, and on avoidance of decomposition of unsaturated acids by oxidation.

With linolenic acid, oxidation problems normally do not appear unless dry lipid or ester samples have been exposed excessively to air, or the solutions of methyl esters in hexane are kept several days before analysis. Lipid extracts kept under inert gas in a freezer usually show little or no degradation within a week or more. Lipids lacking fatty acids with the diallylic methylene (C=C—CH_2—C=C) grouping can be stored for long periods. Conversely, more highly unsaturated acyl groups (e.g., pentaenoic) are proportionately more susceptible to autooxidation than linolenic and should be handled with greater caution, such as developing the chromatogram in chambers filled with nitrogen and analyzing esters soon after preparation.

The quantitation will not give accurate lipid molar ratios if lipid exists in the plasmalogen form unless dimethyl acetal peaks are included with the methyl ester quantitation. Fortunately, plasmalogens have not been reported in photosynthetic tissue. Ether analogs of lipids would also introduce errors which cannot be conveniently corrected for. Again, such material is not likely to be encountered in photosynthetic tissue; for further discussion of such lipids, more extensive publications on lipid analysis are available.[12]

Amide-linked acyl groups as in cerebrosides require longer trans-esterification periods for conversion to ester. Overnight at 68° is adequate. The presence of hydroxy acids in such lipids introduces another series of esters, which are more difficult to quantitate by the methods described here because of possible loss during esterification and chromatography. Although cerebroside does occur in plants, it has not been reported in photosynthetic tissue.

Major advantages of the procedure presented here are its applicability to all acyl lipids without the need for working curves prepared with pure lipid samples, high sensitivity, and the accumulation of acyl group composition data during the analysis.

In the authors' experience the method described above has been the most generally useful for analysis of plant lipids. Several other methods have unique advantages that will make them preferable in some situations.

Lipids can be separated by column chromatography, or by a combination of column and thin-layer chromatography and quantitated by weight. Larger quantities of purified lipids can be prepared in this way, but clear-cut quantitive separation of the sample is generally too difficult to be practical.[11] Detailed general procedures for application of such techniques have been published.[12,19] Analytical procedures based on preliminary separation of complex lipid mixtures into, say, anionic and neutral fractions on ion exchange cellulose prior to thin-layer chromatography can be quite helpful, especially if no solvent system can be found for adequate separation.

Lipids separated by TLC can be quantitated by sugar,[20] or phosphorus analyses.[19] Such methods are equally applicable to acyl, plasmalogen and ether forms, but they do require working curves. Generally these analytical techniques are not as sensitive as determination of acyl

[19] G. Rouser and S. Fleischer, this series, Vol. X, 385 (1967).

[20] P. G. Roughan and A. D. Batt, *Anal. Biochem.* **22,** 74 (1968).

groups by GPC, and reports of poor reproducibility are not uncommon with some of the published methods.

If ^{14}C-, ^{32}P-, or ^{35}S-labeled lipids are available, the elegant techniques of radiochemistry can be applied to lipids separated by any means.[21] Radioautography[22] and scintillation[23] counting are especially sensitive and useful techniques for detection and quantitation.

Lipids can also be separated[22] and quantitated[21,24] after deacylation with alcoholic alkali[22] (Dawson[12]). With this technique lysolipids and their parent lipids give the same deacylation products. Plasmalogen and ether forms are not deacylated and may be partially degraded beyond their simple deacylation products (e.g., lecithin forms some cyclic glycerophosphate).

Quantitation on TLC plates by spot area, densitometry after charring, and other techniques[25] can be applied to lipids, but suffer from being less accurate and less sensitive than the internal standard and radiometric techniques.

Preparative Separation of Lipids

The preparation of up to 50 mg or more of a particular lipid from extracts can easily be accomplished by preparative thin-layer chromatography if the lipid is a major component of the mixture and a suitable solvent system can be found to give a clean-cut separation in one-dimensional chromatography. For minor components (less than 5% of the mixture), purification of more than a few milligrams by this technique alone becomes tedious because of the large number of plates which must be handled. In such cases a preliminary concentration of the component may be possible. For example, the anionic lipids phosphatidylglycerol and sulfolipid can be removed from other components of spinach extract by elution with chloroform–methanol from a DEAE–cellulose column in the acetate form.[11,16] The anionic lipids are then eluted by addition of ammonium acetate or ammonia to the eluting solvent. De-

[21] R. A. Ferrari and A. A. Benson, *Arch. Biochem. Biophys.* **93,** 185 (1961).

[22] A. A. Benson, J. A. Bassham, M. Calvin, T. C. Goodale, V. A. Haas, and W. Stepka [*J. Amer. Chem. Soc.* **72,** 1710 (1950)] present methodology for chromatography and radioautography. A. A. Benson and B. Maruo [*Biochim. Biophys. Acta* **27** 189 (1958); see also other papers by Benson and co-workers] apply such methodology to deacylated lipids.

[23] F. Snyder and N. Stephens, *Anal. Biochem.* **4,** 128 (1962).

[24] J. F. G. M. Wintermans, *Biochim. Biophys. Acta* **44,** 49 (1960).

[25] H. Gänshirt, *in* "Thin Layer Chromatography" (E. Stahl, ed.), p. 44. Academic Press, New York, 1965; reviews quantitative evaluation of thin-layer chromatograms.

tailed instructions for preparation of such columns have been published in this series[19] and elsewhere.[12]

For purification of larger amounts of lipid, DEAE–cellulose column chromatography followed by silicic acid columns will be more convenient. Details of such separations with plant lipids have been published.[11]

Materials

Silicic acid TLC plates 20 × 5 cm, 0.25-mm-thick layer without fluorescent indicator; 2.0-mm-thick layer (without fluorescent indicator preferred)
TLC solvent systems (see above TLC procedures)

Procedure. Trial TLC separations with 5-cm analytical plates aid in selection of an appropriate solvent system for preparative TLC.

The mixture to be separated is brought to a concentration of about 10 mg/ml in chloroform and methanol (9:1), chloroform, or other appropriate volatile organic solvent, and about 10 μl is applied 1 inch from the bottom of a 5 × 20 cm silicic acid plate. This is best done as a 1-cm-long bar formed by four or five overlapping applications of smaller amounts from a capillary tube. One of the above solvent mixtures may give adequate separation, or others may be preferable.[26]

Once an appropriate system has been selected, the lipid mixture is applied as a stripe along the entire length of a 20 × 20 cm plate. If an analytical plate is used (0.25-mm silicic acid layer) 10–15 mg or less lipid is usually the maximum useful load. With the thicker preparative plates roughly 10 times as much sample can be handled. Application of lipid solution by hand is tedious; a commercial TLC sample streaker gives a more uniform application with less effort (Applied Science Laboratories, State College, Pennsylvania).

The plate is developed in the usual manner. Lipid areas are brought out with the Rhodamine 6G–sodium hydroxide spray applied to 1-cm wide strips at the edges and the center of the plate. Glass plates can be used to mask off the remainder of the plate, or a template which clips onto the plate can be prepared from a sheet of plastic. Areas containing the lipid are marked and scraped off with a spatula or razor blade, omitting the sprayed regions. This material is placed in a sintered-glass funnel of appropriate size, and lipid is eluted with solvent, e.g., chloroform–methanol–water (65:25:4) (the water can be omitted if lipid is not strongly adsorbed by the silica gel). A volume of solvent ten to twenty times that of the silica gel will usually suffice if allowed to filter

[26] E. Stahl (ed.), "Thin Layer Chromatography." Academic Press, New York, 1965; this is an excellent comprehensive reference on TLC techniques.

slowly through the adsorbent in several portions. The extract can be brought to dryness with gentle warming under a stream of nitrogen, weighed and taken up in a suitable solvent for future use.

If the lipid is highly sensitive to oxidation, it should be kept under nitrogen at all times, dissolved immediately, and stored in the cold. With such material it is best to weigh the residue from evaporation of an aliquot that can be discarded rather than determining weight by evaporating the entire solution.

The purity of lipid samples prepared as above should be checked by analytical TLC. Frequently further purification by rechromatography in the same or a different solvent system will be necessary to attain complete separation of contaminants. However, this greatly increases the risk of loss with lipids that are easily oxidized. In such cases, preliminary separation by batch elution from a DEAE–cellulose column may be advisable.

Identification of Lipids

Most current lipid analysis is concerned with the individual lipid classes, such as monogalactosyl diglyceride or plant sulfolipid, and the gross acyl group distribution within the class rather than the concentration of individual molecular species, of which there are likely to be many within each class. Physical techniques for identification of lipids are therefore restricted to methods dependent primarily on the gross structural features of the lipid, such as R_f values and infrared spectra. In most work with plant tissues, the lipid composition is already known or can be predicted. In these cases identification is reduced to correlation of R_f values on thin-layer chromatograms with standards. At least two solvent systems of different character, such as the neutral and basic ones, should be used. The lipid TLC patterns of photosynthetic plants and bacteria (Fig. 2) should permit tentative identification of the major lipids likely to be encountered.

As R_f values are sensitive to conditions of chamber saturation, solvent composition, temperature, and adsorbent layer composition, the separations obtained in a given laboratory situation should be checked with a tissue extract or other lipid mixture of known composition. Commercial lipids can also be used, but these have often given more than one spot sometimes because of decomposition in the interval between preparation and use by the customer. It is often more satisfactory to prepare small quantities of standards by preparative TLC.

Further confirmation of lipid assignments should be made with spray reagents, and sometimes by determination of molar ratios of acyl groups, phosphorus, nitrogen, sugar, or other lipid components. In cases of continued uncertainty, products of mild alkaline deacylation,[12,21] or of more extensive hydrolysis[12] can be determined.

Spray Reagents. Numerous procedures for detection of specific functional groups[26] have been published. Of those we have tried, the following have proved useful and reliable:

For NH_2 (purple color with phosphatidylethanolamine, phosphatidylserine, *O*-ornithylphosphatidylglycerol)

Reagent

Ninhydrin, 0.3 g, in 100 ml of *n*-butanol and 3 ml of acetic acid.

Use. Spray the plate lightly with reagent. Cover adsorbent layer with a glass plate and leave at 100° for up to 10 minutes. This spray can be followed by Rhodamine 6G for the other lipids; the ninhydrin-positive lipid spots will not fluoresce with the Rhodamine, and other lipid spots are somewhat weakened.

For P (blue spot with phospholipids)

Reagent.[27] Dissolve 16 g of ammonium molybdate in 120 ml of water. Shake 80 ml of this solution with 10 ml of mercury and 40 ml of concentrated hydrochloric acid. Filter, and add 200 ml of concentrated sulfuric acid and the remainder of the ammonium molybdate solution to the filtrate. Dilute to 1 liter with water.

Use. Moderately sprayed plate gives blue spots on a white background within a few minutes at room temperature. Requires at least 2 μg of phospholipid in a 5-mm diameter spot. Phosphoric acid does not give a positive test. The reagent is stable on storage without refrigeration.

For $—NH—\overset{\overset{\displaystyle O}{\|}}{C}—$ (blue spot with cerebroside and other spingolipids).

Reagents[28]

I. Add 5 ml of acetic acid to a mixture of 5 ml of fresh Clorox brand bleach and 50 ml of benzene. Use immediately after preparation.

II. Benzidine, 0.5 g, and a small crystal of potassium iodide dissolved in 50 ml of ethanol and filtered. Keep out of light and use within 2 hours of preparation.

[27] V. E. Vaskovsky and E. Y. Kostetsky, *J. Lipid. Res.* **9,** 396 (1968).

[28] M. D. Bischel and J. H. Austin, *Biochim. Biophys. Acta* **70,** 598 (1963).

Use. Dry chromatogram to remove ammonia. Spray lightly with reagent I until uniformly wet, allow chromatogram to dry in the hood to remove Cl_2, and spray with reagent II. Blue spots appear with about 10 μg or more of sphingolipid. It has been our experience that phosphatidylethanolamine and other amino lipids also give blue spots.

For sugars (bluish-purple spots with galactolipids, plant sulfolipid)

Reagents[19]

α-Naphthol, 0.5 g (recrystallized from hexane–chloroform) in 100 ml of 50% aqueous methanol
Concentrated sulfuric acid

Use. Spray with naphthol reagent as a mist until chromatogram is damp. Air dry plate, and spray very lightly with concentrated sulfuric acid. Cover adsorbant layer with a glass plate and heat at 120° until color develops.

Infrared spectra are occasionally useful in identification. Spectra of the major lipids of spinach leaf have been published.[11]

[49] Preparation and Assay of Chloroplast Coupling Factor CF_1

By Stephen Lien and Efraim Racker

Treatment of chloroplasts with a dilute EDTA solution was found to liberate a soluble coupling factor.[1,2] The active factor in the EDTA extract is identical[3] to coupling factor CF_1, which contains a latent Ca^{2+}-dependent ATPase.[4]

Assay of Coupling Activity of CF_1

An assay for CF_1 with EDTA-treated chloroplasts or with subchloroplast particles obtained by sonication of chloroplasts in the presence of phospholipids (CP particles) has been described.[5] Since EDTA-treated

[1] A. T. Jagendorf and M. Smith, *Plant Physiol.* **37**, 135 (1962).
[2] M. Avron, *Biochim. Biophys. Acta* **77**, 699 (1963).
[3] R. E. McCarty and E. Racker, *Brookhaven Symp. Biol.* **19**, 202 (1966).
[4] V. K. Vambutas and E. Racker, *J. Biol. Chem.* **240**, 2660 (1965).
[5] See this volume [23].

chloroplasts cannot be stored and the degree of resolution of CP particles is somewhat variable, an assay with stable subchloroplast particles which are completely resolved with respect to CF_1 is used. The procedure for preparing such particles is as follows:

To 20 ml of suspension of phosphorylating subchloroplast particles[6] containing 4 mg of chlorophyll/ml in STN buffer (0.4 *M* sucrose, 0.02 *M* Tricine–NaOH, 0.01 *M* NaCl, pH 7.8), 3.8 ml of 5% silicotungstic acid (Hopkin and Williams Ltd., Chadwell Heath, Essex, England, the pH of the freshly prepared solution is adjusted to 5.5 with 0.5 *N* NaOH) is added slowly at 0° and then stirred gently for 5 minutes. After centrifugation at 144,000 *g* for 30 minutes in a No. 50 rotor in a Spinco ultracentrifuge Model L-2, the supernatant solution is discarded and the pellet resuspended in 20 ml of STN buffer by gentle homogenization. After centrifugation at 144,000 *g* for 30 minutes, the supernatant fluid is discarded and the pellet washed twice with the same amount of STN buffer and finally suspended in 10 ml of STN buffer. The chlorophyll concentration of the suspension is adjusted to about 4 mg of chlorophyll/ml by addition of buffer. Aliquots of 0.5 ml of the suspension are distributed in small tubes which are flushed for 30 seconds with argon, tightly stoppered and stored at −80°. The particles thus obtained are called STA particles and are stable for several months.

Over 99% of the CF_1 in subchloroplast particles is removed by this treatment. In the presence of phenazine methosulfate the rate of phosphorylation is less than 0.1 μmole of ATP formed per milligram of chlorophyll per hour. On addition of CF_1, about 20–30 μmoles of ATP are formed per milligram of chlorophyll per hour. Pyocyanine does not replace phenazine methosulfate.[7] The reconstitution of particles with CF_1 and measurement of photophosphorylation of the reconstituted particles in the presence of phenazine methosulfate are carried out as described.[5]

Assay of Ca^{2+}–ATPase Activity of CF_1

In addition to coupling activity, CF_1 has also a latent Ca^{2+}-dependent ATPase activity which can be activated by heat treatment[4,8] or by trypsin digestion,[4] or by prolonged incubation at high concentrations of dithiothreitol.[9] The ATPase activity provides a rapid and convenient assay of the enzyme and is especially useful in the course of purification.

[6] See this volume [30].

[7] G. Hauska, R. E. McCarty, and E. Racker, *Biochim. Biophys. Acta* **197**, 206 (1970).

[8] F. Farron, *Biochemistry* **9**, 3823 (1970).

[9] R. E. McCarty and E. Racker, *J. Biol. Chem.* **243**, 129 (1968).

Activation of the Latent Ca^{2+}-Dependent ATPase Activity

Activation by Heat in the Presence of Low Concentration of Dithiothreitol

A reproducible and convenient method for activation of the enzyme preparation containing more than 0.5 mg of CF_1/ml was described by Farron.[3] At lower concentration of CF_1 loss of activity occurs.

Reagents

- Tricine–NaOH–EDTA buffer, contains Tricine–NaOH 40 m*M*, EDTA, 2 m*M*, pH 8.0
- ATP 0.2 *M*, pH 7.2
- EDTA 0.2 *M*, pH 7.6
- Dithiothreitol 0.1 *M*
- Tricine–NaOH 1 *M* pH 8.0
- Trypsin, 5 mg/ml in 1 m*M* H_2SO_4
- Trypsin inhibitor, 5 mg/ml in 5 m*M* Tricine–NaOH, pH 7.0

Procedure. An aliquot of enzyme [stored at 4° in 2 *M* $(NH_4)_2SO_4$, containing about 1.5–2 mg of CF_1] is centrifuged for 10 minutes at 5000 *g*, dissolved with 0.6 ml of the Tricine–NaOH–EDTA buffer and desalted by passing through a Sephadex G-50 column (0.8 cm × 15 cm) equilibrated with the same buffer. To 0.2 ml of the desalted enzyme (containing 150–200 μg of CF_1) the following solutions are added: 0.05 ml of ATP, 0.015 ml of dithiothreitol, and 0.03 ml of the buffer. The mixture is placed in a water bath at 60° for 4 minutes; then 1.7 ml of the buffer kept at 20° are added, and the test tube is transferred to a water bath at room temperature.

Activation by Trypsin

This method[4] is especially useful for activation of dilute solutions of CF_1, e.g., in a crude EDTA extract.

Procedure. To an aliquot of desalted CF_1 (as described above) or crude EDTA extract containing 30–50 μg of CF_1, 0.01 ml of ATP, 0.02 ml of EDTA, 0.03 ml of Tricine–NaOH, and H_2O are added to a final volume of 0.93 ml in a test tube. After addition of 0.02 ml of trypsin, the mixture is incubated at room temperature for 6 minutes. The digestion is stopped by adding 0.05 ml of trypsin inhibitor.

Assay of the Activated Enzyme

Principle. The activity of the enzyme is assayed by determining the amount of P_i liberated by the enzymatic hydrolysis of ATP.

Reagents

Tricine–NaOH buffer, 0.02 *M*, pH 8.0
Solution A consists of 40 m*M* Tricine–NaOH, 10 m*M* ATP, 10 m*M* $CaCl_2$, pH 8.0 (prepared immediately before use)
Trichloroacetic acid, 0.5 *M*

Procedure. Aliquots of the activated enzyme containing 2.5 μg of CF_1 are pipetted into test tubes and brought up to a total volume of 0.2 ml with buffer. To each tube 0.8 ml of solution A is added. The tubes are incubated for 10 minutes at 37°, and the reaction is stopped by adding 1.0 ml of ice-cold trichloroacetic acid. After centrifugation at 3000 *g* for 10 minutes at 0°, 1.6 ml of the supernatant is analyzed for P_i.[10]

Units of Activity. One unit of activity is defined as that amount of enzyme which catalyzes the hydrolysis of 1 μmole of ATP/minute under specified assay conditions.

Specific Activity. This is expressed as units per milligram of protein.

Application of Assay to Chloroplasts or Subchloroplast Preparations

The above method can be used to estimate the Ca^{2+}–ATPase activity of various chloroplast and subchloroplast preparations. Trypsin digestion is used for activation of the enzyme. For maximal activation, about 400 μg of trypsin/100 μg of chlorophyll are used. The optimal duration of exposure to trypsin varies with different particle preparations and should be determined in each case. To do this, we recommend a schedule of incubation starting from 12 minutes, with 4-minute increments, up to 32 minutes. Aliquots containing about 8 μg of chlorophyll of the activated mixtures after adding trypsin inhibitor are assayed for Ca^{2+}–ATPase activity as described above.

Preparation of Coupling Factor CF_1

Step 1. Preparation of Crude Extract. Deveined spinach leaves (800 g) are homogenized at 0°–4° in 200-g portions with 250 ml of STN medium (0.4 *M* sucrose, 20 m*M* Tricine–NaOH, and 10 m*M* NaCl, pH 7.8) in a Waring Blendor, for 15–20 seconds. The leaf homogenate is strained through eight layers of cheesecloth and centrifuged at 300 *g* for 2 minutes in a Serval GSA rotor. The pellet is discarded and the supernatant solution is centrifuged at 3000 *g* for 10 minutes. The supernatant fluid is discarded, and the dark green pellet is resuspended gently in 500 ml of STN and centrifuged at 3000 *g* for 10 minutes. The pale green supernatant solution is discarded, and the pellet, containing ap-

[10] H. Taussky and E. Shorr, *J. Biol. Chem.* **202, 675 (1953)**.

proximately 200 mg of chlorophyll, is suspended in 10 mM NaCl to a final volume of 500 ml and is gently stirred with a magnetic stirrer for 10 minutes at 4°. After centrifugation for 20 minutes at 20,000 g in the same rotor, the supernatant solution is discarded and the pellet is evenly resuspended in a minimal amount of 10 mM NaCl ($\approx$ 1 ml/3–4 mg of chlorophyll). After the actual chlorophyll concentration has been determined, the suspension is diluted at room temperature with 0.75 mM EDTA, pH 7.6, to 0.1 mg of chlorophyll/per ml and stirred with a magnetic stirrer for 10 minutes. After centrifugation for 30 minutes at 20,000 g at 20°, the supernatant solution which contains CF_1 is collected. Since CF_1 is cold labile, all operations in the following steps are performed at room temperature.

Step 2. Concentration and Partial Purification on DEAE–Sephadex A-50. To each liter of the crude extract, 5 ml of 0.2 M ATP (pH 7.2), 20 ml of 1 M Tris·SO_4 (pH 7.3), 6 ml of 0.2 M EDTA (pH 7.2), and 40 ml of 2 M $(NH_4)_2SO_4$ are added very slowly with gentle stirring. The whole mixture is passed through a DEAE–Sephadex A-50 column (4 cm × 20 cm) prepared from gel which has been treated as follows: After swelling for 2 hours in distilled water, the gel is washed on a Büchner funnel with 0.5 N HCl (1.5 liters/10 g of dry gel), distilled water (three times the volume of the acid), and 0.5 N NaOH (1.5 liters/10 g of dry gel) followed by a large quantity of distilled water until the pH falls to 8. The gel is equilibrated with two times the gel volume of buffer containing 20 mM Tris·SO_4, 80 mM $(NH_4)_2SO_4$, 2 mM EDTA, pH 7.1. After standing for 30 minutes, the excess buffer is removed by suction and the gel is further equilibrated three times with fresh buffer. Before pouring the column, the gel suspension is kept under vacuum for 2 hours with gentle stirring using a magnetic stirrer to remove any trapped air. After packing the gel into the column, it is washed with 200 ml of the buffer containing 20 mM Tris·SO_4, 2 mM EDTA, 1 mM ATP, and 80 mM $(NH_4)_2SO_4$. The crude extract of CF_1 is then passed through the column. After washing with 600 ml of buffer containing 20 mM Tris·SO_4, 2 mM EDTA, 1 mM ATP, and 0.1 M $(NH_2)_4SO_4$, pH 7.1, CF_1 is eluted with 600 ml of the same buffer containing 0.28 M $(NH_4)_2SO_4$, pH 7.1, and collected in 10-ml fractions. To each fraction an equal volume of saturated $(NH_4)_2SO_4$ solution containing 20 mM Tris·SO_4, 2 mM EDTA, and 2 mM ATP, pH 7.1, is added. After gentle mixing and standing for 10 minutes, fractions that turned turbid are pooled. The combined fractions contain partially purified CF_1, and can be stored at 4° for months without losing activity.

Step 3. Purification with a Second DEAE–Sephadex A-50 Column. The partially purified CF_1 is collected by centrifugation at 20,000 g for

10 minutes in a Serval SS-34 rotor, and the pellet is dissolved in a minimal amount of buffer containing 20 mM Tris·SO_4, 1 mM EDTA, pH 7.1, and clarified by centrifugation at 30,000 g for 20 minutes. The total protein content and $(NH_4)_2SO_4$ concentration of the supernatant are determined,[11,12] and an appropriate amount of the dissolving buffer is added to bring the supernatant to a final $(NH_4)_2SO_4$ concentration of 0.08 M. The whole mixture is passed through a second DEAE–Sephadex A-50 column (2.5 cm × 25 cm) prepared as described in step 2, followed by washing with 200 ml of 0.12 M $(NH_4)_2SO_4$ in the same buffer. CF_1 is eluted by a linear gradient generated by feeding 450 ml of 0.3 M $(NH_4)_2SO_4$, 20 mM Tris·SO_4, 1 mM ATP, 2 mM EDTA, pH 7.1, into a mixing chamber containing 450 ml of the same buffer except that 0.12 M $(NH_4)_2SO_4$ is present. Fractions (5 ml) are collected and a 20-μl sample of each fraction is analyzed for protein.[11] CF_1 appears as a rather sharp peak between 0.16 M and 0.2 M $(NH_4)_2SO_4$ concentration of the linear gradient. The peak fractions containing about 40 mg of protein with an average specific activity of 34, are combined. ATP and neutralized saturated $(NH_4)_2SO_4$ are added to a final concentration of 2 mM ATP and 2 M $(NH_4)_2SO_4$. This preparation is stable at 4° for several months. The pH of the suspension is maintained at 7.1 by periodic addition of dilute NH_4OH. A summary of yield and specific activity of the various fractions is given in the table.

Remarks on the Preparation Procedure. An alternative procedure for the preparation of CF_1 from chloroplasts that have been treated with

PURIFICATION OF CF_1

Purification steps	Total protein (mg)	Specific activity[a]	Total units	% Recovery
Crude extract[b]	162	10.5	1700	100
1st DEAE–Sephadex A-50 column (step elution)	62.5	24.3	1510	89
2nd DEAE–Sephadex A-50 column (linear gradient elution)	39.2	34	1330	78

[a] Ca^{2+}–ATPase is activated by trypsin. Specific activity is defined as micromoles of ATP hydrolyzed per milligram of protein per minute.

[b] The crude extract (2 liters) obtained from chloroplasts (200 mg of chlorophyll) contains 81 μg of protein/ml.

[11] O. H. Lowry, N. J. Rosebrough, A. L. Farr, and R. J. Randall, *J. Biol. Chem.* **193**, 265 (1951).

[12] P. B. Hawk, B. L. Oser, and W. H. Summerson, *in* "Practical Physiological Chemistry," 13th ed., p. 878. McGraw-Hill (Blakiston), New York, 1954.

acetone has been described[8,13] which yields preparations of somewhat lower specific activity. On the other hand, the large volume in step 1 of the present procedure is not convenient for large-scale preparation (i.e., over 200 mg of chlorophyll) unless a Sharples-type centrifuge is available. Chloroplast suspensions containing 600–800 mg of chlorophyll, which yield 6–8 liters of crude extract and about 120–160 mg of purified CF_1 with a specific activity of 30–40 (Ca^{2+}-dependent ATPase) are processed as follows:

1. The diluted chloroplast suspension (0.1 mg of chlorophyll/ml) in 0.75 mM EDTA obtained in step 1 is fed to a refrigerated Sharples centrifuge at 50,000 rpm at a flow rate of 50 ml/minute. The supernatant fluid from the first Sharples is fed to a second Sharples centrifuge at the same flow rate to remove the residual chloroplast fragments, and the resulting pale green supernatant fluid is collected in 1-liter portions; Tris·SO_4 buffer, ATP, EDTA, and $(NH_4)_2SO_4$ are added as described in step 2.

2. In step 2 a larger column (5 × 30 cm) is used. When properly prepared the column should give a flow rate of 400–450 ml per hour under 50–60 cm of hydrostatic pressure (higher pressure results in compression of the gel bed and a reduced average flow rate). The crude extract (ca. 8 liters) can be processed within 20 hours. At room temperature, no deterioration of crude CF_1 preparation is observed. The column is washed with 800 ml of 0.1 M $(NH_4)_2SO_4$ in buffer containing 20 mM Tris·SO_4 1 mM ATP, 2 mM EDTA, pH 7.1. CF_1 is eluted with 1 liter of 0.28 M $(NH_4)_2SO_4$ in the same buffer, pH 7.1, as described in step 2.

3. Final purification on a second DEAE–Sephadex column is carried out as in step 3 except that the dimensions of the gel bed are 2.5 × 40 cm and that after loading of CF_1 the column is washed with 400 ml of buffer containing 0.12 M $(NH_4)_2SO_4$ before eluting with the linear gradient formed by mixing 700 ml of each of the specified buffers.

Purification of CF_1 to Homogeneity by Sucrose Density Gradient Centrifugation

Electrophoretic analysis on polyacrylamide gel showed that preparations of CF_1 obtained from step 3 of the above described procedure contained about 5% contaminating proteins. To obtain a homogeneous preparation, the following procedure was successfully employed:

Reagents

Buffer containing Tris·SO_4, 20 mM; EDTA, 2 mM; and ATP, 1 mM, pH 7.1

[13] A. Bennun and E. Racker, *J. Biol. Chem.* **244,** 1325 (1969).

8% sucrose buffer, 100 ml containing 8 g of sucrose (ultrapure, enzyme grade, Mann), Tris·SO_4, 20 mM; EDTA, 2 mM; and ATP, 1 mM, pH 7.1

25% sucrose buffer, as above except with 25 g of sucrose

Preparation of Sucrose Density Gradient Tubes. Two 36-ml linear (8–25% w/v) sucrose density gradients were set up in two cellulose nitrate centrifuge tubes (Beckman No. 302237) by mixing the 25% sucrose buffer with an equal volume of 8% sucrose buffer in a linear gradient generator at room temperature.

Preparation of Desalted CF_1 Sample and Density Gradient Centrifugation. CF_1 suspension (containing about 40 mg of protein) in 2 M $(NH_4)_2SO_4$ from step 3 was centrifuged at 10,000 g for 10 minutes at 20°. After removing the supernatant, the pellet was carefully dissolved in 1.0 ml of buffer. The solution was clarified by centrifugation at 25,000 g for 20 minutes at 20° and then passed through a column of 1 cm × 20 cm bed volume made up with Sephadex G-50 fine, equilibrated with the same buffer. Desalted CF_1 was collected in a total volume of 3 ml. The desalted sample (1.5 ml) was carefully layered on top of each gradient tube and then centrifuged at 26,000 rpm in a Spinco SW27 rotor for 28 hours at 20°. Fractions (30 drops) were collected by puncturing the tube bottom at room temperature.

Analysis of Fractions. Each fraction (15-μl sample) was assayed for protein.[11] The fractions (50-μg samples) around the major protein peak which appeared at about two-thirds down from the top of the gradient were checked for purity by analytical disc gel electrophoresis[14] on 6% polyacrylamide gel. Fractions, free from contaminating proteins were combined and diluted with an equal volume of buffer. After addition of ammonium sulfate as described in step 3 of the purification procedure, the suspension was stored at 4°. About 50–60% of the starting CF_1 was recovered as a homogeneous protein. The remaining inhomogeneous but highly purified fractions can be stored in 2 M $(NH_4)_2SO_4$ and saved for further purification to homogeneity by the same procedure.

Since the analysis of purity by disc gel electrophoresis are both inconvenient and time consuming, a simple and fast fluorometric assay was developed, based on the fact that CF_1 does not contain tryptophan residues.[8] In contrast to tryptophan-containing proteins which exhibit an emission peak around 350 nm, the fluorescence of homogeneous CF_1 arises from tyrosine residues and peaks at 302 nm. The ratio of fluorescence intensity at 300 and 350 nm was found to correlate with the purity of

[14] B. J. Davis, *Ann. N.Y. Acad. Sci.* **121,** 404 (1964), and L. Ornstein, *Ann. N.Y. Acad. Sci.* **121,** 321 (1964).

CF_1 preparations. The following procedure yielded very reliable results in our laboratory: Samples (25 μl) from the sucrose density gradient fractions were diluted with 1 ml of 20 m*M* Tris·SO_4, pH 7.1, and transferred to a quartz fluorometric cuvette. The relative fluorescence intensities at 300 and 350 nm were measured at room temperature in a Aminco-Bowman spectrophotofluorometer[15] using 280-nm excitation light with slits arranged to give 12.5-nm half-band width for both monochromators. Under the described fluorophotometric conditions, all fractions having a fluorescence intensity ratio at 300:350 nm exceeding 1.85 were consistently found to be homogeneous by polyacrylamide gel electrophoresis. It should be mentioned that fluorescence assay is a convenient way for analyzing fractions from DEAE-Sephadex columns in the earlier steps of CF_1 preparation.

Properties of CF_1

Stability. The soluble enzyme is cold labile, but relatively stable at room temperature. The loss of activity at 0° is accelerated by salts and low pH[3], but if CF_1 is suspended in 2 *M* $(NH_4)_2SO_4$ in the presence of 20 m*M* Tris·SO_4, 2 m*M* ATP, and 2 m*M* EDTA at pH 7.1, the enzyme is stable at 4° for several months. The ATPase is protected against cold inactivation by ATP and several other membrane components.[3,13,16] Heating and tryptic digestion as described for the activation of ATPase, however, destroy coupling activity.[4]

Effect of Cations on ATPase Activity. The activated enzyme shows a specific Ca^{2+} requirement for ATPase activity (optimal concentration of Ca^{2+} is about 10–15 m*M*). Other divalent cations, Ni^{2+}, Mg^{2+}, Co^{2+}, and Sr^{2+} (at 10 m*M*) are less than 3% as effective Ca^{2+}. Furthermore, Mg^{2+}, which is required for the coupling activity of CF_1 inhibits the Ca^{2+}-dependent ATPase. At 0.3 m*M* Mg^{2+}, the inhibition is 50%. The rate of ATP hydrolysis varies with the ATP:Ca^{2+} ratio. The optimal ratio is 1.[4]

Substrate Specificity and Product Inhibition of Ca^{2+}–ATPase. ATP is hydrolyzed four times as fast as GTP or ITP. UTP and CTP are ineffective; ADP is inhibitory.[4]

Inhibitors. In addition to ADP, significant inhibition of ATPase activity is obtained by β-hydroxymercuribenzoate, *n*-butyl-3, 5-diiodo-4-hydroxybenzoate, and NH_4^+ ions. Other compounds, such as *N,N'*-dicyclohexyl carbodiimide and carbonyl cyanide-*m*-chlorophenylhydrazone are without effect at concentrations that give complete inhibition of photophosphorylation in chloroplasts.[4]

[15] American Instrument Company, Silver Spring, Md. 20910.

[16] A. Livne and E. Racker, *J. Biol. Chem.* **244,** 1332 (1969).

[50] Partial Resolution of the Photophosphorylating System of *Rhodopseudomonas capsulata*

By ASSUNTA BACCARINI-MELANDRI and BRUNO A. MELANDRI

Pigmented membrane particles prepared from the facultative phototroph *Rhodopseudomonas capsulata* catalyze light-induced phosphorylation of ADP coupled to cyclic electron transport. This system can be resolved by sonication, in the presence of EDTA, into two components: nonphotophosphorylating particles and a protein factor present in the sonic extract fluid (or in chromatophore acetonic powder) that can restore completely the photophosphorylating activity of the resolved particles.[1,2]

Fractionation Procedures

Growth of Cells: Rhodopseudomonas capsulata (strain St. Louis, American Type Culture Collection) is grown anaerobically in a synthetic medium (initial pH 6.8) containing 0.4% DL-malic acid, 0.1% $(NH_4)_2SO_4$, 0.0001% thiamine hydrochloride, and additional inorganic salts, as specified by Ormerod *et al.*[3] The medium is inoculated with about a 1% (v/v) inoculum of a culture in early stationary phase. Cultures are incubated at 34° and illuminated with banks of 60-W tubular incandescent lamps (light intensity 500 foot candles).

Photophosphorylating Particles: Cells are harvested by centrifugation after 15 hours growth, washed once with 0.05 *M* glycylglycine buffer, pH 7.2, resuspended in the same buffer, and disrupted in a French pressure cell at 16,000 psi. Following a low-speed centrifugation (20,000 *g* for 20 minutes), the supernatant solution is centrifuged for 90 minutes at 200,000 *g* in a No. 50 Titanium rotor (for large scale preparations, centrifugation is for 180 minutes at 100,000 *g* in a No. 30 rotor). The pellets are washed once in the same buffer and recentrifuged. The final particle pellet is suspended in a mixture of 0.1 *M* glycylglycine buffer, pH 7.2, and glycerol (1:1, v/v) at a concentration of approximately 1.5 mg of BChl/ml. The preparation is stored at −20° and is quite stable for two or three weeks.

With these particles, the light-induced phosphorylation ranges from

[1] A. Baccarini-Melandri, H. Gest, and A. San Pietro, *J. Biol. Chem.* **245**, 1224 (1970).
[2] B. A. Melandri, A. Baccarini-Melandri, A. San Pietro, and H. Gest, manuscript in preparation.
[3] J. G. Ormerod, K. S. Ormerod, and H. Gest, *Arch. Biochem. Biophys.* **94**, 449 (1961).

50 to 150 μmoles ATP/hour/mg of BChl. Higher rates can however be achieved by including 2.5 mM $MgCl_2$ in all the buffers used throughout. These particles catalyze a Mg^{2+}-dependent ATPase activity, and the rate is between 40 and 80 μmoles P_i hydrolyzed/hour/mg of BChl. This latter activity is stimulated by 2,4-dinitrophenol (10^{-3} M) as well as FCCP (10^{-6} M) about two- and threefold, respectively, whereas oligomycin (10 μg/ml) inhibits. ATP-$^{32}P_i$ exchange activity measured in dark and in light is in the range 15–50 μmoles $^{32}P_i$ exchanged/hour/mg of BChl, respectively. Light-induced proton uptake has been measured in this preparation.[4] The final extent of H^+ uptake, which is normally reached after about 2 minutes of illumination, is strongly dependent on the initial pH. Typical values in neq. of H^+/mg of BChl, at initial pH's of 6.0, 6.7, and 7.7 are 350–400, 250–280, and 150–180, respectively.

Nonphotophosphorylating Particles: Fresh particles prepared in the absence of Mg^{2+} are diluted with 1 mM EDTA, pH 7.2, to a concentration equivalent to 0.25–0.4 mg of BChl/ml. The suspension is sonicated three times in the cold, each for 30 seconds, with a Biosonik oscillator (Bronwill Scientific) set at 80% of the maximum output, and thereafter it is stirred for 1 hour at room temperature in the dark. Particles are sedimented by centrifugation at 200,000 g in a No. 50 Titanium rotor for 90 minutes; the supernatant solution is decanted, and the particles are carefully resuspended in 1 mM EDTA and recentrifuged. The membrane fragments so obtained can be stored in 0.1 M glycylglycine, pH 7.2, and glycerol (1:1, v/v) at $-20°$ and are stable for more than a month with respect to reactivation. Photophosphorylation capability of these uncoupled membrane fragments is 5–15% in comparison to the original particles; ATPase and ATP-$^{32}P_i$ exchange activities are 20–25% of the original activity. All three of these functions can be restored to the original levels by incubation in the presence of crude supernatant solution or partially purified coupling factor (see below).

Proton uptake, measured in the absence of phosphorylation substrates, is not significantly affected by the EDTA-sonication treatment.

Purification of the Coupling Factor

Although an active component could be precipitated by $(NH_4)_2SO_4$ fractionation of the concentrated sonic extract fluid, a more convenient source of the factor is an acetone powder of phosphorylating particles.

Step 1—Chromatophore Extract: Phosphorylating particles (200 mg of BChl), prepared as described but in the presence of 2.5 mM Mg^{2+}, are

[4] B. A. Melandri, A. Baccarini-Melandri, A. San Pietro, and H. Gest, *Proc. Natl. Acad. Sci.* **67**, 477 (1970).

added dropwise to 20 volumes of cold acetone (—20°). This operation normally is performed rapidly at room temperature. The mixture is filtered (Whatman No. 1 paper), and the partially dried, sticky powder is extracted with 30 ml of 0.05 *M* Tris·HCl, pH 7.6, containing 4 m*M* ATP. After 30 minutes at room temperature, the suspension is centrifuged for 30 minutes at 20,000 *g*. The supernatant solution is collected, and the pellet is extracted again with the same buffer. After a second centrifugation, the two supernatant solutions are combined.

Step 2—High-Speed Centrifugation: The combined supernatant solutions are centrifuged for 1 hour at 200,000 *g* in order to discard membrane fragments and ribosomes. This step gives a twofold increase in specific activity without any loss in total units.

Step 3—Ammonium Sulphate Fractionation: The solution (5 mg of protein/ml) is brought to 0.3 saturation with $(NH_4)_2SO_4$, and the precipitate is removed by centrifugation. Additional $(NH_4)_2SO_4$ is added (127 mg for each milliliter of supernatant solution), and the suspension is allowed to stand in the cold for 1 hour. The pellet is collected by centrifugation and suspended in a small volume of 0.05 *M* Tris·HCl, pH 7.4, containing 4 m*M* ATP. To measure activity, an aliquot of the solution is passed through a Sephadex G-50 column equilibrated with the same buffer. A twofold increase in specific activity could be achieved with 70–80% yield.

Step 4—Column Chromatography: The protein solution (60–80 mg) is applied to a Sepharose 6B column (2.5 × 40 cm) equilibrated with Tris·ATP buffer. Fractions (5 ml) are collected, and the protein is estimated by the Lowry method. During the chromatography, two protein peaks which exhibit coupling factor activity are separated. These two peaks can be further purified by gel filtration on Sephadex G-200. However, the high molecular weight species gives rise again to two peaks, one of high and one of lower molecular weight. Based on gel disc electrophoresis patterns and other criteria, such as sensitivity to heating, protection by ATP, and inactivation properties, it appears possible that the two active peaks represent indeed different states of aggregation of the same protein. The molecular weight of the faster moving species has been estimated, by gel filtration on Sephadex G-200, to be 280,000 daltons whereas the slower moving peak is 78,000.

Assay and Properties of the Coupling Factor

In order to achieve maximum recoupling, a 20-minute contact time is required between nonphosphorylating particles and coupling factor in the presence of Mg^{2+}. The preincubation is currently performed in the dark at the temperature of the assay (usually 30°) and in the presence of the reagents and buffer required, except the ones used to start the reaction. These are: $^{32}P_i$ for photophosphorylation; ATP, phosphoenol pyruvate and pyruvate kinase for ATPase; and ATP and $^{32}P_i$ for ATP-$^{32}P_i$ exchange.

Photophosphorylation Assay:

REAGENTS:

Glycylglycine·NaOH buffer, pH 8.0, 0.8 *M*
$MgCl_2$, 0.5 *M*
D-Glucose, 0.5 *M*
ADP, 0.05 *M*
Na succinate, 0.02 *M*
Sulfate-free hexokinase (Sigma, St. Louis), 250 U/ml
Na_2HPO_4, 0.15 *M* containing ^{32}P (2×10^5 cpm/μmole)

Each reaction mixture contains 0.125 ml of glycylglycine, 0.03 ml of $MgCl_2$, 0.05 ml of glucose, 0.06 ml of ADP, 0.011 ml of succinate, 0.02 ml of hexokinase, 0.05 ml of particles (30–50 μg of BChl), coupling factor (10–100 μg of protein), and distilled water to a final volume of 1.4 ml. The tubes are incubated 20 minutes in the dark at 30°, after which time 0.1 ml of NaH_2PO_4 containing ^{32}P is added to each tube, and the reaction is initiated by turning on the light. A zero time and a dark control are routinely performed. After 10 minutes in the light, the reaction is terminated by adding 0.2 ml of ice-chilled 10% TCA to each tube. The clear supernatant solution obtained by low-speed centrifugation is used to assay organic radioactive phosphate according to Avron.[5]

Since the particles lack any appreciable adenylate kinase activity, photophosphorylation can also be measured spectrophotometrically at 340 nm as the rate of $NADP^+$ reduction in the presence of glucose, hexokinase, and glucose-6-P dehydrogenase.

ATPase and ATP-$^{32}P_i$ Exchange Assays: ATPase activity is measured as the release of inorganic phosphate from ATP at 30° at pH 8.0 in the presence of 5 m*M* $MgCl_2$ and an ATP regenerating system. The reaction is not completely specific for Mg^{2+}, which can be replaced by Mn^{2+} or Ca^{2+}. Inorganic phosphate is measured by the method of Taussky and Shorr.[6]

The reaction mixture to measure ATP-$^{32}P_i$ exchange is essentially that for photophosphorylation, except that glucose, hexokinase, and ADP are omitted and 5 m*M* ATP is included. Since less than 15% of the ATP

[5] M. Avron, *Biochim. Biophys. Acta* **40,** 257 (1960).
[6] H. Taussky and E. Shorr, *J. Biol. Chem.* **202,** 675 (1953).

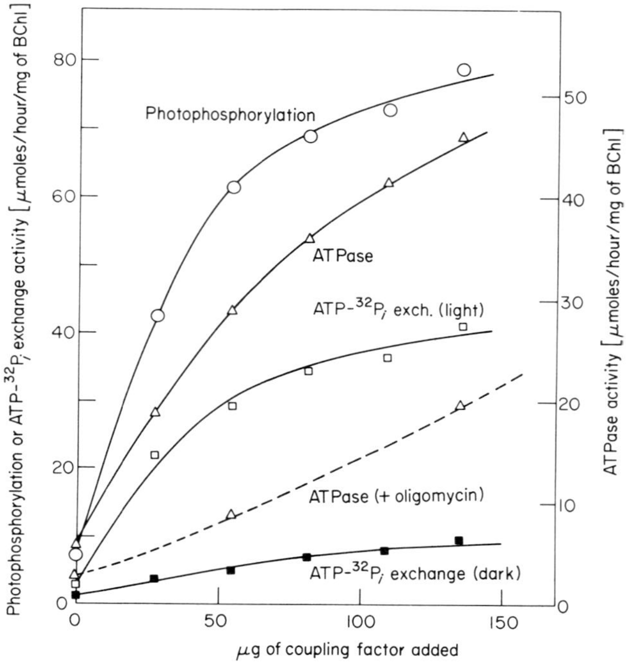

FIG. 1. Restoration of photophosphorylation, ATPase, and ATP-^{32}P exchange in uncoupled membranes (42 μg of BChl) by purified coupling factor (Step 4). Oligomycin (5 μg/ml) was added after preincubation of the coupling factor with the particles and before the addition of ATP, phosphoenol pyruvate, and pyruvate kinase.

present is hydrolyzed during the assay, the specific activity of AT^{32}P formed was assumed to be the same as that of $^{32}P_i$ added.

Properties: The reconstitution of the photophosphorylation, ATPase, and ATP-$^{32}P_i$ exchange (in light and in dark) activities by purified coupling factor (Step 4) is shown in Fig. 1. The reconstituted activities are all sensitive to the energy transfer inhibitor, oligomycin, as are the native ones; reconstituted ATPase is stimulated by DNP (10^{-3} *M*) and FCCP (10^{-6}*M*).

The purified coupling factor from *Rps. capsulata* can be stored at —70° in the presence of 4 m*M* ATP, and it is considerably stable for some weeks. However, in the absence of ATP or in the presence of only 1 m*M* ATP, the activity decreases over a period of 5–8 hours and more rapidly at 37° than at 0°.

The ATPase activity of the soluble factor is rather low, but it is purified, along with the coupling factor activity, reaching values of about

7–8 μmoles/hour/mg of protein. However, although both peaks, obtained after the Sepharose column chromatography, are active in stimulating photophosphorylation and ATPase activity in uncoupled particles, only the high molecular weight peak shows an ATPase activity with an increased specific activity comparable to that of the coupling factor. This could suggest that the low molecular weight peak is an inactive subunit form of the other and that its activity is manifested only when the protein binds to the membranes and probably reaggregates. This ATPase activity was found to be insensitive to oligomycin.

[51] ATP–ADP Exchange Enzyme from Spinach Chloroplasts

By JOSEPH S. KAHN

Assay Method

Principle. The enzyme catalyzes the transfer of the terminal (γ) phosphate from ATP to ADP.[1] When ^{14}C-labeled ADP is used as the acceptor, the rate of the reaction can be calculated from the distribution of label between ATP and ADP as a function of time, provided the reaction has not proceeded to equilibrium (uniform labeling). The method is a simplification of that described by Wadkins and Glaze[2] for mitochondrial exchange enzyme, in that the reaction mixture does not have to be chromatographed quantitatively since only the ratio of label in ATP to that in ADP is required.

The use of labeled ADP is preferred over that of labeled ATP since any breakdown products of the former do not interfere with the assay, while the latter gives rise upon storage to labeled ADP. This would give a large zero-time value for ADP–^{14}C and requires a correction in the rate calculation.

Reagents for Assay

ADP–^{14}C, 10 mM, pH 8.0, containing 0.2–0.4 μCi^{14}C/μmole
ATP, 10 mM, pH 8.0
$MgCl_2$, 50 mM
Tris·HCl, 5 mM, pH 8.0

Enzyme preparation equivalent to 0.5 μg of purified enzyme/0.01 ml.

[1] J. S. Kahn and A. T. Jagendorf, *J. Biol. Chem.* **236**, 940 (1961).
[2] C. L. Wadkins and R. P. Glaze, this series, Vol. X [86].

Procedure. The reaction mixture contains 0.005 ml of 50 mM $MgCl_2$, 0.01 ml of 10 mM ATP, 0.005 ml of 10 mM ADP–^{14}C, and enough 5 mM Tris buffer to give, together with the enzyme preparation, a final volume of 0.08 ml.

Short, wide test tubes (15 × 85 mm) are the most convenient since it is easy to pipette the small volumes into the bottom without touching the sides. The enzyme preparations are pipetted first into the tubes which are kept in an ice bath. After adding the reaction mixture, the tubes are transferred to a 36° water bath and incubated for 4 minutes. The tubes are returned to the ice bath; a few drops of ethanol are added immediately; and then the tubes are placed in a boiling water bath for 2–3 minutes.

Chromatography. Nucleotides are separated most readily by chromatography on Whatman No. 1 paper, using Pabst solvent system No. 1 (isobutyric acid:conc. ammonia:water, 66:1:33).[3] If whole chloroplasts or chloroplast fragments are assayed, applied spots should be washed with dry acetone before developing. This will prevent streaking caused by the high amount of lipid-soluble material.

Application of the reaction mixture to the paper does *not* have to be quantitative. With the aid of disposable Pasteur pipettes, just enough is applied to be clearly visible by UV. After chromatographic development, the paper is dried, and the spots corresponding to ATP and ADP are marked, cut out, and counted directly.

Calculations. The reaction rate, ν, can be calculated as follows:

$$\nu = \frac{(\text{ATP})(\text{ADP})}{(\text{ATP}) + (\text{ADP})} \times \frac{2.3}{t}\left[\log \frac{1}{1 - (n_t/n_\infty)} - \log \frac{1}{1 - (n_0/n_\infty)}\right] \quad (1)$$

where

$$n_\infty, \text{ label distribution in ATP at equilibrium} = \frac{(\text{ATP})}{(\text{ATP}) + (\text{ADP})}$$

$$n_t, \text{ label in ATP at time } t = \frac{\text{cpm ATP}_t}{\text{cpm ATP}_t + \text{cpm ADP}_t}$$

and

$$n_0, \text{ label in ATP at time zero (if any)} = \frac{\text{cpm ATP}_0}{\text{cpm ATP}_0 + \text{cpm ADP}_0}$$

For a given stock solution of ATP and ADP–^{14}C, the following are calculated constants:

[3] P-L Laboratories, Milwaukee, Wisconsin, 1956 (Circular OR-10).

$$R_1 = \frac{(\text{ATP})(\text{ADP})}{(\text{ATP}) + (\text{ADP})} \times 2.3 \tag{2}$$

$$R_2 = \log \frac{1}{1 - (n_0/n_\infty)} \tag{3}$$

The specific enzymatic activity can be calculated as follows:

$$\text{sp. act.} = \frac{R_1}{t} \times \frac{\text{vol (ml)}}{\text{mg of protein}} \times \left[\log \frac{1}{1 - (n_t/n_\infty)} - R_2\right] \tag{4}$$

and is expressed as μmoles per milligram of protein per minute.

Purification of the ATP–ADP Exchange Enzyme

Step 1. Preparation of Extract. Spinach chloroplasts from 3 kg of leaves are isolated according to Jagendorf and Avron,[4] using a grinding medium containing 0.4 *M* sucrose, 25 m*M* Tris·HCl, pH 8.0, and 10 m*M* NaCl. The pellet is resuspended in 2 m*M* Tris buffer, pH 8.0, to a chlorophyll concentration of 0.5–1 mg/ml. The suspension is transferred to a blendor jar, two ice cubes of distilled water are added, and the mixture is blended for 2 minutes at full speed. The suspension is left to stand in the cold with occasional stirring for 2 hours. It is then centrifuged at 100,000 *g* for 1 hour. The pellet is discarded, and the extract is pooled and dialyzed overnight against 2 m*M* Tris buffer, pH 8.0.

The discarded pellet retains about 20–30% of the enzyme, and most of this can be recovered by a second extraction with 2 m*M* Tris buffer.

Step 2. Acetone Precipitation. The dialyzed extract is clarified by centrifugation for 20 minutes at 35,000 *g*. Ice-cold acetone is then added to a final concentration of 90% (v/v) and left to stand in the cold overnight. If no precipitate appears within 5–6 hours, one or two drops of a concentrated sucrose solution are added to the acetone with vigorous stirring, causing the precipitate to appear immediately. The precipitate is collected by centrifugation, resuspended in 200 ml of 2 m*M* Tris buffer, pH 8.0, and dialyzed overnight against the same buffer. The extract is then centrifuged for 20 minutes at 35,000 *g*, and the pellet is discarded.

Step 3. DEAE Chromatography. The supernatant solution is applied to a 25 × 100 mm DEAE–cellulose column, equilibrated with 10 m*M* potassium phosphate, pH 7.2. It is eluted with a linear gradient of potassium phosphate (pH 7.2) from 10 m*M* to 0.3 *M*; total volume, 500 ml. The enzyme elutes between 0.08 and 0.12 *M*.

Step 4. First Sepharose Column. The pooled activity peak is concentrated in an ultrafiltration cell equipped with a membrane of 10,000-MW exclusion limit to a volume of <10 ml. It is then passed through a 25 ×

[4] A. T. Jagendorf and M. Avron, *J. Biol. Chem.* **231**, 277 (1958).

1000 mm agarose gel (Sepharose 6B)[5] column equilibrated with 0.1 *M* NaCl + 5 m*M* Tris·HCl, pH 8.0. The enzyme activity is eluted just prior to a peak of visibly yellow protein.

Step 5. Second Sepharose Column. The active peak is pooled, concentrated to a volume of <3 ml and reapplied to the agarose column as above. A summary of the purification achieved is given in the table. The

PURIFICATION OF THE ATP–ADP EXCHANGE ENZYME FROM SPINACH CHLOROPLASTS[a]

Step	Total activity (μmole/min)	Specific activity[b]	% Yield
1. Blended chloroplasts	936	0.31	100
100,000 *g* supernatant	845	0.72	90
2. Extract from 90% acetone precipitate	636	1.03	68
3. Peak off DEAE–cellulose	282	4.59	30
4. Peak off first Sepharose 6B column	174	5.87	19
5. Peak off second Sepharose 6B column	151	11.8	16

[a] Chloroplasts from 3 kg of spinach leaves.
[b] Micromoles per milligram of protein per minute.

apparent loss of activity on DEAE–cellulose is due to the fact that only the peak tubes were taken for further purification.

The purified enzyme contains four or five minor impurities, as revealed by electrophoresis on acrylamide disc gels. Most of these can be removed by a second passage of the enzyme through DEAE–cellulose. This second passage, however, entails again a large loss of total enzyme activity.

Properties of the Enzyme

Molecular Weight. As estimated from its migration in the agarose column, the Stokes radius of the exchange enzyme corresponds to a molecular weight of 146,000 ± 5000.

Stability. The enzyme is unstable in the presence of concentrated salt solutions, such as 0.5 *M* NaCl or 0.3 *M* K_2HPO_4; consequently, it cannot be fractionated with ammonium sulfate. It is stable in acetone or butanol, provided the salt concentration is low. It is stable to freezing around pH 8, but not at pH <7.

Substrate Specificity. ATP and ADP are the preferred donors and acceptors, the rate of phosphate transfer between other nucleotide di- and triphosphates being <20% that between the adenine nucleotides.[1]

[5] Pharmacia Fine Chemicals, Inc., Piscataway, New Jersey 08854.

The enzyme will not react with pyrophosphate or mononucleotides, nor do these species inhibit the enzyme-catalyzed ATP–ADP exchange. No ATP–P_i exchange, myokinase, 3-phosphoglycerate kinase,[6] pyrophosphatase, ATPase, or other phosphatase or kinase activities are present in the purified enzyme.[1,6]

Metal Requirements. The enzyme has an absolute requirement for a divalent cation, with Mg^{2+}, Mn^{2+}, and Fe^{2+} being equally active. The activity with Co^{2+} and Ca^{2+} is about 60% of that with Mg^{2+}. Optimal concentration for the cations is 2–5 m*M*.

pH Optimum. The enzyme has a broad pH optimum, which peaks between pH 7.5 and 8.0.

Inhibitors. The enzyme is strongly inhibited by GDP, but only slightly by other nucleotide diphosphates.[1] *p*-Hydroxymercuribenzoate and hydrosulfite inhibit the enzyme, while iodoacetate has no effect. The enzyme is inhibited about 50% by 1 m*M* ammonia, butylamine, or ethylamine.

Substrate Concentration. The nature of the assay and the fact that both substrates are also apparently mutual inhibitors makes it difficult to determine the exact optimal concentrations of ATP and ADP for the reaction. The concentrations suggested in the assay (0.6 m*M* ADP, 1.2 m*M* ATP) appear to give maximal activity as well as good resolution by the assay method.

Note. This method of purification is applicable to isolation of the analogous enzyme from spinach, romaine lettuce, Boston ivy, Swiss chard, parsley, collards, kale, pine, and soybeans. The enzyme from chloroplasts of *Euglena gracilis* cannot be purified by acetone precipitation but can be fractionated with ammonium sulfate, followed by chromatography on DEAE-cellulose.

Chloroplast extracts of higher initial specific activity (1.0–1.8 μmoles/mg of protein/minute) can be obtained by lysing the chloroplasts in 2 m*M* Tris buffer without blending. In this case, however, the yield is only 40–50% as compared to 90% with blending.

[6] J. S. Kahn, *Biochim. Biophys. Acta* **79,** 421 (1964).

[52] Adenosine Diphosphoribose Phosphorylase from *Euglena gracilis*[1]

By WILLIAM R. EVANS

$$\text{Ad-R-P-P-R}^{2-} + \text{HPO}_4^{2-} \rightarrow \text{Ad-R-P-P}^{3-} + \text{R-5-P}^{2-} + \text{H}^+ \quad (1)$$

$$\text{Ad-R-P-P}^{3-} + \text{H}^{32}\text{PO}_4^{2-} \rightleftarrows \text{Ad-R-P-}^{32}\text{P}^{3-} + \text{HPO}_4^{2-} \quad (2)$$

Assay Method

Principles. Adenosine diphosphoribose phosphorylase catalyzes the irreversible phosphorolysis of ADPR to ADP and R-5-P.[1a,2] The enzyme also catalyzes an exchange reaction between ADP and P_i. The activity of either the phosphorolytic or exchange reaction can be determined by following the increase in radioactivity associated with the organic phosphate fraction using $^{32}P_i$. A more rapid spectrophotometric assay can be utilized in the phosphorolytic reaction by coupling the ADP produced with myokinase, hexokinase, and glucose-6-phosphate dehydrogenase and measuring the rate of NADPH formation.

Reagents

Tris buffer (Cl^-), 0.5 M, pH 7.8
EDTA ($2Na^+$), 10 mM, pH 7.8
ADP (Na^+), 30 mM
$KH_2{}^{32}PO_4$, 30 mM (approximately 2–5 $\times 10^5$ cpm/μmole)
ADPR, 10 mM
$HClO_4$, 4 N
$(NH_4)_6Mo_7O_{24} \cdot 4\ H_2O$, 80 m$M$
Triethylamine·HCl, 0.8 M, pH 5
Glucose, 10 mM
$MgCl_2$, 30 mM
NADP, 2.5 mM
Myokinase, Sigma, suspension in $(NH_4)_2SO_4$
Hexokinase, Sigma, suspension in $(NH_4)_2SO_4$
Glucose-6-phosphate dehydrogenase, Sigma, suspension in $(NH_4)_2SO_4$

Procedures for ^{32}P–PO_4 Assay. Both the phosphorolytic and exchange reactions result in the formation of β-labeled ^{32}P–ADP. The reaction

[1] Contribution No. 418 from the Charles F. Kettering Research Laboratory.
[1a] W. R. Evans and A. San Pietro, *Arch. Biochem. Biophys.* **113,** 236 (1966).
[2] A. I. Stern and M. Avron, *Biochim. Biophys. Acta* **118,** 577 (1966).

velocity is followed by determining the rate of incorporation of $^{32}P_i$ into the organic phosphate fraction at 30°. The exchange reaction mixture contains the following components in micromoles: Tris(Cl^-), pH 7.8, 50; EDTA, 1; ADP, 3; and $KH_2{}^{32}PO_4$, pH 7.8, 3 in a final volume of 1.0 ml. The components of the phosphorolytic reaction mixture are identical to those of the exchange reaction mixture except that 1 μmole of ADPR is substituted for the 3 μmoles of ADP. The reaction is initiated by the addition of the enzyme and terminated by the addition of 0.05 ml of 4 *N* $HClO_4$. Inorganic phosphate is precipitated by the addition of 0.25 ml of 80 m*M* ammonium molybdate and 0.05 ml of 0.8 *M* triethylamine hydrochloride, pH 5, as described by Sugino and Miyoshi[3] After removal of the triethylamine phosphomolybdate complex by centrifugation, an aliquot of the clear supernatant fraction is plated and the radioactivity determined. Avron's[4] method for determining the radioactivity incorporated into the organic phosphate fraction can also be employed to measure the enzymatic activity.

The reaction mixture for the optical assay contains in micromoles: Tris buffer, pH 7.8, 50; glucose, 10; $MgCl_2$, 3; KH_2PO_4, 4; ADPR, 0.1; NADP, 0.25; and excess myokinase, hexokinase, and glucose-6-phosphate dehydrogenase in 1 ml. The increase in absorbancy at 340 nm is determined at 30°.

It is also possible to measure the phosphorolytic reaction by determining the decrease in P_i during the reaction by any standard assay for P_i.

Purification Procedure

Euglena gracilis strain Z was grown heterotrophically in Hunter's[5] medium in the dark at 25°, or autotrophically in the light in the same medium by omitting the carbon sources. CO_2 (5%) in air was circulated through the cultures for autotrophic growth which were illuminated under 300 foot-candles of daylight fluorescent light. Cells were harvested by low speed centrifugation and washed once with 50 m*M* Tris (Cl^-) buffer, pH 7.5, containing 1 m*M* EDTA.

Step 1. The frozen pellet of whole cells was thawed, refrozen in dry ice–acetone and allowed to thaw again. The cell paste was extracted with 10 m*M* Tris, pH 8, containing 1 m*M* EDTA for 30 minutes at room temperature. If necessary, the pH was maintained at 8 with 1 *N* NH_4OH during extraction. All subsequent manipulations were carried out at 0°–5°.

[3] Y. Sugino and Y. Miyoshi, *J. Biol. Chem.* **239,** 2360 (1964).

[4] M. Avron, *Biochim. Biophys. Acta* **40,** 257 (1960).

[5] S. H. Hunter, M. K. Bach, and G. I. Ross, *J. Protozool.* **3,** 101 (1956).

Step 2. The suspension was centrifuged at 3500 *g* for 15 minutes and the supernatant fluid was saved. The precipitate was suspended in 10 m*M* Tris, pH 8, containing 1 m*M* EDTA, and centrifuged as before. The supernatant fluid was combined with the first supernatant solution and the final precipitate was discarded. One-tenth volume of a 5% solution of streptomycin sulfate was added with stirring to the combined supernatant solution. After 15 minutes, the suspension was centrifuged at 10,000 *g* for 10 minutes. The supernatant fluid was saved, and the precipitate was discarded.

Step 3. The pH of the supernatant fluid was adjusted to 8 if necessary with 1 *N* NH_4OH and solid ammonium sulfate was added to 45% saturation. After 10 minutes, the suspension was centrifuged at 10,000 *g* for 10 minutes and the precipitate was discarded. Additional ammonium sulfate was added to the supernatant fluid to bring it to 65% saturation. The resulting precipitate was collected by centrifugation at 10,000 *g* for 10 minutes and dissolved in 10 m*M* Tris buffer, pH 8.5, containing 1 m*M* EDTA, and the solution was dialyzed against 8 liters of the same buffer.

Step 4. Up to 500 mg of the dialyzed fraction (45–65% ammonium sulfate) was applied to a DEAE–cellulose column (60 × 2 cm) which had been previously equilibrated with 10 m*M* Tris, pH 8.5, containing 1 m*M* EDTA. The enzyme was eluted from the column with 0.1 *N* NaCl in 10 m*M* Tris, pH 8, containing 1 m*M* EDTA. The fractions possessing a high specific activity were combined, dialyzed overnight against 4 liters of 10 m*M* Tris, pH 8, containing 1 m*M* EDTA, and applied to another DEAE column (20 × 2 cm) previously equilibrated with the same buffer. The enzyme was eluted from the column with a linear gradient of increasing NaCl concentration (0–0.5 *M* NaCl in 10 m*M* Tris, pH 8, containing 1 m*M* EDTA).

Step 5. The fractions with the highest specific activity were combined and solid ammonium sulfate was added to 50% saturation. After 30 minutes, the suspension was centrifuged at 10,000 *g* for 10 minutes, and the precipitate was discarded. Additional ammonium sulfate was added to 65% saturation and the suspension was centrifuged as before. The precipitate was dissolved in a minimal amount of 10 m*M* Tris buffer, pH 8, containing 1 m*M* EDTA.

The summary of a typical purification is given in Table I. In general, the purification achieved is about 50-fold, although, in some cases, purification of 100-fold has been achieved. The purified enzyme is fairly stable when frozen so long as EDTA is present.

Properties

Specificity. In the phosphorolytic reaction both ADPR and IDPR are utilized as substrates; no activity was observed with adenosine tri-

TABLE I
PURIFICATION OF ADENOSINE DIPHOSPHORIBOSE PHOSPHORYLASE FROM *Euglena gracilis*

Step	Total protein[a] (mg)	Total activity[b] (units)	Specific activity (units/mg of protein)	Yield (%)
1. Cell extract	4760	88	0.02	(100)
2. 45–60% $(NH_4)_2SO_4$ fraction	455	67	0.15	76
3. First DEAE column eluate	49	14	0.29	16
4. Second DEAE column eluate	7	6	0.89	8
5. 50–60% $(NH_4)_2SO_4$ fraction	3	4	1.3	4.5

[a] Protein determined by a biuret method as described by Zamenhof (Vol. III, p. 696).
[b] One unit is defined as that amount of enzyme which catalyzes the incorporation of 1.0 μmole of P_i minute under the conditions of the assay.

phosphoribose, ADPG, or UDPG. Only ADP serves as a suitable substrate in the exchange reaction; IDP reacts at only 16% of the rate observed with ADP and other nucleoside diphosphates were inactive.

pH Optima. Maximum activities are observed for both the phosphorolytic and exchange reactions at pH 7.8–8.0. The acid slopes of the pH curves appear to be identical while the alkaline slope for the exchange reaction is much steeper than for the phosphorolytic reaction. The slow decline in activity noted on the alkaline side of the pH curve

TABLE II
K_m AND K_i VALUES FOR SUBSTRATES AND INHIBITORS OF ADPR PHOSPHORYLASE[1a,2]

	Phosphorolysis	Exchange
	K_m's	
ADPR	$4–5 \times 10^{-5}$	—
ADP	—	6×10^{-4}
P_i	$4–5 \times 10^{-4}$	$4–5 \times 10^{-4}$
	K_i's	
R-5-P[a]	9×10^{-4}	1.1×10^{-4}
AsO_4[a]	4.1×10^{-4}	5.7×10^{-4}
5′-AMP[b]	1.0×10^{-3}	1.5×10^{-3}
ATP[b]	2.9×10^{-3}	1.3×10^{-3}
R-5-P[b]	4×10^{-3}	

[a] P_i concentration varied.
[b] Either ADPR or ADP concentration varied.

for the phosphorolytic reaction is perhaps due to the liberation of a secondary phosphate group during the reaction.

Kinetic Properties, Inhibitors, and Activators. The K_m and K_i values for the substrates and inhibitors of the enzyme are shown in Table II. In the presence of arsenate and ADPR the reaction products are 5′-AMP and R-5-P. 5′-AMP is the reaction product in the presence of arsenate in the exchange reaction. These observations would seem to suggest the intermediate formation of an enzyme-5′-AMP complex. Kinetic analysis at nonsaturating levels of ADPR and P_i indicate that a "ping-pong" type of mechanism,[6] in which the first substrate adds to, and the first product leaves, the enzyme surface before the second substrate is bound, is involved in the enzymatic reaction. ADPR acts as a competitive inhibitor of P_i in the phosphorolytic reaction but ADP has no effect on the K_m of P_i in the exchange reaction.

The enzyme is reversibly inhibited by *p*-CMB. Ribose,[1a] ribose-1-phosphate,[1a] ribulose-1,5-diphosphate,[1a] glucose-1-phosphate,[1a] glucose-6-phosphate,[1a] adenosine,[2] deoxyadenosine,[2] 3′-AMP,[2] CMP,[2] CDP,[2] UDP,[2] NADH,[2] NADP,[2] and NAD[2] did not inhibit the enzyme. Deoxyribose-5-phosphate[1a] is inhibitory.

There is apparently no metal requirement for enzymatic activity as no inhibition was observed in the presence of EDTA, cyanide, *o*-phenanthroline, or α,α-dipyridyl.

Distribution. There was some slight activity observable in extracts of *Ochromonas danica,* but no activity could be demonstrated in extracts of *Chlorella, P. luridum, P. persicunum, P. aerugineum, Gloeocapsa,* spinach, or peas.

[6] W. W. Cleland, *Biochim. Biophys. Acta* **67,** 104, 173, 188 (1963).

[53] Ribulose Diphosphate Carboxylase from Spinach Leaves

By MARCIA WISHNICK and M. DANIEL LANE

$$\text{D-Ribulose 1,5-diphosphate} + CO_2 + H_2O \xrightarrow{Mg^{2+}} 2\ \text{D-3-phosphoglycerate} + 2H^+$$

Assay Method

Principle. Ribulose diphosphate carboxylase catalyzes the irreversible carboxylation of ribulose 1,5-diphosphate to form 3-phosphoglycerate with the stoichiometry shown in the reaction above. Carboxylase activity is conveniently determined[1] by following the rate of incorporation of $H^{14}CO_3^-$ into phosphoglycerate (acid–stable ^{14}C activity). Alternatively, activity can be determined in the presence of 3-phosphoglycerate kinase, ATP, glyceraldehyde-3-phosphate dehydrogenase, and NADH by following the rate of NADH oxidation spectrophotometrically.[2] This enzyme is widely distributed in plant tissues[3,4] and microorganisms.[5,6]

Reagents

Tris (Cl^-) buffer, 1.0 *M*, pH 7.8 at 25°

$KH^{14}CO_3$, 0.5 *M* (approximately 10^5 cpm/μmole; the specific activity must be accurately determined)

D-Ribulose 1,5-diphosphate-Na_4, 5 m*M*, synthesized enzymatically from D-ribose-5-P[7,8] and assayed by its quantitative conversion to 3-phosphoglycerate in the presence of the constituents of the bicarbonate-^{14}C or spectrophotometric assays and excess ribulose diphosphate carboxylase. The quantity of 3-phosphoglycerate formed is determined either from acid–stable ^{14}C activity formed or NADH oxidized.[2] Ribulose diphosphate can be chromatographically purified as described by Wishnick and Lane[9]

$MgCl_2$, 0.5 *M*

GSH, 0.1 *M*

EDTA (Na), 0.6 m*M*, pH 6.5

Liquid scintillator, 0.25 g of 1,4-bis [2(5-phenyloxazolyl)] benzene (POPOP), 10 g of 2,5-diphenyloxazole (PPO), and 100 g of recrystallized napthalene per liter of dioxane.

Procedure: Bicarbonate-^{14}C Fixation Assay. The complete reaction mixture contains the following components (in micromoles except as indicated): Tris (Cl^-), 100; D-ribulose-1,5-diphosphate, 0.25; $KH^{14}CO_3$ (approximately 2 μCi), 25; $MgCl_2$, 5; EDTA, 0.03; GSH, 3; and ribulose diphosphate carboxylase in a total volume of 0.5 ml. Prior to use, the carboxylase is dialyzed overnight at 4° against 10 m*M* Tris (Cl^-), pH 7.6, containing 0.1 m*M* EDTA and 10 m*M* 2-mercaptoethanol. The final

[1] J. M. Paulsen and M. D. Lane, *Biochemistry* **5**, 2350 (1966).
[2] E. Racker, *in* "Methods of Enzymatic Analysis" (H. U. Bergmeyer, ed.), p. 188. Academic Press, New York, 1963.
[3] C. R. Slack and M. D. Hatch, *Biochem. J.* **103**, 660 (1967).
[4] F. J. Kieras and R. Haselkorn, *Plant Physiol.* **43**, 1264 (1968).
[5] L. E. Anderson, G. B. Price, and R. C. Fuller, *Science* **161**, 482 (1968).
[6] T. Akazawa, K. Sato, and T. Sugiyama, *Arch. Biochem. Biophys.* **132**, 255 (1969).
[7] B. L. Horecker, J. Hurwitz, and P. K. Stumpf, this series, Vol. 3, p. 193.
[8] B. L. Horecker, J. Hurwitz, and A. Weissbach, *Biochem. Prep.* **6**, 83 (1958).
[9] M. Wishnick and M. D. Lane, *J. Biol. Chem.* **244**, 55 (1969).

pH of the reaction mixture is 7.9 at 25°. After a 6-minute incubation at 30°, the reaction is terminated by the addition of 0.2 ml of 6 N HCl. A 0.1-ml aliquot is pipetted into a 15 × 45 mm glass vial (flat bottom shell vial, Demuth Glass Division, catalog number 15045 AW) and is dried in a forced-draft oven for 1 hour at 95°. Water (0.3 ml) is added to each vial followed by 3 ml of liquid scintillator. The vial is capped, inserted into a standard 25-ml polyethylene scintillation counting bottle, and acid–stable ^{14}C activity (as phosphoglycerate-^{14}C-3) determined with a liquid scintillation spectrometer.[10] The carboxylation rate is proportional to enzyme concentration up to a level of approximately 0.012 unit; proportionality to enzyme concentration is maintained until at least 36% of the added ribulose diphosphate has been consumed. The carboxylation reaction follows zero-order kinetics, under the conditions described, after a 1 minute lag period.

Definition of Unit and Specific Activity. One unit of ribulose diphosphate carboxylase is defined as that amount of enzyme which catalyzes the carboxylation of 1.0 μmole of ribulose diphosphate per minute under the assay conditions described. Protein is determined spectrophotometrically[1]: an absorbance (1-cm light path) of 1.00 at 280 mm for the pure enzyme is equivalent to 0.61 mg of protein (see molecular properties). Specific activity is expressed as units per milligram of protein.

Purification Procedure

The purification procedure described is based on that of Paulsen and Lane[1] with some modifications. All the operations are carried out at 4°. The results of the procedure are summarized in the table.

Initial Extract. Fresh spinach, purchased locally, is destemmed and washed with cold tap water. Six hundred grams of leaves are homogenized with 2 liters of 10 mM potassium phosphate buffer, pH 7.6, containing 0.1 mM EDTA, in a 4-liter capacity Waring Blendor for 1 minute at medium speed. The resulting suspension is filtered through Schleicher and Schuell (Keene, New Hampshire) No. 588 fluted filter paper. The filtrate, referred to as the initial extract, is adjusted if necessary to pH 7 with 5 N NH_4OH.

Ammonium Sulfate Fractionation. The initial extract is brought to 37% saturation with solid ammonium sulfate (226 g/liter of extract) and after standing for 1 hour is centrifuged at 9000 g for 30 minutes. The dark green precipitate is discarded, and the supernatant solution is brought to 50% saturation with solid ammonium sulfate (92.5 g/liter). After centrifugation as described above, the precipitate is dissolved in

[10] The inner vials may be discarded and the polyethylene bottles reused.

PURIFICATION OF RIBULOSE DIPHOSPHATE CARBOXYLASE FROM SPINACH LEAVES

Step of purification	Total Protein (g)	Total activity (units[c])	Specific activity (units[c]/mg of protein)	Yield (%)
1. Initial extract from 600 g of leaves	10.12[a]	3542	0.35	100
2. Ammonium sulfate I	5.17[a]	2740	0.53	77
3. Ammonium sulfate II	2.50[a]	1975	0.79	56
4. DEAE-cellulose chromatography	0.97[b]	1067	1.10	30
5. Hydroxyapatite chromatography	0.59[b]	873	1.48	25

[a] Determined by the method of O. H. Lowry, N. J. Rosebrough, A. L. Farr, and R. J. Randall, *J. Biol. Chem.* **193,** 265 (1951).
[b] Determined from absorbance at 280 nm × 0.61.
[c] Determined by the bicarbonate-^{14}C fixation assay.

0.1 *M* potassium phosphate buffer, pH 7.6, containing 0.1 m*M* EDTA and 10 m*M* 2-mercaptoethanol (100 ml of buffer/2 liters of initial extract). This solution (37–50% saturated ammonium sulfate fraction), referred to as ammonium sulfate I, is stored at 4° overnight. It is then diluted with an equal volume of glass-distilled water and saturated ammonium sulfate[11] (pH 7.0, 41 ml/100 ml of diluted ammonium sulfate I) is added. EDTA and 2-mercaptoethanol are added to bring their concentrations to 0.1 m*M* and 10 m*M*, respectively. After centrifugation as described above, the precipitate is discarded and saturated ammonium sulfate[11] (pH 7.0, 8.6 ml/100 ml) is added to the supernatant solution. Following centrifugation, the precipitate is again discarded and additional saturated ammonium sulfate[11] (pH 7.0, 9.5 ml/100 ml) is added to the supernatant solution. The precipitate recovered after centrifugation is dissolved in 0.1 *M* potassium phosphate buffer, pH 7.6, containing 0.1 m*M* EDTA, 10 m*M* 2-mercaptoethanol, and 1 m*M* dithiothreitol (27 ml/100 ml of ammonium sulfate I before dilution with water). This solution, which constitutes a 38–45% saturated ammonium sulfate fraction, is divided into four equal fractions and can be stored at −20° for 6–12 months with relatively little loss of activity.

Although the following steps are given for the purification of the total 38–45% saturated ammonium sulfate solution, the procedures can be readily "scaled down," permitting the preparation of smaller quanti-

[11] Ammonium sulfate is saturated at room temperature and neutralized with NH_4OH so that, when diluted 5-fold, the pH is that indicated.

ties of enzyme from each of the four equal fractions described above. The four fractions are thawed and centrifuged at 15,000 *g* for 10 minutes, and the supernatant solution is applied to a Sephadex G-25 column (4.5 × 42 cm), previously equilibrated with 5 m*M* potassium phosphate buffer, pH 7.6, containing 0.1 m*M* EDTA and 10 m*M* 2-mercaptoethanol. The enzyme is eluted with the same buffer. The light brown fastest moving band, which is clearly resolved from a bright yellow, slower moving band, is collected visually and contains about 90% of the protein applied to the column. This fraction, referred to as ammonium sulfate II, is immediately subjected to DEAE–cellulose chromatography.

DEAE–Cellulose Chromatography. A DEAE–cellulose column (4.5 × 42 cm; exchange capacity 0.8 meq/g; type 20 diethylaminoethyl cellulose, Schleicher and Schuell) is equilibrated with 0.5 *M* potassium phosphate buffer, pH 7.6, then washed with 5 m*M* potassium phosphate buffer, pH 7.6. Prior to use, the column is further washed with 5 m*M* potassium phosphate buffer, pH 7.6, containing 0.1 m*M* EDTA and 10 m*M* 2-mercaptoethanol. The enzyme solution (approximately 2.5 g of protein) from the preceding step is diluted with an equal volume of 5 m*M* potassium phosphate buffer, pH 7.6, containing 0.1 m*M* EDTA and 10 m*M* 2-mercaptoethanol and applied to the DEAE–cellulose column. After the application of an additional 200 ml of the same 5 m*M* phosphate buffer, elution is accomplished with a 4-liter linear potassium phosphate gradient (5 m*M* to 250 m*M*, pH 7.6, containing 0.1 m*M* EDTA and 10 m*M* 2-mercaptoethanol). The column effluent is continuously monitored for protein and is collected fractionally. Fractions exhibiting peak ribulose diphosphate carboxylase activity (using the bicarbonate-^{14}C fixation assay) and containing approximately 50% of the total activity applied to the column are pooled. The pooled enzyme is referred to as the DEAE–cellulose fraction. The enzyme is precipitated by bringing the saturation to 55% with saturated ammonium sulfate[11] (pH 7.0, 122 ml/100 ml of pooled enzyme). Additional EDTA and 2-mercaptoethanol are added to maintain their concentrations at 0.1 m*M* and 10 m*M*, respectively. In preparation for the next step, the enzyme suspension is centrifuged and the precipitate is redissolved in 50 ml of 5 m*M* potassium phosphate buffer, pH 7.6, containing 0.1 m*M* EDTA and 10 m*M* 2-mercaptoethanol. This solution is subjected to gel filtration on a Sephadex G-25 column (4.5 × 42 cm) previously equilibrated with the same 5 m*M* phosphate buffer. The column effluent is monitored from protein, and the protein peak is collected as a single fraction.

Hydroxyapatite Chromatography. The gel-filtered enzyme is applied to a hydroxyapatite column (4.5 × 42 cm; BioRad Bio-Gel HT) previously equilibrated with the same 5 m*M* potassium phosphate buffer.

After application of an additional 500 ml of the 5 m*M* potassium phosphate buffer, the carboxylase is eluted with 25 m*M* potassium phosphate buffer, pH 7.6, containing 0.1 m*M* EDTA and 10 m*M* 2-mercaptoethanol. The column effluent is monitored for protein and ribulose diphosphate carboxylase activity. The enzyme elutes as a distinct peak, and those fractions containing maximum carboxylase activity are pooled (greater than 80% of the total activity applied). These fractions are brought to 55% saturation with saturated ammonium sulfate[11] (pH 7.0, 122 ml/100 ml of pooled volume); EDTA and 2-mercaptoethanol are added to produce final concentrations of 0.1 m*M* and 10 m*M*, resspectively. Ribulose diphosphate carboxylase is stored as a suspension at 0°–2° and retains nearly full activity for about 1 month. The specific activity of the purified carboxylase is 1.4–1.6 units/mg, and these preparations are free of detectable quantities of 5-phosphoriboisomerase and 5-phosphoribulokinase activities.[1]

Properties

Homogeneity and Molecular Characteristics. Ribulose diphosphate carboxylase prepared as described above is homogeneous in the analytical ultracentrifuge[1] and by disc electrophoresis in polyacrylamide gels[12] The carboxylase has a sedimentation coefficient ($s^0_{20,w}$) of 21.0 S.[1] The relation between absorbancy at 280 nm and the refractometrically determined protein concentration is given by the equation $c = 0.61 \ (A^{1\,\text{cm}}_{280\,\text{nm}})$ where c is protein concentration in milligrams per milliliter and A is absorbancy at 280 nm (1.0 cm light path).[1] The absorbancy ratio $A_{280\,\text{nm}}/A_{260\,\text{nm}}$ of the pure enzyme is 1.9. To convert protein concentration determined by the method of Lowry *et al.*[13] using bovine serum albumin as standard to refractometrically determined protein concentration, the former should be multiplied by a factor of 0.79. The molecular weight of the carboxylase determined by the sedimentation equilibrium method[14] is 557,000.[1] The frictional coefficient ratio (f/f_0) of the enzyme calculated from the appropriate molecular parameters is 1.11.[1] This ratio is in the usual range for a protein molecule of spherical shape having typical hydration characteristics (i.e., about 0.2 g of water/g of protein). The spherical shape of the molecule has been verified by electron microscopy.[15] The homogeneous carboxylase is composed of two different kinds

[12] A. C. Rutner and M. D. Lane, *Biochem. Biophys. Res. Commun.* **28,** 531 (1967).
[13] O. H. Lowry, N. J. Rosebrough, A. L. Farr, and R. J. Randall, *J. Biol. Chem.* **193,** 265 (1951).
[14] D. A. Yphantis, *Biochemistry* **3,** 297 (1964).
[15] A. K. Kleinschmidt, M. Wishnick, and M. D. Lane, unpublished observations (1969).

of subunits. The aminoethylated enzyme can be resolved on Sephadex G-100 or G-200 in the presence of 0.5% sodium dodecyl sulfate or 6 *M* guanidine·HCl into two distinct polypeptides which have $s_{20,w}$'s of 3.0 S and 1.8 S respectively, and differing amino acid compositions.[12]

Copper Content. Tightly bound copper has been detected in homogeneous preparations of ribulose diphosphate carboxylase.[16] The enzyme has a typical Cu(II) EPR spectrum ($g\perp \sim g_{max} = 2.09$). Analysis by atomic absorption spectrophotometry, EPR, and neutron activation indicate the presence of 1 gram atom of copper per mole (560,000 *g*) of carboxylase. The bound copper appears to be present as Cu(II). These observations correlate well with the finding that the carboxylase-catalyzed reaction is inhibited by cyanide and that cyanide binds tightly to the enzyme under certain conditions (see Kinetic Properties and Inhibitors). The precise functional role of the copper has not yet been elucidated.

Kinetic Properties and Inhibitors. The pH optimum of the ribulose diphosphate carboxylase-catalyzed reaction is approximately 7.9. The K_m values determined for ribulose 1,5-diphosphate, Mg^{2+}, total "CO_2" ($CO_2 + HCO_3^-$) at pH 7.9 are 0.12 m*M*, 1.1 m*M*, and 22 m*M*, respectively.[1] Mn^{2+}, which can replace Mg^{2+} in the reaction, has a K_m of 39 μM and a V_{max} 56% that with Mg^{2+}.

The active species in the carboxylation of ribulose diphosphate has been shown to be CO_2.[17] Therefore, correction of the Michaelis constant to the concentration of the active species of CO_2 results in a K_m for CO_2 of approximately 0.45 m*M*.[17] The enzyme catalyzes the carboxylation of 1340 moles of ribulose diphosphate per minute per mole of enzyme, or the formation of 2680 moles of 3-phosphoglycerate per minute per mole of enzyme under standard assay conditions.

Ribulose diphosphate becomes inhibitory at concentrations exceeding 0.7 m*M*. Orthophosphate and $(NH_4)_2SO_4$ are competitive inhibitors with respect to ribulose diphosphate and have K_i's of 4.2 m*M* and 3.1 m*M*, respectively.[1] 3-Phosphoglycerate is a noncompetitive inhibitor (K_i = 8.3 m*M*) with respect to ribulose diphosphate and a competitive inhibitor (K_i = 9.5 m*M*) with respect to HCO_3^-.[1] Glyceraldehyde 3-phosphate shows a similar pattern of inhibition; it is a competitive inhibitor (K_i = 22 m*M*) with respect to HCO_3^- and a noncompetitive inhibitor (K_i = 19 m*M*) with respect to ribulose diphosphate.[18] The carboxylase is in-

[16] M. Wishnick, M. D. Lane, M. Scrutton, and A. S. Mildvan, *J. Biol. Chem.* **244**, 5761 (1969).

[17] T. G. Cooper, D. Filmer, M. Wishnick, and M. D. Lane, *J. Biol. Chem.* **244**, 1081 (1969).

[18] M. Wishnick and M. D. Lane, unpublished observations (1969).

hibited 63% by 0.5 mM arsenite in the presence of 0.4 mM BAL. In the absence of BAL, the observed inhibition is 19% while BAL alone causes no inhibition.[18] At levels as high as 10 mM, azide, hydroxylamine, and oxalate are without effect on the carboxylase reaction.[18] The enzyme is inhibited by $HgCl_2$ and *p*-chloromercuribenzoate,[19] as well as by iodoacetamide.[20] Hamamelonic acid 1,5-diphosphate[21] inhibits the carboxylation reaction at low concentrations; 1 μM and 0.1 μM hamamelonic acid 1,5-diphosphate inhibit 90% and 33%, respectively. The inhibition (K_i approximately 0.14 μM) appears to be of mixed character with respect to ribulose diphosphate.[18] The enzyme is reversibly inhibited by cyanide at low concentrations; 10 μM and 0.1 mM cyanide inhibit the carboxylation reaction 51% and 91%, respectively.[9] Inhibition by cyanide ($K_i = 16\ \mu M$) is uncompetitive with respect to ribulose diphosphate and is of mixed character with respect to Mg^{2+} and HCO_3^-. Cyanide binds stoichiometrically to the carboxylase (1 mole of cyanide per mole of enzyme), but only in the presence of ribulose 1,5-diphosphate suggesting the formation of a catalytically inactive ternary complex. One of several possible explanations for these results is the formation of a stable complex between cyanide and a tightly bound metal activator at a site on the enzyme which becomes accessible when ribulose 1,5-diphosphate is bound by the enzyme. This is consistent with the observation that spinach ribulose diphosphate carboxylase contains tightly bound copper (see Copper Content).

Acknowledgments

The methods described in this report were developed with the support of research grants (AM-09117 and AM-09116) from the National Institutes of Health, U.S.P.H.S. M. Wishnick is a Predoctoral Trainee of the U.S.P.H.S. (GM-01234) and M. D. Lane is a Career Development Awardee of the U.S.P.H.S. (K3-AM-1847)

[19] A. Weissbach, B. L. Horecker, and J. Hurwitz, *J. Biol. Chem.* **218,** 795 (1956).
[20] B. R. Rabin and P. W. Trown, *Proc. Nat. Acad. Sci. U.S.* **51,** 501 (1964).
[21] B. R. Rabin, D. F. Shaw, N G. Pon, J. M. Anderson, and M. Calvin, *J. Amer. Chem. Soc.* **80,** 2528 (1958).

[54] Protochlorophyllide Holochrome

By H. W. SIEGELMAN[1] and P. SCHOPFER

$$\text{Protochlorophyllide holochrome} + 2\text{H} \xrightarrow{h\nu} \text{chlorophyllide } a \text{ holochrome}$$

Protochlorophyllide holochrome (PCH) is the protochlorophyllide–protein complex which is photoreduced to chlorophyllide holochrome.[2] It has been obtained primarily from dark-grown bean leaves. An improved isolation and purification procedure was devised to provide reasonable amounts of PCH for experimental studies.[3]

Materials

Bean seedlings (*Phaseolus vulgaris* L. cv. light red kidney; Michigan Bean Corporation, Saginaw, Michigan) were grown in the dark at 24° on vermiculite previously soaked with Kratz–Myers D medium[4] diluted 1:4 with tap water. The primary leaves were harvested under green safelight after 9–11 days of growth (see Vol. XXIV).

Calcium phosphate (brushite) was prepared[5] by dissolving 5 lb of K_2HPO_4 in distilled water and making up to 6.5 liters, and dissolving 4 lb of $CaCl_2 \cdot 2\ H_2O$ in distilled water and making up to 6.16 liters. Into a large vessel (about 20-liter volume), 2 liters of distilled water and 500 ml of the $CaCl_2$ solution were placed. The $CaCl_2$ and K_2HPO_4 solutions were pumped independently and simultaneously at 10 ml/minute into the large vessel which are kept well stirred. Flow rates were checked periodically to ensure that there was always a slight excess of $CaCl_2$ in the large vessel. The product of this reaction is brushite ($CaHPO_4$ 2 H_2O). The brushite was converted to hydroxyapatite [$Ca_{10}(PO_4)_6(OH_2)$] by titration at room temperature with 2 *M* KOH to pH 11 with stirring. Titration was continued until the pH did not drop below 9 on standing. Before use, the hydroxylapatite was washed free of fines by repeated decantation with the buffer to be used to equilibrate the hydroxyapatite column. Fresh hydroxyapatite should be prepared monthly. DEAE–cellulose (DE-11, Whatman) was precycled with 0.5 *N* HCl and 0.5 *N*

[1] Research carried out at Brookhaven National Laboratory under the auspices of the U.S. Atomic Energy Commission.

[2] B. K. Boardman, *in* "The Chlorophylls" (L. P. Vernon and G. R. Seely, eds.), p. 437. Academic Press, New York, 1966.

[3] P. Schopfer and H. W. Siegelman, *Plant Physiol.* **43,** 990 (1968).

[4] W. A. Kratz and J. Myers, *Amer. J. Bot.* **42,** 282 (1955).

[5] H. W. Siegelman, G. A. Wieczorek, and B. C. Turner, *Anal. Biochem.* **13,** 402 (1965).

KOH and then washed with water. It was then titrated with 1 *M* KOH to pH 8.0 in the presence of 0.25 *M* KCl.

The Tricine buffer contained 10 m*M* Tricine, 8 m*M* KOH, 2 m*M* $MgSO_4$ and 1 m*M* Na_2 EDTA. Chromatographic column effluents, except agarose, were monitored with visible and ultraviolet flow monitors and a flow conductivity monitor. Flow rates were maintained with peristaltic pumps. At high flow rates, the flow monitor lamps did not phototransform PCH.

Tricine was prepared by the method of Good.[6] Other chemicals were obtained from the following sources: Triton X-100 from Ruger Chemical Co., polyethylene glycol-6000 from Baker Chemical Co., Polyclar-AT from General Aniline and Film Corp., and agarose: Bio-Gel A 1.5 m, 200–400 mesh from Bio-Rad Laboratory.

Assay Method

The absorption spectra of solutions were measured before and after a saturating red irradiation in 1 cm cells at 10° with a Cary Model 14 spectrophotometer. The spectrophotometric cell was irradiated in the center of a petri dish (6 × 12 cm) filled with a water–ice mixture and covered externally with aluminum foil. There was a 2-cm aperture on one side of the petri dish for the beam of a 150-W lamp (Sylvania DFA). A 3-second irradiation saturated the phototransformation. The relative amount of PCH, subsequently called activity, was calculated from the absorption spectrum before irradiation. A straight line connecting absorbance values at 595 and 670 nm was the baseline. The activity of a preparation was defined as the volume in milliliters times the absorbance at 639 nm measured from the baseline. A sample of 1 ml with $A_{639} = 1.0$ in a 1.0 cm cell has 1.0 unit of activity. The relative amount of chlorophyllide *a* holochrome was calculated from the increase of the absorbance at 678 nm measured after irradiation (A_{678}). The degree of transformation is expressed as the ratio $A_{678}:A_{639}$. Protein determinations were made by the biuret method[7] following trichloroacetic acid precipitation and ethanol washing of PCH.

Extraction and Purification

The PCH is not very stable in 25% glycerol solution, and a rapid purification procedure is essential. All steps of the purification are carried out at 4° under green safelight.

Step 1. Samples of 40 g of leaves were extracted in an Omni-Mixer

[6] N. E. Good, *Arch. Biochem. Biophys.* **96**, 653 (1962).
[7] A. G. Gornall, C. J. Bardawill, and M. M. David, *J. Biol. Chem.* **177**, 751 (1949).

(type OM, Ivan Sorvall) as full speed for 4 minutes with 20 g of Polyclar AT and 160 ml of extraction medium (50 mM Tricine, 50 mM KOH, 2 mM $MgSO_4$, 1 mM Na_2 EDTA, 0.06% (v/v) Triton X-100, 25% glycerol; pH 8.6). The temperature was maintained below 5° by immersing the grinding container in an ice–solid CO_2 bath. The homogenate was clarified by filtration through a 15-cm milk filter disk and centrifuged at 78,000 g at 0° for 1 hour. A lipid layer was aspirated from the top of the centrifuge tubes, and the supernatant solution was retained. It can be stored at −15°.

Step 2. A 50% polyethylene glycol-6000 solution[8] was added to the supernatant solution from step 1 to a final concentration of 15%. The mixture was stirred for 5 minutes, allowed to stand for 30 minutes, and then centrifuged at 48,000 g for 1 hour. The precipitate was dissolved in Tricine buffer and centrifuged at 48,000 g for 30 minutes.

Step 3. A hydroxyapatite column (7.5 × 15 cm) was prepared the day before use and equilibrated with Tricine buffer containing 0.2 M KCl. The PCH solution from step 2 was made 0.2 M with respect to KCl and applied to the column. The column was washed with 2.5 liters of Tricine buffer–0.2 M KCl and the PCH was eluted with Tricine buffer–0.25 M potassium phosphate (pH 8.0). The flow rate was maintained at 20 ml/minute, and all fractions containing PCH were combined.

Step 4. The PCH solution from step 3 was desalted on a Sephadex G-25 column (7.5 × 25 cm) equilibrated with Tricine buffer and then applied to a DEAE-cellulose column (5 × 55 cm) equilibrated with Tricine buffer. The PCH was eluted with a convex exponential gradient using 0.3 liter of 50 mM KCl and 2 liters of 0.25 M KCl in the Tricine buffer. The flow rate was maintained at 20 ml/min and 20-ml fractions were collected. Fractions with an $A_{280}:A_{639}$ ratio between 25 and 28 were pooled and concentrated by ultrafiltration.

Step 5. A portion of the concentrated solution (2 ml) containing about 2 units of PCH was applied to an 8% agarose column (1.75 × 72 cm) equilibrated with the Tricine buffer containing 0.1 M KCl. The same buffer solution was used for elution. The flow rate was maintained at 10 ml/hour with a Mariotte flask, and 2.0-ml fractions were collected.

Yields and specific activities at various stages of purification are given in the table.

Properties

The yield, specific activity, the transformability of PCH is considerably enhanced by extracting the tissue with low concentrations,

[8] A. Polson, G. M. Potgieter, J. F. Largier, G. E. F. Means, and F. J. Joubert, *Biochim. Biophys. Acta* **82,** 463 (1964).

PURIFICATION OF PROTOCHLOROPHYLL HOLOCHROME

Stage of purification	Concentration (units/ml[a])	Transformation ($A_{678}:A_{639}$[b])	Protein (mg/ml)	Specific activity (units/g of protein)	Yield (% units)
1. Crude extract	0.272	1.7	15.3	17.8	100
2. Polyethylene glycol precipitation	0.285	1.8	10.5	27.2	70
3. Hydroxyapatite chromatography	0.110	1.8	3.5	31.5	50
4. DEAE-cellulose chromatography	0.090	2.0	1.6–1.9	47–56	5
5. Agarose chromatography	0.050	1.8	0.43	116	4

[a] One unit/ml = A_{639}/cm before irradiation.

[b] A_{639} before, A_{678} after irradiation.

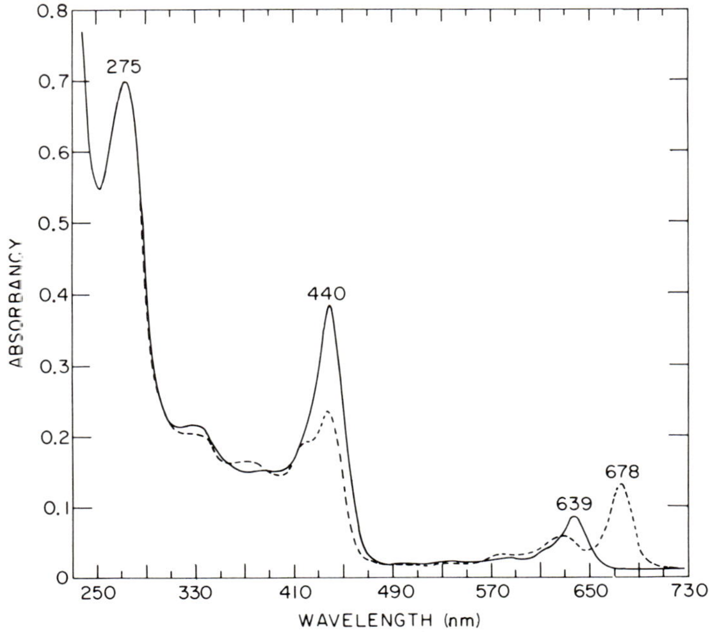

FIG. 1. Absorbance spectra of protochlorophyllide holochrome following agarose chromatography before (———) and after (- - - - - -) phototransformation.

0.05–0.1%, of Triton X-100. Extracted PCH requires the protection afforded by either glycerol or sucrose. High concentrations of sucrose (2 M) markedly increase PCH stability, but the viscosity of the solution is a deterrent to the usual protein purification procedures. The PCH obtained after agarose gel filtration chromatography was judged to be homogeneous based on the constancy of the $A_{280}:A_{639}$ ratio and the presence of only one band on disc electrophoresis at several acrylamide gel concentrations.

The purified PCH has absorbance maxima at 278, 332, 382, 418, 440, and 639 nm. Following exposure to light, the absorbance maxima of chlorophyllide *a* holochrome are at 278, 332, 382, 418, 438, 630, and 678 nm as shown in Figure 1. The 678-nm peak of the newly formed chlorophyllide *a* holochrome shifts slowly toward shorter wavelengths at all stages of purification. There is no spectroscopic evidence for any other chromophore in purified PCH than protochlorophyllide.

Agarose gel filtration resolves PCH into two components of approximately 550,000 and 300,000 daltons each with the same $A_{280}:A_{639}$ ratio. The larger component contains two protochlorophyllide molecules per molecule of protein and the smaller component only one. Apparently PCH is a dissociating system.

Chlorophyllide *a* holochrome and PCH are only weakly red fluorescent under ultraviolet light. Treatment of PCH with 2-mercaptoethanol (50 mM) results in loss of phototransformability in a few hours at 2°, and the solution became strongly light-red fluorescent under ultraviolet light.

[55] Phosphodoxin

By Clanton C. Black, Jr.

$$\text{ADP} + \text{P}_i \xrightarrow[\text{Phosphodoxin + Light}]{\text{Chromatophores or Chloroplasts}} \text{ATP}$$

Photophosphorylation activity usually is determined with chloroplasts or chromatophores in the presence of an exogenously added electron acceptor. The present procedure will describe the isolation and activity of a factor, given the tentative name of phosphodoxin, from photosynthetic organisms. Phosphodoxin is a naturally occurring, water-soluble, heat-stable substance which catalyzes photophosphorylation either by chloroplasts or chromatophores in the absence of other exogenous redox substances.

Assay Method

Principle: Photophosphorylation is assayed by measuring the photochemical formation of $AT^{32}P$ as described below:

Reagents

- Chloroplast or chromatophore suspension, approximately 0.5–1 mg of chlorophyll or bacteriochorophyll per milliliter
- $MgCl_2$, 0.01 *M*
- ADP, 0.01 *M*
- P_i, 0.01 *M*, containing 0.5–1.0 mCi of $^{32}P_i$
- Tris·HCl buffer, pH 7.8
- Phosphodoxin

Procedure: The reaction mixture in a container suitable for illuminating contains in a final volume of 1 ml; 2 μmoles of $MgCl_2$, 1 μmole of ADP, 1 μmole of phosphate + $^{32}P_i$, 50 μmoles of buffer, 10–20 μg of bacteriochlorophyll or chlorophyll as chromatophores or chloroplasts, and various quantities of phosphodoxin. The control reaction mixtures contain all of the components listed above and are kept in the dark. The endogenous photophosphorylation activity of chromatophores and chloroplasts also should be assayed in the reaction mixture given by omitting the phosphodoxin.

The reaction mixtures are irradiated at an intensity of 2000–3000 fc[1] for 1 minute. After illumination, 0.1 ml of 20% trichloroacetic acid is added to all reaction mixtures, including the dark controls. Reaction mixtures are centrifuged at low speed, approximately 1000 *g* for 5 minutes, and the amount of $AT^{32}P$ in the clear supernatant is determined (see [8], Vol. 24). The endogenous photophosphorylation and the dark control values are subtracted from those of the reaction mixture containing phosphodoxin to obtain the photophosphorylation catalyzed by phosphodoxin.

Definition of Specific Activity: A unit of phosphodoxin is defined as that amount which produces a micromole of ATP per milligram of chlorophyll or bacteriochlorophyll per hour in the reaction mixtures given above, corrected for endogenous and dark ATP formation in chloroplasts or chromatophores. Specific activity is expressed as units per milligram dry weight.

Application of Assay Method to Crude Homogenate: The phosphodoxin activity in crude homogenates can be determined; however, sev-

[1] A convenient light source for obtaining such intensities is a 300-W tungsten bulb immersed in a 1500-ml beaker containing a solution of 2% w/v $CuSO_4$. The samples are placed about 3–4 cm from the bulb for irradiation.

eral errors are easily made. First aqueous crude homogenates often have unidentified materials which will partially or completely uncouple phosphorylation. Second, as the amount of crude homogenate is increased, an inhibition of electron transport and the resultant phosphorylation occurs. Thus, difficulty may be experienced in obtaining a linear response between photophosphorylation and the amount of crude homogenate.

Purification Procedure

Some of the phosphodoxin purification procedures presented below have been described by Black *et al.*[2] Purification steps are at room temperature or the temperature noted since phosphodoxin is heat stable.

Step 1—Preparation of Crude Homogenate: Two methods are described for preparing a crude homogenate. First a suspension of chloroplasts or chromatophores (see Reagent section) is treated with 10 volumes of cold acetone (—20°). After 30 minutes, the precipitate is collected and dried at room temperature. The acetone powder is suspended in distilled water and heated to 100°. The suspension is cooled to room temperatures, then centrifuged at 1000 *g* for 10 minutes, and the clear, straw-yellow colored supernatant solution is collected. The pellet is discarded.

In the second method, whole leaves or whole cells are extracted with boiling distilled water.[3] The whole tissue is quickly immersed in water at 100° and heated for 5–10 minutes. The suspension is cooled to room temperature, and the precipitate is collected by centrifugation at 1000 *g* for 10 minutes and discarded. The dark-yellow supernatant contains crude phosphodoxin.

Step 2—Acetone Fractionation: To the aqueous supernatant acetone (1.5 volumes) is added at room temperature. The suspension is centrifuged at 35,000 *g* for 10 minutes, and the precipitate is discarded. The supernatant solution is concentrated in vacuo to remove the acetone.

Step 3—Chromatography and Precipitation: The solution then is streaked on Whatman 3MM paper and chromatographed using *n*-propanol:H_2O:1% NH_4OH (3:1:2) as the solvent. The phosphodoxin, which has an R_f of 0.47 in this solvent system, is eluted with water. The eluted material is taken to dryness *in vacuo,* and the residue is extracted with ethanol. Storage of the ethanol solution overnight at —20° results in a precipitate which is washed with —20° ethanol and fiinally dissolved in distilled water.

[2] C. C. Black, A. San Pietro, D. Limbach, and G. Norris, *Proc. Nat. Acad. Sci.* **50,** 37 (1963).

[3] C. C. Black and A. San Pietro, *in* "Photosynthetic Mechanisms in Green Plants" B. Kok and A. T. Jagendorf, eds.), p. 228. National Academy of Sciences, Washington, D.C., 1963.

The resulting material is slightly yellow in color at neutral or alkaline pH values.

Properties

Identification: A definitive identification of phosphodoxin has not been achieved, although there are reports equating phosphodoxin to a pteridine.[4] The properties of spinach phosphodoxin are described in more detail since it has been studied more exhaustively than phosphodoxin from other organisms. Negative results were obtained in assays for protein and functional metals in spinach phosphodoxin.[3,5]

Absorption, Fluorescence, and ESR Spectra: There is a pronounced pH dependent shift in the absorption spectrum.[2] At acid pH values, i.e., 0.1 *N* HCl, peaks are observed at about 320 nm and 260–265 nm, while at neutral or alkaline pH values, i.e., 0.1 *N* NaOH, a peak is observed at 335 to 360 nm and shoulders at approximately 270 and 250 nm.

A fluorescence emission peak is observed at 440 nm that is activated maximally by irradiation at 358 nm at alkaline pH values. At acid pH values the intensity of both the fluorescence and activation spectra are lower in a manner analogous to the absorption spectra.[2] Spinach phosphodoxin also exhibits a light induced ESR signal which is pH dependant.[5]

Stability: Aqueous solutions near neutrality are stable indefinitely when stored in a refrigerator near 4°. Active solutions have been kept in a refrigerator as long as a year. Boiling phosphodoxin for 10 minutes in 1 *N* HCl does not affect its activity; however, boiling 10 minutes in 1 *N* NaOH inactivates the phosphodoxin. Ultraviolet irradiation for as long as 1 hour at room temperature of an aqueous solution does not alter the activity.[2]

Specificity: In every case thus far studied, spinach chloroplasts will catalyze photophosphorylation with phosphodoxin isolated from any other photosynthetic organism. The same also is true for *Rhodospirillium rubrum* chromatophores.[2,3,6]

Effects of Anaerobic and Other Experimental Conditions on Photophosphorylation Activity: With chromatophores, little effect (slight stimulation) of an argon atmosphere is observed.[2] However, with spinach chloroplasts a pronounced 60–90 inhibition under anaerobic conditions is observed, although complete inhibition is not observed.[2]

The pH optimum for photophosphorylation with spinach chloroplasts is between 7.4 and 7.8.[6] Addition of spinach ferredoxin, ferredoxin–

[4] F. I. Maclean, Y. Fujita, H. S. Forrest, and J. Myers, *Science* **149,** 636 (1965).
[5] C. C. Black, J. J. Heise, and A. San Pietro, *Biochim. Biophys. Acta* **88,** 57 (1964).
[6] C. C. Black, A. San Pietro, G. Norris, and D. Limbach, *Plant Physiol.* **39,** 279 (1964).

NADP reductase, or its antibody do not affect the activity.[6] The normal inhibitors of the Hill reaction, i.e., DCMU, CMU, and uncouplers, i.e., NH_4^+ and CCCP, are effective in the presence of phosphodoxin.[6]

Addition of other Hill oxidants with phosphodoxin gives unexplained results. NADP catalyzed photophosphorylation is not effected, while the ferricyanide catalyzed activity is inhibited and cyclic phosphorylation with PMS is stimulated.[6]

Organisms Containing Phosphodoxin: A water-soluble, heat-stable substance with characteristics similar to spinach phosphodoxin is present in extracts from leaves and chloroplasts of numerous higher plants, from algae, from flagellates, and from *Rhodospirillium rubrum* and *Chromatium* whole cells and chromatophores.[2,7]

[7] C. C. Black and A. San Pietro, *in* "Bacterial Photosynthesis" (H. Gest *et al.*, eds.), p. **223**. The Antioch Press, Yellow Springs, Ohio, 1963.

[56] Quantitative Determination of Carotenoids in Photosynthetic Tissues

By SYNNØVE LIAAEN-JENSEN and ARNE JENSEN

The present chapter is written for the biologist and the biochemist who without specialized training in organic chemistry wants to carry out analytical work in the carotenoid field. It may also be used by the organic chemist as an introduction to quantitative studies on carotenoid pigments. The methods described are routinely used in the authors laboratories.

The treatment is confined to the micro methods required in analysis of carotenoid mixtures in the 1- to 20-mg scale, and does not aim at the isolation of crystalline compounds. Spectroscopic methods like infrared, proton magnetic resonance, and mass spectrometry, indispensable as they are for structural work, are not included, since interpretation of the observations generally requires considerable experience in structural chemistry. Interested readers are referred to the pertinent literature.[1-3]

[1] M. S. Barber, J. B. Davis, L. M. Jackman, and B. C. L. Weedon, *J. Chem. Soc.* **1960**, 2870.
[2] U. Schwieter, H. R. Bolliger, L. H. Chopard-dit-Jean, G. Englert, M. Kofler, A. König, C. v. Planta, R. Rüegg, W. Vetter, and O. Isler, *Chimia* **19**, 294 (1965).
[3] C. R. Enzell, G. W. Francis, and S. Liaaen-Jensen, *Acta Chem. Scand.* **23**, 727 (1969).

General Precautions

Because of the long chain of conjugated carbon–carbon double bonds the carotenoids are sensitive to oxygen, heat, light, and acids. A restricted number (carotenol esters, α-ketols, and some allenic carotenoids) are altered by treatment with weak alkali, whereas hydrolysis of esters of carotenoic acids require strong alkaline conditions.

Experimental work on carotenoids should be carried out under following conditions:

Inert Atmosphere. Extracts or solutions containing carotenoids must be handled and stored *in vacuo* (<10 mm Hg) or in a nitrogen or carbon dioxide atmosphere. The vacuum should always be replaced with inert gas after concentrating solutions. All solutions should likewise be flushed with nitrogen through a capillary for some minutes prior to storage. Rapid transfers in separatory funnels as well as paper and thin-layer chromatography may be carried out in ordinary atmosphere.

Low Temperature. Extractions, manipulations, and reactions should be carried out at room temperature or lower. Pigment solutions must be stored in the deep-freeze (—20°) or in the refrigerator (5°). Concentrations should be carried out *in vacuo* with water baths up to 40°, care being taken to avoid overheating when the concentrate approaches dryness.

Darkness. All solutions should be stored in darkness and covered by black cloth during manipulations. Chromatographic columns are conveniently wrapped in aluminum foil permitting periodic inspection.

Acid-Free Conditions. Do not carry out work on carotenoids in rooms where acidic reagents are used. When reactions involving acids are carried out, the pH must be carefully checked.

Alkaline Conditions. Most carotenoids, except the groups mentioned above, are stable toward treatment with 5–10% methanolic KOH as employed during standard saponification. They are also usually stable toward alkaline adsorbents in chromatographic procedures.

Solvents and Reagents. Peroxide-free (absolute diethyl ether and tetrahydrofuran) and freshly distilled solvents must be used. Neutrality of reagents is imperative.

Extraction

Extraction is preferably carried out on fresh (wet) material with acetone or methanol or with a mixture of both. Treatment of the fresh material in a Waring Blendor with acetone and dry ice constitutes a rapid and gentle method of extraction.

Smaller samples of wet material are best extracted in a mortar with acetone. The material is cut in pieces and ground with a portion of acetone (10–25 ml). The extract is carefully removed with a pipette and filtered through a glass sintered filter (G1). The residue is ground again with another portion of acetone, and the extract is removed. The process is repeated until no more pigment is extracted.

When dry material is used, small-scale tests (by paper chromatography) are undertaken to check that no changes in carotenoid content or composition have occurred during drying. Freeze-drying is the method of choice. Dry material may be extracted with benzene, petroleum ether (for nonpolar carotenoids), diethyl ether, acetone, and methanol Extraction is carried out at ambient or lower temperature. The finely ground material is allowed to stand (in darkness) with the solvent under an inert atmosphere. In some cases moistening of the dry material with water prior to extraction with a polar solvent will facilitate complete extraction.

Phase Separation

A useful separation of carotenoid pigments into nonpolar carotenes and carotenoid esters in one group, and polar xanthophylls (containing hydroxyl groups) in the other may be accomplished by partitioning the pigments between petroleum ether and aqueous methanol. This can be done by dissolving the extract in aqueous methanol in a separatory funnel and adding an equal amount of petroleum ether. After gentle shaking and separation of the phases, carotenes and carotenoid esters are in the upper phase (epiphase) together with chlorophylls *a* and *b* and the bacteriochlorophylls, and the xanthophylls in the lower phase (hypophase) together with chlorophyll *c*. Further data on partition ratios are given below.

Saponification

In cases where the carotenoid components are stable toward alkali it is an advantage to include saponification in the isolation procedure in order to remove chlorophyll and other saponifiable matter.

The pigment is taken to dryness (oil pump); all acetone must be removed from the extract to avoid base-catalyzed formation of condensation products of acetone. The residue is dissolved in a small volume of ether, and the same volume of 10% methanolic KOH solution is added ($\approx$10 ml final solution for each 1 mg of pigment). The mixture is allowed to stand at room temperature for 1–2 hours; then the reaction is interrupted by transfer to ether in a separatory funnel by the addition of a 5% aqueous sodium chloride solution. After separation, the aqueous phase is extracted approximately three times with ether, and the combined ether extracts are washed three times with aqueous sodium chloride solution to remove methanol. The salt facilitates the separation of the phases and prevents formation of emulsions. Should emulsions form, addition of concentrated aqueous magnesium sulfate solution is recommended.

The ether extract may be dried over anhydrous sodium sulfate, which may adsorb some pigment, or better by azeotropic distillation with benzene. The ether extract is concentrated *in vacuo* to approximately 10 ml and $\approx$1 ml of acetone and 50 ml of benzene is added. The resulting solution is taken to dryness. The process is repeated until all the water has been removed. The residue is dissolved in an appropriate solvent for chromatography.

Total Carotenoid Content

The approximate content of total carotenoids in an extract may be determined by measuring the optical density of a sample suitably diluted. The total volume of the extract (v) is measured, a sample is diluted with acetone to give a reading in the spectrophotometer between 0.1 and 0.8 at the wavelength of the middle main absorption maximum of the extract, and the total amount of carotenoids is calculated according to Eq. (1)

$$c = D \cdot v \cdot f \cdot 10/2500 \tag{1}$$

where c is total carotenoids in milligrams, D optical density at the middle main maximum, v total volume in milliliters, and f the dilution factor for the sample actually measured ($f = 10$ if 1 ml is diluted to 10 ml). 2500 is an average extinction coefficient for carotenoids.

Chromatography

Some sort of chromatography is usually needed to separate the individual components of carotenoids extracted from natural sources. Three techniques of choice—paper, thin-layer, and column chromatography—are treated below. The two former methods are suitable for small-scale separations (0.1 mg). Separation of larger quantities must be carried out on columns. However, paper or thin-layer chromatography should always be used in combination with column chromatography for introductory examination of the extract and for checking the purity of column fractions since these methods are more discriminating and rapid.

In the carotenoid field chromatographic separations are usually produced by adsorption effects,[4] the partition technique being less frequently

[4] H. H. Strain, "Chromatographic Adsorption Analysis." Wiley (Interscience), New York, 1942.

applied. Less polar pigments, like carotenes, monohydroxyxanthophylls, methyl ethers, and ketocarotenoids, as well as esters, are best separated on aluminum oxide or calcium hydroxide, while the more polar xanthophylls require less strongly adsorbing materials for their separation. In the latter case magnesium oxide, calcium carbonate, sugar, or cellulose are frequently used.

As a rule the pigment mixture is dissolved in a solvent of the lowest possible polarity (petroleum ether, carbon disulfide), and the separation is carried out on the adsorbent with a solvent series of increasing polarity. It is important that the adsorptivity of the pigment correlates with the strength of the adsorbent so that the separation can take place with solvents with relatively low polarity. A "ranking list" of adsorbents and eluents is given in Table I.

Paper Chromatography. Among the various systems available, two seem to be particularly useful. One uses alumina-containing paper and

TABLE I
ADSORBENTS AND ELUENTS FOR CHROMATOGRAPHY, ARRANGED ACCORDING TO INCREASING ACTIVITY

Adsorbents[a] Weak	Medium	Strong
Cellulose	Calcium carbonate	Magnesium silicate
Sugar	Calcium phosphate	Aluminum oxide
Starch	Magnesium phosphate	Charcoal
Chalk	Magnesium oxide	Kaolin
Sodium carbonate	Calcium hydroxide	

Eluents (the elutropic series)[b]
Petroleum ether, b.r. 40°–50°
Petroleum ether, b.r. 70°–100°
Carbon tetrachloride
Cyclohexane
Carbon disulfide
Benzene
Diethyl ether
Acetone
n-Propanol
Ethanol
Methanol
Water
Pyridine
Organic acids

[a] Listed from top to bottom in order of increasing activity.
[b] Listed in order of increasing power of elution.

petroleum ether–diethyl ether mixtures and is superior for less polar pigments. The other system makes use of kieselguhr paper and petroleum ether–acetone mixtures and is particularly useful for the separation of polar pigments and stereoisomeric sets. (The papers are available from Carl Schleicher & Schüll, Dassel, West Germany, under the designations SS No. 668 and SS No. 287, respectively; circular papers, 18 cm in diameter, are very useful.) A wick (≈1 cm wide) is cut from the periphery to the center of the paper, and a small amount of the extract (10–50 μl containing 20–50 μg of carotenoids) is applied by a capillary in a small, round spot to the center of the paper, the solvent being simultaneously removed by a jet of nitrogen blown in from below. The paper is then placed in the bottom half of a petri dish (16 cm) containing the solvent and covered with a glass plate or the upper half of the petri dish. The speed of the migrating solvent may be regulated to some extent by varying the immersion of the wick into the solvent. When the solvent front has reached the rim of the petri dish, the chromatogram is removed, the front is quickly marked with a pencil, and the distances which the solvent and the various colored bands have migrated are measured. The pigmented rings are cut out, packed tightly in glass tubes (8–10 mm i.d.) which have been drawn out to a capillary at one end, and the carotenoids are eluted with acetone or another suitable solvent directly into small volumetric flasks.

R_f values (ratio of the distance migrated by the pigment to that of the solvent front) for a number of common carotenoids in the two systems mentioned above are given in Table II. Further details are to be found in the original papers.[5,6]

The R_f values will vary somewhat with the moisture content of the papers, which should be stored in desiccators before use. Identification of pigments should be carried out by cochromatography with authentic material and should never be based on R_f values only.

Thin-Layer Chromatography. With the introduction of "Fertigplatten" (factory made thin-layer plates) thin-layer chromatography has become very convenient also for the carotenoid chemist. One may choose between aluminum oxide-, silica gel-, and cellulose-coated plates of either glass, plastic, or aluminum.

The general technique is described in the many textbooks available.[7,8]

[5] A. Jensen and S. Liaaen-Jensen, *Acta Chem. Scand.* **13**, 180 (1959).

[6] A. Jensen, *in* "Carotin und Carotinoide" (K. Lang, ed.), p. 119. Steinkopff, Darmstadt, 1963.

[7] "Thin-Layer Chromatography, A Laboratory Handbook" (E. Stahl, ed.). Springer, New York, 1969.

[8] J. M. Bobbit, "Thin-Layer Chromatography." Reinhold, New York, 1963.

TABLE II
R_f VALUES OF SOME CAROTENOIDS

Pigment	R_f value × 100	
	668P[a]	287A10[b]
Phytofluene	77	—
α-Carotene	43	—
β-Carotene	38	100
Neurosporene	15	—
ζ-Carotene	5	—
Lutein	0	78
Zeaxanthin	—	72
Diadinoxanthin	—	68
Diatoxanthin	—	62
Violaxanthin	—	53
Fucoxanthin	—	48
Chlorophyll *a*	—	35
Neoxanthin	—	26

[a] Schleicher & Schüll No. 668 paper with petroleum ether.
[b] Schleicher & Schüll No. 287 paper with 10% acetone in petroleum ether.

The sample to be tested is applied as a small spot in one corner of the plate (for two-dimensional development) or as a streak along the edge (for subsequent spectroscopic examination) or as a number of spots along one edge (for parallel development of several samples in a cochromatography test). The development is usually carried out by standing the plate directly in the solvent which covers the bottom of the chromatographic chamber. For two-dimensional separation, the plate is dried at room temperature after the first run, turned 90°, and returned to the chamber for a second separation in a different solvent system. Measurements of the distances migrated by the pigments and the solvents allow the calculation of R_f values, which may be kept within reasonable limits provided care is taken to standardize the conditions. R_f values of a number of common carotenoids are given in the pertinent literature.[9]

Colored spots or bands may be scraped off the plates, the carotenoid extracted with acetone, and its spectroscopic properties determined. In quantitative studies great care must be exerted to avoid losses by oxidation and irreversible adsorption.

Column Chromatography. The amount of carotenoid to be separated

[9] H. R. Bolliger and A. König, *in* "Thin-Layer Chromatography, A Laboratory Handbook" (E. Stahl, ed.). Springer, New York, 1969.

decides the size of the column. Roughly, a 25 $\times$ 1 cm (diameter) column is suitable for up to 1 mg of pigment, and a 30 $\times$ 4 cm column can handle up to 20 mg.

The choice of adsorbent should be based on the behavior of the pigment mixture on paper or thin-layer chromatograms. Low polarity of the pigments requires high activity of the adsorbents (see Table I and Ref. 4). The columns may be packed wet (for aluminum oxide and cellulose) or dry. When extrusion of the adsorbent is not foreseen, columns provided with permanent glass-sintered plates are preferred. Otherwise a pad of glass wool is used.

For wet packing the adsorbent is suspended in petroleum ether, the slurry is poured into the glass column and allowed to settle by gravity (aluminum oxide) or pressed hard with a suitable plunger (cellulose powder).

Dry packing may be carried out with suction, or satisfactory packing is achieved by using a plunger of inner diameter approaching that of the inner column diameter (for $CaCO_3$). Add a 1-cm layer of quartz sand on top of the column to prevent disruption of the top of the column and as filter aid in case part of the concentrate should precipitate. Dry packing has the advantage of allowing bed volumes (volume of solvent required to moisten the column material) to be measured. Wet packing is often more satisfactory, and bed volumes can be estimated from a parallel "blank" column packed dry to the same height. Check that the flow rate is satisfactory before the pigment solution is added ($\approx$1 drop/second is usually preferable).

The pigment mixture should be dissolved in the smallest possible volume (not exceeding $\approx 1/10$ of the bed volume) of a solvent of low polarity. If the pigments are not sufficiently soluble in petroleum ether, carbon disulfide may be tried. The pigment concentrate is added to the top of the column with a pipette and washed into the adsorbent with small portions of petroleum ether until all the pigment has entered the column. The chromatogram is then developed with solvents or solvent mixtures of increasing polarity (see Table I). Do not change the solvent system as long as the zones migrate with considerable speed. When the pigments move too slowly in one solvent, add increasing quantities of the next member of the elutropic series (Table I) in steps or continuously. (Make careful notes of the development, including sketches of the various stages, and keep a quantitative record of the solvents used to elute the various zones.) A convenient way of tabulating the data is shown in Table III, which records an actual separation of five pigments of the carotene–monohydroxyxanthophyll type on aluminum oxide. It is

TABLE III
RECOMMENDED PROCEDURE FOR TABULATING THE CHROMATOGRAPHIC EVENT

Journal: p. 3736 Date: 8/18/1969

Extract: 1.45 mg of carotenoid in 2 ml of petroleum ether/CS_2 1:1

Column: Al_2O_3 Woelm neutral, activity grade 2, 1 × 18 cm, packed wet sea sand.

Bed volume: ≈10 ml

Time of chromatography: 1 hour 10 minutes

Pressure: No

Milliliters added	Eluent	Fraction No.[a]	Zone	Color of fr.	Volume (ml)	Notes
2	O. Extract	—	—	—	2	
30	I. Pet. ether	—	—	—	30	
30	II. 3% acetone[a]	—	—	—	5	Development starts
		1	a	Yellow	10	
		—	—	—	5	
		2	b	Orange	10 } 20	
20	III. 5% acetone			Orange	10	
		—	—	—	10	
30	IV. 7% acetone	3	c	Orange	10	
		—	—	—	3	
		4	d	Pink	17	
30	V. 9% acetone	5	d	Pink	30	
30	VI. 10% acetone	6	d/e	Orange	30	
30	VI. 12% acetone	7	e	Orange	20	
		—	—	—	10	
30	VI. 1% methanol	—	—	—	—	

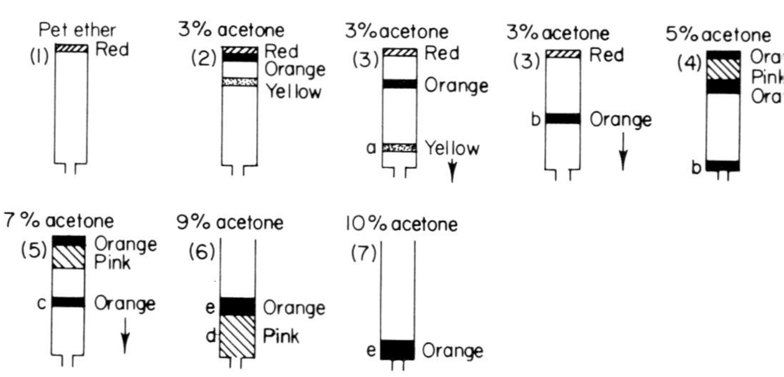

[a] In petroleum ether.

recommended that several fractions be cut from each colored zone on the chromatogram to minimize the size of mixed fractions, which may occur in cases of overlapping. The volumes of the fractions must be measured and the solutions kept in stoppered containers.

Should precipitation of pigment occur on top of the column, add a small amount of carbon disulfide and stir up the sand layer. No chromatographic separation should require more than 1 day.

Quantitative Determination

Absorption spectra in the visible region down to ≈330 nm of aliquots of each fraction are recorded at known dilutions. There is no need to transfer into a single solvent, but the solvent mixture used must be considered when deciding on the chromophore present (see below). Absorption maxima and optical density of the main maximum are tabulated, as recommended in Table IV. The carotenoid content is determined according to the formula given earlier. The choice of extinction coefficient is based on the position of the absorption maxima and the fine structure of the spectrum. The extinction coefficients tabulated in Table V may be used as a guide. Spectra with good fine-structure (e.g., lycopene, Fig. 1) are characteristic of aliphatic chromophores. Medium fine structure (γ-carotene, Fig. 1) and low fine structure (β-carotene, Fig. 1) are indicative of monocyclic or bicyclic chromophores or conjugated carbonyl compounds. *cis*-Isomerization (see below) lowers the extinction coefficient, as does introduction of oxygen functions.

In the example (Table IV), the substances eluted from zones a and b have spectral characteristics corresponding to a *cis* and *trans* undecaene (*cf.* lycopene) chromophore. Zones c and d similarly contained carotenoids with *cis* and *trans* tridecaene (*cf.* spirilloxanthin) chromophores, whereas zone e again contained a carotenoid with undecaene chromophore. The spectrum of fraction 6 showed several peaks, ascribed to a mixture of two compounds with different chromophores. The estimation of the optical density (D_{eff}) of each component in such a mixed fraction is explained in Fig. 2.

The total amount of carotenoid recovered from the chromatogram in the example (Tables III and IV) corresponded to 93%. Recoveries should be no less than 85%.

The composition of the carotenoid mixture of the example is now found to be the one given in Table VI. Note that the uncertainty of the extinction coefficients sets a limit for the accuracy of the analysis.

The composition of each fraction should be checked by circular chromatography on paper to confirm the tentative identification of the various pigments in the column fractions.

TABLE IV

RECOMMENDED PROCEDURE FOR CALCULATING THE CAROTENOID COMPOSITION AFTER CHROMATOGRAPHY (*cf.* TABLE III)

Fraction No.	Eluent, acetone	Zone	Volume (ml)	Dilution	λ_{max}	D_{eff}	$E^{1\%}_{1\,cm}$	Assignment	Carotenoid (mg)
1	3%	a	10	1	345				
					362				
					(440)				
					466	0.370	3100	*cis*-A	0.012
					497				
2	3–5%	b	20	10	345				
					362				
					445				
					472	0.630	3400	A	0.370
					506				
3	7%	c	10	2	365				
					385				
					(460)				
					490	0.410	2400	*cis* B	0.034
					522				
4	7%	d	17	10	365				
					385				
					465				
					493	0.720	2600	B	0.470
					525				
5	9%	d	30	5	365				
					385				
					465				
					495	0.350	2600	B	0.199
					525				
6	10%	d–e	30	1	345				
					362				
					(440)				
					470	0.350	2600	B	0.041
					495	0.250	3100	C	0.024
					525				
7	12%	e	20	5	345				
					362				
					445				
					470	0.670	3400	C	0.198
					499				
								Total:	1.348

Pigment recovery 1.35/1.45 = 93%.

TABLE V

ABSORPTION MAXIMA AND EXTINCTION COEFFICIENTS IN PETROLEUM ETHER[a] FOR SOME CHARACTERISTIC CAROTENOID CHROMOPHORES

Chromophore		Spectral fine structure[b]	Characteristic representative	Abs. max. (nm)[c]					Approx. $E^{1\%}_{1\,cm}$
Aliphatic	3F[d]	0	Phytoene			275	285	296	900
	5F	Medium	Phytofluene			332	347	366	1500
	7F	High	ζ-Carotene			379	400	424	2400
	9F	High	Neurosporene	318	332	415	439	469	2900
	11F	High	Lycopene	345	365	445	472	504	3400
	13F	Medium	Spirilloxanthin	365	384	465	493	527	2500
	10F + C = 0	Medium	Spheroidenone	355	370	(460)	483	515	2200
Monocyclic	11F	Medium	γ-Carotene		349	437	461	493	3100
Bicyclic	11F	Low	β-Carotene		338		450	478	2500
	11F + 2C = 0	0	Cantaxanthin		356		466		2200

[a] In ether mixtures the maxima are insignificantly displaced. In acetone mixtures a bathochromic displacement of 0–8 nm and a ≈5% drop (pure acetone) in extinction coefficient result.

[b] Expressed as intensity of the longest wavelength band divided by intensity of the main middle band, using minimum between these bands as zero line.

[c] The maxima in the first two columns give position of *cis* peak.

[d] F denotes conjugated double bonds.

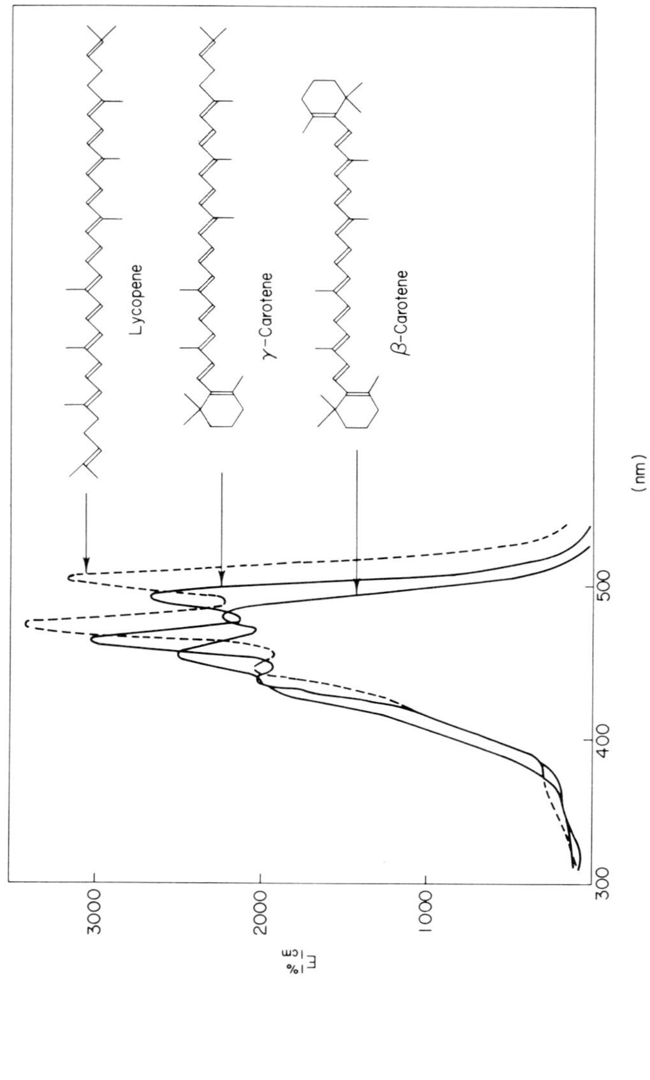

FIG. 1. Absorption spectra in petroleum ether of carotenes with undecaene chromophore demonstrating the spectral effect of cyclization.

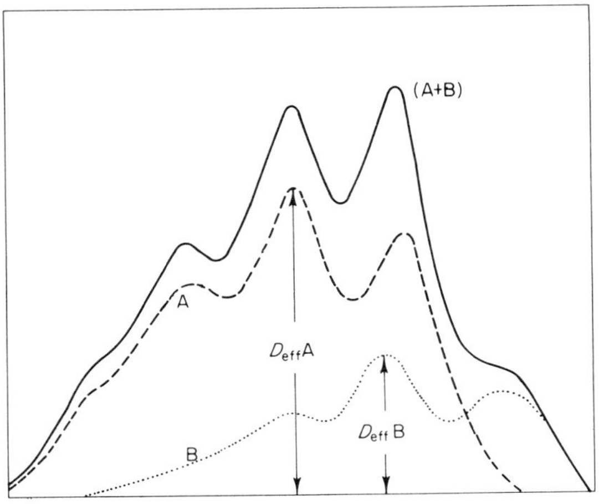

FIG. 2. Absorption spectrum of a mixed fraction—determination of D_{eff} for each component.

Identification of Individual Components

The Chromophore. The absorption spectrum in visible light (see Table V) gives valuable information about the chromophore present in the pigment. No change in the chromophore upon hydride reduction of the compound, indicates the presence of a pure polyene system. However, allenic and acetylenic bonds may be present.

Polarity, R_f Value and Partition Ratio. The polarity of a carotenoid is mainly determined by the number of hydroxyl groups present. More subtle effects are caused by carbonyl, ester, ether, and epoxidic functions. The chromatographic behavior on columns, the R_f values,[2,5,9] the

TABLE VI
COMPOSITION OF CAROTENOID SAMPLE ANALYZED IN EXAMPLE (*cf.* TABLES III AND IV)

Preliminary identification	Amount (mg)		Percent of total	Final identification
cis compound A	0.012	0.382	28	Lycopene
trans compound A	0.370			
cis compound B	0.034	0.744	55	Spirilloxanthin
trans compound B	0.710			
Compound C		0.222	17	Rhodopin
		1.348	100	

partition ratios[10] and the M_{50} value[11] are all very dependent upon the polarity of the pigment and they may give valuable information as to the number and nature of the functional groups involved. Having obtained the above data for an unknown carotenoid, a comparison with published values often reveals which carotenoids should be tried in cochromatographic tests.

cis Isomers. Even when the precautions mentioned above are observed, isomerization of carbon–carbon double bonds inevitably occurs. *cis* Isomers give rise to more or less well separated chromatographic zones. They may be adsorbed above (neo U, neo V, etc., isomers) or below (neo A, neo B, etc., isomers) the *trans* compound. Whereas this complicates the separation and quantitative determination of the carotenoid mixture, it is a useful phenomenon for the direct chromatographic comparison of the pigments (see below).

The phenomenon of *cis* isomerism is described in detail by Zechmeister.[12] *cis* Carotenoids normally show increased absorption in the "*cis*-peak" region with a single or double *cis* peak located at a wavelength 142 ± 2 nm shorter than the maximum at the longest wavelength of the *trans* isomer (in hexane). The position of the *cis* peak thus permits identification of the parent *trans* isomer (*cf.* Table IV). *cis* Isomers usually exhibit main maxima at lower wavelength (shift generally 3–10 nm) than the *trans* isomer, less spectral fine structure, and lower extinction coefficients.

Cochromatography. Cochromatography of the unknown pigment with authentic material is frequently used in carotenoid analysis. Rather than comparing only the *trans* isomers the test should be carried out with the stereoisomeric mixtures of the two pigments. Approximately 0.1 mg of the carotenoid is dissolved in 10 ml of petroleum ether or benzene, one drop of a pale pink solution of iodine in petroleum ether is added, and the solution is left to stand for 2 hours in indirect daylight. Samples of the two stereoisomeric sets are then compared by thin-layer or paper chromatography. Qualitative (R_f value and absorption spectrum of the various isomers) and quantitative agreement in composition of the two sets is a strong indication of the identity of the pigments in question. It is not a definite proof, however. In our experience this test is best carried out by circular chromatography on kieselguhr paper which separates *cis–trans* isomers well.

Preparation of Derivatives. In the absence of crystalline material for further spectroscopic investigations (infrared, proton magnetic reso-

[10] F. J. Petracek and L. Zechmeister, *Anal. Chem.* **28**, 1484 (1956).

[11] N. I. Krinsky, *Anal. Biochem.* **6**, 293 (1963).

[12] L. Zechmeister, "*Cis-trans* Isomeric Carotenoids, Vitamins A and Arylpolynes." Springer, Vienna, 1962.

nance, and mass spectroscopy) preparation of various derivatives on the micro scale should be carried out.

The reactions described below are particularly informative and easy to carry out.

ACETYLATION. Under the conditions described below, primary and secondary hydroxyl groups are acetylated, whereas tertiary hydroxyl groups are not attacked. To the carotenoid (0.1–1 mg) in 1 ml of dry pyridine (dried over BaO) is added acetic anhydride (0.1 ml). The reaction is carried out at room temperature and the reaction course followed by examination of aliquots of the reaction mixture by paper chromatography (e.g., 0 min, 5 min, 10 min, 15 min, 30 min, 1 hour, 3 hours, 6 hours). The number of intermediary products are counted. If only one hydroxyl group is acetylated no intermediates are observed. With two groups (in an unsymmetrical molecule), two monoacetates and a final diacetate are produced. For three groups, three intermediary monoacetates and three diacetates are theoretically expected in addition to the final triacetate. The method thus allows one to decide the number of primary or secondary hydroxyl groups provided the number does not exceed three (thereafter the picture becomes too complicated and the system generally degenerates).

The final product is isolated in the usual manner by transferring to ether in a separatory funnel on dilution with aqueous NaCl solution, and washing the combined ether extract with water.

SILYLATION. Tertiary hydroxyl groups form trimethylsilyl ethers. Hence the number of tertiary hydroxyl groups (except those in the 6 position in cyclic carotenoids) can be determined by submitting the fully acetylated compound to silylation conditions[13]: To 0.1–1 mg of carotenoid in dry pyridine at −35° (cooled in acetone–dry ice) is added hexamethyldisilazane (0.2 ml) and trimethylchlorosilane (0.1 ml) mixed together in pyridine (0.5 ml). The very fast reaction is followed by paper-chromatographic inspection (0 sec, 5 sec, 15 sec, 30 sec, 1 min, 3 min, 6 min, 10 min, etc.) until the final product is quantitatively formed. The number of intermediates observed prior to formation of the final product decides how many tertiary hydroxyl groups are being silylated.

The reaction mixture may be worked up in the usual manner.

REDUCTION WITH LITHIUM ALUMINUM HYDRIDE. Conjugated carbonyl functions are revealed by reduction with lithium aluminum hydride. The resulting alcohol absorbs light at shorter wavelength and is more polar than the parent compound.

The reduction is carried out by dissolving 0.1–1 mg of the pigment

[13] A. McCormick and S. Liaaen-Jensen, *Acta Chem. Scand.* **20**, 1989 (1966).

in anhydrous, peroxide-free diethyl ether or tetrahydrofuran and adding a suspension of finely divided $LiAlH_4$ in anhydrous ether previously filtered through a pad of glass wool. The mixture is shaken a few minutes until a hypsochromic color shift is observed. The reaction is interrupted by pouring the mixture into moist ether in a separatory funnel, and the carotenoid is isolated in the usual manner. Great care must be exerted when working with pigments which are unstable toward basic reagents.

EPOXIDE–FURANOID OXIDE REARRANGEMENT. Epoxides often give a blue color reaction when an ether solution of the pigment is treated with 25% hydrochloric acid in water.[14] Furanoid oxides behave similarly.

Epoxides are rearranged to furanoid oxides upon treatment with weak organic acids or on active adsorbents like alumina. This results in a large hypsochromic shift toward shorter wavelengths in the light absorption spectrum. To the solution of the pigment in ether in a cuvette, 1–5 drops of formic acid are added, the solution is mixed, and the changes in the absorption spectrum are measured at intervals. A hypsochromic shift of the absorption maxima of $\approx$25 nm indicates the presence of one, and a shift of approximately 40 nm indicates two, epoxy groups in the original pigment.

OTHER REACTIONS. Other reactions, e.g., elimination of allylic hydroxyl or ether groups with $HCl–CHCl_3$; methylation of allylic hydroxyl groups with $HCl–CH_3OH$; many other reduction and oxidation reactions are less easily interpreted and are reviewed elsewhere.[15,16]

Conclusion. For identification of a carotenoid of known structure the tests described above, including cochromatography of each derivative will usually be sufficient, provided an authentic sample for direct comparison is available. However, in recent years many minor structural modifications of the classical carotenoid have been elucidated (e.g. carotenoids with C_{45} and C_{50} skeletons) which may alter the R_f values only very slightly. Preparation of as many derivatives as possible reduces the possibility of misidentification. However, where possible, confirmation by mass spectrometry (crystalline material not required) of the conclusions drawn, at least of the molecular weight, should be sought.

[14] P. Karrer and E. Jucker, "Carotenoids." Elsevier, New York, 1950.

[15] S. Liaaen-Jensen and A. Jensen, *in* "Progress in the Chemistry of Fats and Other Lipids" (R. Holman, ed.), Vol. VIII, Part 2, p. 33. Pergamon, New York, 1965.

[16] S. Liaaen-Jensen, *Pure Appl. Chem.* **14**, 227 (1967).

[57] Chlorophyll–Protein Complexes

By ATUSI TAKAMIYA

Because of widespread interest in the states of chlorophyll pigments in the photosynthetic apparatus of green plants and certain bacteria, efforts have been made to isolate the pigments in a dispersed state as possibly native as they occur *in vivo*.[1,2] In this article, only the naturally occurring water-soluble forms of chlorophyll proteins constituting well-defined molecular entities are described.

Chenopodium Chlorophyll Protein CP668

The conditions for extraction and purification of Chenopodium chlorophyll protein are not severely limited; it may be extracted with any buffer solution of neutral to moderately alkaline pH. Addition of low concentrations of salts, e.g., NaCl, may favor the yield of extraction. The purification procedure described is that devised by Yakushiji *et al.*[3]

Preparation

Step 1—Crude Extract: Fresh leaves of *Chenopodium album* (5 kg) are homogenized in 3 liters of 0.01 *M* disodium phosphate solution. Rough fragments are removed by filtration through gauze, and the green juice thus obtained is centrifuged at 9000 *g* in a continuous flow centrifuge to remove the bulk of chloroplasts in the suspension. Acidification, at this step, to pH 5.0 by addition of about 15 ml of acetic acid, facilitates sedimentation without any deletorious effect on the final product.

Step 2—Ammonium Sulfate Precipitation: The somewhat turbid solution is brought to 0.3 saturation with ammonium sulfate and filtered on a Buchner funnel with a thin layer of kieselguhr as filter aid. The filtrate is made 0.6 saturated with ammonium sulfate and filtered as above to collect the precipitate. The filtrate is discarded and the precipitate is dissolved in 250 ml of 0.02 *M* phosphate buffer, pH 7.8. This solution, brownish in color, is dialyzed against 0.01 *M* phosphate buffer, pH 7.8, for 24 hours in a cold room prior to chromatography.

Step 3—Amberlite Chromatography: A column of Amberlite CG 50,

[1] C. S. French, *in* "Encyclopedia of Plant Physiology" (W. Ruhland, ed.), p. 252. Springer-Verlag, Berlin, 1960.

[2] D. W. Kupke and C. S. French, *in* "Encyclopedia of Plant Physiology" (W. Ruhland, ed.), p. 298. Springer-Verlag, Berlin, 1960.

[3] E. Yakushiji, K. Uchino, Y. Sugiura, I. Shiratori, and F. Takamiya, *Biochim. Biophys. Acta* **75,** 293 (1963).

equilibrated with 0.01 *M* phosphate buffer, pH 7.8, is charged with the dialyzed preparation. Beneath the uppermost brown zone, a green zone is formed which is washed in turn with 0.01 *M* and 0.02 *M* phosphate buffer solutions, pH 7.8, and then eluted with 200 ml of 0.5 *M* solution of the same buffer. This green solution has, on 50-fold dilution, an absorbance of 0.7–1.0 at 668 nm (1 cm light path).

Step 4—Second Ammonium Sulfate Precipitation: The ammonium sulfate fractionation is repeated, and the fraction precipitating between 0.3 and 0.6 saturation is dissolved in 0.05 *M* phosphate buffer, pH 7.8, and stored in the dark in the frozen state. To obtain the chlorophyll protein in its native state, CP668, absorbing at 667–668 nm, care must be taken to work under dim light during the course of preparation. Although the substance absorbs less light in the green range of the spectrum, the use of white (dim) light is recommended. (The eye can operate at an extremely low light intensity of white light especially in locating the green band of the substance on a chromatographic column.) The light-converted form of the protein, CP743, may be prepared in light. No preferential enrichment of CP668 or of CP743 occurs during purification of a partially converted mixture of the chlorophyll protein as described.

Terpstra[4] has modified the procedure as follows: The leaves of *Chenopodium album* (fresh or stored frozen at −40°) are ground in cold 0.01 *M* Na_2HPO_4 (150–200 ml/100 g of leaves) in a Waring blendor for 30 seconds. Large fragments are eliminated by centrifugation in a continuous flow centrifuge equipped with filter paper. The suspension is then centrifuged for 20 minutes at 7000 *g* at 0°. The supernatant fluid is acidified to pH 5.8–6.0 with dilute acetic acid and ammonium sulfate added to give successively 30 and 60% saturation. The precipitate formed between 30 and 60% saturation is collected by centrifugation and dissolved in 0.1 *M* phosphate buffer, pH 7.0 (usually 10–25 ml of buffer per 100 g of leaves). The solution is centrifuged for 30 minutes at 40,000 *g* and then for 60 minutes at 144,000 *g* to remove insoluble material which is discarded. The final supernatant solution contains, besides CP668, as a contamination a nonphotoconvertible chlorophyll–protein-like substance absorbing at 668 nm. One gram of "Carbowax 4000" (polyethylene glycol with a molecular weight of 3000–3700) is added per 2 ml of solution of CP668. After standing with occasional stirring for 30 minutes at room temperature, the mixture is centrifuged for 10 minutes at 40,000 *g* at 15° to collect the precipitate of purified CP668 which is dissolved in 0.1 *M* phosphate buffer, pH 7.0.

[4] W. Terpstra, *Biochim. Biophys. Acta* **120**, 317 (1966).

Alternatively, the nonphotoconvertible form occurring in Chenopodium leaves can be removed by first extracting the leaves with a solution of "Carbowax 4000" (1 g per 2 ml of 0.1 *M* phosphate buffer, pH 7.0). Subsequent extraction of the Carbowax-pretreated leaves gives a preparation of CP668 which does not contain the nonphotoconvertible form.

Purity Index

The purity of the chlorophyll protein preparation may be expressed conveniently by a purity index 668/277 defined as the ratio of the absorbance at 668 nm, the red peak, to that at 277 nm, the protein peak. When contaminated with nonphotoconvertible form(s) which also absorb at 668 nm, the ratio of absorbance at 743 nm to that at 277 nm may be used; the absorbance at 743 nm is measured after a sufficiently long illumination of the preparation at the final stage of purification. Values of approximately 0.75 and 0.40 are obtained for purity index 668/277 and purity index 743/277, respectively.

Properties[3,5]

The purified CP668 shows a single peak sedimentation pattern of ultracentrifugation, corresponding to $S_{20} = 2.70$.

The molecular weight of CP668 is 78,000 as measured by the gel filtration method with Sephadex G-100. No change in molecular weight is detected between CP668 and CP743.[6]

CP668 contains both chlorophyll *a* and *b* in an approximate ratio of 6:1–2 per molecule of the chlorophyll protein; no other pigments, such as carotenoid, are present.

The photoconvertibility from the native form, CP668, to CP743 is the most characteristic property of this chlorophyll protein. The main absorption bands in CP668, the native form, are at 277 (protein peak), 429 and 668 nm; in CP743, the photoconverted form, at 277, 362, 399, 429, 567 and 743 nm [Fig. 1(a)]. The difference spectrum is shown in Fig. 1(b).

The light-induced change of CP668 to CP743 is an oxidation. The presence of oxygen (air) or some oxidant (e.g., ferricyanide or mammalian ferricytochrome *c*) is necessary for the photoconversion. Illumination of the photoconverted form, CP743, in the presence of a reducing reagent (e.g., sodium hydrosulfite or ascorbate) causes a complete rever-

[5] A. Takamiya, H. Obata, and E. Yakushiji, *in* "Photosynthetic Mechanisms of Green Plants," p. 479. National Academy of Sciences, Washington, D.C., 1963.

[6] Y. Kamimura and A. Takamiya, unpublished data.

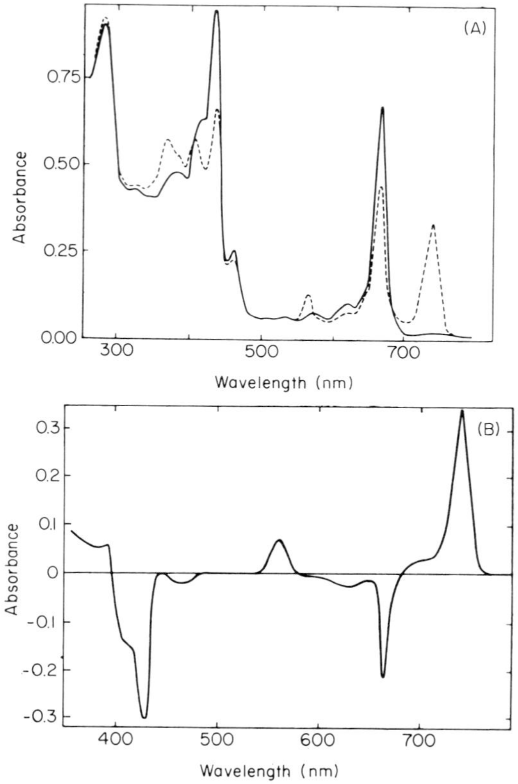

FIG. 1. (a) Absorption spectra of *Chenopodium* chlorophyll protein. Solid line, CP668, native form; dashed line, CP743, photoconverted form. (b) Difference spectrum of photoconversion [from Yakushiji *et al.*[3]].

sion to the native form, CP668. No reversion of CP743 to CP668 occurs either in the dark or upon illumination in the absence of a reducing agent.

Preparation of the Apoprotein[7]

Four volumes of methylethyl ketone are added to the solution of CP668 in 0.01 *M* phosphate buffer, pH 7.2 (temperature, 5°). After shaking in a separatory funnel, the water layer is removed and saved; the extraction is repeated until the water layer is colorless. The apoprotein

[7] T. Murata, Y. Odaka, K. Uchino, and E. Yakushiji, *in* "Comparative Biochemistry and Biophysics of Photosynthesis" (K. Shibata, A. Takamiya, A. T. Jagendorf, and R. C. Fuller, eds.), p. 222. Univ. of Tokyo Press, Tokyo, 1968.

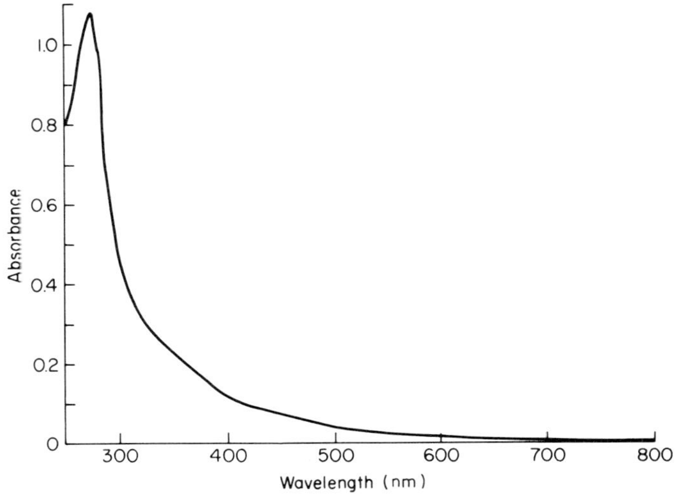

FIG. 2. Absorption spectrum of apoprotein CP668 [from Murata *et al.*[7]].

in the extracted water layer is purified by repeated precipitation with ammonium sulfate at 0.8 saturation (Fig. 2).

Reconstitution of Photoconvertible CP668

An appropriate amount of an alcoholic solution of chlorophyll is added to the apoprotein solution with mild shaking (final alcohol concentration, 10–15%). After standing for about 20 minutes, the solution is brought to saturation with ammonium sulfate and filtered through a thin layer of talcum.

The precipitate of the reconstituted chlorophyll protein is dissolved in 0.05 *M* phosphate buffer, pH 7.2, leaving free chlorophyll adsorbed on the talcum. The ammonium sulfate precipitation is repeated, and the reconstituted chlorophyll protein is purified as described above for the original isolation. The reconstituted product is identical with the original CP668 in photoconvertibility and absorption spectrum as well as the relative content of chlorophyll *a* and *b* regardless of the ratio of chlorophyll *a* and *b* used in reconstitution (Fig. 3).

Similar products can be reconstituted with either chlorophyll *a* or *b* alone. The reconstituted chlorophyll *a* protein (absorption peaks at 429 and 668 nm) is photoconvertible; in contrast, the chlorophyll *b* protein (absorption peaks at 438 and 645 nm) is light insensitive. The phaeophytin ($a + b$) protein reconstituted in a similar way (absorption peaks at 414, 508, 537, and 675 nm) is photoconvertible and exhibits new

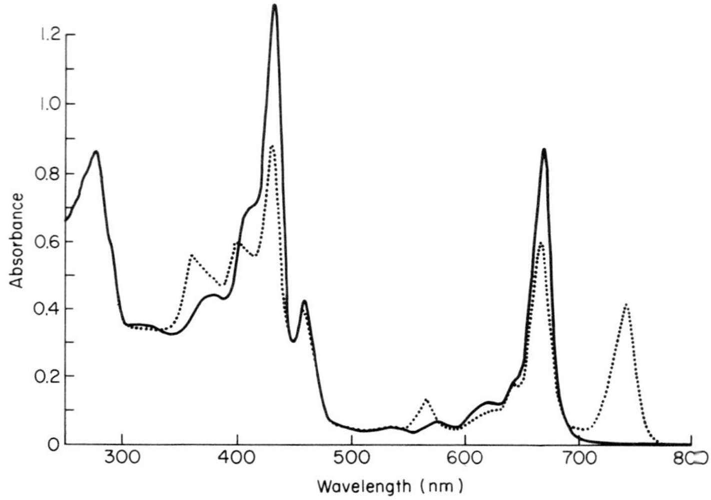

FIG. 3. Absorption spectra of the reconstituted CP668 before (solid line) and after (dotted line) illumination [from Murata *et al.*[7]].

absorption peaks at 391, 522, and 716 nm. A photoconvertible Zn-phaeophytin protein can also be reconstituted. While Cu-, Mn-, and Co-phaeophytin proteins can also be reconstituted, they are not photoconvertible; Ag-phaeophytin does not combine with the apoprotein.

An apoprotein can be prepared similarly from the photoconverted form of the chlorophyll protein, CP743. This reconstituted chlorophyll protein is indistinguishable in the absorption spectrum from the native form CP668, but is insensitive towards light.

Transient Form of CP668 Formed during Photoconversion[8]

On illumination of CP668 with an intense flash of light (flash photolysis) under anaerobic conditions, a transient form of the chlorophyll protein (probably a triplet state intermediate; half life, 2.1 mseconds) is formed (Fig. 4), which is capable of reacting with molecular oxygen.

Distribution

The chlorophyll protein CP668 has a quite limited distribution. It has been discovered in the leaves of *Chenopodium album, C. acuminatum*, var. *ovatum*, and *C. tricolor*. Negative results were obtained with other plants of the same genus, e.g., *C. hybridum* and *C. anthermінticum*.

[8] A. Takamiya, Y. Kamimura, and A. Kira, *in* "Comparative Biochemistry and Biophysics of Photosynthesis" (K. Shibata *et al.*, eds.), p. 229. University of Tokyo Press, Tokyo, 1968.

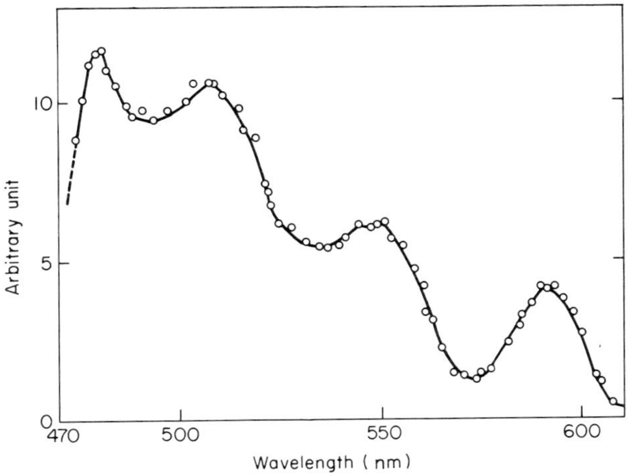

Fig. 4. Absorption spectrum of a transient form of CP668 (Flash photolysis). The time lapse between the main and the measuring flash was 50 μseconds [from Takamiya *et al.*[8]].

In *Chenopodium album,* the yield of the chlorophyll protein markedly depends on the age of the plant and time of the year. Seeds and young cotyledons do not contain the substance; fully grown leaves are the best material. The yield is highest from early to midsummer in the natural habitat. Deep-frozen leaves may serve as the source of the chlorophyll protein, although the yield is sometimes lower than with the fresh material.

Water-Soluble Chlorophyll Proteins of Cruciferae Plants

A group of water-soluble chlorophyll proteins of a closely allied nature have been isolated by Yakushiji[9] from the leaves and green tissues of various Cruciferae plants.

Cauliflower Chlorophyll Protein CP674

Extraction and Purification: The influorescence of cauliflower (1.25 kg) is homogenized in a Waring blendor with addition of 1.5 liters of 0.1 *M* phosphate buffer, pH 7. (Addition of 5% NaCl to the buffer favors the yield of extraction.) Coarse debris is removed by filtration through gauze. To the green juice obtained (2.4 liters), ammonium sulfate (273 *g*) is added to make 0.2 saturation. After standing for 10 minutes, acrinol solution (2%, 125 ml) is added, and 30 minutes later the fluid is filtered through a Buchner funnel layered with talcum. The mucous substance contained in the crude extract is removed by this procedure, and a clear

[9] Y. Yakushiji, *Plant Physiol.* **8,** 73 (1970).

yellow filtrate is obtained. Afterwards, 816 g of ammonium sulfate is added to make 0.7 saturation. The precipitate of the chlorophyll protein is collected by filtration through a fresh talcum layer on the Buchner funnel. The chlorophyll protein is dissolved by rinsing with 0.05 *M* phosphate buffer (Crude solution of CP674). At this stage of purification, a yield of A_{674} (1 cm) $= 1.3 \times 100$ ml is obtained.

The crude sample of CP674 dissolved in 0.01 *M* phosphate buffer, pH 7, is charged on a DEAE–cellulose column. After washing with the phosphate buffer, chlorophyll protein is eluted with higher concentrations (0.05–0.1 *M*) of phosphate buffer, pH 7.

Properties

The molecular weight is 78,000, as determined by the gel filtration method using a column of Sephadex G-100 in phosphate buffer, pH 7 ($\mu = 0.5$). The sedimentation constant, $S_{20,w}$ is 4.7 and the isoelectric point is 4.6, as determined by electrophoresis in a column of 1% ampholine in a sucrose density gradient for 72 hours at 2°.

The absorption spectrum exhibits maxima at 273, 340, 384, 420, 440, 470, 628, and 674 nm (Fig. 5). Occasionally a shoulder appears at 690–703 nm, especially in aged samples. This is due to some deterioration of the chlorophyll in the complex. CP674 contains chlorophyll *a* and *b* in a ratio of 6:1.

In contrast to the water-soluble chlorophyll protein CP668 isolated from the leaves of Chenopodium, the cauliflower chlorophyll protein is

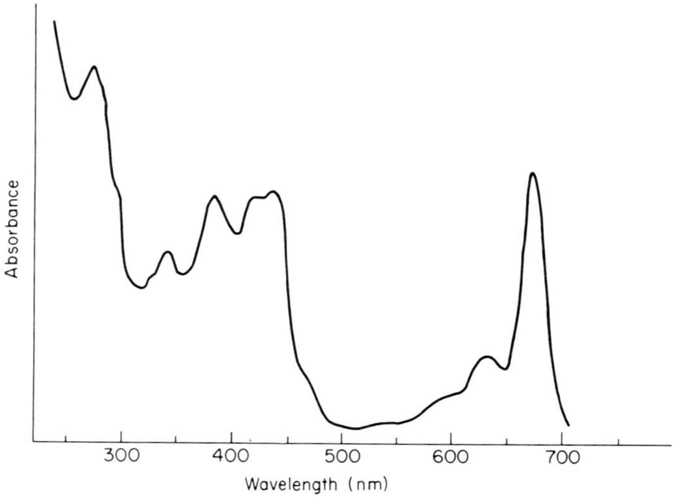

FIG. 5. Absorption spectrum of cauliflower chlorophyll protein CP674 [from Yakushiji[9]].

light-insensitive. Therefore, the purification procedures can be carried out in the day light.

Reconstitution from the Apoprotein

Apoprotein can be prepared from CP674 by extraction of the chlorophyll moiety in a manner similar to that described for CP668. The chlorophyll *a* protein reconstituted by addition of chlorophyll *a* to the apoprotein shows an absorption spectrum similar in general pattern to that of the original chlorophyll protein CP674 (Fig. 6), whereas, the re-

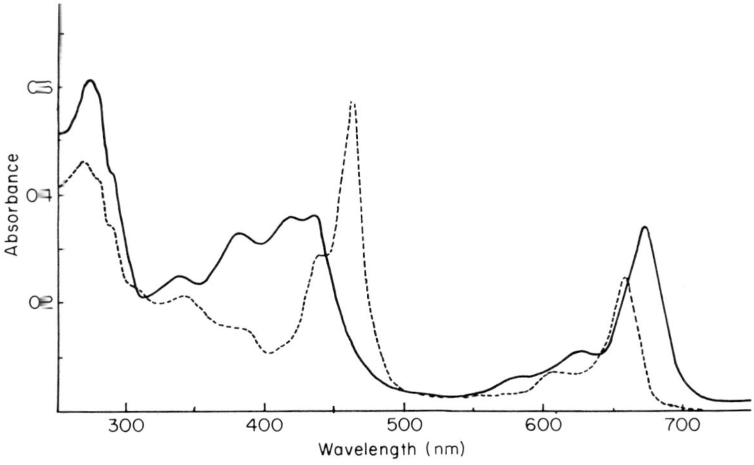

FIG. 6. Absorption spectra of reconstituted cauliflower chlorophyll protein. Solid line chlorophyll *a* protein; dashed line, chlorophyll *b* protein [from Yakushiji[9]].

constituted chlorophyll *b* protein shows an absorption spectrum resembling that of the free chlorophyll *b* in organic solvents (e.g., alcohol). Absorption maxima for the reconstituted chlorophyll *a*-protein are at 342, 383 420, 435, and 672 nm; for the corresponding *b*-protein, at 346, 438, 463 and 657 nm. Reconstitution by addition of chlorophyll $a + b$ to the apoprotein yields products in which the ratio of *a* to *b* is the same as that in the original CP674 (i.e., about 6:1), independent of the variously varied ratios of *a* to *b* in the chlorophyll solution used.

Lepidium Chlorophyll Protein CP662

Extraction and Purification: The whole plants or leaves of *Lepidium virginicum* are homogenized in a Waring blendor, and the chlorophyll protein is extracted and purified in the same way as described above for the cauliflower chlorophyll protein, except that the acrinol treatment is omitted because of the absence of any mucous substance in the *Lepidium* extract. (This is also the cases with other members of the Cruciferae chlorophyll proteins—see below.) After the final steps of purification by DEAE–cellulose column chromatography and Sephadex G-100 gel filtration, the purified sample of the chlorophyll protein is made 0.7 saturated with ammonium sulfate, and it is kept standing in a refrigerator. After some time the chlorophyll protein crystallizes in minute platelets, deep green in color. On microscopic examination, the crystals are almost rectangular in shape—actually parallelograms having the corner angles of about 81° and 99°, respectively. The crystals readily dissolve on dilution with water or buffer solution.

Properties

The molecular weight is 68,000 and the isoelectric point is [illegible].3. The absorption spectrum has maxima at 270, 338, 381, 419, 438, 469, 617, and 662 nm (Fig. 7). The ratio of chlorophyll *a* to *b* is 3:2. The prominent peak at 469 nm corresponds to the high content of chlorophyll *b* in this chlorophyll protein. Carotenoids are not present.

Reconstitution from Apoprotein

Essentially the same are the properties of this chlorophyll protein as those of the cauliflower chlorophyll protein with respect to reversible removal and recombination of the chlorophyll *a* and *b*.

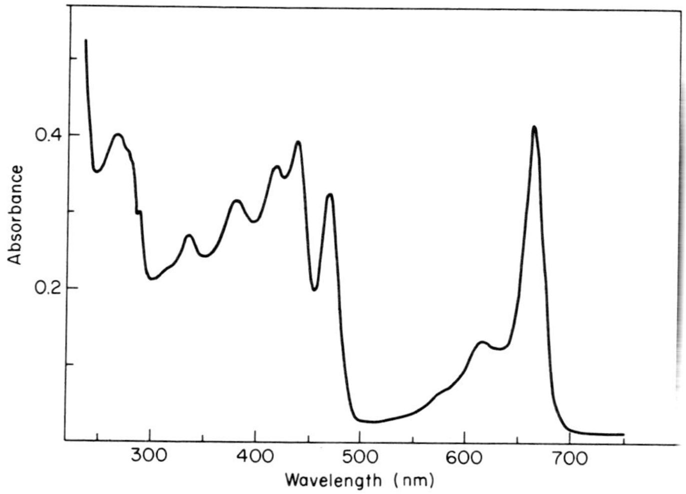

FIG. 7. Absorption spectrum of *Lepidium* chlorophyll protein CP662 [from Yakushiji[9]].

Distribution of Cruciferae Chlorophyll Proteins

The distribution of this group of chlorophyll proteins seems to be rather limited and, moreover, related to the genetic relationship of the plants. Thus far, similar chlorophyll proteins have been isolated from *Brassica nigra* ($n = 8$; B genome); *B. oleracea*, var. *botrys* (Cauliflower; $n = 9$, C genome); *B. oleracea*, var. *capitata* ($n = 9$, C genome); *B. oleracea*, var. *gemnifera* ($n = 9$, C genome); *Capsella Brusa pastoris*; *Cardaria Draba* (syn. *Lepidium Draba*); and *L. virginicum*. Among the various forms of the genus *Brassica*, only those belonging to the B and C genomes with the chromosome numbers 8 and 9 yielded a chlorophyll protein of this type. In contrast, *B. napus* ($n = 10$, A genome) and *B. juncea* ($n = 18 = 8 + 10$, A-B genome) gave negative results.

Although some photochemical activities, e.g., photooxidation of algal cytochrome *c* (553) in the presence of cauliflower chlorophyll protein, are known, the physiological role of this group of water-soluble chlorophyll proteins is still obscure.

[58] Cytochrome Reducing Substance

By YOSHIHIKO FUJITA and JACK MYERS

Photoreactive lamellar fragments from blue-green, red, and green algae and spinach chloroplasts contain a redox substance tentatively named the cytochrome *c* reducing substance (CRS).[1,2] It is reduced by photochemical system I and reoxidized by ferricytochrome *c* or by molecular oxygen.[1,3,4] The CRS content is generally 0.05–0.1 equiv./chlorophyll *a*.[2,3] It is tightly bound and is solubilized only after treatment with polar solvents (as 60% acetone). Solubilized CRS is water soluble, heat stable, and of large (but not determined) molecular weight, but does not contain protein. Solubilized and purified CRS stimulates noncyclic photophosphorylation in spinach chloroplasts as an electron carrier between photochemical system I and O_2.[5]

[1] Y. Fujita and J. Myers, *Arch. Biochem. Biophys.* **113,** 730 (1966).

[2] Y. Fujita and J. Myers, *Arch. Biochem. Biophys.* **113,** 738 (1966).

[3] Y. Fijita and J. Myers, *Arch. Biochem. Biophys.* **119,** 8 (1967).

[4] Y. Fujita and F. Murano, *in* "Comparative Biochemistry and Biophysics of Photosynthesis" (K. Shibata, A. Takamiya, A. T. Jagendorf, and R. C. Fuller, eds.), p. 161. Univ. of Tokyo Press, Tokyo, 1968.

[5] Y. Fujita and J. Myers, *Plant Cell Physiol.* **7,** 599 (1966).

Assay Methods

CRS activity can be assayed by two methods depending upon the kinetics of photochemical oxidation and dark reduction of added cytochrome *c* (method 1) and by stimulation of aerobic photophosphorylation (method 2). The second method is simpler but less specific.

Method 1

Under anaerobic conditions, sonicated and washed lamellae will mediate a photochemical oxidation and dark reduction of cytochrome *c*:

$$\text{Cyt}(\text{Fe}^{2+}) + \text{CRS} \xrightarrow{\text{light}} \text{Cyt}(\text{Fe}^{3+}) + \text{CRS}^- \quad (1)$$

$$\text{Cyt}(\text{Fe}^{3+}) + \text{CRS}^- \xrightarrow{\text{dark}} \text{Cyt}(\text{Fe}^{2+}) + \text{CRS} \quad (2)$$

Thus, cytochrome *c* is oxidized to a certain steady-state level in light and reduced back again exactly to its original level in a following dark

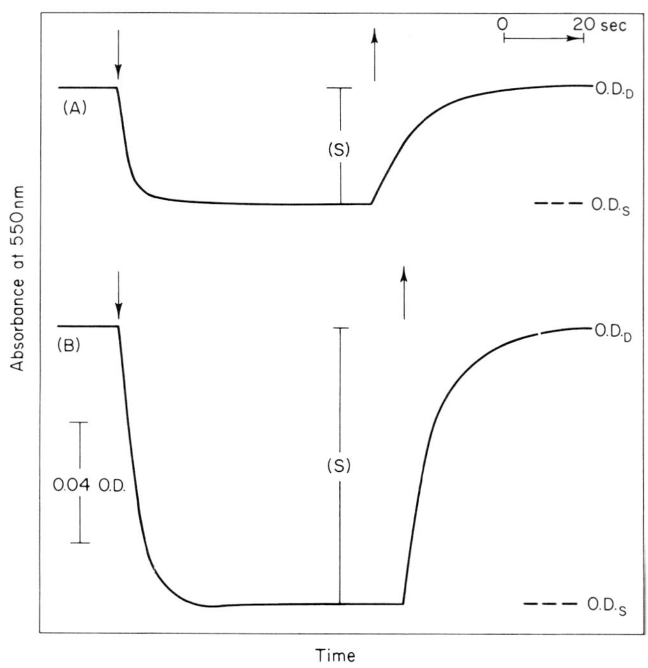

FIG. 1. Time course of anaerobic oxidoreduction reaction of mammalian cytochrome *c* by *Anabaena* lamellae. Curve A, without added CRS; curve B, with added CRS. Downward arrows, light on; upward arrows, light off. [S] measures the absorbance difference between the dark level ($O.D._D$) and the steady-state light level ($O.D._S$).

period (Fig. 1). The CRS content of the reaction mixture can be estimated from the observed kinetics.

Reagents

Sonicated and washed lamellae of *Anabaena cylindrica* (100 μg/Chl/ml) in 0.1 *M* Tris buffer, pH 7.6
Mammalian cytochrome *c*, 80% reduced, 15 mg/ml, in 0.1 *M* Tris buffer, pH 7.6
Sample to be tested in 0.1 *M* Tris buffer
Tris buffer, 0.1 *M*, pH 7.6

Procedure. Sonicated and washed lamellae of *Anabaena cylindrica* are prepared as described in the section on preparation. Cytochrome *c* is reduced with sodium dithionite, and excess dithionite is removed by Sephadex gel filtration. Lamellae (0.4 ml), the sample (1.0 ml), and buffer (2.2 ml) are placed in the main compartment of a Thunberg-type cuvette, and cytochrome *c* (0.4 ml) is placed in the side arm. The gas phase is replaced very carefully with purified N_2 or other inactive gas. A remaining trace of oxygen causes a considerable loss of reduced cytochrome and gives an erroneous result. The reagents are mixed in the dark; then the reaction is followed spectrophotometrically by measuring the absorption changes at 550 nm in the light and dark periods. Recording of optical density requires that the spectrophotometer be modified to provide an added actinic beam to the sample cuvette. A shutter on the actinic beam mechanically synchronized with a shutter on the exit measuring beam will permit brief interruptions of the actinic beam for measurements of optical density. More elegantly, a 550-nm interference filter in the exit measuring beam and a red cutoff filter (as Corning 2-60, >600 nm) in the actinic beam will allow continuous recording. The actinic intensity needed is about 2 mW/cm^2 or an apparent 2000 lux for red light. The preparation is stable and the reactions (Fig. 1) can be run repetitively for 2 hours at room temperature. When lamellar preparations are obtained from spinach, addition of a catalytic amount of plastocyanin is necessary.

Calculation. When concentrations of both oxidized and reduced cytochrome *c* are large compared to CRS, then both light oxidation and dark reduction follow first-order kinetics. The amount of CRS in the reaction mixture can be calculated

$$[\mathrm{CRS}] = \frac{k_L}{k_L - k_D}[\mathrm{S}] \qquad (3)$$

where [S] is the amount of cytochrome *c* oxidized in light and reduced in dark (Fig. 1). The apparent rate constants can be estimated from plots of ln (O.D. — O.D.$_s$) vs time (k_L) and ln (O.D.$_D$ — O.D.) vs time (k_D). The amount of CRS is expressed in terms of electron or cytochrome *c* equivalents. Since lamellar fragments contain appreciable CRS, the amount in the sample tested must be obtained by subtraction.

Method 2

Solubilized CRS stimulates O_2-dependent photophosphorylation of spinach chloroplasts by providing an additional electron carrier between photochemical system 1 and O_2.[5] The stimulation can be used to measure the solubilized CRS content of a sample.

Reagents

1. Spinach chloroplasts (150 μg of Chl/ml) in 0.1 *M* Tris buffer, pH 8.0
2. Na_2HPO_4 containing 3 μCi/ml of ^{32}P (10^{-2} *M*)
3. ADP (10^{-2} *M*) in 10 m*M* Tris buffer, pH 8.0
4. Sample in 10 m*M* Tris buffer, pH 8.0
5. Tris buffer, 10 m*M*, pH 8.0

Procedure. Spinach chloroplasts are prepared by the usual method. The reaction mixture contains 0.1 ml each of reagents 2, 3, and 4 and 0.6 ml of buffer and is prepared in a small glass tube for centrifugation. Before the reaction is started, 0.1 ml of chloroplasts is added. Illumination with more than 5×10^4 lux white light is necessary to give light-saturating conditions. After 5–10 minute illumination at 25°, trichloroacetic acid (20%, 0.1 ml) is added. Chloroplasts are removed by centrifugation, and ^{32}P in the organic phosphate is measured by the method of Avron.[6] The reaction rate is proportional to the amount of CRS added up to 10^{-6} electron equivalent per liter. Many heat-stable substances other than CRS have been reported to have the activity for stimulation of the O_2-dependent photophosphorylation. Therefore, the present assay method is not specific for CRS, especially in case of crude preparation. A check by assay method 1 is necessary.

Preparation of Solubilized CRS[1]

The method has been developed for the isolation of CRS from two sources. Slight modifications are necessary when applied to the other organisms.

Step 1. Algal cells of *Anabaena cylindrica* (late log phase) or spinach chloroplasts (prepared by usual methods) are suspended in Tris buffer

[6] M. Avron, *Biochim. Biophys. Acta* **40**, 257 (1960).

(0.1 *M*, pH 7.6) containing 0.35 *M* NaCl[7] and 1 m*M* EDTA. The suspension is sonicated at 10 kHz for 20 minutes in the cold. After removal of unbroken materials by centrifugation at 3000 *g* for 10 minutes, the supernatant solutions are centrifuged at 30,000 *g* for 90 minutes or 104,000 *g* for 60 minutes. Chlorophyll-containing precipitates are resuspended in the same medium, and centrifuged again. The washing is repeated several times to remove soluble materials.

Step 2. Washed lamellae are suspended in 0.1 *M* Tris buffer (pH 7.6) containing 1 m*M* EDTA at the concentration of 1 mg/Chl/ml. Four volumes of the suspension are added to 6 volumes of cold acetone (—20°) and held for 20 minutes at —20°. Occasional stirring is preferable. The lamellae are collected by centrifugation, and remaining traces of acetone are removed by evacuation in the cold. The material is suspended in 0.1 *M* Tris buffer containing 1 m*M* EDTA and incubated under anaerobic conditions at 2°–5° for 24 hours. A continuous stirring is necessary. CRS is solubilized; precipitates may be extracted repetitively. The combined extracts are heated at 90° for 5 minutes and centrifuged to remove proteins. After dialysis against 10 volumes of water for

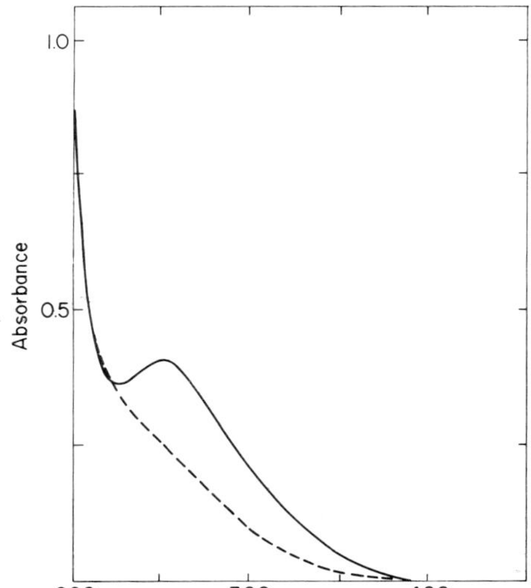

FIG. 2. Ultraviolet absorption spectrum of *Anabaena* CRS in 10 m*M* Tris buffer, pH 7.6. Dotted line shows the absorption spectrum in the presence of borohydride.

[7] NaCl at high concentration gives an aggregation of sonicated lamellae, which helps to effect precipitation at lower centrifugal force. If a 104,000 *g* centrifugation is available, NaCl can be omitted from the medium.

20 hours, the extracts are freeze-dried. The dried material is dissolved in a small volume of water, and any insoluble material is removed by centrifugation.

Step. 3. Further purification is carried out by a Sephadex gel-filtration. The crude CRS solution (20 ml) is applied to a Sephadex G-25 column (3 cm in diameter and 40 cm in length). Before use the column is equilibrated with 0.1 *M* Tris buffer (pH 7.6) containing 0.1 m*M* EDTA. The development is carried out by the same solution in the cold. CRS runs just after the protein position. Fractions containing CRS are combined, dialyzed and freeze-dried. Further gel-filtration is carried out by a column of larger scale compared with the same volume, e.g., 5×50 cm for 5 ml of sample. The CRS solution thus obtained shows the spectral characteristics presented in Fig. 2.

It is evident that the preparation methods so far developed are incomplete.

Properties

Isolated CRS is heat stable and water soluble, but not soluble in organic solvents such as acetone, petroleum ether, and chloroform.[1] From patterns of Sephadex gel filtration it is estimated that the molecular weight of CRS is in the range of 5000–10,000.

The most purified preparation so far obtained showed no light absorption in the visible region. At neutral pH there is a single light absorption peak at 250 nm which is removed by reduction with excess borohydride. The spectral change has no isosbestic point. The preparation is positive to the chemical color reactions specific to quinones and phenols (alkaline phloroglucinol test, acid *o*-phthalaldehyde test). It is evident that characterization is incomplete.

[59] ADP-Glucose Pyrophosphorylase from Spinach Leaf

By GILLES RIBÉREAU-GAYON and JACK PREISS

$$\text{ATP} + \alpha\text{-Glucose-1-P} \rightleftharpoons \text{ADP-glucose} + \text{PP}_i$$

Assay Method

Principle. Enzymatic activity was determined by measuring the synthesis of ATP–^{32}P from ADP–glucose and P–$^{32}P_i$ in presence or absence of the activator 3-phosphoglycerate (3PGA). The ATP is isolated by adsorption onto Norit A and estimated by measuring the radioactivity contained in the Norit. The procedure given below corresponds to the maximum of activity in presence of activator.

Reagents

ADP–glucose, 25 m*M*
Sodium pyrophosphate–^{32}P, 10 m*M* (specific activity $5\text{–}40 \times 10^5$ cpm/μmole)
Glycylglycine, 1 *M*, pH 7.5
$MgCl_2$, 0.1 *M*
Bovine plasma albumin (BPA), 10 mg/ml
3-Phosphoglycerate (3PGA), 10 m*M*
Trichloroacetic acid, 5%
Sodium pyrophosphate, 0.1 *M*, pH 8.0
Norit suspension, 150 mg/ml of water

Procedure. The ADP–glucose (0.01 ml), glycylglycine buffer (0.02 ml), pyrophosphate–^{32}P (0.05 ml), $MgCl_2$ (0.015 ml), BPA (0.01 ml), and 3-PGA (0.02 ml) solutions are mixed together, and the reaction is initiated (in a volume of 0.25 ml) with ADP–glucose pyrophosphorylase (0.0015–0.015 unit). After incubation at 37° for 10 minutes, the reaction is terminated by addition of 3 ml of cold 5% trichloroacetic acid containing 0.033 *M* pyrophosphate. The nucleotides are then adsorbed by the addition of 0.1 ml of the Norit A suspension. The suspension is centrifuged, and the supernatant fluid discarded. The charcoal is washed twice with 3-ml portions of 5% cold trichloroacetic acid and once with 3 ml of cold distilled water. The washed Norit is suspended in 2 ml of an aqueous solution of 0.1% NH_3 and 50% ethanol. One milliliter portions are pipetted into planchets and counted in a gas flow counter.

One unit of enzyme activity is defined as that amount of enzyme catalyzing the formation of 1 μmole of ATP in 10 minutes at 37° under the above conditions. Specific activity is designed as units per milligram of protein. Protein concentration is determined by the method of Lowry *et al.*[1]

Purification Procedure

Step 1. Extraction of the Enzyme. Fresh spinach leaves obtained from local supermarkets are deveined, washed, and then homogenized in a Waring Blendor at top speed for 3 minutes with an equal volume of

[1] O. H. Lowry, N. J. Rosebrough, A. L. Farr, and R. J. Randall, *J. Biol. Chem.* **193**, 265 (1951).

50 m*M* Tris·HCl buffer, pH 7.5, containing 2 m*M* reduced glutathione and 2 m*M* EDTA. The temperature is kept between 0° and 5°. The resultant suspension is then filtered through a Büchner funnel, and the filtrate is centrifuged at 16,000 *g* for 30 minutes. The supernatant fraction is used as the source of the enzyme. The purification described here and summarized in Table I corresponds to a 1370-g weight of deveined leaves.

Step 2. Heat Treatment and Ammonium Sulfate Precipitation. To the opalescent supernatant fraction is added sufficient 1 *M* phosphate buffer, pH 70, to give a final phosphate concentration of 25 m*M*. The resultant solution is then heated with stirring (in 1-liter portions in a 3-liter Erlenmeyer flask) in a water bath maintained at 80°–85°. When the temperature rises to 60°–62° (this occurred in about 5 minutes), the flask is transferred into another water bath maintained at 63°–65° and kept there for 5 minutes. The solutions are then quickly chilled in an ice bath. The precipitated proteins are removed by centrifugation (30 minutes at 16,000 *g*). The supernatant solution, which contained all the enzymatic activity, is then fractionated with solid ammonium sulfate. The fraction precipitating between 35 and 60% ammonium sulfate saturation contained practically all the enzymatic activity. This fraction is dissolved in 0.1 *M* Tris–succinate buffer, pH 7.2, which contained 20% sucrose, 1 m*M* GSH, and 1 m*M* EDTA, and then dialyzed overnight against 20 m*M* Tris–succinate buffer, pH 7.2, containing 20% sucrose, 1 m*M* GSH, and 1 m*M* EDTA.

Step 3. Adsorption of Pigments in a Column of Polyvinylpyrrolidone (PVP). The pigments which give a strong color to the dialyzed solution, (brown when the protein is concentrated and bright yellow when it is diluted) are adsorbed onto a column of insoluble polyvinylpyrrolidone (PVP). The PVP is prepared as described by Loomis and Battaile[2] and the fine particles are eliminated by decantation. Twenty-five milliliters of the PVP suspension is used for each 100 g of fresh spinach leaves. The column is equilibrated with the above buffer in which the proteins were dissolved. The same buffer is used to elute the protein. By following the optical density of this eluate at 280 nm it is possible to obtain most of the protein in about twice the volume added to the column. This step does not remove all the color and does not give any purification but aids in the purification of the enzyme in the latter steps.

Step 4. Chromatography on DEAE–Cellulose Column. The fraction obtained from the PVP column is adsorbed onto a DEAE–cellulose DE 52 column (350 ml of resin bed volume) which is equilibrated with 10 m*M* phosphate buffer, pH 7.4. The enzyme is eluted with a linear gra-

[2] W. D. Loomis and J. Battaile, *Phytochemistry* **5**, 423 (1966).

dient consisting of 10 mM phosphate, pH 7.4, in the mixing chamber (4 liters) and 50 mM phosphate, pH 6.0, containing 0.4 M NaCl in the reservoir chamber (4 liters). Fractions of 20 ml are collected and assayed for enzyme activity. The fractions containing the activity are pooled and the enzyme is precipitated by the addition of solid ammonium sulfate to a concentration of 60% saturation. After centrifugation the precipitate is dissolved in 0.1 M Tris–succinate buffer, pH 7.2, containing 20% sucrose, 1 mM GSH and 1 mM EDTA.

Step 5. Preparative Polyacrylamide Gel Electrophoresis. The concentrated enzyme solution obtained from the DEAE–cellulose chromatography step is dialyzed overnight against 58.7 mM Tris, 3.2 mM phosphate, sucrose 10%, 1 mM GSH, pH 7.2, at 25° and subjected to zone electrophoresis in a Poly-Prep apparatus manufactured by Buchler Instruments, Fort Lee, New Jersey, which incorporates the design features of the system developed by Jovin *et al.*[3] The Tris–glycine polyacrylamide gel buffer system was used.[4] Migration is toward the anode at alkaline pH (pH 10.3 at 0°). The resolving gel is chemically polymerized with persulfate and the concentrating gel is photopolymerized with riboflavin. For the resolving gel, an acrylamide monomer concentration of 7% and a *N,N′*-methylene bisacrylamide concentration of 0.25% are used. For the concentrating gel, the acrylamide concentration is 2.5% and the *N,N′*-methylene bisacrylamide concentration is 0.62%. The length of the resolving gel is 2.5 cm and that of the concentrating gel 0.5 cm The various buffer systems are used exactly as described in the Buchler instruction manual Polymerization of the gels and the fractionation procedure is carried out at 0° by the circulation of coolant from a Buchler refrigerated fraction collector. Concentration and resolution of the applied sample is carried out at a current of 50 mA with a power supply with current regulation. The flow rate is 0.7 ml/minute, and 5-ml fractions are collected. The peak fractions (68 ml) are pooled and concentrated with a Diaflo ultrafiltration cell. This step resulted in an overall purification of 1000-fold over the crude fraction with a 33% yield (Table I). The concentrated enzyme is dialyzed against 20 mM Tris·succinate buffer, pH 7.2, containing 20% sucrose, 1 mM GSH and 1 mM EDTA; it is stored in this buffer at —20°.

Properties of the Enzyme

Homogeneity and Molecular Weight. Analysis of the ADP–glucose pyrophosphorylase on analytical disc gel electrophoresis in several gel

[3] T. Jovin, R. A. Chrambach, and M. Naughton, *Anal. Biochem.* **9**, 351 (1964).

[4] L. Ornstein and B. Davis, *Ann. N.Y. Acad. Sci.* **121**, 321 (1964).

TABLE I
PURIFICATION OF ADP–GLUCOSE PYROPHOSPHORYLASE

Step	Volume (ml)	Protein (mg/ml)	Specific activity (units/mg)	Total activity (units)
1. Supernatant fluid 16,000 *g*	2130	6	0.96	12,230
2. Heat treatment and ammonium sulfate	61.4	69	2.5	10,600
3. PVP chromatography	120	29	2.5	8,900
4. DEAE–cellulose chromatography	7.5	19.6	54.4	8,000
5. Preparative disc gel electrophoresis	3	1.5	935	4,200

concentrations (5–9%) using the Tris–glycine buffer system and Coomassie blue[5] as the protein stain indicated only one protein band being present. A very light band which moved with the dye front in all gel concentrations which may be due to oxidized glutathione present in the enzyme preparation was also observed. After a few days of storage, a new light band with a migration rate slower to the major band appeared. This band seemed to be related to a dimer of the enzyme as it corresponded to a molecular weight about double the native enzyme (see below). The major band was shown to contain all the enzyme activity as determined by an activity stain.[6] The molecular weight of the enzyme was determined by disc gel electrophoresis by the method of Hedrick and Smith[7] employing the Tris–glycine buffer systems. The major band containing the enzyme activity was estimated to have a molecular weight of 210,000. The molecular weight of the very faint slower moving band corresponded to a value of 435,000 suggesting that on standing a dimer of the enzyme may be formed.

The molecular weight by sedimentation equilibrium was determined by carrying out a meniscus depletion run at three different concentrations of enzyme (0.2, 0.4, and 0.8 mg/ml) as described by Yphantis.[8] The molecular weight determined by this method ranged between 195,000 for an assumed partial volume $\bar{v}$ of 0.70 and 240,000 for an assumed $\bar{v}$ of 0.75. The values of the effective reduced molecular weight were similar for the three protein concentrations (3.70 cm^{-2} for 0.2 and 0.4 mg/ml

[5] R. A. Chrambach, M. Reisfeld, M. Wyckoff, and J. Zaccari, *Anal. Biochem.* **20,** 150 (1967).
[6] N. Gentner and J. Preiss, unpublished results.
[7] J. L. Hedrick and A. J. Smith, *Arch. Biochem. Biophys.* **126,** 155 (1968).
[8] D. A. Yphantis, *Biochemistry* **3,** 297 (1964).

and 3.74 cm^{-2} for 0.8 mg/ml). These similar values and the fact that the plot of the log of the blank corrected fringe displacement against the x coordinate, read directly from the comparator gave a straight line is good evidence of the homogeneity of the enzyme.

The addition of sucrose stabilized the enzyme. With no sucrose the activity was decreased 80% in 1 week even after the DEAE–cellulose column step. In 20% sucrose the activity of the purest preparation decreased 25% in 2 weeks. This diminution of activity was related to the formation of a new protein of a molecular weight double of that of the native enzyme (see above). Glycerol protected against inactivation in the same way as sucrose. The stability was found to be similar if the enzyme was kept in ice or at —20°.

Kinetic Properties. The enzyme from spinach leaf is activated by 3-phosphoglycerate and to lesser extents by other glycolytic intermediates (fructose 6-P, fructose di-P, and phosphoenolpyruvate).[9,10] The extent of the 3-PGA activation is dependent on pH; at pH 7.5 the stimulation of the maximal velocity of synthesis is 8- to 9-fold (Table II) and at pH 8.5 it is 60-fold.[9] The effect of 3-PGA on the apparent affinities of the substrates is shown in Table II. The presence of activator reduced the K_m of the substrates, ATP, ADP–glucose, and pyrophosphate 6- to 12-fold. Phosphate has been shown to be a potent inhibitor of the enzyme with 50% inhibition occuring at 60 μM. Phosphoglycerate, however, can antagonize this inhibition, and in the presence of 1 mM 3-PGA, 1.2 mM phosphate is now required to show 50% inhibition. The antagonism of phosphate inhibition by 3-PGA is a common property of many leaf

TABLE II
KINETIC PARAMETERS OF SPINACH LEAF ADP–GLUCOSE PYROPHOSPHORYLASE IN GLYCYLGLYCINE BUFFER pH 7.5

	K_m (mM)		V_{max} (μmoles/mg/min)	
Substrate/effector	−3PGA	+3PGA	−3PGA	+3PGA
ADP–glucose	0.93	0.15	35.5	93.5
Pyrophosphate	0.50	0.04	—	—
ATP	0.45	0.04	8.8	71
Glucose 1-phosphate	0.07	0.04	—	—
$MgCl_2$	1.4	1.6	—	—
3-Phosphoglycerate	0.02		—	

[9] H. P. Ghosh and J. Preiss, *J. Biol. Chem.* **241,** 4491 (1966).
[10] J. Preiss, H. P. Ghosh, and J. Witkop, *Biochem. Chloroplasts* **2,** 131 (1967).

ADP–glucose pyrophosphorylases[11] and of the ADP–glucose pyrophosphorylases from green algae.[12] These phenomena have formed the basis for a hypothesis on the regulation of the biosynthesis of leaf starches during photosynthetic activity.[9–11]

Acknowledgments

Research was supported by a fellowship from the Délégation Générale à la Recherche Scientifique et Technique and by a research grant from the United States Public Health Service (AI 05520).

[11] G. G. Sanwal, E. Greenberg, J. Hardie, E. C. Cameron, and J. Preiss, *Plant Physiol.* **43,** 417 (1968).

[12] G. G. Sanwal and J. Preiss, *Arch. Biochem. Biophys.* **119,** 454–469 (1967).

[60] Carotenoproteins[1]

By Bacon Ke

Despite the common occurrence of carotenoids in the photosynthetic apparatus of green plants and photosynthetic bacteria, little is known about their physiological function. Various roles have been suggested for carotenoids in photosynthesis, among them are energy transfer to chlorophyll,[2] electron transfer,[3,4] oxygen evolution,[5] and protection of chlorophyll against photooxidation.[6]

Certain purple bacteria can be made carotenoid-free either by mutation or by growing in the presence of diphenylamine. Under anaerobic conditions the carotenoid-deficient cells can carry out photosynthesis[6] and photophosphorylation.[7] But under aerobic conditions, illumination leads to photobleaching of the bacteriochlorophylls.[6] These results clearly demonstrate the protective function of carotenoids and at the same time preclude an obligatory role for carotenoids in photosynthesis.

[1] Contribution Number 369 from the Charles F. Kettering Research Laboratory, Yellow Springs, Ohio 45387. This work was supported in part by a National Science Foundation Grant 8460.

[2] L. N. M. Duysens, Doctoral Dissertation, University of Utrecht, 1952.

[3] J. R. Platt, *Science* **129,** 372 (1959).

[4] K. Yamashita, K. Konishi, M. Itoh, and K. Shibata, *Biochim. Biophys. Acta* **172,** 511 (1969).

[5] G. D. Dorough and M. Calvin, *J. Amer. Chem. Soc.* **73,** 2362 (1951).

[6] M. Griffiths, W. R. Sistrom, G. Cohen-Bazire, and R. Y. Stanier, *Nature (London)* **176,** 1211 (1955).

[7] I. C. Anderson and R. C. Fuller, *Nature (London)* **181,** 250 (1958).

Light-induced absorption-spectrum changes due to carotenoids have been observed for some time in photosynthetic bacteria,[8] and more recently in chloroplasts[4,9] and a number of algae.[10-12] The high quantum yield of these changes in photosynthetic bacteria indicate that they may be brought about by environmental changes around the carotenoid pigments[13] and may be related to photophosphorylation.[14,15] It is especially significant that identical spectral changes can be brought about by an electrical potential generated across the membrane by ionic gradients.[16]

The disposition or environment of carotenoids in the photosynthetic apparatus is even less understood. The nature of the *in vivo* environment of carotenoids is reflected by their absorption spectrum. Unlike in photosynthetic bacteria, the overlapping absorption of carotenoids with chlorophylls in green plants makes difficult a direct determination of the *in vivo* carotenoid absorption spectrum. By measuring the difference spectrum between the untreated chloroplasts and chloroplasts lyophilized and extracted with either heptane or isoctane, Ji *et al.*[17,18] showed that β-carotene in chloroplast lamellae absorbs at 493 and 468 nm. These absorption maxima represent a red shift of about 15 nm *in vivo* from those of the free pigments in *n*-hexane. From the further red-shifted absorption spectrum of the isolated β-carotene protein (see below), Ji *et al.* inferred that β-carotene molecules *in vivo* are probably solvated by the phytol chains of chlorophyll and are shielded from the direct effect of the polar groups of lamellar proteins. This is consistent with Butler's previous suggestion, as deduced from energy-transfer measurements on greening plastids, that β-carotene may be in direct contact with the phytol part of chlorophyll.[19]

Griffiths *et al.*[6] and Bergeron and Fuller[20] also suggested a possible

[8] L. Smith and J. Ramirez, *J. Biol. Chem.* **235,** 218 (1960).
[9] K. Sauer and M. Calvin, *Biochim. Biophys. Acta* **64,** 324 (1962).
[10] B. Ke, *Biochim. Biophys. Acta* **88,** 1 (1964).
[11] D. C. Fork and J. Amesz, *Photochem. Photobiol.* **6,** 913 (1967).
[12] Annual Report, Dept. of Plant Biology, Carnegie Institution, Stanford, California, p. 496 (1967–1968).
[13] J. Amesz and W. J. Vredenberg, *in* "Currents in Photosynthesis" (J. B. Thomas and J. C. Goedheer, eds.). Donker Press, Rotterdam, 1965.
[14] D. E. Fleischman and R. K. Clayton, *Photochem. Photobiol.* **8,** 287 (1968).
[15] M. Baltscheffsky, *Arch. Biochem. Biophys.* **130,** 646 (1969).
[16] J. B. Jackson and A. R. Crofts, *Fed. Eur. Biochem. Soc. Lett.* **4,** 185 (1969).
[17] T. H. Ji, J. L. Hess, and A. A. Benson, *in* "Comparative Biochemistry and Biophysics of Photosynthesis" (K. Shibata, A. Takamiya, A. T. Jagendorf, and R. C. Fuller, eds.), p. 36. Univ. of Tokyo Press, Tokyo, 1968.
[18] T. H. Ji, J. L. Hess, and A. A. Benson, *Biochim. Biophys. Acta* **150,** 676 (1968).
[19] W. L. Butler, *Arch. Biochem. Biophys.* **92,** 287 (1965).
[20] J. A. Bergeron and R. C. Fuller, *Nature (London)* **184,** 1340 (1959).

interaction between bacteriochlorophyll and carotenoid pigments as an explanation for the variability in the near-infrared absorption bands in normal and carotenoid-deficient bacterial chromatophores. These notions were corroborated by low-temperature spectroscopic evidence. Clayton and Arnold[21] observed that cooling the chromatophore film changes the near-infrared absorption bands of bacteriochlorophyll in a manner suggesting a suppression of bacteriochlorophyll–carotenoid interactions.

A large number of carotenoproteins have been isolated from a wide variety of animal organisms.[22] Carotenoid–protein association in phytophagous larvae was suggested as early as 1883 by Poulton,[23] and in 1926 Toumanoff suggested a role for lipid in the binding of carotene and protein in holothurians.[24] Among the many invertebrate carotenoproteins, only three have well-known stoichiometric relationships and have been subjected to extensive chemical and physicochemical investigations. These are crustacyanin, a blue protein of lobster carapace; ovoverdin, the green storage protein of lobster eggs, whose lipid content has only been recorded in 1964[22]; and ovorubin, a red glycoprotein. The carotenoids present in carotenoproteins seem to be restricted to astaxanthin (and its ester) and canthaxanthin, both of which have free carbonyl groups in the 4- and 4′-positions on the terminal ionone rings. The ease with which the carotenoids are extracted by acetone or ethanol suggests that covalent bonds are not involved in the binding. In some cases, the carotenoids and protein in the complex are found to have a mutual stabilizing effect, i.e., against photooxidation of the pigment and denaturation of the protein.

One characteristic of the absorption spectrum of the invertebrate carotenoproteins is the red shift of the fundamental absorption bands from those of the free pigments. The majority of the invertebrate carotenoproteins are blue or green, with absorption maxima lying between 560 and 680 nm, and the others are red to purple, with absorption maxima between 490 and 530 nm. Some carotenoproteins are found to have absorption at shorter wavelengths (390–410 nm) than those of the free prosthetic group. The absorption spectra usually show a greater red shift in the canthaxanthin proteins than in astaxanthin proteins. The red shift in absorption spectrum has been attributed to some polarizing effect by the protein on the conjugated double-bond polyene chains.

The β-carotene–protein with red-shifted absorption spectrum and iso-

[21] R. K. Clayton and W. Arnold, *Biochim. Biophys. Acta* **48,** 319 (1961).
[22] D. F. Cheesman, W. L. Lee, and P. F. Zagalsky, *Biol. Rev. Cambridge Phil. Soc.* **42,** 131 (1967).
[23] E. B. Poulton, *Proc. Roy. Soc.* **38,** 269 (1885).
[24] K. Toumanoff, *Publ. Staz. Zool. Napoli* **7,** 479 (1926).

lated from green-plant chloroplasts as well as the carotenoid complex with blue-shifted absorption spectrum and isolated from *R. rubrum* chromatophores are apparently not the natural entities originally present in the photosynthetic apparatus. The complexes are apparently affected by the isolation procedures used, as suggested by the modified absorption spectrum of the complexes. However, the fact that two kinds of complexes have been obtained by different workers, using slightly different extraction methods, indicates that the interactions involving the carotenoid pigments may be very specific.

Preparation Procedures

Since only a few preparation procedures have been reported for β-carotene protein, they are enumerated here. The common requirement among the different procedures for the formation of β-carotene protein appears to be the removal of chlorophyll and lipids.

1. Methanol Extraction of Chloroplasts.[25] The first preparation of a β-carotene protein complex was reported by Menke in 1940.[25] When chloroplasts were extracted with methanol to remove chlorophyll and most of the lipids, a brick-red β-carotene protein residue remained. Water or methanol suspensions of the extracted complex showed absorption maxima at 540 and 490 nm. The red-shifted spectrum was attributed to pigment–protein binding.

2. Ammoniacal Acetone Extraction followed by Acetone Fractionation and Ammonium Sulfate Fractionation.[26] Spinach or parsley leaves (4 kg) were dipped in 3.4 liters of chilled acetone containing 50 ml of 28% ammonia. The mixture was further ground in a mortar with sand and then centrifuged at 1700 *g* for 8 minutes.

To the supernatant solution was added an equal volume of cold acetone and 0.5 g of cellulose pulp. The mixture was kept at 0° for 30 minutes. The dark green precipitate was collected on a Büchner funnel by suction, washed with 1 liter of (3:1) acetone:water mixture containing 5 ml of 28% ammonia.

The pink precipitate remaining on the funnel was successively washed with 1 liter of two-thirds and one-half saturated ammonium sulfate solution (pH 8). The orange precipitate was then extracted with 50 m*M* NaH_2PO_4 solution and centrifuged at 7000 *g* for 15 minutes to precipitate the insoluble portion. The extraction was repeated three times. The combined extract was then made 10% saturated with ammonium sulfate and centrifuged to remove the initial precipitate. The supernatant solution

[25] W. Menke, *Naturwissenschaften* **28**, 31 (1940).
[26] M. Nishimura and K. Takamatsu, *Nature* (*London*) **180**, 699 (1957).

was then made 30% saturated with ammonium sulfate and centrifuged at 7000 *g* for 15 minutes to collect the orange-red precipitate. The precipitate was dissolved in a minimum amount of water and dialyzed against water. The dialyzate was centrifuged at 34,000 *g* for 1 hour and the supernatant fluid was discarded. The precipitate was washed again with minimum amount of water and recentrifuged.

3. Acetone Extraction and Ammonium Sulfate Fractionation of Broken Chloroplasts.[27] Purified chloroplasts were suspended in water at near 0° for 20 minutes and ruptured. To the broken chloroplasts were added 4 volumes of chilled acetone. The mixture was filtered on a Büchner funnel by suction to remove free pigments. The remaining pale-green chloroplasts were then washed with water, ground in a mortar with 0.1 *M* pH 7 phosphate buffer, and centrifuged at 10,000 *g* for 15 minutes.

The red supernatant solution was made 10% saturated with ammonium sulfate (pH 8) and centrifuged to remove the precipitate. The supernatant solution was then brought to 30% saturation with ammonium sulfate. The orange-red precipitate was collected by centrifuging at 3000 *g* for 10 minutes. The ammonium sulfate fractionation was repeated twice more. The precipitate was dissolved in a minimum amount of water and dialyzed against water. The dialyzate was centrifuged at 24,000 *g* for 1 hour to collect the β-carotene–protein complex.

4. Cholate Extraction of Acetone-Treated Chloroplasts.[27] To the acetone-treated chloroplasts (see procedure 3 above), cholate was added to give a concentration of 2%. The chloroplasts were extracted in a refrigerator overnight. The mixture was centrifuged at 7000 *g* for 10 minutes, and the supernatant solution was dialyzed against water to remove cholate. The red solution was purified by ammonium sulfate fractionation as above.

5. Sequential Acetone Extraction of Chloroplast Fragments.[7,18] Deribbed spinach leaves were macerated in a Waring Blendor for 15 seconds at maximum speed with 20 m*M* Tricine (or Tris[28]) buffer, pH 7.9, containing 10 m*M* NaCl and 0.8 *M* sucrose. A volume equal to the leaf weight was used. The homogenate was filtered and chloroplasts were collected between 250 and 650 *g* (10 minutes). The chloroplasts were resuspended in 20 m*M* Tricine (pH 7.9) containing 10 m*M* NaCl and sonicated at 20 kHz for 60 seconds. The sonicated solution was centrifuged at low speed to remove whole chloroplasts. The supernatant solution was then centrifuged at 35,000 *g* for 40 minutes. Hess and

[27] M. Nishimura and K. Takamatsu, *Plant Cell Physiol.* (*Tokyo*) **1**, 305 (1960).

[28] M. W. Banschbach and J. L. Hess, *Abstr. Biol. 046, 158th Nat. Meeting, Amer. Chem. Soc., 1969;* J. L. Hess, personal communication.

Banschbach[28] have also substituted the sonication step by fragmenting the chloroplasts in a glass–Teflon tissue grinder.

The green pellet of chloroplast fragments obtained above was resuspended sequentially in 45, 70, and 80% acetone, and after each resuspension it was centrifuged at 40,000 *g* for 10 minutes. The pellet obtained by centrifuging after 80% acetone extraction was a "red" protein (the β-carotene protein). When the red protein was extracted with 100% acetone, all the β-carotene, quinone, and lipids were removed, and a "gray" protein (the lamellar protein) remained.

This isolation procedure has been applied to spinach as well as oats, wheat, soybean, and *Euglena* to obtain the β-carotene protein complex.[28]

6. Formation during Triton Fractionation of Chloroplasts.[29] In chloroplast fragmentation and fractionation by the action of Triton X-100, subchloroplast particles representing the two photosystems were obtained (see this volume [27] for more details). In the centrifugation step for collecting the photosystem II particles, an orange-red layer of fluid material was observed above the green pellet. This material, after purification by repeated centrifugation, shows the typical absorption spectrum of a β-carotene protein complex (see the table and Fig. 1). No photochemical activity was detected for this complex.

7. Others. Bishop[30] reported the isolation of a β-carotene protein complex from photosynthetic tissues. However, no details were given on the procedure used.

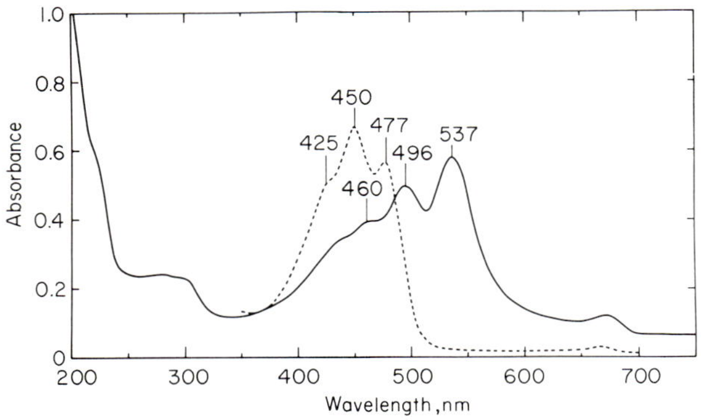

FIG. 1. Absorption spectra of an aqueous suspension of a β-carotene protein (solid curve) and the same material dissolved in *n*-hexane (dashed curve) (see Vernon *et al.*[29]).

[29] L. P. Vernon, B. Ke, H. H. Mollenhauer, and E. R. Shaw, *in* "Progress in Photosynthesis Research" (H. Metzner, ed.), Vol. 1, p. 137. Tubingen, Germany, 1969.

[30] N. I. Bishop, *Biophys. Soc. Meeting, Washington, D.C., 1962,* Abstract FC-10.

ABSORPTION MAXIMA OF β-CAROTENE

Medium	Wavelength (nm)			Reference
n-Hexane	478	450	425	a,b
Phytol	493	468	—	a,b
Isolated chloroplasts	493	468	—	a,b
Protein complexes	540	490	—	c
	537	496	460	d,e
	550	498	465	f
	538	498	460	a,b
	537	496	460	g

[a] T. H. Ji, J. L. Hess, and A. A. Benson, *in* "Comparative Biochemistry and Biophysics of Photosynthesis" (K. Shibata, A. Takamiya, A. T. Jagendorf, and R. C. Fuller, eds.). Univ. of Tokyo Press, Tokyo, 1968.
[b] T. H. Ji, J. L. Hess, and A. A. Benson, *Biochim. Biophys. Acta* **150,** 676 (1968).
[c] W. Menke, *Naturwissenschaften* **28,** 31 (1940).
[d] M. Nishimura and K. Takamatsu, *Nature (London)* **180,** 699 (1957).
[e] M. Nishimura and K. Takamatsu, *Plant Cell Physiol.* (*Tokyo*) **1,** 305 (1960).
[f] N. I. Bishop, *Biophys. Soc. Meeting, Washington, D.C.*, 1962, Abstract FC-10.
[g] Vernon *et al.*[29]

Kahn[31] reported a carotene–protein complex extracted from a chlorophyll–protein complex. The chlorophyll–protein complex was prepared by fractional solubilization of chloroplasts with Triton and DEAE–cellulose column chromatography. The chlorophyll–protein complex was extracted five times with equal volumes of 25% diethyl ether in petroleum ether. The combined extracts were washed with water, dried, and resuspended in 2 m*M* Tris buffer (pH 8) and clarified by centrifugation. The opalescent yellow solution contained the carotene–protein. The absorption spectrum of the carotene in this complex was very similar to that of the free pigment in *n*-hexane and was presumably different from the others described above.

Physical and Chemical Properties

All the carotenoproteins reported for green plants involve β-carotene, and there is a fairly good agreement on the wavelengths of maximum absorption of the complexed β-carotene reported by different workers. However, many of the properties discussed are descriptions and observations made on the complex isolated by given investigators and do not necessarily apply to complexes isolated by others.

The β-carotene protein complex has variably been described as orange-red or red. The complex isolated by Nishimura and Takamatsu[26]

[31] J. S. Kahn, *in* "Photosynthetic Mechanism in Green Plants," p. 496. Nat. Acad. Sci.—Nat. Res. Council, Publ. 1145, Washington, D.C., 1964.

appears under the electron microscope as particles of uniform size, with a mean radius of 26 nm. The weight-average particle weight estimated from the electron micrograph and from sedimentation measurements are 5.7 and 8.3×10^7, respectively. Light-scattering measurements yielded a higher value.[27]

According to Nishimura and Takamatsu,[26] the complex shows positive biuret reaction characteristic of proteins and the Molish reaction for carotenoids. After extraction with hot methanol–acetone mixture (1:1) followed with ether, 33% of its weight was lost, which suggests that the complex may be a lipoprotein. When the complex was first treated with ethanol and then with ether or acetone, a yellow pigment and an insoluble protein were separated. A similar observation was reported earlier by Menke.[25] The complex coagulates irreversibly at lower pH (25). Bishop[30] also reported that the complex is insoluble at low pH and low salt concentration, but becomes soluble at high pH and high salt concentration. The absorption spectrum of the complex was not altered by the addition of ascorbic acid or sodium dithionite.[30]

The β-carotene protein isolated by Nishimura and Takamatsu[26] contained 9.5% nitrogen and 2% β-carotene on a dry weight basis. This would correspond to 3×10^3 β-carotene molecules per particle. The amount of β-carotene present in the isolated protein complex was less than 1% of the total β-carotene.

The absorption spectrum of an aqueous suspension of the β-carotene protein and β-carotene separated and dissolved in *n*-hexane are reproduced in Fig. 1 (from Vernon *et al.*[29]). The wavelengths of maximum absorption reported by various authors are listed in the table. Three sets of spectral values are practically identical. The wavelengths of maximum absorption reported by Bishop[30] are slightly but consistently higher, which could reflect a slightly different state of binding. The absorption maxima of β-carotene in *n*-hexane, phytol, as well as in isolated chloroplasts are also listed for comparison.

Nishimura and Takamatsu[27] described their β-carotene protein as being yellow fluorescent, but no precise emission spectrum was given. Both Nishimura and Takamatsu[27] and Bishop[30] reported that their complex showed streaming birefringence, suggesting an asymmetry in particle shape.

Based on the amount of β-carotene present in the isolated protein complex, Nishimura and Takamatsu[26] early suggested that the complex may represent one of the natural states of β-carotene *in vivo* and have some physiological role. In this case, the red-shifted absorption spectrum of this minute fraction of the complex would be masked by the bulk absorption spectrum.

A similar suggestion that an absorption spectrum with wavelength shift of this magnitude may represent the natural state of the bulk carotenoid *in vivo* was made by Menke[25] earlier for a brown alga. When the brown alga *Laminaria* containing fucoxanthin was heated briefly at 70°, it instantly changed from brown to green. The brown alga had absorption bands at 678, 638, 589, 545, and 499 nm, whereas in the green form the last two bands disappeared, and only an absorption tail at 510 nm remained. On the other hand, the methanol extracts of either the brown or the green form were spectroscopically identical.

Recent work by Ji *et al.* showed that the bulk β-carotene can assume the red-shifted absorption spectrum when most of the chlorophyll and lipids were removed. When the chloroplast fragments were extracted with 45% acetone, 50% of the lipids and all the xanthophylls were removed. The difference spectrum between the 45% acetone-extracted chloroplasts and that subsequently extracted with heptane showed that the wavelengths of maximum absorption of β-carotene were not altered. However, when the chloroplasts were initially extracted with 70% acetone, the amount of lipids and xanthophylls removed was similar to that by 45% acetone, but the amount of chlorophylls removed was approximately 10 times that obtained with 45% acetone. The β-carotene in 70% acetone-extracted chloroplasts showed a red-shifted spectrum, as revealed by difference spectroscopy. Extraction with 80% acetone removed most lipids and all of the pigments except β-carotene and quinones. The residual "red" protein shows absorption peaks at 538, 498, and 460 nm (see the table) and an additional peak at 418 nm due to cytochromes. The cytochromes may be separated from the β-carotene protein by centrifugation in 0.5 *M* sucrose.[28] The purified β-carotene protein contains 40–50 μg of β-carotene per milligram of protein, compared with only 6 μg per milligram of protein in the unpurified preparation. The purified β-carotene protein contains 5–6% of the total protein and 50–60% of the total β-carotene.[28] When β-carotene protein is extracted with 100% acetone and then heptane, a "gray" lamellar protein remains.

According to Ji *et al.*[18] resuspension of the "red" protein in a detergent solution (e.g., 5% sodium dodecyl benzene sulfonate) does not alter the absorption spectrum. The absorption spectrum of detergent suspensions of β-carotene is quite different from that of the "red" protein. By mixing the detergent solutions of β-carotene and the "gray" protein, the absorption spectrum of the free β-carotene is still maintained.

By means of a "double dialysis" technique, Ji *et al.* were able to reassociate free β-carotene and the "gray" lamellar protein into a reconstituted complex with absorption maxima similar to those of the red protein.[17,18] The protein binds approximately 1 mole of β-carotene per mole of lamellar protein assuming a molecular weight of 23,000 for the latter. Formation of the complex appears to be specific since bovine serum albumin does not reassociate with β-carotene.

The shift of the absorption maxima of β-carotene from the free-pigment state to that of the protein complex is about 2×10^3/cm in wavenumbers. From the spectral alterations in the β-carotene absorption regions during sequential extraction, and from the similarity in the absorption spectra of β-carotene in isolated chloroplasts and phytol, Ji *et al.*[17,18] suggested that β-carotene molecules *in vivo* are probably solvated by the phytol chains of chlorophyll, and are shielded from the direct effects of the polar groups of the lamellar protein. As a result of the removal of chlorophylls, the polarization of the surrounding medium—the lamellar protein—comes into play, and consequently resulted in the red shift of the β-carotene absorption spectrum. Thus, the *in vivo* environment of β-carotene in isolated chloroplasts is neither purely hydrophobic nor extremely polar. Interesting, as shown by Colmano, β-carotene in a mixed monolayer with chlorophyll, and in the absence of a protein, absorbs at slightly lower wavelengths than those in the *in vivo* systems.[32]

Carotenoid "Complex" with a Blue-Shifted Absorption Spectrum

Another kind of carotenoid "complex" with a strongly blue-shifted absorption spectrum has recently been reported by several workers.[33-35] Thus far the source of this complex has been *Rhodospirillum rubrum* chromatophores. The first observation was made by Izawa *et al.*[33] When they treated *R. rubrum* chromatophores with 0.1% sodium dodecyl benzene sulfonate (SDBS) at pH 7 for 3 hours, the major near-infrared band at 878 nm due to bacteriochlorophyll disappeared and a new band appeared at 778 nm. In the carotenoid absorption region, the triple-peaked band (475, 504, and 545 nm) also disappeared. However, when the SDBS concentration was increased to 1%, the carotenoid band reappeared but at slightly shorter wavelengths (470, 500, and 530 nm). The absorption maxima of carotenoid in the methanol extract were at 455, 485, and 517 nm. Izawa *et al.*[33] attributed the disappearance of the carotenoid absorption band to an aggregation of carotenoid molecules which were detached from either protein or bacteriochlorophyll by the

[32] G. Colmano, *Nature (London)* **193,** 1287 (1962).

[33] S. Izawa, M. Itoh, T. Ogawa, and K. Shibata, *in* "Studies on Microalgae and Photosynthetic Bacteria" (Japanese Soc. Plant Physiology, ed.), p. 413. Univ. Tokyo Press, Tokyo, 1963.

[34] L. P. Vernon and A. F. Garcia, *Biochim. Biophys. Acta* **143,** 144 (1967).

[35] E. Fujimori, *Biochim. Biophys. Acta* **180,** 360 (1969).

action of SDBS. At higher detergent concentration, the carotenoid aggregates were dispersed into monomers, and thus the absorption band reappeared, but at slightly shorter wavelengths.

Enzymatic digestion of *R. rubrum* chromatophores with pancreatin or α-chymotrypsin in the presence of 0.5% Triton X-100 at 37° for 24 hours produced three pigment–protein complexes which could be separated by sucrose density-gradient centrifugation. In addition to a green and a blue band which contain mainly bacteriochlorophyll, a brown band containing practically all of the carotenoids was also obtained.[34] An aqueous suspension of the brown complex has a major absorption band at 368 nm, but no absorption in the 400 to 500-nm region. Extraction of the complex by acetone yielded the usual carotenoid absorption spectrum (see Fig. 2 top). Vernon and Garcia[34] reported that the transition between the two states is reversible, and addition of more water to the acetone suspension causes a reversal to the blue-shifted spectrum of the carotenoid complex.

Subsequently a carotenoid complex with almost identical absorption spectrum was reported by Fujimori.[35] When *R. rubrum* chromatophores were treated with 3% Triton at pH 2 and centrifuged in a sucrose density gradient, a reddish-brown fraction with an absorption peak at 370 nm and a bacteriopheophytin aggregate fraction absorbing at 850 nm were obtained.

Ke *et al.*[36] observed a marked extrinsic optical rotatory dispersion and circular dichroism associated with the blue-shifted absorption band, as shown in the lower half of Fig. 2. It was suggested that the brown complex may have been formed as a result of the enzymatic digestion in the presence of the detergent. It is likely that the carotenoid molecules were removed from the lipoprotein matrix during the enzymatic digestion step, and the free carotenoid molecules may have reaggregated into a complex with the modified absorption spectrum and optical rotatory properties. This suggestion was independently confirmed by Hager,[37] who reported a similar spectral shift accompanying the aggregation of several carotenoids dissolved in organic solvents upon the addition of water.

Aggregates with similar spectral properties have recently been found among other pigments. Emerson *et al.*[38] have reported on an aggregate of a cyanine dye, 3,3′-bis(β-carboxyethyl)-5,5′-dichloro-9-methylthiacarbocyanine, a photographic sensitizer, and have identified its blue-

[36] B. Ke, M. Green, L. P. Vernon, and A. F. Garcia, *Biochim. Biophys. Acta* **162,** 467 (1968).

[37] A. Hager, *Planta* **91,** 38 (1970).

[38] E. S. Emerson, M. A. Conlin, A. E. Rosenoff, K. S. Norland, H Rodriguez, D. Chin, and G. R. Bird, *J. Phys. Chem.* **71,** 2396 (1967).

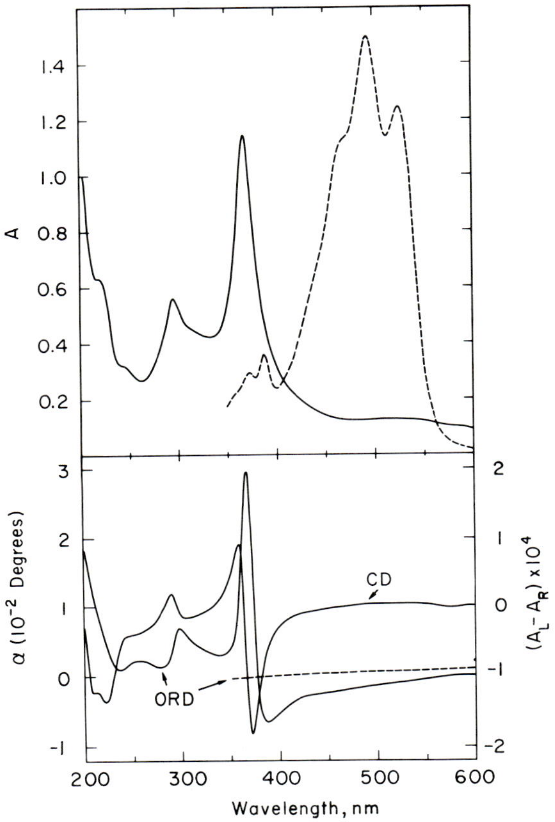

FIG. 2. Absorption spectra (top) and optical rotatory dispersion (ORD) and circular dichroism (CD) spectra (bottom) of the "carotenoid complex" in an aqueous suspension (solid curve) and in acetone (dashed curve).

shifted absorption spectrum by the use of a molecular exciton model. These aggregates are presumed to have a deck-of-cards structure, with the long axes of the molecules perpendicular to the axis of the aggregate. The observed perturbation of the absorption spectrum of the pigment aggregate is caused by the interaction arising from strongly coupled transition dipoles arranged in such a card-deck assembly.

It is of interest to note that a carotenoid aggregate with optical properties very similar to the carotenoid complex described here has more recently been found in a marine organism. Buchwald and Jencks[39] reported that aqueous suspension of the aggregate of astaxanthin, the

[39] M. Buchwald and W. P. Jencks, *Biochemistry* **7**, 834, 844 (1968).

chromophore group of the lobster shell pigment, crustacyanin, shows a large blue shift in its absorption spectrum and also a markedly enhanced optical rotation. The modified optical properties have been attributed to the π-electron interaction between the carotenoid molecules.

[61] Bacteriochlorophyll–Protein of Green Photosynthetic Bacteria

By JOHN M. OLSON[1]

The water-soluble blue-green complex has a molecular weight of 152,000 and contains 20 molecules of bacteriochlorophyll *a*.[2] *In vivo* the bacteriochlorophyll (Bchl) accepts excitation energy from chlorobium chlorophyll and transfers this energy to the reaction center chlorophyll, P840, which may or may not be a part of the complex *in vivo*.

Preparation[2]

Chloropseudomonas ethylica is grown anaerobically[3] in 20-liter carboys with illumination provided by two 60-W incandescent lamps (18-inch Lumiline type) and culture temperature maintained between 30° and 34°. The culture is stirred continually with a magnetic bar. The cells are harvested by precipitation with alum ($KAl(SO_4)_2 \cdot 12\ H_2O$, approximately 2 g/liter of culture medium), followed by low speed centrifugation of the aggregated cells.

Step 1. The packed cells (480-g wet weight) are suspended in 5 volumes (2.5 liters) of distilled water. Solid Na_2CO_3 is added to give a 0.2 *M* solution (pH ~ 10), and the cells are allowed to stand for an hour or more in the cold, before sonication for 10 minutes at 1.2 a in a Raytheon 10-kHz oscillator. The sonicated suspension is centrifuged at 41,000 *g* for 20 minutes.

Step 2. Solid ammonium sulfate (30 g/100 ml) is added to the supernatant solution, and the resulting green precipitate containing both bacteriochlorophyll and chlorobium chlorophyll is mixed with diatomaceous earth (0.5 kg of Celite 545) along with sufficient ammonium sulfate solution (35 g per 100 ml of 10 m*M* Tris, pH 8.0) to make a slurry.

[1] Research carried out at Brookhaven National Laboratory under the auspices of the U.S. Atomic Energy Commission.

[2] J. P. Thornber and J. M. Olson, *Biochemistry* **7,** 2242 (1968).

[3] S. K. Bose, *in* "Bacterial Photosynthesis" (H. Gest, A. San Pietro, and L. P. Vernon, eds.), p. 501. Antioch Press, Yellow Springs, Ohio, 1963.

Step 3. The slurry is poured into a column (60 × 14 cm) and eluted with a constant gradient of decreasing ammonium sulfate (35–0 g/100 ml) in 10 m*M* Tris, pH 8.0 (15 liters). Fractions are collected according to the color of the eluate. The first fraction is colorless, the second straw colored (cytochromes[4]), the third blue-green (bacteriochlorophyll–protein), and the fourth yellow-green. The blue-green fraction (6–8 liters) is concentrated to 150 ml by ultrafiltration and then dialyzed against 10 m*M* Tris, pH 8.0, to remove ammonium sulfate.

Step 4. The concentrated, salt-free Bchl–protein solution is loaded onto a DEAE–cellulose column (40 × 4 cm), which is eluted with a constant gradient of increasing NaCl (0 to 0.3 *M*) in 10 m*M* Tris, pH 8.0 (3 liters). The eluate may be frozen and stored at —10° at this stage.

Step 5. The eluted Bchl–protein solution is made 1 *M* in NaCl and is concentrated to about 10 mg/ml by ultrafiltration. Crystallization is effected by slow dialysis against 5–10 g of ammonium sulfate/100 ml of 1 *M* NaCl in 10 m*M* Tris buffer at 4°. For storage as a noncrystalline precipitate, 30 g ammonium sulfate is added per 100 ml of solution. Since the Bchl–protein is slowly oxidized in the presence of light and oxygen, it should be isolated under green light (F40 green fluorescent lamps behind green celluloid or green 2092 Plexiglas) and stored in the dark.

Properties

Purity. After completion of step 4, the Bchl–protein is at least 99% pure by the criterion of polyacrylamide gel electrophoresis.

Stability.[5-7] In solution the Bchl–protein is most stable in the presence of a salt such as NaCl (0.2–1.0 *M*) buffered between pH 7 and 8. In this pH range, the protein appears to be unaffected by the nonionic detergent Triton X-100 (5%) or by 8 *M* urea. However, sodium dodecyl sulfate (10–40 m*M*) causes a slow denaturation to a blue intermediate which eventually becomes pink due to pheophytinization of the Bchl. Although the Bchl can be extracted quantitatively from the protein by 90% methanol at 40°, the complex appears to be stable in the presence of alcohol if the temperature is low enough. The complex is stable in 50% methanol at 4°–5° and is soluble in a mixture of 60% methanol, 30% ethanol, and 10% water (vol. %) at —44°. At pH 7.7 the complex is stable at 40° for 24 hours, but it denatures within an hour at 72°. At 20°–25° the complex is relatively stable in the pH interval 3–12, although the absorptivity ratio $\epsilon_{809}/\epsilon_{371}$ drops approximately 10% as the pH is dropped

[4] J. M. Olson and E. K. Shaw, *Photosynthetica* **3,** 288 (1969).
[5] A. K. Ghosh and J. M. Olson, *Biochim. Biophys. Acta* **162,** 135 (1968).
[6] Y. D. Kim and B. Ke, *Arch. Biochem. Biophys.* **140,** 341 (1970).
[7] Y. D. Kim, *Arch. Biochem. Biophys.* **140,** 354 (1970).

from 7.8 to 4.0. Below pH 3 the Bchl–protein is converted to the pink bacteriopheophytin protein. Below pH 1.5 a blue intermediate accumulates prior to the formation of pheophytin. Similar blue intermediates are formed above pH 12.

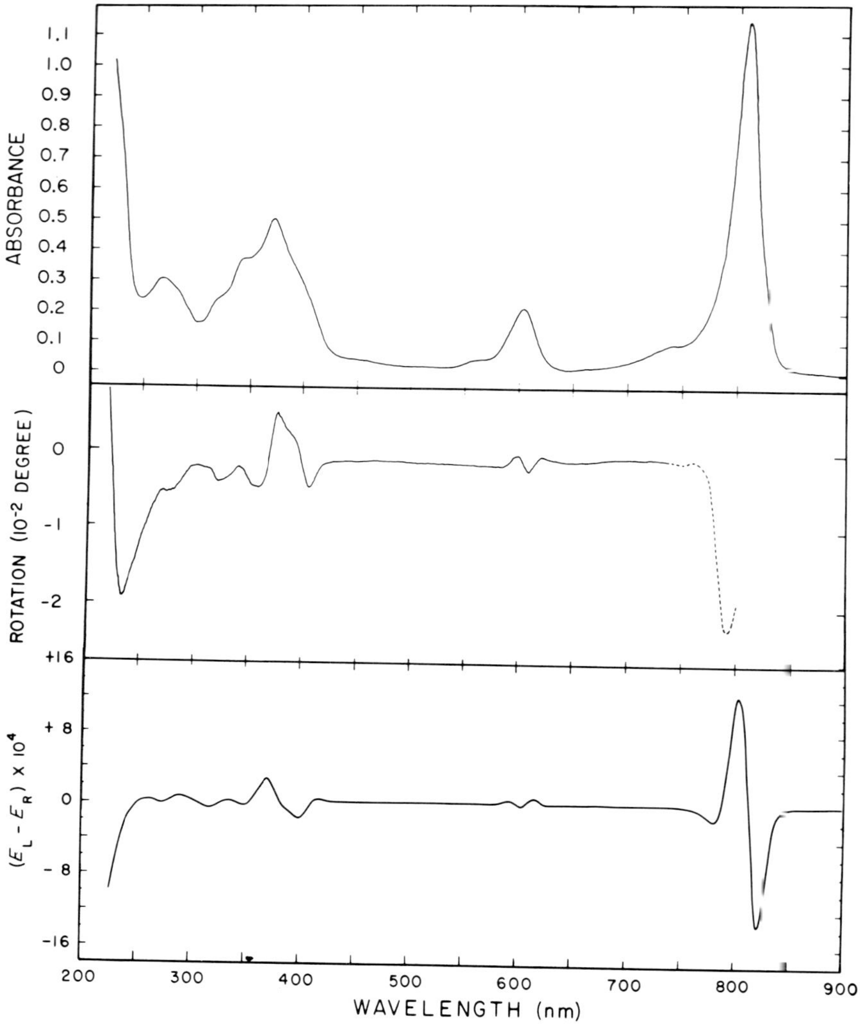

FIG. 1. Absorption and ORD spectra of Bchl–protein dissolved in 0.2 *M* NaCl in 10 m*M* phosphate, pH 7.8. [Bchl] = 7.5 μM [From B. Ke, in "The Chlorophylls" (L. P. Vernon and G. R. Seely, eds.), p. 427. Academic Press, New York, 1966.] Circular dichroism (CD) spectrum of Bchl–protein dissolved in 0.2 *M* NaCl in 10 m*M* Tris, pH 7.4. [Bchl] = 9.7 μM. Data of B. Ke.

Molecular Weight and Volume. The effective molecular weight, $M_{eff,20,w}$ is $3.6 \pm 0.1 \times 10^4$ by equilibrium sedimentation.[8] From the chemical composition the actual molecular weight is computed to be $1.52 \pm 0.04 \times 10^5$ g/mole. Thus the specific volume is calculated to be 0.764 ± 0.014 cm^3/g, and the molecular volume $1.93 \pm 0.08 \times 10^5$ Å^3. The sedimentation coefficient[9] $s_{20,w}$ is 7.3 S.

Chemical Composition.[2] The complex contains bacteriochlorophyll *a* and protein organized in four probably identical subunits of molecular weight 38,000 each containing five Bchl molecules. No carotenoid is present. The amino acid composition (moles per subunit) is Asp (33–34), Glu (28–29), Lys (16), His (7), Arg (18), Thr (13), Ser (23), Gly (34–35), Ala (18–19), Val (29–30), Ile (20), Leu (17–18), Pro (15), Met (3), Cys (2), Tyr (8), Phe (15), Trp 5, Amide (26–29). The N-terminal amino acid is alanine.

Spectral Properties. The absorption spectrum of the Bchl–protein in solution at room temperature is shown in Fig. 1. The absorption characteristics[8] are summarized in Table I. For quantitative measurement of

TABLE I
MOLAR ABSORPTIVITY VALUES BASED ON BACTERIOCHLOROPHYLL

λ (nm)	267	343	371	603	745	809
ϵ (mM^{-1} cm^{-1})[a]	37	49	67	28.4	13.4	154[b]
ϵ/ϵ_{371}[c]	0.55	0.73	1.00	0.42	0.20	2.30[b]

[a] Limit of error estimated to be ±4%.
[b] Variable depending on pH, temperature, and ionic strength.
[c] These are average values. The standard deviation in each case is about 2%.

concentration, the absorbance value at 371 nm is recommended. The absorptivity at 809 nm is variable, depending on ionic strength, pH, temperature, and history of the sample. The absorbance ratio A_{809}/A_{371} is a useful criterion of the state of the complex; values of 2.3–2.4 indicate a sample in excellent condition, whereas values below 2.0 indicate partial denaturation. The fluorescence emission peak at room temperature (20°–25°) is at 818 nm.[8] At 77°K the 809-nm absorption band is resolved into three sharp bands at 825, 814, and 805 nm and a broad band at ~793 nm, as shown in Figs. 2 and 3; the 603-nm absorption band is partially resolved into at least three components with peaks at 599 and 607 nm and a shoulder at 616 nm. The fluorescence emission peak sharpens and shifts to 831 nm as shown in Fig. 3. Fluorescence yields are 0.19 and

[8] J. M. Olson, *in* "The Chlorophylls" (L. P. Vernon and G. R. Seely, eds.), p. 413. Academic Press, New York, 1966.
[9] J. M. Olson, D. F. Koenig, and M. C. Ledbetter, *Arch. Biochem. Biophys.* **129**, 42 (1969).

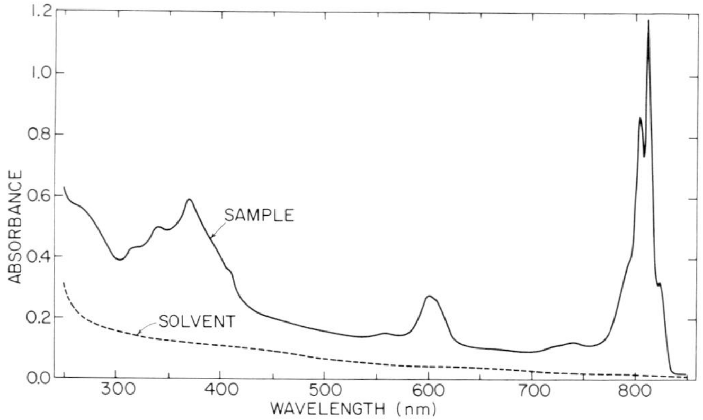

FIG. 2. Absorption spectrum of Bchl–protein dissolved in methanol–ethanol–water mixture at 77°K. Essentially the same spectrum is obtained for solutions in 50% glycerol at 77°K.

0.29 at 293° and 77°K, respectively.[8] Also shown in Fig. 1 is the optical rotatory dispersion (ORD) spectrum[10] at room temperature. This spectrum and the circular dichroism (CD) shown in Fig. 1 (*cf.* Kim and Ke[6]), and elsewhere[11] in this volume, constitute strong evidence of chloro-

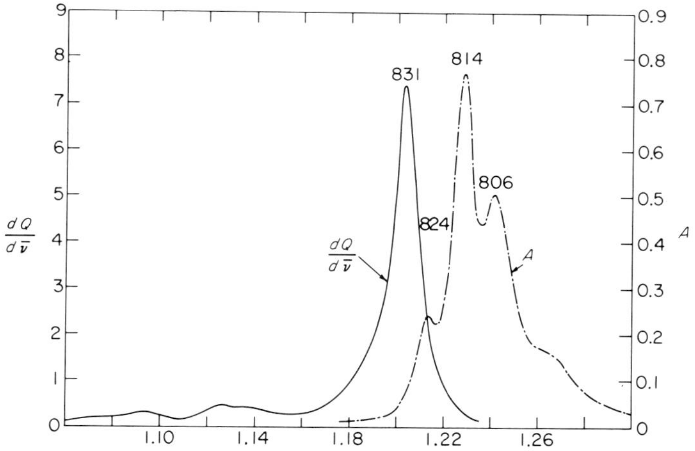

FIG. 3. Far-red absorption and fluorescence emission spectra at 77°K plotted vs wavenumber. Excitation wavelength for fluorescence was 366 nm.

[10] B. Ke, *in* "The Chlorophylls" (L. P. Vernon and G. R. Seely, eds.) p. 427. Academic Press, New York, 1966.
[11] See Vol. XXIV.

TABLE II

MOLECULAR ROTATION VALUES[a] BASED ON BACTERIOCHLOROPHYLL

λ (nm)	233	341	360	377	407	608.5	~792
[φ] (deg/mM/m)	−260[b]	−32	−70	+59	−67	−38	−310

[a] Data of B. Ke.[10]

[b] Specific rotation [α] is 3870° based on protein alone or 3400° based on protein plus Bchl.

phyll–chlorophyll interaction in the Bchl–protein. ORD characteristics are listed in Table II.

Crystal Properties.[9,12,13] Dimensions of the unit cell (hexagonal space group $P6_3$) are $a = b = 195 \pm 1$ Å and $c = 98.4 \pm 0.5$ Å. Each unit cell contains six macromolecules.[7] A shape approximated by an incompletely filled sphere ≈80 Å in diameter, is consistent with the packing arrange-

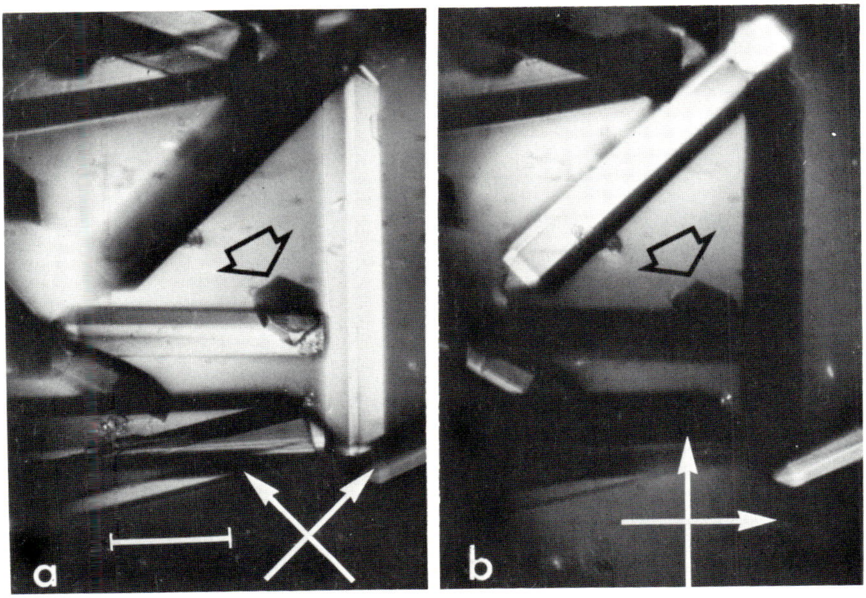

FIG. 4. Birefringence of crystals. The long arrows indicate the orientation of crossed Nicol prisms. Scale marker indicates 0.1 mm. Near the center of each figure, indicated by the blunt arrow, is a crystal habit in which the *c* axis is unusually short. From R. A. Olson, W. H. Jennings, and J. M. Olson, *Arch. Biochem. Biophys.* **129**, 30 (1969).

[12] L. W. Labaw and R. A. Olson, *J. Ultrastruct. Res.* **31**, 456 (1970).

[13] R. A. Olson, W. H. Jennings, and J. M. Olson, *Arch. Biochem. Biophys.* **129**, 30 (1969).

ment in the crystalline state.[12] Crystals exhibit a weak birefringence (2.5×10^{-3} at 550 nm) illustrated in Fig. 4.[13] In white light the polarization color is orange. The absorption spectrum of a crystal is essentially the same as that of a solution. Crystals are weakly dichroic with ratios of $1/(1.21 \pm 0.02)$ at 603 nm and 1.30 ± 0.04 at 809 nm. Selective dispersion of birefringence is negative at 603 nm and positive at 809 nm with respect to the optic axis which is identical to the crystallographic *c* axis.

Paracrystalline Aggregates.[12,14,15] If, after crystallization according to step 5, additional ammonium sulfate is added to the remaining mother liquor, more Bchl–protein comes out of solution as a paracrystalline aggregate consisting of more or less parallel tubules as shown in Fig. 5. Cross sections of tubules show a hexagonal array of electron dense elements surrounding an electron transparent channel. The channel diameter is about 130 Å, and the outside diameter (including cusps) is about 420 Å. The proposed packing arrangement of Bchl–protein macromolecules in the tubules is shown in Fig. 6 compared to the packing in true

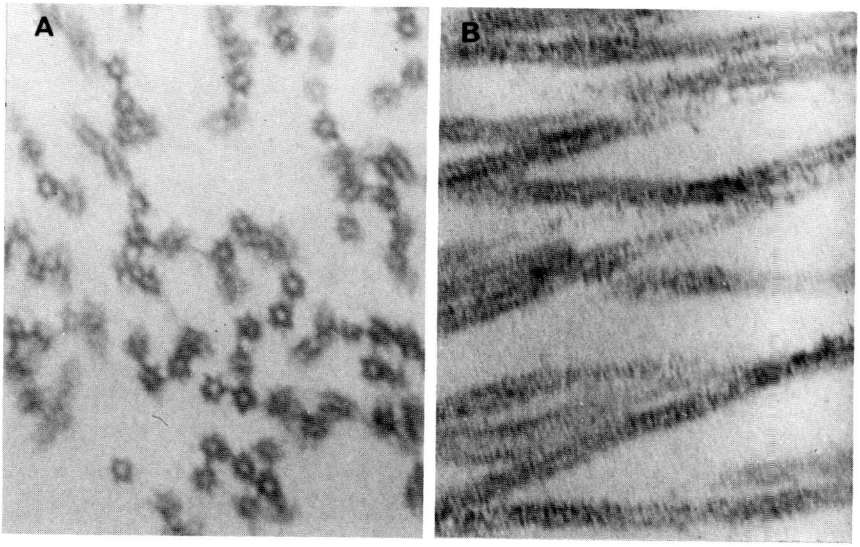

FIG. 5. Electron micrographs of tubules in cross section (A) and in longitudinal section (B). From R. A. Olson, W. H. Jennings, and C. H. Hanna, *Arch. Biochem. Biophys.* **130,** 140 (1969).

[14] R. A. Olson, W. H. Jennings, and C. H. Hanna, *Arch. Biochem. Biophys.* **130,** 140 (1969).

[15] R. A. Olson, *Science* **169,** 81 (1970).

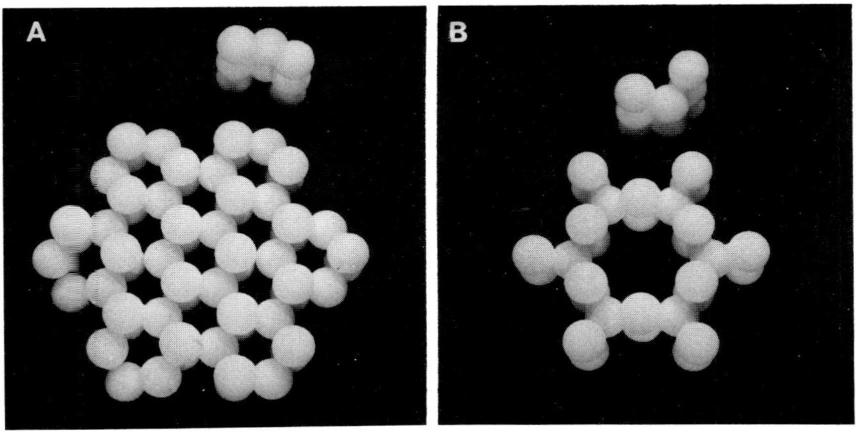

FIG. 6. Actual arrangement in the crystal lattice (A) compared to proposed arrangement in a cross section of tubule in a paracrystalline aggregate (B). From R. A. Olson, W. H. Jennings, and C. H. Hanna, *Arch. Biochem. Biophys.* **130,** 140 (1969).

crystals. In the model shown the channel diameter is 155 Å, and the outside diameter (including cusps) is 437 Å. The absorption spectrum of paracrystalline aggregates is similar to that of crystals and solutions. The dichroic ratio at 809 nm is +1.50 with preferential absorption parallel to the filamentous texture of the aggregates; at 603 nm the dichroic ratio is 1/1.09. The birefringence is of the same order and sign as in the crystals.

Similarity to Other Chlorophyll–Protein Complexes. Chlorophyll *a*–proteins of similar size and subunit structure have been isolated from blue-green algae and higher plants by the use of the detergent sodium dodecyl sulfate.[16]

Note added in proof: Spectral Properties. Since this article was originally written, there have been several measurements of the CD spectrum of the Bchl–protein at room temperature. The spectra appear to depend both upon the particular instrument used and upon the state of the sample examined. There is considerable variation in the quantitative aspects of the various CD spectra. Based on a comparison of five spectra obtained with three different instruments, the author would like to venture the following tentative opinions regarding the relative reliability of the published spectra. In the near UV region, most spectra, including the one

[16] See this volume [66].

shown in Fig. 1, indicate a value for $\Delta\epsilon_{370} - \Delta\epsilon_{400}$ of $50 \pm 5\ M^{-1}$ cm^{-1}. However, in the spectrum shown in Fig. 6 of Sauer,[11] the value is approximately 94 M^{-1} cm^{-1}. The author believes that this value may be twice the actual value measured. In the region of the orange band, all spectra show the same general shape, but there is little quantitative agreement among the various spectra. In the far-red region most spectra including the one shown by Sauer,[11] indicate a value of approximately $240 \pm 20\ M^{-1}$ cm^{-1} for $\Delta\epsilon_{800}$. In the spectrum shown in Fig. 1, however, a value of approximately 130 M^{-1} cm^{-1} is indicated. The author believes that this value may be only half the actual value at this wavelength. The value for the trough ($\Delta\epsilon_{820}$) is extremely sensitive to the condition of the sample and shows a wide range of values: —240 to —340 M^{-1} cm^{-1}.

[62] High Potential Iron Proteins: Bacterial

By ROBERT G. BARTSCH

Iron sulfide type proteins with redox potentials, $E_{m,7} \simeq 0.34$ V have been isolated from the photosynthetic bacteria *Chromatium* strain D[1] and *Rhodopseudomonas gelatinosa* (ATCC 11169, the same as van Niel strain No. 2.2.1).[2] No biochemical function for these iron proteins has yet been recognized.

Assay Method

Because no enzymatic reaction specifically involving high potential iron proteins (HiPIP) is available, it is necessary to depend upon absorption spectrum measurements to detect and to assay the proteins after separation from the bulk of the bacterial extracts. Absorption spectra of reduced and oxidized forms plus the oxidized-minus-reduced difference spectra of the two purified proteins are shown in Fig. 1. The spectroscopic properties of the two isolated HiPIP's are summarized in the table. The ratio $A_{283}:A_{388}$ is used as a purity index, the optimum ratio for the reduced proteins being 2.52 for *Chromatium* and 2.31 for *Rhodopseudomonas gelatinosa* HiPIP's, respectively. Reactions in which the proteins may undergo oxidation or reduction may be monitored by measuring the difference, $A_{480\,(ox)} - A_{480\,(red)}$. In both absolute and difference spectra the absorption bands are broad and of low intensity,

[1] R. G. Bartsch, *in* "Bacterial Photosynthesis" (H. Gest, A. San Pietro, and L. P. Vernon, eds.), p. 315. Antioch Press, Yellow Springs, Ohio, 1963.

[2] H. De Klerk and M. D. Kamen, *Biochim. Biophys. Acta* **2,** 175 (1966).

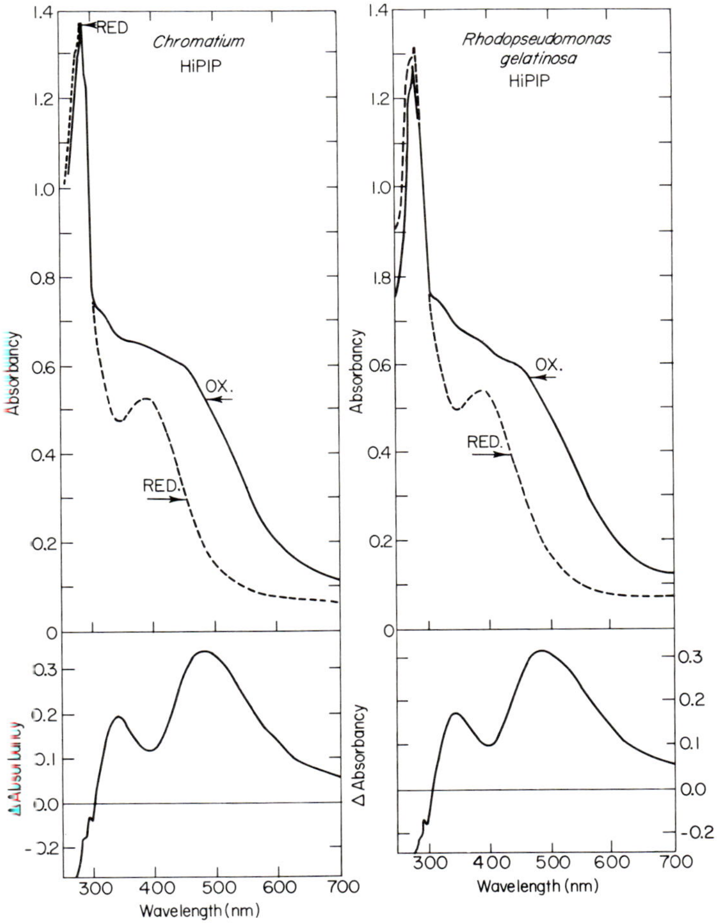

FIG. 1. Absorption spectra of *Chromatium* and *Rhodopseudomonas gelatinosa* high potential iron proteins (HiPIP). *Chromatium* HiPIP, 33 μM, and *R. gelatinosa* HiPIP, 35 μM, in 50 mM phosphate buffer, pH 7.0. The proteins were initially oxidized with Dowex 2-ferricyanide, then reduced with 15 mM 2-mercaptoethanol. From K Dus, H. De Klerk, K. Sletten, and R. G. Bartsch, *Biochim. Biophys. Acta* **140,** 291 (1967). (Reprinted by permission.)

consequently it is impractical to detect the proteins in the presence of highly absorbing pigments such as cytochromes or carotenoids present in the intact cell and in crude extracts.

Mass Culture of Bacterial Cells. Chromatium cells which contain HiPIP may be grown photoautotrophically with H_2 plus CO_2 or thiosulfate plus CO_2 as the major metabolites or photoheterotrophically

CHEMICAL PROPERTIES OF HIPIP'S[a]

Property	*Chromatium*	*Rhodopseudomonas gelatinosa*
Formula weight	10,074	9579
Iron	4	4
Labile sulfur	4	4
Cysteine	4	4
NH_2 terminal residue	Serine	Alanine
Carboxyl terminal residue	Glycine	Alanine
$E_{m,7}$	+0.35 V	+0.33 V
pI (0°)		
Oxidized	3.88	>10
Reduced	3.68	9.5
ϵ_{mM}, 283 nm, red	41.3	35.4
ϵ_{mM}, 388 nm, red	16.1	15.3
$\Delta\epsilon_{mM}$, 480 nm, ox–red	~10	~8.6

[a] K. Dus, H. De Klerk, K. Sletten, and R. G. Bartsch, *Biochim. Biophys. Acta* **140,** 291 (1967).

with malate or succinate as the metabolite. The greatest cell yield per unit volume of medium is obtained with the latter growth mode.[3]

Rhodopseudomonas gelatinosa cells contain HiPIP whether grown photoheterotrophically in a medium[3] containing succinate or malate or aerobically in the dark in a 100-liter fermentor in the same medium. In the latter cells the HiPIP level is reduced to approximately one-third the level in photosynthetically grown cells.

Purification Procedure

The iron proteins can be extracted from both *Chromatium* and *R. gelatinosa* by a variety of means including freeze–thaw lysing, extracting lyophilized cells or acetone powders, extracting cells with detergents (e.g., 1% v/v Triton X-100), and by disrupting the cells by sonication or by high-pressure extrusion.

Suspend 100 g (wet weight) of cells in 400 ml of 0.2 *M* Tris·HCl, pH 7.3, and process with the Sorvall-Ribi cell fractionator or a French press operated at 20,000 psi and at such a rate as to maintain the extruded sample temperature below 20°. Centrifuge the extract at 30,000 *g* for 10 minutes to remove cellular debris. Wash the precipitate by suspending it in 200 ml of 0.2 *M* Tris·HCl, pH 7.3, and centrifuging as before. The combined supernatant solutions are next centrifuged at

[3] See this volume [1]. Also S. K. Bose, *in* "Bacterial Photosynthesis" (H Gest, A. San Pietro, and L. P. Vernon, eds.), p. 501. Antioch Press, Yellow Springs, Ohio, 1963.

100,000 *g* for 90–120 minutes to deposit bacterial membrane fragments (chromatophores).

DEAE Chromatography. The clear yellow-brown supernatant solutions are then passed through a DEAE–cellulose column (Brown Co., Selectacel-Standard, 75-ml bed volume) equilibrated with the extraction buffer. The unadsorbed proteins are washed off the column with 400 ml of the same buffer. The ferredoxin which remains adsorbed may be eluted with 0.32–0.36 *M* NaCl in 20 m*M* Tris·HCl, pH 7.3, after preliminary washing with 0.3 *M* NaCl 20 m*M* Tris·HCl, pH 7.3, to remove proteins less acidic than the ferredoxin.

The combined unadsorbed solutions are desalted by passage through a Sephadex G-25 column equilibrated with 1 m*M* Tris·HCl, pH 8.0, for the *Chromatium* preparation or 1 m*M* phosphate buffer, pH 6.0, for the *R. gelatinosa* preparation. The pH of the desalted solution is checked and adjusted if necessary. Relatively little of the highly basic proteins in the crude *R. gelatinosa* extract are retarded on the Sephadex column. None of the acidic *Chromatium* proteins is retarded.

The *Chromatium* extract is passed into a DEAE–cellulose column (Brown Co., Selectacel Type 20, 100 ml bed volume) equilibrated with 1.0 m*M* Tris·HCl, pH 8, at 4°, to adsorb the colored proteins. The *R. gelatinosa* extract is adsorbed on a CM-cellulose column (Selectacel Standard, Brown Co., 100 ml bed volume) equilibrated with 1 m*M* phosphate buffer, pH 6.0. The columns are then rinsed with 400 ml of the application buffer to remove any unadsorbed proteins.

Chromatium HiPIP is eluted with 400 ml of 40 m*M* NaCl in 20 m*M* Tris·HCl, pH 8.0, after the column is washed with 400 ml of 20 m*M* NaCl in 20 m*M* Tris·HCl, pH 8.0, which primarily elutes cytochrome $c_{553(550)}$. The HiPIP cannot ordinarily be recognized on the column against the highly colored background, but is easy to recognize as a greenish-brown colored substance after elution from the column. At this stage the HiPIP solution is sufficiently pure to be assayed spectrophotometrically; generally the only colored contaminant is cytochrome $c_{553(550)}$ plus chromatophores which are apparently dislodged nonspecifically from the column. The solution is desalted with Sephadex G-25 equilibrated with demineralized water, then adjusted to pH 8.0 with 1 *M* NaOH solution and adsorbed on a DEAE–cellulose column (Type 20, 5-ml bed volume). The protein is eluted in highly concentrated solution with 0.5 *M* NaCl in 20 m*M* Tris·HCl, pH 7.3. Next, solid ammonium sulfate is added to the solution to 60% saturation to precipitate contaminating proteins including the cytochrome $c_{553(550)}$ plus chromatophores. After at least 10 minutes equilibration, the suspension is centrifuged at 30,000 *g* for 10 minutes. The HiPIP is precipitated almost completely from the supernatant solution by raising the ammonium sulfate concentration to 90% saturation. The precipitated protein is dissolved in 3 ml Tris·HCl, pH 8.0, containing 50 μM dithiothreitol to reduce any oxidized HiPIP, and then desalted with Sephadex G-25 equilibrated with demineralized water. The pH of the desalted solution is adjusted to 8 and the protein is adsorbed on a DEAE–cellulose column (Type 20, 20-ml bed volume) equilibrated with 20 mM Tris·HCl, pH 8.0. The column is rinsed with 200 ml of the application buffer, then with 200 ml of 30 mM NaCl in the same buffer to nearly elute the green-brown band off the column. Finally the band is eluted with 2–300 ml of 40 mM NaCl in the application buffer. The purity index of representative samples is measured, and those fractions with $A_{283}:A_{388} = 2.52$ are pooled and concentrated on a small DEAE–cellulose column as described before. The protein can be crystallized from the concentrated solution by adding ammonium sulfate to $\approx$70% saturation.

Chromatium HiPIP is slowly degraded during prolonged (6 months) storage at −20°. The degradation products are unknown, but increased $A_{283}:A_{388}$ ratios develop progressively. Perhaps oxidation of labile sulfur ligands which link to the iron results in destruction of the primary chromophore responsible for the 388-nm absorption band.

Rhodopseudomonas gelatinosa HiPIP is eluted from the CM–cellulose column with 600 ml of 10 mM phosphate, pH 6.5, after a preliminary wash with 400 ml of 2 mM phosphate, pH 6.5, to remove unadsorbed proteins. Cytochrome *cc′* with nearly an identical isoelectric pH contaminates the HiPIP which consists of a mixture of the oxidized plus reduced forms of two chromatographically different species of HiPIP. At this stage the HiPIP is sufficiently free of other colored substances to be assayed spectrophotometrically. The mixture is desalted with Sephadex G-25 equilibrated with 1 mM phosphate, pH 6.5. The more basic cytochrome *cc′* may be preferentially adsorbed by the slightly acidic Sephadex and consequently lag behind the main desalted zone of HiPIP. The HiPIP solution is made 1 mM with 2-mercaptoethanol or 100 μM with dithiothreitol to reduce the oxidized HiPIP. The reduction may require 1–2 hours to reach completion at room temperature. The HiPIP solution is charged into a CM–cellulose column (Selectacel Standard, 25-ml bed volume) equilibrated with the application buffer. The less abundant reduced HiPIP is eluted with 3–400 ml of 1 mM phosphate, pH 6.5, containing 1 mM 2-mercaptoethanol. The predominant HiPIP species is next eluted with 5 mM phosphate, pH 6.5, again containing 1 mM 2-mercaptoethanol. Fractions are collected and assayed, and those of similar purity are pooled. Finally the cytochrome *cc′*, which remains oxidized and therefore relatively strongly adsorbed, is eluted with 500 ml of 20 mM phosphate, pH 6.5. The two HiPIP fractions are treated separately. Indeed only the dominant form has been characterized. The protein solutions are desalted with Sephadex G-25, equilibrated with 1 mM phosphate buffer, pH 7. Some protein may be lost by adsorption on the Sephadex. Alternatively, a fresh undegraded Bio-Gel P-2 column equilibrated with demineralized water may be used, with little loss of HiPIP due to ion exchange adsorption. The desalted solution is concentrated on a small CM–cellulose column (5-ml bed volume) equilibrated with 1 mM phosphate buffer, pH 6.5, and desorbed by 0.2 M phosphate buffer, pH 7. Precipitation of the HiPIP with ammonium sulfate over the range 60–90% saturation may serve to improve slightly the purity index of the sample; crystallization from ammonium sulfate solution has not yet been accomplished.

Both proteins as isolated are essentially completely reduced. The proteins can be completely oxidized without appreciable contamination or dilution of the sample by passage through a small column of anion exchange resin such as Dowex 1-Cl^-, previously charged to 50–70% of capacity with ferricyanide ion. If the concentration of multivalent buffer ion is kept low (1–2 mM) there is no appreciable elution of ferri- or ferrocyanide from the column. The oxidized proteins are very rapidly reduced with sodium dithionite, but with sodium ascorbate or 2-mercaptoethanol or dithiothreitol the proteins are only slowly reduced; with 10 mM 2-mercaptoethanol approximately 15 minutes are required to fully reduce 50 μM HiPIP.

Properties

The properties of the two HiPIP's are summarized in the table. As determined by amino acid analysis, the formula weights for *Chromatium* and for *R. gelatinosa* HiPIP were determined as the sum of the residue weights of component amino acids plus the molecular weights of four iron atoms and four sulfur atoms per mole. Isoelectric pH's were measured by the isoelectric focusing technique of Vesterberg and Svensson[4] using LKB ampholine electrolytes and apparatus.

[4] O. Vesterberg and H. Svensson, *Acta Chem. Scand.* **20,** 820 (1966).

[63] Adenosine Triphosphatase: Bacterial

By T. Horio, K. Nishikawa, and Y. Horiuti

Membrane fragments prepared from a variety of photosynthetic bacteria, when illuminated, can synthesize ATP from ADP and P_i.[1] Chromatophores prepared from the facultative photoheterotroph, *Rhodospirillum rubrum,* catalyze the hydrolysis of ATP into ADP and P_i (ATPase activity); this activity is correlated at least in part with the reversibility of the energy conservation system leading to ATP formation.[2,3]

The enzyme, ATPase, has been extracted and purified[4] from mitochondria. In contrast, all attempts to extract such an enzyme from chromatophores or cells have been unsuccessful.

Preparations

Growth of Bacteria. Rhodospirillum rubrum cells are grown in 1-liter flat bottles at 30° under continuous illumination (approximately 1000 foot-candles) from a bank of tungsten lamps. The components of culture medium are the same as those described previously.[5] After incubation for 3–4 days, cells are harvested and washed two times with 0.1 *M* glycylglycine–NaOH buffer (pH 8.0) containing 10% (w/v) sucrose (glycylglycine–sucrose buffer) by centrifugation. Either the wild-type strain or the carotenoidless mutant strain (G-9; blue-green mutant strain)[6] can be used; essentially the same results are obtained with the two strains. Under conditions as mentioned above, approximately 10 g of cells in wet weight per liter can be obtained.

Preparation of Chromatophores. The washed cells are suspended in approximately 3 volumes (v/w) of 0.1 *M* glycylglycine–sucrose buffer, and ruptured by sonication in a 10-kHz oscillator for approximately 2 minutes at 0°–4°. Other methods, such as grinding cells with aluminum oxide or extrusion of a cell suspension (French pressure cell) are also applicable. Chromatophores prepared by these various methods do not

[1] See Vol. VI [38].

[2] T. Horio, K. Nishikawa, M. Katsumata, and J. Yamashita, *Biochim. Biophys. Acta* **94,** 371 (1965).

[3] Y. Horiuti, K. Nishikawa, and T. Horio, *J. Biochem.* **64,** 577 (1968).

[4] See this series, Vol. X [82].

[5] G. Cohen-Bazire, W. R. Sistrom, and R. Y. Stanier, *J. Cell. Comp. Physiol.* **49,** 25 (1957).

[6] Obtained from Dr. Jack W. Newton.

show conspicuous differences with regard to photosynthetic ATP formation, ATPase, exchange reactions,[7] and electron transport. The resultant suspension of disrupted cells is centrifuged, and the fraction sedimented between 20,000 *g* for 30 minutes and 105,000 *g* for 1 hour is collected. The precipitate is washed two times with buffer, and then suspended in a volume of buffer such that the resultant suspension has an $A_{880\,\text{nm}}$ per milliliter of 50. One unit of $A_{880\,\text{nm}}$ corresponds to 6.5 mμmoles of bacteriochlorophyll.[8] With chromatophores from the mutant strain (G-9), the absorption peak is centered at 873 nm instead of 880 nm. One unit of $A_{873\,\text{nm}}$ of chromatophores from the mutant strain corresponds to 7.1 mμmoles of bacteriochlorophyll. Chromatophores thus prepared can be stored at 0°–4° for at least 2 weeks without serious loss of photosynthetic ATP formation, ATPase, and exchange reactions. Approximately 50 units of $A_{880\,\text{nm}}$ of chromatophores can be obtained per gram wet weight of cells.

Synthesis of γ-^{32}P–ATP

ATP labeled only in the terminal phosphate group is synthesized by means of the ATP–P_i exchange reaction by illuminated chromatophores.[7]

Reagents

- Glycylglycine–NaOH buffer, 0.2 *M*, pH 8.0, 2.5 ml
- $MgCl_2$, 0.1 *M*, 0.5 ml
- ATP, 10 m*M*, 0.5 ml
- $^{32}P_i$ (approximately 10 mCi), 1 m*M*, 0.5 ml
- Phenazine methosulfate, 6 m*M*, 0.5 ml
- Chromatophore suspension ($A_{880\,\text{nm}}$/ml, approximately 50), 0.5 ml
- Water, 2.5 ml

Procedures. The reaction mixture (above) without chromatophores is divided equally and put into five small test tubes (1 × 10 cm). The reaction is started by adding chromatophores and carried out at 30° for 20 minutes under continuous illumination (approximately 1000 foot-candles). The reaction is terminated by adding 0.5 ml of cold 30% trichloroacetic acid to each tube, followed by centrifugation. The resulting supernatant solutions are mixed and passed through a charcoal column (2 × 2 cm). The charged column is washed with 100 ml each of 0.1 *N* and then 0.01 *N* HCl. After the phosphate is removed, labeled ATP is eluted with 60% ethanol containing 1% NH_3. The eluates showing ap-

[7] See this volume [64].

[8] R. K. Clayton, *in* "Bacterial Photosynthesis" (H. Gest, A. San Pietro, and L. P. Vernon, eds.), p. 495. Antioch Press, Yellow Springs, Ohio, 1963.

preciable radioactivity are combined and concentrated with the use of a lyophilization apparatus. All procedures mentioned above, unless otherwise noted, are carried out in a cold room. The radioisotopic yield recovered in ATP is approximately 70%. The sample of γ-^{32}P–ATP thus obtained is free of pyrophosphate, α- or β-^{32}P–ATP, ADP, and AMP.

Assay Methods

Principle. The assay is based on the measurement of $^{32}P_i$ liberated from γ-^{32}P–ATP. The procedures are similar to those of Nielsen and Lehninger[9] as modified by Avron.[10]

Reagents

- NaH_2PO_4, 20 mM
- Ascorbic acid, 1 M
- Ammonium molybdate, 5% containing 1 M H_2SO_4 (molybdate reagent)
- Isobutanol–benzene 1:1 (v/v) saturated with water
- Water saturated with isobutanol–benzene 1:1 (v/v)

Procedure. Standard components of the reaction mixture are: 0.2 M glycylglycine buffer (pH 8.0) containing 10% sucrose, 0.5 ml; 0.1 M $MgCl_2$, 0.1 ml; 1.5 mM γ-^{32}P–ATP (approximately 1×10^6 cpm) 0.1 ml; chromatophore suspension ($A_{880\,nm}$/ml, approximately 50), 0.1 ml; water, to make the total volume 1.5 ml. The reaction mixture is placed in test tubes of 1.0-cm diameter. The reaction is started by adding chromatophores, carried out at 30° for 4 minutes in darkness, and terminated by adding 0.5 ml of 30% trichloroacetic acid, previously cooled in ice water. After standing for at least 10 minutes in an ice-water bath, the test tubes are centrifuged and the supernatant fluid decanted. To 2 × 15 cm tubes containing 0.02 ml of 20 mM NaH_2PO_4 and 0.02 ml of 1 M ascorbic acid, 1-ml aliquots of the clear supernatant liquids are added, followed by 1.5 ml of water saturated with isobutanol–benzene and 0.3 ml of molybdate reagent. The tubes are shaken to mix the contents and allowed to stand for at least 5 minutes. Isobutanol–benzene, 4 ml, saturated with water is added to the mixtures, which are then shaken vigorously for 30 seconds. The two phases separate in about 1 minute; the blue material is transferred into the upper organic phase. One milliliter of the upper layer is removed, placed on a planchet, and evaporated to dryness; the radioactivity is then counted. The amount of $^{32}P_i$ liberated from γ-^{32}P–ATP

[9] S. O. Nielsen and A. T. Lehninger, *J. Biol. Chem.* **215**, 555 (1955).
[10] M. Avron, *Biochim. Biophys. Acta* **40**, 257 (1960).

is calculated by dividing the total radioactivity of the organic phase by the specific radioactivity of γ-^{32}P–ATP of the reaction mixture. The total radioactivity of the organic phase must be corrected by deducting the radioactivity obtained in the absence of chromatophores.

Distribution of ATPase Activity in *R. rubrum* Cells

A cell suspension of the blue-green mutant strain (G-9) of *R. rubrum* (80-mg/ml wet weight) is disrupted by sonication and fractionated as described above. Protein is determined by the biuret method; bacteriochlorophyll is estimated spectrophotometrically. The suspension is extracted with 10 volumes of a mixture of acetone and methanol (7:2, v/v). The molar extinction at 772 nm of extracted bacteriochlorophyll is taken to be 72×10^3.[8] Ubiquinone-10 is estimated by the method as described elsewhere.[11]

FRACTIONATION OF *Rhodospirillum rubrum* CELLS AND DISTRIBUTION OF ATPASE ACTIVITY

	Components or activities/80 mg of cells, wet weight				
				μMoles P_i liberated per hour[a]	
Fraction	Protein (mg)	Bacterio-chlorophyll (mμmoles)	Ubi-quinone-10 (mμmoles)	+ None	+ Oligomycin, 3.3 μg/ml
Cells	11	100	108	0.65	0
Cell debris	5.3	68	77	3.4	0.7
Chromatophores	4.4	35	39	1.7	0.6
Supernatant solution	3.1	0	0	0	0

[a] Experimental conditions are the same as in the text, except that 0.2 *M* Tris·HCl buffer (pH 7.5) and 50 m*M* γ-^{32}P–ATP are added instead of glycylglycine buffer and 1.5 m*M* ATP.

As shown in the table, ATPase activity can be detected in the chromatophores and cell debris fractions, but not in the supernatant solution. The lower activity with whole cells may be the result of impermeability of the substrate through the cell membrane. ATPase activities of both cell debris and chromatophores are significantly inhibited by oligomycin, but not by ouabain (33 μM). In the presence of both 0.13 *M* NaCl and 1.3 m*M* KCl, the ATPase activities are hardly stimulated. Specific activities of cell debris fractions and of chromatophores are al-

[11] See T. Horio *et al.*, Vol. XXIV.

most the same, whether the values are expressed per bacteriochlorophyll or per ubiquinone-10.

Properties of ATPase Activity Associated with Chromatophores

Accelerators and Inhibitors. The addition of divalent cations is required for the ATPase activity: Mg^{2+} and Mn^{2+} are effective; Co^{2+} is less effective; Ca^{2+} is not. The optimum concentration for Mg^{2+} and Mn^{2+} is around 1 m*M*. 2,4-Dinitrophenol (DNP) and arsenate accelerate the activity at a concentration of about 1 m*M* approximately 3- and 1.5-fold, respectively; at higher concentrations, the stimulatory effects are reduced. The effects of redox dyes such as 2,6-dichlorophenolindophenol, phenazine methosulfate, and ascorbate, and of light are described elsewhere.[11]

pH Optimum. The rate of ATPase activity is maximal at pH 7.5; activities at pH 7 and pH 8 are 85% and 95%, respectively, of that at pH 7.5. The activity stimulated by DNP is also maximal at pH 7.5.

Effect of Sonication. When a chromatophore suspension ($A_{88\ nm}$/ml, approximately 50) at pH 8 is sonicated, the ATPase activity gradually decreases as sonication time increases; the decrease of the ATPase activity parallels the decrease of photosynthetic ATP formation and ATP-P_i exchange. All the activities mentioned above are completely inactivated by sonication for 30 minutes.

Specific Activity. In the presence of the standard components of the reaction mixture and under the standard experimental conditions mentioned above, chromatophores hydrolyze 0.6–1.0 μmole of ATP/$A_{880\ nm}$/hour. The K_m for ATP is 0.13 m*M*.

[64] Enzymes Catalyzing Exchange Reactions: Bacterial

By T. Horio, N. Yamamoto, Y. Horiuti, and K. Nishikawa

ADP–ATP Exchange Reaction

An ADP–ATP exchange enzyme can be isolated and purified from a cell-free extract of the photosynthetic bacterium *Rhodospirillum rubrum.* The exchange activity of chromatophores can be completely removed by repeated centrifugal washing, without any loss of photosynthetic ATP formation, ATP–P_i exchange, and ATPase. However, the well-washed chromatophores still possess an ADP–ATP exchange enzyme, which is latent in the bound form; when the chromatophores are disrupted by sonication, the ADP–ATP exchange enzyme is liberated from the particles. The emergence of the activity is accompanied by loss of the other chromatophore activities.

Assay Methods

Two assay methods are applicable; one makes use of ^{32}P-labeled ADP, and the other is spectrophotometric. The former method is more popular and has been used in many laboratories.[1-3] The spectrophotometric method is based on the nucleoside diphosphokinase activity (GTP–ADP transphosphorylase activity) that is dependent on the ADP–ATP exchange enzyme.

Radioisotopic Method. The general principles and procedures described in a previous volume[1] are adopted, except for the following modifications. The components of the standard reaction mixture are 0.1 *M* Tris·HCl buffer (pH 8.0), 25 μl; 0.1 *M* $MgCl_2$, 5 μl; 10 m*M* ATP, 5 μl; 10 m*M* β-^{32}P–ADP (50,000–100,000 cpm),[4] 5 μl; enzyme solution, 5 μl; water, to make the total volume 75 μl. The components are placed in small test tubes (6 × 60 mm) with microliter syringes (Hamilton Co., California). The reaction is initiated by adding the enzyme solution and is carried out at 30° for 5 minutes under aerobic conditions. The reaction is terminated by adding 15 μl of 5 *N* perchloric acid. The acidified solution is neutralized with KOH, the precipitate thus formed is removed, and 10 μl of the resultant clear solution is spotted on filter paper for electrophoresis. On the same sheet of filter paper, 5 μl of a solution containing 10 m*M* each of ATP, ADP, and AMP is spotted. Electrophoresis is carried out at approximately 40 V/cm at 0°–1° for 2 hours using the Flat Plate apparatus (Savants Inc., New York). Spots corresponding to ATP, ADP, and AMP, which are detectable under illumination by ultraviolet light, are cut out and counted.

Adenylate kinase activity is assayed by the same method as described above, except that ATP is omitted from the reaction mixture.

Preparation of β-^{32}P–ADP.[5] The components of the reaction mixture are 10 m*M* AMP, 0.5 ml; 0.1 *M* $MgCl_2$, 0.5 ml; 0.1 *M* Tris·HCl buffer (pH 8.0), 2.5 ml; 5 m*M* γ-^{32}P–ATP (5 mCi), 0.5 ml; 0.26 mg/ml adenylate kinase,[6] 0.5 ml. The reaction is initiated by adding the enzyme solution, and is carried out at 30° for 30 minutes. The reaction is termi-

[1] See this series, Vol. X [86].

[2] See Vol. X [9].

[3] See Vol. VI [32].

[4] In some cases ADP–^{14}C (random labeling) obtained from Schwartz BioResearch Inc. is used.

[5] Alternative method is described in Vol. IV [34].

[6] Commercial preparations of the best quality obtained from Sigma Chemical Co.

nated by immersing the test tubes in a boiling water bath for 5 minutes. After cooling, 0.25 ml of hexokinase solution (5 mg/ml, 200 units/ml)[6] and 0.25 ml of 1 *M* glucose are added, and the mixture is incubated for 10 minutes at 30°, immediately after which the preparation is diluted 10-fold with cold water. The subsequent operations are carried out in a cold room. The resulting solution is passed through a column of Dowex 1 (2 × 2 cm, Cl^- form), and the column is washed with 100 ml of 0.01 *N* HCl. ^{32}P-labeled glucose-6-phosphate and AMP are eluted with 0.01 *N* HCl. The product is then eluted with 0.03 *N* HCl containing 20 m*M* NaCl, and the eluate is collected in 10-ml fractions. The fractions showing radioactivity are mixed; the mixture is adjusted to pH 7.5 by adding Tris (final concentration, approximately 10 m*M*) and NaOH and lyophilized. The lyophilized material is dissolved in 2 ml of water. Approximately 70% of the initial γ-^{32}P–ATP is recovered as β-^{32}P–ADP and is free of ATP, AMP, P_i, glucose-6-phosphate, and α-^{32}P–ADP.

Spectrophotometric Method. The ADP–ATP exchange enzyme can exhibit GTP–ADP transphosphorylase activity (nucleoside diphosphokinase). With added hexokinase, the terminal phosphate of the ATP formed is transferred to glucose to form glucose-6-phosphate, which serves to reduce NADP in the presence of glucose-6-phosphate dehydrogenase; the reaction from GTP is markedly slower in rate than that from ATP. The components of the reaction mixture are 0.2 *M* Tris·HCl buffer (pH 8.0), 0.5 ml; 0.1 *M* $MgCl_2$, 0.1 ml; 5 m*M* NADP, 0.1 ml; 0.2 *M* glucose, 0.1 ml; 20 units/ml of hexokinase,[6] 0.1 ml; 10 m*M* GTP, 0.1 ml; 1 unit/ml of glucose-6-phosphate dehydrogenase,[6] 0.1 ml; enzyme solution, 0.1 ml; water, to make the total volume 1.5 ml. The reaction is carried out in a cuvette of 1-cm optical path at room temperature (approximately 20°), and the reduction of NADP is determined by measuring the difference in absorbancy at 340 nm between the sample and control cuvettes. The control cuvette contains the same reaction mixture except that hexokinase and glucose-6-phosphate dehydrogenase are omitted. The reaction is started by adding the enzyme solution. When the rate of increase of the absorbancy difference at 340 nm is between 0.03 and 0.10 per 100 seconds, the rate is proportional to the amount of ADP–ATP exchange enzyme. Part of the absorbance change at 340 nm results from the adenylate kinase present in enzyme preparation and the transphosphorylase activity of commercial hexokinase. The adenylate kinase activity can be measured by omitting GTP from the reaction mixture. The GTP–ADP transphosphorylase activity of the hexokinase preparation can be measured by omitting the enzyme solution. Under the experimental conditions, the transphosphorylase activity of the hexokinase samples used is less than 0.02 per 100 seconds. The actual ATP–ADP transphosphorylase activity of the enzyme preparation is the value obtained with the complete reaction mixture *minus* the sum of the control values obtained with the adenylate kinase and hexokinase's transphosphorylase assays.

ADP–ATP Exchange Enzyme Associated with Chromatophores

A suspension of wild-type *R. rubrum* cells is ruptured by extrusion from a French pressure cell, and the fraction sedimented between 20,000 *g* for 30 minutes and 105,000 *g* for 1 hour is collected.[7] The fraction is suspended in 0.1 *M* Tris·HCl buffer (pH 8.0) so that the resulting suspension would show an $A_{880\,nm}$/ml of approximately 50. The chromatophore preparation at this stage is represented in Table I by zero washings. The suspension is homogenized for 2 minutes with a Potter-Elvehjem homogenizer driven by a motor, followed by centrifugation at 105,000 *g* for 1 hour. The precipitate is suspended in 0.1 *M* Tris·HCl buffer (pH 8.0); this washing procedure is repeated seven times. The ADP–ATP exchange activity of chromatophores is lowered gradually as the number of washing times is increased; after seven washings, chromatophores are devoid of the ADP–ATP exchange activity, and the ADP–ATP exchange enzyme is recovered in the washings. On the other hand, photosynthetic ATP formation, ATPase and ATP–P_i exchange by

TABLE I
EFFECTS OF WASHING AND SONICATION OF CHROMATOPHORES ON THE ADP–ATP EXCHANGE ACTIVITY

	ADP–ATP exchange[a]	
	Sonication[b]	
Number of washings	Before	After
0	1200	—
1	340	—
2	41	—
3	10	44
4	11	45
5	5	43
6	1	37
7	0	37

[a] Assayed by the spectrophotometric method; mμmoles of NADP reduced/$A_{880\,nm}$/hour.
[b] Chromatophore suspension ($A_{880\,nm}$/ml, approximately 50) is sonicated in a 10-kHz Kubota oscillator for 30 minutes in the cold (0°–4°).

[7] See this volume [63].

chromatophores are not affected by washing. When washed chromatophores are sonicated, another ADP–ATP exchange activity emerges, indicating that chromatophores have the latent activity of an ADP–ATP exchange enzyme which is tightly associated with chromatophores (Table I). The latent enzyme is no longer associated with chromatophores when the activity appears, and it is recovered in a soluble fraction (the supernatant obtained by centrifugation at 105,000 *g* for 1 hour). The enzyme solubilized by sonication appears to be the same as that solubilized by simple washing (see below).

Purification Procedure

Source Material. Cells of the carotenoidless *R. rubrum* mutant (G-9) are grown and harvested[7]; mass culture is carried out in 5-liter reagent bottles under continuous illumination by tungsten lamps at approximately 30°. The cells are washed with water and then lyophilized. All the procedures described below are performed in a cold room. Essentially the same result can be obtained with wild-type cells.

Extraction from Cells and First Ammonium Sulfate Fractionation. The ADP–ATP exchange enzyme can be easily extracted into aqueous solution by disrupting cells. For this purpose, sonication, grinding with aluminum oxide, extrusion from a French pressure cell, and lyophilization are equally effective for the extraction; lyophilization seems most convenient for treatment of a large quantity of cells. Thus, 600 g of lyophilized cells are suspended in 6 liters of 0.1 *M* Tris·HCl buffer (pH 8.0) containing 1 *M* NaCl and extracted overnight with continuous stirring. The suspension is centrifuged at approximately 10,000 *g* for 20 minutes. The precipitate is reextracted with the same volume of buffer. The two extracts are mixed, and ammonium sulfate is added slowly to the mixture with continuous stirring; the final concentration of ammonium sulfate is 176 g/liter (30% saturation). To remove the resulting precipitate, the solution is filtered with suction through filter paper with the aid of Celite. The resulting clear filtrate is desalted by passage through a column of Sephadex G-25 which has been equilibrated with 10 m*M* Tris·HCl buffer (pH 8.0); thus, the desalted solution is in 10 m*M* the buffer

First Application to DEAE–Cellulose Column. The desalted solution is absorbed on a column of DEAE–cellulose (9 × 70 cm), previously equilibrated with 10 m*M* Tris·HCl buffer (pH 8.0). The charged column is washed with 0.1 *M* Tris·HCl buffer (pH 8.0), and then eluted with 0.1 *M* Tris·HCl buffer (pH 8.0) containing 0.2 *M* NaCl. The fractions containing the ADP–ATP exchange enzyme are mixed, and the mixture is desalted on Sephadex G-25 by the same method as mentioned above.

Second Application to DEAE–Cellulose Column. The active fraction is charged on a column of DEAE–cellulose (3 × 40 cm), previously equilibrated with 10 mM Tris·HCl buffer (pH 8.0). The charged column is then eluted with a linear gradient, which is generated with 1 liter of 0.1 M Tris·HCl buffer (pH 8.0) in the mixing vessel and 1 liter of the same buffer containing 0.3 M NaCl in the reservoir. The eluate is collected in 10-ml fractions; the ADP–ATP exchange enzyme is in fractions containing NaCl concentrations of 0.1 M to 0.23 M. The NaCl concentrations are determined simply by measuring the conductivities with a Radiometer type CDM 2d conductivity meter and type CDC114 conductivity cell (Radiometer, Copenhagen). Most of the other proteins present in the starting solution are eluted with the buffer containing NaCl concentrations of 30 mM to 90 mM. The fractions containing the ADP–ATP exchange enzyme are combined and desalted on Sephadex G-25 as mentioned above.

Second Ammonium Sulfate Fractionation. To the desalted solution, 313 g/liter of ammonium sulfate (50% saturation) is added gradually with continuous stirring. After standing for 30 minutes or longer, the solution is centrifuged at approximately 10,000 g for 20 minutes. To the resulting supernatant solution, 176 g/liter of ammonium sulfate (75% saturation) is further added, followed by centrifugation. The precipitate is collected and dissolved in 1 ml of 50 mM Tris·HCl buffer (pH 7.5) containing 0.1 M KCl.

Molecular Sieve Fractionation of Sephadex G-75. The concentrated solution is applied to a column of Sephadex G-75 (3 × 50 cm), which has been equilibrated with 50 mM Tris·HCl buffer (pH 7.5) containing 0.1 M KCl. The same buffer is used for elution. The flow rate is adjusted to 30 ml/hour and the eluate is collected in 3-ml fractions. The proteins present in the sample solution are eluted from the column as three main peaks (assayed as $A_{280\,\mathrm{nm}}$). The first peak is eluted in the void volume. The second peak, which contains the ADP–ATP exchange enzyme, is included in fractions 40–50. The molecular weight of the enzyme is approximately 30,000.

The purification procedure and results are summarized in Table II. In contrast to mammalian sources,[1] crude extracts from *R. rubrum* cells show a low adenylate kinase activity relative to the ADP–ATP exchange activity. When the ADP–GTP exchange rate assayed spectrophotometrically (NADP reduced per minute) is approximately one-tenth the ADP–ATP exchange rate assayed by the radioisotopic method; the ratio remains fairly constant throughout the purification steps. It, therefore, seems unlikely that cell extracts contain another nucleotide diphosphokinase which is more specific for GTP than for ATP.

TABLE II
PURIFICATION PROCEDURE FOR ADP–ATP EXCHANGE ENZYME

Fraction	Total protein (mg)	ADP–ATP exchange[a] Specific activity[b]	ADP–ATP exchange[a] Total activity[c]	Adenylate kinase[a] Specific activity[b]	Adenylate kinase[a] Total activity[c]
1st $(NH_4)_2SO_4$ supernatant	150,000	1.40	210,000	0.070	3500
1st DEAE eluate	11,000	11.5	126,000	0.190	2090
2nd DEAE eluate	1,800	36.4	65,500	0.175	315
2nd $(NH_4)_2SO_4$ precipitate	320	105	33,600	0.112	36
Sephadex G-75 eluate	33	490	16,300	0.036	1.3

[a] Activity is assayed by isotopic method.
[b] Micromoles of P_i transferred per minute per milligram of protein.
[c] Micromoles of P_i transferred per minute.

Properties

Isoelectric Fractionation with Ampholine Carrier Ampholytes. The ADP–ATP exchange enzyme, which has been partially purified by chromatography on DEAE–cellulose column, is subjected to isoelectric fractionation on an electrofocusing column carried out according to the method of Vesterberg and Svensson.[8] The carrier ampholyte used to produce electrophoretically a constant pH gradient from the top to the bottom of a vertical column is a mixture of various derivatives of aliphatic polyaminopolycarboxylic acids of molecular weight as low as 300–600 (obtained commercially as Ampholine carrier ampholytes from LKB Produkter AB., Stockholm-Broma). By the isoelectric fractionation, ampholytes such as proteins can be focused at the values of their respective isoelectric points (pI) on the pH gradient. The pI value of an ampholyte is, therefore, regarded as the pH value of the fraction eluted from the column that contains the maximum concentration of the ampholyte. Ampholine carrier ampholytes capable of producing a gradient from pH 4 to 6 is used at a concentration of 1% (w/v). To obtain stable focused zones of proteins, a stepwise sucrose density gradient (50 steps) from 0–50% (w/v) is produced beforehand. Electrolysis is carried out at 0°–1° for 48 hours in a 100-ml Electrofocusing Column (LKB-Produkter AB.). After electrolysis (about 700 V and 1.3 mA at the final stage), the column content is eluted from the bottom of column, and collected in 2-ml fractions. The ADP–ATP exchange activity is focused in only one zone, the pI value of which is approximately 4.9. When the latent ADP–ATP exchange enzyme associated with well-washed chromatophores is solubilized by sonicating the washed chromatophores and subjected to iso-

[8] O. Vesterberg and H. Svensson, *Acta Chem. Scand.* **20**, 820 (1966).

electric fractionation, it is also focused in one zone of a pI value of 4.9, the same as that for the enzyme solubilized simply by washing.

pH Optimum. The ADP–ATP exchange reaction catalyzed by the purified enzyme is optimum at pH 8.5–9.0.

Heat Stability. When a solution of the purified enzyme (pH 8.0) is heated at 60° for 3 minutes, the activity is almost completely destroyed; the activity is completely protected from heat inactivation if 1 m*M* ADP or ATP is present during heating. Mg^{2+} is not required for the protective effect of ADP and ATP. Neither AMP nor pyrophosphate is effective for protection.

ATP–P_i Exchange Reaction

Besides the photosynthetic ATP formation and the ATPase reaction, chromatophores can catalyze an ATP–P_i exchange reaction.[9] This latter activity is tightly associated with chromatophores; enzymes capable of catalyzing the exchange reaction have not been solubilized.

Preparation of Chromatophores

The culture of *R. rubrum* cells and preparation of chromatophores are performed in the same manner as those described elsewhere in this volume.[7]

Assay Methods

The assay is based on the measurement of $^{32}P_i$ incorporated into ATP. The procedure is similar to that of Nielsen and Lehninger,[10] as modified by Avron.[11]

Reagents. The reagents are the same as those described.[7]

Procedures. The experimental conditions are the same as those described elsewhere,[7] except that 0.1 ml of 50 m*M* ATP and 0.1 ml of 0.1 *M* $^{32}P_i$ (pH 8) (approximately 5×10^6 cpm) are added instead of γ-^{32}P–ATP. The clear supernatant solutions, obtained by centrifugation of the acidified reaction mixtures, are pipetted into 2×14 cm test tubes containing 0.02 ml of 1 *M* ascorbic acid, 1.5 ml of water saturated with isobutanol–benzene and 1.2 ml of acetone. After the tube contents are mixed, 7 ml of isobutanol–benzene saturated with water is added, and then shaken vigorously. After phase separation, 0.8 ml of molybdate reagent is added gently to the lower water phase (along the side wall of test tubes). The water layer is mixed gently with a glass rod. After standing for at least 5 minutes, the test tubes are vigorously shaken for

[9] T. Horio, K. Nishikawa, M. Katsumata, and J. Yamashita, *Biochim. Biophys. Acta* **94**, 371 (1965).

[10] S. O. Nielsen and A. T. Lehninger, *J. Biol. Chem.* **215**, 555 (1955).

[11] M. Avron, *Biochim. Biophys. Acta* **40**, 257 (1960).

30 seconds. The upper organic layer is removed by means of aspiration. The lower water layer is supplemented with 0.02 ml of 20 m*M* phosphate solution and extracted three times in the manner described above. Finally, 1 ml of the water layer is placed on a planchet and evaporated to dryness under a heat lamp. The radioactivity of the dried material is counted. The amount of phosphate exchanged is calculated by dividing the total radioactivity of the water phase by the specific radioactivity of the phosphate of the original reaction mixture. Correction of the calculation by using such an equation as described in a previous volume[1] is not necessary, because the total radioactivity incorporated into ATP is lower than approximately 5% of that of phosphate present in the original reaction mixture.

Properties

Effect of Illumination. The ATP–P_i exchange activity is stimulated by white light (1000 foot-candles) in the presence of either ascorbate or phenazine methosulfate (PMS); the optimal concentration is approxi-

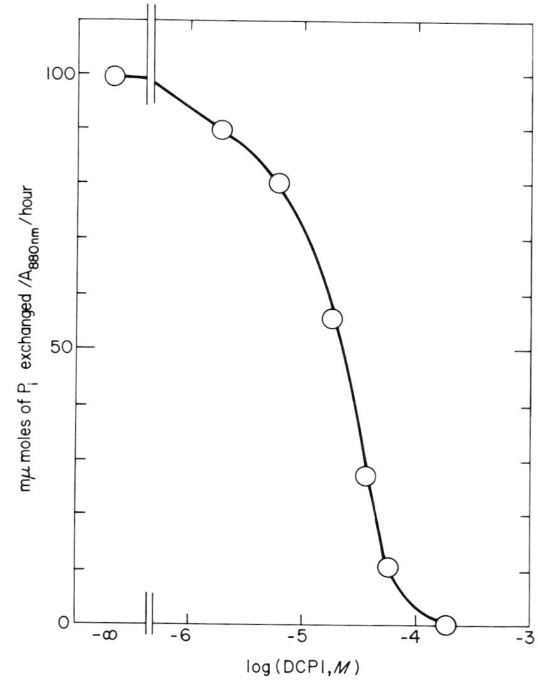

FIG. 1. Effect of DCPI on ATP–P_i exchange activity in darkness. Experimental conditions are as described in the text, except that DCPI is added as indicated.

mately 67 mM for ascorbate and 0.4 mM for PMS. The rate in the light is about four times as high as that in darkness; the rate in darkness is approximately 0.15 μmole of P_i exchanged/$A_{880\,nm}$/hour. Neither ascorbate nor PMS is effective in darkness. The ATP–P_i exchange activity with ascorbate, but not with PMS, in the light is decreased to the level of the activity in darkness, when 0.33 μg/ml of antimycin A is added.

pH Optimum. The ATP–P_i exchange activities in the light and in darkness are optimal at around pH 8.

Inhibitors. The ATP–P_i exchange activities in the light and in darkness are almost completely inhibited by oligomycin (3.3 μg/ml), by 2,4-dinitrophenol (4 mM), and by quinacrine hydrochloride (0.67 mM).

Effect of Sonication. The ATP–P_i exchange activities in the light and in darkness are inactivated by sonicating chromatophores for a prolonged time; the inactivation occurs in parallel with the inactivations of photosynthetic ATP formation and ATPase.[7]

E_h and Ubiquinone-10 Dependency. The rate of ATP–P_i exchange decreases with increasing concentrations of 2,6-dichlorophenolindophenol (DCPIP), as shown in Fig. 1. At 67 μM, the rate of ATP–P_i exchange is negligible, whereas the rate for ATPase activity is stimulated to a maximum extent.[12] The ATP–P_i exchange activity that has been de-

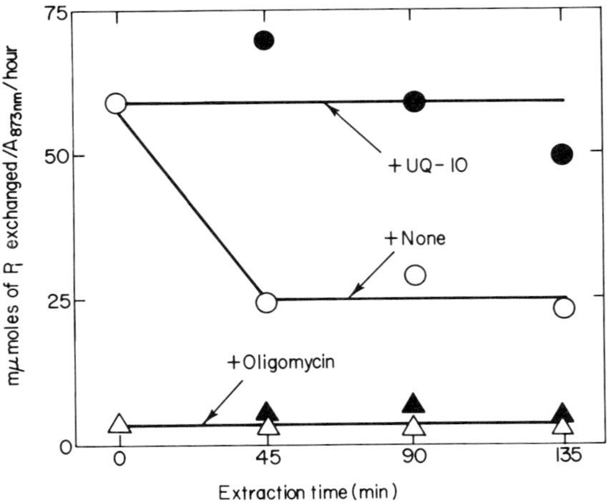

FIG. 2. Effect of quinone extraction and restoration of ubiquinone-10 on ATP–P_i exchange activity. Experimental conditions are as described in the text except for the following details: Chromatophores which were extracted with isooctane for various times are reconstructed by addition of 4 mμmoles/$A_{873\,nm}$ of ubiquinone-10; circles, no addition; triangles, +oligomycin; open symbols, extracted chromatophores; filled symbols, reconstructed chromatophores.

[12] See T. Horio *et al.*, Vol. XXIV.

pressed by 0.67 mM DCPI is restored if ascorbate is added; the maximum restoration is obtained at 6.7 mM ascorbate, where the E_h value of the reaction mixture is approximately +0.1 V.[12] In contrast to the ATPase activity, the ATP–P_i exchange activity is hardly affected at higher concentrations of ascorbate. The ATP–P_i exchange and ATPase activities are not altered by repeated washings of chromatophores with solutions of DCPI and/or ascorbate, indicating that the effect of these redox dyes is due to a reversible reaction. One possible explanation for the ATPase activity may be as follows[12]: Two different oxidation–reduction components are requisite for the ATPase and ATP–P_i exchange activities, and the ATP–P_i exchange activity emerges when both of the components are in the reduced form.

ATP–P_i exchange activity by lyophilized chromatophores is approximately 30% of that of original chromatophores. As shown in Fig. 2, the ATP–P_i exchange activity in darkness is significantly depressed when quinones are extracted from lyophilized chromatophores; however, the activity is restored to the original level when ubiquinone-10 is added back to the extracted chromatophores. The procedure for these experiments are described elsewhere in this series.[12] Approximately one-third of the ATP–P_i exchange activity remains after complete extraction of the associated quinones; the ATP–P_i exchange activity is distinguishable as two types, one dependent and the other independent of the quinones, in the same manner as for the ATPase activity. With both lyophilized chromatophores and ubiquinone-10-reconstructed chromatophores, the ATP–P_i exchange activity is almost completely inhibited in the presence of 3.3 μg/ml of oligomycin.

Of various quinones (4 mμmoles/$A_{873\,\text{nm}}$) added to quinone-free chromatophores, ubiquinone-9, -7, -6, hexahydroquinone-4, rhodoquinone, plastoquinone, α-tocopherolquinone, and phylloquinone are as effective in restoration of ATP–P_i exchange activity as ubiquinone-10. Menadione, 2,3-dimethoxy-5,6-dimethylbenzoquinone, 4-amino-1,2-naphthoquinone, 2-amino-1,4-naphthquinone, 2-acetoamino-1,4-naphthoquinone, and S-(2-methyl-1,4-naphthoquinonyl-3)-β-mercaptopropionic acid are not effective.

Neither addition of serum albumin nor repeated washing with albumin solutions reactivates the ATP–P_i exchange by quinone-free chromatophores, indicating that the restoration of ATP–P_i exchange activity by adding quinones to quinone-free chromatophores is not the result of removal of a trace of isooctane which may have remained.

[65] Isolation of Leaf Peroxisomes

By N. E. TOLBERT

Leaf peroxisomes are respiratory microbodies which are present in the leaves of all plants.[1-3] In appearance, size, and enzymatic composition, leaf peroxisomes belong to a class of microbodies that are similar to peroxisomes from liver and kidney,[4] and they are similar but not identical with glyoxysomes from germinating seed endosperm.[5] They are characterized by a single membrane and a granular matrix without lamellae,[2,3,6] and sometimes they contain a dense core which may be crystalline.[3] This type of subcellular particles is characterized by catalase and specific flavin oxidases,[2,4] and the leaf peroxisomes contain other enzymes of the glycolate pathway.[7-9] In spinach leaves they represent 1–1.5% of the total leaf protein,[2] and in tobacco leaves they are about one-third as numerous as mitochondria.[3] Leaf peroxisomes function in glycolate metabolism, and photorespiration or peroxisomal respiration[7] is characterized by H_2O_2 generation and loss of energy. Photorespiration in plants occurs only in the light when the substrate, glycolate, is photosynthetically biosynthesized in the chloroplasts. Speculation on function of peroxisomal respiration includes regulation of photosynthesis,[7] protection against oxygen toxicity,[4,10] and involvement with gluconeogenesis.[4]

Procedures for isolation of leaf peroxisomes are based upon the pioneering work from de Duve's group on mammalian peroxisomes.[4] Special problems, particularly grinding, confront the plant biochemist, so that leaf peroxisomes were not previously isolated. A complete compilation of the numerous references on microbodies from microscopic and anatomical studies is now available.[11] Considering the great agronomic

[1] N. E. Tolbert, A. Oeser, R. K. Yamazaki, R. H. Hageman, and T. Kisaki, *Plant Physiol* **44,** 135 (1969).

[2] N. E. Tolbert, A. Oeser, T. Kisaki, R. H. Hageman, and R. K. Yamazaki, *J. Biol. Chem.* **243,** 5179 (1968).

[3] S. E. Frederick and E. Newcomb, *Science* **163,** 1353 (1969).

[4] C. de Duve, *Physiol. Rev.* **46,** 323 (1966).

[5] R. W. Breidenbach and H. Beevers, *Plant Physiol.* **43,** 705 (1968).

[6] S. E. Frederick, E. Newcomb, E. Vigil, and W. Wergin, *Planta* **81,** 229 (1968).

[7] N. E. Tolbert and R. K. Yamazaki, *Ann. N.Y. Acad. Sci.* **168,** 325 (1969).

[8] R. Rabson, N. E. Tolbert, and P. C. Kearney, *Arch. Biochem. Biophys.* **98,** 154 (1962).

[9] N. E. Tolbert, *in* "Photosynthetic Mechanism in Green Plants," pp. 648–662. NSF-NRC Publ. 1145, 1962.

[10] C. de Duve, *Proc. Roy. Soc. Ser. B* **173,** 71 (1969).

[11] Z. Hruban and M. Rechcigl, Jr., "Microbodies and Related Particles." Academic Press, New York, 1969.

interest in photosynthesis and photorespiration, rapid further characterization of these particles is expected.

General Procedures

Principle. Fractionation of leaf particles by differential centrifugation does not successfully isolate peroxisomes nor remove contaminating peroxisomes from the other particles, such as chloroplasts or mitochondria. The peroxisomes, about 1 μ in diameter with a specific density similar to that of mitochondria, constitute a contaminant in mitochondrial preparations. Since the peroxisomes are freely permeable to sugars, their specific density increases in a sucrose sugar grinding medium, and they are mostly sedimented by 6000 g in 10 minutes along with the bulk of the broken chloroplasts. Even whole chloroplasts, sedimented at 100 or 600 g, contain significant amounts of peroxisomes. By two or three resuspensions and recentrifugations of whole chloroplasts the bulk of the peroxisomal activity can be removed from the chloroplasts by the combined action of selective breakage of the much more fragile peroxisomes and differential centrifugation. When this happens absorption of the solubilized peroxisomal enzymes on the other particles occurs to some extent. Activity of the peroxisomal marker enzymes, catalase and glycolate oxidase, can be used as a criteria of the degree of contamination in chloroplasts and mitochondria.

The main steps in the procedure are (a) grinding of the leaves; (b) differential centrifugation between 100 and 6000 g, for partial separation and for enrichment of the particulate fraction; (c) an isopycnic, nonlinear sucrose density gradient centrifugation at 39,000 g; and (d) a linear sucrose density gradient centrifugation. The first nonlinear sucrose density gradient acts as a filtration process in which the chloroplasts and mitochondria are stopped by layers of sucrose between 1.3 and 1.75 M. The peroxisomes are trapped at an interface below 1.75 M sucrose generally on top of 2.3 M sucrose.

In isolating leaf peroxisomes, a major difference encountered from work with liver or seed endosperm tissue is the immense bulk of chloroplasts from which the peroxisomes must be separated. In the first gradient the multilayers of sucrose between 1.3 and 1.75 M are almost a necessity to spread out the chloroplasts over a broad section of the gradient. Otherwise, the chloroplast bands are packed in a dense layer through which the peroxisomes do not readily sediment. Some green particles always contaminate the mitochondrial and peroxisomal bands after the first gradient centrifugation. The few chloroplasts in the mitochondrial band appear to be whole chloroplasts in contrast to the bulk of broken chloroplasts at lower densities. In the peroxisomal band the presence of less than 0.1% of the total chlorophyll is, nevertheless, a significant protein and lipid contamination, except perhaps when measuring enzymatic distributions on the gradients. It has been possible and necessary for some assays to recentrifuge the peroxisome band from a discontinuous gradient on a second linear sucrose gradient in order to obtain peroxisomes free of chlorophyll by spectrophotometric assays. In preparation for running the continuous gradient, the peroxisome fraction from the discontinuous gradient may be used, after dilution to about 1.7 *M* sucrose, or they may be so diluted and pelleted.

Sucrose Gradients. The use of sucrose density gradients for isopycnic equilibrium sedimentation of mammalian peroxisomes has been described,[4] and the principles and procedures have been adapted for leaf peroxisomes.[1,2] The sucrose density gradient is used for buoyant density centrifugation in contrast with sucrose gradient separation on the basis of sedimentation velocity.

When grinding medium of Ficol and Ficol gradients were used in lieu of sucrose, peroxisomal enzymatic activity banded at specific densities between the broken chloroplast (1.18–1.20 g/cc) and mitochondria (1.21–1.23 g/cc). Since this location placed the peroxisomes among the many broken chloroplast fractions, the use of Ficol has not been continued. de Duve[4] has explained that the freely permeable single membrane of the liver peroxisomes permits a rapid exchange of sucrose, and thus the density of the particle in sucrose increases. Although leaf peroxisomes may be somewhat less permeable than liver particles because Triton X-100 increases enzymatic activity, the leaf particle is still relatively permeable and does not compartmentalize bound water on a sugar gradient as do other particles. As a consequence during the centrifugational development of a sucrose gradient, the chloroplasts and mitochondria rapidly move to sucrose bands similar to their specific densities, while the peroxisomes absorb sucrose, increase in specific density with time and ultimately stop at a density of around 1.25 to 1.26 g/cc or in layers of 1.9 to 2.0 *M* sucrose. This fortuitous property places the peroxisomal band, after isopycnic centrifugation, far below the other particulate bands. For this reason the peroxisomes are relatively easy to separate from the other major particles. By draining the gradients from the bottom of the tube the peroxisomal band is recovered first and the rest of the bands need not be collected.

The techniques of pelleting peroxisomes to the bottom of the centrifuge tube through sucrose gradients (up to 1.9 *M* sucrose) to retain chloroplasts and mitochondria have been successfully used for enrichment and recovery of these particles. This procedure is unsatisfactory for the first sucrose gradient fractionation of the 6000 *g* pellet from differential centrifugation, because there are many starch grains and a few other dense particles, some of which appear as cell clumps, which sediment to the bottom of the first sucrose gradient even through 2.3 and 2.5 *M* sucrose.

Grinding Procedures. Of special concern in the isolation of leaf peroxisomes are the vigorous grinding procedures, which at this time appear to be a limiting factor. Procedures to date have been by Waring Blendor, mortar and pestle, or mills to break the cellulose cell walls. Investigators of liver peroxisomes and plant endosperm can use a much milder procedure, such as the Potter-Elvehjem homogenizer. After necessary vigorous grinding of leaf tissue, most peroxisomal particles, either do not survive or appear broken. That a peroxisomal band is found on the sucrose gradient suggests that the protein in broken peroxisomes remain partly in a macromolecular gel which does not spill out into the medium. However, these broken peroxisomes cannot be dialyzed and are sensitive to osmotic shock, in contrast to whole liver peroxisomes. The broken particles are also solubilized if a Potter-Elvehjem homogenizer is used to resuspend pelleted fractions. In each gradient run, 25–50% of the peroxisomal enzymatic activity becomes solubilized, in that it floats to the top of the gradient. This is attributed to broken particles. If whole intact peroxisomes could be obtained from leaves, they might be as stable as similar particles from other tissue. We have tried various mills and roller grinders without noteworthy success for obtaining higher yields of leaf peroxisomes. The use of cellulase and pectinase to digest cell walls may provide a better means to obtain leaf peroxisomes.[12]

For each plant it has been necessary to select by trial and error a grinding time for maximum yield of peroxisomal enzymes in the particulate fractions. Very short (5–10 seconds) grinding times in a Waring Blendor have been the most successful. Total enzyme recovery is low because all the cells are not broken. Longer grinding times result in lower recovery of particles but higher total enzymatic activity. After grinding spinach leaves in a Waring Blendor for 7 seconds, 39% of the total glycolate oxidase in the cell free homogenate remained in the particulate fraction, after 20 seconds, 25% and after 60 seconds only 8% On the other hand, failure to completely break all the cells in short grinding periods has resulted in serious underestimation of total enzymatic activity in the plant tissue. Survey experiments should employ two grinding times for each tissue; a long and hard grind for total en-

[12] J. B. Power and E. C. Cocking, *Biochem. J.* **111**, 33 (1969).

zymatic activity, and a short grinding time for isolating peroxisomal particles. At best the percentage of the total enzymatic activity recovered in leaf peroxisomes is low. With spinach and sunflower leaves, we have achieved a 50% yield of peroxisomal activity in the combined particulate fractions after differential centrifugation, but nearly half of this activity was further lost in subsequent steps involving sucrose gradient centrifugation. With corn and sugarcane leaves only a barely significant 1–4% of the peroxisomal enzymes are in the surviving particulate fractions.

Grinding media of varying sucrose concentrations between 0.25 to 1.0 *M* have given similar results. A buffered medium without sucrose gave low recovery of pelleted peroxisomal activity, but higher recovery of total glycolate oxidase due to more complete breakage of the tissue. With 1.25 *M* sucrose in the grinding medium both the percent in the pellet and the total enzyme activity were reduced, perhaps because breakage of the cells was more difficult.

No publications exist on the isolation of peroxisomes by the nonaqueous procedures used for separating chloroplasts and mitochondria.[13,14] In exploratory attempts with this technique we found glycolate oxidase and catalase in the chloroplast fraction. Since Stocking (personal communication) has not observed by electron microscopy intact peroxisomes in chloroplast fraction isolated by nonaqueous procedures, it is possible that peroxisomes do not survive this procedure and the peroxisomal protein is bound to the chloroplasts.

Peroxisome Preparations of Different Scale

Small preparations with 30–60 g of tissue are used to prepare as clean a peroxisomal fraction as possible after one isopycnic nonlinear sucrose gradient centrifugation on a 50-ml gradient. Small gradients could reduce the size of this run, but since the yield of peroxisomes in many leaves is small, we have preferred to use this amount of tissue. These small gradients also contain separate mitochondrial and chloroplast bands, and distribution of enzymes on the gradient among the various particles can be studied.[1,2]

A large-scale preparation of peroxisomes from 150 g of leaf tissue per 50-ml gradient tube has been used for studying peroxisomal components. These preparations start with 450 g of tissue for one centrifuge run of three tubes. This scale gives as good a yield of peroxisomes, but no separation of bands of chloroplasts and mitochondria.

Procedures for separation of liver peroxisomes by zonal centrifugation[15] have been applied to plant plastids.[16] The increasing specific den-

[13] C. R. Stocking, *Plant Physiol.* **34**, 56 (1959).
[14] U. Heber, N. G. Pon, and M. Heber, *Plant Physiol.* **38**, 355 (1963).

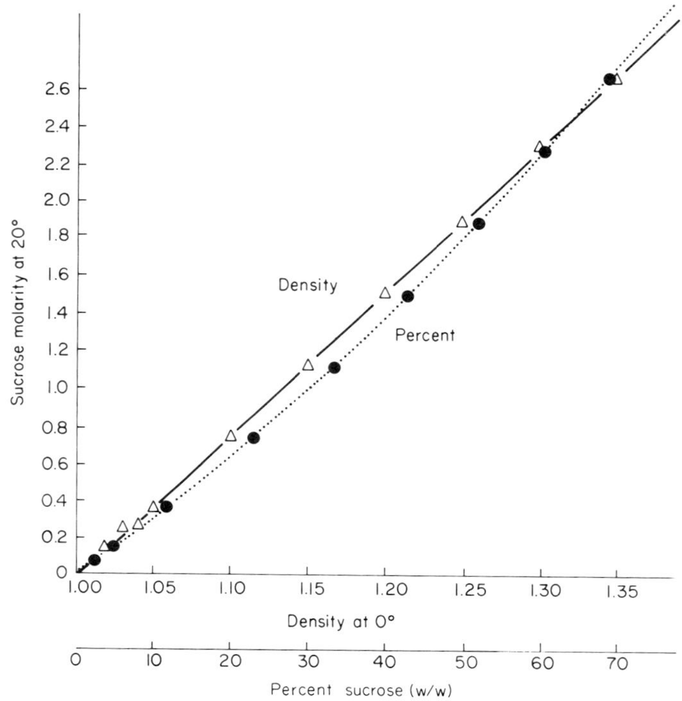

FIG. 1. Relationships between sucrose molarity, percentage of sucrose, and density.

sity of the peroxisomes on a sucrose gradient may restrict the application based on sedimentation velocity, but isopycnic separation in zonal rotors can be expected to provide very large preparations of leaf peroxisomes.

In the procedures described w/v of sucrose at 22° has been used in these large-scale preparations since preparation of solutions is rapid. Preparation of sucrose solution on w/w or molarity basis has been used for gradients[5,15] because it is accurate and desirable for density measurements. Relationships between sucrose molarity, percentage of sucrose, and density are plotted in Fig. 1.

Reagents

Sucrose, 2.5 *M* (stock solution): In a 1-liter calibrated flask dissolve 855 g of reagent grade sucrose in 400 ml of 20 m*M* glycylglycine buffer. The sucrose is added slowly to the buffer at

[15] F. Leighton, B. Poole, H. Beaufay, P. Baudhuin, J. W. Coffey, S. Fowler, and C. de Duve, *J. Cell Biol.* **37**, 482 (1968).

[16] C. A. Price and A. P. Hirvonen, *Biochim. Biophys. Acta* **148**, 531 (1967).

room temperature with stirring by a magnetic stirrer. After final solution, adjust to the 1-liter mark with water. In the 2.5 *M* sucrose the final concentration of glycylglycine is about 10 m*M*. Since the pH cannot be measured at this high sucrose concentration, the pH is assumed to be that of the buffer. This stock is stored at 4° and is stable.

Glycylglycine, 20 m*M*, pH 7.5: The pH of this buffer increases during storage even at 4°. After homogenizing the leaves, the pH of the homogenate decreases from plant acids and needs to be readjusted to pH 7.5. Thus a higher pH of the buffer in the grinding medium is not critical within broad limits.

Grinding medium of 0.83 *M* sucrose in glycylglycine buffer, pH 7.5: Mix well at 4° one part of 2.5 *M* sucrose stock with 2 parts of 20 m*M* glycylglycine buffer. In starting a preparation enough grinding medium should be prepared for also resuspending the pellets. Grinding medium of 0.5 *M* sucrose has been used with similar results. In exploratory experiments it was felt that poorer peroxisome recovery was obtained with grinding media of less than 0.5 *M* sucrose. Grinding media of greater than 0.83 *M* sucrose are more difficult to handle.

Solutions for the sucrose gradient: Aliquots of 50 ml of 2.5 *M* sucrose are diluted in a 100-ml graduated cylinder with 20 m*M* glycylglycine (pH 7.5) to the final volume indicated in Table I. Since the 2.5 *M* sucrose is prepared at room temperature, the final molarities will not be exact if dilutions are made at 4° for operational convience, but this is of minor concern with step gradients. The use of a pipette or dropper during volume adjustment is recommended. The graduated cylinder is capped with a piece of polyethylene and mixed by shaking before removing aliquots.

TABLE I

FINAL VOLUME TO DILUTE 50 ML OF 2.5 *M* SUCROSE FOR GRADIENT FRACTIONS

Final molarity	Final volume (ml)	Final molarity	Final volume (ml)
2.5	50	1.75	71.5
2.3	54.5	1.6	78.0
2.2	57.0	1.5	83.5
2.0	62.5	1.4	89.5
1.9	66.0	1.3	96.0
1.8	69.5	1.25	100.0
1.7	73.5	0.83	150.0

Preparation of Plant Tissues. Freshly harvested leaves have generally been used. Spinach or sunflower leaves have given the highest yields of peroxisomes and bean, corn, and sugarcane leaves the lowest. For recovery of particulate peroxisomal enzymes after grinding, Long Standing Bloomsdale spinach is a somewhat better variety (yields of 40–60%) than Virginia Savoya (yields of 30–50%). Spinach (varieties unknown) from the grocery store is also satisfactory, but lower yields are generally obtained. Whole spinach plants have been stored up to 4 weeks at 4° with no decrease in peroxisomal recovery. In fact in some experiments, better peroxisomal yields from stored leaves were obtained than from freshly harvested leaves. Whole spinach plants cut off just below the crown can be stored at 4° loosely packed in a water-saturated atmosphere in polyethylene bags or covered baskets. Before use, the best leaves are washed in water, slapped dry against paper towels, the midrib cut out, the leaf tissue weighed, and then chilled at 4°.

Grinding Procedure. Procedures for a small-scale or large-scale preparation, are based upon the amount of starting material and upon the analyses to be performed on the gradient. All solutions and material are kept at 4° by working in a cold room or cold chest. One weight, 40–60 g or 150 g, of leaf tissue is cut or torn into shreds as it is placed in a quart size Waring Blendor jar with 1.5 ml/g (90 ml or 225 ml) of 0.83 *M* sucrose grinding medium. Tough leaves, such as sugarcane, corn, or wheat, are cut into pieces about 0.5 cm on a side and ground in 2 volumes of medium. Sixty grams of leaves will produce material for one gradient tube, but three gradients can be run at once in a swinging-bucket rotor. Thus from one 150-g batch of leaves one can run three small-scale gradients at once. For three large-scale gradients, 450 g of leaves are needed. With the quart size Waring Blendor jar only 150 g of spinach leaves can be ground efficiently for peroxisome recovery. This is repeated three times for a large scale run rather than using a larger Blendor jar.

The Waring Blendor is run at first in 1- to 2-second bursts at low speed to chop the tissue into pieces which can be whipped into the grinding medium. Grinding time at maximum speed is judged in part by when the pieces appear homogenized or are no longer visible. For spinach and sunflower this is between 7 and 10 seconds, and for corn, 30 seconds. The grinding time should be as short as possible for the yields of peroxisomes rapidly decrease. The homogenate is squeezed through eight layers of cheesecloth, the pH adjusted to 7.5, the volume recorded, and a sample saved for assay of total enzyme activities.

Differential Centrifugation. This is done at 4° in clear plastic centrifuge tubes since complete resuspension of the pellets requires visual observation for clumps. Whole cells, whole chloroplasts and some starch grains are sedimented at 100 g for 20 or 30 minutes. Shorter periods of centrifugation do not pack the sediment enough for decanting. Shorter periods (10 minutes) of centrifugation at 600 g may also be used, but more peroxisomes are lost in the first pellet.

A 6000 g pellet is obtained after a second centrifugation for 20 minutes, and it contains mainly broken chloroplasts as well as most (two-thirds) of the peroxisomes and about half of the mitochondria. Longer periods of centrifugation packs the pellet so that resuspension is more difficult. A 39,000 g pellet, rich in mitochondria, is obtained by a third centrifugation of the homogenate for 20 minutes. The final supernatant solution must be assayed in studies on distribution of total activity in fractions obtained by differential centrifugation steps. For preparation of peroxisomes only the material sedimented between 100 g and 6000 g needs to be saved.

Each group of particles destined for one sucrose gradient, is resuspended in 4–6 ml of grinding medium and transferred to a graduated tube, if total activity is to be measured in an aliquot. The pellets must be carefully resuspended by repeated suction and extrusion through the tip of a disposable dropper or 5-ml pipette with a small bore until the suspension is uniform. The success of a good sucrose gradient depends upon complete resuspension of the 6000 g pellet. A Potter-Elvehjem homogenizer is not recommended because it appears to break up the peroxisomes.

Nonlinear Sucrose Gradient Centrifugation. Procedures for running the ultracentrifuges and techniques for making the gradient are similar to that used in other particle fractionation procedures. In all cases 50-ml cellulose nitrate tubes for the SW 25.2 rotor are used. A representative discontinuous gradient for a small- and large-scale separation are prepared with different sucrose concentrations in glycylglycine buffer as indicated in Table II.

The gradients are developed for 3 hours at 25,000 rpm (39,000 g) (maximum speed) at 4° in the SW 25.2 rotor for the Spinco Model L or L 2 Centrifuge. Prolongation of centrifugation does not result in a shift in position of the particles since they are being stopped at interfaces of a higher sucrose density than they can penetrate. However, in shorter runs the peroxisomes, which are permeable to sucrose, may not have all reached their final density, because their density is changing during the centrifugation as they move into the more concentrated sucrose.

Handling viscous 2.5 M sucrose at 4° presents operational problems. A 10-ml Mohr pipette with a broken or enlarged tip facilitates flow. In adding each new layer, the tip of the pipette, after touching the previous layer, is placed against the tube wall a few millimeters above the previ-

TABLE II
SMALL-SCALE AND LARGE-SCALE GRADIENTS

Small-scale gradient	Large-scale gradient
4 ml 2.5 *M* sucrose	8 ml 2.3 *M* sucrose
8 ml 2.3 *M* sucrose	10 ml 1.75 *M* sucrose[b]
10 ml 1.8 *M* sucrose[a]	7 ml 1.7 *M* sucrose
15 ml 1.5 *M* sucrose	5 ml 1.6 *M* sucrose
15 ml 1.3 *M* sucrose	5 ml 1.4 *M* sucrose
1–10 ml 6000 *g* pellet	5 ml 1.3 *M* sucrose
	10–15 ml 6000 *g* pellet

[a] Sometimes 1.85 or 1.9 *M* sucrose.
[b] Sometimes 1.8 or 1.85 *M* sucrose.

ous layer. As the sucrose solution is added slowly it flows onto the top of the previous layer. When draining the pipette, blow out the excess gently; otherwise moisture will condense in the cold pipette. One pipette can be used for preparing an entire group of gradients. Between changes of sucrose concentration it is rinsed with the next, more dilute, solution. Kleenex is better for wiping off tips of pipettes than Kimwipes, which do not readily absorb sucrose solutions.

Removal of Sucrose Gradients. This has been done at 4° in a large reach-in vertical cold chest with glass doors. The apparatus for forcing a puncture in the bottom of the centrifuge tube has been described[17,18] and is essentially a permanently mounted needle onto which the gradient tube can be set on dead center and pressured by a screw cap. The needle tip is surrounded by a layer of grease, but the tip must not be blocked. Gradient fractions are drained off through the needle with a short piece of polyethylene tubing into graduated conical test tubes. The collection tubes are placed inside a 500-ml suction flask which is attached to a vacuum pump to facilitate flow of sucrose solutions at 4°.

Samples from the gradient are removed and numbered as described in Table III. The slower the fractions are removed the better; about an hour should be taken to collect the fractions. The operator makes the cuts at the appropriate times as determined by observation of the sucrose bands against a light mounted behind the elution apparatus. A vacuum of 10 lb is sufficient to draw 2.5 and 2.3 *M* sucrose through the needle. The vacuum is reduced to about 5 lb for 1.8 *M* sucrose, and to 1 lb for 1.5 *M* sucrose. If the drainage is too rapid a vortex is formed above the needle and the cuts are not sharp. Determining when to change receivers is based upon the moment a band visibly reaches the exit of the needle

[17] E. H. McConkey, this series, Vol. XII, p. 620.
[18] R. J. Britten and R. B. Roberts, *Science* **131**, 32 (1960).

TABLE III
FRACTIONS FROM DISCONTINUOUS SUCROSE GRADIENTS

Number	Approximate volume (ml)	Sucrose concentration (M)	Component	Appearance
1	3.5–4.0	2.5	Starch grains	Pellet and clear layer
2	4–5	2.3	None	Clear; plus a light green band on top
3	2–3	Interface of 2.3–1.8	Peroxisomes	Hazy with suspension
4	6–7	1.8 or 1.75	None	Clear
5	3–5	Interface of 1.8–1.7	Mitochondria	Hazy and green
6	8–9	1.7	None	Nearly clear
7	10–12	1.5	Chloroplast band	Dark green
8	9–12	1.3	Chloroplast band	Dark green
9	Variable	0.8	Soluble	Yellowish

and is not based upon volume collected. The interfacial bands, No. 3 with peroxisomes and No. 5 with mitochondria, are hazy and readily visible. In general there are two or more major chloroplast bands separated by clear sucrose solutions. We have collected these bands (Nos. 7 and 8) separately but not the clear band in between, in order to reduce the number of subsequent assays.

For large preparations only the peroxisomal band is collected in one receiver from several gradients. In this speeded-up process restraint must be exercised not to drain off the fractions too rapidly.

The peroxisomes are mainly in fraction 3 at the interface on top of 2.3 M sucrose, but they also extend to varying extent into 1.8 or 1.9 M sucrose or fraction 4. For maximum yield, the lower hazy part of fraction 4 can be collected in fraction 3. On the large-scale gradients, the change in sucrose concentrations from 1.75–2.3 M effectively concentrates most of the peroxisomes in a small volume at the sharp interface.

Linear Gradients. Initial separation of peroxisomes from the 6000 g pellet by a linear sucrose gradient has so far not been practical. A wide range of sucrose concentrations must be used to accommodate chloroplasts at 1.5 M and peroxisomes at 1.9–2.0 M sucrose. Because of the immense quantity of chloroplasts in this pellet only small samples can be used giving low yields of peroxisomes. On linear gradients of the 6000 g pellet, peroxisomal enzymes considerably overlapped with the mitochondrial band and even into part of the chlorophyll bands.

Linear gradients are used to further purify the peroxisome fractions

(No. 3) from the nonlinear gradient. These linear gradients range from 1.7 or 1.8 to 2.1 *M* sucrose. The gradients are prepared in a gradient mixing apparatus, of design and operation already described.[17,18] An equal volume (14 ml) of the two sucrose concentrations are used. The lower sucrose concentration is placed in the inner chamber of the gradient maker. The connecting tube must be filled with this lower sucrose concentration to limit a flow back of the denser sucrose when the connecting stop cock is opened. The high sucrose concentration (2.1 *M*) is placed in the outer chamber of the apparatus, which contains the magnetic stirrer and which also drains off into the centrifuge tube. In making the gradient, the stirrer is started, the outlet drain opened, and then the connection between the two changers is opened. The speed of stirring is regulated as the gradient drains out. The linearity of gradients should be checked by absorbancy readings on gradient fractions where a dye is added to one chamber. Also measuring sucrose concentration in trial runs by a hand refractometer is recommended.

The linear gradients are run overnight (12–16 hours) at 39,000 *g* to allow the particles to reach their buoyant density. Fractions from the gradients can be collected on the basis of volume unless bands are visible. With most preparations the peroxisomal band is visible as a white-yellow haze. The whitish peroxisomal band is always accompanied by an adjacent yellow-green band of unknown composition, perhaps proplastids, and the two bands can be visually separated when fractionating the gradient. The peroxisomal band from the linear gradient is devoid of chlorophyll and contains only traces of mitochondrial activity.

Alterations. Little published work has yet appeared on methods for isolating peroxisomes. To date sucrose has been used in grinding media and gradients. Substitutions of Ficol, mannitol, NaCl, and Carbowax 4000 or addition of polyvinylpyrrolidone with the sucrose were no better than sucrose alone and generally gave poorer recoveries. Addition of cations such as Ca^{2+} or K^{+} or anions such as citrate made no difference in yield of particles after grinding in these altered media. Variation in pH has not been studied systematically, but exact control does not seem to be necessary. A pH of 7.5 has been chosen as an average value for the pH optimum of the enzymes in the peroxisomes. Amine buffers have been avoided because Tris inhibits glycolate oxidase, and phosphate buffers have not been used in order to facilitate studies of phosphorylation reactions in the particles. In a limited investigation with other buffers, such as cacodylate, none was found better than glycylglycine. As a result of these preliminary experiments, a simple sucrose–glycylglycine buffer grinding medium and gradient are recommended.

Peroxisomal Properties. Some properties of peroxisomes are mentioned in the introduction. Generally the small-scale preparations contained 0.1–0.3 mg of protein/ml and the large-scale preparations up to 1 mg of protein/ml. Eight enzymes have so far been found in peroxisomes from spinach leaves.[7] The specific activities for catalase was 11.6 mmole/min/mg peroxisomal protein; for NAD–malate dehydrogenase, 87 μmoles; for NAD–hydroxypyruvate reductase, 47 μmoles; for glycolate oxidase, 9.6 μmoles; for aspartate–α-ketoglutarate aminotransferase, 7.3 μmoles; for serine–pyruvate aminotransferase, 4.1 μmoles; for glutamate–glyoxylate aminotransferase, 2.8 μmoles; and for NAD–isocitrate dehydrogenase, 0.12 μmole.

Substrate Specificity. Peroxisomal respiration in leaves appears specific for glycolate. Glycolate oxidase will also oxidize lactate at 40% the rate of glycolate,[19] but in leaves lactate metabolism is not considered significant. The other enzymes of the peroxisome are highly specific and appear to function as dictated by the glycolate pathway.[7–9] NADH oxidation by the peroxisomal isozyme of malate dehydrogenase is a major reaction which characterizes peroxisomal respiration.[20] NADPH oxidation is limited by the low isocitrate dehydrogenase specific activity.

Storage and Stability of Peroxisomes. Good stability of the enzymatic activities of peroxisomes in the isolating sucrose solution is to be emphasized. Peroxisomal fraction 3 in about 2 *M* sucrose has been stored up to 3–4 weeks at 4° without visible spoilage or deterioration of enzymatic activities. Inability to reisolate the stored particles suggest that they do not remain intact. Because of stability at 4°, storage at −18° has not been attempted.

Peroxisomes in fraction 3 from the discontinuous gradient can be diluted to about 1.5 *M* sucrose with 20 m*M* glycylglycine buffer and placed directly on top of a second gradient for further purification. The peroxisomal fraction can be diluted also to 0.8 *M* sucrose and sedimented by centrifugation at 39,000 *g*. This procedure is used to concentrate the particles. After these dilutions, however, part (20–35%) of the enzymatic activity does not sediment as if the particles were further broken or leaked protein.

Latency. An increase of about 35% in each enzymatic activity of the peroxisomes except catalase has been observed upon addition of 0.01% final concentration of Triton X-100. A 0.01% concentration was found to be optimal, and some other detergents tried were not as effective. With liver peroxisomes this latency has not been observed.[4] Triton X-100 is routinely included in all leaf peroxisomal enzymatic assays.

[19] C. O. Clagett, N. E. Tolbert, and R. H. Burris, *J. Biol. Chem.* **178,** 977 (1949).
[20] R. K. Yamazaki and N. E. Tolbert, *Biochim. Biophys. Acta* **178,** 11 (1969).

Inhibitors. Glycolate oxidase is inhibited by hydroxymethy sulfonates,[21] and so is the peroxisomal respiration as assayed manometrically. The anaerobic DCPIP assay cannot be used in the presence of the sulfonates as the two compounds react with each other. Catalase is inhibited by aminotriazole, and so is leaf peroxisomal catalase. Inhibitors for the other peroxisomal enzymes have not yet been described.

Electron Microscopy. A comprehensive summary of electron microscopic investigations of microbodies has required a book length essay,[11] and elegant electron microscopy of leaf peroxisomes are available.[3,6] Leaf peroxisomes, similar to other microbodies, are characterized by electron microscopy as having a single membrane, no lamallae, a granular matrix, and often a dense core or crystalloid, and they specifically stain *in situ* for catalase. In the procedure used by Newcomb's group,[3,6] the epidermis of the leaf is removed and the tissue is fixed in 3% glutaraldehyde containing 50 m*M* phosphate at pH 6.8 for 1.5 hours. The tissue is then rinsed several times with phosphate buffer, postfixed in 2% osmium tetroxide for 2 hours, dehydrated in an acetone series followed by propylene oxide, and embedded in Araldite–Epon. After sections are cut and mounted on copper grids, they are stained for 10 minutes in aqueous 2% uranyl acetate, followed by lead citrate for 5 minutes. By a similar procedure, the leaf peroxisomal fraction from a nonlinear sucrose gradient has been examined by electron microscopy,[2] and the isolated particles appeared similar to those seen *in situ.* So far no detailed electron microscopic study on isolated particles has been done with regard to structure or as an aid to improve isolation procedures. Newcomb's group[22,23] has incorporated a 3,3′-diaminobenzidine stain for catalase activity into these procedures after the glutaraldehyde fixation. The incubation mixture of Beard and Novikoff,[24] prepared immediately before use contains 10 mg of diaminobenzidine, 5 ml of 50 m*M* 2-amino-2-methyl,1,3-propandiol buffer at pH 10, and 0.1 ml of 3% H_2O_2. The pH is adjusted to 9 prior to addition of tissue. Catalase activity in the leaf after glutaraldehyde treatment is found only in peroxisomes, and none is in chloroplasts. Catalase is distributed throughout the peroxisomal granular matrix, but when the core or crystalloid is present, most of the catalase is in this structure within the particle. Because of the unique distribution of catalase in peroxisomes, the use of this stain promises to be a quick and specific method for examining tissue for these microbodies.

[21] I. Zelitch, *J. Biol. Chem.* **224,** 251 (1957).

[22] S. E. Frederick and E. H. Newcomb, *J. Cell Biol.* **43,** 343 (1969).

[23] E. L. Vigil, *J. Histochem. Cytochem.* **17,** 425 (1969).

[24] A. B. Novikoff and S. Goldfischer, *J. Histochem. Cytochem.* **16,** 507 (1968).

Enzymatic Assays

Assays for marker enzymes have been used to identify the various particles in the gradient fractions. For peroxisomes, glycolate oxidase has been used most often as the marker, but catalase and hydroxypyruvate reductase serve equally well.[1,2] For mitochondria, cytochrome *c* oxidase[1,25] is used, and for chloroplasts, chlorophyll determination or NADP–diaphorase is used.[26] Only assays for leaf peroxisomal enzymes are detailed here.

Glycolate Oxidase. Glycolate–O_2 oxidoreductase (EC 1.1.3.1) is assayed anaerobically by 2,6-dichlorophenolindophenol (DCPIP) reduction.[1,2] To a 3-ml Thunberg Beckman cuvette (10 mm in diameter) are added in order: 2 ml of 0.1 M pyrophosphate, pH 8.5, containing 1.5×10^{-4} M DCPIP; 0.05 ml of 5×10^{-3} M FMN (final concentration, 1.0×10^{-4} M) 0.05 ml of 0.5% Triton X-100 (0.01% final concentration); an appropriate amount of enzyme and water so that the final volume will be 2.5 ml; in the side arm 0.1 ml of 0.125 M sodium glycolate (final concentration, 5×10^{-3} M). The cuvette is then evacuated and flushed 10 times with N_2 which has passed through Fieser's solution[27] to remove traces of O_2. The pyrophosphate buffer with dye must be prepared fresh weekly or stored at $-18°$ to avoid an initial lag in the assay.

Dye reduction at 25° is measured at 600 nm with an automatic recording spectrophotometer. The amount of dye (1.2×10^{-4} M final concentration) that can be used in the assay is limited by its absorption. The K_m (dye) is 0.38 mM as determined by extrapolation of a Lineweaver–Burk plot. From the ratio of V_m to V at the designated dye concentration, a correction multiple of 3.95 is obtained for maximum activity, if saturating dye concentration were possible. Based on the extinction coefficient for DCPIP, an absorbancy change of 1.000 is equivalent to 4.78 nmoles of dye and under the assay conditions is equal to 18.88 nmoles of potential glycolate oxidation at V_m. If different dye concentrations were used, a different correction factor for V_m should be calculated.

FMN is generally included in the assay for this flavin oxidase as a precaution, although there is no exogenous flavin requirement for the peroxisomes. Exhaustive removal of oxygen is necessary since diaphorase or any H_2O_2 generated by the oxidase is used by contaminating peroxidases to reoxidize the reduced dye. Addition of KCN to inhibit the

[25] E. W. Simon, *Biochem. J.* **69,** 67 (1958).

[26] M. Avron and A. T. Jagendorf, *Arch. Biochem. Biophys.* **65,** 475 (1956).

[27] Fieser's solution is 20 g of KOH dissolved in 100 ml of water plus 2 g of sodium anthraquinone β-sulfonate and 15 g of $Na_2S_2O_4$. After the N_2 gas has passed through Fieser's solution, it is also passed through a bottle of saturated lead acetate to remove H_2S.

peroxidases does not give maximum rates by this assay with leaf homogenates. Isolated peroxisomes contain no peroxidases, except for the peroxidative activity of catalase. As a result, isolated peroxisomes can be assayed by the dye method aerobically, or after a few nitrogen flushes. However, other particulate fractions and supernatant fractions are so contaminated with peroxidase that the recommended 10 flushings of the cuvettes with N_2 is necessary and is used throughout in order to provide a uniform procedure.

The dye assay is favored because of its sensitivity and because the rate can be recorded automatically. Peroxisomes in 10–30 μl of gradient fraction 3 are generally sufficient. The DCPIP assay is limited by reactions of the dye with other reducing agents, such as cysteine or dithiothreitol, which one might wish to add.

The manometric or Warburg procedure for glycolate oxidase is not used for routine peroxisome assays, because of the larger quantity of peroxisomes needed. Saturating amounts of oxygen are not achieved so that rates vary depending upon the percentage of O_2 in the gas phase, the shaking rate, and the amount of sucrose carried with the enzyme or added to the flasks. Sucrose reduces the rate in the Warburg assay probably by increasing the viscosity of the solution and thus lowering the available O_2. Even with 0.1% Triton X-100 in the Warburg assay to overcome latency, a final concentration of 1.5 *M* sucrose resulted in an 80% decrease in rate of glycolate oxidase activity and 1 *M* sucrose gave a 50% decrease when assayed with an atmosphere of air. These complications also prevent estimations of enzymatic activity in the intact particle in sucrose.

The phenylhydrazine assay for glycolate oxidase[28] has not been evaluated with isolated peroxisomes. This assay will function in the presence of other reducing substances like cysteine. The glyoxylate produced is converted in part to glycine by the particle even in the absence of added glutamate.

Catalase (EC 1.11.1.6). The disappearance of H_2O_2 is measured spectrophotometrically at 240 nm.[1,29] An absorbancy change of 1.000 is equal to 2.76 μmoles of H_2O_2.

Hydroxypyruvate Reductase or D-*Glycerate Dehydrogenase (EC 1.1.1.29)*. Reductase activity is measured spectrophotometrically at 25° in a 1-ml cuvette.[2,30] The reaction mixture contains 0.2 ml of 20 m*M*

[28] A. L. Baker and N. E. Tolbert, this series, Vol. IX, p. 338.

[29] H. Luck, *in* "Methods of Enzymatic Analysis" (H. U. Bergmeyer, ed.), p. 886. Academic Press, New York, 1965.

[30] I. Zelitch, *J. Biol. Chem.* **201,** 719 (1953).

phosphate buffer at pH 6.2, 0.05 ml of 4 mM NADH, 0.03 ml of 0.5% Triton X-100, and 0.62 ml of water and enzyme (generally about 20 μl of gradient fraction 3). After a 5-minute period for measurement of endogenous rate, the reaction is initiated with 0.1 ml of 0.75 M glyoxylate or 10 mM hydroxypyruvate. An absorbancy change of 1.000 is equal to 32 nmoles. The properties of this peroxisomal enzyme from leaf[31] and liver[32] particles have been investigated. In the leaf peroxisomes this reductase probably functions for the reduction of hydroxypyruvate to glycerate. Although the activity with hydroxypyruvate is about 5-fold greater than with glyoxylate and the K_m (hydroxypyruvate) is 0.12 mM compared to a K_m (glyoxylate) of 15 mM, glyoxylate is generally used as substrate, because it is readily available and relatively stable in solution. Because there is a NADPH–glyoxylate reductase in chloroplasts, caution must be used in interpreting these assays on crude homogenates.

NAD–Malate Dehydrogenase (EC 1.1.1.37). This dehydrogenase is assayed spectrophotometrically at 25° by following the oxidation of NADH at 340 mm with oxaloacetate.[20] The assay mixture contains 0.67 ml of 0.1 M HEPES buffer at pH 7.4, 0.03 ml of 0.5% Triton X-100, 0.04 ml of 10 mM oxaloacetate at pH 7.4, 0.02 ml of 2.81 mM NADH and enzyme plus water to give a total volume of 1 ml in the cuvette. Generally, 1–5 μl of peroxisomal fraction 3 is used.

Glutamate–Glyoxylate Aminotransferase (EC 2.6.1.4). This enzyme is assayed by following the formation of glycine–^{14}C.[33] In a final volume of 1.25 ml are 20 μmoles of glyoxylate-1,2-^{14}C (30,000 cpm), 25 μmoles of glutamate, 0.1 μmole of pyridoxal phosphate, the peroxisomes (about 50 μl of fraction 3), and 20 mM phosphate buffer (pH 7.5). Reactions are run at 30° for 10 or 15 minutes and terminated by boiling. Unreacted glyoxylate-^{14}C is removed by a Dowex 1 acetate column (6 $\times$ 50 mm) which is washed with 2 ml of water. Since glycine–^{14}C is the only labeled component of the effluent, aliquots of it are counted.

Aspartate Aminotransferase is measured by linking oxaloacetate formation to malate dehydrogenase.[34] The reaction mixture contains 0.5 ml of 0.1 M HEPES (pH 7.3), 0.05 ml of stabilized malate dehydrogenase (1000 units, Calbiochem), 0.35 ml of enzyme plus water, 0.02 ml of 5 mM pyridoxal 5-phosphate, 0.05 ml of 0.1 M α-ketoglutarate (pH 7), and 0.02 ml of 3.8 mM NADH. The reaction is initiated with 0.05 ml of 40 mM L-aspartate.

[31] N. E. Tolbert, R. K. Yamazaki, and A. Oeser, *J. Biol. Chem.* **245,** 5129 (1970).
[32] S. L. Vandor and N. E. Tolbert, *Biochim. Biophys. Acta* **215,** 449 (1970).
[33] T. Kisaki and N. E. Tolbert, *Plant Physiol.* **44,** 242 (1969).
[34] R. K. Yamazaki and N. E. Tolbert, *J. Biol. Chem.* **245,** 5137 (1970).

Serine–Pyruvate Aminotransferase is measured by linking hydroxypyruvate formation to D-glycerate dehydrogenase.[34] The reaction mixture contains 0.5 ml of 1.0 *M* HEPES (pH 7.3), 0.01 ml of crystalline glyoxylate reductase (or D-glycerate dehydrogenase) (Bohringer), 0.35 ml of enzyme plus water, 0.02 ml of 5 m*M* pyridoxal 5-phosphate (pH 7), 0.05 ml of potassium pyruvate, and 0.02 ml of 3.8 m*M* NADH. The reaction is initiated with 0.05 ml of 0.4 *M* DL-serine.

D_s-threo-Isocitrate Dehydrogenase (NADP) (EC 1.1.1.42). The assay mixture contains 0.67 ml of 0.1 *M* *N*-tris(hydroxymethyl)methyl-2-aminoethanesulfonic acid (TES) (pH 7.4), 0.03 ml of 0.5% Triton X-100, 0.03 ml of 0.1 *M* $MgCl_2$, 0.04 ml of 2.5 m*M* NADP, and 0.13 ml of enzyme plus water. The reaction is initiated by 0.1 ml of 0.2 *M* DL-isocitrate trisodium salt (*allo*-free), and the increase in absorbance at 340 nm is measured.

Chlorophyll Determination. Chlorophyll analyses of peroxisomes is not recommended, for it is inaccurate and unnecessary. Recently we have used the NADP-diaphorase assay as a sensitive indication of chloroplast contamination in peroxisomal fractions. Chlorophyll may be determined by its absorption at 652 nm after extraction into 80% acetone.[35] In peroxisomal fractions there is only a small amount of contaminating chloroplasts, so that large samples must be taken. However, in the high sucrose concentration of the peroxisomal fraction, when the acetone is added, the water is extracted and a sucrose and chlorophyll phase will oil out. Because sucrose is more soluble in cold acetone, these samples can be extracted with acetone at −18° overnight in order to remove the chlorophyll. Since pure peroxisomes contain no chlorophyll, expression of specific activity on a chlorophyll basis is not justified.

[35] D. I. Arnon, *Plant Physiol.* **24,** 1 (1949).

[66] Chlorophyll *a*–Protein Complex of Blue-Green Algae

By J. PHILIP THORNBER[1]

The complex is water insoluble and requires the use of a detergent to dissociate it from the pigmented lipoprotein structure of the cells and to maintain it in solution during its isolation. Excess detergent is detri-

[1] Research carried out at Bookhaven National Laboratory under the auspices of the U.S. Atomic Energy Commission.

mental to the complex, and therefore the time that the chlorophyll–protein is in contact with excess detergent during the purification procedure must be minimized to prevent degradation of the product. If degradation occurs, it is observed upon examination of the absorption spectrum of the isolated complex; the 437-nm peak should have a greater absorbance than the 420-nm shoulder, and the red wavelength maximum should be at 677 nm (see Fig. 2 and the table) if the complex is undegraded. Contact between the complex and detergent is kept minimal by using a peristaltic pump to maintain flow rates through columns and by working rapidly whenever excess detergent is present. Starting with broken cells, the time taken to reach the end of Step 3 should be 3–4 hours.

Purification Procedure

The protein is isolated at room temperature.

Material. Seventy-five grams of packed *Phormidium luridum* cells are mixed with an equal volume of 50 m*M* Tris–HCl buffer, pH 7.8, and broken by freezing to —17°; the cells are allowed to thaw at room temperature, and the supernatant solution is decanted. The procedure is repeated.

Step 1. The broken cells are suspended in Tris buffer and mixed with diatomaceous earth (Celite 545) (10 g of Celite/100 ml of suspension), and the slurry is poured into a column (30 × 10 cm). The packed column is eluted with 50 m*M* Tris, pH 7.8, until the eluate is colorless; in this initial stage a blue biliprotein, phycocyanin, is eluted. The column is subsequently washed with 50 m*M* Tris–0.5% (w/v) sodium dodecyl sulfate (pH 8.0, 400 ml), and the chlorophyll-containing eluate is collected. The rate of flow of solutions through the column is maintained at 10 ml/min.

Step 2. Solid ammonium sulfate is added to the detergent extract until precipitation of the chlorophyll is visible (10 g of ammonium sulfate/100 ml of solution). Celite (≈20 g/100 ml of solution) is then added until the solution becomes a slurry, which is poured into a column (12 × 4 cm). The packed column is eluted with 35% (w/v) ammonium sulfate solution (100 ml) to remove any excess detergent, and then with 35% ammonium sulfate–methanol (100:30 v/v); this latter solution (300 ml) removes detergent-complexed free chlorophyll and carotenoid from the column, but not the chlorophyll–protein complex. Fifty millimolar Tris (pH 7.8, 100 ml) is used to wash the ammonium sulfate and methanol from the column. Finally, 50 m*M* Tris–0.5% sodium dodecyl sulfate, pH 7.8 (about 200 ml) elutes all the chlorophyll–protein complex from the column. The rationale behind each of these eluting solutions is explained in the original report[2] of the isolation of the complex. The rate of flow of solvent through the column is maintained at 10 ml/minute.

Step 3. Immediately after the final chlorophyll-containing eluate has been obtained, it is run into a column (5 × 10 cm) of hydroxylapatite, which has been equilibrated with 10 m*M* sodium phosphate–0.2 *M* sodium chloride, pH 7.0. The chlorophyll-containing solution is washed into the column with 100 ml of the equilibration buffer, and a linear gradient (0.01 to 0.7 *M*) of sodium phosphate in 0.2 *M* NaCl (total volume, 600 ml) is then applied.[2a] The portion of the eluate that contains chlorophyll is collected. The complex of chlorophyll and protein is concentrated by addition of ammonium sulfate to the eluate, centrifugation, and dissolution of the precipitate in 50 m*M* Tris, pH 7.8. The precipitate is best dissolved by gentle mixing with buffer in a Potter homogenizer. A flow rate of 5 ml/minute is used during elution of the column.

The material obtained at this stage is pure with respect to pigment content and can be used as such for photosynthetic studies. However, two colorless proteins are present in trace amounts, and for studies of the composition of the complex these are removed.

Step 4. The concentrated solution is chromatographed on Sephadex G-200 (100 cm × 3 cm) in 50 m*M* Tris, pH 7.8, and the chlorophyll-containing component is collected.

Function

Contrary to the original report[2] the isolated complex contains the reaction center (P700) for photosystem I of the organism; the light-driven reversible absorbance changes between 650 and 750 nm are shown in Fig. 1. The concentration of P697 (P700) in the isolated material is about 1 mole/80 moles of chlorophyll *a*. The complex functions therefore as the light-harvesting antenna for the photosystem I reaction center,[2] and carries out the primary photochemical reaction associated with that system.

Properties

Some 75% of the chlorophyll *a* in the whole cells of the organism is represented by the isolated complex.

Spectral Properties. (i) ABSORPTION. The absorption spectra of the complex at room temperature and at liquid-N_2 temperature are shown in Fig. 2. The millimolar absorptivities at the peak wavelengths are given

[2] J. P. Thornber, *Biochim. Biophys. Acta* **172,** 230 (1969).

[2a] *Note added in proof:* Improved yields and purity can be obtained by using a stepwise elution of the column; 100 ml each of 0.1 *M*, 0.2 *M*, and 0.3 *M* sodium phosphate are used, and the 0.3 *M* phosphate eluate is collected.

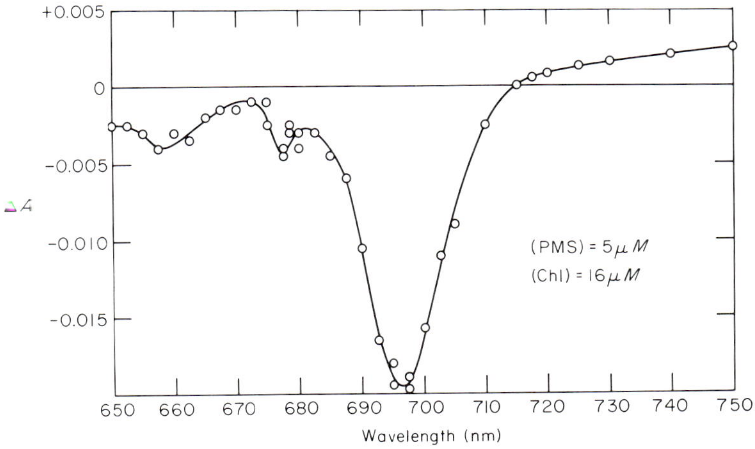

FIG. 1. Light-induced reversible absorbance changes in the chlorophyll *a*–protein complex. Light of 435 nm was used to illuminate the sample.

in the table. (ii) EMISSION. The fluorescence spectra of the chlorophyll–protein at room temperature and at the temperature of liquid nitrogen are shown in Fig. 3.

Stability. The isolated complex can be lyophilized and stored in-

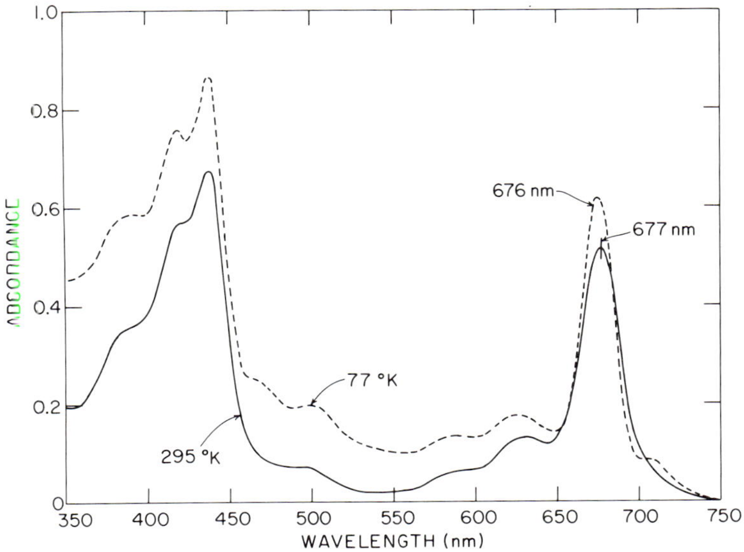

FIG. 2. Absorption spectra of the chlorophyll–protein in 50% glycerol solution at room temperature (——) and at the temperature of liquid nitrogen (- - - -).

Molar Absorptivity Values of the Complex Based on Chlorophyll[a,b]

λ (nm)	275	342	420	437	629	677
$E(\text{m}M^{-1}\ \text{cm}^{-1})$	44	35	75	78	15	60

[a] From J. P. Thornber, *Biochim. Biophys. Acta* **172,** 230 (1969).
[b] Limit of error estimated as ±3%. Calculated from the room temperature spectra of the complex dissolved in 50 m*M* Tris, pH 8.0.

definitely in a dry state; it can also be stored at 4° in the dark for several days without showing any signs of degradation.

Purity. Polyacrylamide gel electrophoresis of the material obtained from Step 4 shows the presence of one protein component which is coincident with a zone of chlorophyll. The component has an electrophoretic mobility of 1.60×10^{-5} $\text{cm}^2/\text{V-sec}$ in an 8% acrylamide gel buffered with 50 m*M* Tris–0.25% sodium dodecyl sulfate, pH 8.0. Ultracentrifugal analysis shows the presence of one boundary of 9 *S* upon examination of the complex in a solution (1 mg of protein/ml) of 50 m*M* Tris–0.2%

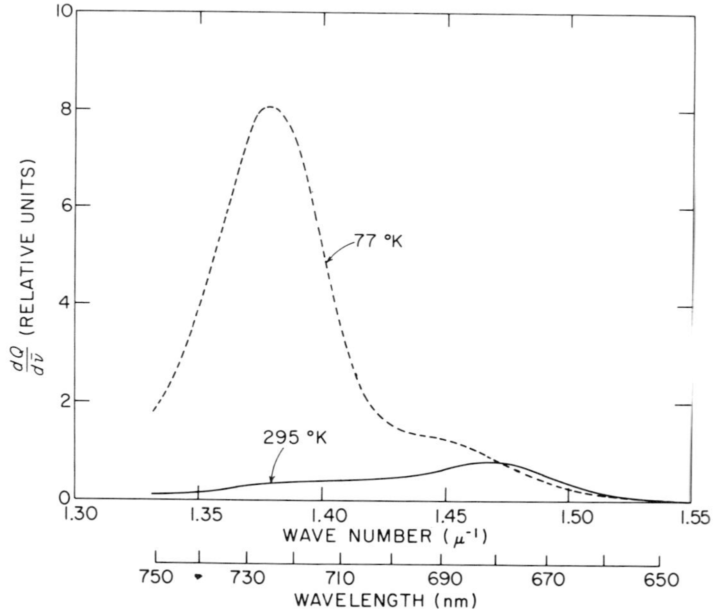

Fig. 3. Emission spectra of the complex dissolved in 50 m*M* Tris, pH 8.0 at room temperature (——) and at the temperature of liquid nitrogen (- - - -). The spectra were recorded for a sample having an absorbance of 0.75 cm^{-1} at 677 nm. The sample thickness was 1 mm. The spectra were corrected for the response of the phototube.

sodium dodecyl sulfate, pH 8.0. The chlorophyll sediments with this boundary.

Molecular Weight. The size of the complex has been determined by comparison of its electrophoretic mobility (see above) in the presence of detergent[3] with that of another complex of chlorophyll and protein of known molecular weight, the bacteriochlorophyll–protein from a green photosynthetic bacterium.[4] A value of 150,000–160,000 g/mole was obtained for the algal complex.[2]

Chemical Composition. The complex contains protein and chlorophyll *a* essentially; traces of carotenoids and a quinone are present. The following molar proportions occur in the isolated complex; chlorophyll *a* (20 ± 1)–β-carotene (0.7 ± 0.1)–echinonone (<0.02)–protein subunits of mol. wt. 35,000 (4). It is believed[2] that the complex contains four identical subunits. The amino acid composition (moles per subunit) is: Asp (27), Glu (21), Lys (11), His (13–14), Arg (11), Thr (19), Ser (19), Gly (33), Ala (32–33), Val (20), Ile (20), Leu (35), Pro (16), Met (6), Cys (3), Tyr (10), Phe (20), Trp (6).

Analogy to Other Chlorophyll–Protein Complexes. Complexes of almost identical molecular size and composition have been isolated from higher plants[5]; a bacteriochlorophyll–protein of similar size and subunit structure has been isolated from green bacteria.[4] Chlorophyll–proteins having the same absorption spectra as the complex described in this article, but of undetermined composition, can be isolated from other blue-green algae by the procedure outlined.

Comment. Every result so far has indicated that the isolated complex is an homogeneous component; it has not proved possible to fractionate the reaction center pigment from the other light-harvesting chlorophylls. Since the complex is a molecule containing 20 chlorophylls and since the reaction center pigment occurs in the ratio of 1 per 80 chlorophylls, then not every isolated molecule of complex can contain a reaction center. Very probably there exist *in vivo* aggregates of the chlorophyll–protein (perhaps tetramers) in which each of the light-harvesting chlorophylls can transfer absorbed energy to P700.

[3] A. L. Shapiro, E. Viñuela, and J. V. Maizel, *Biochem. Biophys. Res. Commun.* **28**, 815 (1967).

[4] See this volume [61].

[5] J. P. Thornber, J. C. Stewart, M. W. C. Hatton, and J. L. Bailey, *Biochemistry* **6**, 2006 (1967).

[67] The Photochemical Reaction Center of *Rhodopseudomonas viridis*

By J. PHILIP THORNBER[1]

The photochemical reaction center of this bacteriochlorophyll *b*-containing organism consists of a photobleachable pigment, P985 (P960 in dried films of lamellae) and a pigment, P830, which shifts to the blue upon illumination.[2] The photochemical reaction center is located in the pigmented membrane system (lamellae) of the organism and requires the use of an anionic detergent to dissociate it from the membrane. The presence of oxidized bacteriochlorophyll *b* ($\lambda_{max} \approx 690$ nm) in the detergent extract of the cells leads to low recoveries of the reaction center material from the hydroxylapatite column chromatography. To prevent large quantities of oxidized bacteriochlorophyll being formed, the preparation is carried out under weak green light; green fluorescent tubes (Sylvania, 40G), in addition to a green Plexiglas filter (2092), are used to prevent oxidation. The preparation should also be done quickly, to minimize oxidation of the chlorophyll, and to this end a peristaltic pump is used to accelerate the flow of eluents through the hydroxylapatite columns. The reaction center is isolated at room temperature.

Preparative Procedure[3]

Material. Photoheterotrophically[4] grown cells are aggregated by the addition of 0.2 *M* potassium aluminum sulfate (400 ml/20 liter of culture medium) and centrifuged. The packed cells (15 g) are suspended in an equal volume of 50 m*M* Tris–HCl, pH 8.0, and ruptured by sonication in a Raytheon 10-kHz oscillator (1.2 A for 10 minutes). The absorbance of the sonicated solution at the far-red wavelength maximum ($\approx$1010 nm) is used to determine the concentration of bacteriochlorophyll *b* ($\epsilon_M = 10^5$ cm^{-1}).

Step 1. The sonicated solution is centrifuged at 15,000 *g* for 30 minutes. Some small pieces of the lamellar system are present in the supernatant solution, but most of the pigment is present in the precipitate;

[1] Research carried out at Brookhaven National Laboratory under the auspices of the U.S. Atomic Energy Commission.

[2] A. S. Holt and R. K. Clayton, *Photochem. Photobiol.* **4,** 829 (1965).

[3] J. P. Thornber, J. M. Olson, D. M. Williams, and M. L. Clayton, *Biochim. Biophys. Acta* **172,** 351 (1969).

[4] K. E. Eimhjellen, O. Aasmundrud, and A. Jensen, *Biochem. Biophys. Res. Commun.* **10,** 232 (1963).

either fraction can be used for the subsequent isolation of the reaction center. The precipitate (or supernatant solution) is homogenized with 50 m*M* Tris–1% sodium dodecyl sulfate, pH 8.0; a sufficient volume of the detergent solution is added to give a final ratio of 12 g of detergent/mmole of bacteriochlorophyll.

Step 2. The homogenate is centrifuged (15,000 *g* for 20 minutes), and the spectrum of the supernatant solution is examined to ensure that the detergent has changed the ≈1010-nm form of the bacteriochlorophyll; the only major absorption peak in the infrared region should be at ≈810 nm as shown in Fig. 1. More detergent solution is added, if necessary, until the requisite spectrum is obtained. The green supernatant is then run into a column of hydroxylapatite (3.5 cm × 8 cm), which is equilibrated in 10 m*M* sodium phosphate–0.2 *M* NaCl, pH 7.0. The homogenate is washed into the column with 100 ml of the equilibration buffer, and then with 100 ml of 0.1 *M* sodium phosphate–0.2 *M* NaCl, pH 7.0. Subsequent elution with 0.2 *M* sodium phosphate–0.2 *M* NaCl removes a small proportion (6–8%) of the bacteriochlorophyll absorbed on the column; this eluate contains the reaction center pigments plus traces of free bacteriochlorophyll *b*.

Step 3. Solid ammonium sulfate is added to the 0.2 *M* phosphate eluate from the hydroxylapatite column to a final concentration of 15%;

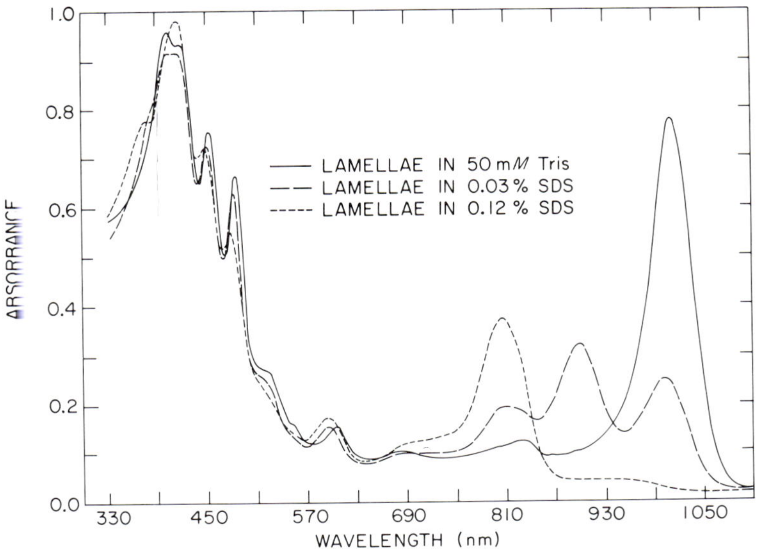

FIG. 1. The effect of sodium dodecyl sulfate on lamellae particles of *Rhodopseudomonas viridis*. (———) lamellae particles, (— — —) lamellae incompletely dissociated by detergent, (------) lamellae completely dissociated by detergent.

all the colored material in the eluate is precipitated. After centrifugation the precipitate is dissolved in a minimal volume of 10 m*M* sodium phosphate–0.2 *M* NaCl, pH 7.0. This concentrated solution is rechromatographed through another hydroxylapatite column (2 cm × 4 cm) using the same conditions as described above (step 2). The final eluate is the reaction center preparation.

Properties

Spectral. The absorption spectrum of the isolated material is shown in Fig. 2, together with the spectrum recorded while the material is exposed to strong white light. The light-induced absorbance changes of the reaction center (bleaching of P960 and the blue shift of P830) are observed. These spectral changes are slowly reversed upon returning the sample to darkness; the presence of phenazine methosulfate (10 μ*M*) or sodium ascorbate (50 μ*M*) permits a rapid recovery of the spectral changes after illumination.

Other Light-Induced Absorbance Changes. In addition to the changes in the reaction center pigments, a cytochrome (C558) becomes oxidized upon exposure of the preparation to light, and in the presence of phenazine methosulfate or sodium ascorbate, the oxidized C558 is rapidly reduced upon returning the sample to darkness. When the redox potential of the preparation is lowered below 0mV, a second cytochrome (C553)

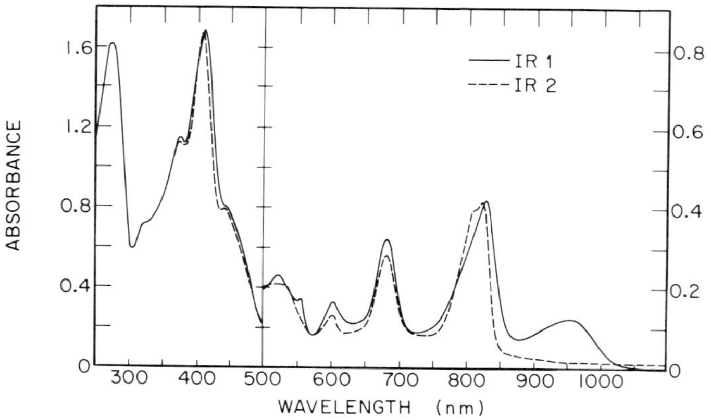

FIG. 2. Absorption spectra of the reaction center preparation, recorded in the IR1 (———) and the IR2 (-----) modes of the Cary 14R spectrophotometer. In the IR1 mode the sample is exposed to a weak beam of monochromatic light, while in the IR2 mode the sample is illuminated by a strong beam of white light. From J. P. Thornber, J. M. Olson, D. M. Williams, and M. L. Clayton, *Biochim. Biophys. Acta* **172,** 351 (1969).

becomes photooxidizable, and its replaces C558 as the electron donor to P960.[5]

Composition. In addition to the reaction center pigments, oxidized bacteriochlorophyll, and protein, other components present in the isolated component are: two cytochromes (C553 and C558) and a carotenoid (dihydrolycopene). The following are the molar proportions of the constituents which have been identified so far: P960 (1); P830 (2); dihydrolycopene (1); C558 (2); C553 (5).

Purity. Polyacrylamide gel electrophoresis reveals that the preparation contains traces of some colorless proteins in addition to the major pigmented complex. The molecular weight of the reaction center complex is about 110,000 g/mole.

Stability. The preparation can be stored in the dark at room temperature, or at 4°, for several weeks without any appreciable loss of photobleaching activity. Alternatively, it can be lyophilized without any loss in activity.

Function. The preparation contains more than just the reaction center pigments in their protein environment, e.g., it contains two cytochromes (C553 and C558), the oxidation of both of which is associated with bleaching of P960,[5] and it contains the primary electron acceptor, whose nature is as yet unknown. The reaction center pigment, the acceptor and C558, and probably other so far unidentified constituents in the preparation participate in a light-driven cyclic electron pathway.[3] Thus, the preparation should be regarded as a photoelectron transport particle rather than as a chlorophyll–protein complex (cf. Thornber[6]; the latter being used for a pigmented complex which contains a single protein species.

[5] G. D. Case, W. W. Parson, and J. P. Thornber, *Biochim. Biophys. Acta* **223**, 112 (1971)

[6] See this volume [66].

[68] Spinach Leaf D-Fructose 1,6-diphosphate 1-Phosphohydrolase (FDPase) EC 3.1.3.11

By Jack Preiss and Elaine Greenberg

Fructose diphosphate + H_2O → fructose 6-phosphate + orthophosphate

Assay Method

Principle. Enzymatic activity is determined by continuous measurement of fructose 6-phosphate formation in the presence of excess phosphohexose isomerase, glucose 6-phosphate dehydrogenase, and NADP. The rate of NADPH formation is equal to the rate of fructose 6-phosphate formation. The reaction may also be followed by measuring the amount of orthophosphate formed with the Fiske–SubbaRow method.[1]

Reagents

Tris·HCl buffer, 1 *M*, pH 8.5
$MgCl_2$, 0.1 *M*
Disodium ethylenediaminetetracetate (EDTA), 0.032 *M*
Bovine plasma albumin (BPA), 10 mg/ml
Fructose diphosphate, sodium salt, 20 m*M*
NADP, 10 m*M*
Glucose 6-phosphate dehydrogenase, 1 mg/ml; diluted from a commercial preparation (specific activity 300 units/mg) with water
Phosphohexoseisomerase from yeast, 1 mg/ml; diluted from commercial preparation (specific activity 400 units/mg) with water

Procedure. The Tris buffer (0.1 ml), $MgCl_2$ (0.05 ml), EDTA (0.05 ml), fructose diphosphate (0.05 ml), NADP (0.05 ml), glucose 6-phosphate dehydrogenase (0.01 ml), and phosphohexoseisomerase (0.01 ml) are mixed together in a volume of 0.9–0.99 ml. The reaction is then initiated with 0.01–0.1 ml (0.001–0.02 unit) of the fructose diphosphatase (final volume of the reaction mixture is 1.0 ml), and absorbance readings are taken at 30-second intervals.

Definition of Unit and Specific Activity. One unit of enzyme is defined as that amount which hydrolyzes 1 micromole of fructose diphosphate per minute at 25° under the conditions of the assay. The specific activity is expressed as units of enzyme activity per milligram of protein. Protein concentration is determined by the method of Lowry *et al.*[2] with crystalline bovine serum albumin as a standard.

Application of Assay Method to Crude Tissue Preparations. The method described above is satisfactory for assay of the crude spinach leaf extracts.

Purification Procedure

Spinach leaves obtained from a local supermarket are deveined and washed with distilled water. The following procedures are then carried out at 0°–4° with 4700 g of spinach leaves.

Step 1. Preparation of Crude Extract. Spinach leaves are ground in 5 liters of cold 0.1 *M* phosphate buffer, pH 7.5, containing 10 m*M* EDTA

[1] C. H. Fiske and Y. SubbaRow, *J. Biol. Chem.* **66**, 375 (1925).
[2] O. H. Lowry, N. J. Rosebrough, A. L. Farr, and R. J. Randall, *J. Biol. Chem.* **193**, 265 (1951).

and 5 mM dithiothreitol (DTE) in a gallon-size Waring Blendor for 3–4 minutes. The resulting suspension is filtered by vacuum through 2 layers of filter aid pads placed on a large Büchner funnel. The filtrate is centrifuged for 10 minutes at 12,000 g.

Step 2. Ammonium Sulfate Fractionation. To each liter of crude extract, 314 g of ammonium sulfate is added. After 20 minutes the suspension is centrifuged for 10 minutes at 12,000 g. To each liter of the clear supernatant solution is added 325 g of ammonium sulfate. After 20 minutes the suspension is centrifuged for 10 minutes at 12,000 g and the precipitate is dissolved in 50 mM phosphate buffer, pH 7.0, and dialyzed against 15 liters of the same buffer overnight.

Step 3. DEAE–Cellulose Chromatography 1. A column (5×13 cm) containing DEAE–Cellulose (Whatman DE-52) is equilibrated with 10 mM phosphate buffer, pH 7.2, and 364 ml of the ammonium sulfate fraction are adsorbed onto the column. The column is washed with 350 ml of 10 mM phosphate buffer, pH 7.2, and then eluted with 0.2 M PO_4 buffer, pH 7.0, until the absorbance at 280 nm decreased below 1. The column is then eluted with 1 M phosphate buffer, pH 7.5, and the fractions containing FDPase activity are pooled. The enzyme is precipitated with solid ammonium sulfate which is added to a concentration of 0.9 saturation. The enzyme is dissolved in 10 mM phosphate buffer, pH 7.2, and dialyzed against 15 liters of the same buffer.

Step 4. DEAE–Cellulose Chromatography 2. Three hundred and twenty-six milliliters of enzyme is adsorbed onto a column (1×13 cm) containing DEAE–cellulose (Whatman DE-52) which is equilibrated with 10 mM phosphate buffer, pH 7.5. A linear gradient that contains 250 ml of 0.2 M phosphate buffer, pH 7.5, in the mixing chamber and 250 ml of 1 M phosphate buffer, pH 7.0, in the reservoir chamber is used to elute the enzyme. Fractions containing the enzyme activity are pooled, and the enzyme is precipitated with ammonium sulfate crystals which are added to a concentration of 0.95 saturation. The enzyme is dissolved in 9.4 mM imidazole buffer, pH 6.7, and dialyzed against this buffer overnight.

Step 5. Preparative Disc Gel Electrophoresis. Thirty milliliters of the enzyme fraction is subjected to zone electrophoresis in a Poly-prep apparatus (Buchler Instruments, Fort Lee, New Jersey) which incorporates the design features of the system developed by Jovin *et al.*[3] An imidazole–borate polyacrylamide gel buffer system is used.[4] Migration is toward the anode at pH 7.5 (0°). Both the resolving and concentrating

[3] T. Jovin, R. A. Chrambach, and M. Naughton, *Anal. Biochem.* **9**, 351 (1964).

[4] The imidazole borate polyacrylamide gel discontinuous buffer system was developed by Dr. George B. Bruening of the Department of Biochemistry and Biophysics, University of California, Davis, California.

gels are polymerized with ammonium persulfate. The length of the resolving gel is 2 cm, and that of the stacking or concentrating gel is 3.2 cm. The polymerizing mixture for the formation of the resolving gel is 1 part of a 30% acrylamide monomer, 1% *N,N'*-methylene bisacrylamide solution; 1 part of a buffer solution containing, per 200 ml, 48 ml of 2 *N* HCl, 17.5 ml *N*-ethylmorpholine, and 60 ml of glycerol (the pH of this solution when diluted 8-fold is 7.3); and 2 parts of a 0.2% ammonium persulfate solution. The polymerizing mixture for the formation of the concentrating gel is 2 parts of a 5% acrylamide monomer–1.25% *N,N'*-methylene bisacrylamide solution; 1 part of a buffer solution containing, per 100 ml, 24 ml of 2 *N* HCl, 5.1 g of imidazole, and 0.2 ml of tetramethylethylenediamine (TEMED); and 1 part of a 0.4% ammonium persulfate solution. The upper or cathode buffer, pH 8.6, contained per liter, 11.2 g of boric acid, 90 ml of 2 *M* triethanolamine, and 10 ml of glycerol. The lower or anode buffer is a 1:12 dilution of the buffer used for polymerizing the resolving gel. The eluting buffer is a 1:3 dilution of the buffer used for the polymerization of the resolving gel.

The polymerization of the gels and the fractionation procedures are carried out at 0° (circulation of a coolant from a Buchler refrigerated fraction collector). The various buffer systems are used exactly as described in the Buchler Instruction Manual. Concentration of the applied sample is carried out at a current of 50 mA with a power supply with current regulation. At the moment that the tracking dye (chlorophenol red) enters the resolving gel, the current is reduced to 40 mA. The flow rate of the eluting buffer is 0.8 ml/minute, and 5-ml fractions are collected. The activity is located and the peak fractions are pooled (35 ml) and precipitated with 23 g of ammonium sulfate. The enzyme is dissolved in 10 m*M* phosphate buffer, pH 7.2, and dialyzed against 500 ml of this buffer overnight.

The yields and specific activities obtained in each step of the purification procedure are summarized in the table.

Purification of FDPase

Fraction	Volume	Protein (mg/ml)	Units/ml	Specific activity	Total units
1. Crude extract	8200	6.2	1.35	0.22	11,070
2. Ammonium sulfate	365	10.1	24.1	2.4	8,800
3. DEAE–cellulose I	327	0.46	11.1	24.2	3,640
4. DEAE–cellulose II	35	1.5	48.4	32.3	1,695
5. Disc gel electrophoresis	3.1	3.7	426	115	1,320

Properties of the Enzyme

Homogeneity, Molecular Weight. Disc electrophoresis of the enzyme fraction from step 5 at three different pH values, 7.5, 8.5, and 9.0, and with varying gel concentrations, showed only one protein band when the gels were stained with Coomassie blue.[5] The enzyme was estimated to have a molecular weight of 195,000 using the disc gel technique of Hedrick and Smith.[6] The enzyme migrated as a single symmetrical peak in the analytical ultracentrifuge at 50,740 rpm; S_{obs} was 7.8 S at a protein concentration of 3.6 mg/ml in 10 m*M* phosphate buffer, pH 7.2.

pH Optima. The pH optimum of the enzyme depends on the concentration of magnesium ion.[7] At 5 m*M* $MgCl_2$ the pH optimum for activity is 8.5 and there is very little activity at 7.0. Increasing concentrations of magnesium ion up to 80 m*M*, however, causes a shift of the pH optimum to 7.5–7.7 either in Tris·chloride or *N*-2-hydroxyethyl piperazine–*N'*-2-ethanesulfonic acid (HEPES) buffers. Appreciable activity is observed at pH 7.0 at the higher magnesium ion concentrations (20–80 m*M*).

Kinetics. The fructose diphosphate concentration vs. velocity curve was found to be sigmoidal at either pH 8.5 or pH 7.5,[7] suggesting that there are multiple interacting sites on the protein. The concentration of fructose diphosphate required for 50% of maximal velocity at pH 8.5 and 5 m*M* $MgCl_2$ is about 0.25 m*M*. At pH 7.5 the concentration of fructose diphosphate required for half-maximal activity is dependent on the concentration of $MgCl_2$ present, varying from 2.5 m*M* fructose disphosphate at low concentrations of Mg^{2+} (10 m*M*) to 0.3 m*M* at high concentrations of Mg^{2+}.

EDTA stimulates fructose diphosphatase activity about 1.5- to 2-fold throughout the pH range. However, EDTA does not shift the pH optimum to a more neutral pH as in the case of the mammalian FDPases. No inhibition of the spinach leaf FDPase by 5'-adenylate has been observed. Other properties of the spinach leaf FDPase have been previously described.[8,9] Activation of the Mg^{2+}-dependent FDPase by reduced ferredoxin has been reported by Buchanan *et al.*[10] These results, although preliminary, are intriguing in view of the *in vivo* studies of

[5] R. A. Chrambach, M. Reisfeld, M. Wyckoff, and J. Zacceri, *Anal. Biochem.* **20,** 150 (1967).

[6] J. L. Hedrick and A. J. Smith, *Arch. Biochem. Biophys.* **126,** 155 (1968).

[7] J. Preiss, M. L. Biggs, and E. Greenberg, *J. Biol. Chem.* **242,** 2292 (1967).

[8] E. Racker and E. A. R. Schroeder, *Arch. Biochem. Biophys.* **74,** 326 (1958).

[9] E. Racker, this series, Vol. V, p. 272.

[10] B. B. Buchanan, P. P. Kalberer, and D. I. Arnon, *Biochem. Biophys. Res. Commun.* **29,** 74 (1967); *Fed. Proc., Fed. Amer. Soc. Exp. Biol.* **27,** 344 (1968).

Pedersen *et al.*[11] that suggest that a light-induced activation occurs at the fructose diphosphatase stage. The *in vitro* effects of reduced ferredoxin observed by Buchanan *et al.*[10] may serve to explain the light induced regulation of photosynthetic carbon assimilation observed at the FDPase stage since reduced ferredoxin is an *in vivo* product of the photochemical reactions.

[11] T. A. Pedersen, M. Kirk, and J. A. Bassham, *Physiol. Plant.* **19**, 219 (1966).

[69] Photochemical Reaction Centers from *Rhodopseudomonas spheroides*

By Roderick K. Clayton and Richard T. Wang

General Properties; Chemical and Spectrophotometric Assays

The hallmark of the known bacterial photosynthetic reaction centers is the presence of a bacteriochlorophyll (BChl) molecule that is specialized to act as a primary photochemical electron donor. This BChl, called P870, P890, etc. after the wavelength of its long-wave absorption maximum, is oxidized by light, while an unspecified electron acceptor becomes reduced. Oxidation of P870 is signaled by loss of the long-wave absorption band ("bleaching"). In living cells the oxidized P870 is reduced by one or more *c*-type cytochromes. In purified reaction centers the source of electrons for the re-reduction of oxidized P870 depends on the environment. In any case the defining assay for reaction centers is the reversible light-induced bleaching of P870. Alternatively the P870 can be oxidized with ferricyanide; the bleaching by chemical oxidation is reversible and shows a midpoint potential between 450 and 500 mV (one-electron titration curve).

In *Rhodopseudomonas spheroides* the extinction coefficient of P870 at its absorption maximum near 865 nm has been estimated[1] to be 113/mM/cm. The differential coefficient, $\Delta\epsilon$, for the bleaching is then 100/mM/cm at the peak of absorption. For the light-harvesting BChl component B870, in blue-green mutant *Rhodopseudomonas spheroides*, $\epsilon =$ 127/mM/cm at the peak.[2]

Observation of the light-induced bleaching of P870 is facilitated by

[1] R. K. Clayton, *Photochem. Photobiol.* **5**, 669 (1966).

[2] The wavelength of the long wave maximum of B870 varies from about 850 to 875 nm in different strains, under different culture conditions, and after different treatments, such as freeze drying or rupture of the cells.

two procedures: removal of the chemically inert light-harvesting BChl and removal or oxidation of cytochromes that would otherwise react very rapidly with oxidized P870, restoring it to its unbleached form. With purified reaction centers, the bleaching can be measured easily with any commercial spectrophotometer than can be adapted for illumination of the sample during measurements of optical density in the near infrared. It is usually possible to cut a hole in the lid of the sample compartment so as to bring in an excitation beam. If the sample is excited with blue light, a red filter in the measuring light path can prevent scattered blue exciting light from striking the detector(s). Alternatively the Cary 14R spectrophotometer can be used to compare spectra recorded with weak light (IR1 mode) and with strong light (IR2 mode) striking the sample (Fig. 1 gives an example of this). The white light that strikes the sample in the IR2 mode is so strong that care must be taken to prevent irreversible damage to the sample during the measurement. The beam can be attenuated before it strikes the sample, with compensating attenuation in the reference beam, and the scan can be kept rapid to minimize the time of exposure.

In reaction centers made from *Rhodopseudomonas spheroides*[3] or *Rhodospirillum rubrum*[4] the bleaching of P870 is accompanied by blue shift of a band near 800 nm due to another specialized BChl, P800. There appear to be two or three P800 for each molecule of P870. The absorption increase at 780 nm due to this blue shift is often easier to measure than the bleaching at 870 nm. Optical interference from the fluorescence of light harvesting BChl can be evaded by measuring at 780 nm, with a 780-nm interference filter between the sample and the measuring beam detector.

Bleaching of the long-wave band of P870 is also accompanied by bleaching at 600 nm (see Fig. 1); this effect can be measured in the "visible" range of a spectrophotometer, using a photomultiplier that is insensitive to the near infrared. Then if the exciting light is confined to wavelengths greater than about 800 nm, there is no interference in the measurement from either fluorescence of BChl or scattered exciting light.

Rhodopseudomonas viridis contains the longer-wave pigment BChl *b*. Reaction centers made from this organism[5] contain P830 and P960, analogous to the P800 and P870 of *Rhodopseudomonas spheroides*. The blue shift of P830 can be measured as an increase in optical density at 810 nm.

[3] D. W. Reed and R. K. Clayton, *Biochem. Biophys. Res. Commun.* **30**, 471 (1968).

[4] G. Gingras and G. Jolchine, *in* "Progress in Photosynthesis Research" (H. Metzner, ed.), Vol. 1, p. 209. Tubingen, Germany, 1969.

[5] See this volume [67].

Source Material

Methods that have yielded highly purified reaction center preparations from *Rhodopseudomonas spheroides* (as well as from *Rhodospirillum rubrum*) have been consistently effective only with mutant strains that lack colored carotenoids. Satisfactory growth of these blue-green mutant strains requires that the combination of light, O_2, and BChl be avoided. Anaerobic bottle cultures can be started from slants grown in the dark so as to have little pigment. In serial bottle culture transfers,

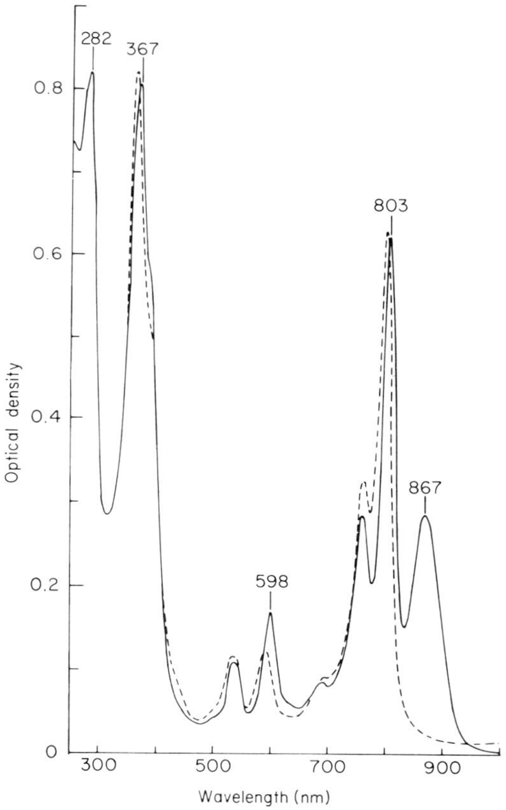

FIG. 1. Solid curve: absorption spectrum of a suspension of reaction centers prepared as described in the text, measured in the IR1, visible and UV modes (sample exposed only to weak light) of a Cary 14R Spectrophotometer. The smaller maxima at 535 and 759 nm are ascribed to bacteriopheophytin. Dashed curve: absorption spectrum of the same material, measured in the IR2 mode (sample exposed to strong light during measurement), showing the bleaching of P870 and a blue-shift of P800 from 803 to 798 nm. Path length, 1 cm.

freshly inoculated cultures should be incubated in the dark for several hours prior to illumination, to allow respiration to deplete O_2 in the medium. Cultures should be harvested at or just before the peak of exponential growth; reaction centers made from older cultures usually contain larger amounts of bacteriopheophytin as a contaminant.

The material shown in Fig. 1 was made from strain 2.4.1/R-26 of *Rhodopseudomonas spheroides,* grown in modified Hutner medium[6] at 30° under illumination by 40-W tungsten lamps at an intensity of about 6 mw/cm^2. The cells were grown in completely filled 2-liter glass-stoppered bottles containing magnetic bars for very gentle stirring.

The culture sometimes develops a sickly grayish color; it should then be discarded and a fresh one started from stock. It is well to keep viable freeze dried stocks of the "normal" bright blue cells against this contingency. Occasional reisolation from single colonies on aerobic culture plates grown in darkness, is also recommended as a safeguard against changes in the pigmentation of the culture.

Purification Method[7]

The method given here is an outgrowth of earlier ways[8,9] to destroy light-harvesting BChl while preserving P870. It is a refinement of a technique developed in this laboratory[3] and subject to continuing improvement. In particular, the value and best use of gel filtration and other chromatographic and electrophoretic methods are not well defined as yet. The use of gel filtration as mentioned earlier[3] and in this article should be regarded only as examples of the potential value of the method. Another matter of choice, largely unexplored, is the selection of one out of many possible detergents. Our best results to date have been obtained with the zwitterionic detergent Ammonyx-LO[10] (1% v/v) rather than Triton X-100 as used earlier.[3]

Except for the incidental fact that the centrifuge rotors are maintained near 0°, all steps in the following procedure are carried out at room temperature. Exposure to light should be minimized from the time that detergent is added until the ammonium sulfate fractionation has been completed (a green safelight is useful). This helps to prevent con-

[6] G. Cohen-Bazire, W. R. Sistrom, and R. Y. Stanier, *J. Cell. Comp. Physiol.* **49,** 25 (1957).

[7] This work has been supported by AEC Contract No. AT-(30-1)-3759.

[8] R. K. Clayton, *Biochim. Biophys. Acta* **75,** 312 (1963).

[9] P. A. Loach, G. M. Androes, A. F. Maksim, and M. Calvin, *Photochem. Photobiol.* **2,** 443 (1963).

[10] Ammonyx-LO is dimethyl lauryl amine oxide, made by Onyx Chemical Company, Div. of Millmaster Onyx Corp., Jersey City, New Jersey. We are indebted to the Research Department of Onyx Chemical Company for a gift of this detergent.

tamination of the final preparation by degradation products of light-harvesting BChl.

The first step is to prepare a chromatophore suspension. Harvest the cells of blue-green mutant *Rhodopseudomonas spheroides* by centrifugation; wash and suspend to about 30% (wet v/v) in 10 m*M* Tris·HCl buffer, pH 7.5. Disrupt the cells by passing the suspension twice through a French pressure cell at about 20,000 psi. Centrifuge the extract at 12,000–15,000 *g* for 30 minutes to remove debris. Recover the supernatant solution as a clear blue extract, sacrificing a little so as to avoid carrying over any of the sedimented material. Then centrifuge the extract at about 200,000 *g* for 60 minutes (100,000 *g* for 90 minutes is optional) to obtain a clear, deep blue jellylike pellet of chromatophores and a brown supernatant liquid. Remove the supernatant fluid and rinse the walls of the tube and the surface of the pellet gently with H_2O. Above the dense pellet there may be a fluffier, paler pellet. If so, remove and discard the latter while rinsing the dense pellet. Redisperse the clear dense pellet in 10 m*M* Tris·HCl, pH 7.5, leaving behind any hard grayish core in the pellet. Wash the resulting chromatophore suspension twice by centrifugation at 200,000 *g* for 60 minutes and redisperse in 10 m*M* Tris·HCl, pH 7.5. Adjust the final suspension to an optical density (1-cm path) of 50 at the long-wave absorption maximum near 870 nm.

To the adjusted chromatophore suspension, stirred in a beaker, add dropwise 1/30 vol of 30% aqueous Ammonyx-LO. Put the mixture in 10- or 12-ml centrifuge tubes (angle rotor; a swinging bucket is not necessary); 5 or 6 ml in each tube. Beneath this, using a syringe, layer the first 3 ml of 0.5 *M* aqueous sucrose solution and then enough 1.0 *M* sucrose, at the bottom, to fill the tube. Centrifugation at about 200,000 *g* for 90 minutes[11] yields a blue band of light-harvesting BChl attached to membrane fragments near the original 0.5 *M*/1 *M* sucrose boundary and a clear pellet of cell wall fragments and other debris at the bottom. The uppermost 5 or 6 ml, originally occupied by the chromatophore–detergent mixture, is deep bluish brown. It contains reaction centers and three principal colored contaminants: detergent-solubilized BChl, bacteriopheophytin, and a BChl degradation product. These substances have absorption maxima at 770, 760, and 680 nm, respectively. A little light-harvesting BChl, with maximum absorption at 850–860 nm (shifted by the detergent), might also be present in addition to the P800 and P870 of the reaction centers. Most of the detergent remains in this upper 5- to

[11] Or longer at lower centrifugal force, but the higher force gives a preparation with less contamination by detergent-solubilized BChl.

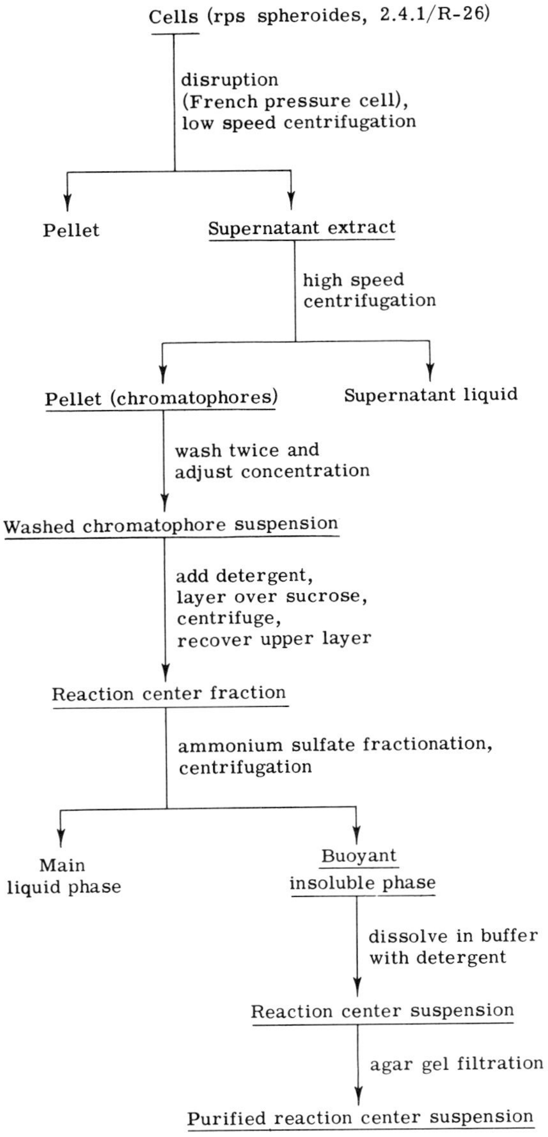

FIG. 2. Outline of a method for preparing reaction centers from *Rhodopseudomonas spheroides*.

6-ml fraction. Transfer this "reaction center fraction" to a beaker for ammonium sulfate fractionation.[12]

For each milliliter of reaction center fraction, add gradually, while stirring, 0.3 g of solid ammonium sulfate. Continue stirring for 5 minutes. An insoluble phase containing the reaction centers rises to the top upon centrifugation (about 7000 *g* for 20 minutes); this "levitate" is a brown oil at room temperature or a green scum at lower temperature. Withdraw the submergent (aqueous) fluid with a syringe, so as to recover the levitate. Dissolve the levitate in one-third the original volume of 10 m*M* Tris·HCl, pH 7.5, containing 2% Ammonyx-LO. After 10 minutes of stirring, add dropwise an equal volume of saturated ammonium sulfate (pH near 7) to achieve 50% of saturation. Continue the stirring for 5 minutes, centrifuge, and once more recover the levitate, dissolving it as before in one-third the original volume of 10 m*M* Tris·HCl, pH 7.5 containing 2% Ammonyx-LO. This time add saturated ammonium sulfate (pH $\approx$ 7) to 30% of saturation. Centrifuge; set aside the levitate, and treat the submergent liquid with more saturated ammonium sulfate, to give 50% of saturation. A new levitate, the "30–50" fraction, can then be harvested by centrifugation and dissolved in a little 10 m*M* Tris·HCl, pH 7.5, containing Ammonyx-LO. This is the desired fraction. It should contain relatively little of the colored contaminants mentioned earlier, as indicated by the spectra of Fig. 1.

Further purification, as by filtration on Sephadex or Bio-Gel (agar gel) in the range 2×10^5–1.5×10^6 particle weight, can be regarded as optional. The material shown in Fig. 1 had been passed through a Sephadex G-200 column, eluting from bottom to top with 10 m*M* Tris·HCl, pH 7.5, containing 0.03% Ammonyx-LO, after the ammonium sulfate fractionation. The only apparent improvement was that the ratio $O.D._{280}:O.D._{800}$ (in the "best" fraction, near the center of the eluted band) diminished from 1.5 to 1.3.

The foregoing procedure is summarized in Fig. 2. Indications of composition and yield are shown in the table. This is currently our best material for optical investigations.

Criteria of Purity, and More Properties

The final preparation, at the concentration shown in Fig. 1, should have a clear ice-blue color. A green or brown tingle would indicate the presence of colored contaminants.

Absence of the light-harvesting BChl component B870 is indicated by

[12] The fractionation method given here was chosen empirically and, like other aspects of this rapidly evolving procedure, can undoubtedly be improved.

RECOVERY AND PURIFICATION (RELATIVE TO OTHER COMPONENTS) OF P870 IN REACTION CENTER PREPARATIONS

Sample	Percent recovery of P870	[B870]:[P870][a]	$O.D._{280}$:O.D.(P870)[b]
Washed chromatophore suspension	100 (nominal)	25	38
Reaction center fraction after centrifugation over sucrose	62	—	—
Reaction center suspension after ammonium sulfate fractionation	50	0.0	3.3
Final reaction center suspension	25	0.0	2.9

[a] Based on extinction coefficients given in the text.
[b] O.D.(P870) means the O.D. attributable to P870 at 867 nm.

the completeness of the reversible bleaching of the 865-nm band when sufficiently strong exciting light is used. However, a long exposure to strong light could cause irreversible bleaching of any B870 that is present, leaving a colorless contaminant in the material. Some irreversible bleaching of P870 and P800 is also caused by prolonged exposure to strong light, but the degradation of B870 is an order of magnitude faster.

The absorbance near 800 nm, due to P800, should be 2.2 times that at 865 nm due to P870. A smaller ratio would suggest that some B870 is present. A larger ratio would prevail if some of the P870 were in its oxidized (bleached) form. To be sure that all the P870 is reduced, let the preparation stand in the dark at least 10 minutes, preferably in the presence of a mild reductant such as 10^{-4} *M* sodium ascorbate, before measuring.

The absorbance near 280 nm is an indicator of the protein content, but it can also reflect other substances, notably BChl and Triton X-100. BChl in ether shows a peak[13] at about 260 nm; $\epsilon \approx 20/mM/cm$. The absorption maximum of Triton X-100 at 275 nm has $\epsilon = 1.3/mM/cm$ (mol. wt. 700). One of the useful traits of Ammonyx-LO is that it does not have an absorption band near 280 nm, so its presence does not interfere with protein estimation. In our experience thus far, every successful effort to remove most of the detergent has caused aggregation of the reaction center particles,[14] giving a turbid suspension. A clear suspension

[13] J. C. Goedheer, *in* "The Chlorophylls" (L. P. Vernon and G. R. Seely, eds.), p. 147. Academic Press, New York, 1966.

[14] The reaction center preparations described earlier[3] probably contained amounts of Triton X-100 that contributed significantly to the absorption at 280 nm, despite some evidence to the contrary.

is regained when detergent is added back. Limiting detergent concentrations for this effect depend strongly on the ionic environment and have not been well characterized.

The purity (or environment) of reaction centers with respect to small molecules can, of course, be adjusted by dialysis. The reaction centers are retained by conventional dialysis membranes and by membrane filters designed to pass particles of molecular weight less than about 20,000.

We do not know what ratio of absorbance at 280 nm to that at 800 or 865 nm will represent "pure" reaction centers. This will depend on the irreducible size of the particle that can support photochemically active P870. The particle size is not yet settled; the reaction centers made with Triton X-100 behaved during gel filtration as though their molecular weight was 650,000, but the ones being made now with Ammonyx-LO behave like particles of molecular weight between 50,000 and 200,000.

The compositions of reaction center preparations made with Triton X-100 have been published, with lists of the ratios of certain components (quinones, cytochromes, metals) to P870. But the preparations made with Ammonyx-LO contain far less of these components, and yet show cyclic electron transport beginning with the photochemical oxidation of P870 and ending with the reduction of oxidized P870. Again it seems premature to assert what components are intrinsic to "pure" reaction centers in the sense that they are indispensable for cyclic electron flow, as contrasted with the direct return of an electron from reduced primary acceptor to oxidized P870:

$$P + A \overset{h\nu}{\rightleftarrows} P^+ + A^-$$

The direct back reaction, which can be induced by lowering the temperature or by adding *o*-phenanthroline,[15] has a time constant of about 50 mseconds. The cyclic route, which predominates in reaction centers suspended in 10 mM Tris·HCl, pH 7.5, at room temperature, requires several seconds or longer.

It is hoped that further analysis will show what meaning (if any) should be attached to the idea of a reaction center particle, and what structural relationship exists between the photochemical reaction centers and the associated systems of the living cell: light harvesting system; electron transport and phosphorylating machinery.

[15] R. K. Clayton, H. Fleming, and E. Z. Szuts, unpublished experiments (1969).

Author Index

Numbers in parentheses are reference numbers and indicate that an author's work is referred to although his name is not cited in the text.

A

B

D

E

H

I

J

K

R

U

V

W

Subject Index

A

B

N

O

P

S

T

U

V